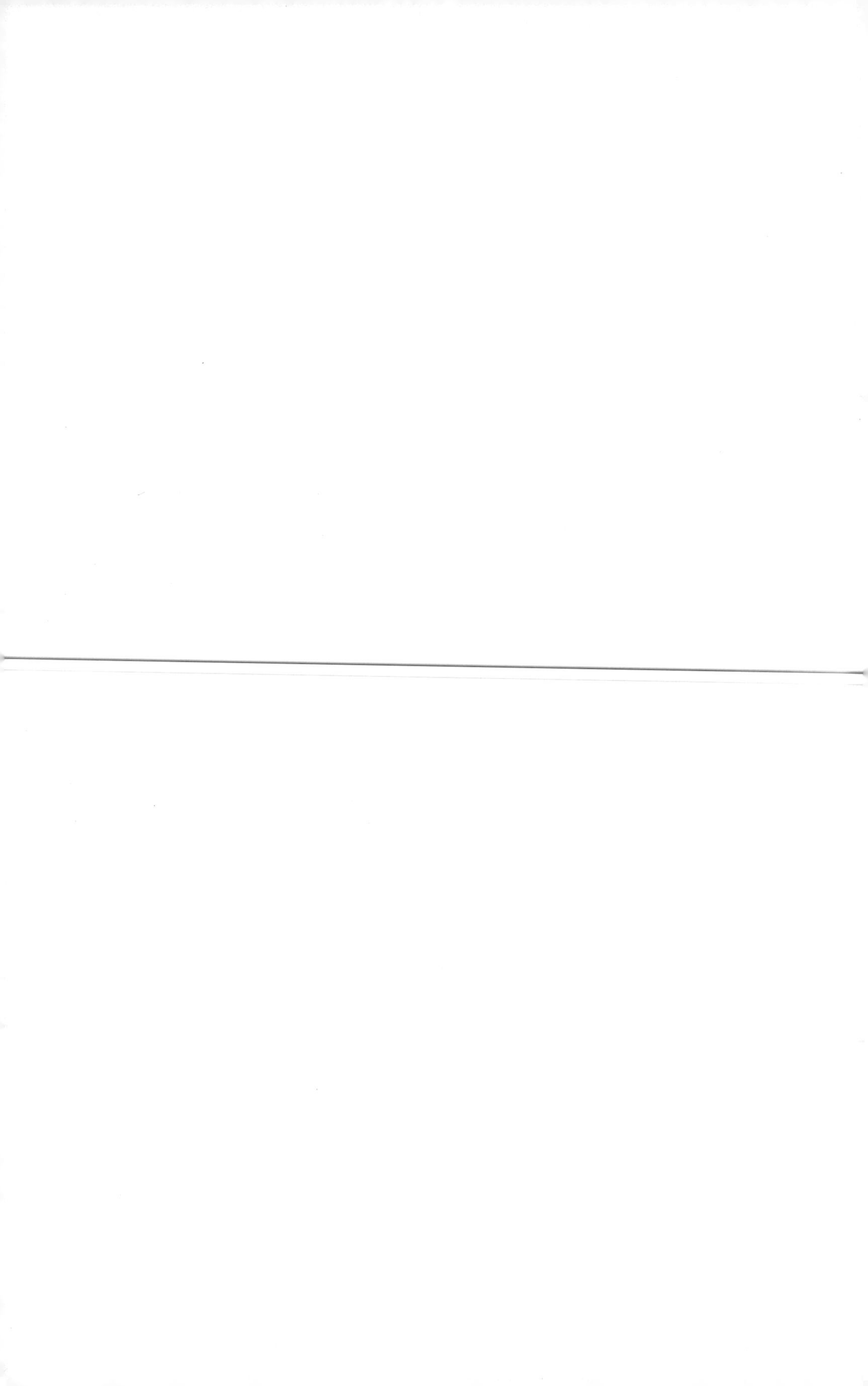

SYNCHROTRON RADIATION AND DYNAMIC PHENOMENA

CONFERENCE PROCEEDINGS NO. **258**

PARTICLES AND FIELDS SERIES 49

SYNCHROTRON RADIATION AND DYNAMIC PHENOMENA

48TH INTERNATIONAL MEETING OF PHYSICAL CHEMISTRY
GRENOBLE, FRANCE 1991

EDITOR:
ALBERTO BESWICK
LURE & UNIVERSITÉ
PARIS-SUD, ORSAY

CO-EDITORS:

H. BAUMGÄRTEL
C. R. A. CATLOW
P. DECLEVA
H. DEXPERT
E. GEISSLER
J. GOULON
M. J. HUBIN-FRANSKIN
J. LAJZEROWICZ
I. NENNER
J. PANNETIER
C. TROYANOWSKY

AIP

American Institute of Physics **New York**

L.C. Catalog Card No. 92-53790
ISBN 1-56396-008-7
DOE CONF-910937

Printed in the United States of America.

CONTENTS

IV. TIME-RESOLVED X-RAY SPECTROSCOPY AND SCATTERING

Contents

II. CORE-HOLE EXCITATION AND RELAXATION

III. TIME-DEPENDENT FLUORESCENCE DECAYS

IV. TIME-RESOLVED X-RAY SPECTROSCOPY AND SCATTERING

Note: The author who presented the talk is designated by an asterisk.

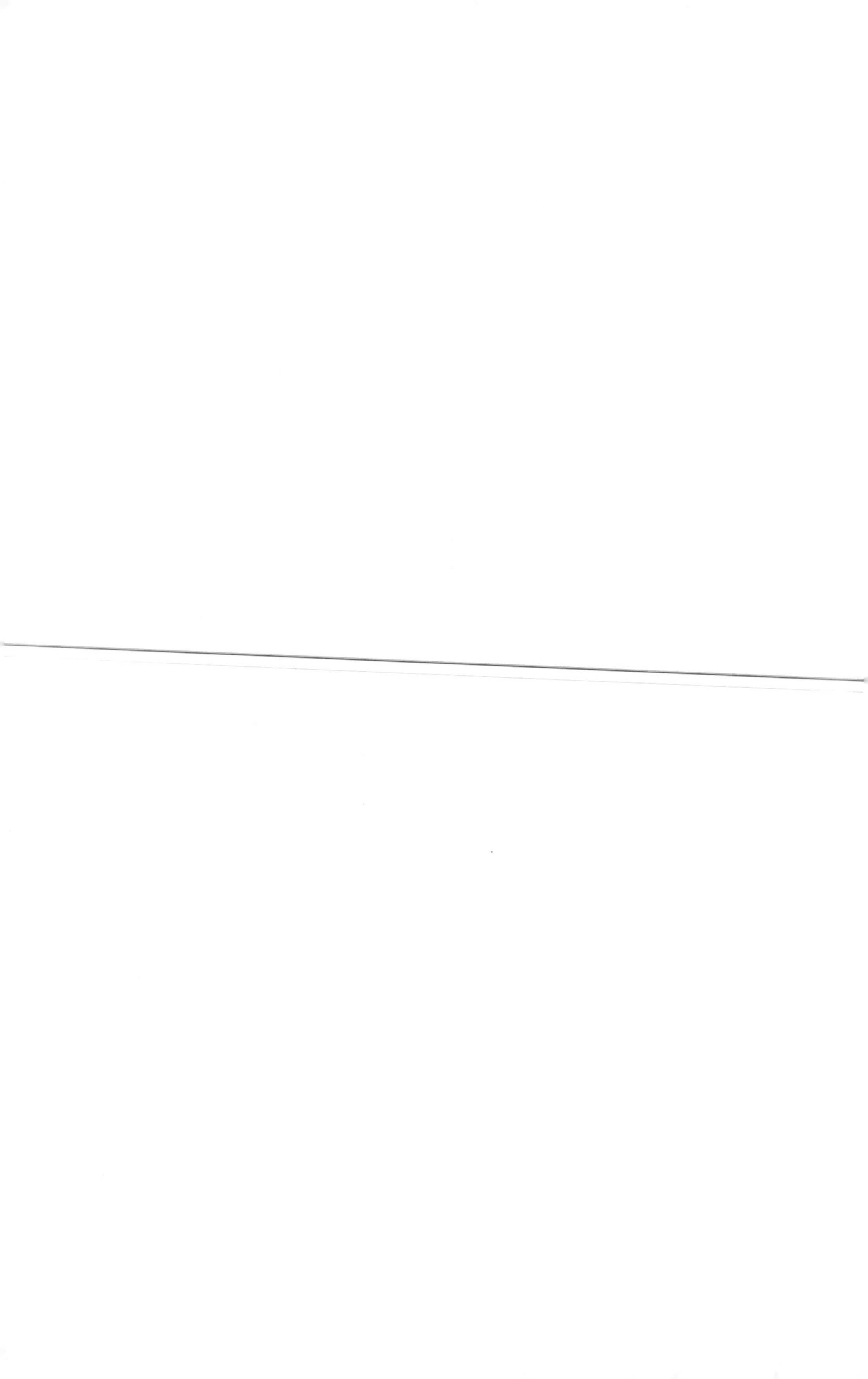

Foreword

Our 1991 international meeting was one of the series the European Physical Chemistry Societies organize jointly, the conference being alternately held in France, Germany, Italy, or the U.K. We had this year the pleasure of being joined in the organization by our Belgian colleagues.

Whenever possible these meetings aim at dealing with an interdisciplinary subject that will bring together people from different but overlapping fields. In this respect the developments of synchrotron radiation research were a perfect target, considering the number of domains SR research covers, and the very promising interacting between biophysics and SR studies. In fact preliminary enquiries showed that the SR community already numbered 8000 to 9000 people, and that 30 SR facilities are at present operating or under construction. The problem, evidently, was not to find which subjects exhibited a sufficient overlap, but which ones should be discarded, so as to avoid an excessive crowd.

Our Committee therefore decided to limit the scope of the conference to time-dependent phenomena. The venue of the meeting stemmed from its theme, and we met in Grenoble, where the ILL (Institut Laue-Langevin, built around a high neutron flux reactor) has been operating for many years and where the European Synchrotron Radiation Facility is now being built. The place offers many additional advantages: a large university where Chemical Physics and Physics are very active, and where many colleagues have the experience of our meetings and were ready to help with practical as well as scientific matters. No other place could have been as appropriate for our purpose.

We selected as main topics: (1) Dynamic phenomena in metallic, semiconductor, and molecular clusters; (2) molecular photodissociation and photoionization; (3) dynamics of chemisorbed and condensed molecules; (4) core-hole excitation and relaxation; (5) time-dependent fluorescence decays; and time-resolved x-ray spectroscopy and scattering.

With such "restricted" topics most areas of chemical physics and biophysics could already be involved: the exceptional properties of SR sources, wavelength tunability, brightness, and pulsed character provide ideal capabilities to study in detail time-dependent processes. SR enables one to prepare excited species and monitor their subsequent decay, and also to probe physical and chemical processes in condensed phase, in a time-scale range extending from a few picoseconds to hours. We thus built a really interdisciplinary conference which brought together physicists, chemists, and biophysicists. The attendance was not very large, owing to the present squeeze on research budgets and travel allowances, but over 50 people came from outside France, providing a very good sampling of what is studied in more than half the existing SR centers. It is true that a small attendance makes it more difficult to balance a budget, but it helps in improving the contacts and in giving additional impetus to the discussions.

If instrumental problems were not at the heart of the conference it was of course important to assess what can be expected from the third generation of SR sources, which will be active in the very near future: these were presented in the opening lecture, dealing with ESRF and its performance, and the round table on "time-dependent experiments: present and future," to be both found at the end of this book. This printing order is deliberate, so as to conclude these Proceedings with the lecture and panel discussion which were essentially aimed at, not reported on recent achievements but at the future and prospects of research.

The reports on recent results were many and provided a very attractive overview of present results and trends. The topics were too numerous to be all mentioned, but the following incomplete list will give a fair idea of their variety: (1) Experimental and theoretical studies on clusters; (2) state-selective chemistry; (3) photoexcitation by the combined use of lasers and SR; (4) ion–ion coincidence experiments leading to the understanding of fragmentation dynamics; (5) high-energy excitation of physisorbed molecules; (6) polarization and orientational effects in photoionization; (7) electronic and core-level excitation of molecular clusters; (8) many-body effects in core-level XPS and resonance photoemission; (9) core-induced photodissociation of surface molecules; (10) core-hole relaxation effects in high-energy spectroscopies; (11) nuclear Bragg diffraction, a new source for solid-state physics and hyperfine spectroscopy; (12) time-resolved fluorescence studies of excited liquids and solids; (13) time-resolved fluorescence studies of biosystems, (14) time-resolved scattering applied to polymers; (15) VUV photochemical etching; and (16) many aspects of time-resolved and energy-dispersive x-ray scattering such as: fast-scanning EXAFS, solid-state combustion studies, EXAFS studies of redox and dehydration reactions, etc.

Many authors brought to Grenoble their latest results and data, ensuring a satisfactory blend of fresh achievements in well-established domains along with very recent breakthroughs that bode well for the future. Many of these results were of an interest broad enough to fill—sometimes to overflow—the ample time we always reserve for discussions. Although discussions are the "raison d'être" of our conferences, we seldom saw such lively debates.

The joint European Physical Chemistry meetings were first organized in 1947, held regularly since 1973, and yearly since 1988. They are an exellent opportunity for working together, and I have pleasure in expressing our thanks to our friends H. Baumgärtel (Deutsche Bunsen Gesellschaft), C.R.A. Catlow (Faraday Division, RSC), P. Decleva (Associazione Italiana di Chimica Fisica), and M. J. Hubin-Franskin (Division de Chimie Physique, Société Royale de Chimie). They were associated to an efficient French group which included A. Beswick, Chairman, H. Dexpert, E. Geissler, J. Goulon, J. Lajzerowicz, I. Nenner, and J. Pannetier. They all won our gratitude by building a good meeting which left all participants with good memories. Special thanks are certainly due to the Grenoble members of the team, who coped with the problems of practical management in addition to the scientific duties they shared with the others. Thanks are also due to the European Synchrotron Radiation Facility: its Director General Professor Haensel gave us financial support, found enough time in his heavy schedule to take care of the opening lecture, and provided our participants with a very open welcome on the ESRF site.

And as usual our thanks go chiefly to the authors who presented us with their newest results and to whom we owe a lively meeting, which we hope will prove useful to the large and ever expanding synchrotron radiation community.

Clément Troyanowsky
Conference Officer
SFC/Division de Chimie Physique

ACKNOWLEDGMENTS

This conference was jointly organized by (1) Division de Chimie Physique of Société Française de Chimie, host society; (2) Associazione Italiana di Chimica Fisica; (3) Deutsche Bunsen Gesellschaft für Physikalische Chemie; (4) Faraday Division of the Royal Society of Chemistry; and (5) Division de Chimie Physique of Société Royale de Chimie.

It has been hosted by the Université Joseph Fourier, Grenoble, whose practical support and help have been most valuable.

Support has also been received from the following organizations: (1) Centre National de la Recherche Scientifique, Département Chimie; (2) Commissariat à l'Energie Atomique, DRECAM; (3) Direction des Recherches, Etudes, et Techniques; (4) European Synchrotron Radiation Facility; (5) LURE, Laboratoire pour l'Utilisation du Rayonnement Electromagnétique; and (6) U.S. Office of Naval Research, London branch.

This support is gratefully acknowleded.

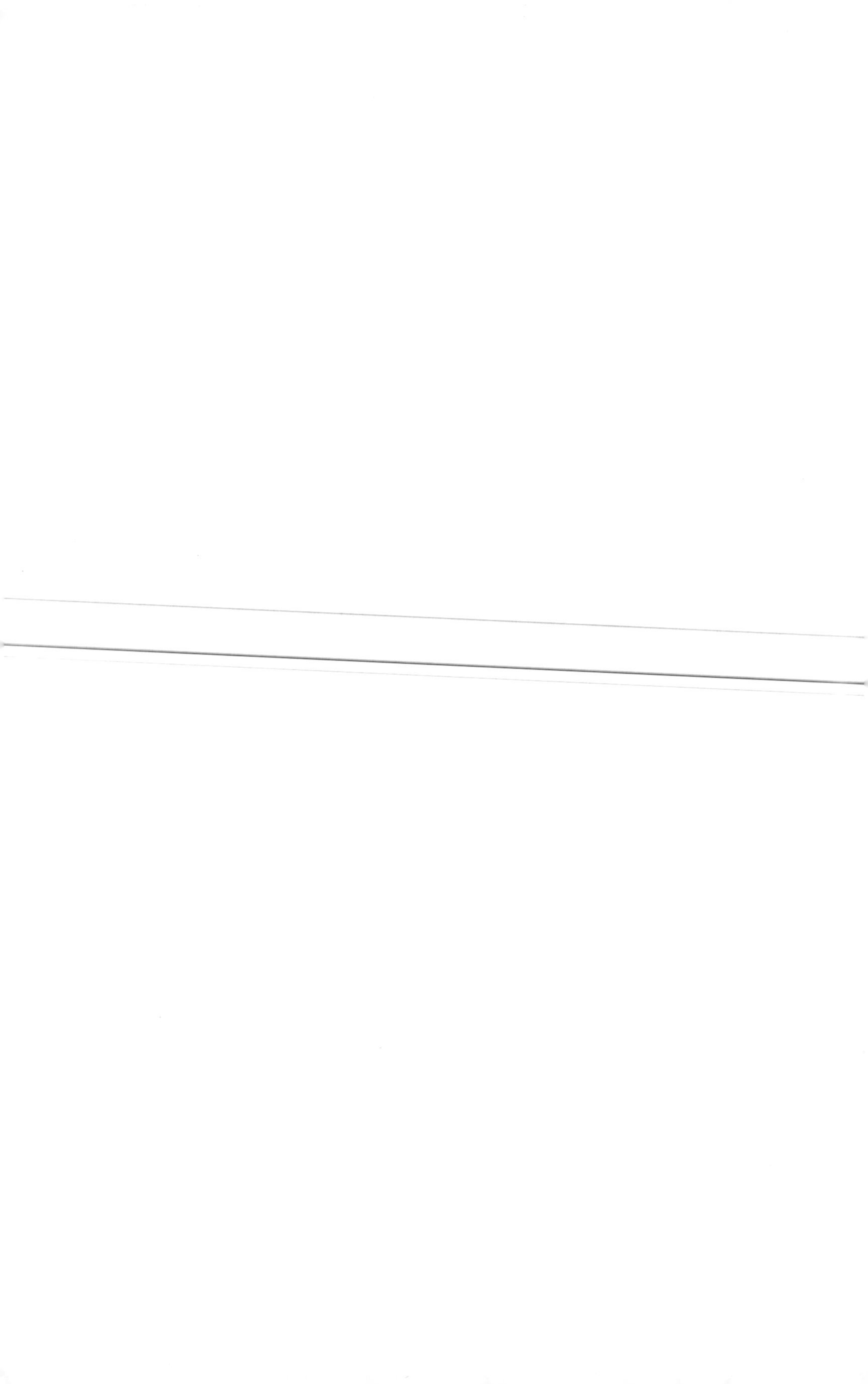

I. PHOTODISSOCIATION AND PHOTOIONIZATION: FROM ISOLATED SPECIES TO CLUSTERS AND ADSORBED MOLECULES

REACTIONS OF STATE SELECTED IONS STUDIED WITH VUV RADIATION

Tomas Baer
Department of Chemistry
University of North Carolina
Chapel Hill, NC 2759-3290

ABSTRACT

Methods for state preparation of ions by photoelectron photoion coincidence (PEPICO) technique are reviewed. The use of the pulsed structure of the photons from synchrotron radiation in collecting threshold electrons is pointed out. Recent results from laboratories in Chapel Hill, and LURE, the synchrotron facility at the University of Paris, Orsay, are highlighted. These include dissociation rate measurements of energy selected butylbenzene and ethylchloride ions. Some recent results on the isotope effect in the H and D loss channels of $CH_3OCD_3^+$ are used to show that the C-H bending modes are still largely intact in the region of the transition state. The bimolecular ion-molecule studies between $H_2^+(v)$ and H_2 or D_2 are presented. Finally, the advantages of using supersonically cooled samples for PEPICO studies are discussed.

INTRODUCTION

Photoelectron photoion coincidence (PEPICO) studies have been carried out for a number of years. Among the first such studies were those of Brehm and von Puttkamer[1] and Eland.[2] More recently, Stockbauer,[3] Meisels,[4] Baer,[5] Guyon,[6], Rosenstock,[7] Koyano,[8] and Ng[9] have contributed to this field. Two recent reviews have focussed on dissociation studies[10] and experimental aspects[11] of PEPICO. In this paper, we highlight some of the important experimental considerations, in particular those relating to the use of pulsed synchrotron radiation, and the use of molecular beams as a source of cold samples.

The PEPICO technique is based on conservation of energy. If a molecule is ionized with a photon of energy $h\nu$, and electron is ejected with an energy, E_e, the ion internal energy will be given by:

$$E_{int}(AB^+) = h\nu - IP + E_{th} - E_e \quad (1)$$

where IP is the ionization energy of the molecule, and E_{th} is the thermal energy of the molecule. If the ion is measured in delayed coincidence with an energy selected electron, then only ions of a specific internal energy will be collected. It is clear that good energy resolution for the ions requires a narrow distribution of thermal energies, E_{th}, and a precise method for measuring the electron kinetic energy. In recent years, the most common approach has been to detect threshold (or zero energy) electrons.

THE DETECTION OF THRESHOLD ELECTRONS

The principal of threshold electron detection is based on two features of low energy electrons, their low angular divergence when extracted by a small electric field,[12] and their time of flight (TOF).[13] The low angular divergence of threshold electrons can be used to advantage by discriminating against energetic electrons, as shown in Figure 1. The resulting electron transmission as a function of the electron energy is also shown in this figure. An analyzer based only on angular discrimination has been termed a steradiancy analyzer[12] and can be made to work quite well if one is willing to correct for the high energy electron tail. This tail comes about because there are always some energetic electrons that are ejected in the direction of the detector. They cannot be stopped by the steradiancy analyzer alone. However, they can be stopped either by an energy analysis if the electric field across the ionization region is not too large, and the photoionization region is very narrow. These conditions sometimes conflict with desired experimental parameters. By far the best way to stop the energetic electrons is to use a pulsed light source and thereby select the threshold electrons by their TOF.[14] The synchrotron radiation source is ideal for this purpose.

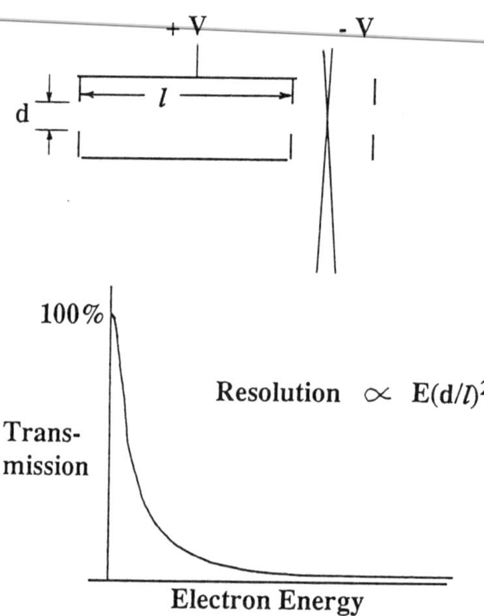

Fig. 1. a) Diagram of the steradiancy analyzer for detecting threshold electrons. Two trajectories are shown, one for a threshold electron, the other for an energetic electron with some off axis velocity. b) the resulting transmission function for the electrons.

What is required for the synchrotron is a time between photon pulses of at least 60 ns. With this time separation an electron drift tube of about 10 mm with an applied electric field of about 2 V/cm can be used. However, it is very clear that the resolution can be increased significantly by increasing the time between bunches. Other factors that enter in are the size of the photon source at the experiment (this determines the aperture sizes that can be utilized) and the time spread of the photon pulse. The narrower it is, the more precisely can the electrons be timed. The advantage of using electron TOF in conjunction with the steradiancy analysis is that both the electron collection efficiency and the electron energy resolution can be improved, each by about an order of magnitude. Resolution of 5 meV can thus be readily achieved.[15]

ION LIFETIME MEASUREMENTS

The measurements of dissociation rates of energy selected ions is shown in Figure 2. Slowly dissociating ions are those that move some distance from their origin in the acceleration region and dissociate in this region. As shown in the Figure, each position of dissociation results in a different fragment ion TOF. The resulting asymmetric TOF distribution can then be analyzed in terms of an exponential (or bi-exponential) decay. A single parameter, the mean ion lifetime is used to model these data.

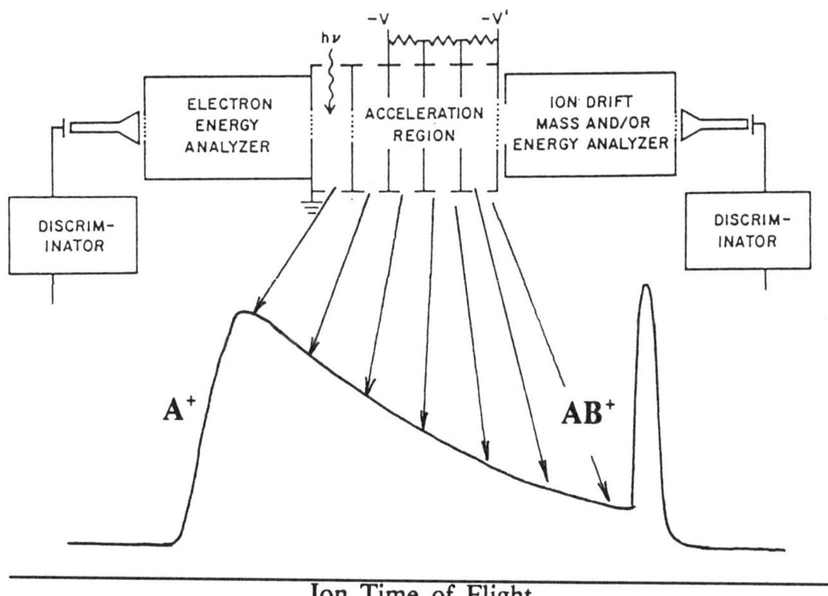

Fig. 2. A diagram of a PEPICO apparatus showing the time of flight distribution of a fragment ion resulting from a slow dissociation taking place in the acceleration region.

TWO EXAMPLES OF SLOW IONIC DISSOCIATIONS

The dissociation of n-butylbenzene ions is a text book example of a reaction in which the low energy path leads to products that involve a rearrangement of the ion in the region of the transition state, while a higher energy path proceeds by a simple bond fission.[16] A product ion TOF distribution at low ion internal energy is shown in Figure 3. The solid line is a fit to the experimental data.

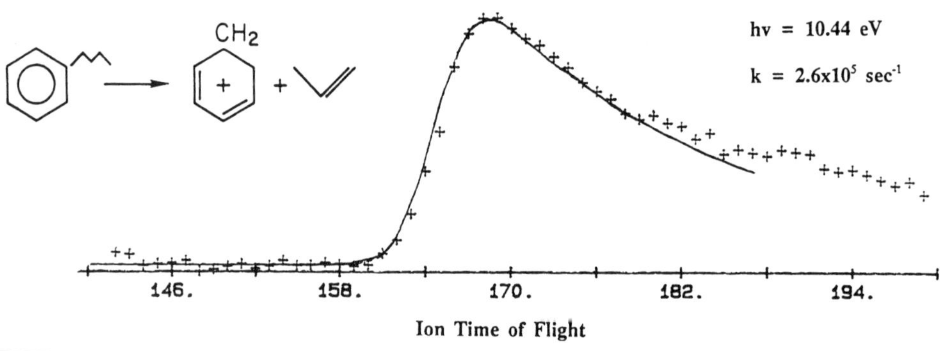

Fig. 3. The TOF distribution for $C_7H_7^+$ product ions from n-butylbenzene ion dissociation. The solid line is a fit to the data from which the dissociation rate constant of 2.6×10^5 sec^{-1} is derived.

The dissociation of butylbenzene can be well explained in terms of the statistical theory of unimolecular decay (RRKM or QET). These calculations can be carried out with a knowledge of the threshold energy for dissociation and a knowledge of the vibrational frequencies of both the ion and the transition state. The latter are approximated based on a knowledge of the neutral molecule's frequencies.

A very different reaction is the loss of HCl from ethylchloride ($C_2H_5Cl^+$) ions. Here the threshold energy for dissociation is so low that no RRKM calculation could reproduce the slow measured dissociation rates.[17] In fact, the calculated rate constants were about five orders of magnitude too fast. In addition, we noticed from our *ab initio* molecular orbital calculations of the potential energy surface, that there is a substantial energy barrier involved for the transfer of an H atom from the terminal carbon group to the Cl atom prior to HCl loss.[18] These results suggest that the slow rate for HCl loss from the ethyl chloride ion may be a result of tunneling through the barrier. We have thus measured the rates for both HCl loss from $C_2H_5Cl^+$ and DCl loss from $C_2D_5Cl^+$ and compared them with calculations including tunneling.

The tunneling calculations utilized an Eckart potential which is parameterized in terms of the barrier heights, forward and reverse, and the curvature at the top of the barrier. These three parameters can be obtained directly from the *ab initio* calculations. In addition, the calculations also provide us with the vibrational frequencies of the ion and the transition state. We can thus calculate the RRKM/tunneling rate, as according to the formalism of Miller,[19] with no adjustable parameters.

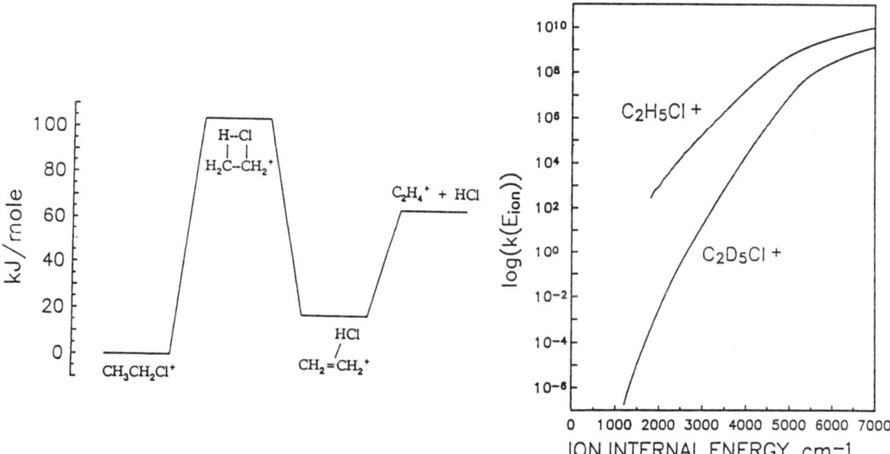

Fig. 4 The potential energy curve for HCl loss from ethylchloride ions. Based on *ab initio* calculations with a 6-31G*/MP2 basis set. Taken from Booze *et al.* (ref. 17) with permission.

Fig.5 The calculated dissociation rates based on an RRKM model with tunneling corrections for the indicated reactions. Taken from Booze *et al.* (ref. 17) with permission.

The calculated rates using the *ab initio* parameters did not fit very well the measured rates. However, by lowering the barrier height (excluding the zero point energy) from 104 kJ/mol to 72 kJ/mol and by reducing the critical frequency (which is related to the potential barrier curvature) from 1499 cm^{-1} to 1300 cm^{-1}, excellent fits were obtained with our measured rates for the normal ethylchloride ion. The rates for the deuterated sample could then be calculated with no further adjustments in the parameters, except those dictated by the replacement of the H by a D atom. As shown in Figure 5, the resulting rates for DCl loss were significantly lower. Yet, these calculated rates fit perfectly the measured rates for DCl loss.

The excellent agreement of the calculated and measured DCl loss rates represents strong support for the tunneling model. Yet, the need for adjusting the barrier height by about 30 kJ/mol is disquieting. Is this adjustment required to offset the approximations inherent in the use of a one-dimensional Eckart potential, or is it necessary because the *ab initio* molecular orbital calculations of transition state energies is not very precise. These questions are difficult ones to answer because there is so little information available about barrier heights. Furthermore, tunneling calculations that include additional dimensions and curvature in the reaction path, require a considerable greater effort than those described here.

THE FATE OF BENDING MODES IN THE TRANSITION STATE

When an ion such as dimethylether looses an H atom, the number of vibrational frequencies goes from 18 to 17 at the transition state, to 15 in the final products. We make the assumption that at the transition state, there is only one reaction coordinate (imaginary frequency) and that all other vibrations are still intact. However, to what extent are the two bending modes, which are associated with the C-H stretch still intact?

We have measured the onset of the H and D loss reaction from $CH_3OCD_3^+$ with the Orsay synchrotron. The high resolution provided by electron TOF was essential for measuring this small isotopic shift. Figure 6 shows the breakdown diagram for this reaction in the vicinity of the dissociation onset. The breakdown diagram is the fractional abundance of

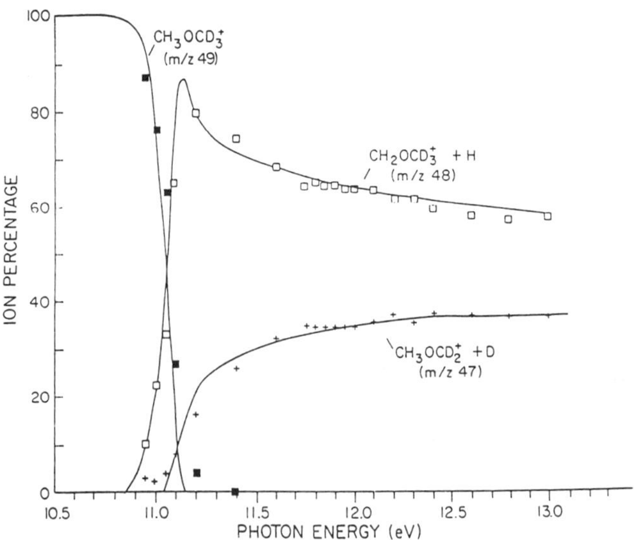

Fig. 6. The breakdown diagram for partially deuterated dimethylether. Taken with permission from Dutuit *et al.* (ref. 20).

the parent and daughter ions as a function of the parent ion internal energy. The solid lines are obtained from RRKM calculations, which in this case are not so sensitive to the absolute rates, but are very sensitive to the relative rates for H and D atom loss. The measured difference in the onset for H and D loss is 460 cm^{-1}.

The breakdown diagram contains information about fate of the zero point energies of the relevant vibrations, which include the three vibrations associated with the C-H and the C-D bonds. The two stretch frequencies are 2914 and 2152 cm^{-1}, while the bending frequencies for the C-H and C-D units are 1470 and 1051 cm^{-1} respectively. One stretch and two bending frequencies are lost in product formation. As a result, the zero point energy difference between product energies for H and D loss is 800 cm^{-1}, the H loss being lower in energy. This would be the measured onset energy difference if the two bending modes were reduced to 0 cm^{-1} at the transition state. From these data and the assumed vibrational frequencies we can determine that the bending modes have been reduced by only 80% to 1176 and 841 cm^{-1}, respectively.

ION MOLECULE REACTIONS

The PEPICO technique can also be used to investigate ion-molecule reactions. A diagram of the set-up used in the Orsay experiment is shown in Figure 7. State selected $H_2^+(v)$ ions are extracted from the ionization cell and transported to the ion-molecule interaction region where the products are extracted and measured by TOF.

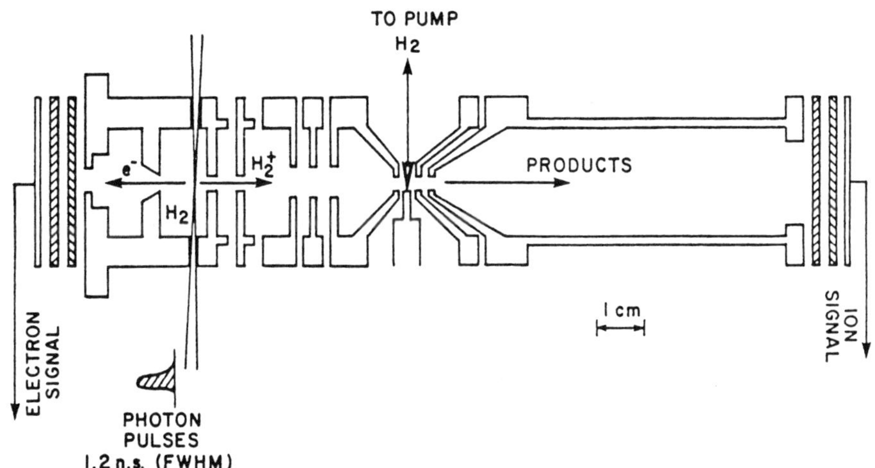

Fig. 7. The double electron ion TOF apparatus at Orsay used to investigate ion-molecule reactions with state selected ions. Taken with permission from Cole *et al* ref. 21.

Figure 8 illustrates the TOF distribution obtained when $H_2^+(v=5)$ reacts with H_2. The primary beam is due to unreacted ions, the peak at long TOF is a result of a symmetric charge transfer reaction, while the small peak at short TOF is a result of a collision induced dissociation (CID). Of particular interest is the fact that the H^+ TOF distribution has considerable structure, which in fact changes dramatically with the ion internal energy. This structure is a result of the kinetic energy of the product ions. We have then information not only about the state of the reactants but also the products. We are thus close to having state-to-state information. These results have been interpreted theoretically by Eaker and Schatz[23] using a surface hopping trajectory calculation. They conclude that some of the reactions leading to CID proceed via a relatively long lived H_3^+ intermediate.

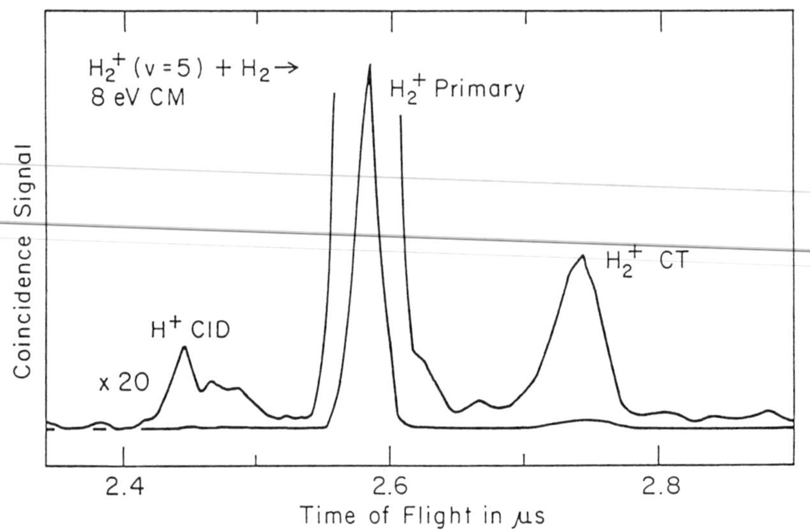

Fig. 8. A typical PEPICO TOF distribution for the products of the $H_2^+(v=5)$ + H_2 reaction at an ion laboratory energy of 16 eV. The collection time was about 20 min. Taken with permission from Guyon *et al.* ref. 22.

PEPICO STUDIES WITH SUPERSONICALLY COOLED SAMPLES

The internal energy distribution of molecules becomes increasingly broad as the molecular size increases. This is a direct result of the many low vibrational frequencies in large molecules. In the investigation of these molecules, the resolution of 5 meV attainable now at the synchrotron facilities with timing of electrons is of little use because the internal energy distribution of the starting neutral molecules ranges up to several hundred meV. However, it is possible to cool these vibrations by expanding the sample through a supersonic nozzle. To what extent can these samples be cooled? A comparison of the breakdown diagrams obtained with a sample at 298 K and a supersonically cooled sample provides a direct measure of the cooling.[24,25]

The dramatic effect of the supersonic expansion can be seen in the ion TOF distributions. Figure 9 shows the TOF distributions when iso-butane is photoionized at 10.95 eV photon energy. The two main features are the much narrower peaks when the gas is cold. From the peak width, we determine that the translational temperature (transverse to the direction of the molecular beam) is 3 K. The other feature is the near disappearance of the fragment ion peak. This is a direct result of the cooling of the internal degrees of freedom. At a photon energy of 10.95 eV, the cold sample has insufficient energy to dissociate, whereas a significant fraction of the molecules in the warm sample do.

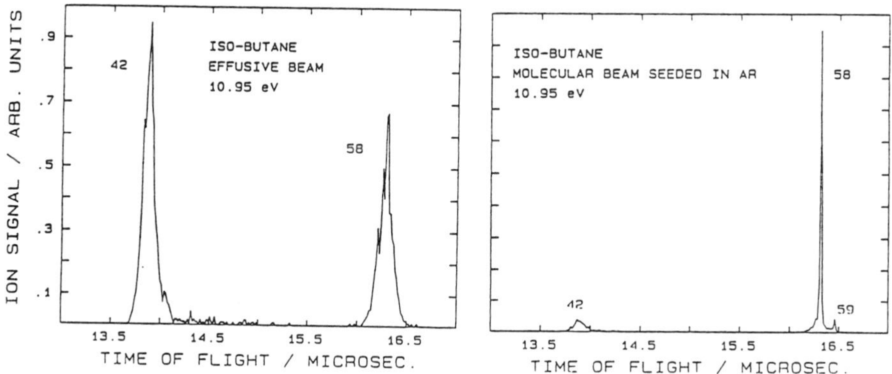

Fig. 9. The TOF distributions of isobutane gas under the two indicated experimental conditions.

When data such as shown in Figure 9 are collected at several ion energies, and the fractional abundances of the parent and daughter ion are plotted in the form of a breakdown diagram (Figure 10), the amount of cooling can be determined quantitatively. The shift in the cross over energy is a direct measure of the internal energy removed by the cooling process. The solid lines show the calculated breakdown diagrams, in which various experimental aspects, such as the contributions of energetic electrons and the thermal background gas (ca 25%) in the molecular beam experiment are taken into account. The analysis shows that the vibrational and rotational energies are reduced from an average of about 100 meV to less than 5 meV. The expansion conditions were 100 torr of sample gas diluted in 300 torr Ar. The nozzle diameter was 180 μm.

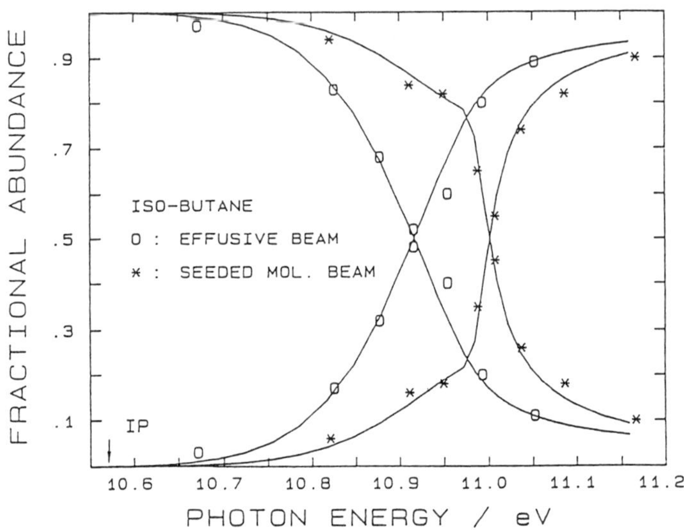

Fig. 10. The breakdown diagram of isobutane at 298 K and in the beam conditions. The solid lines are calculated breakdown diagrams. Taken with permission from Weitzel *et al.* ref. 24.

One of the interesting features of the TOF distributions in Figure 9 is that the parent ion in the beam experiment has an extremely narrow TOF distribution (ca 12 ns) compared to the thermal sample (ca 120 ns). Similarly, the fragment ions produced from cold parent ions have broad TOF distributions because of the inevitable release of translational energy. This feature makes it thus possible to distinguish readily the difference between a parent ion and a daughter ion simply from the width of the TOF distribution.

One of the big problems in the study of clusters is the uncertainty concerning the parentage of various cluster ions observed in the mass spectrometer. For instance, how can one distinguish the dimer ion, A_2^+ produced by the direct photoionization of the dimer neutral, A_2, from a similar mass dimer ion produced by the dissociative photoionization of A_3 + $h\nu \dashrightarrow A_2^+ + A$? The answer lies in the kinetic energy of the dimer ion. This can thus be readily determined in a TOF experiment if the TOF resolution is sufficient to distinguish them. The TOF distribution in Figure 11 shows an example for the case of CH_3OH dimers and protonated dimers. The latter peak is broad because it resulted from a dissociative reaction, while the very small, but narrow peak is the dimer ion which is clearly formed by direct photoionization of the dimer neutral. Figure 12 shows data for the ethylchloride dimer[23] ion which is very broad. The narrow peaks in the middle show what the parent ion should look like if it had been produced from the photoionization of ethylchloride neutral.

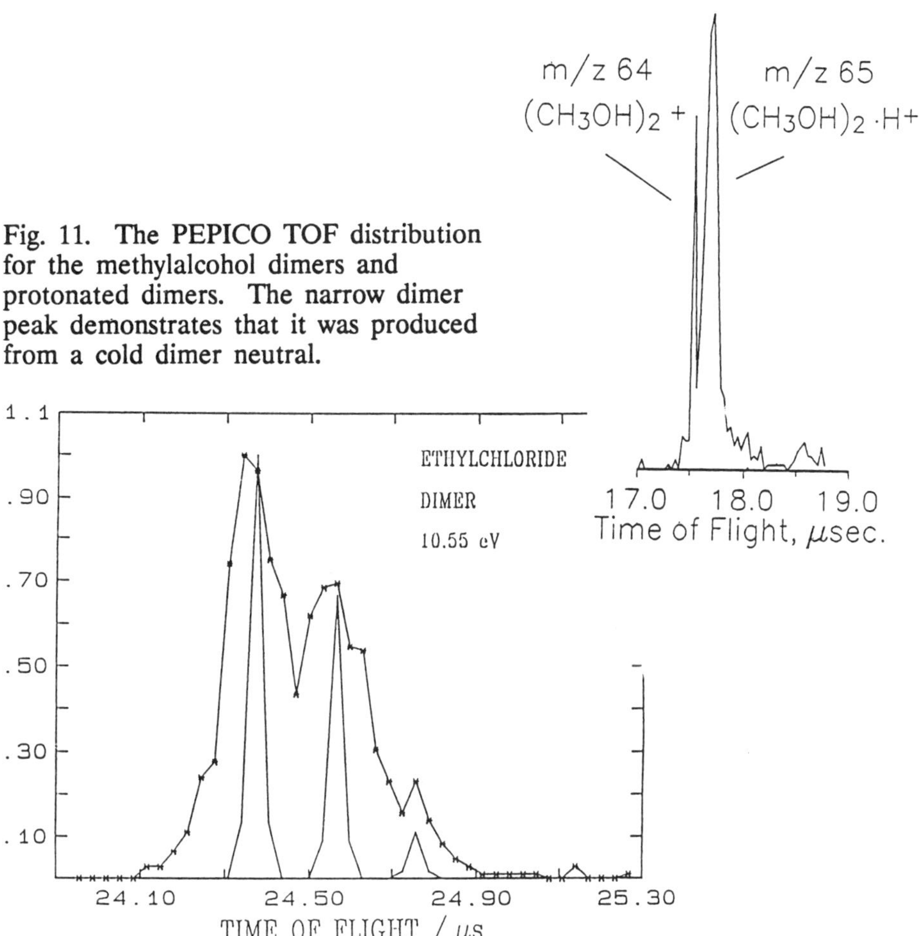

Fig. 11. The PEPICO TOF distribution for the methylalcohol dimers and protonated dimers. The narrow dimer peak demonstrates that it was produced from a cold dimer neutral.

Fig. 12. The PEPICO TOF distribution of the ethylchloride dimer ion. The peaks corresponding to the various Cl isotopes and broadened by kinetic energy. The narrow peaks in the middle are calculated to correspond to a cold dimer ion TOF distribution. The broad experimental peaks demonstrate that these dimer ions were produced from a dissociated ionization of a higher order cluster. Taken with permission from Weitzel *et al.* ref. 26.

ACKNOWLEDGEMENTS

I wish to acknowledge all of my co-workers who took part in the experimental work as well as the many exciting scientific discussions. Among these are my colleagues at LURE, Paul Marie Guyon, Odile Dutuit, and Tom Govers, and my collaborators in Chapel Hill, among them Jon Booze and Karl-Michael Weitzel.

REFERENCES

1. B. Brehm and E. von Puttkamer, Z. Naturforsch. Teil A **22** 8 (1967)
2. J.H.D. Eland, Int. J. Mass Spectrom. Ion Proc. **8** 143; **9** 397 (1972)
3. R. Stockbauer, J. Chem. Phys. **58** 3800 (1973)
4. P.R. Das, T. Nishimura, and G.G. Meisels, J. Phys. Chem. **89** 2808 (1985)
5. T. Baer, B.P. Tsai, D. Smith, and P.T. Murray, J. Chem. Phys. **64** 2460 (1976)
6. I. Nenner, P.M. Guyon, T. Baer, and T.R. Govers, J. Chem. Phys. **72** 6587 (1980)
7. H.M. Rosenstock, R. Stockbauer, and A.C. Parr, J. Chem. Phys. **73** 773 (1980)
8. T. Kato, K. Tanaka, and I. Koyano, J. Chem. Phys. **79** 5969 (1983)
9. K. Norwood, G. Luo, and C.Y. Ng, J. Chem. Phys. **91** 849 (1989)
10. T. Baer, Adv. Chem. Phys. **64** 111 (1986)
11. T. Baer, K.M. Weitzel, and J. Booze, in "Vacuum Ultraviolet Photoionization and Photodissociation of Molecules and Clusters" C.Y. Ng, Ed. World Scientific P 259
12. T. Baer, W.B. Peatman, and E.W. Schlag, Chem. Phys. Lett. **4** 243 (1969)
13. B.P. Tsai, T. Baer, and M.L. Horowitz, Rev. Sci. Instrum. **45** 494 (1974)
14. T. Baer, P.M. Guyon, I. Nenner, A.Tabche-Fouhaille, R. Botter, L.R. A. Ferreira, and T.R. Govers, J. Chem. Phys. **70** 1585 (1979)
15. P.M. Guyon, J. Hepburn, T. Weng, F. Heiser, and D. Reynolds, Phys. Rev. Lett. (1991) in press
16. T. Baer, O. Dutuit, H. Mestdagh, and C. Rolando, J. Phys. Chem. **92** 5674 (1988)
17. J.A. Booze, K.M. Weitzel, and T. Baer, J. Chem. Phys. **94** 3649 (1991)
18. J.C. Morrow and T. Baer, J. Phys. Chem. **92** 6567 (1988)
19. W.H. Miller, J. Am. Chem. Soc. **101** 6810 (1979)
20. O. Dutuit, T. Baer, C. Metayer, and J. Lemaire, Int. J. Mass Spectrom. Ion Proc. (1991) in press
21. S.K. Cole, T. Baer, P.M. Guyon, and T.R. Govers, Chem. Phys. Lett. **109** 285 (1984)
22. P.M. Guyon, T. Baer, S.K. Cole, and T.R. Govers, Chem. Phys. **119** 145 (1988)
23. C.W. Eaker and G.C. Schatz, J. Chem. Phys. **89** 6713 (1988)
24. K.M. Weitzel, J.A. Booze, and T. Baer, Chem. Phys. **150** 263 (1991)
25. K.M. Weitzel, J.A. Booze, and T. Baer, Int. J. Mass Spectrom. Ion Proc. **107** 301 (1991)
26. K.M. Weitzel, J.A. Booze, and T. Baer, Z. Phys. D Atoms, Mol. Clusters **18** 383 (1991)

DISCUSSION

RUHL - Was the breakdown graph of iso-butene corrected for contributions of hot electrons ?

BAER - The breakdown curve in the figure is exactly as collected. However, the hot electrons were taken into account in constructing the solid line.

AVALDI - Have you never considered the possibility to use the penetrating field technique to improve the collection of the threshold electrons and the rejection of "hot" electrons in your experiment?

BAER - This is a very good method for optimizing the detection of threshold electrons. On the other hand, we take great pains to have a reasonably large electric field (ca 10.V/cm) which is very constant in order to interpret the ion TOF distribution. For this reason, we sacrifice some resolution in the electron energy selection.

BRUTSCHY - Did you check in the results of iso-butene (I.B) the possibility of fragmentation of heterogeneous clusters (I.B).-Ar, by using for example He gas ?

BAER - The possibility of the participation of a "hidden" mixed cluster which would give rise to our ion signal is an intriguing one. I believe that our sharp signals (in TOF) for the parent iso-butene ion rules out this possibility for them. However, it is possible that the fragment ions were formed that way. We cannot rule if out, but I suspect that in these systems, these processes did not take place. However in the cluster study on CH_3OH and $C_2H_5C\ell$ such processes may well take part.

LEACH - (A) A few years ago you had studied C—H dissociation in $CH_3C\ell^+$ cations. The yields of $CH_3C\ell^+ \rightarrow CH_2C\ell^+ + H$ were very different for energy selected photon as compared with electron impact sources. Are these differences now understood?

(B) Concerning the effets of rotation on ion dissociation, it is not only the total energy that is important but also the particular J and K level distributions. This is clearly seen in the Kühlewind et al. experiments on benzene ion dissociation. With deuterium substitution a different range of J and K levels is populated in these experiments, as compared with $C_6H_6^+$, for an equivalent total internal energy, giving rise to different dissociation yields for isotopic benzenes.

BAER - (A) The problem of H loss from $CH_3C\ell^+$ still remains a mystery. By electron impact, H loss is seen at the thermochemical onset, whereas in photoionization, no such dissociation path is observed. Another example of a difference between EI and PI is in CH_3NO_2. By EI a strong metastable ion is observed, whereas in PI only fast dissociations are observed. There is no simple explanation for this.

(B) The role of rotations in dissociation is still largely unstudied. The Kühlewind results appear to raise as many questions as they answer. We need more studies of this sort. At the present time our experiments do not address this problem except in a very crude manner. By cooling the sample in a molecular beam, we cool both the vibration as well as the rotations. It is thus difficult to test the effect of rotational cooling on the dissociation rates.

BAUMGARTEL - Why is only one vibration taken to be important in the dynamic behaviour of the transition state ?

BAER - The transition state is precisely the ion configuration for which N-1 vibrational frequencies are real (that is the potential energy surface is concave upward) and 1 frequency is imaginary (the surface is concave downward). This is the definition of the saddle point. However, if you ask which frequencies in the molecular ion change in going to the transition state, the answer is that several different frequencies are important. So it really depends on your basis set. In calculating the transition state, we choose the normal modes of the transition state. With this basis set, only one frequency is the reaction coordinate.

DUJARDIN - In Iso-butene you showed that the fragmentation probability curves are shifted to higher energies when going from room temperature molecules to coded molecules. As you said this shift is most probably due to the thermal internal energy of room temperature molecules. However there may exist some cases (see for example the case of $Fe(CO_5)^+$ studied by A. Ding et al., in Berlin) where the shift is much higher than the thermal internal energy. This latter case may occur if the ion fragmentation is specifically induced by the rotational excitation of the room temperature molecule.

BAER - The situation in which rotational energy changes the reaction path so that the E_0 is lowered seems rather unlikely to me. As we discussed before, the observed shift in the $Fe(CO_5)^+$ onset may be larger than the average thermal energy because instead of using the crossover energy, the onset was chosen. These onsets are rather ill defined and depend very much on the signal to noise ratio.

WEITZEL - You showed data on the H-loss from di-methyl-ether. As a first guess one would expect that this dissociation is characterized by a loose transition state. You indicated that the transition state might not be entirely loose. Did anybody search from the transition state by variational transition state theory and what is known about it's "looseness".

BAER - This is a very interesting question. We have observed that H-loss onsets are often characterized by tight transition states. For instance in benzene the H-loss onset has a tighter transition state than C_3H_3 loss, even though the latter involves a rearrangement. More experimental and more theoretical work needs to be done on such systems.

ATOMIC SPECTROSCOPY WITH LASER AND SYNCHROTRON RADIATION

P. van der Meulen[a,b], E. de Beer[a], C.A. de Lange[a], N.P.C. Westwood[a,c] and M.O. Krause[b]

[a]Laboratory for Physical Chemistry, University of Amsterdam, Nieuwe Achtergracht 127, 1018 WS Amsterdam, The Netherlands.

[b]Oak Ridge National Laboratory, Oak Ridge, Tennessee 37831-6201, U.S.A.

ABSTRACT

Oxygen and nitrogen atoms which play a crucial role in many chemical processes are studied with photoelectron spectroscopy. In the case of oxygen synchrotron radiation is employed to study the inner-valence excitation region between 20 and 30 eV and detailed information about bound-continuum interactions between discrete Rydberg levels and various ionization continua is obtained. In the case of nitrogen laser multiphoton ionization is employed to study extensive novel Rydberg series below the lowest ionization energy which show appreciable local perturbations arising from bound-bound interactions.

INTRODUCTION

The study of atomic spectroscopy and dynamics of energy levels around and above the lowest ionization energy (IE) is an important area of research. Historically the experimental method for accessing this energy region has been through vacuum ultraviolet (VUV) absorption spectroscopy. Photons were generated in discharge lamps and were subsequently monochromatized, but experimentally the problems have always remained appreciable and VUV photoabsorption has never developed into a routine tool. At present much more versatile radiation sources such as lasers and synchrotrons are available which can be used to advantage in a wide range of spectroscopic techniques. In practice, these modern photon sources are employed in a complementary, rather than in a competitive fashion. In the energy range where pulsed lasers are routinely available (visible, UV) the radiation produced is characterized by a large degree of tunability and monochromaticity, often in combination with a large photon flux. Synchrotron radiation sources generate a lower photon flux per unit band

[c]On leave from the Guelph-Waterloo Centre for Graduate Work in Chemistry, University of Guelph, Guelph, Ontario, N1G 2W1, Canada.

width, but the energy range available (essentially from the X-ray to the infrared region) is completely out of bounds for present laser techniques. Also, photoabsorption is no longer the unique experimental tool it once was for atomic photophysics and photochemistry. Especially when quantum states located above the ionization limit come into play, photoelectron spectroscopy (PES) with excitation at continuously tunable photon energies and with kinetic-energy resolved and preferably also angle-resolved electron detection has proved to be a very powerful method for elucidating atomic photoionization processes which are a unique probe in obtaining fundamental information about the interaction between photons and matter. In this paper we shall show how photoelectron spectroscopy, on the one hand in combination with one-photon ionization processes initiated by synchrotron radiation, on the other hand in conjunction with resonance-enhanced multiphoton ionization (REMPI) processes stimulated by lasers with high pulse energies, can be employed to obtain detailed and novel information about simple first-row atoms.

For reasons of experimental convenience photoelectron spectroscopy of atoms for many years has concentrated on simple closed-shell systems. When the attention is focused on open-shell atoms, both experimentally and theoretically photoionization studies are far from straightforward. The first experimental difficulty one has to contend with is the generation of usually highly reactive open-shell atoms in a concentration sufficient for detailed spectroscopic study. Secondly, the contamination and corrosion problems associated with the handling of most open-shell atoms should not be taken lightly. Theoretically the calculation of reliable photoionization parameters such as cross sections (σ) and asymmetry parameters (β) is severely complicated by the large number of strongly coupled open channels for these systems. Hence, quantum chemical methods of a high degree of sophistication are required from the start. However, both experimentalist and theoretician should not yield too easily to the various difficulties associated with photoionization studies of simple open-shell atomic species. When research of this type can be conducted successfully, important information on the central issue of atomic science, *viz.* electron correlation, will be brought to light.

The present study is concerned with atomic oxygen and nitrogen. Both are open-shell species and highly reactive, and both play a central role as intermediates in a large variety of chemical reactions. Such reactions can take place in the earth's upper atmosphere (where O is particularly abundant) under the influence of solar radiation, or in a multitude of combustion processes where both atoms figure emphatically. Their occurrence in stellar outer layers is also common. Under our laboratory conditions we generate atoms either by means of a microwave discharge (oxygen) or *via* photodissociation of a suitable precursor (nitrogen). In view of their chemical and physical importance it may appear surprising that relatively little is known about the photoionization spectroscopy and dynamics of these key atoms.

Atomic oxygen is studied with angle-resolved photoelectron spectroscopy using synchrotron radiation. As an example we shall discuss a one-photon inner-valence 2s electron excitation into states in the energy range between 20 and 30 eV which subsequently decay through autoionization. It is well known that under these circumstances much of the photoionization dynamics is reflected in the line shapes which arise from an interaction between discrete Rydberg levels and the various photoionization continua into which these Rydberg levels can autoionize. Theoretically the subject of *bound-continuum* interactions is fraught with subtleties which have to be considered before reliable photoionization parameters can be extracted from the line shapes. The situation for oxygen atoms is treated in some detail.
Atomic nitrogen is studied with laser photoelectron spectroscopy by means of a (2+1) REMPI process. In the photodissociation process which was employed, and whose principal aim was to generate the interesting diatomic radical NH, N atoms were produced abundantly in the metastable $2p^3\ ^2P^o$ or $^2D^o$, rather than in the $2p^3\ ^4S^o$ electronic ground state. *Via* two-photon absorption from these metastable states novel 2p→np,nf Rydberg series converging to the lowest $2p^2\ ^3P^e_{2,1,0}$ ionic limits are reached which are difficult to access otherwise. In addition, an isolated interloper which is a member of a Rydberg series converging to the $2p^2\ ^1D^e_2$ ionic limit is seen to interact with the other Rydberg series. This perturbation reflects a *bound-bound* interaction between quantum states below the lowest IE, but still has profound effects on the photoelectron spectra which provide useful probes of the compound state character of the levels from which photoionization takes place. In this "atomic spectroscopy without really trying" the previous knowledge about Rydberg series in atomic nitrogen is extended considerably.

EXPERIMENTAL

The electron spectrometry with synchrotron radiation studies were carried out at the Aladdin storage ring in Madison, WI, USA. A detailed account of the experimental procedures has already been published elsewhere [1,2]. Only a brief summary will be given here.
Within the dipole approximation the photoionization process can be characterized by two parameters : the partial photoionization cross section σ_i, to which the measured intensity is proportional, and the asymmetry parameter β_i, which describes the angular distribution of the photoelectrons. Our electron spectrometry apparatus contains three electrostatic analysers mounted at right angles to each other on a platform that can be rotated in a plane perpendicular to the incoming photon beam, thus allowing ready determination of σ_i and β_i [1,2]. Experiments were performed in the Constant Ionic State (CIS) mode [1,2], combined with conventional photoelectron (PE) spectra for normalization and calibration purposes. Our relative partial cross sections were put on

an absolute scale using previously obtained electron spectroscopy modulation data at 21.22 eV [3].

A 1m Seya-Namioka monochromator equipped with a 1440 l/mm Os coated grating was used in the O(I) studies presented here. The monochromator bandpass was generally set to 0.7 Å. Both the monochromator and the electron spectrometer were calibrated using standard procedures [1,2].

Atomic oxygen was produced in a 2.45 GHz microwave discharge. Because only partial dissociation (20 %) was achieved, two sets of experiments were performed, *viz.* one set with the microwave discharge off and another with the discharge on. Interfering molecular signals could then easily be removed by a suitable subtraction procedure [1,2].

The experimental arrangement for the laser photoelectron spectroscopy studies has been described previously[4]. Briefly, the experiments were carried out using a frequency doubled, excimer pumped dye laser (Lumonics Inc.) with a pulse width of ca. 10 ns at a repetition rate of 30 Hz. The dye laser was operated with the Coumarin dyes 540, 500, 480 and 460. Frequency doubling, using a β-BaB_2O_4 (BBO) crystal, was employed to generate tunable radiation between 290 and 226 nm with a spectral width of ~0.5 cm^{-1}. The laser output was focused into the ionization region of a "magnetic bottle" spectrometer. The original design[5] has been modified to allow the study of reactive and transient species. With this spectrometer electron kinetic energies can be measured by means of a time-of-flight technique with 50% collection efficiency. Hence, this spectrometer provides much greater sensitivity than an electrostatic electron analyser, but by using the "magnetic bottle", information on the angular distribution of the photoelectrons is lost. In order to obtain equal energy resolution over a wide range of electron energies, a retarding voltage is applied in the flight tube and a conversion from time-of-flight to energy is performed only for the high-resolution part of the time spectrum. By scanning the retarding voltage the entire photoelectron spectrum is swept through this time-of-flight window. This procedure results in a constant resolution of 6-8 meV over a wide range of energies (~0.05 to 5 eV).

Metastable N atoms are produced from photodissociation of NH in its electronically excited $a^1\Delta$ state. NH ($a^1\Delta$) is, in turn, produced from photodissociation of HN_3, which is formed by gently heating a solid mixture of NaN_3 and stearic acid to ~380K.

ANGLE-RESOLVED PHOTOELECTRON SPECTROMETRY OF THE OXYGEN ATOM USING SYNCHROTRON RADIATION

It has long been known[6-7] that the broad and asymmetric line shapes of absorption lines associated with Rydberg levels above the first ionization energy are due to autoionization phenomena. The problem of configuration interaction between a single discrete Rydberg state and one or more ionization continua has been solved analytically by Fano and Cooper[8-10]. They showed that the photoionization cross section across an isolated resonance is given by :

$$\sigma(E) = \sigma_b + \sigma_a (\varepsilon + q)^2 / (1 + \varepsilon^2) \quad (1)$$

$$\varepsilon = (E - \zeta) / (\Gamma / 2) \quad (2)$$

In Eq. (1) $\sigma(E)$ is the energy dependent photoionization cross section, σ_a and σ_b represent the resonant and non-resonant part of the cross section, q is the line profile index and ε is the departure from the resonance energy ζ expressed in units of the half-width $\Gamma/2$. All five parameters are assumed to be independent of the energy E.

The configuration interaction theory of Fano and Cooper has been extended by Mies[11] to include the effect of interference between individual resonances through coupling to the same continuum. To describe the interaction between a discrete level n and a set of continua $\{\psi_{\beta E}\}$, we may introduce an effective continuum[8,11] Ψ_{nE} which is given by a linear superposition of continuum states $\psi_{\beta E}$:

$$\Psi_{nE} = (2\pi / \Gamma_n)^{1/2} \Sigma_\beta \vartheta_{n\beta} \psi_{\beta E} \quad (3)$$

where $\vartheta_{n\beta}$ is the configuration interaction matrix element between the (shifted) discrete state n and the continuum β, and Γ_n represents the width (FWHM) of autoionizing level n. The total interaction between n and the set $\{\psi_{\beta E}\}$is embodied in Ψ_{nE}. The remaining continua can be chosen to be orthogonal to both n and Ψ_{nE}, and are responsible for the background cross section (compare σ_b in Eq. (1)[8-10]).

If the effective continua Ψ_{mE} for all discrete states m are mutually orthogonal each resonance may be treated as a single isolated resonance[11]. However, strong interference effects may be expected when the effective continua associated with different bound states overlap. Therefore, the resonances are best classified according to the overlap matrix between their effective continua defined by :

$$\Theta_{nm} = \langle \Psi_{nE} | \Psi_{mE} \rangle = (4\pi^2 / \Gamma_n \Gamma_m)^{1/2} \Sigma_\beta \vartheta^*_{n\beta} \vartheta_{m\beta} \quad (4)$$

Γ_n can be expressed as a sum of partial widths according to :

$$\Gamma_n = \Sigma_\beta \Gamma_{n\beta} = 2\pi \Sigma_\beta \vartheta^*_{n\beta} \vartheta_{n\beta} \quad (5)$$

In the Fano notation each resonance state can be associated with a q_n factor determined by :

$$q_n = t^b_n / (\pi \Sigma_\beta \vartheta_{n\beta} t^c_\beta) \quad (6)$$

in which t_n^b and t_β^c stand for the transition matrix elements between the initial state and the bound state n and the continuum ß respectively. The energies ζ_n of the discrete levels can be obtained by diagonalizing the matrix which describes the couplings between the unperturbed bound levels due to their mutual interactions with the continua.

Often a superposition of isolated resonances (Eq. (1)) is used to parametrically represent the photoionization cross section across an autoionization manifold :

$$\sigma(E) = \sigma_b + \Sigma_n \, \sigma_{an} \, (\varepsilon_n + q_n)^2 \, / \, (1 + \varepsilon_n^2) \qquad (7)$$

$$\varepsilon_n = (E - \zeta_n) \, / \, (\Gamma_n \, / \, 2) \qquad (8)$$

However, it should be emphasized that Eq. (7) is only applicable to those cases where there is negligible interference between the resonances, i.e. if $\Theta_{nm} = \delta_{nm}$. When the continua do overlap the apparent widths, q parameters and resonance energies obtained by fitting an autoionization manifold to Eq. (7) may have no immediate relationship to the intrinsic atomic properties Γ_n, q_n, and ζ_n defined above, even though the quality of the fit, expressed as a least squares error, can be quite good. Hence, parametrizations as in Eq. (7) should be viewed with great caution.

A mathematically convenient parametrization of the cross section profile across an autoionization complex has been given by Shore[12-14] :

$$\sigma(E) = C + \Sigma_n \, (a_n \varepsilon_n + b_n) \, / \, (1 + \varepsilon_n^2) \qquad (9)$$

$$\varepsilon_n = (E - \zeta_n) \, / \, (\Gamma_n \, / \, 2) \qquad (10)$$

Although the scattering formalism employed by Shore in the derivation of Eq. (9) has been shown[11,15] to be in principle capable of correctly describing the interference effects discussed by Mies[11], Eq. (9) is only valid in the limit of no interaction between the discrete states[16]. Thus, like Eq. (7), Eq. (9) can only be used to describe the cross section across several, not necessarily well separated, bound levels which autoionize into orthogonal effective continua. The parameters C, a_n and b_n are related to the Fano parameters in Eq. (7) by :

$$C = \sigma_b + \Sigma_n \, \sigma_{an} \qquad (11)$$

$$a_n = 2 q_n \sigma_{an} \qquad (12)$$

$$b_n = (q_n^2 - 1) \sigma_{an} \qquad (13)$$

Alternative parametrizations can be obtained on the basis of the Multichannel Quantum Defect Theory (MQDT)[17-18]. Formula's similar to Eq. (1) have been derived[18] for the case of a single Rydberg series which interacts with one or more continua. In the limit of an isolated resonance Fano's formula, Eq. (1), is recovered[19].

Next we shall apply the theory of autoionization profiles outlined above to the O(I) $2s^22p^4\,(^3P^e) \rightarrow 2s2p^4(^4P^e)np$ autoionizing Rydberg series. In contrast to the O(I) $2s^22p^4\,(^3P^e) \rightarrow 2s^22p^3(^4S^o, {}^2D^o, {}^2P^o)ns,nd$ Rydberg series for which the basic properties were determined in a photoabsorption experiment back in 1967[20], experimental studies addressing the $2s \rightarrow np$ Rydberg series in atomic oxygen have been published only recently[1,2,21]. Here we shall show that the n=3 manifold of the $2s \rightarrow np$ Rydberg series can, in a good approximation, be considered as a superposition of isolated resonances and therefore Eqs. (7)-(13) can be used to obtain accurate values for ζ_n, Γ_n, and q_n.

Assuming L-S coupling is valid, in exciting a 2s electron only np states of $^3S^o$, $^3P^o$, and $^3D^o$ symmetry can be reached from the $^3P^e$ ground state. From these states the $np(^3S^o, {}^3D^o)$ levels are allowed to autoionize into the $^4S^o$ $\varepsilon s/\varepsilon d$ continua, while autoionization into the $^2P^o$ $\varepsilon s/\varepsilon d$ continua is limited to the $np(^3P^o, {}^3D^o)$ levels. All three $np(^3S^o, {}^3P^o, {}^3D^o)$ states are allowed to decay into the $^2D^o$ $\varepsilon s/\varepsilon d$ continua. Thus, omitting fine structure, two autoionizing Rydberg series can be expected in each of the $^4S^o$ and $^2P^o$ channels, whilst in the $^2D^o$ channel three series are anticipated. From the point of view of the resonance states we find that the $np(^3S^o, {}^3P^o, {}^3D^o)$ levels may autoionize into two, three, and four continua respectively.

The absolute photoionization cross sections for the $^4S^o$ and $^2D^o$ ionic states across the $2s \rightarrow np$ Rydberg series are shown in Fig. 1. As pointed out above, the first (n=3) member of the series shows evidence for the $^3S^o$ and the $^3D^o$ resonances in the $^4S^o$ channel, whereas in the $^2D^o$ channel all three $^3S^o$, $^3P^o$, and $^3D^o$ resonances are observed. It should be noted that the two resonances in the $^4S^o$ channel are much more pronounced in the spectrum of the asymmetry parameter across the n=3 resonance[1]. The individual members are no longer resolved for the higher members (n>3) of the series.

In the discussion of the O(I) $2s \rightarrow np$ autoionizing Rydberg series it is important to realize that resonances of different L-S symmetry do not interfere due to their interactions with the continua, or, expressed in terms of Eq. (4), $\Theta_{nm} = 0$ if n and m belong to different L-S terms. This means that the $^3S^o$, $^3P^o$, and $^3D^o$ resonances of a single $2s2p^4np$ configuration, in spite of their small separation, are merely superimposed. This is generally not so for resonances of identical symmetry belonging to different configurations, e.g. $np(^3S^o)$ and $(n+1)p(^3S^o)$. However, when the separation between those two states is much larger than the width of the individual resonances their interaction can, in good approximation, be neglected[11]. As can be seen

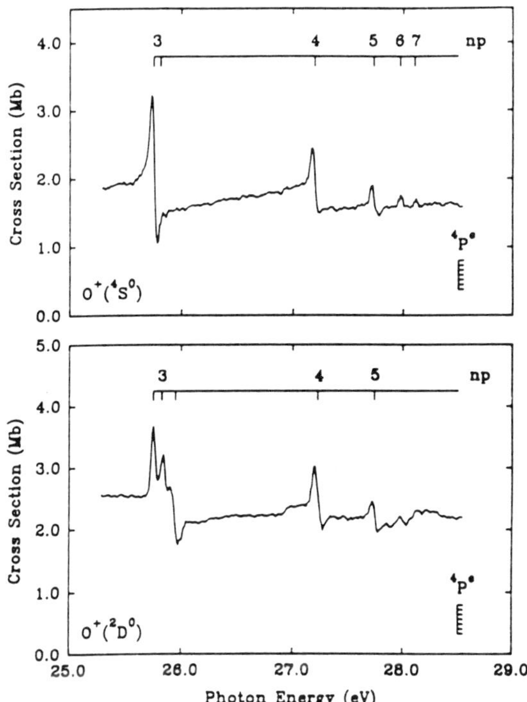

Fig. 1. Absolute photoionization cross sections for the $^4S^o$ (upper panel) and $^2D^o$ (lower panel) ionic states across the 2s → np autoionizing Rydberg series in atomic oxygen.

from Fig. 1 this is indeed the case for the n=3 and n=4 manifolds, certainly if one considers that the resonances in Fig. 1 are broadened by the finite bandpass of the monochromator. Thus, the n=3 manifold of the 2s → np Rydberg series can be accurately described by a superposition of two or three (depending on the ionic state studied) isolated resonances, and can, hence, be fitted using Eq. (9). The corresponding Fano parameters were obtained from Eqs. (11)-(13). The fit procedure employed a Simplex[22] algorithm, and the broadening of the lines by the monochromator slit function was accounted for by convoluting the profile calculated on the basis of Eq. (9) with a Gaussian of 0.71Å FWHM. Details have been described elsewhere[2]. The results of the fit are shown in Fig. 2 and in Table I.

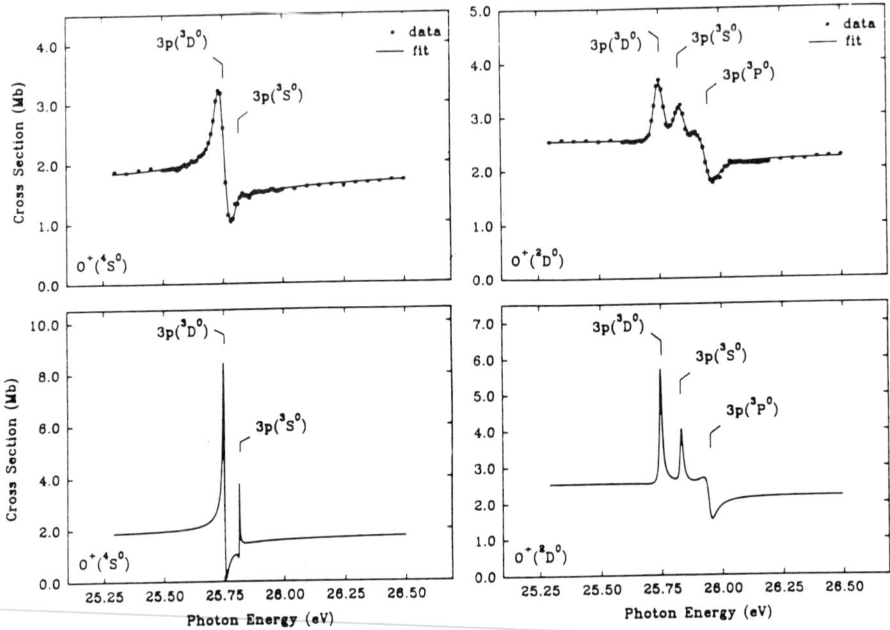

Fig. 2. Absolute photoionization cross sections for the $^4S^o$ (left) and $^2D^o$ (right) ionic states across the 2s → 3p ($^3S^o$, $^3P^o$, $^3D^o$) autoionizing resonances. The lower panels contain the profiles as calculated using Eqs. (9)-(10) and the parameters in Table I. The upper panels show the same profiles convoluted with a Gaussian of 0.71 Å, which represents the monochromator slit function, together with the experimental data points.

Table I. Resonance parameters for the 2s → 3p ($^3S^o$, $^3P^o$, $^3D^o$) autoionizing Rydberg states derived from the experimental profile. The numbers in parentheses are estimated probable errors which include the uncertainty in the description of the monochromator slit function, and, for the resonance energy, the energy calibration of the monochromator.

Ionic State	Resonance State	Energy (eV)	FWHM (meV)	q
$^4S^o$	$2s2p^4(^4P^e)3p(^3D^o)$	25.75(0.03)	8(1)	-2.1(0.1)
	$2s2p^4(^4P^e)3p(^3S^o)$	25.82(0.03)	2(1)	2.7(0.1)
$^2D^o$	$2s2p^4(^4P^e)3p(^3D^o)$	25.75(0.03)	13(1)	10(2)
	$2s2p^4(^4P^e)3p(^3S^o)$	25.83(0.03)	15(1)	4(1)
	$2s2p^4(^4P^e)3p(^3P^o)$	25.95(0.03)	32(1)	-0.7(0.1)

An interesting feature in Fig. 2 is the very small cross section in the minimum of the $3p(^3D^o)$ resonance at 25.755 eV for the $^4S^o$ ionic state. Because the only contributors to the $^4S^o$ partial cross section are the $^4S^o\varepsilon s$ ($^3S^o$) and the $^4S^o\varepsilon d$ ($^3D^o$) channels, and the cross secion for photoionization into the $^4S^o\varepsilon d$ ($^3D^o$) channel vanishes at the bottom of the $3p(^3D^o)$ resonance ($\varepsilon(^3D^o) = -q(^3D^o)$), we find that the cross section for photoionization into the $^4S^o\varepsilon s$ ($^3S^o$) is quite small. Thus, by far the largest contribution to the $^4S^o$ partial cross section comes from the $^4S^o\varepsilon d$ channel.

When our experimental results are compared[2] with the results of a calculation by Taylor and Burke[23], it appears that the calculated widths are about an order of magnitude too small. This discrepancy is even more pronounced when it is realized that the calculation of Taylor and Burke[23] pertains to the total photoionization cross section, so the calculated widths correspond to the total widths Γ_n in Eq. (5). On the other hand, our experimental widths should be identified with the partial widths $\Gamma_{n\beta}$ (or the sum of at most two partial widths). In general, information about the *partial* widths $\Gamma_{n\beta}$ and the *partial* cross sections $|t^c_\beta|^2$ and $|t^b_n|^2$ can only be obtained from photoelectron spectrometry experiments. Furthermore, the study of the angular distribution of the photoelectrons across an autoionization resonance[1] can provide important additional insight into these quantities. Photoabsorption or photoionization measurements will only yield the *total* width and the *total* photoionization cross section. As is clearly expressed in Eqs. (4)-(6), in order to completely characterize an autoionization complex, both the (partial) photoionization cross sections and the asymmetry parameters should be studied in all possible channels.

PHOTOELECTRON SPECTROSCOPY OF THE NITROGEN ATOM USING LASER RADIATION

High-lying excited states of the N atom have been studied by (2+1) REMPI from the $^2D^o$ and $^2P^o$ metastable states. Branching ratios for ionization into the energetically accessible ionic states have been investigated employing PES. A specific objective of these studies is to establish whether the state-selective production of N ions in their electronic ground or excited states is feasible.

N atoms are produced by photodissociation of NH ($a^1\Delta$) exclusively in their $2p^3$ $^2D^o$ and $^2P^o$ metastable states. No N atoms are detected in the $2p^3$ $^4S^o$ ground state as this is a spin forbidden transition.

Electronically excited Rydberg states of the N atom below the first IE have been studied by (2+1) REMPI-PES from both the $^2D^o$ and $^2P^o$ initial states in the wavelength region 290 to 226 nm. The photodissociation of the precursors occurs with the same laserpulse as the subsequent (2+1) REMPI process. The exact dissociation mechanism that leads to the formation of N atoms in their $^2P^o$ and $^2D^o$ states is unknown. However, it is clear from the present experiments that N atoms are produced effectively throughout the energy range studied, and that the photodissociation cross section does not vary drastically with wavelength.

Assuming L-S coupling is valid, the selection rules for two-photon absorption are *even-even, odd-odd,* $\Delta S=0$, $\Delta l=0,\pm 2$ and $\Delta J=0,\pm 1,\pm 2$; thus from our initial states two-photon transitions to doublet states of odd parity can be observed.

The J=3/2 and 5/2 levels of the N(I) $2s^2 2p^3\ {}^2D^o_{3/2,5/2}$ states are separated by 8.713 cm^{-1} [24], and transitions from the two J components are thus observed as separate lines in the REMPI spectrum. From the ${}^2D^o_{3/2,5/2}$ states the only accessible states within the available energy range and with the correct symmetry are the $({}^3P^e)3p$ and $({}^3P^e)4p$ states, where $({}^3P^e)$ refers to the ion core. These transitions are indeed observed in our REMPI spectrum.

The J=1/2 and J=3/2 components of the N(I) $2s^2 2p^3\ {}^2P^o_{1/2,3/2}$ metastable states are separated by only 0.386 cm^{-1} [24] and cannot be resolved in the present experiments. From the ${}^2P^o_{1/2,3/2}$ state the np and nf Rydberg series converging to the $2s^2 2p^2\ {}^3P^e_{0,1,2}$ limit are accessible by two-photon absorption. In the REMPI spectrum, the highest observed principal quantum number for the np series is n=25. The nf series are much weaker and can be observed to n=10. Many of these energy levels are observed for the first time. In the literature n=5 is the highest observed principal quantum number of the doublet $({}^3P^e)np$ Rydberg series[24]. This is, in part, due to the fact that doublet states of even parity are not accessible by one-photon absorption from the ${}^4S^o$ ground state.

In the region below the first IE (14.534 eV above the ground state[24]) there are only few states which belong to Rydberg series converging to an excited ionic state, namely the ${}^2F^o$, ${}^2P^o$ and ${}^2D^o$ states arising from the $({}^1D^e)3p$ configuration[24]. Higher members of these series have not been reported in the literature, but will lie above the first IE and outside the energy range studied. These $({}^1D^e)3p$ interlopers strongly perturb the $({}^3P^e)np$ Rydberg series around n=5. A limited Multichannel Quantum Defect Theory (MQDT) analysis has been performed to assess the degree of mixing and hence the relative contributions of $({}^3P^e)np$ and $({}^1D^e)3p$ character in the wavefunctions. To limit the number of interacting channels, it is assumed that around n=5, L and S are still good quantum numbers and that only states of the same L, S, and J can interact. This reduces the problem to a series of two channel MQDT analyses, resulting in wavefunctions of the following type:

$$\Psi_{L,S,J} = c\ \psi(({}^3P^e)np)_{L,S,J} + d\ \psi(({}^1D^e)3p)_{L,S,J} \qquad (14)$$

Photoelectron spectra have been recorded for many of the resonances observed in the REMPI spectrum. Depending on the initial state (${}^2D^o$ or ${}^2P^o$) and the photon energy one (${}^3P^e$) to three (${}^3P^e$, ${}^1D^e$ and ${}^1S^e$) electronic states of the ion are energetically accessible in (2+1) REMPI. The branching ratios for transitions to these ionic states can be determined by photoelectron spectroscopy and can, in principle, be related to the coefficients squared in the wavefunctions of the Rydberg states determined from the MQDT analysis. A quantitative comparison, however, between the wavefunction composition and the photoelectron branching ratios is not straightforward. For transitions to the $^3P^e$ and $^1D^e$ ionic states the intensities are proportional to

$$|< \Psi(^1D^e)\varepsilon l \,|\sum_j \mathbf{r}_j|\, c\,\psi\,((^3P^e)np) \;+ d\,\psi\,((^1D^e)3p) >|^2 \qquad (15a)$$

and

$$|< \Psi(^3P^e)\varepsilon l \,|\sum_j \mathbf{r}_j|\, c\,\psi\,((^3P^e)np) \;+ d\,\psi\,((^1D^e)3p) >|^2 \qquad (15b)$$

Cooper minima are known to play a role in the (2+1) REMPI of $N(^2D^o)$ atoms *via* the $(^3P^e)3p$ Rydberg state[25,26]. As similar states and electron energies are accessed in the present experiments, Cooper minima are also likely to play a role here. Therefore, the energy dependence of the cross sections is not a trivial matter and without detailed calculations only a very crude comparison between the wave functions and the photoelectron branching ratios is possible.

As an example, the branching ratios for ionization *via* the $2p^24p \leftarrow 2p^3(^2P^o)$, $2p^25p \leftarrow 2p^3(^2P^o)$ and the $2p^210p \leftarrow 2p^3(^2P^o)$ two-photon transitions will be discussed. The PE spectrum for ionization *via* the $(^3P^e)4p\ ^2P^o{}_{3/2}$ state is shown in Fig. 3A. In a (2+1) ionization scheme, the two lowest electronic states of the ion, the $^3P^e$ and $^1D^e$ states, are energetically accessible. The transition to the $^3P^e$ state is the strongest feature in the PE spectrum, but the transiton to the $^1D^e$ state is also observed. The intensity ratio is approximately 1 : 0.13. As can be expected for direct ionization of an unperturbed Rydberg state, the transition in which the ionic core is preserved, dominates. The splitting between the J = 0, 1 and 2 components of the $^3P^e$ ionic states is 6 and 10 meV, respectively. These components should thus, in principle, be resolvable in the PE spectrum, our instrumental resolution being 6-8 meV. The N atoms, however, get a significant degree of translational excitation in the photodissociation process which leads to a Doppler-type broadening of the PE signals of ~15 meV. Therefore, the three components of the $^3P^e$ level cannot be resolved, although the line is clearly broad.

The PE spectrum for ionization *via* the $(^3P^e)5p\ ^2P^o{}_{3/2}$ state is shown in Fig. 3B. The three lowest electronic states of the ion, the $^3P^e$, $^1D^e$ and $^1S^e$ states, are now energetically accessible. In the PE spectrum the transition to $^1S^e$ is not observed. Transitions to the $^3P^e$ and $^1D^e$ ionic states are observed as in the ionization via the $(^3P^e)4p\ ^2P^o_{3/2}$ state, but the branching ratios are very different. The $^3P^e$: $^1D^e$ intensity ratio of 0.94 : 1 is due to a strong perturbation by the $(^1D^e)3p$ state. This bound-bound interaction leads to appreciable energy shifts and strong mixing of wavefunctions.

From a two-channel MQDT analysis the mixing coefficients |c|=0.82 and |d|=0.57 (Eq.(14)) are obtained. Due to this mixing with the wavefunction of the interloper, which possesses a $^1D^e$ core, ionization into the $^1D^e$ ionic state is strongly enhanced.

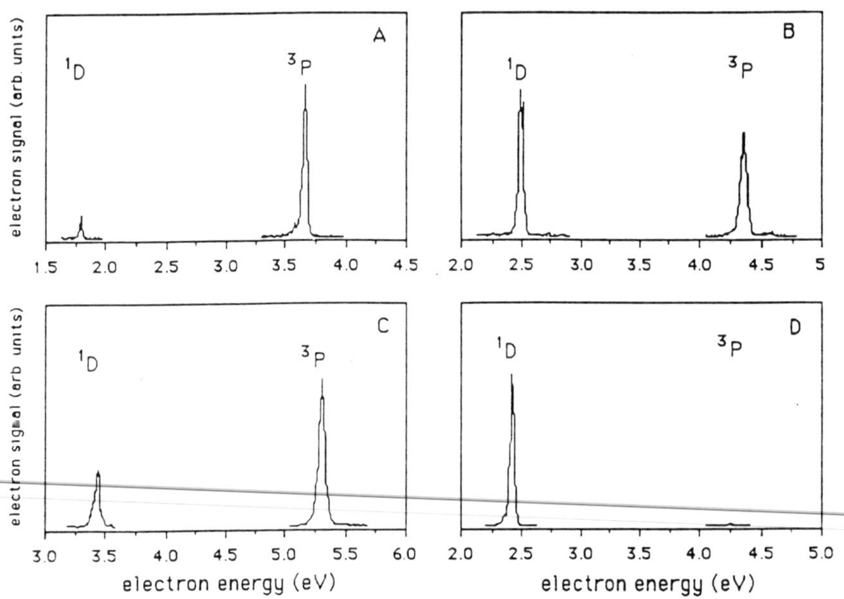

Fig. 3. Photoelectron spectra of the N atom (initial state $^2P^o$) indicating final $^3P^e$ ground and $^1D^e$ excited ionic states. A. *via* the intermediate $(^3P^e)4p\ ^2P^o_{3/2}$ state; B. *via* the intermediate $(^3P^e)5p\ ^2P^o_{3/2}$ state; C. *via* the intermediate $(^3P^e)10p\ ^2P^o_{3/2}$ state; D. *via* the intermediate $(^1D^e)3p\ ^2F^o_{5/2,7/2}$ states.

The PE spectrum for ionization via the $(^3P^e)10p\ ^2P^o_{3/2}$ state is shown in Fig. 3C. The three lowest electronic states of the ion, the $^3P^e$, $^1D^e$ and $^1S^e$ states, are energetically accessible. In the PE spectrum the transition to $^1S^e$ is not observed. The branching ratio for ionization into the $^3P^e$ and $^1D^e$ states is 1: 0.22.

As can be seen from the present PE results, photoionization of a Rydberg state cannot generally be described as removal of the Rydberg electron with complete core preservation. Even in the absence of perturbations, e.g. the $(^3P^e)10p$ Rydberg state, photoionization not only takes place to the $^3P^e$ ionic state, but core switching, leading to a sizable transition probability to the $^1D^e$ ionic state, is also apparent. This demonstrates the fact that for a Rydberg electron the photoionization probability is small during a large part of the orbit where the electron is far removed from the core, and only becomes appreciable when the electron interacts closely with the core. The degree of core interaction, exchange of angular momentum, and hence the amount of core switching, is strongly dependent on which Rydberg state is considered. When photoionization of the N atom occurs via the $(^1D^e)3p\ ^2F^o$ Rydberg state, whose PE spectrum is depicted in Fig. 3D, a high degree of state selectivity is observed. Following this route, N^+ ions are generated with high specificity (>95%) in their electronically excited $^1D^e$ state.

ACKNOWLEDGEMENTS

This work was supported in part by the U.S. Department of Energy, under Contract No. DE–AC05–840R21400 with the Martin Marietta Energy Systems, Inc. Financial support from the Netherlands Organization for Scientific Research and the North Atlantic Treaty Organization under Grant No. 0404/87 is also gratefully acknowledged. The Synchrotron Radiation Center is operated under National Science Foundation Grant No. DMR-8821625. The authors would also like to thank dr. W. Vassen for assistance in performing the MQDT analysis.

REFERENCES

1. P. v.d. Meulen, C. A. de Lange, M. O. Krause, and D. C. Mancini, Phys. Scr. **41**, 837 (1990).
2. P. v.d. Meulen, M. O. Krause, and C. A. de Lange, Phys. Rev. A **43**, 5997 (1991).
3. W. J. v.d. Meer, P. v.d. Meulen, M. Volmer, and C. A. de Lange, Chem. Phys. **126**, 385 (1988).
4. B. G. Koenders, D. M. Wieringa, Karel E. Drabe, and C. A. de Lange, Chem. Phys. **118**, 113 (1987).
5. P. Kruit and F. H. Read, J. Phys. E **16**, 313 (1983).
6. J. A. R. Samson, *Advances in Atomic and Molecular Physics* (Academic, New York, 1966), Vol. 2, p. 177.
7. G. V. Marr, *Photoionization processes in gases* (Academic, New York, 1967).
8. U. Fano, Phys. Rev. **124**, 1866 (1961).
9. U. Fano and J.W. Cooper, Phys. Rev. **137**, A1364 (1965).
10. U. Fano and J.W. Cooper, Rev. Mod. Phys. **40**, 441 (1968).
11. F. H. Mies, Phys. Rev. **175**, 164 (1968).
12. B. W. Shore, Rev. Mod. Phys. **39**, 439 (1967).
13. B. W. Shore, J. Opt. Soc. Am. **57**, 881 (1967).
14. B. W. Shore, Phys. Rev. **171**, 43 (1968).
15. H. Feshbach, Ann. Phys. **43**, 410 (1967).
16. G. V. Marr and J. M. Austin, J. Phys. B : At. Mol. Phys. **2**, 107 (1969).
17. M. J. Seaton, Rep. Prog. Phys. **46**, 167 (1983).
18. K. Ueda, Phys. Rev. A **35**, 2484 (1987).
19. J. P. Connerade, J. Phys. B : At. Mol. Phys. **16**, L329 (1983).

20. R. E. Huffmann, J. C. Larrabee, and Y. Tanaka, J. Chem. Phys. **46**, 2213 (1967).
21. G. C. Angel and J. A. R. Samson, Phys. Rev. A **38**, 5578 (1988).
22. J. A. Nelder and R. Mead, Comput. J. **7**, 308 (1964).
23. K. T. Taylor and P. G. Burke, J. Phys. B : At. Mol. Phys. **9**, L353 (1976).
24. S. Bashkin and J. A. Stoner, *Atomic Energy Levels and Grotrian Diagrams* (North-Holland, Amsterdam 1975) Vol.1.
25. S. T. Pratt, J. L. Dehmer, and P. M. Dehmer, Phys. Rev. A **36**, 1702 (1987).
26. S. T. Manson, Phys. Rev. A **38**, 126 (1988).

DISCUSSION

NAHON - a. Could you give us the dissociation ratio you can achieve in the dissociation of O_2 by microwave discharge ?

b. Have you tried to study the variations of $\frac{\sigma(O)}{\sigma(O_2)}$ vs $h\nu_{SR}$?

DE LANGE - a. For O_2 the degree of dissociation observed in the ionization region, which is located about 30cm downstream from the microwave discharge, is approximately 20%.

b. The ratio of the cross sections for atomic and molecular oxygen, i.e. $\frac{\sigma(O)}{\sigma(O_2)}$ has been determined in an electron spectroscopy modulation experiment at the He$I\alpha$ wavelength[1]. This value has been used to put our synchrotron results for atomic oxygen on an absolute scale. We have not explicitly measured this ratio as a function of photon eergy. However, some information about this ratio may be obtained from absolute atomic[2,3,4] and molecular[5,65,7,8] cross section data available in the literature.

1. W.J. van der Meer, P. van der Meulen, M. Volmer, and C.A. de Lange, Chem. Phys. **126** 385 (1986).
2. P. van der Meulen, M.O. Krause, and C.A. de Lange, Phys. Rev. A **43** 5997 (1991).
3. J.A.R. Samson and P.N. Pareek, Phys. Rev. A **31** 1470 (1985).
4. G.C. Angel and J.A.R. Samson, Phys. rev. A **38** 5578 (1988).
5. J.A.R. Samson, J.L. Gardner, and G.N. Haddad, J. Electr. Spectrosc. **12** 281 (1977).
6. C.E. Brion, K.H. Tan, M.J. van der Wiel, and Ph.E. van der Leeuw, J. Electr. Spectrosc. **17** 101 (1979).
7. K. Kimura, Y. Achiba, M. Morishita, and T. Yamazaki, J. Electr. Spectrosc. **15** 269 (1979).
8. J.A.R. Samson, G.H. Rayborn, and P.N. Pareek, J. Chem. Phys. **76** 393 (1982).

BESWICK - When using photodissociation to produce radicals should we consider possible alignment effects when analyzing the rotationally resolved photoelectron spectrum ?

DE LANGE - Yes, the relative population of the M_J levels does affect the ionic rotational distribution. In contrast to stable species, fragments formed by photo-dissociation will generally not have a statistical population of M_J levels. Both for $H_2O_2 \rightarrow 2$ OH and $HN_3 \rightarrow N_2 +$ NH photodissociation this distribution has been measured by sub-Doppler LIF measurements using a photolysis beam of 266 nm.[1,2] From these studies it is known that the alignment is low in both OH and NH. The alignment effects have, nevertheless, been taken into account in the calculations[3]. However, it should be realized that a magnetic bottle analyser is hardly suitable for angular resolved electron detection, which prohibits an independent measurement of the role of alignment effects.

1. K.H. Gericke, R. Theinl and F.J. Comes, J. Chem. Phys. **92** 6548 (1990).
2. S. Klee, K.H. Gericke, and F.J. Comes, Ber. Bunsenges, Phys. Chem. **92** 429 (1988).
3. E. de Beer, C.A. de Lange, J.A. Stephens, K. Wang, V. McKoy, J. Chem. Phys. **95** 714 (1991).

MORIN - Can you give evidence of two electron effects by analysis of Rydberg series converging to $4s^{-1}$ level as it is the case in Krypton for instance ?

DE LANGE - Two-electron effects appear to play an important role in the spectrum of atomic bromine just below the $4s^{-1}\,P^{\circ}_{2,1,0}$ thesholds. In particular, autoionization structure due to the $4s^2 4p^5 \rightarrow 4s^2 4p^3 nln'l'$ two-electron excitations was observed in the same spectral region as the $4s^2 4p^5 \rightarrow 4s^1 4p^5(^3P^{\circ}_{2,1,0})np$ Rydberg series. The autoionization profiles of the doubly excited states were found to depend strongly on the continuum channel. The quantum defect and the shape of the $4s^1 4p^5(^3P^{\circ}_{2,1,0})np$ resonances are very similar to that of the equivalent series in Kr. The results of our experiments on Br will be published in a forthcoming paper[1].

1. P. van der Meulen, M.O. Krause, and C.A. de Lange, J. Phys. B., 1991, in press.

LEACH - 1. The oxygen atom has some superexcited states, not far above the first ionization limit, which autoionize only very slowly. Indeed fluorescence is observed to be competitive with autoionization in these cases. Did you observe these weakly autoionizing states in your CIS type studies ?

2a. The photolysis of $NH(a^1\Delta)$ with photons of about 5eV should occur via excitation to the $C'\Pi$ state of NH which is only weakly bound and which dissociates to excited Nitrogen atoms ($N\ ^2D$) and $H\ ^2S$. However, the $C'\Pi$ state of NH may be partially mixed with the lower lying $A^3\Pi$ state. This could give rise to a partial exit channel to $N\ ^4S + H\ ^2S$. Did you observe this channel ? If so, it would be interesting to vary the wavelength of the photons used to dissociate $NH(a'\Delta)$; this would enable one to follow the coupling between the $c'\Pi$ and $A^3\Pi$ states of NH in the region above the dissociation limit of $NH(c'\Pi)$.

2b. Is the $b'\Sigma^+$ state of NH also formed in the photolysis of N_3H ? If so, since it is a metastable state, it could also possibly be photolyzed by $\approx$ 5eV photons if a suitable excited state existed. This might lead to $N^2P + H^2S$ products.

DE LANGE - 1. The $2s^2 2p^3(^2P^{\circ})ns(^3P^{\circ})$, $2s^2 2p^3(^2D^{\circ})nd(^3P^{\circ})$ and $2s2p^5(^3P^{\circ})$ resonances, which, assuming L-S coupling is valid, are forbidden to autoionize, were nevertheless clearly observed in our CIS spectrum[1]. However, their intensity seemed to be much reduced when compared with the high-resolution ion-yield measurement of Dehmer *et al*[2,3,4]. As a possible explanation we suggested[1] the presence of traces of molecular oxygen in their He discharge lamp which has been shown[5,6] to selectively enhance the population of the $^3P^{\circ}$ states.

For these $^3P^\circ$ states the autoionization rate is much reduced and fluoerscence competes effectively with autoionization. Consequently, the resonances can be observed in both fluorescence and photoionization experiments

1. P. van der Meulen, M.O. Krause, and C.A. de Lange, Phys. Rev. A **43** 5997 (1991).
2. P.M. Dehmer, J. Berkowitz, and W.A. Chupka, J. Chem. Phys. **59** 5777 (1973).
3. P.M. Dehmer and W.A. Chupka, J. Chem. Phys. **62** 584 (1975).
4. P.M. Dehmer, W.L. Luken, and W.A. Chupka, J. Chem. Phys. **67** 195 (1977).
5. J.A.R. Samson and V.E. Petrosky, J. Electr. Spectrosc. **3** 461 (1974).
6. D.M. de Leeuw and C.A. de Lange, Chem. Phys. **54** 123 (1980).

2a. No resonances associated with REMPI from the $N(^4S)$ ground state have been observed in the present study. It should, however, be noted that because one laser is used for both photolysis and subsequent REMPI, the $N(^4S)$ production can only be probed at very specific wavelengths. It is, in principle, possible to vary the dissociation wavelength over the region of interest while continuously monitoring the $N(^4S)$ production with a second laser. This would allow a study of the coupling between the $c^1\Pi$ and $a^3\Pi$states.

2b. Only NH in its a $^1\Delta$ state has been observed in our study. Rohrer and Stuhl have studied the $b(^1\Sigma^+)$ quantum yield at photolysis wavelengths of 193 and 248nm and found it to be very small in both cases[1]. A possible mechanism for the formation of $N(^2D^\circ)$ and $N(^2P^\circ)$ is a one- or two-photon absorption from the a $^1\Delta$ state accessing dissociative singlet states that lead to ground state $H(^2S^\circ)$ and nitrogen in its lowest doublet states $N(^2D^\circ)$ and $N(^2P^\circ)$.

1. F. Rohrer and F. Stuhl, J. Chem. Phys. **88** 4788 (1988).

<u>CAEULETTI</u> - Can you give us more details about the magnetic bottle spectrometer, precisely with regard to its resolution ?

<u>DELANGE</u> - With the magnetic bottle spectrometer electron kinetic energies are measured by means of a time-of-flight technique with ~50% collection efficiency. The constant energy resolution of 6-8 meV is achieved by applying retarding voltages. The original design[1] has been modified to allow the study of reactive and transient species. First, the pumping has been improved by differentially pumping the ionization region and the flight tube. Secondly, the ionization region has been made chemically resistant to volatile species by coating it with colloidal graphite. Also, such a coating prevents local charging effects of critical surfaces.

1. P. Kruit and F.H. Read, J. Phys. E **16** 313 (1983).

<u>WONG</u> - 1. What are he intrinsic lifetimes of O, N, Cl and Br ?
2. What is a typical measurement time of your PE spectrum ?

<u>DE LANGE</u> - 1. The transport time between the discharge and the ionization region is of the order of 50msec thereby putting a lower limit on the lifetime of the O, Cl and Br atoms generated in the microwave discharge.

2. The typical measurement time of a PE spectrum is about 600 seconds.

PHOTOIONIZATION AND PHOTODISSOCIATION STUDIES OF SMALL MOLECULES USING THE MAX SYNCHROTRON RADIATION

L.-E. Berg, P. Erman, E. Källne, S. L. Sorensen and G. Sundström
Department of Physics I, Royal Institute of Technology,
S-100 44 Stockholm, Sweden

ABSTRACT

The Swedish synchrotron facility MAX in Lund has a high flux (10^{12} ph/s · mA · mrad·0.1% band width) in the region 400-600 Å and is therefore well suited for photoionization studies of free atoms and molecules up to about 35 eV. Accordingly we have initiated a series of studies of properties of small molecules using this facility using in the first place mass spectroscopic tools. In the present report we will give some examples of results from this series of measurements including photoionization of molecular nitrogen and oxygen where the ions $N_2^{\pm}$, N^+ and $O_2^{\pm}$, $O^{\pm}$ have been studied in the range 350 - 2000 Å. Very rich line structures are observed in the mass spectra following autoionization and dissociation of singly and doubly excited Rydberg states in N_2 and O_2 as well as ion pair formation in the case of O_2.

INTRODUCTION

A great amount of information concerning the structure and dynamics of molecules may be collected from studies of mass selected ions and ionic fragments formed via photoionization using tunable photon sources. In the region $\lambda > 600$ Å where helium Hopfield continuum sources may be used, photoionization of small molecules have been extensively studied. However, for higher energies where synchrotron radiation has to be used, the experimental photoionization data are considerably more scarce and frequently restricted to narrow wavelength region. Since our present equipment at the MAX synchrotron enables mass spectroscopy at a photon resolution of 0.4 Å FWHM at target pressures of about 10^{-5} Torr (i. e. collision free conditions), it was decided to perform photoionization studies of small fundamental molecules such as N_2, O_2, CO_2 and NO in the first place in the range 350 - 2000 Å.

EXPERIMENTAL

The new normal incidence beam line at MAX[1] includes a 1 m monochromator with a 1200 ℓ/mm grating (Bausch and Lomb) blazed at 800 Å. At a slit width of 50 μm a line width of 0.4 Å is obtained in second order. A 50 cm long movable glass capillary tube (inner diameter 3 mm) is used to guide the synchrotron light from the exit slit of the monochromator to the target chamber. Together with an efficient pump system this arrangement allows to maintain a pressure of about 10^{-9} Torr in the monochromator at a target pressure of 10^{-5} Torr. The ions formed in the target region are conducted to the entrance of a quadrupole mass spectrometer (VG SXP 300) by an electrostatic mirror. Standard detectors (a GaAsP diode for the photons and a channeltron for the ions) are used together with pulse counting electrics, combined with a mini computer to acquire and store the data. Typical count rates after the capillary tube are 10^9 photons/s using a 50 μm slit yielding about 10^4 molecular ions/s. These high count rates allow recordings of spectra in second order of diffraction in a short time (typically 10 Å/min).

PHOTOIONIZATION AND PHOTODISSOCIATION OF N_2

The ionization potential (I.P.) of N_2 is 15.58 eV (795,8 Å) and the first structures in the N_2^+ mass spectrum appear above this limit and originate from autoionization of the lowest Rydberg levels in N_2 above the I.P. which converge to the X (v'=1) and A (v'=0-4) states in N_2^+. At 18.75 eV (661 Å) the B (v'=0) state in N_2^+ is reached (Fig. 1) and the N_2^+ spectrum[2] in the region 660 - 700 Å (Fig. 2) is dominated by autoionization of N_2 levels in the three series converging to the N_2^+ B state, i. e. the Hopfield "apparent" emission series and the Ogawa Tanaka absorption series. As seen from Fig. 2 the emission series is a beautiful example of a Fano-Beutler resonance where a destructive interference between the different ionization channels causes a deep minimum in the photoionization cross section. Such resonances have been well studied in particular in

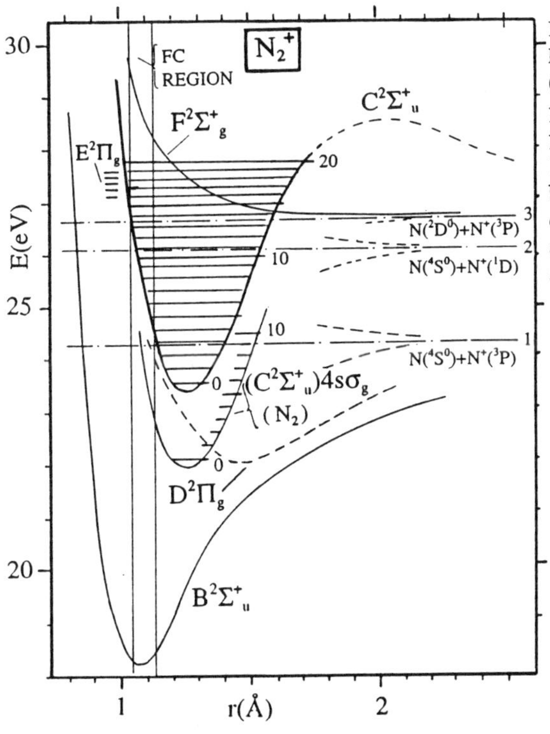

Fig. 1. Potential curves of N_2^+ in the vicinity of the C state constructed from photoelectron data. The potential curve of the doubly excited $(C\ ^2\Sigma_u^+)4s\sigma_g$ Rydberg state in N_2 has been constructed assuming the same shape as the N_2^+ C state. (from ref[2])

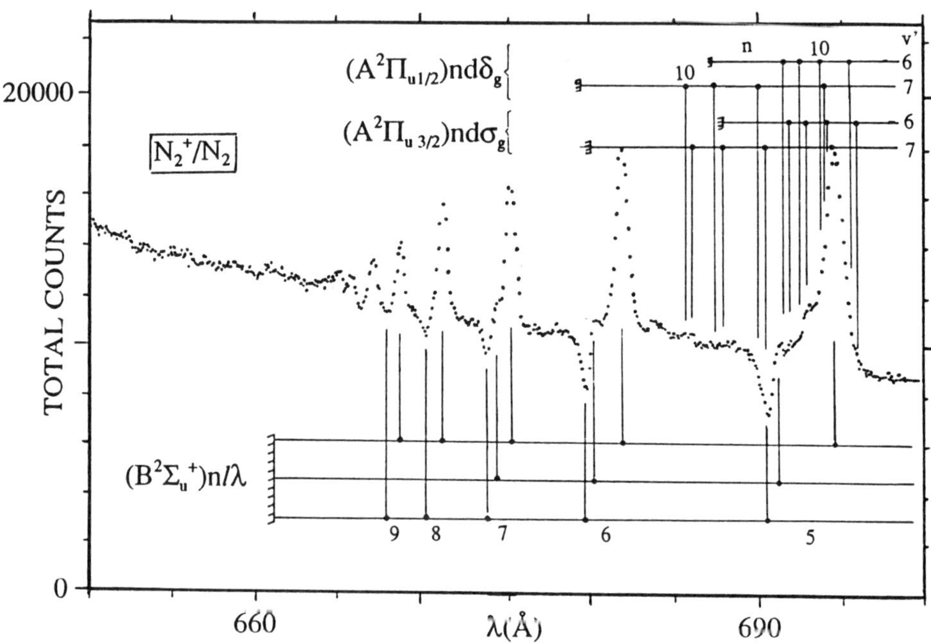

Fig. 2. Photon induced production rate of N_2^+ ions from N_2 in the range 650 - 700 Å. (from ref[2])

the case of rare gases and also for highly excited states in small molecules. In the example displayed in Fig. 2, the n = 5 - 8 members of the emission series may be used to evaluate the Fano parameters[3] in the line profile formula

$$\sigma(E) = \tau_c + \tau_r(q^2 - 1 + 2q\varepsilon)/(1 + \varepsilon^2) \quad (1)$$

where $\varepsilon = (E - E_0)/\Gamma/2$ (Γ = half-width) is the reduced energy, σ_c the continuum strength, and τ_r that of the resonance. A fit to Fig. 2 gives $q^2 \sim 0.25$ and $\Gamma \sim 0.024$ eV.

Fig. 3 shows a section of our N^+ fragment spectrum[2] formed from photon excitation of N_2. The production of N^+ starts at the first dissociation limit at 510.4 Å and a progression of peaks appear at energies corresponding to the N_2^+ C (v'>3) levels (cf Fig. 1). These peaks should be formed in the well-known predissociation of the N_2^+ C state (cf ref[4]). However, Fig. 3 contains at least one more progression with spacings similar to those of the above mentioned fragments from the C state. Since the strongest of the peaks coincide energetically with the doubly excited $(C^2\Sigma_u^+)4s\sigma_g$ "Codling" levels (Fig. 1) they should originate from photodissociation of these levels. Our N_2^+ spectrum[2] in the region 500-540 Å shows that the "Codling" levels may also decay via autoionization.

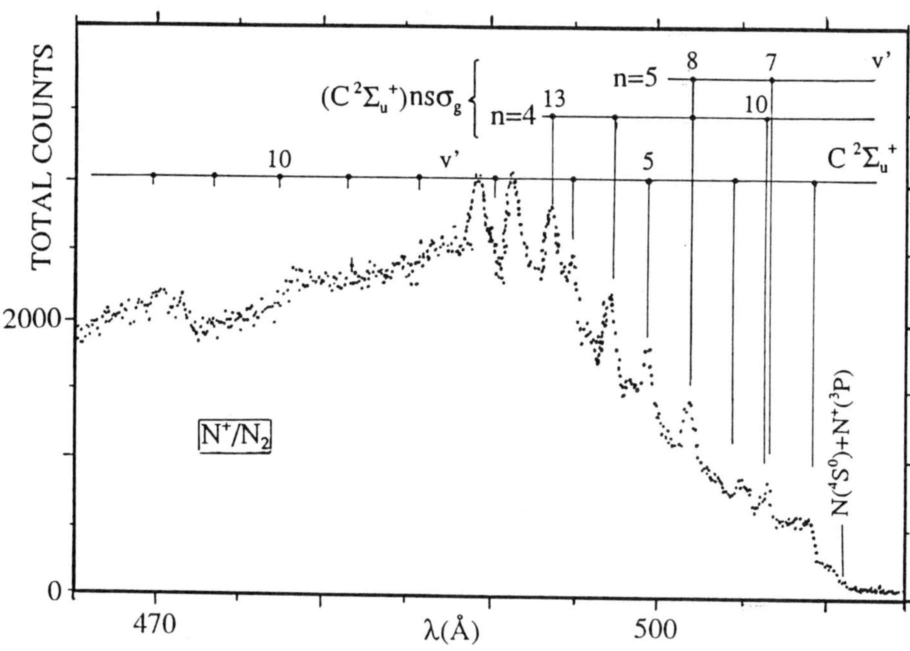

Fig. 3. Photon induced production rate of N^+ fragment ions from N_2 in the range 465 - 515 Å. (from ref[2])

PHOTOIONIZATION AND PHOTODISSOCIATION OF O_2

The first structures in the O_2^+ mass spectrum above the first ionization limit at 1028 Å follow from autoionization of Rydberg series in O_2 converging to the lowest doublet (A, B) and quartet (a, b) states in O_2^+ (Fig. 4). In the region 500 - 600 Å autoionization is observed of two Rydberg series which converge to the C state in O_2^+. As seen from our spectrum[5] (Fig. 5) one of these appears as an apparent emission window series with strongly broadened peaks which might be fitted to the Fano formula (1). For n = 3 is obtained $q^2 \sim 0.55$ and $\Gamma \sim 0.081$ eV. The Fano resonances mentioned above are seen even more clearly in the O^+ fragment spectrum (Fig. 6) which should be a consequence of predissociations of the actual Rydberg states converging to the $c(O_2^+)$ state. Concerning the strong, broad peak around 558 Å (Fig. 6) we tentatively assume that this originates from photodissociation of a Rydberg series in O_2 converging to the highest $(1\pi_u)^3\ (1\pi_g)^2\ {}^2\Pi_u$ state in O_2^+.

As follows from Fig. 4 there are well known low lying ion pair states in O_2. In fact we find the first traces of O^+ and O^- fragments at 17.27 eV well below the first dissociation limit of O_2^+ (= 18.73 eV). A rich line spectrum of $O^\pm$ is observed which indicates that the fragmentation takes place via discrete O_2 Rydberg states rather than by a transition from the O_2 ground state via the continuum of the actual ion pair state. It could be added that analogous ion pair states in N_2 were searched for but not found. Further improvements on the normal incidence beam line at MAX are in progress including the installation of 2400 ℓ/mm grating and a photon detection channel as well as an electron spectrometer thus making possible a simultaneous recording of ions, fluorescent photons and photoelectrons.

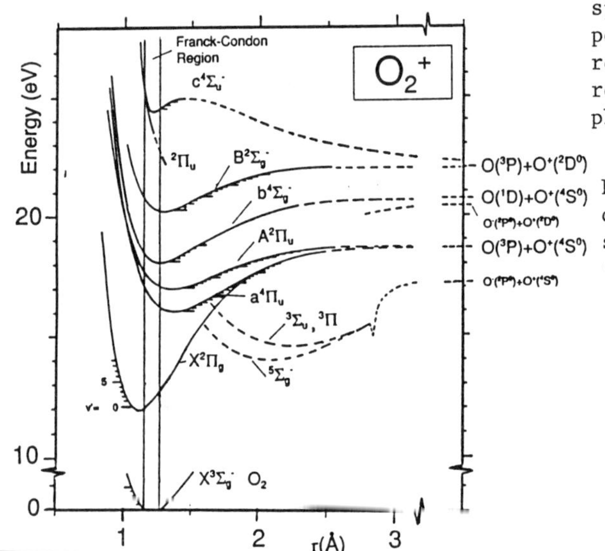

Fig. 4. Potential curves of low lying states in O_2^+. (from ref[5]).

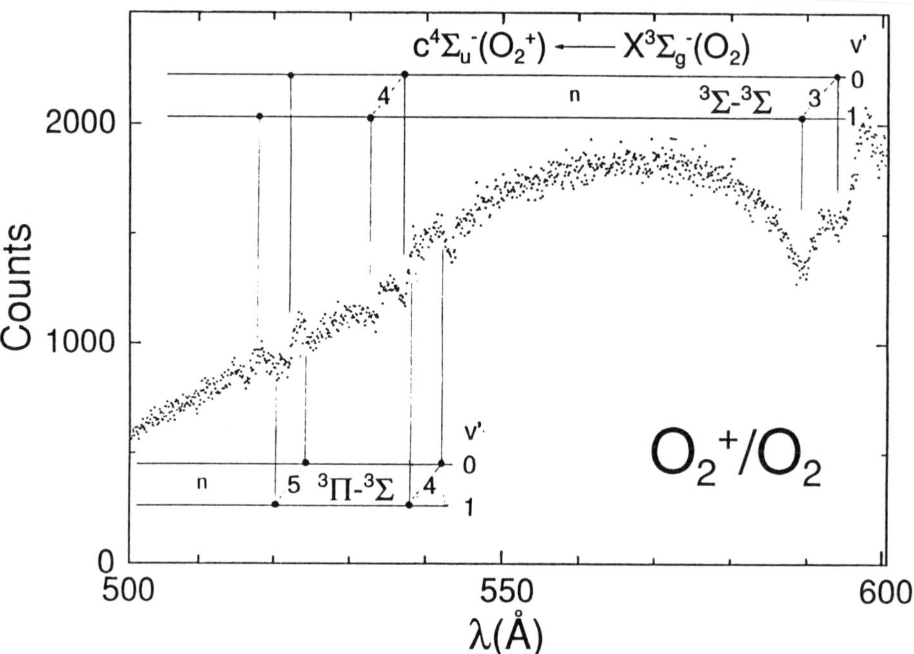

Fig. 5. Photon induced production rate of O_2^+ ions from O_2 in the range 500 - 600 Å. (from ref[5])

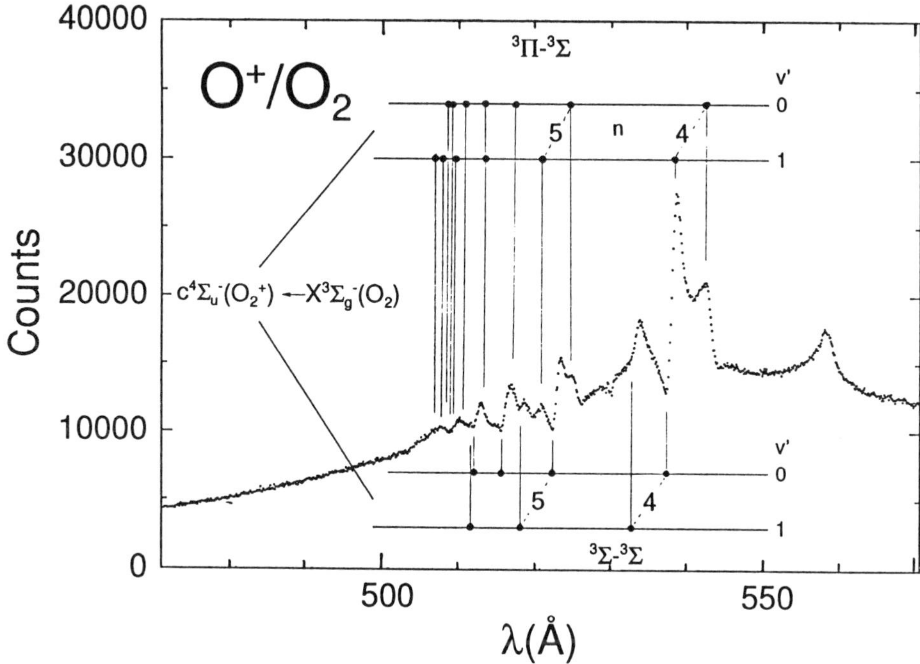

Fig. 6. Photon induced production rate of O^+ ions from O_2 in the range 500 - 600 Å. (from ref[5])

REFERENCES

1. S. L. Sorensen, B. J. Olsson, O. Widlund, S. Huldt, S.-E. Johansson, E. Källne, A. Nilsson, R. Hutton, U. Litzén and A. Svensson, Nucl. Instr. Meth. A 297, 296 (1990).
2. L.-E. Berg, P. Erman, E. Källne, S. L. Sorensen and G. Sundström, Physica Scripta 44, 131 (1991).
3. U. Fano, Phys. Rev. 124, 1866 (1961).
4. P. Erman, Physica Scripta 14, 51 (1976).
5. L.-E. Berg, P. Erman, E. Källne, S. L. Sorensen and G. Sundström, Physica Scripta (in press).

SINGLE AND DOUBLE PHOTOIONIZATION OF AROMATICS : A COMPARATIVE STUDY OF C_6H_6 ISOMERS BETWEEN 25 AND 70 eV

Ori BRAITBART[1], Seiji TOBITA[2,3], Sydney LEACH[2,4], Pascale ROY[5] and Irène NENNER[1,5]

1 DRECAM/SPAM, C.E.A. Saclay, 91191 Gif sur Yvette, France
2 Laboratoire de Photophysique Moléculaire du C.N.R.S., Université Paris-Sud, 91405-Orsay, France
3 Gunma College of Technology, 580 Toriba, Maebashi, Gunma, Japan
4 DAMAP, Observatoire de Paris-Meudon, 92195-Meudon, France
5 LURE, Université Paris-Sud, 91405-Orsay, France

Abstract

Monochromatized synchrotron radiation from the LURE-ACO storage ring was used to excite and ionize three C_6H_6 isomers in the 25 - 70 eV photon energy range. The compounds studied were the cyclic species benzene and the two-open chain molecules 1,5-hexadiyne and 2,4-hexadiyne. Four types of measurements are described which involve a high resolution reflectron TOF mass analyzer. Half-integer and ^{13}C isotope mass peaks in the PEPICO TOF-MS provided information on parent and fragment dications. From the PEPIPICO spectra, the correlated times of flight values for the two monocations originating in the Coulomb dissociation of a dication gave information on the principal fragmentation pathways. Similarities and differences in the various spectra provide insight into dynamic processes, including structural rearrangements, which follow single and double ionization in these species, but require more analysis and also modelisation of energy deposition for further clarification.

INTRODUCTION

The principal questions raised by previous studies of the fragmentation of monocations[1-3] and dications[2,4-7] of C_6H_6 isomers concern (i) the structure of the parent mono- and di-cations, (ii) possible isomerization to common ionic structures before fragmentation, and (iii) the mechanisms of metastable ion dissociations. As compared with monocations, additional problems arise in dications from the existence of two types of fragmentation: Coulomb dissociation into two monocations, and covalent dissociation into a dication fragment plus neutral(s)[8]. Interest in the photostability of aromatic hydrocarbon mono- and di-cations has also arisen from possible astrophysical relevance[8,9].

Most laboratory studies on the isomeric $C_6H_6^{n+}$ (n=1,2) ions have been carried out by electron impact methods and various mass spectrometric techniques. Collisional activation and charge transfer methods have also been used. Only a few studies have used photon impact as the means of ion formation. Results from the various techniques have often given different answers to the key structural and dynamic problems enumerated above, both for mono- and di-cations of isomeric C_6H_6. Conflicting results and interpretations most probably arise from changes in the ionic structures produced at the different excitation conditions in the various experiments. The different distributions of energy deposited in the newly formed mono- or di-cation may lead to modified distributions of ionic structures each of which could have its own fragmentation pattern.

To clarify some of these questions requires varying the excitation energy and starting with different isomers. The present study on the cyclic compound benzene, and the open chain species 2,4-hexadiyne and 1,5-hexadiyne uses monochromatized synchrotron radiation to that end, over the excitation range 25-70 eV. The experimental techniques used include time-of-flight mass spectrometry (TOFMS), photoelectron-photoion-photoion coïncidences (PEPIPICO), and photoelectron-neutral-photoion coïncidences (PENPICO), described in the next section.

EXPERIMENTAL

Synchrotron radiation over the energy range 25-70 eV was provided by the super-ACO-LURE facility. A toroidal grating monochromator, resolution 0.1 nm, was used to disperse the light.

The sample vapour was introduced into the reaction chamber as an effusive jet. Crossed beam interaction with the exciting radiation occurs at the centre of a strong, uniform field (fig. 1). This serves as the ionization source of a reflectron-type time-of-

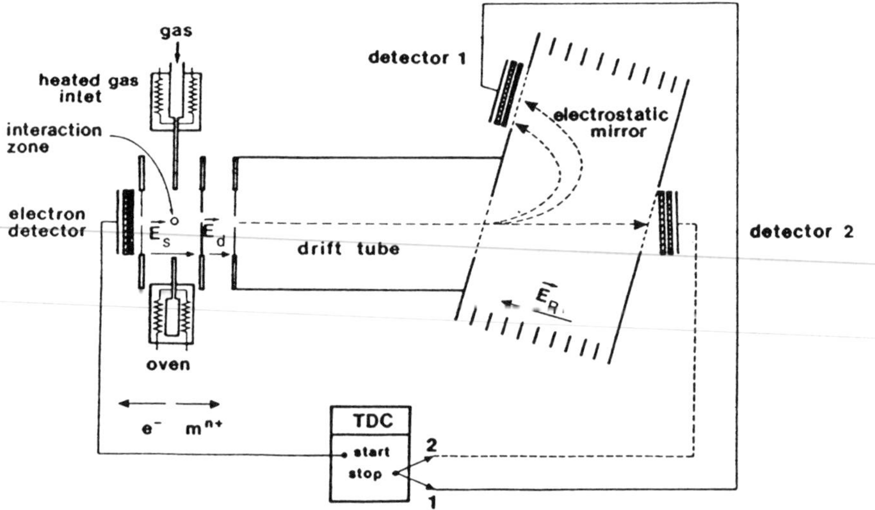

Figure 1. Schematic view of the reflectron-type time-of-flight mass spectrometer

flight mass spectrometer(10) whose principles are described elsewhere(11). This instrument not only has good ion mass resolution at the reflex detector (≈ 600 in our experiments), but it is also capable of detecting neutral fragments striking the straight-through detector. Both detectors are microchannel plates. The length of the ionization zone is 10 mm and the distance between the ion acceleration plate and the entrance to the drift tube is 5 mm. The length of the drift tube is 460 mm and the width of the electrostatic mirror is 153 mm. The reflector angle is 7.73°. Typical potentials are

+1600/+1700 V on the electron plate, -1600 V on the ion plate, -3000 V on the drift tube and +2000/+2300 V on the electrostatic mirror. It should be noted that the detected electrons are not energy analyzed and are used as a zero marker for measuring absolute times of flight for ions detected in (delayed) coïncidence.

The instrument can be operated in the following multi-coïncidence[12,13] modes:

(i) Photoelectron-photoion coïncidences (PEPICO). This provides a "conventional" time-of-flight mass spectrum i.e. of both single and multiply-charged cations.

(ii) Photoion-photoion coïncidences (PIPICO), which give time-of-flight spectra of correlated pairs of monocations issuing from Coulomb dissociation of molecular dications.

(iii) Photoelectron-photoion-photoion coïncidences (PEPIPICO). This triple coïncidence mode measures absolute TOFs for each ion and, in principle, provides identification of correlated ion pairs, as well as information on the mechanisms and dynamics of charge separation reactions, including metastable decompositions of molecular dications.

(iv) Photoelectron-neutral-photoion coïncidences (PENPICO). Metastable molecular ions (mono- or di-cations) may fragment in the drift tube, or in the reflex part of the spectrometer, to produce a fragment ion and a neutral ($AB^{2+} \rightarrow A^{2+} + B$; $AB^{+} \rightarrow A^{+} + B$). The daughter ion, detected by the reflex detector, and the neutral, are both born with the velocity of the precursor ion. The momentum acquired by the neutral is sufficient for a signal to be produced on striking the straight-through detector.

Using different field conditions, modes (i), (ii) and (iii) can be used either with the reflex detector or the straight-through detector. The latter variant enables one to measure the kinetic energy release associated with a particular dissociation. Such measurements cannot be done easily with the reflex detector since the time-of-flight of ions in this mode is a complex function of their kinetic energy.

In the PENPICO mode the electric fields are normally adjusted so as to fulfill the Wiley-McLaren time focussing conditions[14] for the ions detected on the reflex detector. This gives optimal mass resolution for the ions but poorer resolution for the neutrals since the neutral particle focussing conditions are then not satisfied. Identification of the neutral species is not directly afforded by the PENPICO spectra but in some cases it can be deduced by detailed analysis of the shapes and centers of gravity of the two-dimensional t_1(neutral),t_2(ion) time-of-flight patterns[15].

The results reported here concern only the PEPICO and PEPIPICO modes. PENPICO measurements of metastable dissociations in benzene are reported elsewhere[15].

RESULTS AND DISCUSSION

Time of flight mass spectra

PEPICO spectra of benzene were measured at six photon excitation energies, respectively 25, 36, 40.8, 45, 50 and 70 eV. The linear isomers were studied at 35 and 70 eV. For 1,5-hexadiyne measurements were also made at 1 eV intervals between 25 and 35 eV. Only a portion of the PEPICO results is reported here.

Figure 2 shows the TOF-MS for benzene (36 eV), 1,5 hexadiyne (35 eV), and 2,4-hexadiyne (35 eV).

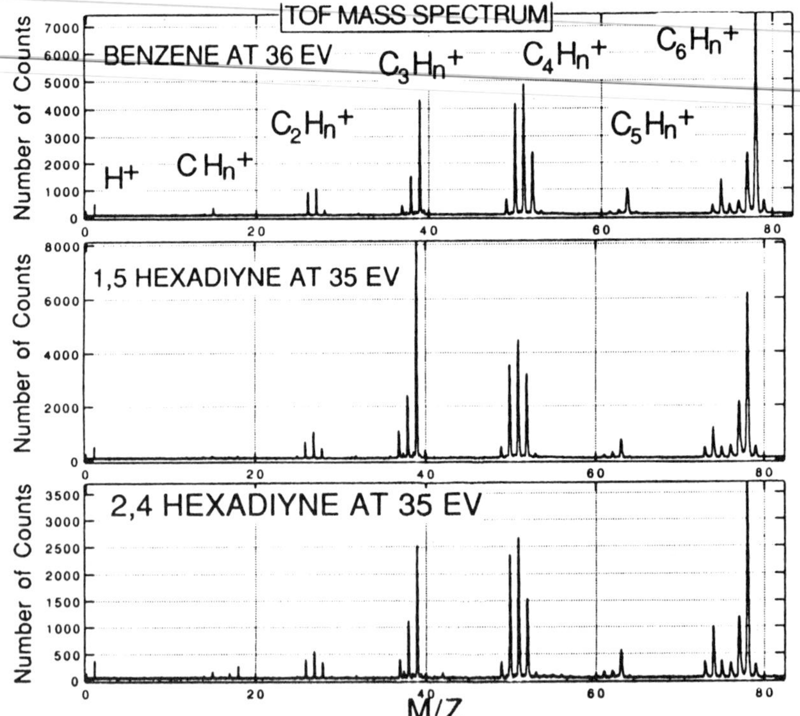

Figure 2. Time-of-flight mass spectra (PEPICO) : Benzene at photon excitation energy E_{exc} = 36 eV; 1,5-hexadiyne at 35 eV; 2,4-hexadiyne at 35 eV

The spectra for benzene and 2,4-hexadiyne are apparently similar, which might indicate common intermediate isomerization. The TOF-MS of 1,5-hexadiyne has some marked differences with respect to the other two spectra, in particular concerning the relative intensities of m/z = 78 and m/z = 39. Examination of half-integer peaks, and isotopic peaks involving ^{13}C provided information on parent and fragment dications and enabled us to determine the relative proportions of dications and monocations given in table 1.

Table 1 : Relative yields of some ions in PEPICO TOF-MS spectra of three C_6H_6 isomers at 35 and 70 eV

yield ratio	E_{exc}/eV	Benzene	2,4-hexadiyne	1,5 hexadiyne
$C_6H_6^{2+}/C_6H_6^+$	35	0.14	0.10	< 0.01
	70	0.73	0.46	0.32
$C_6H_6^{2+}/C_3H_3^+$	35	0.6	0.29	< 0.01
	70	1.7	0.80	0.32
$C_6H_5^{2+}/C_6H_6^{2+}$	35	0.03	0.09	< 0.01
	70	0.02	0.04	0.03

PEPIPICO spectra

Figure 3 shows a detail of the benzene PEPIPICO spectrum at 70 eV. The ion pairs formed are indicated along the t_1 and t_2 axes. Branching ratios of the charge separation processus at E_{exc} = 35(36) eV are given in figures 4 and 5.

Figure 4 shows that charge separation from $C_6H_n^{2+}$ precursor ions predominate, but that charge separation from $C_4H_n^{2+}$ also occurs. The importance of the $C_3H_n^+/C_3H_m^+$ charge separation branching ratio indicates that central C—C bond breaking is favored over terminal C—C bond rupture in both hexadiynes. This branching ratio is greater than that for $CH_n^+/C_5H_m^+$ for the two hexadiynes, whereas the $C_2H_n^+/C_4H_m^+$ ratio is largest for benzene. The high energy H^+ loss reactions are of relatively little importance, as was found for naphthalene and azulene by PEPIPICO experiments(16).

Looking more closely at the individual reactions, we find that the reaction $C_6H_6^{2+} \rightarrow CH_3^+ + C_5H_3^+$ is the most important charge separation channel in benzene and 2,4-hexadiyne, whereas in 1,5-hexadiyne the dominant charge separation pathway

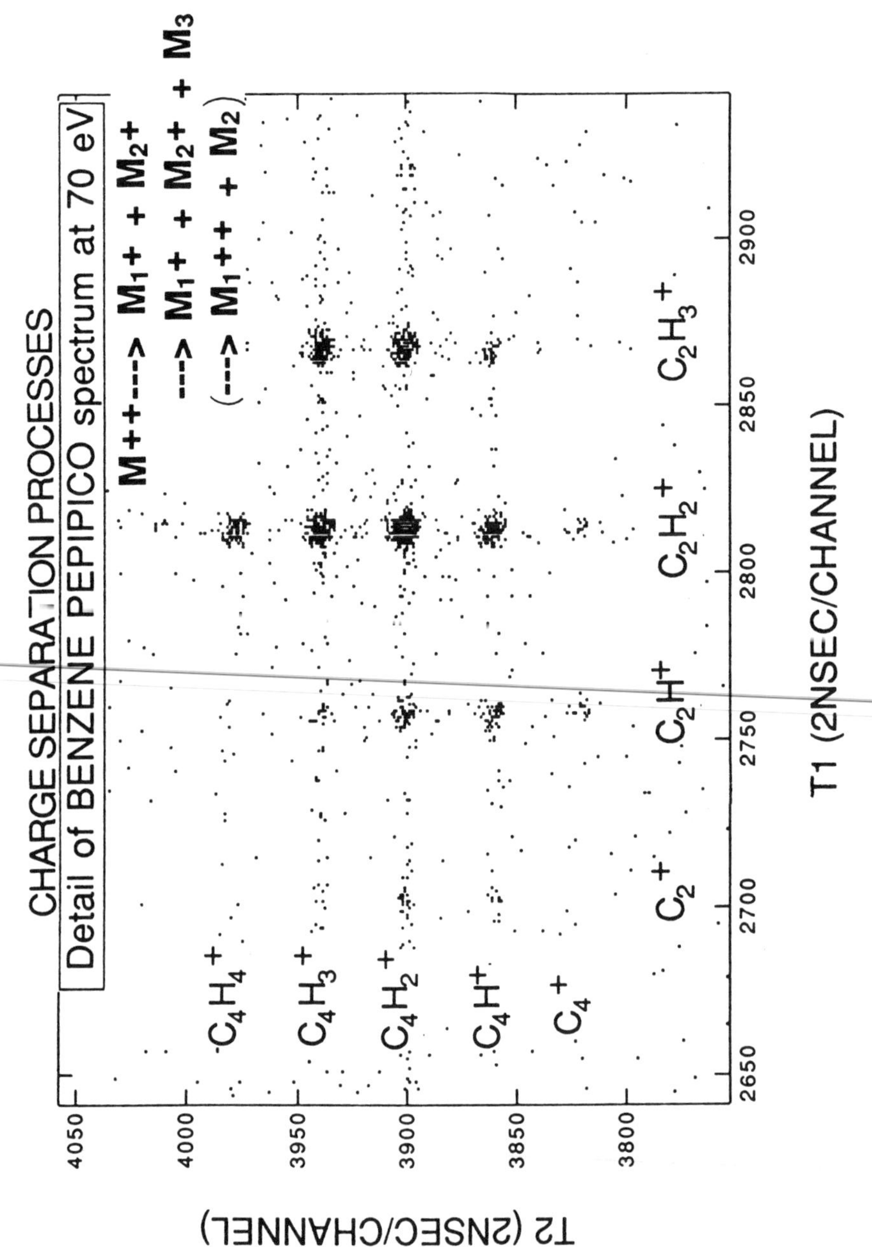

Figure 3. Part of Benzene PEPIPICO spectrum at E_{exc} = 70 eV

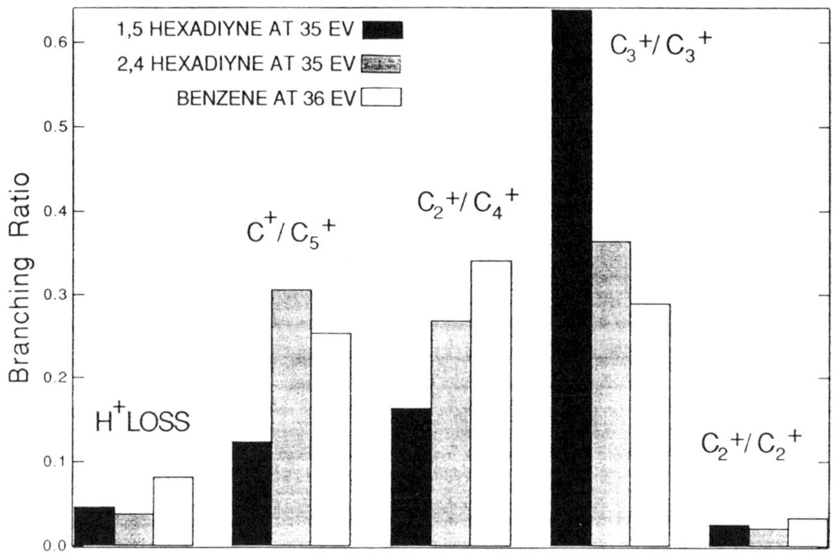

Figure 4. Branching ratios of principal groups of charge separation reactions for three $C_6H_6^{2+}$ isomers at E_{exc} = 35 and 36 eV.

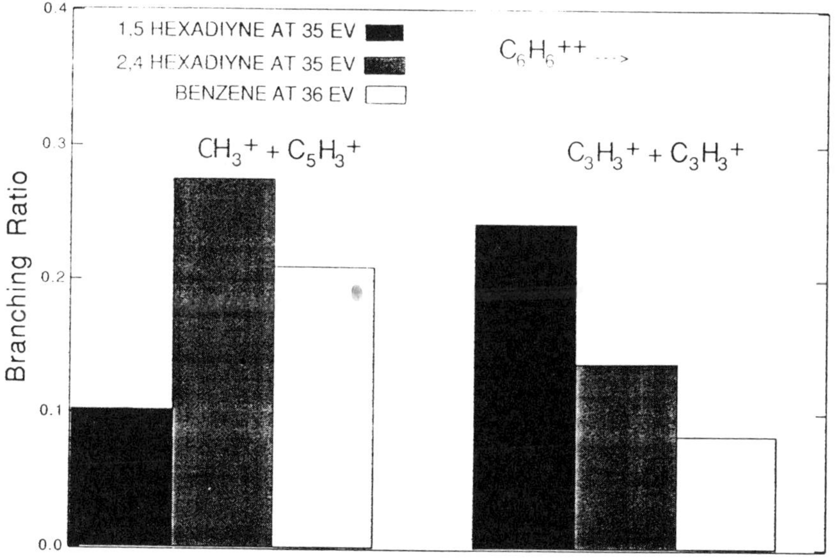

Figure 5. Branching ratios of $C_6H_6^{2+} \rightarrow CH_3^+ + C_5H_3^+$ and $C_6H_6^{2+} \rightarrow C_3H_3^+ + C_3H_3^+$ charge separation reactions for three $C_6H_6^{2+}$ isomers at E_{exc} = 35 and 36 eV.

is $C_6H_6^{2+} \rightarrow C_3H_3^+ + C_3H_3^+$. This again indicates that central C— C bond breaking is predominant in 1,5 hexadiyne^{2+}. Richardson et al[(6)] had already observed from PIPICO studies, that $C_6H_6^{2+} \rightarrow CH_3^+ + C_5H_3^+$ is the chief charge separation channel in benzene at 40.8 eV.

The TOF-MS and PEPIPICO results both show significant differences in relative product yields from the three isomers. This is in contrast to the case of the $C_{10}H_8$ isomers naphthalene and azulene for which the relative yields are closely similar, indicating the involvement of a common isomer[(16)].

The C_6H_6 isomeric differences can indicate that different precursor ionic structures are formed, for both monocation and dication, and/or the existence of a variety of internal energy dependent dissociation pathways. Deeper analysis of the whole of these spectra, and closer examination of energy deposition in the three isomers, are currently underway in order to clarify some of these issues. We have already modelised the energy deposition and studied the effect of internal energy modifications on slow and fast dissociation processes of the benzene monocation, as reported elsewhere[(15)].

Suggestive of the role of internal energy is the potential energy diagram given in fig.6 for low energy processes of the three $C_6H_6^{2+}$ isomeric ions. This diagram, constructed from observed or inferred energies, is consistent with the relative yields of the two principal charge separation reactions in the TOF-MS and PEPIPICO spectra of the three C_6H_6 isomers. Our proposed further analysis of results on the three isomeric $C_6H_6^+$ and $C_6H_6^{2+}$ ions should enable us to clarify some of the issues raised by electron impact and collisional excitation work[(1-5,7)], and also to integrate some of the results of theoretical calculations of the ionic isomers and their dissociation products[(17,18)].

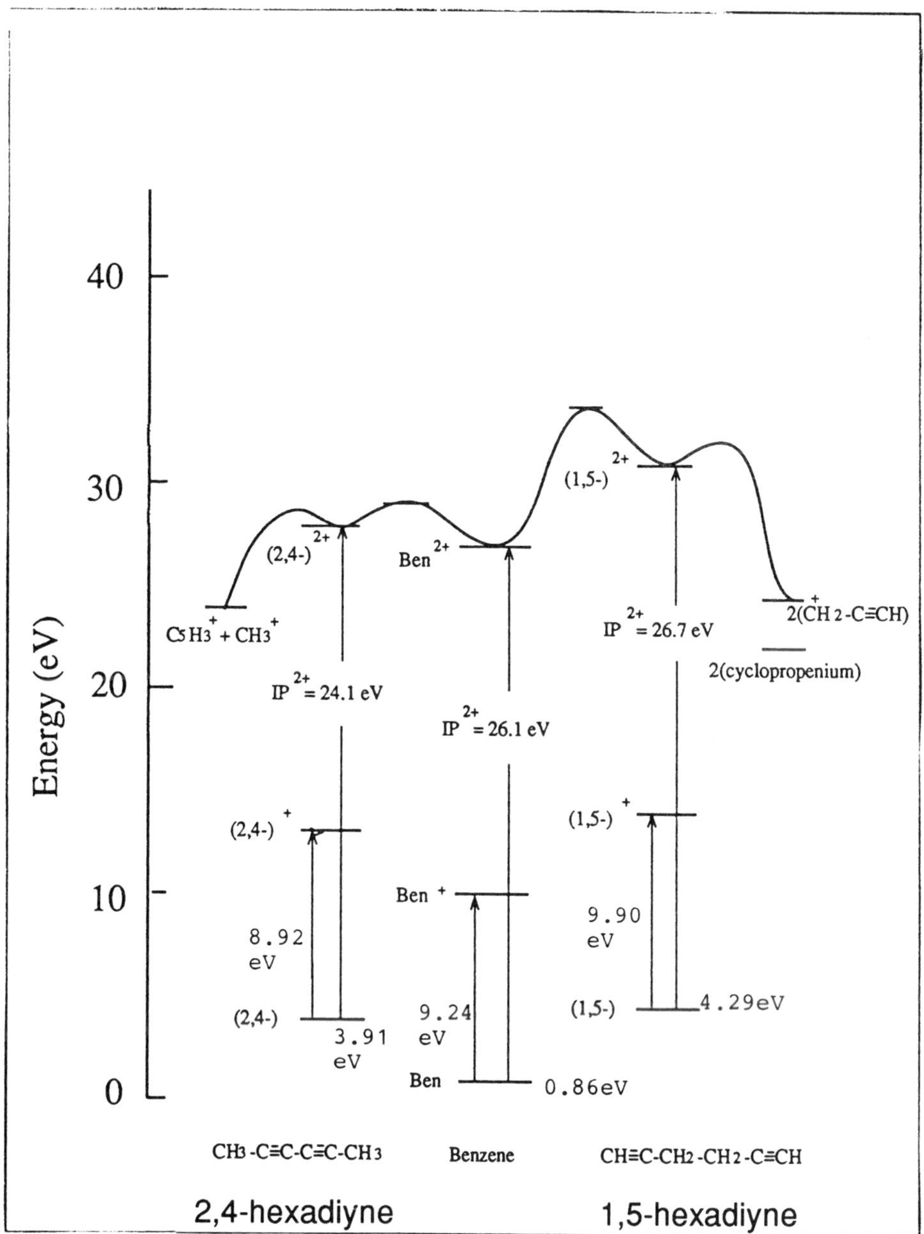

Figure 6. Schematic potential energy diagram for the low energy dissociation processes of three isomeric $C_6H_6^{2+}$ dications.

REFERENCES

(1) H.M. Rosenstock, J. Dannacher and J.F. Liebman, Radiat. Phys. Chem. 20, 7 (1982).

(2) M.J. Hayward, Md.A. Mabud and R.G. Cooks, J. Am. Chem. Soc. 110, 1343 (1988).

(3) N. Ohmichi, Y. Malinovich, J.P. Ziesel and C. Lifshitz, J. Phys. Chem. 93, 2491 (1989) and references therein.

(4) T. Ast, Adv. Mass Spectrom. 8A, 555 (1980).

(5) R.G. Kingston, M. Guilhaus, A.G. Brenton and J.H. Beynon, Org. Mass Spectrom., 20, 406 (1985).

(6) P.J. Richardson, J.H.D. Eland and P. Lablanquie, Org. Mass Spectrom. 21, 289 (1986).

(7) K. Vékey, A.G. Brenton, and J.H. Beynon, Org. Mass Spectrom. 24, 31 (1989).

(8) S. Leach, J. Elect. Spectros. Rel. Phen. 41, 427 (1986).

(9) S. Leach, in Interstellar Dust (L.J. Allamandola and A.G.G.M. Tielens, eds), I.A.U. Symp. 135, Kluwer, Dordrecht (1989), p. 155.

(10) B.A. Mamyrin, K.I. Karatev, D.V. Shmikk and V.A. Zagulin, Sov. Phys. JETP 37, 45 (1973).

(11) M.J. Besnard-Ramage, P. Morin, T. Lebrun, I. Nenner, M.J. Hubin-Franskin, J. Delwiche, P. Lablanquie and J.H.D. Eland, Rev. Sci. Instrum. 60, 2182 (1989).

(12) J.H.D. Eland, F.S.Wort and R.N. Royds, J. Elect. Spectrosc. Rel. Phen. 41, 297 (1986).

(13) M. Simon, T. Lebrun, P. Morin, M. Lavollée and J.L. Maréchal, Nucl. Inst. Meth. 62, 167 (1991).

(14) W.C. Wiley and I.H. McLaren, Rev.Sci.Instrum. 26, 1150 (1955).

(15) O. Braitbart, S. Tobita, P. Roy, I. Nenner et S. Leach, to be submitted.

(16) S. Leach, J.H.D. Eland and S.D. Price, J. Phys. Chem. 93, 7575 (1989).

(17) K. Lammertsma and P. von Ragué Schleyer, J. Am. Chem. Soc. 105, 1049 (1983)

(18) K. Krogh-Jespersen, J. Am. Chem. Soc. 113, 417 (1991).

SPLINE BASIS IN VARIATIONAL DESCRIPTION OF DISCRETE AND CONTINUUM STATES

M.Brosolo and P.Decleva
Dipartimento di Scienze Chimiche, Università di Trieste
Via A.Valerio, 38 I-34127 Trieste (Italy)

ABSTRACT

A variational approach for the determination of continuum orbitals recently proposed in the one dimensional case is generalized to the multicenter molecular problem. A test application is performed on the H_2^+ molecule in the one-centre approximation, with a radial basis of B-spline functions. The resulting eigenvalue equation is solved by inverse subspace iteration giving a full set of K-matrix normalized solutions at each prefixed energy.
Expansion up to l=20 gives very satisfactory agreement with exact results. The potential of a multicenter LCAO type implementation is pointed out.

1. INTRODUCTION

A large amount of experimental data on molecular photoionization, such as total and partial cross sections and angular distributions, is becoming increasingly available through the use of syncrothron radiation. By means of an adequate theoretical description and interpretation these observables can provide a lot of informations both on collision processes and intrinsic properties of initial and final target states.

The first purpose in photoionization properties calculation is a good representation of the continuum wavefunction. This can be performed by direct numerical integration of the differential equation or by expansion of the continuum functions as finite linear combinations of basis orbitals. Altough a variety of approaches are available for small systems, only a few appear to be widely applicable to the multicenter problem posed by complex poliatomic targets.

Recently a variational algorithm,based on a pure L^2 formulation, has been proposed for the one dimensional atomic case, and a successful application to resonances in Helium photoionization has been performed[1, 2]. This approach is based on the use of a spline function basis, pioneered by Shore[3-6], and on a new variational principle, which involves the diagonalization of the matrix A=H-ES, obtained in the given basis. Splines vanish identically outside a closed volume, so the method amounts to solve the Schrödinger equation within a finite volume, large enough to reach the asymptotic region, and to extract the full K-matrix normalized wavefunction by fitting the asymptotic part to a linear combination of the regular and irregular asymptotic solutions. The main features of this approach are: it is variational, it does not need imposing any condition on the boundary, it is purely algebraic and is extremely simple.

In this work we have considered an application to the H_2^+ molecule, in a simple one-centre expansion both for the bound and the continuum states. The coupling of different partial waves and the need of extracting a full set of independent solutions at each energy have been solved by applying the subspace inverse iteration approach. The results obtained show excellent agreement with Richards and Larkins[7] calculations, indicating the potential of present approach in the molecular case.

2. METHOD

Recently a generalization of the Rayleigh-Ritz-Galerkin approach for bound states to the determination of continuum wavefunctions has been proposed[1]. For the bound state problem this procedure, by expansion in basis set $\Phi = \Sigma_i c_i B_i$, leads to the Galerkin equations

$$b_i = \langle B_i | H-E | \Phi \rangle = 0 \qquad i=1,\ldots\ldots,n \tag{1}$$

and to the algebraic eigenvalue problem $(H-ES)c=0$, determining eigenvalues and corresponding eigenvectors.

In the continous spectrum the energy is fixed, so that the equation $Ac=0c$, $A(E)=H-ES$, in general does not admit non trivial solutions $c\neq 0$. In this case it is proposed to take as approximate solution the eigenvector c relative to the eigenvalue closest to zero (i.e. of minimum modulus), so that the expectation value $\langle\Phi|H-E|\Phi\rangle$ will be minimized, with the constraint $\Sigma_i |c_i|^2 = 1$. This condition is intuitively related to the minimization of the residual $\Sigma_i |b_i|^2$, which takes to the eigenvalue equation $A^+Ac=\lambda c$. Altough the two equations are different, $Ac=ac$ implies $\langle c,A^+Ac\rangle=a^*a$, so that if the minimum modulus eigenvalue of A is well separated from the others (as we have always verified when the basis is flexible enough to provide an accurate solution), then the corresponding eigenvector should be close to the eigenvector of A^+A relative to its minimum eigenvalue. Other formulations, corresponding to the choice of different norms, are currently investigated.

The required vectors have been extracted by applying inverse iteration method to the relative subspace, which proved to be very effective, thanks to the large separation between the moduli of the n lowest eigenvalues and the rest.

Basis functions employed for the H_2^+ problem are products of radial B-splines times spherical harmonics:

$$\Phi_{ilm} = 1/r \; B_i(r) Y_{lm}(\theta,\varphi) \tag{2}$$

Splines are piecewise polynomials S(x) chosen to satisfy fixed constraints[8]. Given an interval [a,b], divided into m subintervals by m-1 internal knots k_i:

$$a < k_1 < k_2 < \ldots < k_{m-1} < b, \tag{3}$$

S(x) is a polynomial of degree n on each subinterval, and it is of class C^{n-1} over all [a,b]. This is realized by requesting continuity of the function and its n-1 derivatives in each knot. We chose maximum continuity at the interior knots, but no constraints at the endpoints, apart from the condition f(0)=0, which is satisfied simply deleting the first spline from the basis. The symmetrical condition f(R)=0 at the end of the interval is employed only in the bound state case, deleting the last spline.
The dimension of the basis is m+n, so it follows that increasing the order or the number of the intervals by one just adds one basis function. The B-spline basis is defined by the requirement of minimal support, which leads to basis functions being non-zero over at most n+1 adjacent intervals.

The flexibility of this kind of basis allows a very good representation of the oscillatory nature of continuum orbitals as well as of the flatter bound states. Another advantage is given by the banded structure of the resulting matrices, because of the zero overlap between functions with disjoint supports.

Calculations were performed at the internuclear distance of R=2.0 au. After some tests we chose splines of order 10, an interval of 20 au, and a uniform mesh with step h=.125. This allows good accuracy to be retained up to a k= $\sqrt{(2E)}$ value around 4, that is photoelectron energies up to 8 au can be studied. In the Coulomb problem we found that the maximum k value obtainable for a preset accuracy is linearly varying with inverse step size. The fixed mesh allowed the matrices S and H to be computed only once for all energies. Radial integrals have been evaluated by gaussian integration of order 12, and are essentially exact.

For the solution of the bound state problem, as well as for tests on the eigenvalue spectrum of the A matrix, standard algorithms have been employed[9, 10], while for the continuum calculations subspace inverse iteration was used. Basically one starts with n independent zero order vectors x_i°; at each iteration the system

$$A y_i^{k+1} = x_i^{k} \tag{4}$$

is solved, and the y_i^{k+1} are symmetrically orthonormalized:

$$x_i^{k+1} = S^{-(1/2)} y_i^{k+1} \tag{5}$$

The solution of the linear system is performed throughout the LU factorization of matrix A. The iteration is stopped when the subspace distance between two successive iterations is less than a prefixed threshold.

Because of the $D_{\infty h}$ symmetry of the H_2^+ molecule only one partial wave for each l value is allowed in each symmetry, with only l values of the same parity allowed. We have employed an expansion up to a maximum of l=20, giving 10 independent channels.

The obtained solutions are fitted to linear combinations of the asymptotic regular and irregular solutions (Coulomb functions f_l and g_l[11]) at the last two interior knots of the interval, and normalized to K-matrix boundary conditions:

$$\Phi'_n = \Sigma_{lm}\ 1/r\ P_{lm,n}(r)\ Y_{lm} \tag{6}$$

$$P_{lm,n} = CS_{lm,n}\ f_l(r) + CC_{lm,n}\ g_l(r) \tag{7}$$

$$K = CC{*}CS^{-1} \qquad \Phi_n = \Sigma_{n'}\ CS_{n',n}^{\ 1}\ \Phi'_{n'} \tag{8}$$

$$\Phi_n(r) \to \Sigma_{lm}\ 1/r\ (\delta_{lm,n}\ f_l + K_{lm,n}\ g_l)\ Y_{lm} \tag{9}$$

The K-matrix so obtained is very nearly symmetric, as it is required for an exact solution[12]. It gives informations about the coupling between different angular momenta due to the non spherical molecular potential; it is a compact way to represent electron-molecule interaction. K-matrix eigenvalues

$$k_i = \tan \eta_i,\ i=1,\dots,n \tag{10}$$

define the partial wave phase shifts η_i.

Photoionization cross sections and angular distribuitions have been calculated by Dill and Dehmer's formulas[13, 14].

3. RESULTS AND DISCUSSION

The quality of the present one centre expansion may be judged by the bound state energies calculated for the H_2^+ molecule.

The convergence towards exact values rapidly increases with angular momentum. The largest differences are founded for the lowest σ_g and σ_u states, probably tied to the cusps present in the LCAO expansion $1\sigma_{g,u} = 1s_1 \pm 1s_2$, which require large l values in a one-centre expansion. For π_g or δ_g symmetries essentially exact results are obtained for the lowest level. The final value (lmax=20) for the σ_g

ground state, 1.10250 au, compared with the exact one, which is 1.10263 au[15], shows an accuracy of 10^{-4} au, which is important because of the strong dependence of the dipole matrix elements on the details of the ground state function[16].

As far as continuum wavefunctions, a test on their goodness is given by the phase shifts. They provide a good representation of the rescattering of the initial electron angular momentum into a range of momenta due to the anysotropy of the molecular potential. The results have been compared with Richards and Larkins' ones[7]. As they extracted phase shifts neglecting coupling between different angular momenta, we reported both full phase shifts from K-matrix eigenvalues and uncoupled diagonal K-matrix elements. At low energy the effect of coupling is small, and the agreement of full results with ref. 7 is very satisfactory. The energy getting higher, the differences between coupled and uncoupled values get larger, and the results of Richards and Larkins have to be compared essentially with uncoupled values.

Because of the coupling, it appears probable that present results are more accurate than previous ones, despite the essentially exact wavefunction of the former study.

As eigenvalues, also K-matrix eigenvectors yield some informations.

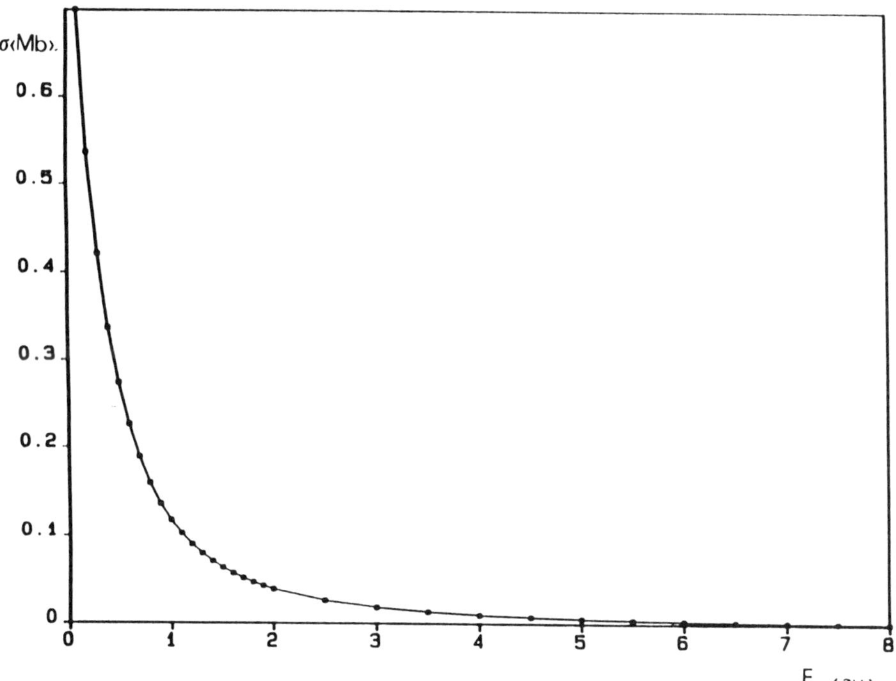

Figure 1. Total cross section for H_2^+ photoionization.

According to Loomba et al.[17] if an eigenchannel is dominated by a particular orbital momentum, then the eigenchannel wavefunction has the characteristic angular pattern of the corresponding spherical harmonic.

Photoionization cross sections show an excellent agreement with exact results of ref. 7, the error being on the fourth significant figure. The monotonic decrease of the total cross section with photoelectron energy is illustrated in Figure 1.

We also studied the relative contributions of each partial wave component, and the behaviour obtained is indeed very similar to that previously reported by Chapman and Hayes[18].

Also the pattern of the asymmetry parameter β in function of photoelectron energy, reported in Figure 2, is in excellent agreement with that reported by Richards and Larkins[7]. Nonetheless, it is significantly different from the one-centre result of Chapman and Hayes, but this disagreement is not to be attributed to the inadequacy of the one-centre expansion, nor to the l-uncoupled approximation of ref. 7, but probably to an incorrect evaluation of the β parameter in ref. 18.

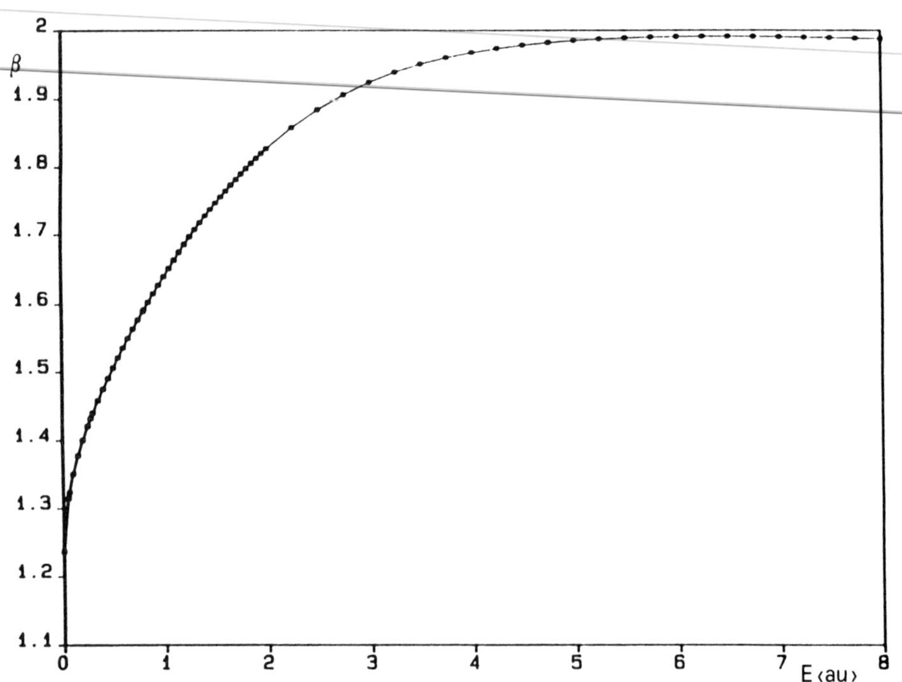

Figure 2. Asymmetry parameter for photoionization of H_2^+ as a function of photoelectron energy.

In conclusion, let us summarize the principal points of this approach.

It is purely L^2, variational and very simple. It does not require boundary conditions, nor supplementary asymptotic solutions. A one centre approximation has been employed, but an LCAO method is entirely feasible, at the expense of multicenter matrix elements.

The spline basis shows several advantages: it leads to rapid convergence with increasing order, which does not inordinately increase the dimension of the basis, and gives rise to very sparse matrices.

The extraction of few eigenvectors of a given matrix relative to eigenvalues closest to zero is the core of the method and it is efficiently performed through inverse subspace iteration.

4. REFERENCES

1. C.Froese Fischer and M.Idrees, Comput. Phys., 3 (1989) 53
2. C.Froese Fischer and M.Idrees, J.Phys.B: At.Mol.Opt.Phys., 23 (1990) 679
3. B.W.Shore, J.Chem.Phys., 58 (1973) 3855
4. B.W.Shore, J.Chem.Phys., 59 (1973) 6450
5. B.W.Shore, J.Phys.B:At.Mol.Phys., 7 (1974) 2502
6. B.W.Shore, J.Chem Phys., 63 (1975) 3835
7. J.A.Richards and F.P.Larkins, J.Phys.B:At.Mol.Phys., 19 (1986) 1945
8. C.de Boor, A Practical Guide To Splines (Springer-Verlag New York 1978)
9. W.H.Press, B.P.Flannery, S.A.Teukolsky and W.T.Vetterling, Numerical Recipes (Cambridge University Press, Cambridge 1986)
10. NAG Fortran Library Manual, Mark 13, (NAG Central Office, 256 Banbury Road, Oxford OX2 7DE, U.K.)
11. A.R.Barnett, Comput.Phys.Comm., 27 (1982) 147
12. R.K.Nesbet, Variational Methods in Electron-Atom Scattering Theory (Plenum New York 1980)
13. J.L.Dehmer and D.Dill in: Electron-Molecule and Photon-Molecule Collisions, T.Rescigno, V.Mc Koy and B.Schneider Eds., (Plenum, New York 1979) p.225
14. D.Dill and J.L.Dehmer, J.Chem.Phys., 61 (1974) 692
15. L.Laaksonen, P.Pyykkö and D.Sundholm, Int.J.Quantum Chem.,23 (1983) 309
16. H.Le Rouzo and G.Raseev, Phys.Rev.A, 29 (1984) 1214
17. D.Loomba, S.Wallace, D.Dill and J.L.Dehmer, J.Chem.Phys., 75 (1981) 4546
18. F.M.Chapman,Jr. and E.F.Hayes, J.Chem.Phys., 67 (1977) 2974

THEORETICAL AND EXPERIMENTAL STUDIES OF THE ELECTRONIC STRUCTURE OF $SnCl_2$ BY PHOTOIONIZATION WITH SYNCHROTRON RADIATION

S. Stranges ° and M. Y. Adam
LURE Université de Paris Sud Orsay Cedex, France

C. Cauletti and M. de Simone
Dip. Chimica, Università di Roma "La Sapienza", Italia

M. N. Piancastelli
Dip. di Scienze e Tecnologie Chimiche, II[a] Università di Roma "Tor Vergata", Italia

ABSTRACT

Molecular effusive beam and dispersed synchrotron radiation at SuperACO positron storage ring allowed an investigation of the complete outer- and inner-valence shells photoionization of gaseous $SnCl_2$ for the first time. New and resonanting features in the inner-valence region were observed in the photoelectron spectrum of this molecule. Auger processes following the relaxation of Sn 4d hole in the free molecule $SnCl_2$ were detected, analogus to those observed in atomic tin.

INTRODUCTION

Very little is reported in literature about photoemission of tin compounds studied in vapour phase by synchrotron radiation [1]. We present here the results of an investigation on the valence electronic structure on tindichloride. This study is particularly interesting due to the presence of this molecule in gas phase as a practically pure molecular species [2] at temperature above the melting point (247° C), despite the covalent nature of the binding in solid state.

Recently some of us have performed pseudo potential ab-initio calculations on this molecule and compared these results with the experimental Ionization Energies (IEs) obtained with laboratory sources (HeI/HeII) [3], obtaining a better qualitative agreement than in the studies previously reported in literature [4]. This preliminary study prompted us to deepen the knowledge of the ionization process with particular regard to the ionization of the inner-valence levels,

°permanent address: Dip. Chimica, Università di Roma "La Sapienza", Italia

including possible many body effects, resonant phenomena and so on. Synchrotron radiation is the ideal source for such purposes, due to its unique characteristic of high flux and tunability.

We measured photoelectron spectra in the photon energy range 25÷52 eV. We were able to observe the ionization of both outer- and inner-valence shells including also Cl 3s- and Sn 4d-based orbitals.

The ionizations of Cl 3s orbitals were observed for the first time for this molecule.

Both Sn 5s and Cl 3s orbitals give rise to a complex spectral structure which can be related to many body effects.

The Sn $4d^{-1}$ ionization is accompained by a clear Auger decay.

Finally a resonant behavior is qualitatively observed in the IE region of the innermost σ_{SnCl} MO (with main Sn 5s character) due to a promotion of Sn 4d electron to an empty MO*.

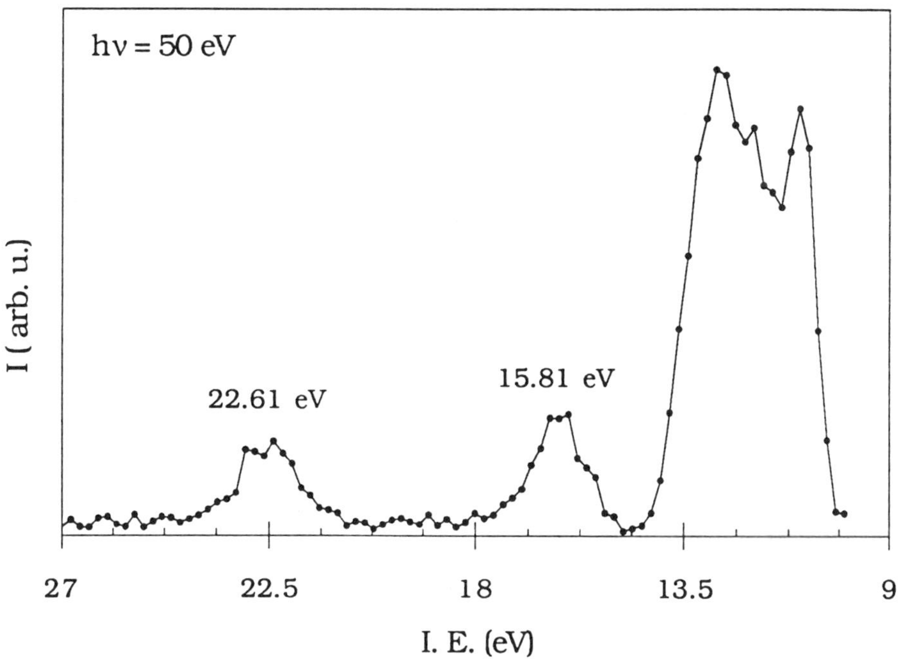

fig. 1 Photoelectron spectra of the outer- and inner-valence shells of $SnCl_2$, with the exclusion of the Sn $4d^{-1}$ region, taken at 50 eV.

EXPERIMENTAL

The sample used was obtained commercially (Aldrich, $SnCl_2$

99.99% purity). Particular care was taken in handling the sample to exclude moisture. Before recording the spectra the powder was purified in situ under vacuum, at first by heating for a few hours to eliminate all traces of water, then for a long time at higher temperature (ca. 180°C) to remove possible traces of $SnCl_4$ [5]. Successively the temperature of the cell was raised very slowly until a usable count rate was reached (T=275÷280°C). The heating apparatus employed in this experiment, with the effusive beam at right angle with respect the propagation direction of the light, is that described in ref.[6] without the part containing the TOF mass spectrometer and the deflector.

The dispersed synchrotron radiation (by means of a TGM monochromator equipped with two interchangeable gratings [7]) at the SA23 beam line of SuperACO storage ring was used. Photoelectron spectra were recorded at right angle with respect to the polarization plane of the synchrotron radiation by a 127° sector electron spectrometer, in the 23÷52 eV photon energy range, with total instrumental resolution ranging from ca. 0.35 to ca. 0.70 eV.

RESULTS AND DISCUSSION

The photoelectron spectra taken at hν = 50 eV, of the outer and inner-valence shells of $SnCl_2$, with the exclusion of Sn $4d^{-1}$ region, is shown in fig. 1. We can distinguish two regions, in order of increasing IEs: i) the outer- valence region, already well analised in the previous studies, including bands related to $4a_1$,$3b_2$,$1a_2$,$1b_1$ and $3a_1$ MOs (see tab. I) for the assignement and the comparison with the above mentioned theoretical calculation; ii) the inner-valence region above 13 eV including two broad and structured spectral features: one with maximum at 15.94 eV and a shoulder on the low IE side, related to the ionization of the $2a_1$ MO, the other with a maximum at 22.61 eV, which has to be ascribed to the ionization $1b_1$ and $1a_1$ MOs, mainly localised on the Cl 3s orbitals. As already reported in many cases [8] for the ionization of the inner-valence levels no one to one correspondence between spectral bands and MOs does exist, due to the breakdown of the one-electron picture. Instead, complex spectral structure deriving from the overlap of many satellites are evident. Furthermore an interesting phenomena accompanying the ionization of some of the satellites in the Sn 5s IE region exhibit a strong cross section effect as photon energy is varied in the range 24÷29 eV: a maximum in the spectral intensity is observed in-between 25÷26.26 eV. This effect is shown in fig. 2. This photon energy at which the maximum of the cross-section is found is consistent with the presence of an autoionization process following a transition Sn 4d -> MO*.

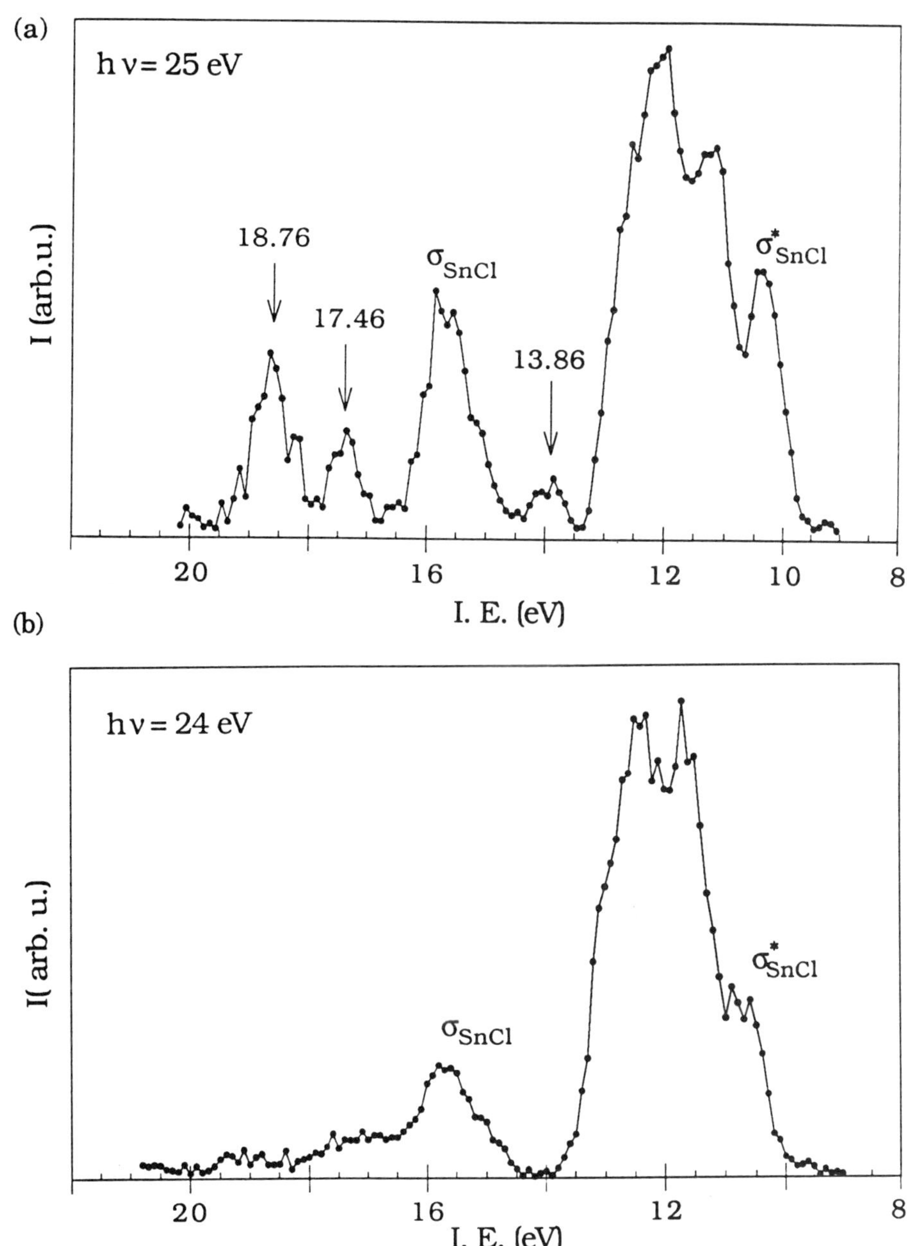

fig. 2 Photoelectron spectra in the outer-valence region taken (a) at the maximum of the resonant effect (hν = 25 eV) and (b) at hν = 24 eV (out of resonance).

MO	$-\varepsilon_{MO}$(eV)	% population a						dominant character	IE (eV)b
		Sn			2Cl				
		s	p	d	s	p	d		
$4a_1$(HOMO)	10.36	36.4	21.6	0.9	0.0	41.2	0.0	σ^*_{SnCl}[c]	10.38
$3b_2$	12.03	0.0	0.6	0.4	0.0	98.9	0.0	n_{Cl}[c]	11.39[e]
$1a_2$	12.32	0.0	0.0	1.3	0.0	98.7	0.1	n_{Cl}[e]	11.39[e]
$1b_1$	12.88	0.0	7.0	0.9	0.0	92.0	0.2	n_{Cl}[e]	12.39[e]
$3a_1$	13.11	2.8	6.2	1.4	0.0	89.5	0.2	n_{Cl}[c]	12.39[e]
$2b_2$	13.85	0.0	13.7	1.4	1.7	82.9	0.4	σ_{SnCl}	12.78
$2a_1$	17.23	48.0	0.9	0.0	5.9	44.6	0.5	σ_{SnCl}	15.94
$1b_2$	28.99	0.0	2.6	0.8	96.2	0.5	0.0	n_{Cl}	22.61[f]
$1a_1$	29.46	5.0	1.6	0.2	92.1	1.1	0.1	n_{Cl}	22.61[f]

a) Population analysis of Mulliken, b) our experimental results obtained with a conventional source HeI /HeII [1] c) on plane molecular orbital d) out of plane molecular orbital e) double band f) from our measurements with synchrotron radiation.

tab. I MO calculation results for $SnCl_2$.

Fig. 3 shows the doublet corresponding to $^2D_{5/2}$ and $^2D_{3/2}$ spin-orbit components of Sn 4d-based levels together with the related Auger structure. It is evident the better quality of the spectrum in this region with respect to the literature data obtained with laboratory sources.

The straightforward interpretation of this intense structure, whose maximum has a kinetic energy of ca. 5.7 eV, is to ascribe it to molecular Auger processes following the formation of Sn 4d ionized molecular states. Atomic analogous Auger decay processes has been recently observed also in the photoionization of atomic tin [9].

Although the deconvolution of the Auger broad peak is not possible, since information concerning to the energies of the electronic ground and excited states of the doubly-charged molecular ion is missing in literature, an extime of the energy range corresponding to these doubly-charged states below the 4d thresholds is possible by analysing the shape of the band and taking into account the width of the single Auger line due to the analyser pass energy (Δ_1 = 0.72 eV at the experimental conditions), the experimental spin-orbit splitting of the 4d^{-1} Sn ionic states (Δ_2 = 1.05 eV [10]) originating the Auger decays, and their experimental intensity ratio ($I(^2D_{5/2})/I(^2D_{3/2})$=1.62 at hv=48.5 eV). From the dotted area reported in fig. 3 it is evident that such a states are localized in the energy range 4.37÷6.22 eV below the 4d^{-1} ($^2D_{5/2}$) threshold (IE=33.48 eV [10]).

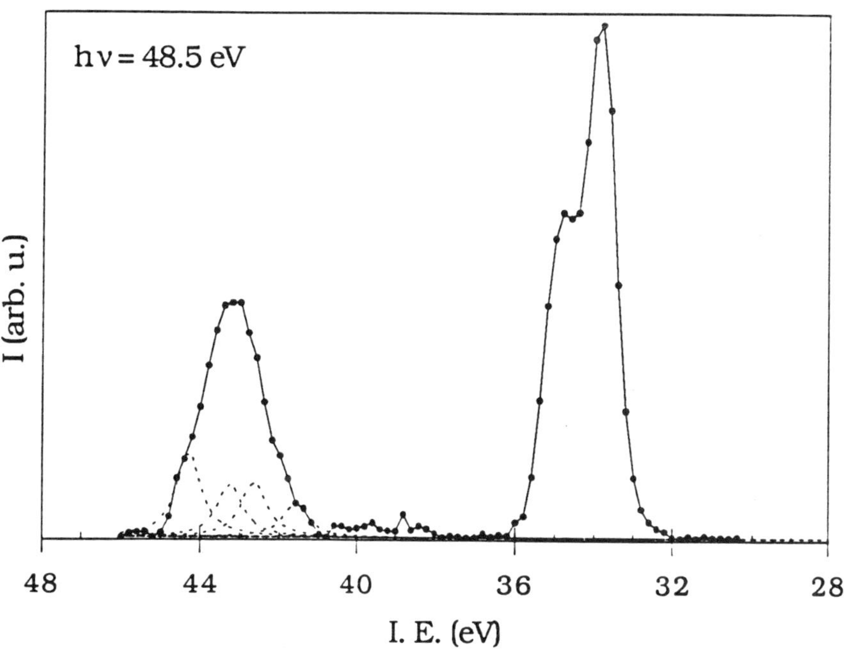

fig. 3 Photoelectron spectra of the Sn 4d based levels together with the related Auger structure.

ACKNOWLEDGMENTS

We are very gratefull to L. Hellener, M. J. Besnard-Ramage and G. Dujardin for letting us use their experimental devices. We wish to thank J.M.Bizau for assistence in managing the monochromator. S.S. gratefully acknowledges a grant from the Italian National Research Council.

REFERENCES

[1] G. M. Bancroft, W. Gudat and D. E. Eastman, J. Electron Spectrosc. Relat. Phenom. 10, (1977), 407; R. Malutzyski and V. Schmidt, Phys.

B 19, (1986), 1035; I. Novak, J. M. Benson, A. Svesson and A. W. Potts, Chem. Phys. Lett. 135, (1987), 471;G. M. Bancroft, K. T. Sham, D. E. Eastman and W. Gudat, J. Am. Chem. Soc.(1977), 1752.

[2] S. Ciach, D. J. Knowles, A. J. C. Nicholson and D. L. Swingler, Inorg. Chem. 12, 1443, (1973).

[3] C. Cauletti, M. de Simone and S. Stranges, J. Electron Spectrosc. Relat. Phenom. to be published

[4] I. Novak and A. W. Potts, J. Electron Spectrosc. Relat. Phenom. 33, 1 (1984),: D. H. Harris, M. F. Lappert, J. B. Pedley and J. Sharp, J. Chem. Soc. Dalton Trans. , 945 (1976); R. T. Poole, J. Electron Spectrosc. Relat. Phenom. 15, 91 (1979),; J. M. Ricart, J. Rubio and F. Illas, Chem. Phys. Lett. 123, 528 (1986).

[5] L. Andrews and D. Frederick, J. Amer. Chem. Soc., 92 (1970) 775.

[6] M.Y. Adam, L. Hellner, G. Dujardin, A. Svensson, P. Martin and F. Combet-Farnoux, J. Phys. B: At. Mol. Opt. Phys., 22 (1989) 2141.

[7] P.K.Larsen,W.A.M. van Bers, J.M. Bizau, F. Wuilleumier, S. Krummacher, V. Schmidt and D. Ederer, Nucl. Instr. and Meth., 195 (1982) 245.

[8] M. Y. Adam, C. Cauletti, M. N. Piancastelli and A. Svesson, J. Electron Spectrosc. Relat. Phenom. 50, 219 (1990) and references therein.

[9] R. Malutzki and V. Schimdt, J. Phys. B: At. Mol. Phys., 19, 1035 (1986).

[10] A. W. Potts and Lyus, J. Electron Spectrosc. Relat. Phenom. 13, 327, (1978).

LINEAR DICHROISM IN THE ANGULAR DISTRIBUTION OF PHOTOELECTRONS FROM ORIENTED MOLECULES

N. A. Cherepkov
Aviation instrument making institute
190 000 Leningrad, USSR

G. Schönhense
Institut für Physik, Johannes Gutenberg–Universität
Staudinger Weg 7, D – 6500 Mainz, FRG

ABSTRACT

This paper presents a theoretical treatment of Linear Dichroism in the Angular Distribution of photoelectrons (LDAD), that is a difference between photoelectron currents ejected at a definite angle by photon beams of two different linear polarizations rotated by 90^0 relative to each other. Apart from a trivial term existing already in the cases of unpolarized atoms and unoriented molecules, for oriented molecules additional terms appear in the electric dipole approximation. Their measurement can give additional information on the molecular structure complementing measurements of the circular dichroism in the angular distribution investigated earlier. As an example, LDAD in a model case of an oriented diatomic molecule is investigated.

1. INTRODUCTION

Since early discussions of the "complete experiment" (see, e.g. Ref. 1) possible ways to realize such experiments have attracted wide interest. In the sense of quantum mechanics the term "complete" refers to the number of basic quantities (i.e. dipole matrix elements and relative phases of partial waves) which can in principle be extracted from a set of experimental data. Already in the case of low–energy photoionization of closed–shell atoms five basic quantities exist so that five independent experimental parameters are required for a complete experiment. In order to reach this challenging goal, special spectroscopic methods have been developed such as the energy–, angle– and spinresolved photoelectron spectroscopy with circularly polarized synchrotron radiation[2], which has been successfully applied to several atomic species.

For molecules, there exists principally an unrestricted number of basic quantities due to the unlimited series of allowed partial waves. In practice, the partial wave expansion series can often be cut off at a certain angular momentum l_{max} because higher terms are negligible. Then, in this approximation, complete experiments become possible for molecules. In this paper we present a new and relatively simple spectroscopic method which provides detailed additional information on molecular photoemission, suitable for a complete experiment. The basic effect is termed *Linear Dichroism in the Angular Distribution of photoelectrons* (LDAD), which means a difference between photoelectron currents ejected at a definite angle by linearly polarized light of two mutually perpendicular polarizations.

The LDAD effect is strongly related with a corresponding phenomenon arising in photoemission with *Circularly* polarized light (CDAD), which was discovered and investigated recently [3-8]. CDAD appears already in electric dipole approximation and occurs for all spatially fixed molecules and for aligned or oriented atoms and molecules.

2. GENERAL DERIVATION OF LDAD

A simple derivation shows that already for unpolarized atoms and rotating molecules the LDAD is different from zero. It is connected with the fact that the angular distributions of photoelectrons in these cases are nonisotropic and are defined by the direction of the polarization vector $\vec{e}$. If $\vec{e}$ is directed along the z axis of a coordinate frame, the angular distribution of photoelectrons has the standard form [9]

$$I_{\vec{e}} = \frac{d\sigma(\omega)}{d\Omega} = \frac{\sigma(\omega)}{4\pi}[1 + \beta(\frac{3}{2}\cos^2\vartheta - \frac{1}{2})] \qquad (1),$$

where σ is the partial photoionization cross section, ω is the photon energy, and β is the angular asymmetry parameter.

We will consider two mutually perpendicular linear polarizations, thus it is convenient to use a coordinate frame with the z axis directed along the photon beam, characterized by the unit vector $\vec{q}$, while two mutually perpendicular linear polarization states are defined by polarization vectors $\vec{e}_x$ and $\vec{e}_y$ directed along the x and y axes, respectively (see Fig. 1). In this coordinate system the angular distribution (1) leads to the following expression for the *LDAD of unoriented molecules or atoms*

$$I_{LDAD} = I_{\vec{e}_y} - I_{\vec{e}_x} = -\frac{3\sigma(\omega)}{8\pi}\beta\sin^2\vartheta\cos 2\varphi \qquad (2)$$

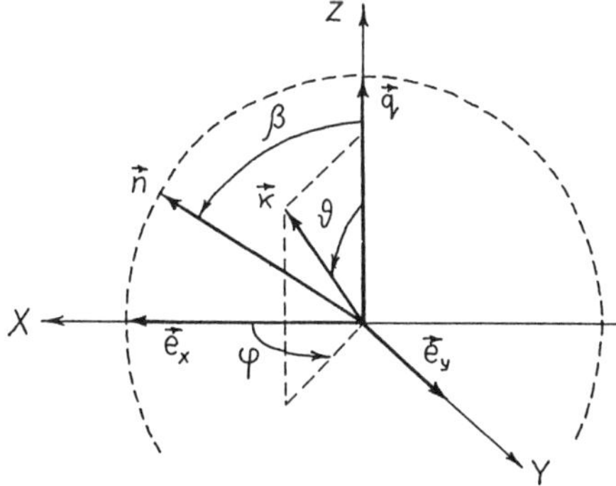

Fig. 1: Geometry of an experiment for LDAD measurements. The unit vectors $\vec{q}$, $\vec{e}_x$ and $\vec{e}_y$ denote the photon propagation and polarization directions, respectively; $\vec{n}$ is the molecular axis and $\vec{\kappa}$ is the unit vector along the photoelectron momentum.

For spatially fixed molecules, the differential cross section and LDAD term are more complicated because the anisotropy of the initial state plays an important role. Explicit derivation of the dipole matrix element for the two perpendicular polarization states of the ionizing photon beam and forming the difference of the corresponding differential cross sections (to be published elsewhere) yields the result for the *LDAD for space–fixed molecules*

$$I^i_{LDAD}=I^i_{\vec{e}_y}(\vec{\kappa},\vec{q})-I^i_{\vec{e}_x}(\vec{\kappa},\vec{q})=\sqrt{\frac{3}{4\pi}}\,\sigma_i(\omega)\sum_{L,M}\sum_{M_J}A^{2M_J}_{LM}\cdot$$

$$(-1)^{M_J}Y_{LM}(\hat{\vec{\kappa}})\left[D^2_{-M_J2}(\alpha,\beta,\gamma)+D^2_{-M_J-2}(\alpha,\beta,\gamma)\right] \qquad (3),$$

where the parameters $A^{JM_J}_{LM}$ are defined by

$$A^{JM_J}_{LM}=\frac{\alpha\,\omega\,p}{\sqrt{3}\sigma_i(\omega)}\sum_{l_1m_1}\sum_{l_2m_2}\sum_{\mu\,m'm''}[l_1,l_2,L,J]^{1/2}(-1)^{m'+m_2+1}\begin{pmatrix}l_1&l_2&L\\0&0&0\end{pmatrix}\cdot$$

$$\begin{pmatrix}l_1&l_2&L\\-m_1&m_2&-M\end{pmatrix}\begin{pmatrix}1&1&J\\-m'&m''&M_J\end{pmatrix}<0|\hat{d}^*_{m'}|il_1m_1\mu>\;<il_2m_2\mu|\hat{d}_{m''}|0> \qquad (4).$$

p is the photoelectron momentum and $\vec{\kappa}=\vec{p}/p$, α is the fine–structure constant, $\sigma_i(\omega)$ is the partial photoionization cross section for the ith subshell; the dipole matrix elements in the molecular frame $<il_2m_2\mu|\hat{d}_{m''}|0>$ are defined in Ref. 5, and $[l]\equiv 2l+1$. The parameters $A^{JM_J}_{LM}$ coincide with the parameters $A^{JM_J}_{LMSM_S}$ introduced in Ref. 5 with $S=M_S=0$. They are normalized by the condition $A^{00}_{00}=1$. $D_{-M_J2(-2)}(\alpha,\beta,\gamma)$ are the Wigner D–functions with σ,β,γ being the Euler angles defining the transformation from the molecular frame to the photon frame[5].

For experimental reasons [7,8] it can be more convenient to determine the normalized LDAD intensity which we will term A_{LDAD}

$$A_{LDAD}(\vec{\kappa},\vec{q}) = \frac{I^{i}_{\vec{e}_y}(\vec{\kappa},\vec{q}) - I^{i}_{\vec{e}_x}(\vec{\kappa},\vec{q})}{I^{i}_{\vec{e}_x}(\vec{\kappa},\vec{q}) + I^{i}_{\vec{e}_y}(\vec{\kappa},\vec{q})} = \frac{I^{i}_{LDAD}}{I^{i}_{+}} \qquad (5)$$

where

$$I^{i}_{+} = 3\sigma_i(\omega)\sum_{L,M} Y_{LM}(\hat{\kappa})\left[\frac{1}{\sqrt{3\pi}}A^{00}_{LM} + \sqrt{\frac{2}{15}}\sum_{M_J} A^{2M_J}_{LM}\,Y_{2M_J}(\hat{q})\right] \qquad (6)$$

Formally, LDAD appears due to the fact that the dipole operators for two orthogonal linear polarizations in the molecular frame contain two terms, one of which is identical for the two polarizations and the other one differs by a sign. The angular distribution for each of these polarizations is proportional to the square modulus of the corresponding matrix element of the operator in the molecular frame, and contains a group of terms which are identical for both polarizations, and a group of terms which differ by sign. This latter group of terms is responsible for the LDAD. So, the origin of LDAD is connected with the existence of a distinguished direction in space defined by the molecular orientation, and with the dipole selection rules.

LDAD is a kind of optical activity. The usual optical activity of unoriented chiral molecules appears due to the dissymmetry of their structure and is described by the electric dipole – magnetic dipole interference terms.[10] The optical activity of oriented molecules is connected with a dissymmetry of the geometry of the experiment[5] and appears already in the electric dipole approximation.

3. LDAD FOR DIATOMIC MOLECULES

As the simplest possible example we will consider the case of *homonuclear diatomic molecules,* for which $A^{JM_J}_{LM} = (-1)^J A^{J-M_J}_{L-M}$ and $M + M_J = 0$ and $A^{JM_J}_{LM} = 0$ for L odd. A possible experimental geometry is shown in Fig. 1, where the molecular axis lies in the XOZ plane so that the Euler angles $\alpha = \gamma = 0$ and $0 \leq \beta \leq \pi$. If furthermore the photoelectron momentum is fixed at $\Theta = \varphi = 45^\circ$, the contribution of eq. (2) for unoriented molecules is zero. Using the explicit expressions for the Wigner D–functions[11], we find from (3)

$$I^i_{LDAD}(\vec{\kappa},\vec{q}) = \frac{\sqrt{3}}{\pi\sqrt{2}}\,\sigma_i(\omega)\,\Big\{\sum_{L=o}^{\infty} A^{20}_{L0}\sqrt{\frac{3\pi}{2}}\,Y_{L0}(\hat{\kappa})\sin^2\beta$$

$$+\sum_{L=1}^{\infty} A^{21}_{L-1}\,\Theta_{L-1}(\vartheta)\sin\beta\cos\beta\cos\varphi$$

$$+\sum_{L=2}^{\infty} A^{22}_{L-2}\,\Theta_{L-2}(\vartheta)\,(1-\frac{1}{2}\sin^2\beta)\cos 2\varphi\Big\} \tag{7}$$

Where $\Theta_{LM}(\vartheta)$ is a part of the spherical harmonic $Y_{LM}(\vartheta,\varphi) = \Theta_{LM}(\vartheta)\frac{e^{iM\varphi}}{\sqrt{2\pi}}$. For homonuclear diatomics only even L terms contribute.

It is interesting to integrate this equation over all electron ejection angles thus deriving the *Linear Dichroism (LD) in photoabsorption of oriented molecules*

$$I^i_{LD}(\vec{q}) = I^i_{\vec{e}_y}(\vec{q}) - I^i_{\vec{e}_x}(\vec{q}) = \frac{3\sigma_i(\omega)}{\sqrt{2}}\,A^{20}_{00}\sin^2\beta\cos 2\gamma \tag{8}$$

This nonzero result for the LD shows that oriented linear molecules (as well as nonlinear molecules) exhibit optical activity in photoabsorption. Note that in the geometry of Fig. 1 $\gamma = 0$ and I_{LD} shows a very simple angular dependence. Of course, for unoriented molecules LD is always equal to zero.

For comparison, we derive the *Circular Dichroism in the Angular Distribution* (CDAD) for linear molecules using the notations introduced above.

$$I^{i}_{CDAD}(\vec{\kappa},\vec{q}) = I^{i}_{+1}(\vec{\kappa},\vec{q}) - I^{i}_{-1}(\vec{\kappa},\vec{q}) =$$

$$\frac{1}{\pi}\sqrt{\frac{3}{2}}\,\sigma_i(\omega)\sin\vartheta\sin(\varphi-\alpha)\sum_{L=1}^{\infty}\Theta_{L-1}(\vartheta)\,\mathrm{Im}A^{11}_{L-1} \qquad (9)$$

Comparison of eqs. (7), (8) and (9) immediately reveals the complementary nature of LDAD, LD and CDAD results. They depend on different dynamical parameters $A^{JM}_{LM}J$.

Finally, we present a numerical example for the model case of an oriented homonuclear diatomic molecule calculated in a special geometry. The angle

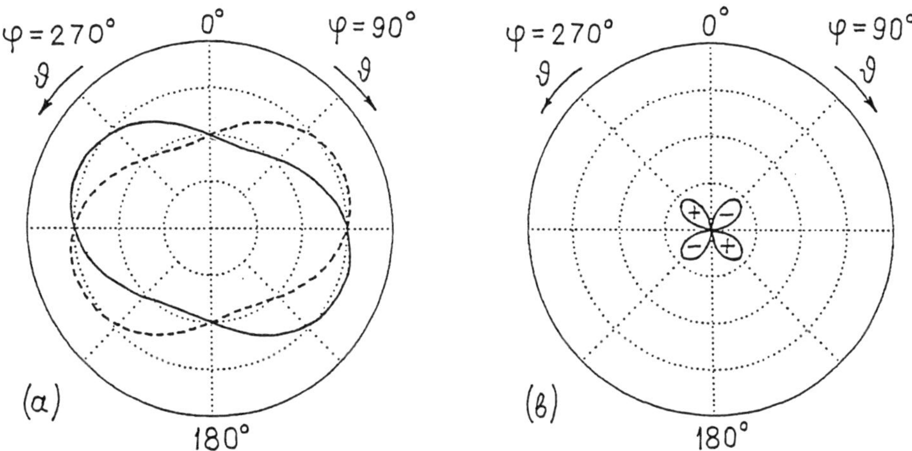

Fig. 2: LDAD for an oriented homonuclear diatomic molecule (geometry see text).

a) $I_{\vec{e}x}(\vartheta)$, dashed curve; and $I_{\vec{e}y}(\vartheta)$, full curve

b) LDAD as a function of the angle ϑ

between molecular axis $\vec{n}$ and photon propagation direction $\vec{q}$ is the magic angle 54.7° (second Legendre polynomial vanishes), electron observation is in the plane perpendicular to the plane spanned by $\vec{n}$ and $\vec{q}$. An initial σ orbital is ionized to sσ, pσ and pπ continuum states. The dipole matrix elements for pσ and pπ are equal to 1 and are two times smaller for the sσ channel. The phase shifts are $\delta_{p\sigma}-\delta_{s\sigma}=\frac{\pi}{4}$, $\delta_{p\pi}-\delta_{p\sigma}=\frac{\pi}{18}$.

The results of our model calculation for this particular case are shown in Fig. 2 as polar plots. When switching from $\vec{e}_x$ to $\vec{e}_y$ polarization, the angular distribution pattern (a) exhibits a significant rotation. The LDAD signal (b) is the intensity difference between the two curves in (a). Experimentally, it can be determind by detecting – at each angle ϑ – the photoelectron intensities for both photon polarizations and by plotting the intensity difference (or asymmetry according to eq. 5).

An especially sensitive signal occurs if the molecular axis $\vec{n}$ is rotated in the geometry of Fig. 1 when $\vartheta = \varphi = 45^0$, so that the contribution of eq. (2) for unoriented molecules is equal to zero.

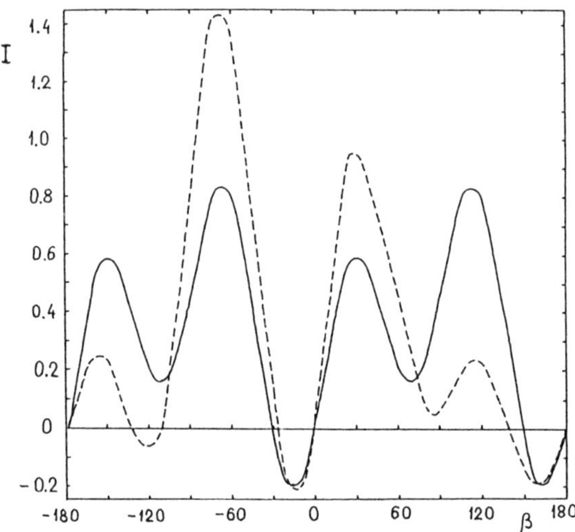

Fig. 3: LDAD for the geometry of experiment presented in Fig. 1 for the model case of a homonuclear (full curve, nonzero parameters are $A^{20}_{00}=A^{20}_{20}=A^{21}_{22-1}=A^{22}_{2-2}=1$) and a heteronuclear (dashed curve, nonzero parameters are $A^{20}_{00}=A^{20}_{10}=A^{20}_{20}=A^{21}_{1-1}=A^{21}_{2-1}=A^{22}_{2-2}=1$) diatomic molecule.

Fig. 3 shows the relative values of I_{LDAD} (expression in brackets of eq. (7)) for this model case for both homonuclear and heteronuclear molecules. The angular dependence of I_{LDAD} is rather complicated indeed, and being measured, enables one to extract simultaneously several parameters.

Performing successively a series of measurements of CD, LD, CDAD, LDAD and of the angular distributions I, one can extract all essential parameters characterizing the process provided that from each measurement only a few parameters are extracted. In the model case considered

PARAMETER	KIND OF MEASUREMENTS				
	CDAD	LD	LDAD case 1[a]	LDAD case 2[b]	I
A_{10}^{00}	–	–	–	–	+
A_{20}^{00}	–	–	–	–	+
A_{00}^{20}	–	+	–	+	–
A_{10}^{20}	–	–	–	+	–
A_{20}^{20}	–	–	–	+	–
A_{11}^{1-1}	+	–	–	–	+
A_{21}^{1-1}	+	–	–	–	+
A_{11}^{2-1}	–	–	+	–	–
A_{21}^{2-1}	–	–	+	–	–
A_{22}^{2-2}	–	–	–	+	+

Table 1. Possible set of experiments from which ten parameters $A_{LM_J}^{JM_J}$ with L $\lesssim$ 2 can be extracted (marked by +).

[a] The geometry of the experiment is defined above.

[b] The geometry of the experiment is defined by the Euler angles $\alpha = 0$, $\beta = 45^0$, $\gamma = 0$, and $0 \lesssim \vartheta \lesssim \pi$, $\varphi = 90^0$ or 270^0.

above, it is easy to find a sequence of measurements which enables one to extract all ten parameters with no more than three parameters being extracted from each measured curve. Table 1 illustrates such a sequence for one particular set of experiments.

CONCLUSION

LDAD and LD are new phenomena in photoemission of space–fixed molecules, appearing in the electric dipole approximation. The kind of information gained through these measurements is complementary to CDAD results [3–8]. The following points are essential:

i) Both LDAD and CDAD do not depend on the spin–orbit (or spin–axis) interaction, and if the fine–structure splitting of either the initial or final ionic state is not resolved, both LDAD and CDAD remain essentially unchanged [7].

ii) CDAD is described by the imaginary part of the $A_{LM}^{JM_J}$ parameters (4) with J=1, while LDAD is given by the parameters with J=2. It means that the information which can be extracted from LDAD measurements is not identical with the information extracted from CDAD measurements. In particular, Im $A_{LM}^{1M_J}$ is proportional to sines of phase shift differences of continuum wave functions $(\delta_{l_1m_1}-\delta_{l_2m_2})$, while the parameters $A_{LM}^{2M_J}$ contain cosines of the phase shift differences. Therefore LDAD and CDAD measurements supplement each other when one needs to extract the dipole matrix elements and phase shifts from the measured quantities.

iii) In principle, LDAD measurements can give more information than CDAD measurements since the general expression for LDAD contains more terms. In the particular case of linear molecules it is evident from the direct comparison of eqs. (7) and (9). Here CDAD is defined by the parameters A_{L-i}^{ii} which contain only interference terms between degenerate photoelectron continua differing by ±1 in their m values, $\Delta m = m_2 - m_i = \pm 1$, while LDAD is described by the parameters A_{L0}^{20}, A_{L-1}^{21}, A_{L-2}^{22}, which contain interference terms between continua differing by $\Delta m = 0, \pm 1$ and ± 2, respectively.

Experiments of LDAD for oriented molecules adsorbed at surfaces are just being performed.

Acknowledgement

One of the authors (N.A.C.) acknowledges the hospitality of Bielefeld University FRG, where this work has been started, financially supported by the Deutsche Forschungsgemeinschaft (SFB 216), and the hospitality of the University of Nebrasca–Lincoln, where it has been completed.

REFERENCES

1. J. Kessler, Comments Atom. Molec. Phys. 10, 47 (1981)
2. Ch. Heckenkamp, F. Schäfers, G. Schönhense and U. Heinzmann, Z. Phys. D 2, 257 (1986).
3. N.A. Cherepkov, Chem. Phys. Lett., 87, 344 (1982).
4. R.L. Dubs, S.N. Dixit and V. McKoy; Phys. Rev. Lett., 54, 1249 (1985);
   J. Chem. Phys., 85, 656 (1986).
5. N.A. Cherepkov and V.V. Kuznetsov. Z. Phys. D 7, 271 (1987); J. Phys. B 22, L405 (1989); J. Chem. Phys., in press (1991).
6. J.R. Appling, M.G. White, T.M. Orlando and S.L. Anderson. J. Chem. Phys. 85, 6803 (1985);
   J. Winniczek, R.L. Dubs, J.R. Appling, V. McKoy and M.G. White, J. Chem. Phys., 90, 949 (1989).
7. C. Westphal, J. Bansmann, M. Getzlaff and G. Schönhense. Phys. Rev. Lett., 63, 151 (1989); Vacuum, 41, 87 (1990); J. Electron Spectrosc. Relat. Phenom. 52, 613 (1990);
   C. Westphal, J. Bansmann, M. Getzlaff, G. Schönhense, N.A. Cherepkov, M. Braunstein, V. McKoy and R.L. Dubs. Surface Sci., 253, 205 (1991).
8. G. Schönhense. Phys. Scripta, T31, 255 (1990).

9. J. Cooper and R.N. Zare. J. Chem. Phys., 48, 942 (1968).
10. D.J. Caldwell and H. Eyring. The Theory of Optical Activity (Wiley: New York 1971).
11. D.A. Varshalovich, A.N. Moskalev and V.K. Khersonskii. Quantum Theory of Angular Momentum, (World Scientific: Singapore 1988).

QUESTION

Q - What are the specific advantages of LDAD and CDAD measurements over angular distribution measurements ?

CHEREPKOV and SCHONHENSE - The main advantage of LDAD and CDAD measurements in comparison with the angular distribution measurements is connected with the fact that the number of terms contributing to LDAD or CDAD is strongly reduced. One can easily extract two or three parameters from the CDAD or LDAD angular dependence, while it is quite problematic to extract five or even more parameters from the angular distribution curve.

CI MODELS FOR THE DESCRIPTION OF MANY-BODY EFFECTS IN THE PHOTOIONIZATION SPECTRA.

P. Decleva, G. De Alti, G. Fronzoni and A. Lisini
Dipartimento di Scienze Chimiche, Universita' di Trieste, Via A. Valerio 38, I-34127 Trieste (Italy)

ABSTRACT

Large basis set CI calculations are performed for the satellite states in the valence and core photoelectron spectra of a series of representative small systems, at various excitation levels, up to 4h-3p. The explicit inclusion of relaxation gives significant improvement at lowest excitation level and is therefore employed in all calculations. A valence space strongly interacting is identified by means of a projection scheme, allowing the use of restricted schemes, like 3h-2v-1p, which are still adequate for a qualitative description of all spectral regions. Improved intensity evaluation in the valence region is afforded by the use of a generalized overlap approach.

1. INTRODUCTION

Many-body effects lead to the appearance of satellite peaks in the photoelectron spectra and to the breakdown of the one-particle picture of ionization[1]. A detailed interpretation of the shake-up structures requires a quantitative reproduction both of the energies and the intensities of the satellites and has been proven to be very difficult because of the density of final states and the strong coupling of highly excited configurations. Their calculation is therefore a severe test for the theoretical models employed, especially when accurate experimental results are available for the comparison.

The shake-up states have usually strong contribution from 2h-1p configurations so an overall picture of the shake-up spectrum can be achieved, whitin a CI framework, by the simple 2h-1p CI model. It has been proven to be useful for a first description of the photoelectron spectra also in large systems just because of its computational simplicity. It is necessary, however, to use an adequate basis set, which should comprise a certain number of Rydberg orbitals, and to take into account the relaxation by the use of SCF relaxed orbitals for the ions[2]. This is important not only for the description of the core holes but also in the case of valence ionizations, as has been verified in recent calculations on the second and third row hydrides[3]. To obtain a quantitative comparison with the experimental data it is however necessary to extend the 2h-1p CI scheme, including higher excitations (3h-2p, 4h-3p) with the consequence of producing increasingly large CI matrices. A useful way to limit the dimension of the problem is the individuation among the virtual orbitals of a valence space in which excitations higher than 2h-1p are induced to take into account the strongest quasi-degeneracy correlation effects, while the rest of correlation is described by single or double excitations in the remaining space. This valence space can be obtained by means of a projection of the entire virtual space of the extended basis set onto the minimal basis set and has proven to be adequate for the description of both primary and excited ionic states[4]. Also a quantitative reproduction of the intensities is often rather difficult, even in the high energy limit where it is expected that the relative intensities are satisfactorily described by the Sudden Approximation (SA). In a bound-state calculation the intensity of a ionization process is characterized by the amplitudes:

$$\chi_{ki} = \langle \Psi_k^{N-1}, a_i \Psi_0^N \rangle \tag{1}$$

where i is the orbital annihilated.

At high energy there are two alternative definitions of SA to describe the intensity ratio: the spectral strength ratio:

$$\frac{I_{k'}}{I_k} = \frac{R_{k'}}{R_k} \qquad R_k = \Sigma_i \left| \chi_{ki} \right|^2 \tag{2}$$

and the generalized overlap ratio:

$$\frac{I_{k'}}{I_k} = \frac{\left| \langle \varphi_{\varepsilon k'} \mid \varphi_{k'} \rangle \right|^2}{\left| \langle \varphi_{\varepsilon k} \mid \varphi_k \rangle \right|^2} \tag{3}$$

with the Dyson orbital $\varphi_k = \Sigma_i \chi_{ki} \varphi_i$ and $\varphi_{\varepsilon k}$ given by a plane wave[2].

Both definition of the intensities are strongly influenced by electron correlation and only in the case of large CI calculations it is possible to reach converged intensities values. In particular, it has been shown in the Ne and Ar valence spectra [2], that in the case of Rydberg excitations the use of the spectral strength definition is inadequate to accurately represent the intensities, requiring at least the use of the SA including a plane wave description of the photoelectron or even an explicit treatment of the continuum. On the contrary such effects are found of minor importance in the core region, which is accurately reproduced by the spectral strengths values.

In order to analyze the capability of the (n+1)h-np CI schemes, coupled with both definitions of SA, to describe the satellite structures in PES, we report selected results for a series of representative small systems for which accurate experimental data exist for comparison.

We refer to previous work [2,3,7] for details of the computational procedures employed and extensive reference to previous investigations.

2. RESULTS AND DISCUSSION

A natural starting point for the discussion is the examination of the spectra of noble gases, for which individually resolved satellites measured with high accuracy are available for comparison[5]. The importance of the use of SCF relaxed orbitals is illustrated in Table I for the valence ionization spectrum of Ar.

Table I. Satellite energies[a] and intensities[b] in Argon.

Ar	EXP.[5]		2h-1p NR[c]		2h-1p R[d]		(2h-2p)	3h-2p		4h-3p		
	E	I	E	I	E	I	I	E	I	E	I	I[e]
$3p^{-1}$	(15.8)	159.	(13.93)	(0.946)	(13.84)	(0.959)	(0.904)	(15.54)	(0.924)	(15.79)	(0.924)	159.
$3s^{-1}$	13.4	100.	13.57	(0.667)	13.42	(0.629)	(0.591)	13.97	(0.675)	13.12	(0.602)	100.
$3p^4 4p$	19.8	0.6	25.81	1.37	19.83	0.24	0.40	21.40	0.63	19.29	0.47	0.57
$3p^4 4s$	20.7	0.6	23.68	0.25	20.67	0.71	0.75	23.18	1.80	20.24	0.86	0.45
$3p^4 4p$	21.4	3.7	26.67	0.29	21.58	1.48	2.00	23.22	2.25	20.98	2.26	3.39
$3p^4 3d$	22.8	18.6	26.79	14.71	23.42	12.28	12.35	24.79	11.97	22.68	19.52	18.70
$3p^4 4p$	23.8	1.5	27.95	0.54	23.68	0.41	0.59	25.30	0.77	23.30	0.79	1.20
$3p^4 4d$	25.4	9.4	29.47	6.09	25.85	5.96	6.04					
$3p^4 5d$	26.9	4.1			27.21	4.60	4.67					
$3p^4 6d$	27.6	1.5			27.87	1.46	1.48					

[a] The absolute energy is reported for the lowest state, relative energies are reported for all other states.

[b] Theoretical spectral strengths are reported for the primary peaks, satellite intensities are normalized to the experimental intensities of the primary peaks.

[c] Unrelaxed orbitals; [d] Relaxed orbitals.

[e] Generalized overlap ratios.

At the 2h-1p unrelaxed level the energies are overestimated while the intensities are too low; the use of relaxed orbitals brings a decisive improvement for the energies. To reach an increase in the intensity values it is necessary to use also a 2h-2p correlated ground state. The presence in the spectrum of the $3s3p^6 \leftrightarrow 3s^2 3p^4 3d$ interaction gives rise to an intense shake-up line. To describe satisfactorily the intensity of this state it is necessary to extend the CI excitations up to 4h-3p level. As concerns the intensity calculations performed with the two different formulations of the SA, it can be seen that the generalized overlap values are in better agreement with the experiment, confirming that the use of spectral strength alone is generally insufficient for a quantitative reproduction of the experimental intensities.

The shake-up spectra of the hydrides HCl, H_2S and PH_3 represent a next step to understand the many-body effects in PES because of their simple electronic structure. In fact, the inner valence region of these molecules shows some similarities with the isoelectronic noble gas Ar so a similar behaviour of the 2h-1p CI calculations is expected.

Table II. Experimental and calculated ionization energies (eV) and intensities of HCl.[a]

EXP[8]		2h-1p		3h-2v-1p		4h-3p		
E	I	E	I	E	I	E	I	I[b]
2π								
(12.8)	100.	(12.78)[c]	100.	(12.78)[c]	100.	(12.78)[c]	100.	100.
				18.09	0.53	16.87	0.55	0.90
				18.29	0.46	17.01	0.47	0.83
				20.38	0.42	17.88	0.40	0.69
						21.38	0.46	0.63
5σ								
3.86	50.	3.96	50.	4.27	50.	3.91	50.	
4σ								
11.10	28.	11.67	61.2	11.35	29.95	10.81	34.34	31.48
13.20	67.	13.99	67.	14.04	67.	13.37	67.	67.
15.87	2.9	16.27	7.0			15.64	5.27	4.70
17.16	11.	17.84	7.0	16.90	3.68	16.85	4.29	5.17
						18.24	6.11	4.15
19.44	13.	19.30	5.3	19.33	7.36	19.44	3.90	3.52
		21.72	2.2	20.81	1.40			
20.89	11.	21.80	6.1	22.14	1.74			
		21.98	1.7	22.33	10.68			
21.87	9.6	22.49	9.7	22.85	2.80			
		23.12	1.7					
24.2	18.	25.96	12.1	25.77	2.04			
		26.45	3.4	26.38	11.39			
		26.57	3.6	26.90	1.58			
		26.72	3.9	27.21	0.92			
		28.56	5.1	28.42	6.31			
		29.24	2.4	29.04	2.10			
		29.95	3.4					

[a] The absolute energy for the lowest state is reported; for all other states relative energies are given; [b] Generalized overlap values.
[c] 4h-3p CI value.

From Table II for HCl and Figure 1 for H_2S and PH_3 it can be seen that this scheme gives a good overall description of the spectra, provided that an adequate basis set is chosen and SCF relaxed orbitals are used. In the HCl molecule there is still a resemblance with the Ar atom, in particular the presence of 3d quasidegeneracy, although a strong shake-down line is apparent. A significant improvement is obtained at 3h-2v-1p CI level and in particular the intensity ratio between the first two lines, which is rather poor in the 2h-1p calculations, is brought in accordance with the experiment.

In the H_2S spectrum we can observe that the main line is split into several components of comparable intensities with a complete break-down of the simple one-particle picture. The 3h-2v-1p CI calculation gives a redistribution of the intensities of the main band with respect to the 2h-1p CI results bringing the calculated spectrum in close resemblance with the experiment.

The shake-up structure of PH_3 is only fairly reproduced by 2h-1p CI calculations; the improvement of the 3h-2v-1p results is clearly seen in the figure 1 although the main band appears too sharp at variance with the experiment.

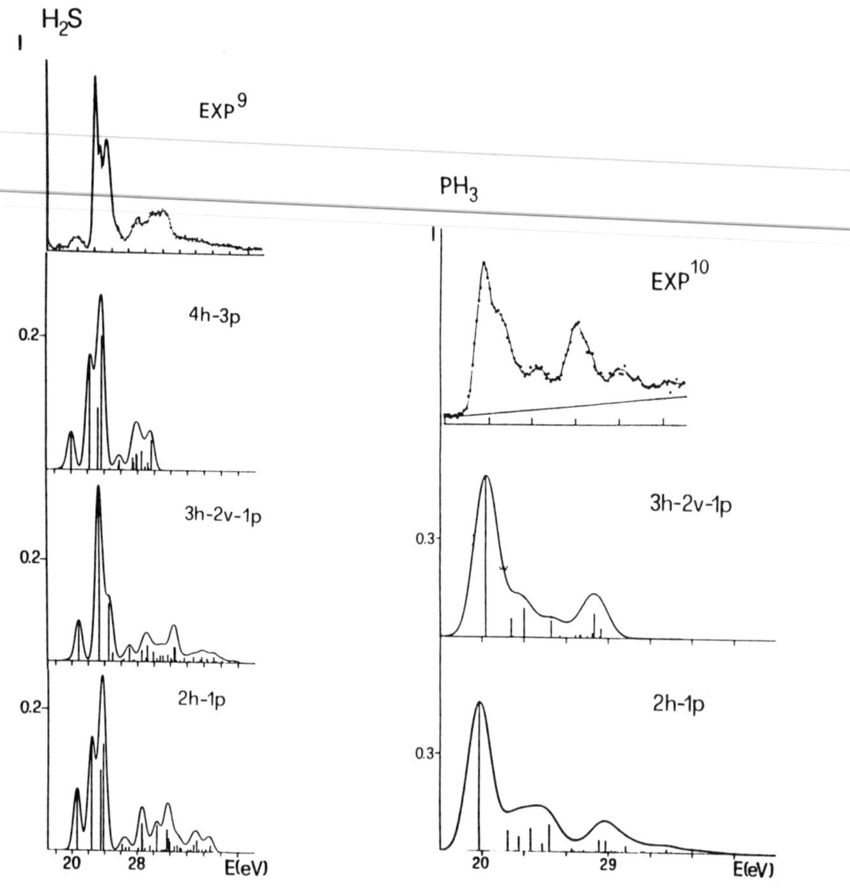

Fig. 1. Experimental and calculated inner valence spectra of H_2S and PH_3.

For HCl and H_2S also 4h-3p CI calculations are performed to get a quantitative comparison with the well resolved experimental spectra available. There is a general improvement as concerns the energies of both molecules. In HCl there is a better reproduction of the low energy part of the spectrum, although some accuracy may be lost for the highest states computed. A further improvement in the intensities is afforded also here by the use of the generalized overlap scheme. In H_2S the rearrangement of the intensities within the main band gives a pattern similar to that obtained at 2h-1p CI level. For both molecules present 4h-3p results are quite close to recent ADC(4) GF results[6], indicating the convergence of the calculations, and pointing out the difficulty of a reproduction of the experimental values for small molecules comparable to that in atoms. Both inadequate evaluation of the intensity at the S.A. level, and nuclear motion effects are probably responsible of the persisting discrepancies.

Table III. Satellite energies (eV) and intensities (%) in the core ionization of CO[a].

2h-1p		3h-2v-1p		4h-3p[7]		ADC(4)[11]		Exp[11]	
O1s									
541.72	0.73	543.82	0.65	543.10	0.64	541.56	0.61	542.57	100.
14.18	0.1	13.66	0.0	13.11	0.0	14.88	0.0	15.9	11.2
17.92	4.4	16.85	10.8	15.21	0.1	15.80	0.0	17.25	1.3
18.43	0.4	17.61	0.2	15.68	5.2	17.00	0.4	18.11	4.2
19.15	0.2	19.00	1.3	16.90	0.5	17.47	0.2	20.06	1.2
19.92	4.9	19.33	0.6	17.20	6.0	18.37	0.9	21.7	
22.92	0.7	20.29	0.1	17.38	1.2	20.32	1.5	23.7	
23.93	0.8	21.50	0.6	19.22	0.8	20.54	1.5	26.9	
25.69	0.3	22.96	0.0	20.19	0.5	21.67	5.9	27.9	
25.97	1.1	24.18	0.3	21.30	0.1	22.11	0.6		
27.61	1.1	24.73	0.2	22.10	0.4	22.50	4.3		
		24.85	0.3			24.19	0.0		
C1s									
297.75	0.78	297.71	0.72	297.00	0.73	296.08	0.68	296.19	100.
6.39	2.2	10.23	1.5	8.20	2.5	9.11	2.2	8.34	2.3
17.12	1.6	18.46	1.6	15.77	2.0	17.12	5.0	14.88	4.8
20.46	1.1	19.73	0.4	17.37	1.1	19.89	1.3	17.85	1.9
21.51	0.8	20.17	0.8	18.24	1.6	20.79	0.4	19.22	1.1
22.16	2.0	21.43	1.8	18.61	0.6	21.18	0.2	20.09	1.6
23.14	0.2	23.34	1.4	19.80	1.1	21.64	1.7	21.1	0.4
23.54	0.3	23.81	0.3	21.12	0.1	22.31	0.7	22.2	0.5
23.83	1.0	24.36	0.5					23.2	2.7
24.59	0.5	27.63	1.9						

[a] Absolute energy and spectral strength is reported for the main ionization.

Also in the case of core holes, despite the simplification offered by the usual hole localization, a quantitative treatment of satellites is not easy even in simple molecule. The strong relaxation often present gives a decided advantage to the use of relaxed orbitals for the ionic states. Moreover the strong field of the core holes pulls valence empty orbitals down in energy, so that doubly excited configurations mix very strongly and may even dominate the physical states.

Selected results for the CO molecule are reported in Table III. Despite having been repeatedly investigated, a detailed assignement of the full spectrum is still unavailable, although the low energy features are reasonably understood. Employing relaxed $1s^{-1}$ orbitals a semiquantitative description of the main features of the spectrum is already available at the 2h-1p level, altough errors in the energy positions are rather large, and the intensities are generally underestimated. Double excitations within the valence space give rise to a 3h-2v-1p scheme which affords a significant improvement for the lower states, notably a good reproduction of the high intensity line relative to the O 1s spectrum, although the agreement deteriorates at higher energies. A highly correlated 4h-3p CI calculation[7] gives satisfactory agreement for the lowest lying satellite energies, although the intensities are still significantly underestimated in some cases. ADC(4) calculations of comparable sophistication probably suffer from inadequate treatment of relaxation, as is especially suggested by wrong intensity distribution obtained in the O $1s^{-1}$ case[11].

4. CONCLUSIONS

The 2h-1p CI scheme with relaxed orbitals gives good results for the satellites, considering the simplicity of the model which can therefore serve as a first description also in large systems. Improved results can be obtained supplementing this scheme with higher excitations in the valence space only, as in the case of 3h-2v-1p CI scheme which can be proposed for the interpretation of the spectra dominated by quasi-degeneracy effects. Very accurate results both for the energies and intensities are afforded at 4h-3p CI level, although a quantitative reproduction of the latter is more difficult and may requires al least the use of the generalized overlap approach.

REFERENCES

1. L.S. Cederbaum, W. Domke, J. Schirmer and W. von Niesse, Adv. Chem. Phys. 65 (1986) 115, and ref. therein.
2. P. Decleva, G. De Alti, G. Fronzoni and A. Lisini, J. Phys. B: At. Mol. Opt. Phys., 23 (1990) 3777.
3. A. Lisini, P. Decleva and G. Fronzoni, J. Mol. Struct. (Theochem) 228 (1991) 97.
4. P. Decleva, G. Fronzoni and A. Lisini, Chem. Phys., 134 (1989) 307 and ref. therein.
5. S. Svensson, B. Eriksson N. Mårtensson, G. Wendin and U. Gelius,

J. Electron Spectry., 47 (1988) 327.
6. W. von Niessen, P. Tomasello., J. Schirmer, L.S. Cederbaum, R. Cambi, F. Tarantelli and A. Sgamellotti, J. Chem. Phys., 92 (1990) 4331; W. von Niessen and P. Tomasello, Phys. Scr., 41 (1990) 841.
7. M. Ohno and P. Decleva, Chem. Phys., 00 (1991) 00.
8. S. Svensson, L. Karlsson, P. Baltzer, B. Wannberg, U. Gelius and M.Y. Adam, J. Chem. Phys., 89 (1988) 7193.
9. M.Y. Adam, A. Naves De Brito, M.P. Keane, S. Svensson, L. Karlsson, E. Källne and N. Correia, J. Electron Spectry., 56 (1991) 241.
10. C. Cauletti, M.N. Piancastelli and M.Y. Adam, J. Mol. Struct., 174 (1988) 135.
11. J. Schirmer, G. Angonoa, S. Svesson, D. Nordfors and U. Gelius, J. Phys. B: At. Mol. Phys., 20 (1987) 6031.

PHOTODISSOCIATION OF H_2 INTO H(1S) + H(n=3) FRAGMENTS: WHICH ATOMIC STATE?

M. Glass-Maujean
Laboratoire de Spectroscopie Hertzienne de l'E.N.S.*
Université P. et M. Curie, 4 pl. Jussieu, 75252 Paris
and L.U.R.E. 91405 Orsay

H. Frohlich
Laboratoire des Collisions Atomiques et Moléculaires*
Université Paris-Sud, 91405 Orsay
and L.U.R.E. 91405 Orsay

ABSTRACT

The relative populations of the 3s, 3p, 3d states of the fragments have been determined for various excitation energies, by a fluorescence time analysis. The branching ratios show strong variations. The possible predissociation mechanisms are discussed.

INTRODUCTION

The photodissociation of H_2 may appear as an overstudied system. H_2 is the simplest molecule, *ab-initio* calculations can reach an amazing precision, nevertheless many experimental results are still puzzeling.

The excited states of atomic hydrogen are degenerated, the Mulliken correlation diagrams are enable to lead to the determination of the final state of the dissociation.

For the first optically allowed dissociation channel: H(1s) + H(n=2), a partition into the H(2s) and H(2p) fragments oscillating with the excess energy was predicted [1] and observed [2] on the continuum. Such a behaviour is due to an interference between the B and B' continua simultaneously excited in the Franck-Condon region and coupled at large internuclear distance.This unusual situation araises from the energy degenescence of the excited levels of the H fragment.

For the second optically allowed dissociation channel: H(1s) + H(n=3), very little is known. It opens at λ=748.5Å. The $(1s\sigma_g,3p\pi_u)$ D $^1\Pi_u$ state is known to correlate to H(1s) + H(3p) [3], the $(1s\sigma_g,4p\sigma_u)$ B"B $^1\Sigma^+_u$ state to H(1s) + H(3p) too [4]. The $(1s\sigma_g,4p\pi_u)$ D' and $(1s\sigma_g,5p\sigma_u)$ [5] are expected to correlate to H(1s) + H(n=4). All the other $^1\Pi_u$ or $^1\Sigma_u$ states correlated to H(1s) + H(3s) or H(3d) are unidentified, they are expected to be doubly excited in the Franck-Condon region.

First to observe the H(n=3) excitation spectrum and then to determine the excited level of the fragment will help in the understanding of such apparently simple system.

EXPERIMENT

The V.U.V. continuum of the synchrotron radiation of Super-ACO storage ring at

* Unités associées au C.N.R.S.

Orsay was dispersed by a 3-m normal incidence monochromator equiped with a 2400 l/mm holographic grating. The spectral range from 700 to 750Å was explored with a 0.2Å bandwidth. The transmitted light passed through a differencially pumped cell, where the pressure, measured by a Baratron-type gauge, was maintained to a few militorrs.
Balmer α fluorescence was detected perpendicularly to the incident synchrotron light beam, through a light pipe, with a refrigerated photomultiplier. The fluorescence photon counts were recorded in a computer file. The transmitted photon counts recorded without gas had been used to numerically normalize the fluorescence signal to give the relative H_α cross section.

We took adventage of the pulse character of the synchrotron radiation to perform a time analysis of the H_α fluorescence. The dissociation and predissociations occur in a time very short compared to the atomic decay times, the fluorescence decay is then just the superposition of the H(3s, 3p and 3d) atomic exponential decays. The atomic lifetimes are quite different: 158, 5.3 and 15.5 ns respectively and can be easily distinguished [6]. When operating with two positron packets, the delay between two incident light pulses is of 115 ns, the analysis time is of 100 ns for 256 channels.

The branching ratios were determined on six predissociation peaks and one point between, with an incident spectral bandwidth of 1Å. For each incident wavelength the decay curve was analysed as a sum of exponentials of definite widths corresponding to the theoretical lifetimes. The populations have been deduced from the integration of the related decay over the time running between two consecutive incident pulses, taking into account that all the 3s and 3d states radiate through Balmer α line whilst only 15% of the 3p do [6].

RESULTS AND DISCUSSION

The H_α excitation spectrum is dispayed in Fig. I. The H_2 absortion spectrum had been assigned for wavelengths longer than 770Å [7,8]. For higher energies, only the ionization spectra (H_2^+, $H^+ + H^-$) observed at very low temperature had been assigned[9,10]. The levels involved in predissociation are more probably not the high Rydberg, low vibrational levels observed in ionization but low Rydberg, high vibrational levels. In H_2 for dissociation to compete successfully against ionization, ionization has to occur with a large vibrational change($\Delta v >> 1$) as vibrational interaction is the most efficient mechanism [9].

To make a temptative assignement, we used the quantum defect approach. The npσ $^1\Sigma^+{}_u$ and npπ $^1\Pi_u$ quantum defect functions had been determined by Jungen and Atabek many years ago [5]. From these values, we calculated the wavelengths of the rotational lines of the first Rydberg series. We used the $(1s\sigma_g,3p\pi_u)$ $^1\Pi_u{}^-$ series as a test, the Q_1 lines had been indentified up to v=17 [11].The $^1\Pi$ potential curves are smooth up to their dissociation limit, the 3pπ Q_1 series can be followed without any accident, the quantum defect function in ref. 5 is able to give reliable results. The $^1\Sigma$ potential curves are more perturbed by doubly excited states, the B" state exhibits a second minimum [12], the smooth quantum defect function of ref 5 is not able to reproduce the very high vibrational levels of such a state. From the *ab-initio* values of the B"B potential [12], we deduced quantum defect values for large internuclear distances, extrapolating the previous values.

Most of the lines may be assigned as Q_1 and Q_2 lines of the 4pπ and 5pπ progressions. Several strong peaks are probably due to the 6pσ progression. The results are summarized in Fig. I.

Figure I: H_α excitation spectrum.

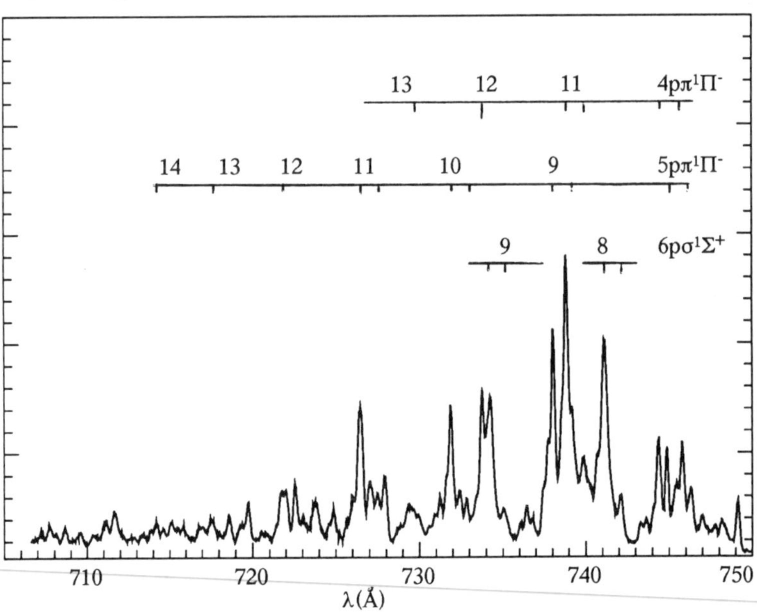

The $^1\Pi_u^-$ levels may be coupled to the 3pπ $^1\Pi_u^-$ continuum through homogenuous perturbation. Such continuum is correlated to H(1s) + H(3p). Only H(3p) atoms are then expected to be detected in a first approximation.

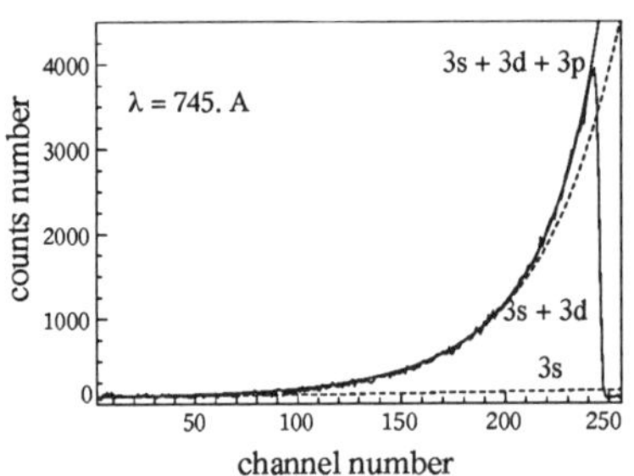

Figure II: H_α decay, and the exponential decomposition.

Such situation is rather different from the H(1s) + H(n=2) dissociation for which all the predissociations were due to couplings to the $^1\Sigma^+_u$ state. The $^1\Pi^-$states are known to be very resistant to ionization[11].

The 6pσ $^1\Sigma^+_u$ may be coupled to the B"B $^1\Sigma^+_u$ continuum through homogenous coupling. Once again the limit is H(1s) + H(3p) [4]. At very large distance (R≈35a_0), the(H(1s) + H(n=3)) $^1\Sigma^+_u$ states are coupled through the ion pair state [13], some transfert probabilities between the n=3 states are then expected[14].

The branching ratios determined on six peaks and one vale are summarised in table I.Obviously, the H(n=3) fragments are not overwhelmingly in the 3p state.

At large distance, the 3pπ $^1\Pi$ and the 3dπ $^1\Pi$ states are denerated and coupled through Van der Waals interaction [15], a 3p-3d mixing is to be expected. This mixing would not depend on the $^1\Pi_u^-$ predissociation peak as far as the direct excitation of the continuum may be neglected [1]. The branching ratios show strong variations for the $^1\Pi_u^-$ predissociation peaks. The direct excitation of the D and B"B continua has to play a role.The importance of the 3s fragments indicates that the B"B continuum must be efficient.

λ(Å)	3s	3p	3d
745.	14%	33%	53%
743.	24%	48%	29%
741.	47%	<5%	53%
740.	38%	<5%	62%
739.	21%	22%	57%
734.	22%	12%	66%
732.	39%	6%	55%
727.	26%	30%	44%

Table I:
3s, 3p, 3d branching ratios

None of the above mechanisms may explain the disappearence of the 3p fragments for some values.

CONCLUSION

The study of the branching ratios shows the high complexity of photodissociation in the case of degenerated states.

Even the simplest molecular system may keep some mystery. The degeneracy of the fragment states with large distance couplings between them yields to an interesting theoretical and yet unresolved problem.

REFERENCES

[1] J.A. Beswick and M.Glass-Maujean, Phys. Rev. A35, 3339 (1987).
[2] M. Glass-Maujean, H. Frohlich and J.A. Beswick, Pys. Rev. Lett. 61,157 (1988).
[3] A. Monfils, Bull. Acad. Roy. Belg. cl. Sci.(5) 47, 816 (1961) and J. Mol. Spectrosc. 15, 265 (1965).
[4] L. Wolniewicz and K. Dressler, J. Mol. Spectrosc. 96, 195 (1982).
[5] Ch. Jungen and O. Atabek, J. Chem. Phys 66, 5584 (1977).
[6] H.A. Bethe and E.E. Salpeter,"Quantum Mechanics of One and Two Electron Atoms", (Springer, Berlin 1957).
[7] S. Takesawa, J. Chem. Phys. 52, 2575 (1970).
[8] G. Herzberg and Ch. Jungen, J. Mol. Spectrosc. 41,425 (1972).
[9] P.M. Dehmer and W.A. Chupka, J. Chem. Phys. 65, 2243 (1976).
[10] W.A. Chupka, P.M. Dehmer and W.T. Jivery, J. Chem. Phys. 63, 3929 (1975).
[11] M. Glass-Maujean, J. Breton and P.M. Guyon, Chem. Phys. Lett. 112, 25 (1984).
[12] W. Kołos, J. Mol. Spectrosc. 62, 429 (1976).
[13] D.R. Bates and J.T. Lewis, Proc. Phys. Soc. A68, 173 (1955).
[14] I.V. Komarov and V.N. Ostrovskii, J.Phys. B12, 2485 (1975).
[15] V.N. Ostrovskii, private communication.

ABSOLUTE MEASUREMENTS OF PHOTOABSORPTION, PHOTOIONIZATION, AND PHOTODISSOCIATION CROSS SECTIONS OF SiH_4 AND Si_2H_6 IN THE 13-22 eV REGION[a]

Kosei Kameta, Masatoshi Ukai, Norihisa Terazawa, Kazunori Nagano, Yuji Chikahiro, Noriyuki Kouchi, and Yoshihiko Hatano
Department of Chemistry, Tokyo Institute of Technology, Meguro-ku, Tokyo 152, Japan

Kenichiro Tanaka
Photon Factory, National Laboratory for High Energy Physics, Tsukuba, Ibaraki 305, Japan

ABSTRACT

The absolute values of the photoabsorption cross sections and photoionization quantum yields of silane and disilane have been measured in the 13-22 eV region. Using these values, we have evaluated the absolute cross sections of the photoionization and neutral dissociation. The individual structures observed in the cross section spectra have been interpreted in terms of the characters of superexcited states with strong emphasis on the neutral dissociation. The $3a_1^{-1}\sigma^*(t_2)$ state of silane presenting a broad peak at around 14.6 eV in an absorption cross section is shown to be strongly dissociative, whereas Rydberg series to the $(3a_1)^{-1}$ limit showing up vibrational structures between 16 and 18 eV predominantly undergo fast autoionization decay. A general importance of competition between autoionization and neutral dissociation is indicated of disilane in much higher energy region than the first ionization potential.

INTRODUCTION

Competition between autoionization and neutral dissociation processes following superexcitation plays an important role in the photon-molecule or electron-molecule interactions.[1,2] Experimental difficulties in absolute cross section measurements of the photoionization and photodissociation in the extreme-UV (EUV) region limited the information on the dynamic aspect of superexcited molecules.

The photoionization quantum yield (η) i.e., the probability of ionization on a single photoabsorption event, which is defined as σ_I/σ_T (σ_I: photoionization cross section, σ_T: photoabsorption cross section), is one of the most fundamental quantities to understand photon-molecule interactions. We underline that η-measurements are quite important in the absolute evaluation of photoionization and photodissociation cross sections. Using synchrotron radiation as a light source and a multistage ionization chamber, we have measured systematically η-values for simple organic molecules in the region lower than 11.8 eV.[3,4] We have recently extended the absolute measurements to the higher energy region by employing metallic films as window materials.[5]

Photoabsorption and photoionization cross sections for silanes and heavier group-IV compounds are of interest in comparison with those for hydrocarbons, which have been relatively well examined. Recent experimental efforts on silane have been mainly devoted to the inner shell ionization[6-11] or the absorption spectra lower than 11.8 eV.[12,13] However, no absolute cross section of the photoabsorption or photoionization was obtained between 11.8 eV and the 2p shell absorption region. Very recently, Cooper et al.[14] have measured the photoabsorption cross sections and partial photoionization cross sections for silane using the dipole (e,e) and (e,e+ion) spectroscopy. A considerable discrepancy has existed in the near threshold behavior of their cross sections with the results of "real" photon experiments.[12,13] Photoexcitation processes of disilane have been much less investigated.[13,15-18]

In this paper we show the first observation of absolute photoabsorption cross sections and photoionization quantum yields of silane (SiH_4) and disilane (Si_2H_6) in the 13-22 eV region. In this energy region, the existence and the dissociation of superexcited states have been suggested by electron impact studies.[18-20] We also present photoionization and photodissociation cross sections and discuss the competitive decay processes of superexcited states.

EXPERIMENTAL

The experimental setup and procedure are almost the same as those in our previous experiments[3,4,21,22] except for an ionization chamber design and the use of window materials.[5] A modified Samson-type multistage ionization chamber with a cylindrical cross section[23]

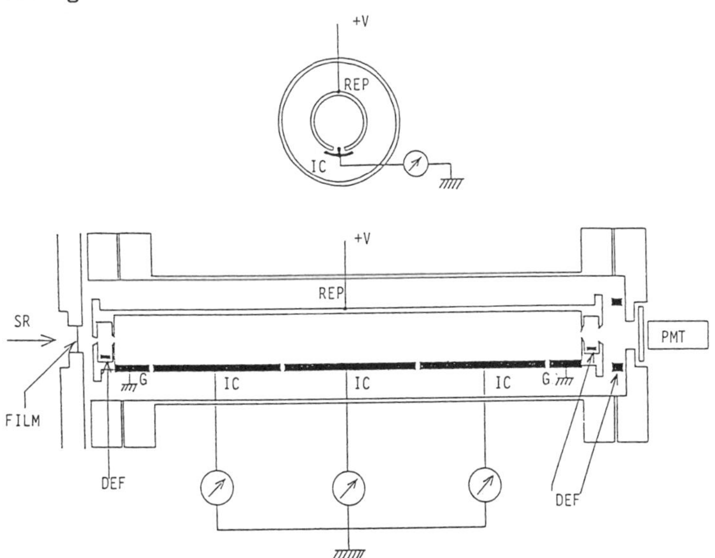

Fig.1 Multistage photoionization chamber with a cylindrical cross section. IC: ion collectors; G: Guard electrodes; REP: ion repellors; DEF: secondary electron deflectors; FILM: metallic thin film window.

was newly adopted in the present experiment (Fig.1) so as to maximize the collection efficiency of the photoion. The chamber consisted of and three pairs of secondary electron deflectors. Synchrotron radiation (SR) from the 2.5 GeV positron storage ring of Photon Factory dispersed with a 1m Seya-Namioka monochromator entered into the ionization chamber through In (for the incident photon energy of 13-17.2 eV), Sn (16.8-24 eV), or, if necessary, Te (22-40 eV) thin film as an entrance window. Use of metallic films enabled exact measurements in this EUV region by eliminating the second order components of the incident light and by preventing sample gas effusion back into the optical path. A typical wavelength scan step was 0.1 nm with a nominal bandpass of 0.1 nm. The absolute wavelength was calibrated using the Ar*3s* and Xe*5s* resonances. Sample gas pressures of 10-40 mTorr were measured with a capacitance manometer. The incident light intensity was monitored as a photo-current from gold mesh in front of the ionization chamber.

The photoabsorption cross sections and photoionization quantum yields (η) were obtained using the equations,[3,31]

$$\sigma_T = \frac{\ln(i_1/i_2)}{nL}, \tag{1}$$

$$\eta = \frac{i_1/e}{I_o \exp(-\sigma_T nl)\{1-\exp(-\sigma_T nL)\}}, \tag{2}$$

where, i_1, i_2 are the collected ionic photo-currents; e, the electronic charge; I_o, the absolute photon flux at the entrance of the chamber; σ_T, the photoabsorption cross section of a sample molecule; n, the number density of sample molecules, l, the distance of the collection plate from the entrance of the chamber, and L, the effective length of the collection plates. The absolute photon flux I_o was determined by photo-currents in Ar and Xe (η =1). An experimental error limit was estimated to be within ±10% for the absorption cross sections, and less than ±5% for the η values, but in the near η =1 region the error was less than 2%.

Multiplying the photoionization quantum yield by the photoabsorption cross section, the absolute photoionization cross sections are obtained. Subtracting the photoionization cross sections from the photoabsorption cross sections we have obtained the total cross section of neutral dissociation σ_D. This is the most reliable method to obtain the absolute photoionization and photodissociation cross sections.

RESULTS AND DISCUSSION

A. Silane

The ground state configuration of silane in T_d symmetry is written as $(1a_1)^2(2a_1)^2(1t_2)^6(3a_1)^2(2t_2)^6$; $\tilde{X}\ ^1A_1$.

The photoabsorption (σ_T), photoionization (σ_I) and photodissociation (σ_D) cross sections together with photoionization quantum yield (η) obtained in this experiment are shown in Fig.2. Some dis-

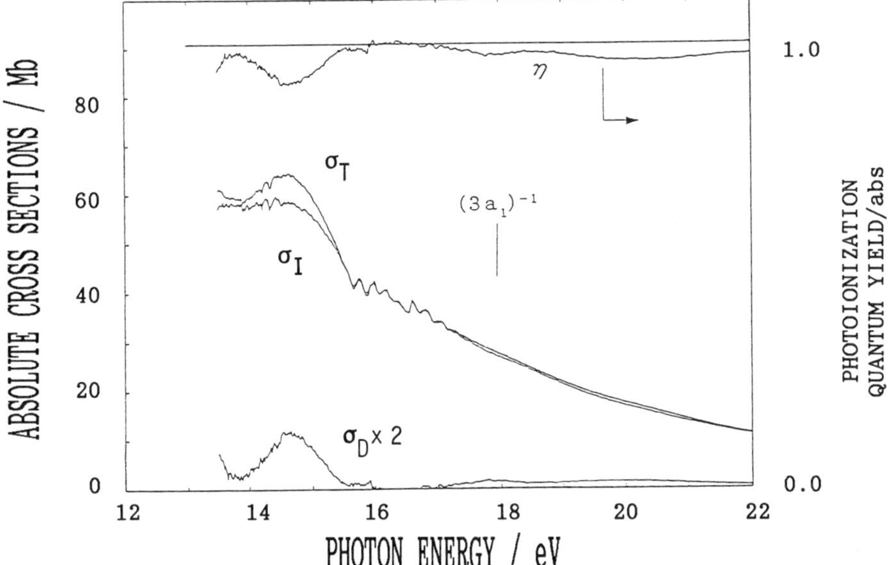

Fig.2 Photoabsorption (σ_T), photoionization (σ_I), and photodissociation (σ_D) cross sections and photoionization quantum yield (η) of silane in the 13.6-22 eV region.

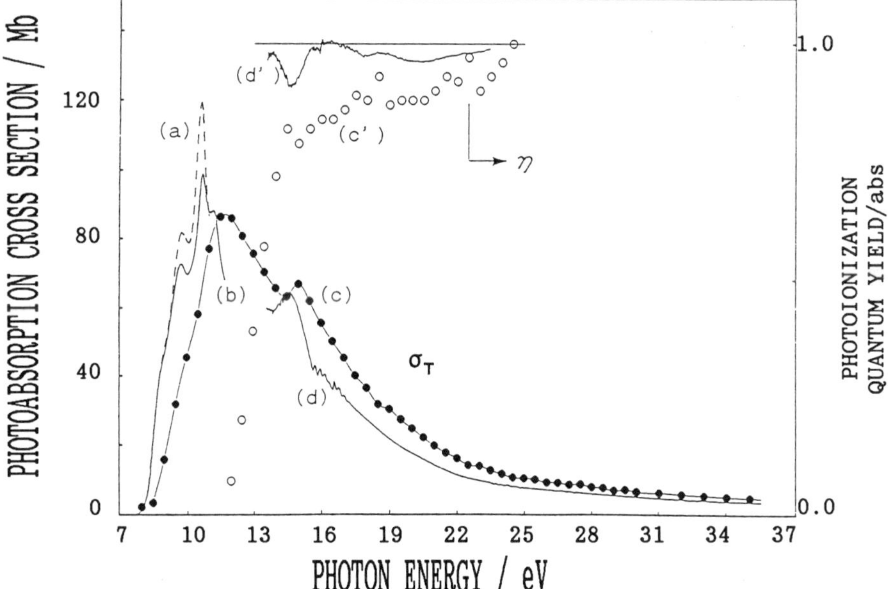

Fig.3 An over view of extreme-UV photoabsorption cross section of silane. (a) Itoh et al.[13]; (b) Suto and Lee[12]; (d) present result. Comparisons of σ_T and η with (c) Cooper et al.[14] are also shown.

tinct structures in the spectrum are observed. The σ_T value of 60 Mb at 13.6 eV increases toward a broad peak maximum, then decreases gradually with increasing the photon energy down to 11 Mb at 22 eV. The general appearance of the absorption spectrum is similar to the small angle electron-energy-loss spectrum.[24]

Figure 3 compiles σ_T results in the present and the previous lower energy studies.[12,13] Although σ_T is not obtained between 11.8 and 13.4 eV because of the lack of window materials, the present σ_T appears to approach continuously to the previous result by Suto and Lee.[12] Also our present result combined with the result by Suto and Lee satisfies the Thomas-Kuhn-Reiche (TKR) sum rule. A broad peak is observed with a peak maximum at 14.6 eV (Fig.2). Dillon et al.[24] assigned this broad band to the optically allowed $3a_1$ electron transition to an antibonding $\sigma^*(t_2)$ orbital. Distinct vibrational structures in 16-18 eV are also revealed which are ascribed to vibration-accompanying Rydberg progressions to the $(3a_1)^{-1}$ limit. The same structures were shown in mass-spectrometric photoionization efficiency curves,[25-27] the electron impact spectrum,[24] and the threshold photoelectron spectrum.[28] However, no evidence on the appearance of the parent ion above the $(3a_1)^{-1}$ limit was reported in mass sepectrometric studies.[25-27] This indicates that the $(3a_1)^{-1}$ ionic state should have a stable molecular configuration in the Franck-Condon region but that somewhat important relaxation processes such as predissociation into ground $\tilde{X}\ ^2T_2$ state should be involved.

The η-spectrum in Fig.2 shows that even in much higher energy region than the ionization threshold (11.6 eV), the η value is not converged to unity. A distinct deviation with a minimum at around 14.6 eV is observed. The position of the minimum corresponds to the peak maximum in σ_T. At 14.6 eV the peak of σ_I is less pronounced than that in σ_T, whereas σ_I is almost identical to σ_T in the vibrational structure region between 16 and 18 eV. The peak position at 14.6 eV of σ_D well corresponds to the peak position of σ_T and the minimum of η. Two kinds of superexcited states with different characters are observed. The superexcited state produced by the $3a_1$ electron transition to the antibonding $\sigma^*(t_2)$ orbital, which is shown as the 14.6 eV broad band, decays through fast neutral-dissociation processes competing with autoionization. A similar contrast in the strongly dissociative superexcited state was seen in the adjacent energy region with regard to the Si*2p* excitation.[10] The superexcited Rydberg states in 16-18 eV converging to the second ionization potential decay mainly through autoionization. Therefore, none of appreciable amount of σ_D is obtained in this region.

In consideration of the growing up behavior of the η-value above the ionization threshold,[3-5] σ_D should have a considerable magnitude in the lower energy region than the present lowest limit of the photon energy but σ_D decreases with increasing the photon energy to almost zero at 16 eV and grows up slightly again above 17 eV. This slight growth is also shown in the η-spectrum as the deviation from unity. The deviation in the energy region higher than the second IP (18.2 eV) is attributed to superexcited states converging to many-body states at 24.2, 26.4, and 30.5 eV (vertical IP).[14] The non-zero values of σ_D above 17 eV should be ascribed to the dissociative nature of the neutral many-body-states (doubly excited states) with the expanded or diffused molecular configuration in the Franck-Condon region.

Finally, Fig. 3 indicates also apparent discrepancies between the present and previous dipole (e,e) results[14] of σ_T and η not only in the spectral features but also in the absolute values and the peak positions of the spectrum. The difference seems to be originated from the low-resolved electron beam and the normalization involved in the experimental procedure of the dipole (e,e) spectroscopy.

B. Disilane

The ground state configuration of disilane in D_{3d} symmetry is written as $(4a_{1g})^2(4a_{2u})^2(2e_u)^4(2e_g)^4(5a_{1g})^2$; $\tilde{X}^1A_{1g}$.

The present σ_T, σ_I, σ_D, and η in the 13.6-22 eV region are shown in Fig.4 . The σ_T value of 107 Mb with a broad peak maximum at 14.2 eV decreases gradually with increasing the photon energy down to 20 Mb at 21.2 eV. The general appearance of the absorption spectrum is also similar to the small angle electron energy loss spectrum.[17]

The broad peaks observed at around 14.2 and 15.3 eV in the present experiment are ascribed to the $4a_{2u}$ electron transitions to 4s and 5s Rydberg levels, respectively. These peaks are superimposed on the spectra of direct transitions to the ionization continuum.

The η-spectrum also shown in Fig.4 again underlines the importance of neutral dissociation even in much higher energy region than the first ionization potential at 10.53 eV;[15,16] the η-value is

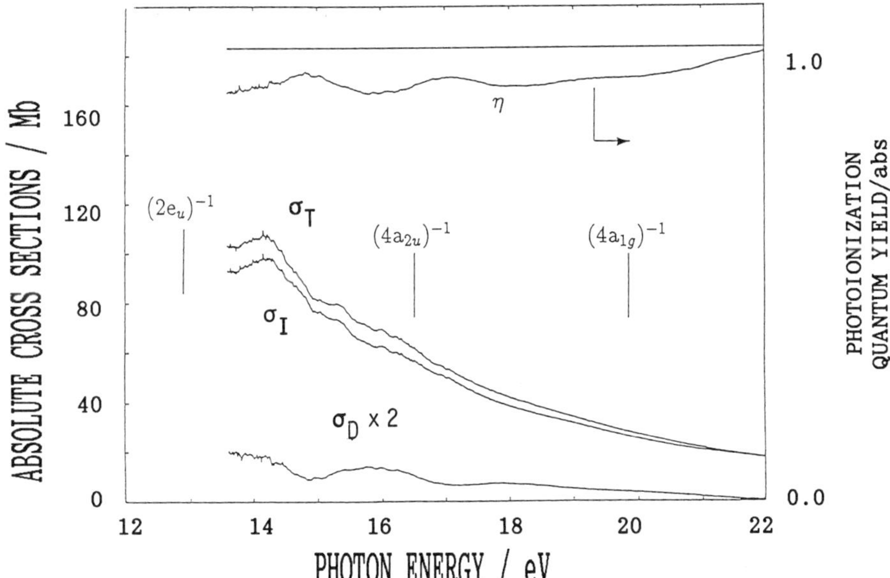

Fig.4 Photoabsorption (σ_T), photoionization (σ_I), and photodissociation (σ_D) cross sections and photoionization quantum yield (η) of disilane in the 13.6-22 eV region.

not converged to unity. This indicates evidence of the existence of superexcited states decaying via neutral dissociation in competing with autoionization. Since experimental errors of η-values are within about 2% in the energy region, the deviation from unity in η-values is real. The minima centered at around 16 and 18 eV might be mainly due to the neutral dissociation of superexcited states converging to the $(4a_{2u})^{-1}$ (16.50 eV) and $(4a_{1g})^{-1}$ (19.84 eV) limits, (Ref.15,16) respectively. The larger deviation from unity than that of silane over the whole energy region examined indicates that the density of superexcited states are relatively larger than silane, or that neutral dissociation processes are preferred to the autoionization processes in the case of disilane.

The σ_I curve shows almost the same structures as those of the σ_T curve, which means that the superexcited states corresponding to each broad peak decay not only through neutral dissociation but also through autoionization. The general appearance of the σ_D curve for disilane is almost the mirror image of the η curve.

The superexcited states converging to unknown ionization satellite levels, i.e. doubly excited states, are expected to lie in this energy region and to contribute to the neutral dissociation processes as mentioned in the case of silane.

In summary, we have presented the absolute photoabsorption, photoionization, and photodissociation cross sections of silane and disilane in the 13.6-22 eV region and demonstrated an importance of the neutral dissociation of superexcited states. This implies that neutral dissociation occurs quite efficiently in much higher energy region than the first ionization thresholds. We also remark the importance of the photoionization quantum yield in the absolute evaluation of photoionization and photodissociation cross sections in the extreme-UV region.

The authors thank Drs. M.Inokuti, T.Hayaishi, T.Ibuki for their helpful discussions. They also thank the staff of Photon Factory for their support. A part of this research has been financially supported by the grant-in-aid of Ministry of Education, Science, and Culture.

REFERENCES

a. K.Kameta, M.Ukai, R.Chiba, K.Nagano, N.Kouchi, Y.Hatano, and K.Tanaka, J.Chem.Phys. 95, 1456 (1991); K.Kameta, M.Ukai, N.Terazawa, K.Nagano, Y.Chikahiro, N.Kouchi, Y.Hatano, and K.Tanaka, ibid. 95 (1991), in press.
1. Y.Hatano. Comments At.Mol.Phys. 13,259(1983).
2. Y.Hatano, Radiochimica Acta 43,119(1988).
3. H.Koizumi, K.Shinsaka, T.Yoshimi, K.Hironaka, S.Arai, M.Ukai, M. Morita, H.Nakazawa, A.Kimura, Y.Hatano, Y.Ito, Y.Zhang, A. Yagishita,K.Ito ,and K.Tanaka, Radiat.Phys.Chem. 32,111(1988).
4. H.Koizumi, K.Shinsaka, and Y.Hatano, Radiat.Phys.Chem. 34,87 (1989).
5. K.Kameta, M.Ukai, T.Kamosaki, R.Chiba, K.Shinsaka, N.Kouchi, Y.Hatano, H.Koizumi, Y.Ito, and K.Tanaka, to be published.

6. W.Hayes and F.C.Brown, Phys.Rev. A6,21(1972).
7. A.Yagishita, S.Arai, C.E.Brion, T.Hayaishi, J.Murakami, Y.Sato, and M.Ukai, Chem.Phys.Lett. 132,437(1986).
8. Y.Sato, K.Ueda, A.Yagishita, T.Sasaki, T.Nagata, T.Hayaishi, M. Yoshino,T.Koizumi, and A.A.MacDowell, Phys.Scripta 41,55(1990).
9. E.Shigemasa, K.Ueda, Y.Sato, A.Yagishita, H.Maezawa, T.Sasaki, M. Ukai, and T.Hayaishi, Phys.Scripta 41,67(1990).
10. G.G.B.de Souza, P.Morin, and I.Nenner, Phys.Rev. A34,4770(1986)
11. S.Bodeur, P.Millié, and I.Nenner, Phys.Rev. A41,252(1990).
12. M.Suto and L.C.Lee, J.Chem.Phys. 84,1160(1986).
13. U.Itoh, Y.Toyoshima, H.Onuki, N.Washida, and T.Ibuki, J.Chem. Phys. 85,4867(1986).
14. G.Cooper, T.Ibuki, C.E.Brion, Chem.Phys. 140,133(1990).
15. H.Bock, W.Ensslin, F.Fehér, and R.Freund, J.Am.Chem.Soc. 98,668 (1976).
16. J.-H.Xu, J.V.Mallow, and M.A.Ratner, J.Phys. B16,3863(1983).
17. M.A.Dillon, D.Spence, L.Boesten, and H.Tanaka, J.Chem.Phys. 88, 4320(1988).
18 J.Perrin and J.F.M.Aarts, Chem.Phys. 80,351(1983).
19. S.Tsurubuchi, *Oyo Buturi*(Appl.Phys.) 59,808(1990).
20. T.Tsuboi, K.Nakashima, and T.Ogawa, Bull.Chem.Soc.Jpn.64,1(1991).
21. H.Koizumi, T.Yoshimi, K.Shinsaka, M.Ukai, M.Morita, Y.Hatano, A. Yagishita,and K.Ito, J.Chem.Phys. 82,4856(1985).
22. H.Koizumi, K.Hironaka, K.Shinsaka, S.Arai, H.Nakazawa, A.Kimura, Y.Hatano, Y.Ito, Y.Zhang, A.Yagishita, K.Ito, and K.Tanaka, J. Chem.Phys. 85,4276(1986).
23. J.A.R.Samson, *"Handbuch der Physik"* vol.31, edited by W.Mehlhorn (Springer, Berlin, 1982) p.123
24. M.A.Dillon, R.-G.Wang, Z.-W.Wang, and D.Spence, J.Chem.Phys. 82, 2909(1985).
25. K.Börlin, T.Heinis, and M.Jungen, Chem.Phys. 113,93(1986).
26. J.Berkowitz, J.P.Greene, H.Cho, and B.Ruscic, J.Chem.Phys. 86, 1235(1987).
27. T.Hayaishi, T.Koizumi, T.Matsuo, T.Nagata, Y.Sato, H.Shibata, and A.Yagishita, Chem.Phys. 116,151(1987).
28. T.Heinis, K.Börlin, and M.Jungen, Chem.Phys.Lett. 110,429(1984)
29. A.W.Potts and W.C.Price,F.R.S., Proc.R.Soc.London, Ser.A 326,185 (1972).

DISSOCIATIVE MULTIIONIZATION IN MOLECULES

J.H.D. Eland and B.J. Treves-Brown
Physical Chemistry Laboratory, Oxford, OX1 3QZ, U.K.

ABSTRACT

The manifestations of dissociative multiionization in two-dimensional time-of-flight mass spectra are described and are illustrated by results on ICN and SF_6. Concerted reactions and two sequential mechanisms can be distinguished on the basis of vector correlations between initial momenta, which determine the peak shapes. Slower sequential reactions produce metastable signatures that give information on ion lifetimes, complementing deductions from the peak shapes. Developments which will enhance this technique for study of reaction dynamics are discussed.

INTRODUCTION

Almost all multiionization of small molecules is dissociative, even double ionization. Although many molecules possess low-lying states in which they are effectively stable, in the great majority of states formed by impact processes all doubly charged small molecules dissociate. By contrast to singly charged ions which can only eject a neutral fragment, the first step of doubly charged ion decay can be one of two processes, covalent dissociation

$$m^{2+} \text{ ----> } m_1^{2+} + m_2$$

or coulombic dissociation, also called charge separation

$$m^{2+} \text{ ----> } m_1^{+} + m_2^{+}$$

After covalent dissociation as a first step, charge separation is likely to follow as a second, and after either process secondary decay of the primary fragments may occur. The time scales may range from femtoseconds to microseconds. These rich mechanistic possibilities of dissociative double ionization (and the easily extrapolated greater richness of higher multiple ionization) are explored effectively by techniques in which all ionic products are detected in coincidence with electrons from the initial ionization event. For double ionization a new three-dimensional mass spectrometry arises[1], providing novel tools to investigate ion reaction mechanisms; there are new varieties of metastable peaks and above all the main ion-pair peak shapes represent vector correlations between initial ion momenta which give intimate clues to the reaction dynamics[2]. In this paper we try to illustrate the power of the PEPIPICO (photoelectron-photoion-photoion coincidence) technique, as far as it is yet understood, concentrating on spectra of small molecules, particularly ICN and SF_6 as examples.

EXPERIMENTAL TECHNIQUE

The PEPIPICO technique is now used by several groups in the world, and details of the different experiments are given in specialist papers[3-6]. Common to all is that photoionization by extreme ultraviolet radiation or soft X-rays takes place in a region of uniform electric field, which accelerates all the ions into the flight path of a time-of-flight (TOF) mass spectrometer and accelerates the electrons towards a multiplier detector. Detection of an electron provides a signal from which the arrivals of two or more ions are timed. The spectra are maps of intensity as a function of the arrival times of both the first and the second detected ions as parameters. To reduce the background "noise" from false coincidences, and to simplify interpretation of peak shapes in the spectra it is desirable that ionization should be confined to a small defined zone; for this reason target molecules are supplied in the form of a gas jet which crosses the beam of ionizing radiation. For studies of reaction dynamics the uniformity of the electric field in the ionization region is crucially important, as is collimation of the target gas beam. Current results point, as we shall see, to the need for a true molecular beam, preferably a beam cooled by supersonic expansion. As in all coincidence experiments high collection efficiency for the particles is needed; this militates against energy analysis of the electrons. Nevertheless heroic experiments with energy analysis to provide initial state selectivity have already been performed[6,7], though not yet under conditions yielding detailed dynamical information. The spectra discussed here have been taken using wavelength-selected vacuum ultraviolet light from atomic discharges, with only weak and unintentional electron energy selectivity.

RESULTS AND DISCUSSION

The PEPIPICO spectra of SF_6 and ICN taken at a wavelength of 25.6 nm are shown in Figures 1 and 2, using two different forms of presentation. The density of printed points, as in the SF_6 spectrum, can represent only a very limited dynamic range of intensity variations, so some contrast enhancement has been used to exhibit weak features. Each ion pair produces a peak in the spectra at coordinates (times) proportional to the square roots of the masses of the two ions, so the identities of the pairs can be read off. In detail most peaks consist of a more or less long and thin line at a definite slope to the axes; in addition certain peaks have weak tails extending to the principal diagonal of the Figure. These tails represent a form of "metastable peak" peculiar to PEPIPICO spectroscopy, for relatively slow charge

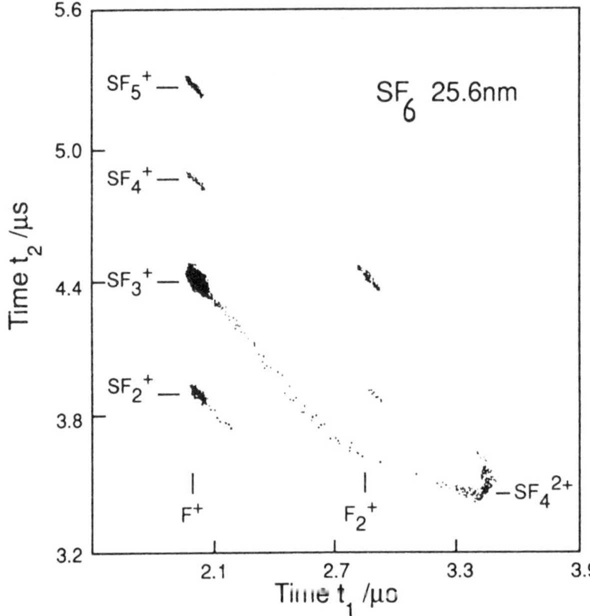

Fig. 1. PEPIPICO spectrum of SF_6 at 25.6 nm as a dot plot, showing metastable tails for slow decays of SF_4^{2+} and SF_3^{2+}.

separation reactions. The tail seen on the main peak in the spectrum of SF_6 corresponds to the fragment ion reaction

$$SF_4^{2+} \text{ ----> } SF_3^{+} + F^{+}$$

occurring in times from a few tens of nanoseconds to several microseconds. The ion pairs contributing to the main peak represent SF_4^{2+} ions which dissociate within less than 10 ns of their formation; those that fall in the tail are longer-lived ions that form a small minority of the population. The distance along the tail away from the peak towards the diagonal is a measure of the actual lifetime; any ion that lives long enough to be fully accelerated before decay contributes to the "V" shape abutting the diagonal at the normal time of flight for SF_4^{2+}. The distance from the main peak to the "V" thus represents a lifetime range equal to the acceleration time of the doubly charged progenitor. The intensity distribution along the tail and in the "V" can thus be used to characterize the true lifetime distribution of the ions; simulations for different mixtures of mean lifetimes are used to match the experimental data. It turns out in this case that to fit the intensity variation along the tail two mean lifetimes of SF_4^{2+} of 2 µs and 40 ns are needed. In the spectrum of ICN there is also a metastable tail on the main peak for I^{+} + CN^{+} pairs, not visible in the contour diagram of Figure 2, showing that some of the ICN^{2+} parent ions are also metastable with a similar lifetime.

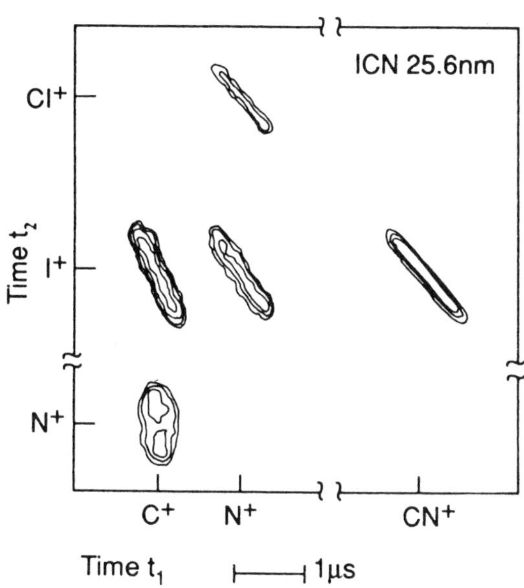

Fig. 2. PEPIPICO spectrum of ICN at 25.6 nm as a contour diagram. Successive contours are at intervals of a factor of two in intensity. Note the scale breaks.

Two-body ion pair processes

The simplest peaks in PEPIPICO spectra represent ion pairs from coulomb dissociation of parent dications producing two bodies only: examples are

$$ICN^{2+} \rightarrow I^{+} + CN^{+},\ ICN^{2+} \rightarrow IC^{+} + N^{+},\ SF_6^{2+} \rightarrow SF_5^{+} + F^{+}$$

Two ions from dissociation of a dication initially at rest have anticorrelated momenta from their mutual coulomb repulsion. Because the time of flight in properly tuned apparatus deviates from its nominal value by an amount proportional to the ion's initial momentum along the spectrometer axis, and the directions are random in space, all two-body pairs should form lines in the spectra with slopes of precisely -1. The real two-body ion pairs do form narrow diagonal peaks, as shown in Figures 1 and 2, with varying lengths and finite widths. To interpret the lengths and widths we must briefly look at the theory in more detail.

Under space-focussing conditions the time-of-flight t for any ion is given to a good approximation by:

$$t = t^0 - p_z/(eE) \qquad (1)$$

where t^0 is the flight time for an ion starting at rest, p_z is the initial ion momentum projection on the line towards the detector, E is the drawout field and e the elemental charge. The actual initial momenta of ions from charge separation can be considered to have three components. First, there is the momentum p along the dissociation coordinate from release of the coulomb repulsion energy. Its projection along the spectrometer axis is $p\cos\theta$, θ being the angle between the two directions. If the dissociation coordinate is assumed to have a random distribution in space it has a sinθ distribution to the spectrometer axis. Secondly, the thermal velocity v_{th} of the molecules before ionization has a Gaussian distribution along the spectrometer axis: each product ion inherits a momentum mv_{th} proportional to its mass. Thirdly, rotation of the molecule is converted into anti-correlated equal momenta of the two fragments perpendicular to the bond axis. To interpret the shapes of two-body peaks it is convenient to take one dimensional distributions of the sums and differences of the flight times for the two ions in each pair.

In the distribution of the sum, $t_1 + t_2$, the effects of the anticorrelated contributions to the initial momenta cancel, leaving only the thermal contribution. The shapes have exactly the same origin and meaning as TOF peak shapes for parent ions, which were described long ago by Franklin, Hierl and Whan[8]. The peaks should have a Gaussian shape corresponding to the one-dimensional Maxwell-Boltzmann distribution of molecular velocities along the spectrometer axis in the target gas before ionization. Our peak shapes for parent ions and for $t_1 + t_2$ distributions from two-body reactions are approximately Gaussian, but their widths imply kinetic temperatures lower than the actual gas temperature. The peak width and shape depend on the distance of the gas nozzle from the light beam. This is because at the pressure in the source there are no more molecular collisions, and the narrowness of the light beam selects the central collimated part of the gas jet for ionization. The measured widths define effective transverse kinetic temperatures of about 70 K for singly charged ions and 120 K for pairs. We do not fully understand the difference, but part is due to the double contribution of instrumental broadening to the ion pairs. The narrowing (by comparison with room-temperature widths) is crucial for good mass resolution, and for detailed study of peak shapes, where thermal velocities provide unwanted broadening.

The distribution of the difference $t_2 - t_1$ for a single PEPIPICO peak is equivalent to the shape of a peak in the simpler PIPICO (photoion-photoion coincidence) technique[9,10]. The distribution contains contributions from all three sources of initial momentum, except in the case of two equal mass fragments where the thermal velocity contribution vanishes. Because the coulomb repulsion energy is very large by comparison with the other contributions, the momentum given to the ions can be considered almost single-valued. A pure single-valued energy release would produce square flat-topped peaks, exactly as for large kinetic energy releases in TOF mass spectrometry[8]; the distributions over $t_2 - t_1$ are indeed broad and flat-topped. Initial intercharge distances in dissociating dications are not single-valued, however, but cover a range corresponding to the width of the Franck-Condon zone in the ions' formation, so the momentum release p is not single-valued either and the $t_2 - t_1$ distributions have rounded ends. The intrinsic coulomb energy release distributions cannot be deduced directly from the observed peak shapes, because thermal velocities and the rotational momentum release smear the TOF distributions. The smearing effect of thermal velocities can be eliminated if necessary, since the deviation of $t_1 + t_2$ for each individual pair from the mean value gives the thermal velocity on axis, which can be used to correct the values of t_1 and t_2 individually before subtraction. The smearing effects of molecular rotation, which cannot be removed in this way have not been calculated, but are certainly significant. The use of a cooled molecular beam would effectively eliminate them.

Kinetic energy releases in two-body charge separations are deduced in practice from the half-height widths of the $t_2 - t_1$ distributions using formulae given before[2]. Results for the ICN and SF_6 spectra are listed in the Table.

Peak slopes

In a two-body dissociation the initial linear momenta of the fragments are exactly anticorrelated, so the true slope of the peak in PEPIPICO is unambiguously -1. The measured slope often differs from this value by more than the apparent standard deviation, however, when it is determined in a least-squares calculation treating the two times in each pair as of equal statistical weight. A major reason is that the thermal velocity of the neutral molecule before ionization, which both

Table. Details of cation pairs from ICN and SF_6 at 25.6 nm

Pair	Intensity[a]	U_{CS}/eV	U_{neut}/eV
$CN^+ + I^+$	64.0	5.0+/-0.4	-
$N^+ + CI^+$	2.7	5.5+/-1	-
$C^+ + I^+$	34.0	5.5+/-1	1[b]
$N^+ + I^+$	18.0	5 +/-1	0.8[c]
$C^+ + N^+$	2.2	5 +/-1	?
$F^+ + SF_5^+$	1.5	4.8+/-0.5	-
$F^+ + SF_4^+$	0.4	4.9+/-1	?
$F_2^+ + SF_3^+$	0.8	4.6+/-0.8	?
$F^+ + SF_3^+$	27.3	4.3+/-0.5	0.5[d]
$F_2^+ + SF_2^+$	0.1	?	?
$F^+ + SF_2^+$	1.9	3.5+/-1	?

[a]Intensities are on a scale where the total singly charged ion intensity is 250. [b]Exponential distribution with this mean, with free rotation of intermediate CN^+. [c]Gaussian distribution with this mean and 0.3 eV half-width, retaining linearity. [d]Gaussian distribution with 2 eV width and this mean, truncated at zero.

fragments inherit, gives each one a random momentum component proportional to its mass. The spread or "error" in the time for the heavier ion is thus proportionally larger than the spread in that for the lighter one: the correct unit slope can be determined if the two variables are properly weighted in the least-squares treatment. When pairs of ions are produced by more complex reactions, perhaps involving several steps, the spreads of momentum carried by each ion are unknown and may be considerably larger than those produced by thermal energies. To circumvent this difficulty, and contrary to earlier practice[2], we now consider that the only reliable method of determining peak slopes is to use a contour diagram, after smoothing if necessary. If the long axis of a peak is marked by straight, parallel contours their slope should be used. For peaks with curved or non-parallel contours the concept of slope is of less value. This is unfortunately also the case when a peak is "hollow" because of the loss of ions with large sideways velocity components.

Three-body peaks with a single slope of -1

Peaks in PEPIPICO spectra for three-body or more complex reactions come with a great variety of shapes, some of which are exhibited in Figure 2. Where single processes are involved the shape is governed mainly by the momentum given to the unseen neutral fragment(s). Peaks with a true slope of -1 can arise either from concerted simultaneous reactions in which neutral fragments receive no correlated impulse, or from deferred charge separation. The deferred charge separation mechanism is one in which a doubly charged parent ion first ejects a neutral fragment, after which the remaining doubly charged fragment splits into two singly charged ions.

$$m^{2+} \dashrightarrow m_1^{2+} + m_2: \qquad m_1^{2+} \dashrightarrow m_3^+ + m_4^+$$

Evidence for the temporal separation of these two steps may come from the existence of the intermediate doubly charged fragment as a distinct ion in the mass spectrum, or from inference that this same dication is the immediate parent of several different ion pairs. More direct evidence is the presence of a "diagonal" metastable tail in the PEPIPICO spectrum, as illustrated in Figure 1, which shows that the charge separation step in formation of $SF_3^+ + F^+$ from SF_6 via SF_4^{2+} can take 10ns or longer. A similar but distinguishable metastable tail could appear for simultaneous formation of three fragments, two ions and one neutral, the whole process being delayed after ionization; it would connect to a point on the diagonal corresponding to the total mass of the doubly charged precursor. In this case the SF_4^{2+} ion, which is definitely the precursor, is a prominent feature of the mass spectrum, making this reaction one of the clearest examples of deferred charge separation. Other examples of neutral ejection before charge separation include:

$CH_3CN^{2+} \rightarrow CH_2CN^{2+} + H$ (Acetonitrile)

$C_2H_4^{2+} \rightarrow C_2H_3^{2+} + H$ (Ethylene)

$CH_3OH^{2+} \rightarrow CHOH^{2+} + H_2$ (Methanol)

$CF_4^{2+} \rightarrow CF_3^{2+} + F$ (Tetrafluorethylene)

$C_6F_6^{2+} \rightarrow C_6F_5^{2+} + F$ (Hexafluorobenzene)

$C_4H_4N_2^{2+} \rightarrow C_4H_4^{2+} + N_2$ (Pyridazine)

Reactions of these types are very common, in fact loss of a neutral F atom as a first step has been observed in the PEPIPICO spectrum of every single perfluorinated compound so far examined.

In deferred charge separation the peak is expected to have a slope of -1; the peak shape is also characteristic of the reaction mechanism. The energy release in the first step, although usually small, is not zero. If the time between the two steps is sufficient for free rotation of the dication fragment before charge separation, the directions of the two momentum releases will be almost completely uncorrelated; simulations by a Monte-Carlo method confirm that the result is a peak of slope -1 with more than thermal width and a lozenge-like shape. The shape for the $SF_3^+ + F^+$ pair is shown in Figure 3, and is echoed in a simulation. The

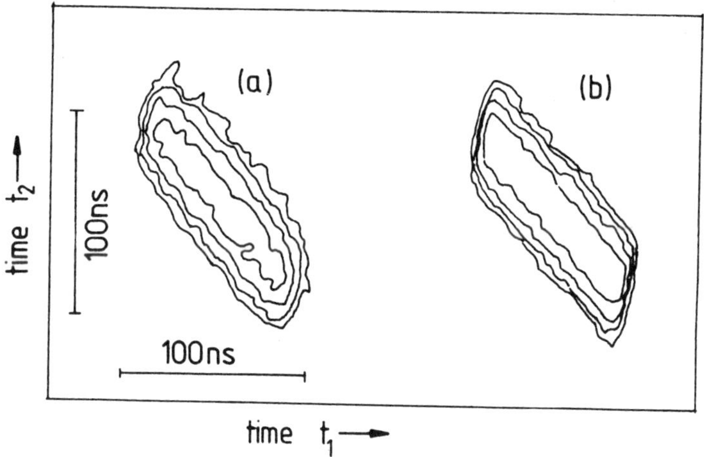

Fig. 3. Contour diagrams (a) for the observed $F^+ + SF_3^+$ peak and (b) from a simulation assuming an F_2 neutral fragment ejected with a total energy release of 0.5 eV with a broad distribution given in the Table.

peak shape is a lozenge of unit slope, rather than a bar, because the momentum spreads of the ions from the first step of the reaction, proportional to the masses, are unequal. The ends of the lozenge should have slopes of m_4/m_3, the ratio of the ions' masses, here 89/19. In cases where the energy release in the first step is small or where the energy released in the second step has a broad spread, a peak of unit slope with rounded ends is seen, and the weak peaks for $SF_4^+ + F^+$ and $SF_3^+ + F_2^+$ in Figure 1 are examples. Information on the energy release in neutral fragment ejection and on its distribution can be obtained from the distribution of $t_1 + t_2$, as for two-body reactions. The $t_1 + t_2$ distribution corresponding to Figure 3 is shown in Figure 4 and indicates a broad energy release distribution in the neutral ejection step; unfortunately it is not known whether two F atoms or an F_2 molecule are ejected, though we suspect that it is molecular fluorine.

In deferred charge separation both ions receive the same velocity from reaction to the departure of the neutral fragment and this velocity is known for each ion pair from the deviation of $t_1 + t_2$ from its mean value. The effect of energy release in the first step can thus be subtracted in calculating the distribution of $t_2 - t_1$ in order to characterize the energy release in the second step of the reaction. There is a definite improvement in sharpness of the $t_2 - t_1$ distribution as demonstrated in Figure 4, when the two ions are of very unequal mass.

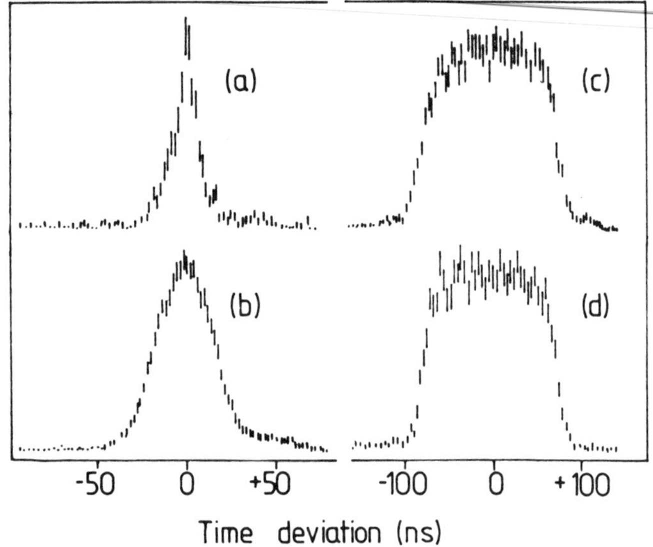

Fig. 4. One dimensional distributions from the spectrum of SF_6 as in Fig. 1. (a): $t_1 + t_2$ distribution for $F^+ + SF_5^+$ (pseudo-thermal) (b): $t_1 + t_2$ distribution for $F^+ + SF_3^+$ showing the effect of neutral ejection. (c): $t_2 - t_1$ distribution for $F^+ + SF_3^+$ as measured and (d): the same as (c) after removal of the effect of momentum release in ejection of the neutral, as explained in the text.

Concerted breaking of two bonds in a dication, the mechanism sometimes called simultaneous coulomb explosion, can also result in singly ionized fragments with anticorrelated momenta and one neutral fragment, all liberated simultaneously. An example is the dissociation of doubly ionized carbon disulphide[11]:

$$^{32}SC^{34}S^{2+} \rightarrow {}^{32}S^{+} + C + {}^{34}S^{+}$$

A simple peak with parallel straight contours and unit slope (as seen in this example) will appear for such a reaction if the momentum given to the neutral is zero, or if it is random in direction relative to the direction of charge separation. More complex peak shapes are produced by concerted dissociations in which the neutral fragment receives a significant correlated impulse, however, and these are discussed later.

Three-body peaks with non-unit single slopes

PEPIPICO peaks of non-unit single-valued slope can arise from simultaneous explosions, from deferred charge separation with the neutral ejected on the same line as the ions, or from secondary decay of the products of charge separation[2]. We believe that the last mechanism is the most common.

The secondary decay mechanism can be written

$$m^{2+} \rightarrow m_1^{+} + m_2^{+} \qquad m_1^{+} \rightarrow m_3^{+} + m_4$$

though the starting point may already be a fragment, and both ions from charge separation may subsequently dissociate. The critical parameter determining both the description of the mechanism and the appearance of the peaks is time, the time between charge separation and loss of the neutral fragment. The deferred charge separation mechanism discussed above corresponds to negative times in this sense, concerted immediate explosions to zero time (within a few femtoseconds) and secondary decays to positive time. Details of the peak shapes respond to magnitudes of the time from microseconds, in metastable peak phenomena, to femtoseconds in secondary decay.

Secondary decay of a primary ion about a picosecond or later after charge separation leads to a PEPIPICO peak whose slope is equal to the mass ratio m_3/m_1 (or the inverse if m_2 is heavier than m_3). For times in this range and for normal rotational temperatures the intermediate ion m_1^{+} can rotate freely before the secondary decay occurs, so the momentum release directions in the two steps are uncorrelated. Since the time is long enough for escape, secondary decay occurs outside the range of coulomb repulsion. If an energy release occurs in the second step it affects only one of the two ions detected, m_3^{+}, so the peak shape will be a lozenge with vertical (or horizontal) ends, showing that one of the ions has a much wider spread of momentum along the axis than the other. Secondary decay of primary ions is just the same process as monocation dissociation studied in normal mass spectra, and it is no surprise that the energy release is not usually larger than a few hundred millielectron volts. As a result the majority of peaks of this type have rounded ends and fairly small perpendicular widths. Some clear examples of secondary decay are[12,13]:

CH_3R^{2+} --> R^+ + CH_3^+: CH_3^+ --> CH_2^+ + H (R = OH, I, SCN, etc.)

$(CH_3)_2SO^{2+}$ --> CH_3^+ + CH_3SO^+: CH_3SO^+ --> SO^+ + CH_3

$C_4H_4O^{2+}$ --> $C_2H_2^+$ + $C_2H_2O^+$: $C_2H_2O^+$ --> CH_2^+ + CO (Furan)

Close agreement between the measured slope and a mass ratio is in itself good evidence for the secondary decay mechanism; the primary ion pair is almost always present in the spectrum, and often the same ions break up in two or more secondary decays. If the range of lifetimes in secondary decay extends into the metastable range (10 ns to 10 µs) a broad ridge is seen in the PEPIPICO spectrum extending along the line of m_2^+ from m_1^+ to m_3^+. In such a case the decay of the singly charged ion involved is likely to be known as a metastable process in normal mass spectrometry, and perhaps to be visible in the photoionization mass spectrum itself.

In the PEPIPICO spectrum of ICN the peak for I^+ + C^+ offers itself as a probable case of secondary decay, as the slope is almost exactly 26/12 and the CN^+ ion is certainly formed by charge separation. This case involves an ambiguity of interpretation, however, because C is the central atom. Even if the C-N bond were broken instantaneously at the moment charge separation began, the C^+ ion would have to push the N atom out of its way before escaping. Such a mechanism[2], "obstructed instantaneous explosion", cannot easily be distinguished from secondary decay. The means of distinction will be to vary the rotational temperature of the molecules; at high rotational temperature the N atom from an obstructed explosion should escape sideways at an earlier time, and the slope of the peak should approach -1.

Some peaks are found to have single-valued slopes that are neither -1 nor close to a recognisable mass ratio. In polyatomic molecules the possibility that both members of a primary pair have undergone secondary decay must be considered; both of the individual secondary reactions should be observable separately for this to be plausible. Experimental artefacts may sometimes disguise the correct slope, but there remain a group of very interesting peaks whose slopes cannot be explained in this way; a prime example is the pair I^+ + N^+ in the spectrum of ICN. The slope of the peak is -1.36±0.04 whereas secondary decay of CN^+ would give -1.86 (26/140) and secondary decay of CI^+ would give -0.92 (127/139). One arbitrary but possible explanation is the release of a single definite kinetic energy in secondary decay along the direction of charge separation, before the diatomic fragment has time to rotate. A formula for the resulting slope in terms of the masses and kinetic energies was given in the earlier paper[2], and provides an excellent starting point for simulations; the best match on this model is given in the Table. Another possibility is that secondary decay takes place at a short distance from the other (undissociated) primary ion, within the coulomb repulsion zone. If the secondary decay involves no energy release the slope for a single-valued critical distance is easily calculable, though simulations using exponential lifetime distributions are probably more realistic. For I^+ + N^+ from ICN at 25.6 nm this model would require a mean CN^+ lifetime of about 30 fs. Up to now it has not been possible to decide between these and other model mechanisms; the finer resolution made possible by a rotationally cold molecular beam may allow a decision to be made.

Energy releases in secondary decays which give peaks of the expected slope can be determined from the widths of the peaks perpendicular to their lengths, or by comparisons of the peak shapes with simulations. There is reason to expect the energy releases and their distributions to be very similar to those exhibited by the same monocations in normal mass spectra, but no systematic investigation of this aspect of the spectra has yet been made.

Peaks with curved contours

In three-body decays where the neutral fragment is ejected within less than a few picoseconds before or after charge separation, directional correlation of all three initial momenta is to be expected. Simulations[11,13] and algebraic analysis[14] indicate that the correlation should show up in the form of peaks with curved contours, and two limiting cases can be delineated clearly. If the neutral fragment is ejected at a significant angle to the line of charge separation, the PEPIPICO peak should be broad at its centre but narrow at the ends. For near-perpendicular neutral ejection the centre of the peak corresponds to charge separation perpendicular to the mass spectrometer axis; in this orientation reaction to the neutral momentum component can lie along the spectrometer axis and so have maximum effect on the flight times. The ends of the peak, by contrast, represent charge separation along the axis, where an additional perpendicular momentum component, perpendicular to the axis, has minimal effect. Several peaks with the resulting "egg" shape have been found and are particularly common in cases where a triply charged ion dissociates into three fragment ions, only two of which are observed. Neutral fragments normally carry much less momentum than ions, even if it is correlated, so ovoid peaks are less often observed in double ionization, where their true shape may be swamped by the thermal velocity distribution. One exceptionally clear and intense example[2] is the reaction

$$SO_2^{2+} \text{ ----> } O^+ + S^+ + O$$

In the spectrum of ICN an example is seen in the very weak peak for C^+ + N^+. In this case the oval has its major axis almost vertical, showing that the I and N^+ fragments travel in nearly opposite directions, leaving very little momentum to be carried by the C^+ ion. The curvature of the contours shows that the angles involved in the dissociation are not random (nor zero), however. The width of the peak suggest a large angle between C^+ and N^+ directions, but because of the low intensity of the peak an exact analysis has not been attempted. The width of the distribution of N^+ times in the peak is almost exactly the same as its width in the N^+ + CI^+ peak, suggesting that the initial process resembles the simple charge separation.

A different peak shape can arise from preferential ejection of a neutral fragment with significant momentum along the line of charge separation. If the momentum release is single-valued (as assumed to explain the I^+ + N^+ peak) the peak should be a thin line at an unexpected slope, but for a momentum release distribution it should have an "hourglass" shape, fat at the ends but narrow in the middle because of the overlap of many lines of different slopes. No satisfactory real example has yet been found. Peak contours are often distorted into the semblance of an hourglass shape by loss of ions with large sideways velocity components. It is almost impossible to compute the effects of discrimination with complete fidelity, and such peaks unfortunately convey little dynamical information beyond the bare magnitude of the coulomb energy release. Angular anisotropy in the initial ionization process, which has hitherto been neglected here, can also produce similar appearances.

Peaks with non-parallel contours

PEPIPICO peaks for a number of secondary decays are found on close examination to have a "twist": the intense part of the peak has a different slope from the less intense part, as illustrated in Figure 5 by

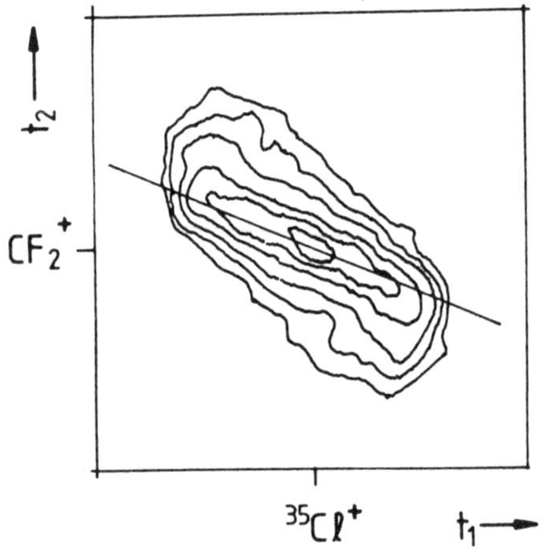

Fig. 5. PEPIPICO peak contours for $^{35}Cl^+$ + CF_2^+ from CF_2ClBr at 30.4 nm, showing a strong "twist". The diagonal line shows the slope calculated for a slow secondary decay of CF_2Br^+.

an example from the spectrum of CF_2ClBr. One or both of the intense or less intense parts usually has a slope either close to -1 or close to the mass ratio for a secondary decay, as in these examples. In some cases the explanation is that a particular ion pair is formed by two concurrent mechanisms with different probabilities and different characteristic slopes, and the two simply overlap. This is certainly not the explanation in all cases, however, and the twisted form of peak generally implies a correlation between the probability of secondary decay and one of the parameters which determine peak slope. The probability must be systematically related to either the momentum given to the neutral along the charge separation direction, to the distance between ions at which secondary fragmentation takes place or to the time of the event after charge separation.

If time is the critical variable, twisted peaks can be explained by secondary decay occurring in a time range of 10fs to 1ps, the range in which the slopes vary markedly as functions of the position in the coulomb field at which the neutral fragment is released. The most frequent form of twisted peak has its greatest intensity near the limiting slope for a secondary reaction, equal to a mass ratio, but is surrounded by an aureole with contours of slopes between the limiting value and -1. The pile-up near the limiting mass ratio is well explained on the time hypothesis, where the lifetime distribution should be approximately exponential, by the non- linear relationship between slope and time elapsed since charge separation. After a time of the order of 300 fs, which depends on the masses and the initial intercharge distance, all later decays will contribute to intensity at slopes close to the limiting value. Only extremely fast reactions, essentially complete in one or few vibration periods, will not have a tail of lifetimes piling up near the limiting slope. Confirmatory evidence for this interpretation is that when the energy of the ionizing light is reduced it is always the part of the peak at the secondary reaction slope that becomes relatively more intense, presumably because only the slowest reactions persist. Detailed tests to validate or contradict this interpretation will probably require the use of cold molecular beams, and initial energy selection of the reacting ions. If they do validate it we shall have a new route to femtosecond molecular reaction dynamics.

CONCLUSIONS

The PEPIPICO technique is shown to be a powerful tool for the study of doubly charged ion dissociations; by obvious extension it can be applied to reactions of more highly charged species. To make the method really effective, two improvements are needed. First, target gas must be provided in the form of a rotationally cold, tightly collimated molecular beam. Secondly, initial state selectivity must be added to the technique instead of the simple wavelength variation available at present, or alternatively the statistics of existing energy-selective forms of the technique must be improved beyond recognition. Developments on both these lines are currently under way.

REFERENCES

1. J.H.D. Eland, Acc. Chem. Res., **22**, 381 (1989).
2. J.H.D. Eland, Mol. Phys., **61**, 725 (1987).
3. L.J. Frazinski, M. Stankiewicz, K.J. Randall, P.A. Hatherley and K. Codling, J. Phys. B. **19**, L819 (1986).
4. P. Lablanquie, I. Nenner, P. Millié, P. Morin, J.H.D. Eland, M.-J. Hubin-Franskin and J. Delwiche, J. Chem. Phys., **82**, 2951 (1985).
5. T. LeBrun, M. Lavollée and P. Morin, AIP Conf. Proc. **215**, 846 (1990).
6. D.M. Hanson, C.I. Ma, K. Lee, D. Lapiano-Smith and D.Y. Kim, J. Chem. Phys., **93**, 9200 (1990).
7. R. Murphy and W. Eberhardt, J. Chem. Phys., **89**, 4054 (1988).
8. J.L. Franklin, P.M. Hierl and D. A. Whan, J. Chem. Phys., **47**, 3148 (1967).
9. G. Dujardin, S. Leach, O. Dutuit, P.-M. Guyon and M. Richard-Viard, Chem. Phys., **88**, 339 (1984).
10. P.M. Curtis and J.H.D. Eland, Int. J. Mass Spectrom. Ion Processes, **63**, 241 (1985).
11. J.H.D. Eland, Laser Chem., 1991, in press.
12. E. Ruhl, S.D. Price, S. Leach and J.H.D. Eland, Int. J. Mass Spectrom. Ion Processes, **97**, 175 (1990).
13. Unpublished work in this laboratory.
14. T. LeBrun, Doctoral Thesis, Université Paris XI, 1991.

QUESTION

BRUTSCHY - 1. Have you studied these fragmentation patterns at different energies, and did you possibly see a change of the pattern, telling you something of the transition state ?

2. The structure of more complicated fragments is or may be still open. Do you believe one may combine this elegant method with a collision induced fragmentation (CID) section ?

ELAND - In answer to the first question, we regularly take PEPIPICO spectra at several wavelengths between 40.7nm and 25.6nm (300eV to 48eV). Where reactions are simple two-step processes of deferred charge separation or secondary decay, the peak shapes do not change. The only concerted reaction studied over a wide range so far is $SO_2^{2+} \rightarrow O^+ + S^+ + O$, where the fragments separate at almost 120°. In this case the angles do not change much with wavelength, but the magnitude of the energy release increases at high energy. The shapes of "twisted" peaks do change strongly with wavelength.
On the second question, I fear that the technique you suggest would be too difficult for technical reasons, unless detectors of true 100% detection efficiency become available.

DE LANGE - When your doubly charged ion separates into two charged fragments, you can only detect both fragments if their direction after formation allows them to reach the detector. If the dissociation takes place far from the detector this situation is not nearly as probable as when it occurs close to the detector. Is it true that there is a large geometrical factor of this nature that one should correct for? As a second point, I want to come back to what you said about dissociation of ICN^{2+}. Of course neutral ICN is linear, but how about ICN^{2+}? Will non-linearity of the doubly charged ion have a significant effect on the line shape?

ELAND - On the first point, we normally operate our apparatus under conditions where all fragments are detected, even if they fly apart in the source exactly perpendicularly to the axis. If they dissociate later, they are certain to hit that active area of the detector, so no conveciion is needed. the possibility of such ion losses is however fully included in the Monte-Carlo modelling with which the experimental data are compared. On the second point ICN^{2+} may be non-linear in excited states where electrons other than the outer π-electrons have been lost. This could certainly affect the peak shape, but will not explain it entirely: particularly it does not explain the large momentum of the neutral I atom.

LEACH - One reason for the sideways ejection of carbon in the ICN dication case could be the large change in the bending force constant expected when one removes two electrons from ICN. Neutral ICN is a 16 valence electron linear species, the outer 4 electrons being in a π orbital. In such species, the electron occupation of this π orbital governs the ease of bending the linear molecule. This is well known for the centrosymmetric series CO_2, CO_2^+, NCN,...C_3 for which the bending frequency diminishes strongly as successive πg electrons are removed. A greater flexibility is to be expected for ICN^{2+} with respect to ICN. The existence of a heavy Iodine atom may also help to tunnel energy into the bending vibration via anharmonic coupling between stretch and bend vibrations. Thus a propensity for sideways ejection of carbon, by strong excitation of the bending vibration, could result from ejection of two valence electrons from the outer π orbital in ICN.

BESWICK - Could you comment on the possible contribution of the zero-point bending energy of ICN in the suprisingly large contribution of the $I + C^+ + N^+$ concerted reactions ?

ELAND - The zero-point bending is probably very important , because it is effectively amplified by this Coulomb repulsion, as we have seen in dissociation of triply charged ions such as OCS^{3+}. The best model of the $ICN^{2+} \rightarrow C^+ + N^+ + I$ reaction seems to be an "initial" separation into $N^+ + CI^+$, followed by $CI^+ \rightarrow C^+ + I$ before rotation of the diatomic fragment from its initial, small, zero-point angle. As CI^+ is so heavy, it will rotate relatively slowly. As C^+ is light, by contrast, a small energy release in CI^+ decay will propel it fast towards N^+, whose repulsion will deflect it to the observed large angle (≅100° from N^+) outside the Coulomb zone.

DUJARDIN - When you analyse a particular peak shape, you assume that there is a single process leading to the corresponding fragments ions. However it may be that several processes are superimposed and give rise to more complicated analysis. Could this explain the "unexplained" peak shapes?

ELAND - You are right that because initial states are not selected, mixtures of mechanism are observed. In many cases we see trivial superimposition of peak shapes where two distinct pathways can produce the same ion pair. An interesting example is the peak for $H_2+ + CH_2^+$ from ethylene oxide C_2H_4O, which arises at 30.4nm from both CH_4^{2+} + CO by deferred charge separation in CH_4^{2+} , and by secondary decay of CH_2CO^+ from the primary pair $H_2+ + CH_2CO^+$. At low photon energy the first component vanishes.

In other cases, such as $S^+ + O^+$ from SO_2, the peak shape remains the same at all wavelengths, but the size changes. We take this to mean that the same general mechanism (here concerted explosion) is operative in decay from a range of initial states, with changing total energy release.

BAER - Can you say anything about why some ions are metastable ? Is it tunneling or a statistical process ? Are strong metastables related to large E_0's for dissociation ?

ELAND - We have good evidence, from detailed RRKM modelling, that in the PEPIPICO spectra of large aliphatic hydrocarbons almost all the primary singly-charged ions from charge separation decay statistically including those that produce metastable peaks. The charge separation step in these cases is fast, but seems to allow a statistical energy distribution among the fragments.

By contrast, the metastable charge separations of such small ions as ICN^{2+}, SF_4^{2+}, N_2O^{2+}, CO_2^{2+} or $C_2H_2^{2}$ seem likely to be due to tunneling processes undergone by that small minority of ions in energy levels near the top of the barriers. Metastable charge separations of aromatic species such as benzene or pyridazine might be explained either way, and more detailed measurements are needed before a distinction can be made.

POLARIZATION AND ORIENTATIONAL EFFECTS IN PHOTOIONIZATION

N.A. Cherepkov

Aviation Instrument Making Institute,
190000 Leningrad, USSR

ABSTRACT

The ways to obtain additional information on molecules as compared to that given by partial photoionization cross section and the angular asymmetry parameter measurements are discussed. Photoelectron spin polarization measurements for unoriented molecules can give three additional parameters for achiral molecules plus five more parameters for chiral molecules. Maximum information can be obtained from investigations of either space-fixed or aligned and oriented molecules. It is shown that measurements of circular and linear dichroism in the angular distribution of photoelectrons for these molecules is the simplest way to extract rather complete information on their structure, as well as on their orientation.

INTRODUCTION

Photoionization of molecules is one of the simplest processes which enables one to investigate the molecular structure and properties. It is always desirable to extract as much information as possible from experimental data. But if in the case of atoms one can perform a "complete" quantum-mechanical experiment and extract from the experimental data all theoretical parameters describing the photoionization process in the electric-dipole approximation[1-3], in the case of molecules the same goal is unattainable. Indeed, the number of theoretical parameters in molecules is in principle infinite, since the orbital angular momentum of electron is not a good quantum number, while we still have to use the partial wave expansion for continuous spectrum wave functions. Therefore the partial wave summation extends up to infinity, giving the infinite number of dipole matrix elements to be extracted from experiment. Of course, for practical reasons one can restrict this summation by finite number of terms, but even in this case the number of parameters will be substantially larger than in atoms.

Nevertheless, there are ways to obtain rather complete information on molecules too, from a series of experiments which will be discussed in this talk. These experiments include

i) measurements of the partial photoionization cross section $\sigma_i(\omega)$, where ω is a photon energy;

ii) the usual measurements of the angular asymmetry parameter $\beta(\omega)$;

iii) angle- and spin-resolved photoelectron spectroscopy of unoriented molecules, that is measurements of Angular Distribution of photoelectrons whith defined Spin Polarization (ADSP);

iv) measurements of ADSP for either space-fixed or aligned and oriented molecules, which in principle give the most complete information on the process, but are too complicated for practical realization;

v) as a simplification of the preceding point, measurements of Circular and Linear Dichroism in the Angular Distribution of photoelectrons (CDAD and LDAD, respectively), from oriented molecules without detecting the spin polarization of photoelectrons, which are much simpler from experimental point of view and still are giving rather complete information on molecules. All these processes take place already in the electric-dipole approximation, which solely will be used here.

UNORIENTED MOLECULES

It is well known that if molecules are unoriented (rotating), the angular distribution of photoelectrons is defined by the angular asymmetry parameter $\beta(\omega)$[4,5] and has the same form as in atoms[6]

$$I^i_{\vec{e}_z}(\vec{\varkappa}) = \frac{\sigma_i(\omega)}{4\pi}\left[1 + \beta\left(\frac{3}{2}\cos^2\vartheta - \frac{1}{2}\right)\right] \qquad (1)$$

where $\vec{\varkappa}$ is the unit vector in the direction of photoelectron momentum, $\sigma_i(\omega)$ is a partial photoionization cross section of the i-subshell, ω is the photon energy, and light is linearly polarized along the z axis. The process under consideration is described by two vectors, one of which is used to define a coordinate frame, while the other one remains free. Therefore, the angular distribution (1) depends on only one angle ϑ between these two vectors and possesses cylindrical symmetry. The coordinate frame is usually connected with the photon beam, so that its z axis coincides with the polarization vector $\vec{e}$ for linearly polarized light or with the direction of the photon beam (which will be characterized in the following by the unit vector $\vec{q}$) for circularly polarized and unpolarized light.

Together with the partial photoionization cross section which is implied to be known, after measuring the angular distribution of photoelectrons one have two measured quantities. The further information can be obtained from measurements of the spin polarization of photoelectrons. Again, if molecules are unoriented, the angular distribution of photoelectrons with defined spin polarization has the same form as in atoms[3], and is defined by three additional parameters[7] (for achiral molecules)

$$I_m^i(\vec{\varkappa},\vec{s},\vec{q}) = \frac{\sigma_i(\omega)}{8\pi}\left\{1+\frac{(2-3m^2)}{2}\beta P_2(\vec{\varkappa}\vec{q})+mA^i(\vec{s}\vec{q})-m\gamma^i\left[\frac{3}{2}(\vec{\varkappa}\vec{q})(\vec{\varkappa}\vec{s}) - \frac{1}{2}(\vec{s}\vec{q})\right]+(2-3m^2)\eta^i(\vec{s}[\vec{\varkappa}\vec{q}])(\vec{\varkappa}\vec{q})\right\} \quad (2)$$

where m=±1 for circularly polarized light and m=0 for linearly polarized one (in this case the unit vector $\vec{q}$ had to be replaced by the polarization vector $\vec{e}$). For unpolarized light m=1 in (2) and terms linear in $\vec{q}$ had to be excluded. $\vec{\varkappa}$ and $\vec{s}$ are the unit vectors in the directions of the photoelectron momentum and spin, respectively. The new parameters A^i, γ^i, η^i, like the angular asymmetry parameter β, are expressed through dipole matrix elements and phase shifts, and contain the dynamical information on the process, while all dependence on the angles (the geometrical part) is explicitly given in (2) and is independent on the photon energy or molecular structure.

The term proportional to A^i in (2) gives the spin polarization of photoelectrons parallel to the photon beam when light is circularly polarized. This term is the only one which survives after integration over the electron ejection angles. The next term proportional to γ^i gives a contribution also for circularly polarized light. It describes the spin polarization in the scattering plane, defined by the vectors $\vec{q}$ and $\vec{\varkappa}$. And the last term in (2), proportional to the parameter η^i, gives the contribution for any polarization of light, as well as for unpolarized light. It describes the spin polarization perpendicular to the scattering plane, which for linearly polarized light is defined by the vectors $\vec{e}$ and $\vec{\varkappa}$.

General expressions for the spin polarization parameters A^i, γ^i, η^i are given in[3] using the partial wave expansion for the continuous spectrum wave function, and in[8] (see also Ref.9) using the expansion over generalized harmonics which incorporate the symmetry properties characteristic for nonlinear molecules. The case of linear molecules was considered in[7].

The spin polarization of molecular photoelectrons appears through the spin-orbit (or spin-axis) interaction, which manifests itself in the multiplet splitting of molecular levels and in the difference between continuos spectrum wave functions corresponding to different projections of the electron spin on the molecular axis. The spin-orbit splitting of molecular levels leads to the degree of polarization of the order of unity.To observe it one have to separate photoelectrons corresponding to different fine-structure components of the molecular ion state. If the fine-structure splitting is not resolved, the spin polarization of photoelectrons will be small, of the order of $(\alpha Z)^2$, where α is the fine-structure constant and Z is the nucleus charge. Indeed, the polarization parameters A^i, γ^i, η^i in the nonrelativistic limit have the opposite sign for different fine structure components and are inversely proportional to the statistical weights of these states, so that after averaging over these states the net result will be zero in the nonrelativistic limit. The nonzero result appears due to the influence of the spin-orbit interaction on the wave functions, which is of order of $(\alpha Z)^2$. In general, the spin polarization of molecular photoelectrons is expected to be lower than in atoms due to a greater number of allowed transitions in molecules[7].

The general predictions of the spin polarization of molecular photoelectrons[3,7] were followed by the calculations for HI molecules in the region between the $^2\Pi_{1/2}$ and $^2\Pi_{3/2}$ ionization thresholds where the spin-orbit autoionization takes place[10]. Calculations in an extended photon energy region for the πnp outer-shell and d inner shell photoionization of HBr and HI molecules have been performed in Ref.11. The first measurements for CO_2, N_2O[13,14], Br_2, I_2, CH_3I[14] supported general predictions of the theory described in[7]. Both total photoelectron current [12,13] and photoelectrons ejected at a definite angle[13,14] have been analysed with respect to spin. The most detailed experimental measurement so far has been performed with circularly polarized synchrotron radiation emitted out of the plane of the storage ring BESSY for the spin polarization parameter A^i of HI molecule[15] shown in Figure 1. A strong variation in both $A^{1/2}$ and $A^{3/2}$ due to autoionization is observed. In general, the experimental data are in accordance with the approximate relationship $A^{1/2}=-A^{3/2}$ valid when the spin-orbit interaction is neglected [7].

A theoretical analysis performed in Ref. 15 shows that the parameter A^i in this energy region has oscillatory structure due to the autoionization, and at the moment there is only qualitative agreement between theory and experiment. Further investigations, both theoretical and experimental, are needed in order to understand completely the observed spectra.

With VUV laser light sources developed recently it is possible now to measure the spin polarization of photoelectrons correspondind to different rotational bran-

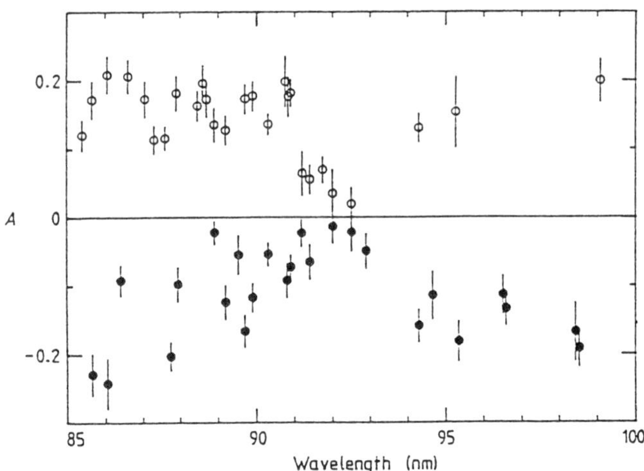

Fig.1. Spin-polarization parameter A^{l} for HBr^{+} $^{2}\Pi_{3/2}$ (v=0) (open circles) and $^{2}\Pi_{1/2}$ (v=0)(full circles) final ionic states [15].

ches of a resonance feature, as has been shown in Ref.16 in the case of HI molecule. The corresponding theory has been also developed [17]. The first calculations[18] of the spin polarization parameters performed using this theory and assuming Hund's case (e), also show the pronounced rotational structure, though the direct correspondence between the theoretical and experimental structures has not been established yet. The key point here is to which Hund's case belongs the part of the spectrum under consideration. From the analysis of the experimental results for a sequence of Rydberg states in the photoabsorption spectrum of HI using a multichannel quantum defect theory it was concluded in[19] that there is a transition from Hund's case (c) to case (e) with increasing the principal quantum number. Unfortunately, the results obtained in[17,18] show that the spin polarization measurements hardly can be used to distinguish between different rotational branches, since the polarization paramemeters only weakly depend on the rotational quantum numbers.

The ADSP for unoriented chiral molecules contains five additional terms as compared to eq.(2)[20]

$$I_m^i(\vec{\varkappa},\vec{s},\vec{q}) = \frac{\sigma_i(\omega)}{8\pi}\Big\{mD^i(\vec{\varkappa}\vec{q}) + mC^i(\vec{s}[\vec{\varkappa}\vec{q}]) + B_1^i(\vec{\varkappa}\vec{s}) + B_2^i(\vec{\varkappa}\vec{q})(\vec{s}\vec{q}) + B_3^i(\vec{\varkappa}\vec{q})^2(\vec{\varkappa}\vec{s})\Big\} \quad (3)$$

Here the first term gives CDAD for unoriented chiral molecules which was predicted at first by Ritchie[21]. The next term gives the transverse polarization of photoelectrons which has a different sign for left and right circularly polarized light. And the last three terms give the longitudinal polarization of photoelectrons, which also appears for absorption of linearly polarized (in this case again the unit vector $\vec{q}$ had to be replaced by $\vec{e}$) and unpolarized light.

Coefficients of all five terms in (3) are proportional to differences of pairs of dipole matrix elements, which differ from each other by signs of all projections of orbital angular momenta and spins. For nonchiral molecules having a plane of symmetry these differences are identically equal to zero, whereas for chiral molecules they are different from zero. A realistic estimation of these differences is quite complicated. They should be proportional to an asymmetry factor η discussed in[22], which depends on the degree of dissymmetry in the structure of the molecule and can be of the order of 10^{-2} (see Refs.23 and 24). Up to now nobody measured the spin polarization of photoelectrons ejected from chiral molecules.

PHOTOIONIZATION OF ORIENTED MOLECULES

Consider now the cases when the initial state of a molecule do not possess the spherical symmetry and can be characterized by at least one vector. It means that rotating molecules had to be aligned or oriented, for example, through excitation by linearly or circularly polarized light, respectively, which produces nonequal population of states with different projections of the total angular momentum including rotation. The alternative possibility is to stop molecular rotation. For example, molecules adsorbed at a surface or molecules in liquid crystals are not rotating. Molecules can be oriented by the external field[25]. It is possible also to select the photoionization processes of oriented-in-space molecules in a gas phase by coincidence measurements of fragment ions and photoelectrons, provided the molecular ion decays rather fast after ionization[26]. Fixed-in-space molecules, depending on their structure, can be characterized by one, two or even three vectors. In the following, both space fixed and aligned or oriented molecules will be called oriented for brevity. In many respects both space-fixed and aligned or oriented molecules behave identically, that is geometrical parts, describing particular processes, are the same for these two cases, while the dynamical parts (the corresponding parameters) can look differently.

It is convenient now to consider the photoionization process in the coordinate frame, defined by the molecular orientation. Then ADSP depends on three vectors and can be presented as an expansion in spherical functions[27]

$$I^i_m(\vec{\varkappa},\vec{\xi},\vec{q})=\sqrt{3\pi}\ \sigma_i(\omega)(-1)^{1-m}\sum_{LM}^{2lmax}\ \sum_{JM_J}^{2}\ \sum_{SM_S}^{1}\begin{pmatrix}1&1&J\\-m&m&0\end{pmatrix}A^{JM_J}_{LMSM_S}Y_{LM}(\hat{\varkappa})$$
$$\times\ Y_{JM_J}(\hat{q})Y_{SM_S}(\hat{s}) \qquad (4)$$

where lmax is the highest orbital momentum retained in the partial wave expansion of the continuous spectrum wave function. The coefficients of this expansion, $A^{JM_J}_{LMSM_S}$, for space-fixed molecules are defined in[27]. They are normalized by the condition $A^{00}_{0000}=1$. Equation (4) again is presented in a very convenient form of a product of geometrical (the spherical functions) and the dynamical (the parameters $A^{JM_J}_{LMSM_S}$, which depend on the photon energy) factors.

Equation (4) is very complicated and has, in principle, infinite number of terms. A general analysis of this expression was performed in[27]. In particular, it was shown there that for linear molecules in Hund's cases (a) and (b) photoelectrons can be polarized only parallel to the molecular axis, while in Hund's case (c) they can be polarized in any direction. Unfortunately, the spin polarization measurements are quite complicated and are connected with the loss of at least three orders of magnitude in intensity[28]. Therefore, though ADSP from oriented molecules (4) contains the most complete information on molecules, at the moment there are no experimental results for it.

To make the experiment much more simple, one have to exclude the electron spin polarization measurement. To this end it is necessary to sum (4) over spin projections

$$I^i_m(\vec{\varkappa},\vec{q})=I^i_m(\vec{\varkappa},\vec{s},\vec{q})+I^i_m(\vec{\varkappa},-\vec{s},\vec{q})=\sqrt{3}\ \sigma_i(\omega)(-1)^{1-m}\sum_{LM}^{2lmax}\ \sum_{JM_J}^{2}\begin{pmatrix}1&1&J\\-m&m&0\end{pmatrix}$$

$$\times A^{JM_J}_{LM00} Y_{LM}(\hat{\vec{\varkappa}}) Y_{JM_J}(\hat{\vec{q}}) \qquad (5)$$

This is the angular distribution of photoelectrons for oriented molecules, which was obtained at first by Dill[29] and was analysed in[30,31] for the case of space-fixed molecules.

From the experimental point of view measurements of the angular distribution (5) is much easier than the measurements of ADSP (4), while the number of terms in (5), in principle, is still infinite, so that measurements of (5) give quite exhaustive information on molecules. But the practical use of these measurements is restricted by the difficulty of extracting of many parameters from one measured curve. Therefore it is worth while to look for the way to diminish the number of terms in (5). It can be achieved by investigating the difference between photoelectron currents ejected at a definite angle by light of two orthogonal polarizations, either circular or linear, which will be called, following Ritchie[32], circular or linear dichroism in the angular distribution of photoelectrons, respectively[33-35].

QUALITATIVE CONSIDERATION OF CDAD

Consider now qualitatively the origin of CDAD. Suppose that the light beam is propagating along the x axis of our laboratory frame defined by molecular orientation. Then the dipole operator for left and right circularly polarized light is given by

$$\hat{d}_{\pm}=(\vec{e}_z \pm i\vec{e}_y)\vec{r}=z\pm iy= \sqrt{4\pi/3}\left\{Y_{10}(\hat{\vec{r}})\pm\frac{1}{\sqrt{2}}\left[Y_{11}(\hat{\vec{r}})+Y_{1-1}(\hat{\vec{r}})\right]\right\} \qquad (6)$$

For simplicity we also suppose that the inital state of an oriented molecule is characterized by a wave function which can be expressed through the spherical function with a given l (summation over l does not alter the answer but makes the equation more complicated)

$$\Psi_i=\Psi_{ilm}(\vec{r})\sim R_{il}(r)Y_{lm}(\hat{\vec{r}}) \qquad (7)$$

The photoelectron wave function $\Psi^-_{\vec{p}}(\vec{r})$, which contains in the asymptotic region the superposition of a plane wave, propagating in the direction of the electron momentum $\vec{p}$, and a converging spherical wave, is presented as a partial wave expansion

$$\Psi^-_{\vec{p}}(\vec{r})\sim\sum_{l'm'} Y_{l'm'}(\hat{\vec{r}})Y^*_{l'm'}(\hat{\vec{\varkappa}})e^{-i\delta_{l'm'}} \qquad (8)$$

With these assumptions we will find for the dipole matrix element after integration over spherical angles

$$\langle\Psi^{-}_{\vec{p}}|\hat{d}_{\pm}|\Psi_{ilm}\rangle \sim \sum_{l'm'}(-1)^{m'} e^{(i\delta_{l'm'})} Y_{l'm'}(\hat{\vec{\varkappa}})\left[\begin{pmatrix} l' & 1 & l \\ -m' & 0 & m\end{pmatrix} \pm \frac{1}{\sqrt{2}}\begin{pmatrix} l' & 1 & l \\ -m' & 1 & m\end{pmatrix} \pm \frac{1}{\sqrt{2}}\begin{pmatrix} l' & 1 & l \\ -m' & -1 & m\end{pmatrix}\right]\langle pl'\|d\|il\rangle \qquad (9)$$

The angular distribution of photoelectrons is proportional to the square modulus of this matrix element. The key point is that in (9) there are both terms having the same sign for two circular polarizations and terms having opposite signs. Due to that the angular distributions for two circular polarizations will also contain a group of terms having the same sign and a group of terms having opposite signs. The latter terms will contribute to CDAD, while the former will cancel. In the case of unoriented molecules the laboratory frame can be always taken to be coincident with the photon frame, in which the dipole operator for left and right circularly polarized light is proportional to only one spherical function, $Y_{11}(\hat{\vec{r}})$ or $Y_{1-1}(\hat{\vec{r}})$, respectively, and CDAD does not appear.

If states with different m are degenerate and equally populated, the angular distribution has to be averaged over m

$$I^{i}_{\pm 1}(\vec{\varkappa},\vec{q}) \sim \frac{1}{2l+1}\sum_{m}\left|\langle\Psi^{-}_{\vec{p}}|\hat{d}_{\pm 1}|\Psi_{ilm}\rangle\right|^{2} \qquad (10)$$

Using equation (9) and the condition that the dependence on m is given solely by the spherical function $Y_{lm}(\hat{\vec{r}})$, one can show that the terms which have different signs for two polarizations disappear from (10). And vice versa, if there is any unequivalence between states with different projections m, these terms will give nonzero contribution and will cause the appearance of CDAD. Unequivalence between states with different m can be connected with unequal population of states with different m[35] (alignment and orientation of molecules), and with the dependence of a radial part of wave function in our laboratory frame on the projection m, as it takes place when molecules are fixed in space, and states with different m are not degenerate[33]. The same arguments hold for polarized (aligned and oriented) atoms, which also reveal CDAD[35,36].

From the consideration presented above it is seen

that CDAD directly follows from the dipole selection rules, and therefore is expected to be of the order of unity[33], that is, of the same order of magnitude as the differential photoionization cross section itself for the same angle. After integration over electron ejection angles one finds that space-fixed (as well as oriented, but not aligned[36]) molecules reveal also Circular Dichroism[27] (CD). In the case of space-fixed molecules CD is different from zero for molecules of rather low symmetries[37], in particular, for molecules, which do not have two mutually perpendicular planes of symmetry.

The usual optical activity of unoriented chiral molecules, and CD in particular, is caused by the dissymmetry in their structure. Namely, chiral molecules possess neither plane of symmetry no centre of inversion. Like circularly polarized light, they can be presented as a helix. The absorption of light is different depending on whether the screw senses of these two helices are the same or the opposite. This leads to the appearance of CD in photoabsorption by chiral molecules[38].

CDAD from oriented molecules appears due to dissymmetry in the geometry of the experiment. Three vectors, describing the experiment, $\vec{\varkappa}$, $\vec{q}$ and $\vec{n}$ ($\vec{n}$ is the unit vector in the direction of the molecular orientation), can form a basis for a left or right coordinate system provided they are noncoplanar.

Suppose that these vectors are mutually perpendicular. Then they can be arranged in a way shown in Fig.2, forming a part of left-handed (as in Fig.2) or right-

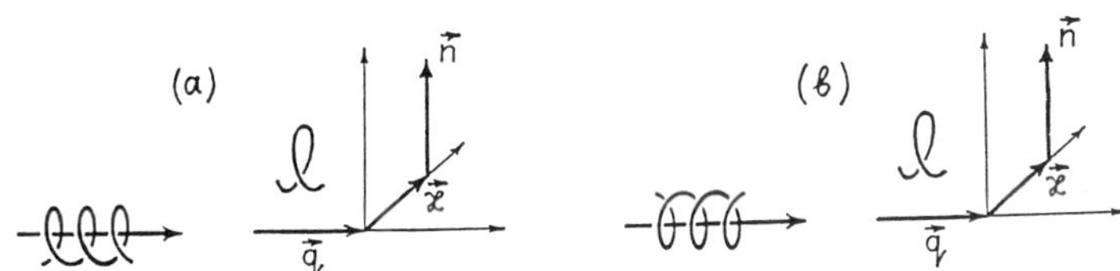

Fig.2. Absorption of light depends on whether the screw senses of light and experimental arrangements are the same or opposite.

handed helix. Again, absorption of light will be different depending on whether the screw senses of light and of the experimental arrangment are the same or opposite. As a result, CDAD will appear.

It is also important to mention that the usual optical activity of unoriented chiral molecules is given by the electric dipole - magnetic dipole interference terms, which are α times smaller than the pure electric dipole terms, responsible for the effects considered here.

The detailed study of LDAD from oriented molecules and its comparison with CDAD is presented in the contributed paper[39] and is not repeated here. It is also shown there, that performing successively a series of measurements of CDAD, LDAD, CD, LD and of angular distributions, one can extract all essential parameters characterizing the process provided only few parameters are extracted from each measurement.

SURVEY OF EXPERIMENTAL RESULTS FOR CDAD

The first prediction of CDAD[33], as well as the first calculations of CDAD for CO molecules[34], have been made for the case of space-fixed molecules. Later it was realized that aligned molecules must reveal the same behaviour[35], and the first experimental evidence of the existense of CDAD was obtained for aligned NO molecules[40,41]. In these experiments NO molecules have been excited to the $A^2\Sigma^+$ state by absorption of one[41] or two[40] linearly polarized photons from the pump laser beam and then ionized by the counterpropagating probe laser beam. The laser beams crossed the molecular beam, which entered the interaction region at right angles to both the propagation direction of the lasers and the detector axis of the electron spectrometer. The angular dependence of CDAD was obtained by rotating the linear polarization vector $\mathbf{e}$ of the pump laser beam. These experiments for the first time demonstrated that CDAD is not a small effect, and that it can be successively used to detect molecular alignment. Afterwards CDAD measurements have been used in[42] to probe the alignment of ground state NO fragments produced by the UV photodissotiation of methyl nitrite, CH_3ONO.

CDAD for space-fixed molecules have been observed for the first time by Westphal et al[43-45]. They investigated both diatomic (CO, NO) and polyatomic (benzene, CH_3I) molecules, oriented by adsorption on Pd(111) or graphite(0001) surface. Adsorption on a single-crystal surface yields a perfectly oriented ensemble of molecules with a rather high target density, about two orders of magnitude larger than in the gas phase, resulting in a huge photoelectron intensities. As a sourse of circularly polarized light they used synchrotron radiation. A disadvantage of these experiments is connected with the fact that photoelectrons are ejected into a hemisphere only, which is further restricted by the systematic effect of refraction of photoelectrons due to the electrostatic surface barrier. Therefore, the grazing emission parallel to the surface is equivalent to the emission angle $\theta^{in}<90^\circ$ (shown by dotted line in Fig.3)

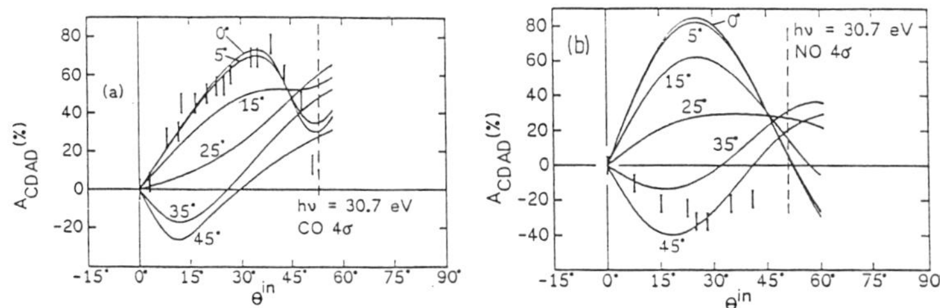

Fig.3. Normalized CDAD spectra for the 4σ levels of CO (a) and NO (b) molecules, calculated for various bent orientations, and measured experimentally in[44] for the geometry when $\theta_q=130^\circ$, $\varphi_q=0$, $\varphi_\varkappa=90^\circ$ and $\theta_\varkappa$ is varied.

for the case of free molecules. Experimentally it was more convenient to measure the normalized CDAD value A_{CDAD} which is a ratio of the difference to the sum of intensities at each angle.

The experimental results obtained in[44] for diatomic molecules NO and CO were compared with calculations performed in the Hartree-Fock approximation. Generally the agreement between the calculated and measured CDAD values is quite encouraging, particularly in view of the very simple model of an isolated molecule assumed in these calculations. In[44] it was also investigated how sensitive are CDAD spectra to the tilt of the molecule. Fig.3 shows CDAD spectra for various angles of tilt, calculated and measured for 4σ photoemission of CO and NO molecules, as a function of the angle θ^{in} within the surface barrier, which corresponds to a free molecule fixed in space. For each angle of tilt the spectra correspond to an orientational average around the surface normal. The 4σ level in these molecules is only weakly influenced by the substrate, therefore calculations performed for free molecules are expected to be reliable. Results of calculations show that the CDAD spectra change significantly with increasing angle of tilt and even change a sign. Comparison of these calculations with experimental data shows that CO molecules are adsorbed normal to the surface, while NO molecules have an average bent orientation between 35° and 45°. These results are consistent with earlier angle-resolved photoemission studies[46] and demonstrate the potential of CDAD measurements as a probe of adsorbate orientation.

LDAD for oriented molecules was predicted theoretically very recently[39], and up to now no experimental observations of LDAD have been reported. CD and optical rotation by oriented molecules were investigated in[37] for molecules of different symmetry groups. It was shown there that these effects are not connected with each other by the Kramers-Kronig relation and are irreversible, like the Faraday effect in a magnetic field. The optical rotation by the nematic liquid crystal of p-azoxyanisole, observed earlier[47], can be explained on the basis of this theory.

CONCLUSION

It was shown that the angle- and spin-resolved photoelectron spectroscopy of unoriented molecules gives an opportunity to find three new parameters in addition to the partial photoionization cross section and the angular asymmetry parameter β obtained from the angle-resolved photoelectron spectroscopy. For unoriented chiral molecules the number of new parameters increases up to 8. In atoms the spin polarization measurements have been used to perform the complete quantum-mechanical experiment[3,48], to identify autoionization resonances[49]. In molecules these investigations, both theoretical and experimental, are only starting, and they seem to be very promising in interpretation of complicated molecular spectra.

The new observable quantities, CDAD and LDAD, can be measured with aligned or oriented, as well as with space-fixed molecules. Measurements of CDAD and LDAD enable one to get rid of many terms giving contribution to the usual angular distributions, and to specify the kind of information extracted from the experiment. Changing the kind of measurement (CDAD or LDAD, CD or LD) and its geometry, it is possible to extract rather many parameters from a set of measurements and to make an essentinal step towards the complete quantum-mechanical experiment for molecules. CDAD measurements have been proposed to use in search of Cooper minima in molecules[50]. The other purpose to use the CDAD and LDAD measurements is to probe molecular orientation, produced by any method, for example, by adsorption on a surface, by scattering processes, by chemical reactions, by external fields, by photon excitation and so on. Therefore CDAD and LDAD measurements have a vast variety of applications. With synchrotron radiation as a source of light, they do not differ much from experiments currently performed in many laboratories. From theoretical point of view, these phenomena are well understood, but there are only few numerical calculations for particular molecules. Therefore further investigations are desired, both theoretical and experimental, and opening of new synchrotron radiation facilities will stimulate new researches in this potentially very rich field.

REFERENCES

1. J. Kessler, Comments At. Mol. Phys. 10, 47 (1981).
2. U. Heinzmann, J. Phys. B 13, 4353, 4367 (1980).
3. N.A. Cherepkov, Adv. At. Mol. Phys. 19, 395 (1983).
4. J.C. Tully, R.S. Berry, and B.J. Dalton, Phys. Rev. 176, 95 (1968).
5. A.D. Buckingham, B.J. Orr, and J.M. Sichel, Philos. Trans. R. Soc. London, Ser. A 268, 147 (1970).
6. J. Cooper, and R.N. Zare, J.Chem. Phys. 48, 942 (1968).
7. N.A. Cherepkov, J. Phys. B 14, 2165 (1981).
8. N. Chandra, Phys. Rev. A 40, 752 (1989).
9. N.A. Cherepkov and V.V. Kuznetsov. Phys. Rev. A 44, no. 3, in press.
10. H. Lefebvre-Brion, A. Giusti-Suzor, and G. Raseev, J. Chem. Phys. 83, 1557 (1985).
11. G. Raseev, F. Keller, and H. Lefebvre-Brion, Phys. Rev. A 36, 4759 (1987).
12. U. Heinzmann, F. Schafers, and B.A. Hess, Chem. Phys. Lett. 69, 284 (1980).
13. U. Heinzmann, B. Osterheld, F. Schafers, and G. Schonhense, J. Phys. B14, L79 (1981).
14. G. Schonhense, V. Dzidzonou, S. Kaesdorf, and U. Heinzmann, Phys. Rev. Lett. 52, 811 (1984).
15. H. Lefebvre-Brion, M. Salzmann, H.-W. Klausing, M. Muller, N. Bowering, and U. Heinzmann, J. Phys. B 22, 3891 (1989).
16. T. Huth-Fehre, A.Mank, M. Drescher, N. Bowering, and U. Heinzmann, Phys. Rev. Lett. 64, 396 (1990); Phys. Scripta, 41, 454 (1990).
17. G. Raseev, and N.A. Cherepkov. Phys. Rev. A 42, 3948 (1990).
18. M. Buchner, G. Raseev, and N.A. Cherepkov. J. Chem. Phys., to be published.
19. A. Mank, M. Drescher, T. Huth-Fehre, N. Bowering, U. Heinzmann, and H. Lefebvre-Brion. J. Chem. Phys., to be published.
20. N.A. Cherepkov. J. Phys. B. 16, 1543 (1983).
21. B. Ritchie. Phys. Rev. A 13, 1411 (1976).
22. A. Rich, J. Van House, and R.A. Hegstrom. Phys. Rev. Lett. 48, 1341 (1982).
23. D.M. Campbell and P.S. Farago. Nature, 318, 52 (1985).
24. K. Blum, R. Fandreyer, and D.Thompson. J. Phys. B 23, 1519 (1990).
25. F. Harren, D.H. Parker, and S. Stolte. Comments At.

Mol. Phys. 26, 109 (1991).
26. A.V.Golovin, V.V. Kuznetsov, and N.A. Cherepkov. Sov. Tech. Phys. Lett. 16, 363 (1990).
27. N.A. Cherepkov and V.V. Kuznetsov. Z. Phys. D 7, 271 (1987).
28. J. Kessler. Polarized Electrons (Springer, Berlin, 1985).
29. D. Dill. J. Chem. Phys. 65, 1130 (1976).
30. D. Dill, J. Siegel, and J.L. Dehmer. J. Chem. Phys. 65, 3158 (1976).
31. J.W. Davenport. Phys. Rev. Lett. 36, 945 (1976).
32. B. Ritchie. Phys. Rev. A 12, 567 (1975).
33. N.A. Cherepkov. Chem. Phys. Lett. 87, 344 (1982).
34. R.L. Dubs, S.N. Dixit, and V. McKoy. Phys. Rev. Lett. 54, 1249 (1985).
35. R.L. Dubs, S.N. Dixit, and V. McKoy. J. Chem. Phys. 85, 656 (1986); 85, 6267 (1986).
36. N.A. Cherepkov, and V.V. Kuznetsov. J. Phys. B 22, L405 (1989).
37. N.A. Cherepkov and V.V. Kuznetsov. J. Chem. Phys., in press (1991).
38. L.D. Barron. Molecular Light Scattering and Optical Activity (Cambridge Univ. Press, Cambridge, 1982).
39. N.A. Cherepkov and G. Schonhense. This volume, p.
40. J.R. Appling, M.G. White, T.M. Orlando, and S.L. Anderson. J. Chem. Phys. 85, 6803 (1986).
41. J.R. Appling, M.G. White, R.L. Dubs, S.N. Dixit, and V. McKoy. J. Chem. Phys. 87, 6927 (1987).
42. J.W. Winniczek, R.L. Dubs, J.R. Appling, V. McKoy, and M.G. White. J. Chem. Phys. 90, 949 (1989).
43. C. Westphal, J. Bansmann, M. Getzlaff, and G. Schonhense. Phys. Rev. Lett. 63, 151 (1989).
44. C. Westphal, J. Bansmann, M. Getzlaff, G. Schonhense, N.A. Cherepkov, M. Braunstein, V. McKoy, and R.L. Dubs. Surf. Science, in press (1991).
45. G. Schonhense, Phys. Scripta, T31, 255 (1990).
46. E. Miyazaki, I.Kojima, M. Orita, K. Sawa, N.Sanada, K. Edamoto, T. Miyahara, and H. Kato. J. Electr. Spectr. Rel. Phen. 43, 139 (1987).
47. R. Williams. J. Chem. Phys. 50, 1342 (1969).
48. U. Heinzmann. Phys. Scripta, T17, 77 (1987).
49. M. Muller, N. Bowering, F. Schafers, and U.Heinzmann. Phys. Scripta, 41, 42 (1990).
50. H. Rudolph, R.L. Dubs, and V. McKoy. J. Chem. Phys. 93, 7513 (1990).

QUESTION

ELAND - I should like to remark that the first electron-ion coincidence measurements showing anisotropic angular distributions from dissociation more rapid than rotation were done at ANL in about 1980, and the results have also been confirmed most recently (J. Meas. Sci. Technol. 1, 1989. A clear forward-backward asymmetry was observed in electron ejection to form one state of NO^+.

CHEREPKOV - I thank you for mentioning these references. Still measurements of Golovin et al. (1990) are more related to the subject of my talk.

GOULON - Could you comment about the possibility to detect CDAD for deep inner shell photoelectrons, i.e. by doing the experiment in the soft, or medium soft X-Ray range.
In conjunction with the idea of carrying out the experiments at higher energy, don't you think that the Electric quadrupolar terms will generate additional CD mechanisms for oriented molecules ?

CHEREPKOV - In principle, CDAD can be observed also in deep inner shell photoionization, though the effect will be the lower, the closer is the molecular wave function to the atomic one.
All the effects, discussed in my talk, appear already in the electric-dipole approximation, which gives the leading contribution at photon energies, say, below 1keV. At higher photon energies higher multipoles (electric quadrupole in particular) will contribute also to CDAD, so that the corresponding equations should be modified.

NAHON - 1. How do you expect CDAD to vary with the kinetic energy of the photoelectron.
2. In the case of inner-shell photoionization, do you expect strong inter-shell coupling effects on CDAD ?

CHEREPKOV - 1. I expect that the variation of CDAD with photon energy qualitatively will coincide with the variation of, say, angular asymmetry parameter ß, namely, it will variate sharply in the region of autoionization resonances, and will variate smoothly outside resonances.
2. From my experience in atomic calculations I would not expect strong influence of intershell correlations ion on the CDAD since it is described by the parameters containing the ratio of dipole matrix elements. Intershell correlations usually influence all the transitions from a given subshell, so that the ratio will be more stable, except for the case when the Cooper minimum is present in one of the transitions. In this case the intershell correlations will move the Cooper minimum to another position and due to that will change all the parameters, CDAD in particular.

MENZEL - For determination of orientation of non-normal adsorbed particles by CDAD: how dependent will the found angle be on :

a) admixture of metal wave functions in ground state,
b) effects of reflexion and refraction of the PE ?

CHEREPKOV - Of course the determination of molecular orientation relative to a surface is an approximate procedure. To make it more exact and to diminish the influence of substrates, one has to measure CDAD not from the outermost subshell, but at least from the next one. Effects of reflection and refraction could not be avoided in this way, but they can be taken into account semi-empirically as it was done by Westphal et al. (1991).

BESWICK - How will the CDAD signals differ when the molecule is sitting rigidly forming an angle different from zero with respect to the case when the molecule is vibrating ?

CHEREPKOV - If the molecule is sitting rigidly forming an angle with the surface normal different from zero, the CDAD signal will be equall to zero in the direction of the molecular axis. If the molecule is vibrating around the direction of the surface normal, the CDAD signal will be equal to zero in the direction of the surface normal.

HIGH RESOLUTION THRESHOLD PHOTOIONISATION OF H_2

R. I. Hall[1,2], L. Avaldi[1,3], K. Ellis[1], G. Dawber[1], A. McConkey[1] and G.C. King[1]

[1] Department of Physics, Manchester University, Manchester M13 9PL, U.K.
[2] Laboratoire de Dynamique Moleculaire et Atomique, Universitè P. et M. Curie, 4 pl. Jussieu B75, 75252 Paris Cedex 05, France
[3]I.M.A.I. del CNR, Area della Ricerca di Roma, CP10 Monterotondo Scalo, Italy

Abstract

A novel spectrometer, which uses the penetrating field technique, has been used to obtain high resolution threshold photoelectron spectra of H_2^+ at both liquid nitrogen and room temperatures. Photoionisation of vibrationally excited H_2, produced by recombinative desorption in a recently developed source, has also been investigated.

Introduction

Photoelectron spectroscopy have added a wealth of information to our knowledge of molecules and ions. Threshold photoelectron spectroscopy is a particular type of photoelectron spectroscopy in which the incident photon energy is varied and only zero energy electrons are detected. While conventional photoelectron spectroscopy is largely limited to studying direct transitions, that is transitions to ionic states with favourable Franck-Condon factors, in threshold experiments both direct and autoionisation transitions are excited. Therefore ionic states which may not be accessible by direct transitions can appear through degenerate auotoionisation processes, since near zero energy electrons are detected. This technique has been pioneered by Peatman in the seventies[1,2].

In this paper a high sensitivity threshold spectrometer[3], which makes use of the penetrating field technique[4], is used for a spectroscopic investigation of molecular hydrogen. A considerable amount of experimental[2,5-7] and theoretical[8-10] work on this molecular ion exists and a high resolution threshold photoelectron spectrum (TPES) up to 725 Å (17.1eV) has been presented by Peatman[2]. In the present case the experimental investigation has been extended up to the dissociation limit and TPES at both liquid N_2 and room temperatures have been measured. Moreover the possibility to detect H_2^+ ions with the same experimental apparatus, just by reversing the polarity of the extracting field and of the electrostatic analyzer, enable a direct correlation of the autoionisation transitions from H_2 Rydberg states with molecular ion states to be established. Moreover a TPES of vibrationally hot H_2, produced by means of a recently developed source[11], has been also measured.

Experimental

The experiments have been performed at the Synchrotron Radiation Source (SRS) of the Daresbury Laboratory, UK. The apparatus consists of a threshold spectrometer used in conjunction with the normal incident 5m McPherson monochromator of the VUV 3.2 beam line at SRS. The monochromator is equipped with gratings which cover the wavelength, λ , range 45-100nm (25-12eV) with a photon flux of about 10^{10}ph/sec at $\Delta\lambda \cong 0.1$Å[12]. The photon beam from the monochromator was trasported to the target via a pyrex capillary tube of internal diameter 1mm . The photoelectron spectrometer (fig. 1) was developed from the one previously described by King et al[3]. It consists of a 127° cylindrical deflector analyzer (CDA) and an input lens system, formed by two three-aperture asymmetric lenses, which collects electrons from the target region and focusses them onto the entrance plane of the CDA. The detector at the exit plane of the CDA has been placed off-axis and a pusher electrode is used to deflect the outgoing particles towards the detector itself. In detecting ions the cone potential is raised to several KV to improve the detection efficiency of the channeltron.

In the threshold mode of operation an extraction electrode generates a penetrating field in the interaction region, which selectively extracts

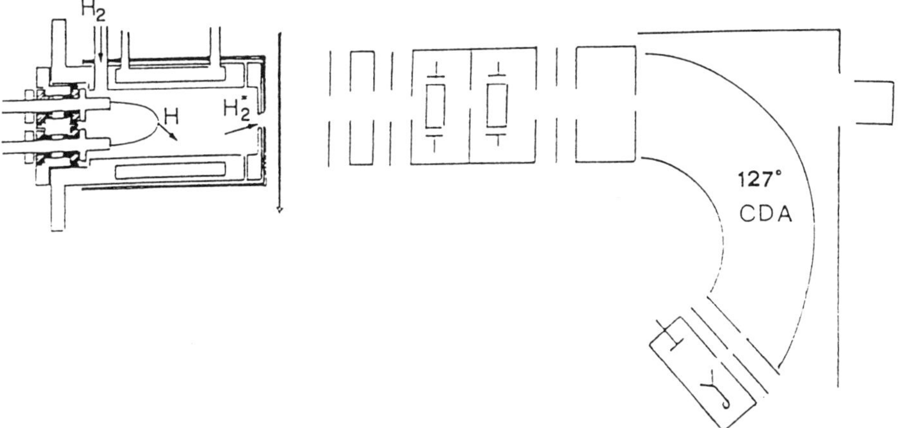

Fig. 1. Schematic diagram of the threshold spectrometer and of the hot H_2 source

low energy particles[4]. This technique results in a very large collected solid angle (≅4π) and also produces a cross-over point in the electron trajectories, which acts as an input to the following lens system further providing a good energy selection. The performances of the spectrometer are illustrated in fig.2, where the threshold electron and total ion yields in the region of the Ar^+ $^2P_{3/2}$ and $^2P_{1/2}$ ion states are reported versus photon energy. The ion spectrum was obtained by reversing the polarity of the voltages of the extraction electrode and of the 127° CDA. The structures in the ion yield correspond to high lying neutral Rydberg states converging to the $^2P_{1/2}$ ionisation potential. The excited neutral states can decay to the lower $^2P_{3/2}$ ion state through autoionising transition with the emission of electrons of kinetic energy between 3 and 150meV. Detection or suppression of these energetic electrons is a good test of the performances of a threshold spectrometer. From figure 2 it is clear that even the 3 meV electrons from the decay of the 11s' Rydberg state are heavily discriminated against. Indeed the $^2P_{3/2}$ photoelectron peak is not broader than that of $^2P_{1/2}$ as would be the case if it contained contributions from the 3meV electrons from the autoionising process. Moreover the $^2P_{3/2}/^2P_{1/2}$ ratio, that in this case is 2.3/1, close to the statistical weights, indicates that the threshold resolution was less than 3meV. This width is essentially that of the incident photon beam, suggesting that the threshold resolution of the analyser is somewhat less than this value.

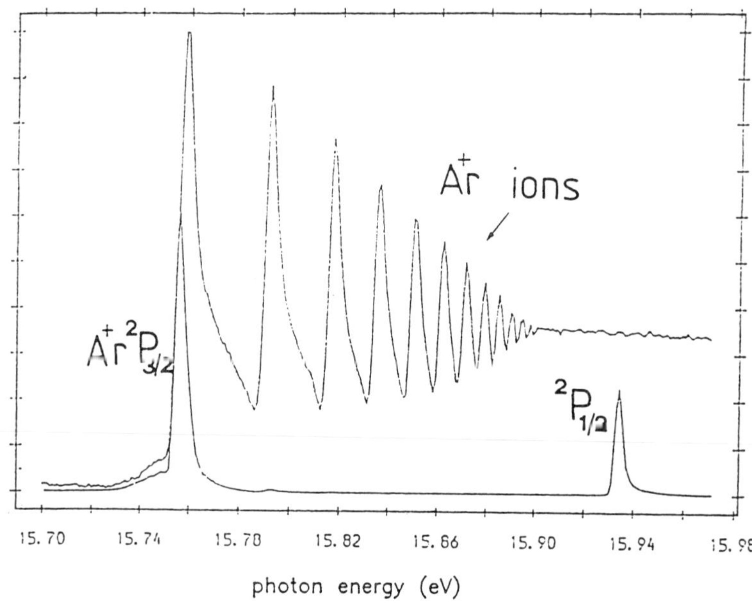

Fig. 2. Threshold photoelectron spectrum in the region of the $^2P_{3/2,1/2}Ar^+$ ionic states, superimposed on an ion yield obtained in the same energy region

The vibrationally excited H_2 molecules are produced by means of a source developed by Hall et al.[4]. The source placed in front of the electron spectrometer (fig. 1) is composed of a cylindrical cell made of stainless steel and cooled by water flowing through an outer jacket. A magnetic shield around the cell prevents the field generated by the tantalum filament to perturbing the threshold electrons generated in the scattering region. The vibrational excitation of H_2 molecules results from recombinative desorption of atomic hydrogen on the cell surfaces following atomization of H_2 on the incandescent filament[11].

The same cell has been also used to produce cold H_2 molecules. In this case liquid nitrogen was flowed through the outer jacket of the cell.

Results and discussion

The H_2 TPES obtained in these experiments at liquid N_2 and room temperatures are reported in figures 3 and 4, respectively. In the inset in fig. 3 an expanded view of the region close to the dissociation limit is also shown. In a preliminary analysis of the data the assignment of the peak has been done using Herzberg and Howe's molecular parameters[13] for $H_2(X\ ^1\Sigma_g^+)$ and the ones determined by Peatman in a previous threshold experiment[2] for $H_2^+(X\ ^2\Sigma_g^+)$. All the peaks assigned by Peatman are also observed in this work, but the higher intensity of the light source as well as the higher efficiency of the spectrometer enable the observations of other peaks. Moreover the present investigation has been extended up to the dissociation limits and features belonging to the v'=18 band are clearly observable in the spectrum.

Despite the high threshold resolution of the present experiment ($\leq$3meV), still the relative intensities of the peaks bear little resemblance to the expected intensities on the basis of the Franck-Condon factors calculated for instance by Villarejo[16]. As previously discussed by Peatman[2] and Stockbauer[7] this is due to the contribution of autoionising states which are degenerate with ionic states and yield the zero or low energy electrons which are detected. Autoionisation, being two order of magnitude more intense than direct ionisation, dominates the spectrum in the region up to about 17eV. This is clearly shown by the lower part of figure 4 where the total H_2^+ ion yield measured with the same experimental apparatus is reported. The most striking aspect of the ion spectrum is the wealth of autoionising features and the large cross section of the autoionising processes, which make the direct ionisation process indiscernible.In this regard it should also be pointed out that the large number of autoionising states makes it likely that various ion states which have only a small cross section by direct ionisation will be brought out by degenerate states . It is possible to identify these peaks by

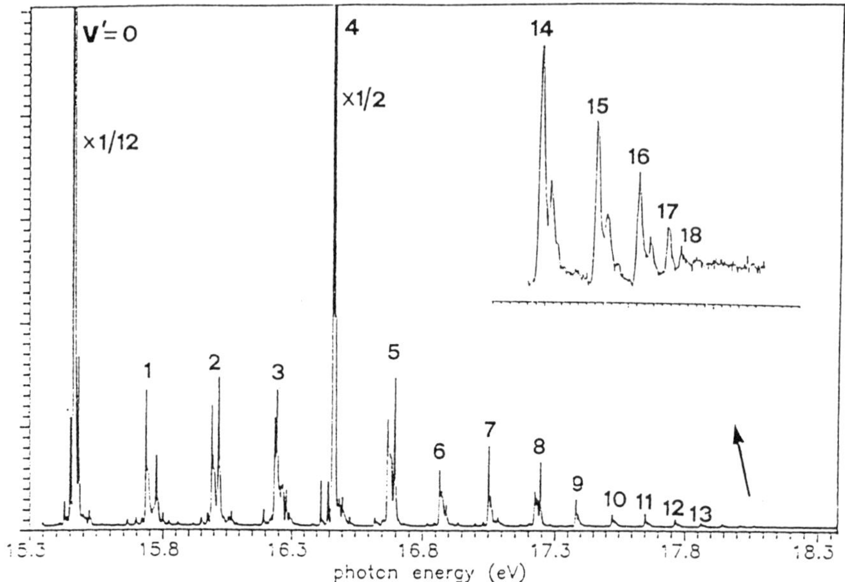

Figure 3: Threshold photoelectron spectrum of H_2 at liquid nitrogren temperature

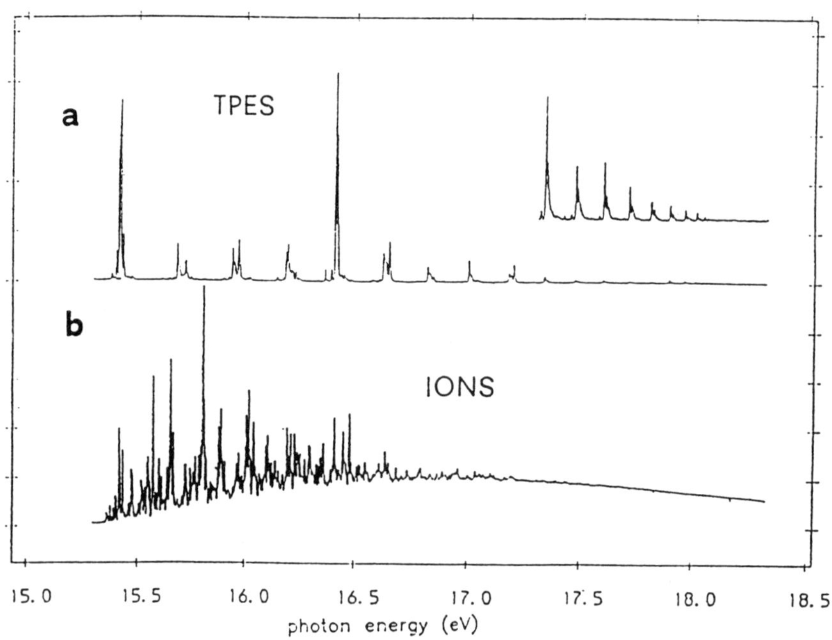

Figure 4: Threshold photoelectron spectrum (a) and total H_2^+ ion yield (b) at room temperature.

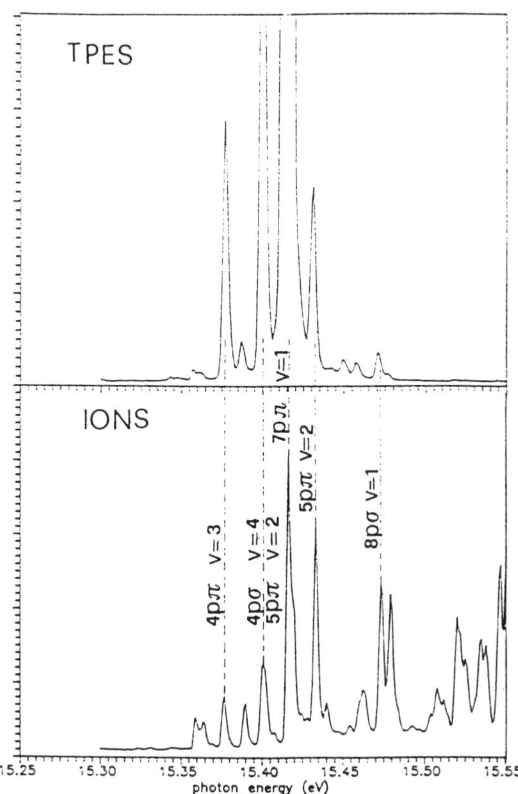

Fig. 5. Threshold photoelectron spectrum and total ion yield in the energy range 15.3-15.5eV. Assignments of some autoionising transitions are proposed by means of the Rydberg spectra of H_2[5,16].

comparing the two spectra in fig.4. An example is given in figures 5 a and b where the TPES and the ion spectrum in the energy region between 15.30eV and 15.55eV are reported.In this latter figure it clearly appears that the threshold peaks with anomalous intensities are degenerate with autoionising state, which can be assigned by means of the Rydberg spectra of H_2[5,15].Some of the proposed assignments are reported in the figure.

Finally in figure 6 a TPES of vibrationally excited H_2 in the energy region between 12.4eV and 15.4eV is shown. This spectrum has been obtained by heating the tantalum filament at a temperature of about 3000°K. The TPES is dominated by the atomic hydrogen peak at 13.605eV, produced from H_2 on the incandescent filament. However in the same spectrum peaks

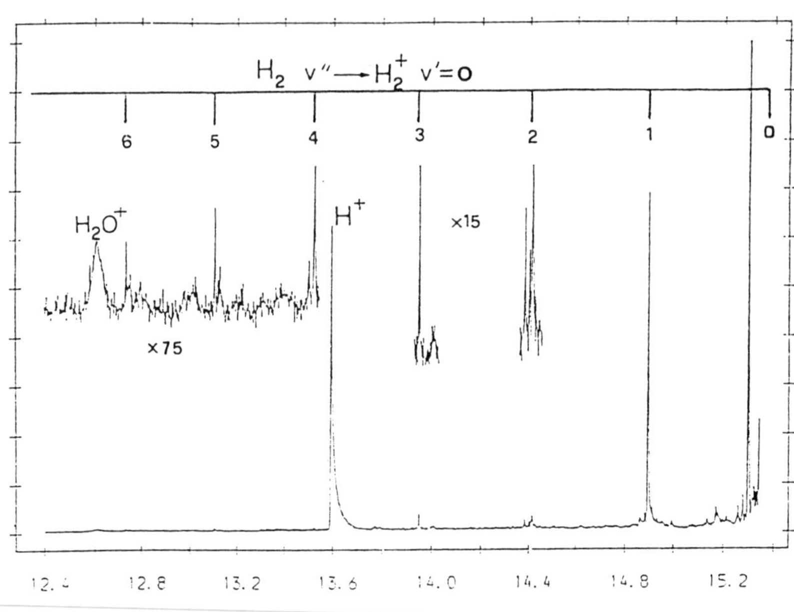

Fig. 6. Threshold photoelectron spectrum of vibrationally excited H_2 in the energy region 12.4–15.3eV

corresponding to transitions from H_2(v''=1–6) to H_2^+ (v'=0) are clearly identified. The assignment of the various transitions has been done by means of the same molecular constants used for the other TPES.

The determination of internal state distributions of molecular hydrogen and its isotopes is of great importance in many branches of physics such as surface science and plasma physics. For this purpose several techniques[11,16–18] have been studied for H_2 diagnostics. From the present results it appears that threshold photoemission is not a competitive technique, despite its sensitivity, compared to resonant enhanced multiphoton ionisation (REMPI)[18] and dissociative attachment[11,16]. Indeed, for example, in the dissociative attachment the increase of the cross section compensates the relatively low populations of the high v' levels present in the hot H_2 gas. This seems not to be the case in photoionisation. Moreover in order to establish the relative population values of the photoionisation cross section, vibrationally-rotationally resolved, at threshold are needed. So far only calculations for v'=1 and 2 at the He I-α radiation (584Å) exist[19,20]. From the experimental point of view it has also to be mentioned that degenerate autoionisation can also alter the relative intensities of the the different peaks at threshold, in this way hampering the extraction of the internal state distributions.

Conclusions

A sensitive electron spectrometer has been used to study threshold photoionisation of H_2. Operation in the threshold mode is a particularly powerful technique because it provides both high resolution and sensitivity. The high sensitivity of the spectrometer together with the tunability and high flux of the SRS at Daresbury enabled the spectroscopic investigation to be extended up to the dissociation limit.

The present investigation confirms that the H_2 threshold spectrum is dominated by the presence of degenerate autoionisation transitions from Rydberg states. These transitions result in an intensity distribution of the different peaks which is completely different from the one expected on the basis of the Franck-Condon factors.

Finally for the first time a threshold photoelectron spectrum of vibrationally hot H_2 is reported.

Acknowledgment

We are most grateful to the Daresbury SRS facility and staff for providing such excellent working conditions. In particular we are much idebted to Dr. M. Hayes for furnishing the data acquisition and instrumental control program and to Dr. M. MacDonald for his generous help during the measurements.

Work partially supported by a CNR(Italy)-Royal Society cooperative agreement.

References :

[1]W.B. Peatman, G.B. Kasting and D.J. Wilson, J. Elec. Spectrosc. 7,233 (1975)

[2] W.B. Peatman, J. Chem. Phys. 64, 4368 (1976)

[3] G.C. King, M. Zubek, P.M. Rutter and F.H. Read , J. Phys. E :Sci. Instrum. 20, 440 (1987)

[4] S. Cvejanovic and Read F.H. J. Phys. B: At. Mol. Phys. 7, 1180 (1974)

[5] W.A. Chupka and J. Berkowitz, J. Chem. Phys. 51, 4244 (1969)

P. M. Dehmer and W. A. Chupka J. Chem Phys. 65, 2243 (1976)

[6] L. Asbrink, Chem. Phys. Lett. 7, 549 (1970)

[7] R. Stockbauer, J. Chem. Phys. 70, 2108 (1979)

[8] H. Wind, J. Chem. Phys. 43, 2956 (1965)

[9] G. Hunter, A.W. Yau and H.O. Pritchard, At. Data Nucl. Data Tables 14, 11 (1974)

[10] T. Murai, J. Phys. Soc. Jpn. 38, 514 (1975)

[11] R. I. Hall, I. Cadez, M. Landau, F. Pichou and C. Schermann, Phys. Rev. Lett. 60,337 (1988)

[12] D.M.P. Holland, J.B. West, A. A. MacDowell, I.H. Munro and A.G. Beckett, Nucl. Instrum. Meth. B44, 233 (1989)

[13] G. Herzberg and L. Howe ,Can. J. Phys. 37,636 (1959)

[14] D. Villarejo, J. Chem. Phys. 49, 2532 (1968)

[15] S. Takezawa, J. Chem. Phys. 52,2575 (1970) and 52, 5793 (1970)

[16] D. Popovic, I. Cadez, M. Landau, F. Pichou,, C. Schermann and R.I. Hall, Meas. Sci. Technol. 1,1041 (1990)

[17] M. Pealat, J. E. Taran, M. Bacal and F. Hillion, J. Chem. Phys. 82, 4943 (1985)

[18] G. D. Kubiat, G. O.Sitz, R. N. Zare and A. H. Kung Chem. Phys. Lett. 83,2538 (1983)

[19] J. Tennyson, J. Phys. B:At. Mol. Phys. 20,L375 (1987)

[20] S. Hara, J. Phys. Soc. Jap. 58, 17 (1989)

Addendum

Very recently the same technique used in the H_2 experiment has been extended to perform coincidence experiments devoted to probing double ionisation processes near their threshold. For this purpose a time of flight analyzer, equipped with an extraction electrode to generate the penetrating field, has been placed in front of the 127° CDA. This apparatus enables several type of coincidence measurements to be performed[1] : threshold photoelectron-photoion (TPEPICO), photoelectron-photoion (PEPICO), photoion-photoion (PIPICO) and finally threshold photoelectron-threshold photoelectron (TPEsCO).
An example of TPEsCO spectrum for the double ionisation of CO is reported in the figure. The coincidence spectrum reveals a sharp onset located at 41.27±0.03 eV and a series of structures forming an irregular envelope that can be attributed to the formation of vibrational levels of what appears to be several states of CO^{2+}. Photo-double ionisation can proceed by either direct or indirect mechanisms. The mutual correlation of the two electrons in the continuum inhibits the direct double escape process and the cross section for such a process in the framework of the Wannier theory[2] is predicted to be vanishingly small. The indirect process proceeds through the formation of singly charged ion in an excited state, member of a Rydberg series converging to a higher state of the double ion, which subsequently autoionises to a lower state of the double ion. The threshold cross section of this process is finite and it can be expected that autoionisation will be the dominant process in the threshold region. According the most recent calculations[3] of the potential curves of CO^{2+} the bound states present in the investigated region are the $^3\Pi$ one, that is metastable due to predissociation by a repulsive $^3\Sigma^-$ state at large distance, and the $^1\Sigma^+$ and the $^1\Pi$ states that lie about 0.25 and 0.52 eV above the $^3\Pi$ ground state. A definite analysis of the TPEsCO spectrum in terms of the above three states is not possible because the expected non Franck-Condon behaviour of the peak intensities is a considerable handicap especially as the resolution is not yet sufficient to isolate the individual vibrational levels. However according the predictions of Larsson et

al.[3] a tentative analysis in terms of the vibrational progression of the $^3\Pi$ and $^1\Sigma^+$ states has been made and the proposed assignments for the two progressions, labelled "p" and "s" respectively, are shown in the figure.

In summary a new technique for probing double ionisation processes has been developed. Its performances in the case of CO show that this technique provides accurate values for the onset energies and enables vibrational structure of the double charged ions to be resolved.

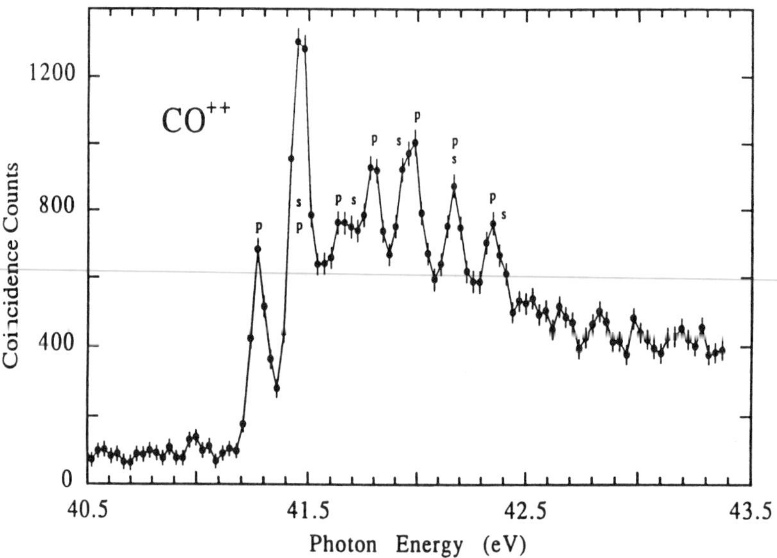

TPEsCO spectrum of CO recorded with a 30 meV channel width. The total dwell time was 7 min/ch.

References :

[1] R.I. Hall, A. McConkey, K. Ellis, G. Dawber, L. Avaldi and G.C. King submitted to Meas. Sci. Technol. (1991)

[2] G.H. Wannier Phys. Rev. **90**,817 (1953)

[3] M. Larsson, B.J. Olsson and P. Sigray Chem. Phys. **139**,457 (1989)

DISCUSSION

BAER - These are very beautiful results. However, I am surprised that your CO^{++} spectrum shows such a strong origin peak. Is this the true adiabatic IP or is the signal at lower energy ?

AVALDI - We think that the strong origin peak is located at the adiabatic IP. The signal at lower energies (if it is real) can be attributed to autoionization from a satellite state of the single ion to the double ion continuum at large nuclear distances (precursor mechanisms). This process was observed by Winkoun et al. (1988) in H_2O and by Lablanquie et al. (1989) in CO as well.

NENNER - Your beautiful results on CO^{2+} by double threshold electron coincidences raise the following question. Have you tried to compare your data with very high resolution Auger data of Siegbhan or Moddeman, in order to estimate the role of autoionization of singly charged ions ?

AVALDI - The data by Siegbhan et al. (1969) show a single vibrational progression with a spacing of about 0.3eV which has ben attributed to the $^1\Sigma^+$ state by Correia et al. (1985). In this work a more rich structure is observed indicating the presence of other states.

ELAND - What time resolution can you use for e^--e^- coincidence with that inhomogenesous field extraction mehod ? Would any differences of electron energy show up as time differences ?

AVALDI - The time resolution (FWHM) in the TAC spectrum was 80 nsec, the main contribution to the time spread coming from the path of the electrons in the low field in the neighbourhood of the interaction region. This spread would mask small differences in the electron energy. Energetic electrons are filtered out by the field configuration and the electron optics.

MORIN - In the case of e^--e^- experiment, how sensitive are you to any misalignment of the photon beam ? What about the size of the photon beam ?

AVALDI - The potential is weakly varying in the interaction region (1mm diameter photon beam). Any slight misalignment can be compensated by suitable tuning of the extraction and needle potentials.

DE LANGE - 1. In the case of vibrationally excited hydrogen not only the cross section σ, but also the asymmetry parameter β is a strong function of vibrational quantum number. Tennyson has performed a calculation of β as a function of V. It would be interesting to see to what extent your results show a β dependence.

2. Could you comment on the very asymmetric line shape of the hydrogen atom signal in your TPES experiment in vibrationally excited H_2 ?

AVALDI - 2. The asymmetric line shape is characteristic of the response function of the threshold analyzer as you can see in the case of the Ar $^2P_{3/2}$ and $^2P_{1/2}$ lines (fig. 2).

LEACH - From your results on rotationally resolved TPES of H_2 it should be possible to extract Fano parameters from the autoionization profiles. The detailed theory has been developed a few years ago by C. Jungen, M. Raoult and A. Giusti. Has such an analysis of your experimental results been carried out ?

AVALDI - In a treshold spectrum a degenerate autoionising state results only in an enhancement of the ionic peak intensity, thus a study of the lineshape to extract Fano parameters is not possible. Such a study would have to be performed in a constant ionic state measurement in which any autoionising state, embedded in the continuum of the ionic state investigated, will appear as a structure with a typical resonance profile.

R-MATRIX CALCULATION OF THE CORE AND VALENCE PHOTOIONIZATION CROSS SECTION OF LITHIUM

A. Lisini
Dipartimento di Scienze Chimiche, Università di Trieste,
Via A. Valerio 38, I-34127 Trieste (Italy)

ABSTRACT

Core and valence photoionization cross section of lithium has been calculated with the R-matrix method, including 11 target states and 17 coupled channels. Good agreement has been obtained with the few available experimental data, and the extreme complexity of the region above the first core ionization threshold calls for a detailed experimental investigation. Branching ratios relative to the shake-up processes on and off resonance have been obtained in a wide energy range.

1. INTRODUCTION

The calculation of the photoionization cross section of atoms and molecules gives important information both on the electronic structure of the target and the dynamics of the process. In fact on one hand the analysis of the resonances gives a balance of the interaction between bound and continuum states and on the other hand the branching ratios relative to the various target states give the entity of both direct and conjugate shake-up processes, which often accompanies the main ionization process and reflect the importance of correlation effects in the target.

To have a clear insight of these important aspects it is necessary to analyze accurately the cross section over a wide energy range, which can be a difficult task not only for molecules but also for atoms, especially if many channels are opening in narrow energy ranges. Among the atoms, helium has been analyzed in detail [1] and it is considered an ideal test case for the theoretical models. The next more complex atom, i.e. lithium, which is even more attractive since it is the simplest open shell system, has been much less investigated theoretically, although it has been studied experimentally both in the valence region and more recently in the core region [2-5].

We present here a study of the photoionization of lithium using the R-matrix method [6-8] which can be applied both at the threshold and at higher energy, also in the presence of several target states and coupled channels, giving a balanced description of the processes involved. In a previous paper [5] we have examined the behaviour of total and partial cross sections and branching ratios over a limited energy range and we extend here the analysis to lower energies.

In particular the following processes, in which the ion+electron system is left in the $^2P^o$ state, have been included in the calculations:

$$h\nu + Li(1s^2 2s\ ^2S) \rightarrow Li^+ (1s^2) + e^- (kp)$$
$$\rightarrow Li^+ (1s2s\ ^{1,3}S) + e^- (kp)$$
$$\rightarrow Li^+ (1s2p\ ^{1,3}P) + e^- (ks+kd)$$
$$\rightarrow Li^+ (1s3s\ ^{1,3}S) + e^- (kp)$$
$$\rightarrow Li^+ (1s3p\ ^{1,3}P) + e^- (ks+kd)$$
$$\rightarrow Li^+ (1s3d\ ^{1,3}D) + e^- (kp+kf)$$

giving a total of 11 target states and 17 coupled channels. As regards the computational details, the same scheme of ref. 5 has been employed.

2. RESULTS AND DISCUSSION

The lowest energy part of the cross section of lithium has been already calculated by Peach et al. up to about 30 eV with the R-matrix method [9]; in Figure 1 we report also the higher energy part which becomes very complex for the presence of many resonances.

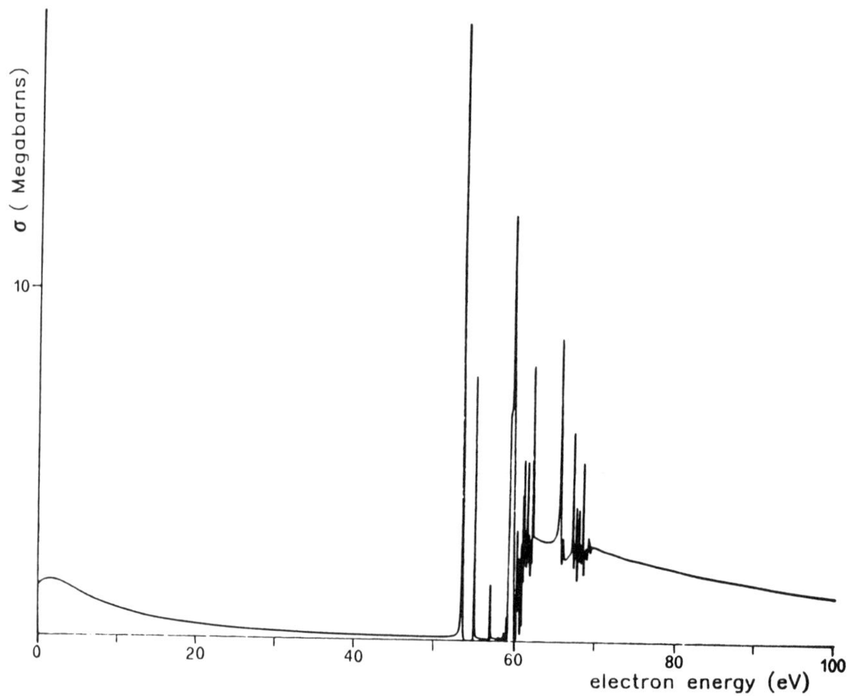

Fig. 1. Calculated total cross section of photoionization of lithium.

Table 1. Experimental and calculated ionization energies of lithium and Rydberg series converging to each threshold.

Rydberg series		Threshold	Energy (eV)	
			Exp.[a]	Calc.
		$1s^2$	5.390	5.393
$(1s2s\ ^3S)$ np	-->	$1s2s\ ^3S$	64.412	64.312
$(1s2s\ ^1S)$ np	-->	$1s2s\ ^1S$	66.152	66.149
$(1s2p\ ^3P)$ ns,nd	-->	$1s2p\ ^3P$	66.672	66.529
$(1s2p\ ^1P)$ ns,nd	-->	$1s2p\ ^1P$	67.608	67.472
$(1s3s\ ^3S)$ np	-->	$1s3s\ ^3S$	74.172	74.014
$(1s3s\ ^1S)$ np	-->	$1s3s\ ^1S$	74.670	74.501
$(1s3p\ ^3P)$ ns,nd	-->	$1s3p\ ^3P$	74.760	74.573
$(1s3d\ ^3D)$ np,nf	-->	$1s3d\ ^3D$	74.976	74.767
$(1s3d\ ^1D)$ np,nf	-->	$1s3d\ ^1D$	74.980	74.769
$(1s3p\ ^1P)$ ns,nd	-->	$1s3p\ ^1P$	75.038	74.854

[a] Ref. 10.

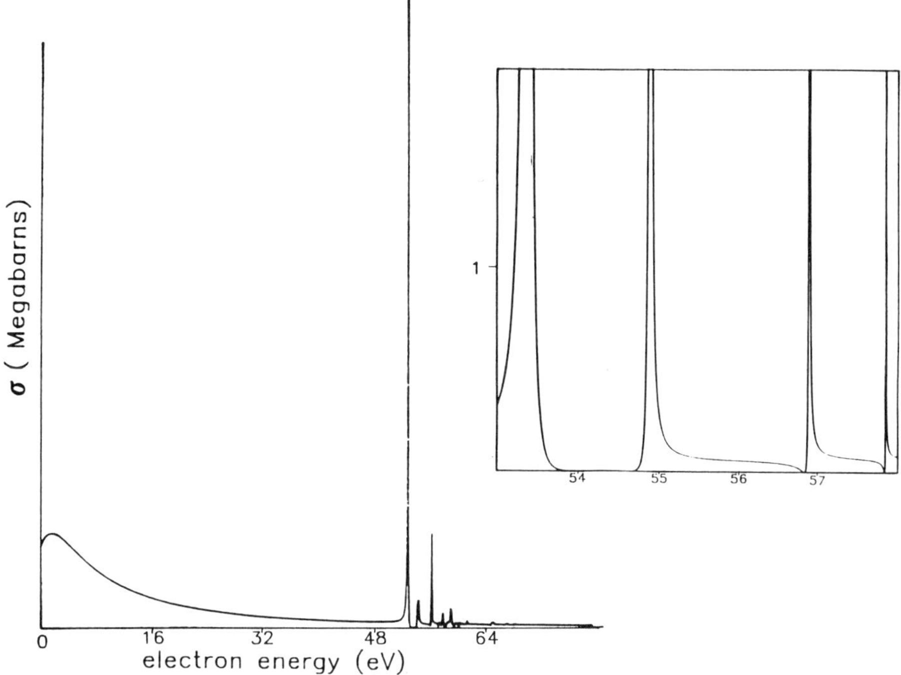

Fig. 2. Calculated partial cross section of photoionization of lithium into the $1s^2$ state. In the inset the first four resonances are shown in detail.

In fact there are several Rydberg series converging to each threshold as shown in Table I, and the relevant structures have been so far only partly identified. The (1s2s ^3S)np series and the lowest members of the (1s2s ^1S)np, (n=2,3) fall below the threshold for core ionizations, so they are weakly coupled only to the valence $1s^2kp$ continuum. These core excited resonances are very sharp, intense, almost discrete states, as can be seen in Figure 2, where the calculated lower energy members are presented in detail.

Besides the experimental data for these series, there are sparse calculations in the literature and a complete analysis is still missing; the present calculations therefore represent a complementary part of previous work. In Table II the results for energies and oscillator strengths relative to the 1s2snp Rydberg series are reported. Excellent agreement is apparent between the present calculations and the experimental data reported in the table. It is to be noted that there are some new calculated values for the (1s2s ^1S)np series, in particular the feature found at 64.05 eV by Ederer et al. [11] is attributed to the (1s2s ^1S)3p resonance. Various theoretical values for the intensity ratio between the 1s(2s2p ^3P) and 1s(2s2p ^1P) resonances have been proposed in the literature, our calculated value of 24 is quite near the photoemission result of 33 [12].

Table II. Experimental and calculated energies and oscillator strengths for the 1s2snp Rydberg series.

	Experimental[a]			Calculated			
	E(eV)	Oscillator strength	Relative value	E(eV)	Oscillator strength	Relative value	Width(eV)
1s(2s2p ^3P)	58.91	0.24	1.000	58.78	0.24	1.000	3.88×10^{-3}
(1s2s ^3S)3p	62.42	0.053	0.230	62.28	0.050	0.208	0.23×10^{-3}
(1s2s ^3S)4p	63.36	0.019	0.077	63.25			
(1s2s ^3S)5p	63.76		0.035	63.67			
1s(2s2p ^1P)	60.40	0.007	0.030	60.28	0.010	0.042	9.73×10^{-3}
(1s2s ^1S)3p	64.05[b]			64.14	0.001	0.004	0.04×10^{-3}
(1s2s ^1S)4p	65.29			65.10			
(1s2s ^1S)5p	65.65		0.0014	65.51			

[a] Ref. 12. [b] Ref.11.

The most complex part of the cross section is between the 1s2s ^3S and the 1s2p ^1P thresholds, as can be seen in Figure 3, where the calculated partial cross sections are presented. In fact all six (or four) Rydberg series interact strongly in this region and only a

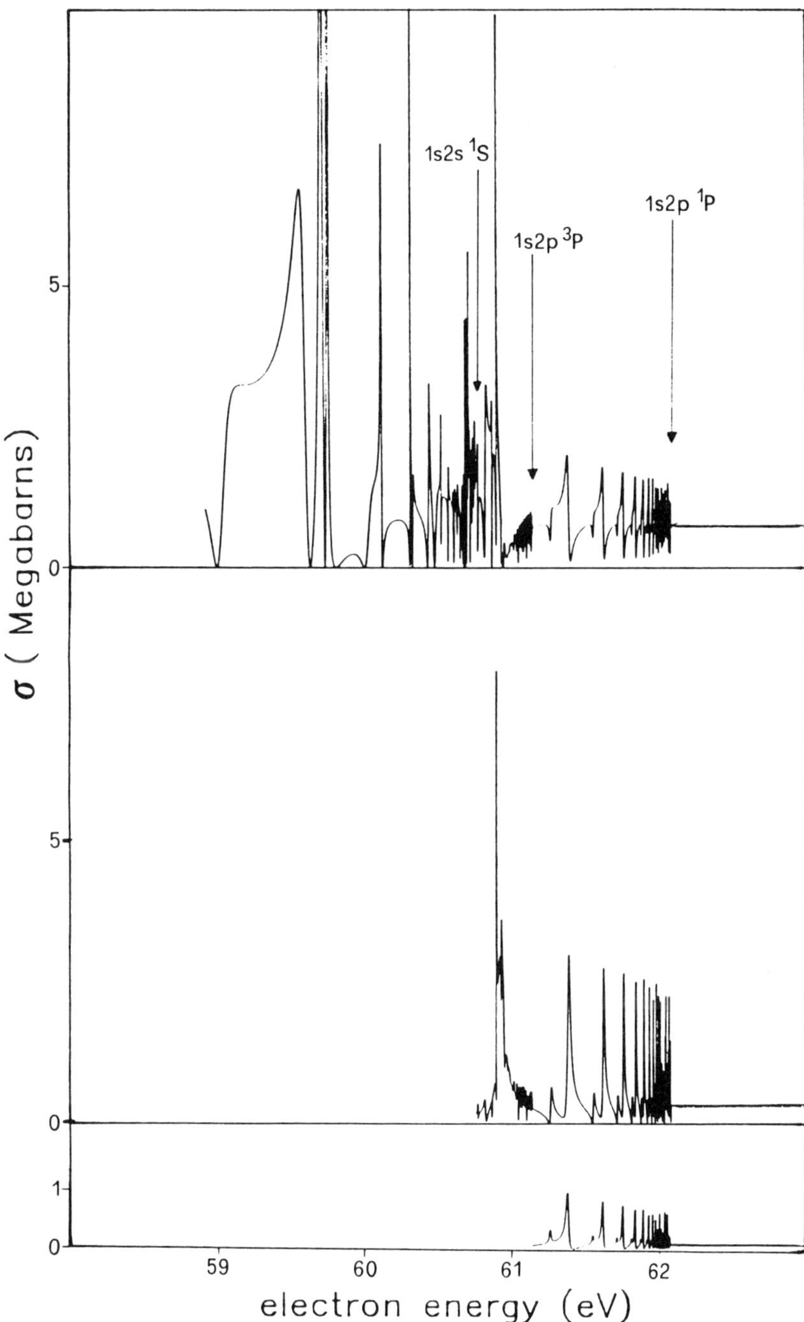

Fig. 3. Calculated partial cross sections of the photoionization of lithium into the 1s2s ^{3}S (top), 1s2s ^{1}S (middle) and 1s2p ^{3}S (bottom) states.

tentative identification is possible.

The ^3P partial cross section is the simplest one, since there are only two series converging to the 1s2p ^1P threshold, i.e. (1s2p ^1P)ns and (1s2p ^1P)nd, but their members are less than one tenth of an eV apart, rendering their attribution still problematic. The lowest members of these series, i.e. (1s2p ^1P)3d and (1s2p ^1P)3s are probably hidden in the broad band between 59 eV and 59.6 eV visible in the ^3S partial cross section, where probably other resonances are present such as the lowest members of the (1s2s ^1S)np and (1s2p $^{1,3}P$)ns series. At higher energy, there are still some relatively isolated resonances, but they become soon very close to each other in proximity of the 1s2s ^1S and 1s2p ^3P thresholds.

It is remarkable the close resemblance between the calculated ^3S partial cross section in Figure 3 and the photoabsorption cross section measured by Mehlman et al. [13]; only some minor differences in the resonance intensities are present because of the lower resolution in the experiment, where sometimes such high intensities cannot be reached.

Going towards the 1s3l thresholds, the resonances appear again well identifiable, in particular the two 1s3s3p resonances, which are isolated from the others, have been studied experimentally and the relevant total and partial cross sections have been accurately reproduced by the R-matrix method [5].

Table III. Experimental and calculated shake-up energies and intensities relative to the core ionization of lithium[a].

	E(eV)		I(%)					
Photon Energy(eV)			70.96	71.30	80	100	150	200
	Exp[b].	Calc.						
1s2s ^3S --> 1s2p ^3P	2.26	2.22	63.21	43.69	14.77	12.55	7.66	6.35
--> 1s2p ^1P	3.20	3.21	185.33	18.86	8.60	3.62	2.20	1.55
--> 1s3s ^3S	9.76	9.70			19.92	24.83	27.19	27.34
--> 1s3s ^1S	10.26	10.19			4.33	4.63	5.56	5.74
--> 1s3p ^3P	10.35	10.26			6.99	4.32	2.37	1.63
--> 1s3d ^3D	10.56	10.45			8.18	4.88	2.45	1.57
--> 1s3d ^1D	10.57	10.46			0.89	0.64	0.60	0.35
--> 1s3p ^1P	10.63	10.54			2.25	1.13	0.70	0.27

[a] Energies and percentage intensities with respect to the 1s2s ^3S state. [b] Ref. 10.

The study of the shake-up processes in the vicinity of the resonances is particularly interesting, since their intensities undergo dramatic changes. In the case of the 1s(3s3p ^3P) and 1s(3s3p ^1P) resonances, these states autoionize to the underlying 1s2p ^3P and 1s2p ^1P ionic states, with the consequent strong enhancement of the 1s2s $^{1,3}S$ --> 1s2p $^{1,3}P$ conjugate shake-up intensities. As can be seen in Table III, where the behaviour of the shake-up processes with the photon energy is reported, both conjugate shake-up intensities are affected at the resonance energies, indicating a considerable mixing between the autoionizing states, then they drop at higher energies, showing, as expected, a different behaviour from the direct shake-up processes, the intensities of which increases slowly, approaching the sudden limit at high energy.

3. CONCLUSIONS

The results obtained illustrate the complexity of the photoionization dynamics of a simple open shell system, and provide a detailed description of the cross section over a wide energy range, showing the full capability of the present R-matrix approach.

Good agreement with the available experimental data is obtained, notably resonance parameters and branching ratios for all channels considered in the case of isolated resonances. Further analysis is needed to fully disentangle the structures above the first ionization threshold.

ACKNOWLEDGEMENTS

Many thanks are due to P.G. Burke, A. Hibbert and K.A. Berrington of the Queen's University of Belfast for providing us with the updated versions of the RMATRIX and CIV3 programs.

REFERENCES

1. K.A. Berrington, P.G. Burke, W.C. Fon and K.T. Taylor, J. Phys. B: At. Mol. Phys., 15 (1982) L603, and references therein.

2. T.A. Ferrett, D.W. Lindle, P.A. Heimann, W.D. Brewer, U. Becker, H.G. Kerkhoff and D.A. Shirley, Phys. Rev. A, 36 (1987) 3172, and references therein.

3. B. Langer, J. Viefhaus, O. Hemmers, A. Menzel, R. Wehlitz and U. Becker, Phys. Rev. A, 43 (1991) 1652.

4. G.B. Armen and F.P. Larkins, J. Phys. B: At. Mol. Opt. Phys., 24 (1991) 2675, and references therein.

5. A. Lisini, P.G. Burke and A. Hibbert, J. Phys. B: At. Mol. Opt. Phys., 23 (1990) 3767.

6. P.G. Burke and W.D. Robb, Adv. At. Mol. Phys., 11 (1975) 143.

7. K.A. Berrington, P.G. Burke, M. Le Dorneuf, W.D. Robb, K.T. Taylor and Vo Ky Lan, Comput. Phys. Commun., 14 (1978) 367.

8. K.A. Berrington, P.G. Burke, K. Butler, M.J. Seaton, P.J. Storey, K.T. Taylor and Yu Yan, J. Phys. B: At. Mol. Phys., 20 (1987) 6379.

9. G. Peach, H.E. Saraph and M.J. Seaton, J. Phys. B: At. Mol. Opt. Phys., 21 (1988) 3669.

10. C.E. Moore, Atomic Energy Levels (NBS Circular 467) (Washington, DC: US Govt Printing Office).

11. D.L. Ederer, T. Lucatorto and R.P. Madden, Phys. Rev. Letters, 25 (1970) 1537.

12. P. Gerard, Ph. D. thesis, Université de Paris-Sud, 1984.

13. G. Mehlman, D.L. Ederer, E.B. Saloman and J.W. Cooper, J. Phys. B: At. Mol. Phys., 11 (1978) L689.

CAGE EFFECT ON THE PHOTODISSOCIATION OF SMALL MOLECULES IN RARE GAS MATRICES

H. Kunz, R. Schriever, M. Chergui, J.G. McCaffrey[1] and N. Schwentner
Institut für Experimentalphysik, Freie Universität Berlin,
Arnimallee 14, D–1000 Berlin 33, Germany

ABSTRACT

Photodissociation of H_2O (D_2O) and Cl_2 trapped in rare gas matrices has been investigated in relation to the cage effect. Barrier energies for the exit of the fragments from the matrix cage have been determined, together with the excess kinetic energy dependance of the photodissociation quantum efficiency. Simple analytical models as well as classical Molecular Dynamics simulations are used to discuss the results.

INTRODUCTION

One of the avenues leading to an understanding of photoreactions in condensed matter involves a microscopic description of the simplest of all such reactions, i.e., the photodissociation of simple molecules embedded in rare gas lattices.

The interest in using such model systems is outlined in Refs. 1 and 2, together with the experimental methodology, which consists in determining energy thresholds for cage exit of the fragments and the dependance of the permanent photodissociation quantum efficiency with excess kinetic energy, using synchrotron radiation. The cases of H_2O (D_2O) and Cl_2 in rare gas matrices have been investigated. Since most of the data have been published or are in press, we will only present the general results in this contribution.

RESULTS AND DISCUSSION

Photodissociation of H_2O (D_2O) in Ar[2,3], Kr[4] and Xe[5] matrices was investigated using fluorescence of OH in Ar and Kr or H in Xe matrices to determine the concentration of fragments for given irradiation doses. Barrier energies for the H atom cage exit of 1.3 eV to ~ 1.8 eV with respect to the H_2O adiabatic dissociation energy were determined. They are rationalised by summing the H–Rg pair potentials at the D_3 site of the fcc lattice through which the H fragment passes when leaving the matrix cage[3]. The photodissociation quantum efficiency was found to increase with increasing excess kinetic energy and was fitted with a near–square power law as a function of energy[4]. The prefactor (which we call the characteristic efficiency) in the power law increases from Ar to Xe, from D_2O to H_2O and with temperature. The temperature effect is weaker in Kr than Ar[4] and vanishes in Xe matrices[5]. The prefactor was rationalised in terms of a model involving the fractional energy loss per elastic collision between the H fragment (which carries most of the excess energy) and the matrix atoms. The model reproduces the trends from matrix to matrix and with isotopic substitution in a satisfactory way.

---

[1]Alexander von Humboldt Fellow 1989–1991
Permanent address; Dept. of Chemistry, St. Patrick's College, Maynooth, Co. Kildare, Ireland.

The temperature effect is interpreted in terms of a continuum model based on heat conductivity equations[6], whereby local melting of the lattice contributes to the overcoming of the barrier by the H fragment. The local melting is due to the combined contributions of the energy transfer per elastic collision between H and the matrix atoms and the decrease of the heat conductivity of the crystal with increasing temperature. Thus, the model[6] predicts a strong temperature effect in Ar matrices, a milder one in Kr and none in Xe, as actually observed[4,5]. In determining the quantum efficiency, the absorption cross–section of matrix isolated H_2O is of great importance. We have recently obtained more reliable absorption spectra of H_2O in Ar, Kr and Xe matrices which will be presented in a forthcoming publication[7]. The differences with the absorption spectra of Refs. 3–5 do not however alter the shape of the photodissociation yield curves.

In order to enable comparison of our results with Molecular Dynamics (MD) calculations, we measured the absolute photodissociation quantum efficiency of H_2O in Ar matrices[8]. Our results are in fair agreement with the MD simulations of H_2O dissociation in Ar matrices[9] and HI in Xe matrices[10].

Wavelength specific measurements of the photodissociation of molecular chlorine in crystalline argon samples[11] showed that a dominant threshold exists in the 130 nm band at 9.2 eV corresponding to absorption into the bound 1 $^1\Sigma_u$ state. The maximum quantum yield for permanent dissociation in the 130 nm band was found to be 0.3. Photoexcitation of the $^3\Sigma_u$ state at 180 nm also results in the permanent dissociation of chlorine but with a quantum yield of 0.02. The dissociation efficiency of this band was found however to be very sample preparation dependent viz., sample crystallinity. In crystalline samples, dissociation efficiencies were typically one order of magnitude less than in non–crystalline samples. No dissociation occurs following excitation into the lower energy C $^1\Pi(_{1u})$ band at 330 nm irrespective of sample preparation conditions. Measurement of the dissociation threshold of molecular chlorine in the 1 $^1\Sigma_u$ state (9 eV) as a function of temperature showed little variation. From spectroscopic data it is concluded[12] that the dissociation mechanism is occuring by an impulsive mechanism involving curve–crossing from the initially populated bound 1 $^1\Sigma_u$ state to repulsive potentials correlating with ground state atomic chlorine. A simple microscopic model, drawn from MD calculations[13] and experimental data is constructed, using pair–wise addition of ArCl potential terms, to estimate the energetics of the steps involved in this dissociation process in the solid lattice.

Isolation of atomic chlorine in two dominant trapping sites is observed[14] following photodissociation of Cl_2 in crystalline Xe throughout the spectral region from the C $^1\Pi_{1u}$ state to the region of Xe exciton absorption. The production efficiency of the thermally stable trapping site[15,16] was most pronounced in the region of the onset of Xe_2Cl excitation feature and was found to increase significantly with temperature. In contrast to Ar[11,12] and Kr[17] matrices, photodissociation of Cl_2 in Xe appears to occur without a pronounced cage effect. This observation is rationalized in terms of the larger lattice parameters of Xe which allows[18] isolation of Cl_2 and Cl at single substitutional and octahedral interestitial sites respectively. From this structural information, simple models are presented for the production of Cl in the two different kinds of trapping sites. Thus production of the thermally unstable site involves the symmetric dissociation of Cl_2 with both Cl atoms occupying octahedral interstitial sites separated by a single lattice constant of Xe. The thermally stable site involves the isolation of one Cl atom at the substitutional site originally occupied by the Cl_2 parent molecule and the other at an octahedral interstitial site. Correlation between the excitation spectrum of Xe_2Cl emission and the onset of efficient production of the thermally stable site of Cl is discussed in terms of the possibility of a charge transfer induced dissociation mechanism.

Measurements are now underway[19] to investigate the cage effect on the photodissociation of HCl in rare gas matrices.

REFERENCES

1. M. Chergui and N. Schwentner, to appear in Research Trends: Chemical Physics, ed.: J. Menon (Trivandum, India).
2. N. Schwentner and M. Chergui, to appear in Optical Processes in Excited States of Solids, ed.: B. Di Bartolo (Plenum Press).
3. R. Schriever, M. Chergui, H. Kunz, V. Stepanenko and N. Schwentner, J. Chem. Phys. 91, 4128 (1989).
4. R. Schriever, M. Chergui, Ö. Ünal, N. Schwentner and V. Stepanenko, J. Chem. Phys. 93, 3245 (1990).
5. R. Schriever, M. Chergui and N. Schwentner, J. Phys. Chem. 95, 6124 (1991).
6. E. T. Tarasova, A. M. Ratner, V. M. Stepanenko, I. Ya. Fugol, R. Schriever, M. Chergui and N. Schwentner, to be published.
7. M. Chergui, R. Schriever and N. Schwentner, to be published.
8. R. Schriever, M. Chergui and N. Schwentner, J. Chem. Phys. 93, 9206 (1990).
9. I. Gersonde and H. Gabriel, private communication.
10. B. Gerber, R. Alimi and V. A. Apkarian, J. Chem. Phys. 89, 1974 (1988).
11. H. Kunz, J.G. McCaffrey, R. Schriever and N. Schwentner, J. Chem. Phys., 94, 1039 (1991). Please note that the pictures appearing as 2,3 and 5 have been interchanged and belong to the captions numbered Fig. 5, Fig. 2 and Fig. 3 respectively.
12. J.G. McCaffrey, H. Kunz and N. Schwentner, J. Chem. Phys., in Press.
13. R. Alimi, R.B. Gerber, J.G. McCaffrey, H. Kunz and N. Schwentner, manuscript in Preparation.
14. J.G. McCaffrey, H. Kunz and N. Schwentner, J. Chem. Phys., submitted August 1991.
15. H. Kunz, J.G. McCaffrey, M. Chergui, R. Schriever, Ö. Ünal and N. Schwentner, J. Luminesc., 48 & 49, 621 (1991).
16. H. Kunz, J.G. McCaffrey, M. Chergui, R. Schriever, Ö. Ünal and N. Schwentner, J. Chem. Phys., 95, 1466 (1991).
17. H. Kunz, J.G. McCaffrey and N. Schwentner, manuscript in preparation July (1991).
18. R. Alimi, A. Brokman and R.B. Gerber, J. Chem. Phys., 91, 1611 (1989).
19. K.–H. Gödderz, J.G. McCaffrey, M.Chergui and N. Schwentner, unpublished results

THE ROLE OF SHAPE RESONANCE IN IONIZING AND NON-IONIZING DECAYS OF SUPEREXCITED ACETYLENE MOLECULES IN THE EUV REGION

Masatoshi Ukai, Kosei Kameta, Ryo Chiba, Kazunori Nagano, Noriyuki Kouchi, Kyoji Shinsaka[a], and Yoshihiko Hatano
Department of Chemistry, Tokyo Institute of Technology,
Meguro-ku, Tokyo 152, Japan

Hironobu Umemoto
Department of Applied Physics, Tokyo Institute of Technology,
Meguro-ku, Tokyo 152, Japan

Yoshiro Ito
Nagaoka University of Technology, Nagaoka, Niigata 940-21, Japan

Kenichiro TANAKA
Photon Factory, National Laboratory for High Energy Physics
Oho, Ibaraki 305, Japan

ABSTRACT

By the absolute measurements of the photoabsorption, photoionization, and photodissociation cross sections of C_2H_2 in the 53-93 nm region we have shown the σ^* and π^* shape resonances in photoionization of C_2H_2. Excitation spectra of $C_2(d^3\Pi_g \to a^3\Pi_u)$, $C_2(C^1\Pi_g \to A^1\Pi_u)$, CH $(A^2\Delta \to X^2\Pi)$, and H(*Lyman-α*) fluorescence have revealed rich density of superexcited states and their competitive neutral dissociation with autoionization. The ionizing and non-ionizing decay processes of superexcited C_2H_2 in competition have been shown and the role of shape resonance in overlapping nature with superexcited Rydberg states have been discussed.

INTRODUCTION

The decay process of acetylene molecules in superexcited states is one of the fundamental examples of the dynamic behavior of molecules excited well above the first ionization potential.[1,2] Previously obtained photoionization quantum yields[3-5] evidenced superexcited states in the 70-100 nm (about 12-18 eV) region in a strong competition of neutral dissociation decays with autoionization. This involved that an absolute cross section for a non-ionizing decay process, namely the neutral dissociation cross section would be appreciably large in the extreme-UV region, where the absorption cross section possesses its maximum value.[3,4,6] Previous studies, however, were mainly concentrated on ionizing decay processes[7-13] and only limited experimental efforts were applied to elucidate the non-ionizing decay channels.[6]

Shape resonance in the photoionization has been investigated of various molecules.[14] Several theoretical calculations on the photoionization of C_2H_2 indicated the shape resonant enhancement.[15,16] Interpretations of experimental results were, however, confused.[12,13,16-18] By the application of a photoelectron spectroscopy (PES) method, Holland et al.[12,13] examined the π^* shape resonance at around 80 nm excitation. However, as theories suggested at least three shape resonances, the problem still remained unresolved.

In the present paper, the absolute photoabsorption cross section and the photoionization quantum yield of C_2H_2 have been measured in detail using continuum monochromatized synchrotron radiation in the 53-93 nm region. The absolute photoionization and photodissociation cross section have also been obtained. We remark an overlapping nature of shape resonances with other superexcited states to result in a peculiar behavior of ionization and non-ionizing decay processes in competition. Excitation spectra of $C_2(d^3\Pi_g \rightarrow a^3\Pi_u)$, $C_2(C^1\Pi_g \rightarrow A^1\Pi_u)$, $CH(A^1\Delta \rightarrow X^1\Pi)$, and H(*Lyman-α*) fluorescence have also been obtained to substantiate selective dissociation processes in the non-ionizing decay channels of superexcited C_2H_2. Effect of the shape resonances have also shown up to depress the symmetry selected dissociation of superexcited C_2H_2.

EXPERIMENTAL

Experiments were performed using an extreme-UV synchrotron radiation (SR) from a 2.5 GeV positron storage ring at the Photon Factory as a tunable light source in the 50-100 nm region. A typical photon flux from a Seya-Namioka monochromator was 1×10^{10} photon/s at 50 nm with the typical photon bandpass of 0.4 nm. The flux was monitored by a photo-current from Au-mesh and a sodium salicylate coated glass window with a photomultiplier (PMT) before and behind the photon impact chamber, respectively.

Measurements of a photoabsorption cross section σ_T and a photoionization quantum yield η were made using a multistage ionization chamber of a parallel plate design.[19-21] Thin Sn and In films of about 100 nm thickness were employed as entrance window materials for the wavelength regions of 53-74 nm and 72-93 nm, respectively, to prevent gas effusion back into the optical path and to eliminate higher order dispersed radiation. A photoionization cross section σ_I and a photodissociation cross section σ_D were given by $\sigma_I = \eta\,\sigma_T$ and $\sigma_D = (1-\eta)\sigma_T$

Experimental setup for fluorescence excitation spectrum measurements was described elsewhere.[23,24] Briefly, UV and visible (VIS) fluorescence emitted by a dissociative photoexcitation in a gas cell was refocused using a pair of quartz lenses and detected by a PMT with a UV-VIS response set perpendicularly with the SR beam. Optical filters were placed between the lenses to select fluorescence from a specific dissociation fragment. The center wavelengths and the bandpass (fwhm) of the filters were 516 (11), 358 (16), and 423 (16) nm to resolve $C_2(d^3\Pi_g \rightarrow a^3\Pi_u, \Delta v=0)$, $C_2(C^1\Pi_g \rightarrow A^1\Pi_u, \Delta v=1)$, and $CH(A^2\Delta \rightarrow X^2\Pi, 0\rightarrow 0)$ emissions,[6] respectively. H(*Lyman-α*) emission was detected by a CsI coated microchannel plate (MCP) through a MgF_2 filter.[24]

RESULTS AND DISCUSSION

A. Shape Resonance in Ionizing Decay

The present photoabsorption σ_T, photoionization σ_I, photodissociation σ_D, cross sections and the photoionization quantum yield η are shown in Fig.1. By the use of metallic thin films, measurements were limited in the 53-93 nm region. A detailed comparison of the present results with the previous ones has already been described

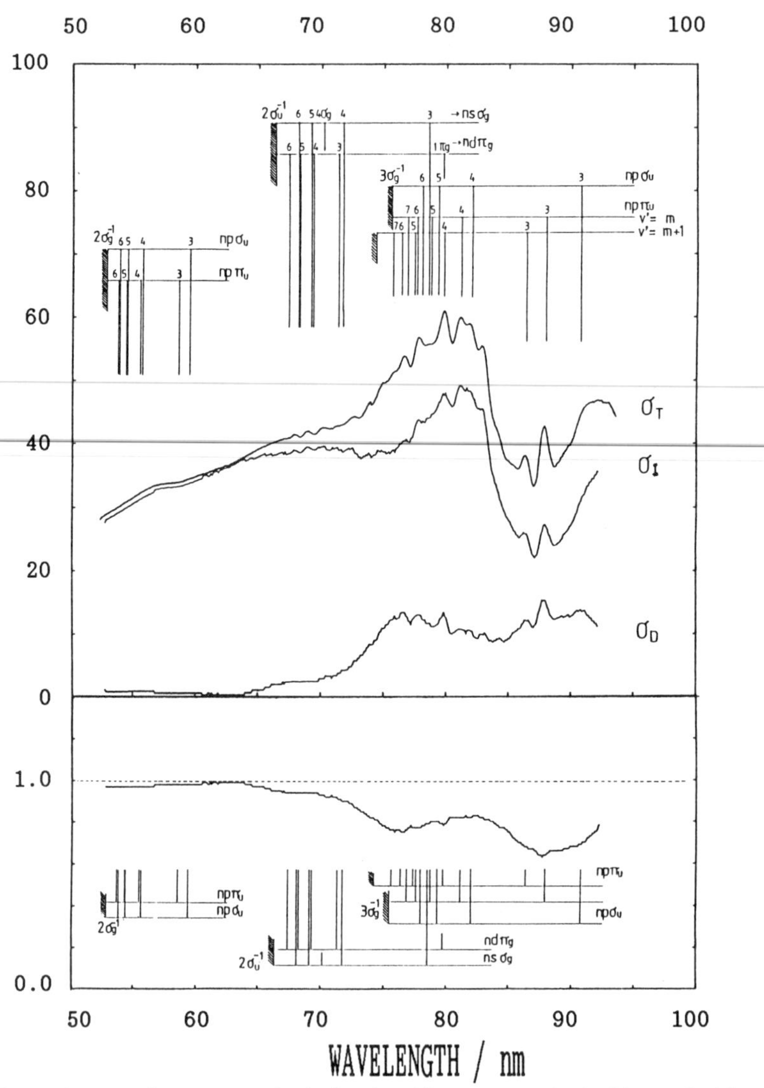

Fig.1 Photoabsorption σ_T, photoionization σ_I, photodissociation σ_D cross sections and photoionization quantum yield η of C_2H_2 obtained with a 0.4 nm photon bandpass. Vertical lines indicates calculated level positions of the Rydberg states.[10,13,17]

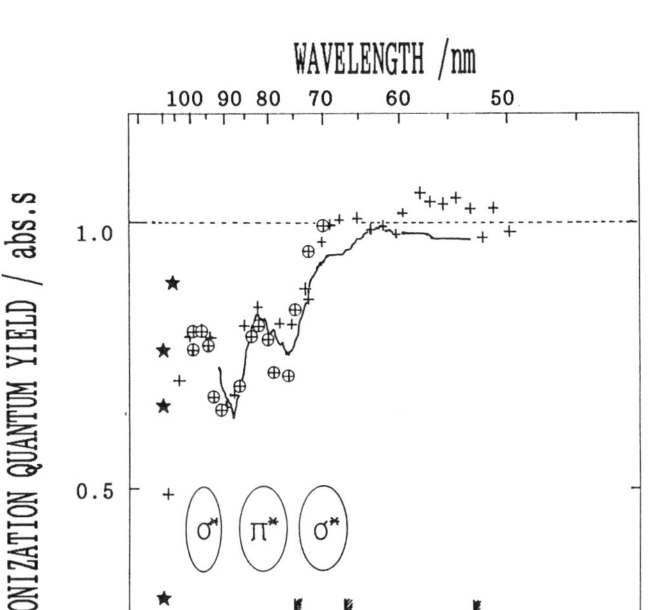

Fig.2 Comparison of photoionization quantum yield; − present work; ⊕, Metzger and Cook[4]; ★, Person and Nocole[5]; and + ,Cooper et al.[26] Positions of ionization limits and the shape resonances are also indicated.

elsewhere.[25]

The σ_T spectrum is characterized by the broad maxima at around 80 nm and 92 nm, the former of which is accompanied by the Rydberg like oscillatory structures in the 66-88 nm region. Vertical lines of calculated $3\sigma_g \rightarrow np\sigma_u, np\pi_u$, $2\sigma_u \rightarrow ns\sigma_g, nd\pi_g$, and $2\sigma_g \rightarrow np\sigma_u, np\pi_u$ Rydberg series by Langhoff et al.[13,17] and Hayaishi et al.[10] relatively explain the small oscillation. Good agreement is obtained of the $3\sigma_g \rightarrow np\pi_u$ series.[10] Quite large oscillator strengths were obtained for the $3\sigma_g \rightarrow 3p\sigma_u$, $2\sigma_u \rightarrow 4\sigma_g$, and $2\sigma_u \rightarrow 1\pi_g$ transitions.[13,17] However, no pronounced intensity is obtained at these level positions.

The η spectrum in the 53-93 nm region shows several characteristic structures. Comparison with the previous results[4,5,26] is shown in Fig.2. The present metallic film transmission prevented the observation of a steep growing-up structure just above the first ionization threshold at 11.4 eV. The η spectrum shows two minima down to 60 and 70 % with large widths at around 76 and 88 nm, respectively. A shallow shoulder exists in the 63-70 nm region. In this region, the quantum yield is nearly 90 %. The η approaches almost unity at around 63 nm. A slight decrease is also exhibited in the region shorter than 60 nm. The σ_D is shown to be almost the mirror image of the η curve.

The σ_I shows two broad peaks at around 70 nm and 81 nm. The latter peak is less pronounced than that in the σ_T, whereas the former is not shown in the σ_T curve. Several small oscillations are superimposed on the continuum ionization cross section. The σ_I again increases with increasing the wavelength to the present longest limit. Below the $3\sigma_g^{-1}\tilde{A}^2\Sigma_g$ second ionization limit, the present σ_I which is distinctively the $\tilde{X}^2\Pi_u$ partial ionization cross section, shows broad enhancements at two regions.

Very simply, a photoabsorption cross section σ_T is expressed by a sum of direct ionization cross section σ_C and an excitation cross section onto the superexcited states σ_S as $\sigma_T = \sigma_C + \sigma_S$. The dissociation quantum yield 1-η is thus expressed as,[25]

$$1-\eta = \frac{\sigma_S}{\sigma_C + \sigma_S} \cdot \frac{\tau_D^{-1}}{\tau_D^{-1} + \Gamma/\hbar}, \tag{1}$$

where τ_D^{-1} and $\Gamma/\hbar$ are the rate of dissociation and that of autoionization respectively. The former term of the product of eq.(1) is the ratio of the excitation processes. This term explains enhancements of σ_D at around ionization thresholds because of the dense superexcited Rydberg states. The latter term estimates the amount of the dissociation to total decay processes. A broad peak in the σ_T could be interpreted as representing a strong repulsive molecular configuration or as indicating a extremely short electronic relaxation lifetime. Concerning to the broad peaks at 80 and 92 nm, the lamped-up enhancements of the η and the depressed σ_D imply that the peaks are originated from the efficiently enhanced σ_I by extremely short ionization lifetimes. This is contrary to the implication of the $2\sigma_u \to 1\pi_g$ and $3\sigma_g \to 3p\sigma_u$ intervalence transitions by Langhoff et al.[13,17] but strongly indicates the shape resonances correlated with the first ionic $1\pi_u^{-1}\tilde{X}^2\Pi_u$ state. Holland et al.[12] mentioned that the peak at 80 nm was the π^* shape resonance but that at 93 nm was the $3\sigma_g \to 3p\sigma_u$ intervalence transition. However, the η result strongly supports an identification of the σ^* shape resonance as suggested by Robin.[18] Finally, the broad peak centered at around 70 nm is due to the σ^* shape resonance correlating with the $3\sigma_g^{-1}\tilde{A}^2\Sigma_g$ ionic state as discussed by Lynch et al,[16] but not to the $2\sigma_u \to 4\sigma_g$ transition.[27] Again, the η spectrum shows a piled-up shoulder in the same region which is in good correspondence with the dug-up shoulder in the σ_D curve.

The shape resonances at 70, 80, and 93 nm, thus contribute to ionization decay and the ionization quantum yield is much enhanced. The sum of the ionization cross sections for each peak area of the shape resonances may explain the large f-values for the $3\sigma_g \to 3p\sigma_u$, $2\sigma_u \to 1\pi_g$, and $2\sigma_u \to 4\sigma_g$ transitions.[13,17]

B. Shape Resonance in Non-ionizing decay

Fluorescence excitation spectra (FES) recorded in the 50-100 nm region in Fig.3 are shown to have individual characteristic behaviors with the incident wavelength. In consideration of lowest dissociative ionization limits to produce $C_2(d^3\Pi_g)$, $C_2(C^1\Pi_g)$, $CH(A^2\Delta)$, and $H(2p)$ and negligible amount of overlapping emission species,[6] every FES in Figs.3 is due purely to neutral dissociation processes of super-excited C_2H_2 to produce the above excited fragments. FES intensities

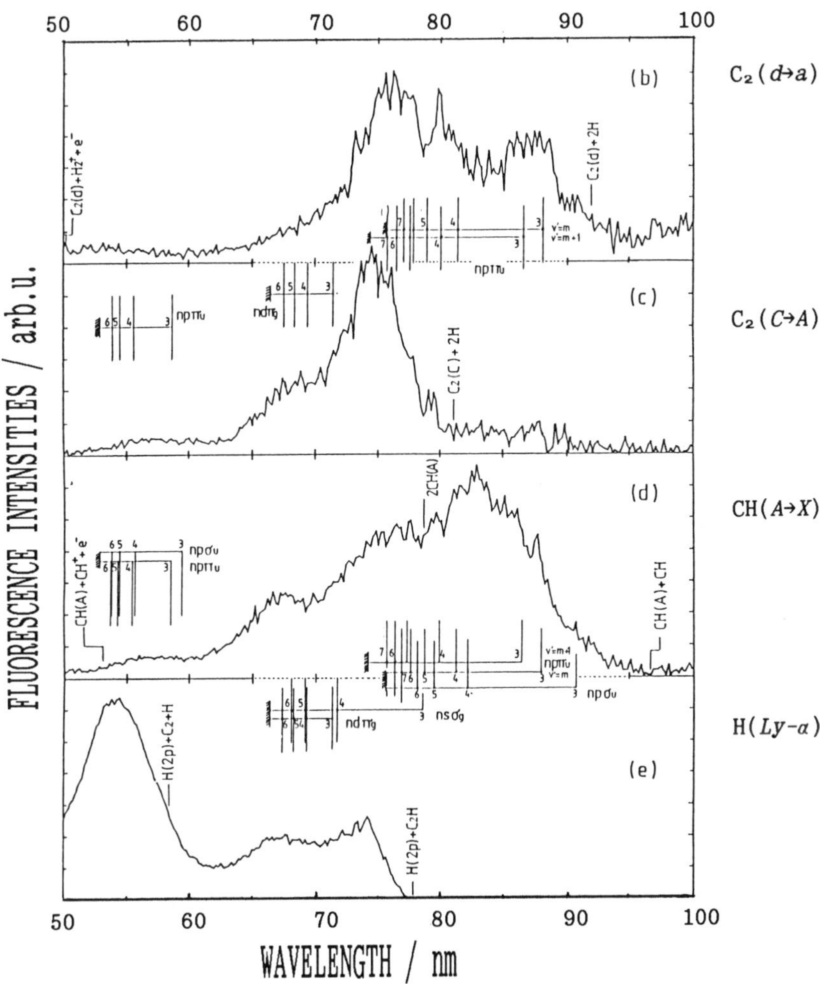

Fig.3 Fluorescence excitation spectra obtained 0.4 nm photon band pass for $C_2(d^3\Pi_g \to a^3\Pi, \Delta v=0)$, $C_2(C^1\Pi_g \to A^1\Pi_u, \Delta v=1)$, $CH(A^2\Delta \to X^2\Pi, 0 \to 0)$, and H(*Lyman-α*). Considering the symmetry conservation principle, calculated levels of the Rydberg series and their series limits by Langhoff et al.[13,17] and by Hayaishi et al.[10] are indicated.

growing up at the indicated dissociation limits in Fig.3 show that the present excited fragments are produced mainly via predissociation of superexcited molecules. The difference of the σ_D spectrum from the FES corresponds to the production of non-emitting fragments.

According to the symmetry conservation principle,[27] superexcited precursor states of the present emitting dissociation fragments is considered. $C_2(d)$ and $C_2(C)$ are originated from the $3\sigma_g^{-1}np\pi_u$, $2\sigma_u^{-1}nd\pi_g$, and $2\sigma_g^{-1}np\pi_u$ Rydberg series in the $^1\Pi_u$ symmetry.[13,17] The difference between the two FES for C_2^* emission is determined only by the difference between dissociation limits for the two channels. The CH(A)+CH(X) dissociation is also ascribed to the $3\sigma_g^{-1}np\pi_u$ Rydberg states. The origins of the CH($A^2\Delta$)+CH($A^2\Delta$) dissociation emerge at 78 nm are $2\sigma_u^{-1}ns\sigma_g, nd\pi_g$, and $2\sigma_g^{-1}np\sigma_u, np\pi_u$ Rydberg series. H($2p$) fragments are also originated both of these series. Spectral features of the FES in Fig.3 should reflect the density of the above Rydberg states.

As seen in the ionization quantum yield, weakened fluorescence intensities for emission from neutral products between 60 and 64 nm are due to a diluted density of the superexcited states. However, peculiar depressed structures are obvious in the FES. The $C_2(d\to a)$ FES shows a minimum in 80-86 nm where the π^* shape resonance is obvious in an ionization process. We propose that the Rydberg state density is depressed by the π^* shape resonance. Because the $3\sigma_g^{-1}np\pi_u$ Rydberg states and the π^* shape resonance are in the same $^1\Pi_u$ symmetry, a preferential transition to the π^* shape resonance with a larger f-value makes the depression of the excitation to the Rydberg states in the same symmetry existing at the same excitation region (symmetry depression of Rydberg transitions). Similar effects are also seen in the FES of CH($A\to X$) and H($2p$) emissions at around 70 nm. Especially the CH($A\to X$) FES originating from the $2\sigma_u^{-1}ns\sigma_g$ series in the $^1\Sigma_u^+$ symmetry is strongly depressed by the σ^* shape resonance. In the threshold region of the CH(A)+CH(X) dissociation in 91-96.5 nm, only a weak increase in the CH($A\to X$) FES is observed. On the contrary, an enhanced FES maximum at 83 nm for CH($A\to X$) emission does not suffer from depression by the π^* shape resonance. These phenomena suggest, against the symmetry conservation, the CH(A)+CH(X) dissociation in the region shorter than 78 nm has a contribution from the $3\sigma_g^{-1}np\sigma_u$ Rydberg series. A strong vibronic coupling of the $3\sigma_g^{-1}np\sigma_u$ states with ν_4 or ν_5 vibrational modes in the Π_g or Π_u symmetry,[27] respectively, may yield Π states undergoing the CH(A)+CH(X) dissociation.

It is of interest that the $3\sigma_g^{-1}4p\pi_u$,v'=m+1 state[10] is much more enhanced than the other members involved in the same converging series. The energy position coincides with the calculated $2\sigma_u^{-1}1\pi_g$ state which is identified not to a valence excited state[13,17] but to the π^* shape resonance[12,15,18] by the above discussion. However, a kind of discrete-continuum configuration mixing between the Rydberg state and the shape resonance may enhance this dissociation peak. This enhancement is also seen in the σ_D spectrum and the CH($A^2\Delta\to X^2\Pi$) FES.

We should here stress again the importance of the coexisting nature of the shape resonances and the Rydberg states to affect the photoionization and photodissociation processes of C_2H_2 in the EUV region. This nature makes the neutral dissociation process, involving dissociative excitation, of the molecule extremely complicated.

REFERENCES

a. Present address: Department of Electronics, Kanazawa Institute of Technology, Nonoichi-machi, Ishikawa 921, Japan

1. Y.Hatano, *Comm.At.Mol.Phys.* 13,259(1983).
2. S.Leach, "*Photophysics and Photochemistry in the Vacuum Ultraviolet*" edited by S.P.McGlynn, G.L.Findley, and R.H.Huebner, (Reidel,Dortrecht,1982) p.297.
3. J.Berkowitz, "*Photoabsorption, Photoionization, and Photoelectron Spectroscopy*", (Academic, NY, 1979)
4. P.H.Metzger and G.R.Cook, *J.Chem.Phys.* 41,642(1964).
5. J.Person and P.Nicole, *J.Chem.Phys.* 53,1767(1979).
6. J.C.Han, C.Ye, M.Suto, and L.C.Lee, *J.Chem.Phys.* 90,4000(1989).
7. R.Botter, V.H.Dibeler, J.A.Walker, and H.M.Rosenstock, *J.Chem. Phys.* 44,1271(1966).
8. V.H.Dibeler, J.A.Walker, and K.E.McCulloh, *J.Chem.Phys.* 59,2264 (1973).
9. Y.Ono, W.A.Osuch, and C.Y.Ng, *J.Chem.Phys.* 76,3905(1982).
10. T.Hayaishi, S.Iwata, M.Sasanuma, E.Ishiguro, Y.Morioka, Y.Iida, and M.Nakamura, *J.Phys.* B15,79(1982).
11. A.C.Parr, D.L.Ederer, J.B.West, D.M.P.Holland, and J.L.Dehmer, *J. Chem.Phys.* 76,4349(1982).
12. D.M.P.Holland, J.B.West, A.C.Parr, D.L.Ederer, R.Stockbauer, R.D. Boff, and J.L.Dehmer, *J.Chem.Phys.* 78,124(1983).
13. P.W.Langhoff, B.V.McKoy, R.Unwin, and A.M.Bradshaw., *Chem.Phys. Lett.* 83,270(1981).
14. J.L.Dehmer,D.Dill, and A.C.Parr, "*Photophysics and Photochemistry in the Vacuum Ultraviolet*" edited by S.P.McGlynn, G.L.Findley and R.H.Huebner, (D.Reidel, Dortrecht, 1982) p.341.
15. R.Unwin, I.Khan, N.V.Richardson, A.M.Bradshaw, L.S.Cederbaum, and W.Domcke, *Chem.Phys.Lett.* 77,242(1981).
16. D.Lynch, M.-T.Lee, R.R.Lucchese, and V.McKoy, *J.Chem.Phys.*80,1907 (1984).
17. L.E.Machado, E.P.Leal, G.Csanak, B.V.McKoy, and P.W.Langhoff, *J. Electron Spectrosc.* 25,1(1982).
18. M.B.Robin, "*Higher Excited States of Polyatomic Molecules*",vol.2 & 3 (Academic, NY, 1975 & 1985)
19. H.Koizumi, T.Yoshimi, K.Shinsaka, M.Ukai, M.Morita, Y.Hatano, A.Yagishita, and K.Ito, *J.Chem.Phys.* 82,4856(1985).
20. H.Koizumi, K.Hironaka, K.Shinsaka, S.Arai, H.Nakazawa, A.Kimura, Y.Hatano, Y.Ito, Y.Zhang, A.Yagishita, K.Ito, and K.Tanaka, *J. Chem.Phys.* 85,4276(1986).
21. H.Koizumi,K.Shinsaka, and Y.Hatano, *Radiat.Phys.Chem.*34,87(1989).
22. H.Koizumi,K.Shinsaka,T.Yoshimi,K.Hironaka,S.Arai,M.Ukai,M.Morita, H.Nakazawa,A.Kimura,Y.Hatano,Y.Ito,Y.Zhang,A.Yagishita,K.Ito, and K.Tanaka, *Radiat.Phys.Chem.* 32,1(1988).
23. S.Arai, T.Yoshimi, M.Morita, K.Hironaka, T.Yoshida, K.Shinsaka, Y.Hatano, A.Yagishita, and K.Ito, *Z.Phys.* D4,65(1986).

24. S.Arai,T.Kamosaki,M.Ukai,K.Shinsaka,Y.Hatano,Y.Ito,H.Koizumi, A. Yagishita,K.Ito, and K.Tanaka, J.Chem.Phys. 88,3016(1988).
25. M.Ukai,K.Kameta,R.Chiba,K.Nagano,N.Kouchi,K.Shinsaka,Y.Hatano, H. Umemoto,Y.Ito, and K.Tanaka, *J.Chem.Phys.* 95(1991) *in press.*
26. G.Cooper,T.Ibuki,Y.Iida,and C.E.Brion, *Chem.Phys.* 125,307(1988).
27. G.Herzberg, *"Electronic Spectra and Electronic Structure of Polyatomic Molecules", vol.3, Molecular Spectra and Molecular Structure*,(Van Nostrand, NY, 1967).

DISCUSSION

GLASS-MAUJEAN - 1. What is the effect of C_2H fluorescence ?

2. Most of the absorption leading to processes other than ionization in the 85-90nm region does not appear in the total fluorescence spectrum. Is it a dissociation giving fundamental-state fragments ?

UKAI - 1. As is well known. C_2H fluorescence appears in a wide wavelength region and overlaps with other significant fluorescence. We also tried to obtain C_2H fluorescence excitation spectrum but could not obtain a meaningful result according to weak signal.

2. I think so. Because we are looking at dissociative excitation, none of the information about fundamental or non-fluorescent fragments is obtained.

Detection of fundamental fragments from superexcited states is now being examined using multi-colour laser technique. However, combination of a laser assisted detection with synchrotron radiation excitation still has a considerable difficulty due to their different time structures.

FLUORESCENCE EXCITATION SPECTRA EXCITED BY PHOTODISSOCIATION AND PHOTOIONIZATION OF CO_2

Masatoshi Ukai, Kosei Kameta, Noriyuki Kouchi, Kazunori Nagano, and Yoshihiko Hatano
Department of Chemistry, Tokyo Institute of Technology, Meguro-ku, Tokyo 152, Japan

Kenichiro Tanaka
Photon Factory, National Laboratory for High Energy Physics, Oho, Tsukuba, Ibaraki 305, Japan

Abstract

Fluorescence excitation spectra of $CO_2^+(\tilde{A}^2\Pi_u \rightarrow \tilde{X}^2\Pi_g)$, $(\tilde{B}^2\Sigma_u^+ \rightarrow \tilde{X}^2\Pi_g)$ and $CO(A^1\Pi \rightarrow X^1\Sigma)$ emissions produced in the photoionization and neutral photodissociation of CO_2 have been obtained in the threshold regions of the $CO_2^+(\tilde{A}^2\Pi_u)$ and $CO_2^+(\tilde{B}^2\Sigma_u^+)$ states. A strong competition between autoionization and neutral dissociation has been observed to yield a preferential enhancement of the neutral dissociation. This preferential enhancements have mainly been explained by the autoionization rate of the individual superexcited Rydberg series. A drastic step-down decrease in the VUV-FES at the ionic thresholds have been ascribed to the dramatic density dilution of the superexcited states into the continuum.

Introduction

In spite of the detailed studies of the photoionization process of CO_2 in the extremely-UV region,[1-19] a special interest was not paid to the competition between autoionization and neutral dissociation of superexcited states. A photoionization quantum yield[9] shows a striking behavior in the 60-83 nm region to indicate a dense concentration of superexcited states and their strongly favored neutral dissociation decay. Understanding of this nature with respect to the competitive decay is of considerable importance, especially at the threshold region of individual ionic states. It is clear that partial ionization cross sections show threshold autoionization behavior of superexcited states into individual final channels. Constant-ionic-state (CIS) PES[7,9,11] and dipole (e,2e)[21] studies were performed to obtain partial ionization cross sections of CO_2^+. However, PES is not favorable in reliable detection of near-zero energy electrons at the ionization thresholds. Furthermore, the overlapping photoelectron peaks of the $CO_2^+(\tilde{A}^2\Pi_u)$ and $(\tilde{B}^2\Sigma_u^+)$ states[16] make partial ionization cross section measurements rather difficult. We stress here an advantage of fluorescence excitation spectrum (FES) measurements from ionic species to obtain information on threshold behavior of partial ionization cross sections. Although a strong intersystem crossing between the $CO_2^+(\tilde{A}^2\Pi_u)$ and $(\tilde{B}^2\Sigma_u)$ states introduced a controversy between the PES[7,8] and FES[12,18] results, it should be noted that the $\tilde{B}$ state FES reproduces the relative partial ionization cross section and that below the $\tilde{B}$ state threshold, $\tilde{A}$ state FES reasonably provides the partial ionization cross section.

We also stress that the neutral dissociation of superexcited CO_2 was not investigated in detail.[14,17,19,22] Mitsuke et al.investigated the ion-pair predissociation.[23]

In this paper, we present FES of $CO_2^+(\tilde{A}^2\Pi_u \rightarrow \tilde{X}^2\Pi_g)$ and $(\tilde{B}^2\Sigma_u^+ \rightarrow \tilde{X}^2\Pi_g)$ emissions and FES from neutral dissociation fragments in the threshold regions of the $CO_2^+(\tilde{A}^2\Pi_u)$ and $(\tilde{B}^2\Sigma_u^+)$ states. The purpose of this investigation is to present a threshold behavior of the competitive-decay feature of superexcited states. A strong competition between autoionization and neutral dissociation has been shown to yield a preferential enhancement of the neutral dissociation. A drastic decrease in the VUV-FES at the ionic thresholds have also been shown to indicate the dramatic density dilution of the superexcited states into the continuum.

Experimental

The experiment was performed using an extreme-UV synchrotron radiation (SR) from a 2.5 GeV positron storage ring at the Photon Factory. The experimental apparatus (Fig.1) was similar to that described previously.[25,26]

A monochromatized synchrotron radiation (SR) with a typical flux of 5×10^9 photon/s at 50 nm and a typical bandpass of 0.1 nm was employed as an excitation light source.

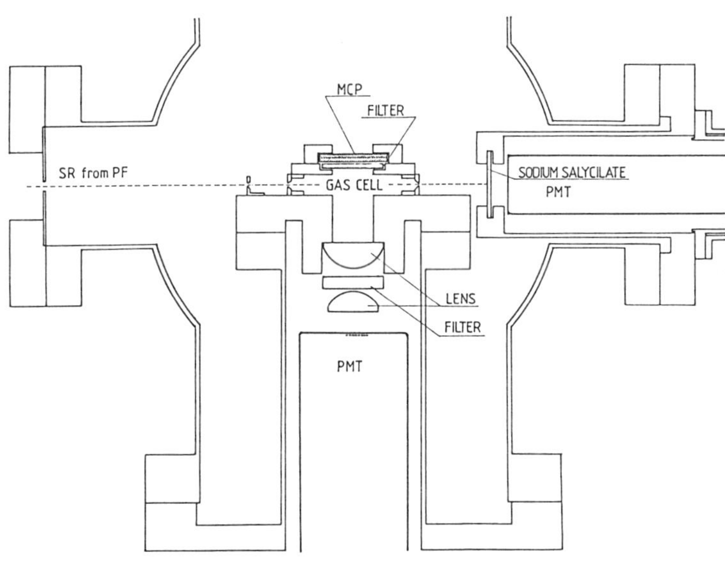

Fig.1
Experimental apparatus for undispersed fluorescence detection.

Undispersed VUV optical emissions in the 105-180 and 134-180 nm regions from sample gas were detected by a CsI coated microchannel plate (MCP) through a LiF or BaF_2 window, respectively. Visible (VIS) and ultraviolet (UV) emissions were extracted to outside the vacuum, refocused using a pair of quartz lenses, and detected by a photomultiplier (PMT). An appropriate optical filter was placed between the lenses to select ionic emission. A UV-VIS sensitive PMT with a 300-530 nm bandpass filter detected $CO_2^+(\tilde{A}^2\Pi_u \rightarrow \tilde{X}^2\Pi_g)$ emission. A UV (160-320 nm) sensitive PMT with a narrow bandpass filter (250 nm center, 25 nm bandpass) detected $CO_2^+(\tilde{B}^2\Sigma_u^+ \rightarrow \tilde{X}^2\Pi_g)$ emission.

Results and Discussion

The neutral dissociation of superexcited CO_2 is viewed by a VUV-FES in Fig.2. The origin of the VUV-FES in the wavelength region longer than 63.5 nm is ascribed to the dissociatively excited $CO(A^1\Pi \rightarrow X^1\Sigma)$ emission.[17,26] Negligibly small contribution of atomic O and C emissions[17] does not give a severe problem in the VUV-FES. The dissociative ionization of $CO(A^1\Pi)+O^+(^4S^o)$ appears at 45.8 nm. The VUV-FES in Fig.2 is thus regarded as a neutral dissociation spectrum.

A general feature of the present FES is approximately in good agreement with the previous results in 68.5-92 nm region with lower resolution.[14,17] Peak structures in this region are explained by Rydberg states in Tanaka-Ogawa series converging to the vibronic CO_2^+ $(\tilde{A}^2\Pi_u)$ states in the 68.7-78.7 nm region and two Henning series (R_B series) to the vibrationally ground $CO_2^+(\tilde{B}^2\Sigma_u^+)$ state in 68.5-76.5 nm. At the $CO_2^+(\tilde{A}^2\Pi_u, v'=0$ and $1)$ thresholds the fluorescence intensity shows striking step-down structures. The intensity also decrease quite steeply at the $CO_2^+(\tilde{B}^2\Sigma_u^+, v'=0)$ threshold. These threshold depressions of the fluorescence intensity indicate a dramatic density-dilution of the superexcited Rydberg states into the respective ionization continua. Among these, the behavior of fluorescence intensity is most pronounced at the $CO_2^+(\tilde{B}^2\Sigma_u^+, v'=0)$ threshold. A step-down structure at the $CO_2^+(\tilde{B}, v'=1)$ series limit is not so prominent as that at the $CO_2^+(\tilde{B}, v'=0)$ limit. The PES results showed very small transition probabilities for the higher vibronic $CO_2^+(\tilde{B}^2\Sigma_u^+)$ states. This also indicates that very small density of higher vibronic Rydberg states to the $CO_2^+(\tilde{B}^2\Sigma_u^+, v')$ limits are plausible above the $CO_2^+(\tilde{B}^2\Sigma_u^+, v'=0)$ threshold. In the region shorter than 68.5 nm, FES is not completely dark but shows a small converging oscillations toward 64 nm. Peak structures due to the Tanaka-Jursha-Leblanc[27](TJL) Rydberg series (R_C series), especially absorption and weak absorption series, converging to the $CO_2^+(\tilde{C}^2\Sigma_g^+, v'=0)$ series limit are obvious in the 64-69.5 nm region. Any considerable enhancement, however, are not obtained at the positions of R_C emission series. Again the FES manifests a drastic dilution of the superexited states above the CO_2^+ $(\tilde{C}^2\Sigma_g^+, v'=0)$ series limit. None of severe difference in the VUV-FES was obtained by alterring the short wavelength cut off from 105 (using LiF) to 134 nm (BaF_2). This denies an peculiar increase in excited O atom emissions. A similar enhancement was obtained in the ion-pair predissociation cross section[23] producing O^-.

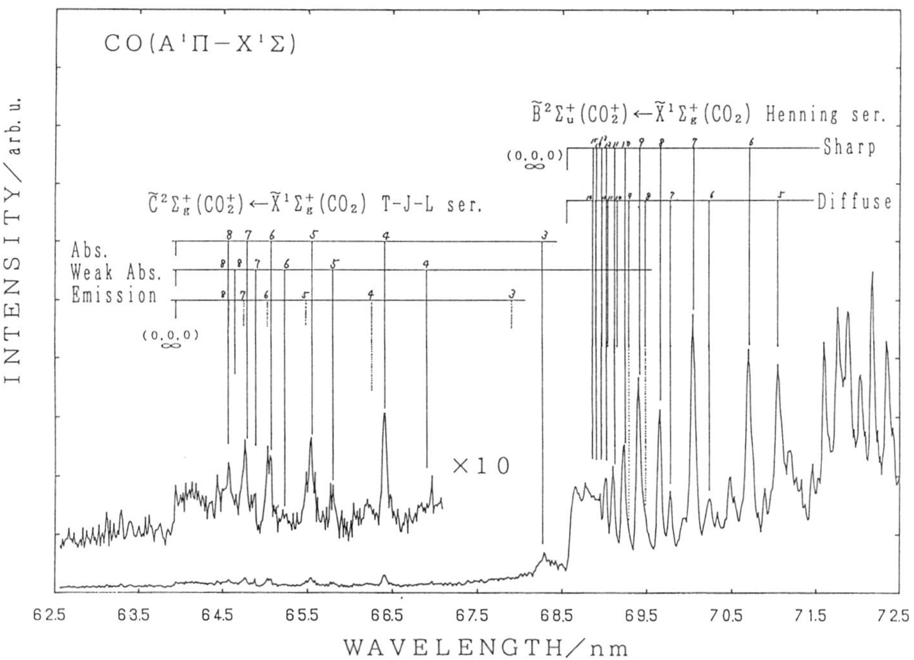

Fig.2 Fluorescence excitation spectrum for VUV emission detected by MCP through LiF window with nominal bandpass of 0.04 nm. Level assignments from Ref.27 are indicated, but those for Henning (R_B) sharp series follow McCulloh.[2]

Figures 3a and 3b present FES of $CO_2^+(\tilde{A}^2\Pi_u \rightarrow \tilde{X}^2\Pi_g)$ and $CO_2^+(\tilde{B}^2\Sigma_u^+ \rightarrow \tilde{X}^2\Pi_g)$ emissions. Spectral features of the two FES are in approximate agreement with the previous results[14,18] with lower resolution.

The threshold feature of the $CO_2^+(\tilde{A} \rightarrow \tilde{X})$ FES is characterized by two R_B Rydberg progressions of Henning sharp and diffuse series to the $CO_2^+(\tilde{B}^2\Sigma_u^+, v'=0)$ limit superimposed on a stepwise growth at the higher vibrational $\tilde{A}^2\Pi_u$ state thresholds of symmetric stretching vibration. These perturbed structures of the R_B series autoionization are also observed in the $\tilde{X}^2\Pi_g$ cross section.[11] It is clear that just above the $CO_2^+(\tilde{A}^2\Pi_u)$ threshold the partial ionization cross section possesses a larger contribution from superexcited states through R_B autoionization in comparison with the direct ionization. Vibrational autoionization structures due to the Tanaka and Ogawa Rydberg series are not obvious. The present spectrum with much higher resolution than the previous ones displays finer R_B profiles. Difference in the peak widths between the sharp and diffuse series progressions is clearly recognized. Cross sections themselves, especially at the sharp series levels possess larger values within the same principal quantum number. Gentieu and Mentall mentioned that the peak heights at the sharp and diffuse levels were inverted in the FES relative to

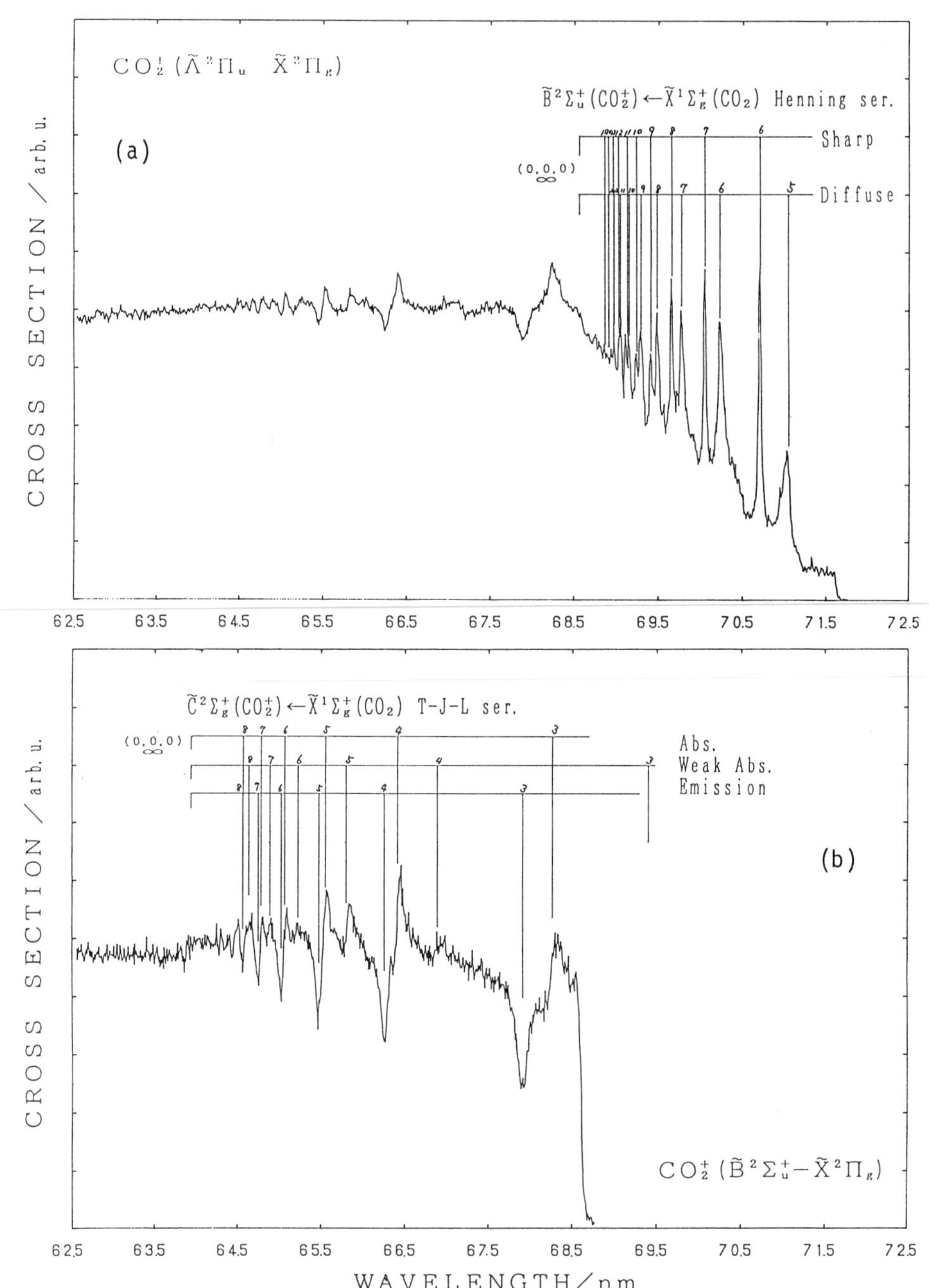

Fig.3 Fluorescence excitation spectrum of (a) $CO_2^+(\tilde{A}^2\Pi_u \rightarrow \tilde{X}^2\Pi_g)$ and (b) $CO_2^+(\tilde{B}^2\Sigma_u^+ \rightarrow \tilde{X}^2\Pi_g)$ emissions with nominal bandpass of 0.04 nm. Level assignments are same as in Fig.2.

the total absorption cross section.[18] In the present higher resolution measurement, although the total cross-section area at the diffuse levels are larger than those at the sharp levels, the increased peak heights with decreased widths at the sharp levels are not inversely obtained in comparison with those remaining constant at the diffuse levels. Three R_C series progressions are observed in the $CO_2^+(\tilde{A}\rightarrow\tilde{X})$ FES.

The FES of $CO_2^+(\tilde{B}^2\Sigma_u^+\rightarrow\tilde{X}^2\Pi_g)$ emission grows up quite steeply at the v'=0 threshold. Prominent structures due to the above three R_C series are shown just above the $\tilde{B}^2\Sigma_u$ state threshold. The R_C absorption and weak absorption series show up with the larger and smaller peak profiles. The window profiles of the R_C emission series indicates a strong interchannel coupling with continuum. It is clear that Resonance widths of the three R_C series are still broader than those of the R_B diffuse series ones. The R_C series structures in the $\tilde{B}\rightarrow\tilde{X}$ FES appear quite similar to those in the $\tilde{A}\rightarrow\tilde{X}$ FES but are more prominent. As mentioned, this similarity was ascribed to a 55 % intersystem-crossing of the $\tilde{B}$ state into the $\tilde{A}$ state.[15,28]

The large amplitudes of the R_C series structures again indicates largely involved resonance contribution of the R_C superexcitation cross sections. These large perturbations of the R_C series autoionization did not show up in the $\tilde{A}^2\Pi_u$ and $\tilde{X}^2\Pi_g$ partial cross sections of a CIS-PES study.[11] This implies that the autoionization structures in an absorption spectrum[27] are ascribed to the strong interchannel coupling of the R_C levels with the $\tilde{B}$ state continuum.

A small step-down decrease in the $\tilde{B}\rightarrow\tilde{X}$ FES at the $CO_2^+(\tilde{C}^2\Sigma_g^+$, v'=0) series limit also suggests a drastic dilution of the largely involved resonance cross section into the $\tilde{C}^2\Sigma_g^+$ state continuum.

Strong competition of autoionization and neutral dissociation is shown in the $\tilde{A}$ and $\tilde{B}$ threshold regions. Discrete R_B and R_C progression profiles indicate predissociation of the R_B and R_C states and very small contributions of a continuum dissociation to yield neutral species.

Both of the two R_B series show up with peak profiles in the $\tilde{A}\rightarrow\tilde{X}$ FES. Almost the same cross sections for the two R_B series in the $\tilde{A}\rightarrow\tilde{X}$ FES involve that there is not a preferential autoionization into the $\tilde{A}$ continuum among the two series. However, the VUV-FES in Fig.2 clearly shows by far larger intensities of the R_B sharp series than those of the diffuse series. This was also be seen for the smaller n (principal quantum number) states in the longer wavelength region.[17] This clearly visualize that not dissociation rates of the two R_B states but autoionization rates themselves determine the peak broadening widths and the R_B sharp states with the smaller autoionization rates yield larger enhancement of the VUV-FES than the R_B diffuse states with the larger rates. This contrast is not seen at the R_B diffuse n=5 state. This can be explained by a coincidence of the level with the $\tilde{A}$,v'=1 threshold increasing the Rydberg state density converging to this limit and the small vibrational multiplicity of the $\tilde{A}$ continuum. Less pronounced peak intensity in the $\tilde{A}\rightarrow\tilde{X}$ FES clearly support this.

Among the three R_C series as shown by the prominent perturbed structures in the $\tilde{B}\to\tilde{X}$ FES, the enhancement of VUV-FES is obtained only at the TJL-absorption and weak absorption levels. The broader profile widths of the series implying the larger autoionization rates explain the less intense relative emission cross section by one order of magnitude in the VUV-FES. An increased multiplicity of the continuum channels does not explain the broader profile widths than those of the R_B series because none of pronounced structures of the $\tilde{X}$ and $\tilde{A}$ cross sections was observed in the CIS-PES result.[11] A preferential autoionization of the R_C states into the $\tilde{B}$ continuum through such as the fast Coster-Kronig type autoionization involves a strong interchannel coupling.

The peak enhancement is obtained of the R_C absorption and weak absorption series. The R_C emission series progression, however, is no more shown in the VUV-FES. Window profiles of the perturbed structures in the $\tilde{B}\to\tilde{X}$ FES indicating an exclusively strong interchannel coupling with the $\tilde{B}$ continuum explain the absence of this progression in the neutral dissociation decay.

In summary, we have shown FES for the ionic $CO_2^+(\tilde{A}^2\Pi_u\to\tilde{X}^2\Pi_g)$ and $(\tilde{B}^2\Sigma_u^+\to\tilde{X}^2\Pi_g)$ emissions and VUV-FES from a dissociatively excited CO fragment. A strong competition between autoionization and neutral dissociation observed in the R_B and R_C progression regions yielding the preferential enhancement of the neutral dissociation have mainly been explained by the autoionization rate of the individual superexcited Rydberg series. A drastic step-down decrease in the VUV-FES has been observed at the $\tilde{A}^2\Pi_u$, $\tilde{B}^2\Sigma_u^+$, and $\tilde{C}^2\Sigma_g^+$ thresholds indicating the dramatic density dilution of the superexcited states into the continuum.

The authors thank the staff of the Photon Factory for support. They also thank Prof.L.C.Lee and Prof.M.Suto for valuable discussion. Financial support of a Grant-in-Aid from the Ministry of Education, Science, and Culture is aknowledged

References

1. N.Wainfan, W.C.Walker, and G.L.Weissler, Phys.Rev.99,542(1955).
2. K.E.McCulloh, J.Chem.Phys. 59,4250(1973).
3. J.H.D.Eland and J.Berkowitz, J.Chem.Phys. 67,2782(1977).
4. P.Roy, I.Nenner, P.Millie, P.Morin, and D.Roy, J.Chem.Phys. 84, 2050(1986).
5. R.Frey, B.Gotchev, O.F.Kalman, W.B.Peatman, H.Pollak, and E.W. Schalg, Chem.Phys. 21,89(1977).
6. T.Bear and P.M.Guyon, J.Chem.Phys. 85,4765(1986).
7. J.L.Bahr, A.J.Blanke, J.H.Caver, J.L.Gardner, and V.Kumer, J. Quant.Spectrosc.Padiat.Transfer, 12,59(1972).
8. J.L.Gardner and J.A.R.Samson,J.Electr.Spectrosc. 8,469(1976).
9. T.Gustafsson, E.W.Pliummer, D.E.Eastman, and W.Gudat, Phys.Rev. 17,175(1978).
10. P.Roy, I.Nenner, M.Y.Adam, J.Delwiche, M.-J.Hubin-Franskin, P. Lablanquie, and D.Roy, Chem.Phys.Lett. 109,607(1984).
11. M.-J.Hubin-Franskin, J.Delwiche, P.Morin, M.Y.Adam, I.Nenner,

and P.Roy, J.Chem.Phys. 81,4246(1984).
12. L.C.Lee and D.L.Judge, J.Chem.Phys. 57,4443(1972).
13. R.W.Carlson,D.L.Judge, and M.Ogawa, J.Geophys.Res.78,3194(1973).
14. L.C.Lee, R.W.Carlson, D.L.Judge, and M.Ogawa, J.Chem.Phys. 63, 3987(1975).
15. J.A.R.Samson and J.L.Gardner, J.Chem.Phys. 58,3771(1973).
16. J.A.R.Samson and J.L.Gardner, J.Geophys.Res. 78,3663(1973).
17. E.P.Gentieu and J.E.Mentall, J.Chem.Phys. 58,4803(1973).
18. E.P.Gentieu and J.E.Mentall, J.Chem.Phys. 64,1376(1976).
19. C.Y.Wu and D.L.Judge, Chem.Phys.Lett. 68,495(1979).
20. J.Berkowitz, "Photoabsorption, photoionization, and photoelectron Spectroscopy" (Academic, NY,1979).
21. A.P.Hitchcock, C.E.Brion, and M.J.Van der Wiel, Chem.Phys. 45,461 (1980).
22. C.Y.Robert Wu, E.Phillips, L.C.Lee, and D.L.Judge, J.Geophys.Res. 83,4869(1978).
23. K.Mitsuke, S.Suzuki, T.Imamura, and I.Koyano, J.Chem.Phys. 93, 1710(1990)
24. S.Arai, T.Kamosaki, M.Ukai, K.Shinsaka, Y.Hatano, Y.Ito, H. Koizumi, A.Yagishita, K.Ito, and K.Tanaka, J.Chem.Phys. 88,3016 (1988).
25. M.Ukai, K.Kameta, R.Chiba, K.Nagano, N.Kouchi, K.Shinsaka, Y. Hatano, H.Umemoto, Y.Ito, and K.Tanaka, J.Chem.Phys. in press.
26. E.Phillips, L.C.Lee, and D.L.Judge, J.Chem.Phys. 65,3118(1976).
27. Y.Tanaka, A.S.Jursha, and F.J.LeBlanc, J.Chem.Phys.32,1199(1960).
28. E.W.Schlag, R.Frey, B.Gotchev, W.P.Peatman, and H.Pollak, Chem. Phys.Lett. 51,406(1977).

DISCUSSION

MEYER - Spectral similarity between $CO_2^+(B \rightarrow X)$ and $CO_2^+(A \rightarrow X)$ FES suggests to me an experimental artifact of secondary fluorescence from optical filters. Have you examined this possibility ?

UKAI - We did not examine the secondary emission from the optical filters. However, as we employed narrow band pass filters and different photomultipliers with the appropriate response, we believe this kind of artifact is removed.
Furthermore, fluorescence intensities of $CO_2^+(B \rightarrow X)$ and $CO_2^+(A \rightarrow X)$ are comparable. The possibility of secondary fluorescence detection is very small because the fluorescence quantum yields of glass filter are very small.
The similarity in the FES structures between $CO_2^+(B \rightarrow X)$ and $CO_2^+(BA \rightarrow X)$ above the $CO_2^+(B)$ state threshold is not caused by the secondary fluorescence from filters but by non-radiative intersystem crossing between the A and B states. This was confirmed by PES and time-resolved threshold electron-fluorescence coincidence. However, the probability of the intersystem crossing should be examined in more detail.

STATE SELECTIVE IONIZATION OF O_2 IN A FRAMEWORK OF VAN DER WAALS MOLECULES[a]

Masatoshi Ukai, Kosei Kameta, Kyoji Shinsaka[b], and Yoshihiko Hatano
Department of Chemistry, Tokyo Institute of Technology, Meguro-ku, Tokyo 152, Japan

Takato Hirayama[c], Shin-ichi Nagaoka[d], and Katsumi Kimura
Institute for Molecular Science, Myodaiji, Okazaki 444, Japan

ABSTRACT

A mass spectrometric study of van der Waals molecules, $(O_2)_2$, $(O_2)_3$, and ArO_2 in the 50-100 nm region is presented. Autoionization structures of O_2 between 70 and 100 nm were completely absent in photoionization efficiency (PIE) curves of $(O_2)_2^+$, $(O_2)_3^+$, and ArO_2^+, whereas the contribution of the autoionization due to the $c^4\Sigma_g^- 3s\sigma_g$ Rydberg states of inner valence excitation remained. In the dissociative ionization yielding O_3^+ and ArO^+ products, molecular rearrangement was shown totally very important. Especially, an excellent agreement of the ArO^+ PIE curve with the $O_2^+(b^4\Sigma_g^-)$ partial photoionization cross section indicated a selective molecular photoionization in a frame of a molecular cluster or a van der Waals molecule.

INTRODUCTION

Photoionization of binary clusters or van der Waals (vdW) molecules gives an important opportunity to investigate ionization processes concerning the intermolecular process within an intramolecular framework of a cluster. Especially, the differences of the individual physical properties of the cluster components give rise to internal energy transfer into discrete and continuum levels.[1-9] Bimolecular reactions involving short-lived active species, such as the collisional deexcitation of optically allowed species,[10-13] are involved under a half-reaction condition. The electronic or atomic rearrangements are involved in most of these cases, which make it possible to investigate intramolecular potentials, coordination effects, and so on. However, only limited works were devoted to extreme-UV photoionization of molecular clusters.

Photoionization spectra for every ionic state and species of O_2 are characterized by numerous autoionization structures appearing from the ionization onset far into the extreme-UV (EUV) region, which was elucidated by extensive photoionization studies for individual ionization processes.[14-22] Linn et al.[8] previously studied the dissociative and non-dissociative photoionization of $(O_2)_m$ clusters (m=2-4) in the outer valence excitation region. Absence of a number of O_2 Rydberg structures in their cluster spectra suggested the ionization of superexcited clusters strongly coupled with dissociation of vdW bond. Atomic rearrangement following photoionization of a cluster was also indicated in the dissociative ionization spectra.

The present paper reports photoionization of the vdW molecules, $(O_2)_2$, $(O_2)_3$, and ArO_2, in the 50-100 nm region, where inner valence excitation as well as outer valence one is involved. The dimension of an excited molecule involved in a cluster should strongly depend on the concerning molecular orbital. Moreover, rearrangement within a cluster should depend on the accompanying molecules. The aim of this paper is to investigate these effects on the molecular photoionization. Autoionization structures associated with the $O_2^+(c^4\Sigma_u^- 3s\sigma_g)$ state have been clearly observed for these vdW molecules. An evidence of the preferential ionization process of the $O_2^+(b^4\Sigma_g^-)$ state has also been observed in dissociative ionization of ArO_2.

EXPERIMENTAL

The experiment was performed at the beam line BL2-B2 of the UVSOR at Institute for Molecular Science.[9,23,24] A monochromatized extreme-UV synchrotron radiation intersected a nozzle-expanded gas beam at a right angle. The ionic products were extracted perpendicularly to both of the photon and the molecular beam, and then analyzed using a quadrupole mass spectrometer. vdW molecules were produced in an expansion of a 1:1 gas mixture of O_2 and Ar at a stagnation pressure of 7 atm (maximum) through a 50μmϕ nozzle aperture at -100°C.

Photoionization efficiency (PIE) curves for individual ionic species were recorded in the 50-100 nm region with 0.04-0.2 nm bandpass. The spectra were normalized by the photon intensity, monitored behind the intersection region. Absolute photon wavelength readings were calibrated for the Ar 3s and 3p Rydberg levels.

RESULTS AND DISCUSSION

A typical mass spectrum pattern is presented in Fig.1, obtained at 60 nm with a back pressure of 5 atm. The spectrum is not corrected for the transmission efficiency of the mass spectrometer. The monomer

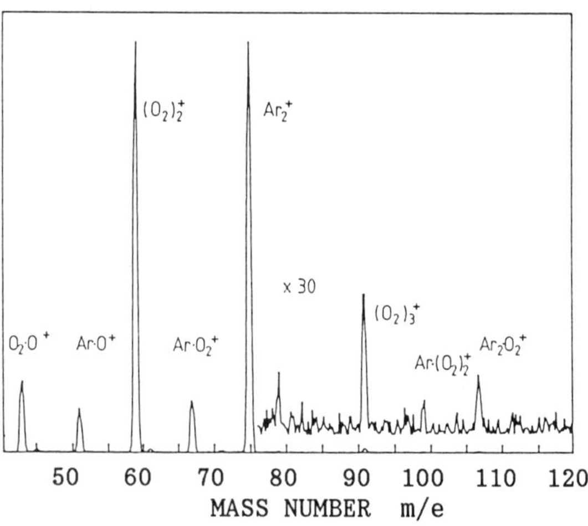

Fig.1
Typical mass pattern at the 60 nm ionization wavelength with the 5 atm back pressure of a 1:1 gas mixture of Ar and O_2 at -100°C nozzle temperature.

region is not shown. The relative intensities of the $(O_2)_2^+$, $(O_2)_3^+$, and Ar_2^+ peaks are larger than those for mixed cluster ions, ArO_2^+, $Ar(O_2)_2^+$, and $Ar_2O_2^+$ by about one order of magnitude.

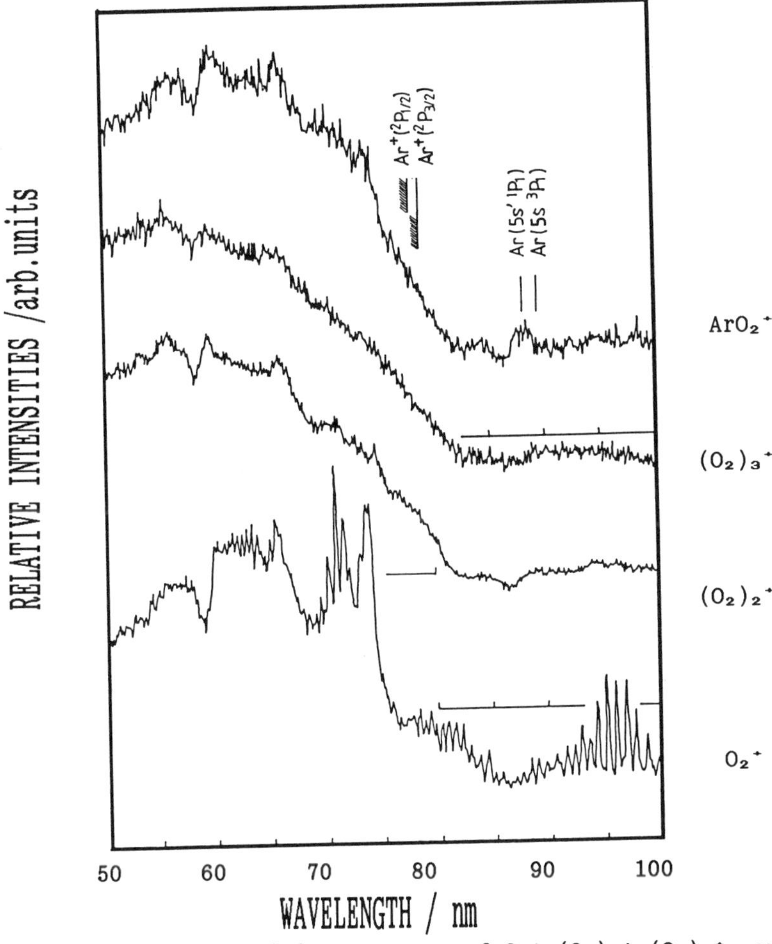

Fig.2 Photoionization efficiency curves of O_2^+,$(O_2)_2^+$,$(O_2)_3^+$, and ArO_2^+ in the 50-100 nm region.

Photoionization efficiency (PIE) curves recorded for O_2^+,$(O_2)_2^+$, $(O_2)_3^+$, and ArO_2^+ are shown in Fig.2. Since the purpose of this research is not to determine the ionization thresholds, the wavelength region studied is restricted to 50-100 nm. The distinct different behaviors O_2^+, $(O_2)_2^+$, $(O_2)_3^+$, and ArO_2^+ PIE curves confirm the reliable single collision condition in a supersonic nozzle beam and negligible secondary reactions following initial photoionization. Judging from the intensity ratios of the individual mass peaks in Fig.1, main contributions to the dimer and trimer signals are not due to fragmentation of the higher mass clusters but directly due to parent ions. As earlier reported by Linn et al.[8] the numerous Rydberg structures in the spectrum of monomer ion were almost completely diminished in the PIE curves of $(O_2)_2^+$ and $(O_2)_3^+$ in the region longer than 66 nm; only several small undulations are observed.

The ground electronic configuration of O_2 is presented by $(1\sigma_g)^2(1\sigma_u)^2(2\sigma_g)^2(2\sigma_u)^2(3\sigma_g)^2(1\pi_u)^4(1\pi_g)^2;X^3\Sigma_g^-$. The electronic structure of O_2^+ below about 75 nm is ascribed to the Rydberg series of the $1\pi_u$ shell excitation. Oscillations in the region from the peak at 73.2 nm to 60 nm is due mainly to the Rydberg series converging to the $b^4\Sigma_g^-$ and $B^2\Sigma_g^-$ limits,[14,18] in which the $3\sigma_g$ electron promotion is involved. These molecular orbitals concern a valence bond; the equilibrium internuclear separation of O_2 molecule with an electron vacancy in one of these orbitals is larger than that in the ground state.[25] Excitation to these superexcited states of O_2 within $(O_2)_n$, whether it is vibrationally excited or not, is substantially coupled to nuclear motion; i.e., the O_2-O_2 bond with about 10 meV well[22] will readily be dissociated. This is why the superexcited $(O_2)_n$ molecules in this wavelength region contribute to the O_2^+ signal, but not to $(O_2)_n^+$ or $(O_2)_{n-1}^+$, after dissociation of the vdW bond.

As seen in Fig.2, humps and shoulders observed in the $(O_2)_2^+$ PIE curve are due to opening of new ionization channels into higher ionic continuum. The gross features of the PIE curves for $(O_2)_2^+$ and $(O_2)_3^+$ are quite similar to the non-oscillatory component of the O_2^+ PIE curve. The intensity of the PIE is given by $I_T=I_C+I_S$, where I_T is the total photoion intensity, I_C, the component of direct ionization, and I_S, the autoionization yield of coexisting superexcited states. The oscillatory structures in the PIE curve are originated from resonant contribution of I_S. The PIE curves in the region longer than 66 nm of $(O_2)_2^+$ and $(O_2)_3^+$ are mainly ascribed to direct ionization. However, the remaining non-oscillatory component of I_C should show the same feature as that observed in Fig.2. In the region shorter than 66 nm, on the other hand, some characteristic features are recognized in the $(O_2)_2^+$ and $(O_2)_3^+$ curves. Especially, the window at around 59 nm is presented very clearly. The window structure is due to the first member of the $c^4\Sigma_u^-ns\sigma_g$ series of an isolated O_2.[18-20] The $c^4\Sigma_u^-ns\sigma_g$ Rydberg molecules with a non-bonding $2\sigma_u$ shell vacancy retains the equilibrium internuclear separation of the neutral ground state. This suggests the absence of vdW-bond dissociation and explains the distict autoionization structure in the $(O_2)_2^+$ and $(O_2)_3^+$ PIE curves.

The general feature of the ArO_2^+ PIE in Fig.2 is very similar to those of $(O_2)_2^+$ and $(O_2)_3^+$. Characteristic structures above 66 nm are observed very clearly. Considering the Ar-O_2 vdW potential,[22-24] the spectrum is interpreted in the same manner as that for $(O_2)_2^+$ and $(O_2)_3^+$. The observation of more distinguished humps and shoulders of O_2^+ structures in the ArO_2^+ PIE than that in the $(O_2)_2^+$ PIE implies that Ar atom in ArO_2^+ is less interactive with the O_2^+ ion than O_2 in $(O_2)_2^+$. No structure concerning the Ar ionization is observed except for humps observed at around 88 nm. The structures are the excitation to the Ar$(3p)^5 5s'\,^1P_1$ and $5s\,^3P_1$ states at 88.0 and 89.4 nm, respectively followed by the autoionization into the ArO_2^+ continuum, i.e., Penning ionization of O_2 by Ar in an intramolecular framework.[1-7] The small hump at 84nm may also be the contribution from the Ar(*6s*)O_2 autoionization. The oscillator strength decreasing rapidly with increasing the principal quantum number makes the structure ambiguous.

The PIE curves of O^+ and O_3^+ ions are shown in Fig.3. The threshold structure at the first $O^+(^4S^o)+O(^3P)$ limit (66.2 nm) and the general appearance of the O^+ PIE curve is quite similar to previous results.[8,15] The small peaks below the threshold are originated from the ion-pair formation into the $O^-(^2P^o)+O^+(^4S^o)$ limit at 71.8 nm. Enhancements of O^+ PIE emerge at dissociation limits.

As presented by Linn et al.[8] the threshold structure of the PIE curve for O_3^+ is not identical to that of O^+. In the region shorter than 65 nm, the two spectra are also completely different. Most of the autoionization structures observed in the O^+ spectrum are diminished in the O_3^+ curve. The energetic stabilization defines the threshold behavior of the O_3^+ curve.

The appearance onset of O_3^+ in much longer wavelength region than that of O^+ is ascribed to the intramolecular rearrangement or the so-called ion-molecule half-reaction[8,29,30]

$$O_2^+(a^4\Pi_u, v'>5)\cdot O_2 \rightarrow O_3^+ + O. \qquad (1)$$

However, the overview of O_3^+ curve is not identical to the partial ionization cross section of the $O_2^+(a^4\Pi_u)$ state.[21,22] The PIE curve of O_3^+ above the threshold is quite similar to the PIE curves of $(O_2)_2^+$ and ArO_2^+ but not to O^+; the broad structures with humps and shoulders are well distinguished. It is suggested that formation of the parent ion $(O_2)_2^+$ by photoabsorption is the first step, and then intramolecular rearrangement occurs to give a main contribution to O_3^+ formation. In the 50-74 nm region the photoionization of O_2 is mainly ascribed to the partial ionization to the $X^2\Pi_g$, $a^4\Pi_u$, $A^2\Pi_u$, $b^4\Sigma_g^-$, and $B^2\Sigma_g^-$ states[22] and the PIE curves of O_2^+ and $(O_2)_2^+$ are made up of the superposition of individual partial ionization processes. On the other hand the dissociation of the $X^2\Pi_g$ state is unlikely. The $a^4\Pi_u$ and $A^2\Pi_u$ ions are endothermic for the first dissociation limit and only intramolecular rearrangement in $(O_2)_2^+$ offers a chance for dissociation. The $b^4\Sigma_g^-(v'>5)$ and $B^2\Sigma_g^-(v'>0)$ ions are exothermic for the first limit. The PIE curves of O_3^+ is, thus, produced by the superposition of the available dissociation and rearrangement.

The feature of the ArO^+ PIE curve in Fig.4 is completely different from those observed in the O^+ and O_3^+ spectra. The onset is located at around 68 nm between those of O^+ and O_3^+. The position of the peak maximum is at a slightly longer wavelength than that observed in the O^+ curve. However, these effects are not originated from the molecular rearrangement in the hetero dimer-ion frame. Figure 4 compares the present result with the partial ionization cross section of the $O_2^+(b^4\Sigma_g^-, v'=0\text{-}3)$ state.[31] The onset at 68 nm, the position of peak maximum at 66 nm, the oscillatory structure in 60-65 nm, and the rapid decrease above 60 nm with windows at 59 and 53 nm in the ArO^+ PIE spectrum are completely identical to those for the photoionization cross section of the $O_2^+(b^4\Sigma_g^-)$ state. The vibro-

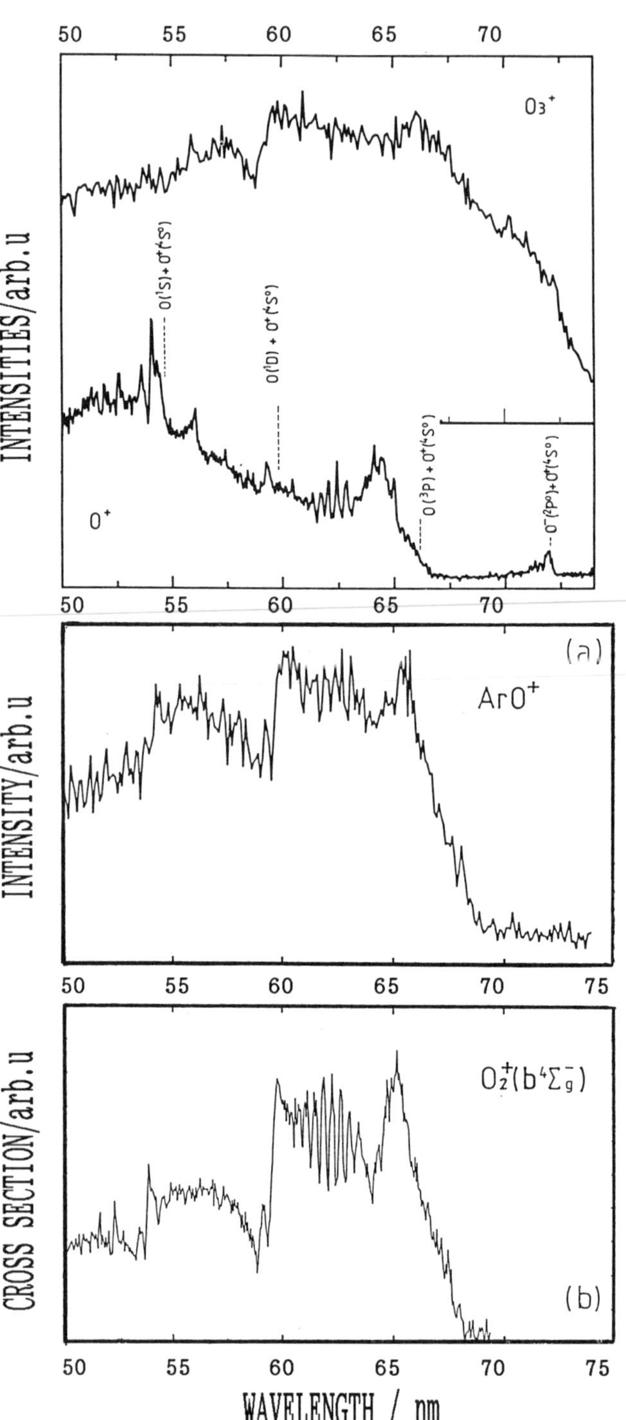

Fig.3 Relative photoionization efficiency curves for O^+ and O_3^+ in the 50-74 nm region.

Fig.4(a)Relative photoionization efficiency curve for ArO^+ in the 50-74 nm region and (b) partial ionization cross section of the $O_2^+(b^4\Sigma_g^- v'=0\text{-}3)$ state.[31]

nic levels of the $O_2^+(b^4\Sigma_g^-, v'=0-3)$ states are slightly below the first dissociation limit by 0.56-0.15eV. The coincidence of the threshold and the spectral similarities indicate that the $ArO_2^+(b^4\Sigma_g^-, v'=0-3)$ states give a significant contribution to formation of ArO^+, i.e.,

$$ArO_2^+(b^4\Sigma_g^-, v'=0-3) \rightarrow ArO^+ + O, \qquad (2)$$

which is also a kind of intramolecular rearrangement or a half-process of ion-molecule reaction. The endothermic energy defect is supplied by the stabilization due to the Ar and O^+ interaction. The $O_2^+(B^2\Sigma_g^-)$ predissociation with a large kinetic energy release[13] hardly populates ArO^+ but rather prefers complete decomposition. This is why the ionization channel of the $O_2^+(b^4\Sigma_g^-)$ state is selectively observed in the presence of Ar atom. Furthermore, it should be noted that ArO^+ was not observed as a product of a "full" O_2^++Ar ion-molecule reaction.[32] Concerning the present case, the short lifetime of the $O_2^+(b^4\Sigma_g^-)$ state prevents us from performing a "full" collision experiment. The single opportunity may be offered in such an unimolecular decomposition.

In summary, photoionization processes of $(O_2)_2$, $(O_2)_3$, and ArO_2 have been studied in the 50-100 nm region. A strong selectivity has been observed in the superexcitation. In the production of parent ions, superexcited states correlated with outer valence orbitals have been ruled out via the vdW-bond dissociation, but those related with the inner valence orbitals have participated in the process. Strong selectivity of the photoionization have also been observed in the production of the dissociative ions, O_3^+ and ArO^+.

The authors wish to thank Dr.H.Shiromaru for his help contribution in the early stage of the experiments. They also thank the staff of UVSOR for their cooperation.

REFERENCES

a M.Ukai, K.Kameta, K.Shinsaka, Y.Hatano, T.Hirayama, S.Nagaoka, and K.Kimura, Chem. Phys. Lett. 167, 334 (1990).

b Present address: Department of Electronics, Kanazawa Institute of Technology, Kanazawa, Ishikawa 921, Japan.

c Present address: Department of Physics, Gakushuin University, Mejiro, Toshima-ku, Tokyo 171, Japan.

d Department of Chemistry, Faculty of Science, Ehime University, Matsuyama, Ehime 790, Japan

1. P.M.Dehmer and S.T.Pratt, "Photophysics and Photochemistry in the Vacuum Ultraviolet, edited by S.P.McGlynn, J.L.Findley, and R.H.Huebner (D.Reidel, Dortrecht,1985) p.467.
2. S.T.Pratt and P.M.Dehmer, J.Chem.Phys. 76,4865(1982).
3. W.Kamke, B.Kamke, H.U.Kiefl, and I.V.Hertel, Chem.Phys.Lett. 122, 356(1985).
4. W.Kamke, B.Kamke, H.U.Kiefl, Z.Wang, and I.V.Hertel, Chem.Phys. Lett. 128,399(1986).

5. E.Rühl,P.Bisling,B.Brutschy,K.Beckmann,O.Leisin, and H.Morgner, Chem.Phys.Lett. 128,512(1986).
6. W.Kamke, B.Kamke, Z.Wang, H.U.Kiefl, and I.V.Hertel, Z.Phys. D2, 159(1986).
7. B.Kamke, W.Kamke, Z.Wang, E.Rühl, and B.Brutschy, J.Chem.Phys.86, 2525(1987).
8. S.H.Linn, Y.Ono, and C.Y.Ng, J.Chem.Phys. 74,3348(1981).
9. H.Shiromaru, H.Suzuki, H.Sato, S.Nagaoka, and K.Kimura, J.Phys. Chem. 93,1832(1989).
10. M.Ukai, H.Koizumi, K.Shinsaka, and Y.Hatano, J.Chem.Phys. 84,3199 (1986).
11. M.Ukai, Y.Tanaka, K.Koizumi, K.Shinsaka, and Y.Hatano, J.Chem. Phys. 84,5575(1986).
12. M.Ukai, H.Nakazawa, K.Shinsaka, and Y.Hatano, J.Chem.Phys. 88, 3623(1986).
13. M.Ukai, H.Yoshida, Y.Morishima, K.Shinsaka, and Y.Hatano, J.Chem. Phys. 90,3199(1989).
14. P.M.Dehmer and W.A.Chupka, J.Chem.Phys. 62,4525(1975).
15. T.Hayaishi,Y.Iida,Y.Morioka,M.Sasanuma,E.Ishiguro,and M.Nakamura, J.Phys.B19,2861(1986).
16. E.Nishitani,I.Tanaka,K.Tanaka,T.Kato, and I.Koyano, J.Chem.Phys. 81,3429(1984).
17. C.Y.R.Wu, E.Phillips, L.C.Lee, and D.L.Judge, J.Chem.Phys. 71,769 (1979).
18. A.Tabche-Fouhaile, I.Nenner, P.M.Guyon, and J.Delwiche, J.Chem. Phys. 75,1129(1981).
19. P.Morin, I.Nenner, M.Y.Adam, M.J.Hubin-Franskin, J.Delwiche, H. Lefebvre-Brion, and A.Giusti-Suzor, Chem.Phys.Lett.92,609(1982).
20. M.Ukai, A.Kimura, S.Arai, P.Lablanquie, K.Ito, and A.Yagishita, Chem.Phys.Lett. 135,51(1987).
21. T.Gustafsson, Chem.Phys.Lett. 75,505(1980).
22. J.W.Gallagher, C.E.Brion, J.A.R.Samson, and P.W.Langhoff, J.Phys. Chem.Ref.Data 17,9(1988).
23. K.Kimura, Y.Achiba, and H.Shiromaru, Rev.Sci.Instrum. 60,2205 (1989).
24. H.Shiromaru, H.Shinohara, N.Washida, H.-S.Yoo, and K.Kimura,Chem. Phys.Lett. 141,7(1987).
25. F.R.Gilmore, J.Quant.Spectrosc.Radiat.Transfer 5,369(1965).
26. A.Radzig and B.M.Smirnov, "Reference Data on Atoms,Molecules, and Ions" (Springer Verlag, Berlin, 1985).
27. G.Henderson and G.E.Ewing, J.Chem.Phys. 59,2280(1973).
28. F.P.Tully and Y.T.Lee, J.Chem.Phys. 57,866(1972).
29. J.M.Ajello, K.D.Pang, and K.M.Manahan, J.Chem.Phys.61,3152(1974).
30. P.M.Dehmer and W.A.Chupka, J.Chem.Phys. 62,2228(1975)
31. M.Ukai, K.Kameta, N.Terazawa, Y.Chikahiro, K.Nagano, N.Kouchi, Y. Hatano, and K.Tanaka, to be published
32. By contrast, ArO^+ was observed in an $Ar^+ + O_2$ charge transfer reaction. The appearance threshold for ArO^+ was at 2.2 eV kinetic energy (c.m.). The formation of ArO^+ was interpreted to proceed via $Ar \cdot O_2^+(a^4\Pi_u)$ charge-transfer complex; S.Scherbarth and D.Gerlich, J.Chem.Phys. 90,1610(1989); G.D.Flesch, S.Nourbakhsh and C.Y.Ng, ibid. 92, 3590 (1990). However, the present result

strongly suggests an important role of the $Ar{\cdot}O_2^+(b^4\Sigma_g^-)$ intermediate state also in the "full" $Ar^+ + O_2$ ion-molecule reaction. The ionization threshold for $O_2^+(b^4\Sigma_g^-, v'=0)$ is almost identical to the appearance potential of ArO^+ within the experimental error of the above ion-molecule reaction.

DISCUSSION

LEACH - Your results on Ar.O_2 complexes make me ask whether there exists similar dissociative ionization processes in the upper atmosphere for complexes between metal atoms and oxygen molecules (M.O_2) ?

UKAI - I am sorry I do not know much above the metal atom-oxygen molecule complex in the upper atmosphere.
(addendum)
Because of the close shell structure of the ground Ar atom, the intermolecular force betrween Ar and O_2 is dispersive rather than chemical, which makes the initial conditions rather different. Furthermore, dissociative ionization of Ar.O_2 complex is correlated to ionization of O_2 whereas that of M.O_2 should have more relation with ionization of M.
However, we can find some similarities in intermolecular force between an excited Ar atom and a O_2 molecule with that of M.O_2. I therefore think there must be a kind of similarity in the processes involving autoionization.

BESWICK - Do you think it could be possible to perform similar experiments on the Ar H_2 complex ? This system is of considerable theoretical interest in relation with the role of non adiabatic transitions in the entrance channel which have been studied in scattering experiments.

UKAI - I think it is possible. However my next target is not Ar H_2 but ArCO_2 which is much easier to produce and more analogous with respect to ArO^+ formation.

DUJARDIN - In the high energy range of the photoionization curves there are very few features observed although there probably exist a lot of Rydberg states in this energy range. A more sensitive way of looking at these Rydberg states would be to look at negative ions produced by ion pair formation in the clusters.

UKAI - Higher resolution results by Dehmer and Chupka (1975) on O_3^+ in the lower energy region showed richer structure of Rydberg states than our result.
It would be possible to extract fine structures in the higher energy side using more sensitive detection technique such as negative ion detection.
However, we had an experimental difficulty in detecting negative ions, we could only succeed in O^- observation so far.

WONG - What are the molecular structures of $(O_2)_2$ and $(O_2)_3$?

UKAI - I think the geometrical structure of $(O_2)_2$ is still on discussion. A floppy structure of ground $(O_2)_2$ with weak potential minima at "T" and dihedral shapes is proposed by the infrared absorption. However, the structures of excited $(O_2)_2$ are not determined. Ar.O_2 has "T" and linear shape structures in the ground state. I have no idea on $(O_2)_3$.

SMALL CLUSTERS AND FULLERENES: NEW INSIGHTS FROM AB-INITIO MOLECULAR DYNAMICS

Wanda Andreoni
IBM Research Division, Zurich Research Laboratory,
8803 Rüschlikon, Switzerland

ABSTRACT

Using the Car-Parrinello method we have investigated theoretically the physical properties of clusters of several materials. A brief survey is given of the major goals and the new insight gained so far into the temperature dependence of the structural pattern and the dynamics of microclusters. The recent application to the study of fullerenes is also discussed.

1. INTRODUCTION AND METHOD

In this paper I present a concise survey of the recent progress made in the study of small clusters, using the Car-Parrinello (CP) method.[1] First I discuss the by now "almost standard" applications of simulated annealing techniques to the search of the low-energy structures, as well as the still rare simulations at finite temperature, for metal and semiconductor clusters. I shall try to emphasize those results which have indeed allowed us to gain new insight into the physics and chemistry of this new "phase of matter". Secondly, I shall present some very recent applications to fullerenes, and underline those aspects for which CP-like techniques can be particularly useful.

For the description of the method, I refer to the original work and to the specific articles which will be cited throughout this paper. The CP method unifies Density Functional Theory[2] and Molecular Dynamics (MD).[3] It is by no means restricted to the use of approximations such as the Local Density Approximation (LDA)[4] for the exchange-correlation functional or the pseudopotential for the approach to the electronic problem. However, in practical applications, these approximations are used and their validity governs the reliability of the results for the electronic structure. In the application to clusters, the original scheme proposed by Car and Parrinello for infinite systems is invariably used, i.e. periodic boundary conditions are retained which imply the use of plane-waves for the expansion of one-electron orbitals. Care must be taken in using a unit cell sufficiently large so as to "minimize" spurious effects due to the interaction between the different images on the quantities of interest. The input parameters that determine the level of accuracy of specific LDA-pseudopotential calculations carried out with this computational technique are the size of the unit cell and the energy cutoff used in the plane-wave expansion. Only very few test calculations showing the accuracy of the computational scheme have appeared so far.[5-7]

The advantages of this method are obvious, i.e. the possibility to treat ionic and electronic variables simultaneously, to calculate structural, electronic and dynamical properties at finite temperature, and to handle low-symmetry systems. However, a remark must be made concerning an important technical point, i.e. the value of the time step used in the simulations. This is generally of the order of 10^{-4} psec, i.e. two orders of magnitude smaller than the typical time step of classical MD. This fact constitutes an obvious limitation with respect to classical MD, thus restricting the size of the cluster and the choice of the dynamical processes one can study. For instance, classical MD simulations have demonstrated that studying the dynamics of chemical reactions on a small 13-atom cluster of Ni[8,9] or the melting of large gold clusters[10] with classical MD requires long observation times and a large statistical basis. It is probably worth emphasizing that such a restriction on the simulation time is not intrinsic to the theoretical method and requires mainly an increase of computer speed in order to be overcome.

2. SODIUM CLUSTERS

Several aspects of the physics of alkali-metal clusters have been investigated successfully prior to the application of the Car-Parrinello method, and in particular the occurrence of specific magic numbers has been explained.[11] In fact, owing to the simple type of bonding, the electronic properties can be understood in terms of jellium models which ignore the influence of the structure, and shell models (spherical and ellipsoidal) were able to account for the relative abundance of clusters of different sizes as well as for trends in ionization potentials. Determination of low-energy structures had already been performed with the pseudopotential-LDA approach,[12,13] and first-principle configuration interaction (CI) calculations[14] had already given the structure of sodium clusters up to nine-atom ones. The new information that a Car-Parrinello-like technique can give concerns the dependence of both structural and electronic properties on temperature and the dynamics. In addition, being more easily extendable to larger sizes, it allows one to gain insight into the evolution of the structural properties with size. This is what has indeed been achieved in the study of sodium microclusters in the range of 6-20 atoms.[7,15] The main new results can be summarized as follows:

i) The low-energy portion of the potential energy surface corresponds to structures with pentagonal rings, analogous to complexes of Lennard-Jones systems and in contrast to metal crystalline subunits. The similarity with the Lennard-Jones systems was pointed out for the first time in Ref. 7 and can be understood on the basis of the non-directionality of the chemical bonding, which also implies that structural compactness is the dominant criterion in determining the relative stability of different isomers.

ii) The electronic structure corresponds fairly well to the predictions of the self-consistent spherical shell model,[16] as can be seen for instance in the behavior of the KS one-electron energy levels for clusters of different sizes

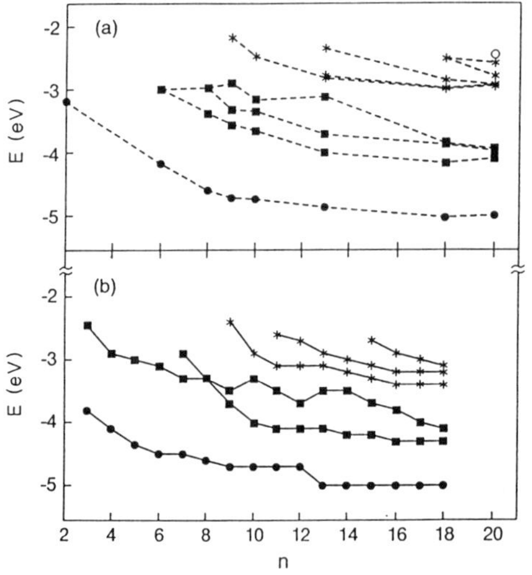

Figure 1. Kohn-Sham one-electron eigenvalues as a function of the number of atoms in the cluster: (a) from calculations in Ref. 7, (b) from the self-consistent spheroidal model of Ref. 16. From the low-energy side upward, the sequence of levels is as follows: ●: $1s$, ■: $1p$, ∗: $1d$, o: $2s$ in the spherical-shell model. Note that this labelling corresponds only to the dominant component of the eigenstate. From Ref. 7.

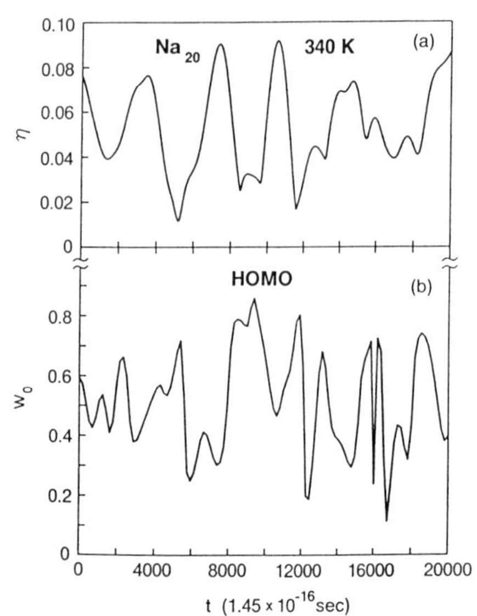

Figure 2. Na_{20} at 340 K: Time evolution of (a) the eccentricity parameter and (b) the weight of the spherical component of the highest-occupied-molecular-orbital (HOMO) w_0. Note that in the spherical shell model for Na_{20}, $w_0 = 1$. From Ref. 7.

illustrated in Fig. 1. In particular, the cluster orbitals can be classified in good approximation with the angular momentum ℓ. Deviations are larger for the non-magic number clusters.

iii) Increasing temperature drives an increase in the eccentricity of the cluster, in contrast to widespread ideas according to which warming up a cluster makes it "more spherical". This can also be understood in terms of the change in the electronic structure with temperature which we calculate, and specifically with the increase of hybridization between states of different angular momentum. Fig. 2 illustrates the time variation of the eccentricity parameter and of the s-component of the highest occupied molecular orbital (HOMO) for the case of a "warm" 20-atom cluster. Although the correlation is not 1:1, since the HOMO hybridization is not the only factor determining the shape variation, the link is clear. This is actually a nice example of the link between electronic and ionic variables at finite temperature, a result unique to the Car-Parrinello method.

iv) Concerning the issue of whether the small clusters are rigid and as to the interplay between isomerization and melting, our simulations, although limited to time intervals of a few psec, can provide useful information. For a small cluster, such as a 6-atom cluster, isomerization is well identified. Interestingly, the dynamics at 600 K can be described as simple oscillation between only two types of isomers, both related to the T=0 isomers (one 3D and one 2D). For larger clusters, such as Na_{18} and Na_{20}, it was not possible to distinguish between isomerization and melting, and the atomic mobility was calculated to be of the same order as that of the bulk system at the same temperature. Na_{10} was found to be "not rigid" in the two simulations, one done at 260 K and the other at 440 K. While all atoms are observed to diffuse at 440 K, the exceptionally high mobility at 260 K is due exclusively to the movement of the two capping atoms on the "surface" of the cluster. The presence of two caps and a core of eight atoms also characterizes the T=0 isomers of the cluster. In particular, Fig. 3 shows the vibrational spectrum at the lower temperature, analyzed in the contribution of the caps and of the core atoms (Fig. 3). The capping atoms contribute mostly to the low-frequency modes. The whole picture is suggestive of a "surface"-driven "melting" of the cluster, similar to what is actually observed in computer experiments on large metal particles.[10] The analogy however is obviously qualitative.

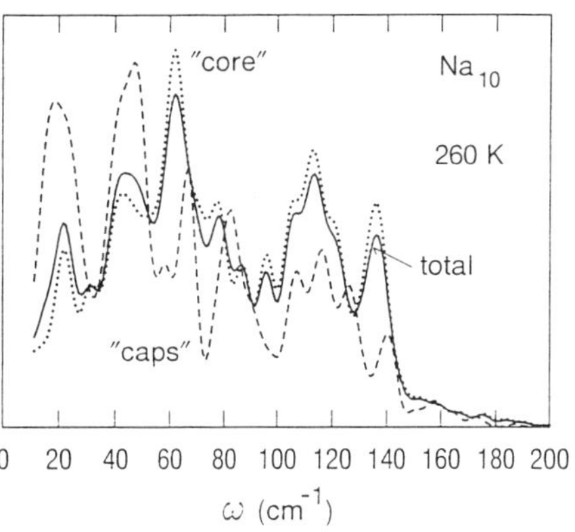

Figure 3. Na_{10}: Vibrational spectrum calculated at ~260 K (solid line). Dashed line = contribution from the "caps." Dotted line = contribution from the "core" atoms. From Ref. 15.

3. SEMICONDUCTOR CLUSTERS

Previous to the application of the Car-Parrinello method, several calculations had been published on microclusters of silicon up to 10 atoms, giving a partial account of low-energy structures. However, a clear picture of the special features of the chemical bonding compared to that in the bulk had already been obtained.[17-19] Likewise, the application of the Car-Parrinello method to silicon clusters is for the moment confined primarily to the structural properties and only a limited account of dynamical properties and temperature dependence of electronic and structural properties has been published.[20-22] Refs. 20 and 21 contain the results for silicon aggregates of up to 10 atoms. Particularly successful has been the dynamical simulated annealing search for the 10-atom cluster for which it has predicted a new structure as the lowest energy one, later confirmed by CI calculations.[23] The 10-atom cluster structure (tetracapped trigonal prism) has now been suggested as the building block of the 30-atom cluster of silicon.[24] Trigonal prisms and antiprisms constitute also the seed of the low-energy structures of the 13-atom cluster of silicon.[25] Two isomers are shown in Fig. 4: (a) can be described as a trigonal antiprism with six caps

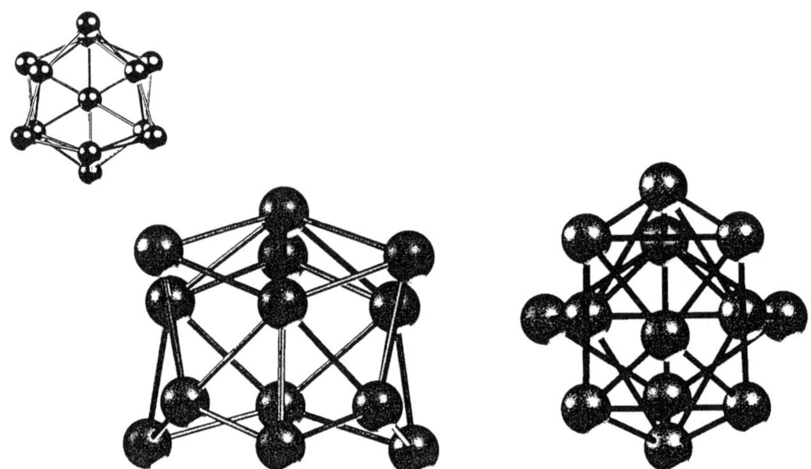

Figure 4. Low-energy isomers of Si_{13} calculated in Ref. 25.

on the three-fold faces and one 6-fold coordinated additional cap, while (b) can be viewed as a fully-capped trigonal prism with two more three-fold caps. The energy difference between (a) and (b) is calculated to be only 0.2 eV. Unfortunately, a direct experimental determination of the structure of small clusters is lacking. However, in agreement with our predictions, two types of low-energy isomers are believed to exist for clusters of $N > 8$.[24] Indeed, our simulations at finite temperature for Si_8[15] confirm that the cluster is rigid, with the atoms oscillating about the positions of the "$T = 0$" lowest energy isomer at least up to 1000 K. The dynamics of Si_{10} is richer.[15] When we heat the cluster up to 1000 K we do not see, within our observation times, any transition to a different isomer. Instead, both at 500 K and at 1000 K, we observe a pseudorotation, whose single step is characterized by the relaxing of two bonds and the formation of two new ones. The result is identical to that of a rotation by 90 degrees. Being that our clusters are inhibited to rotate, we know that what we observe is a pseudorotation. This is indeed characteristic of this type of structure and we find it also for Mg_{10} whose lowest energy structure is likewise a tetra-capped trigonal prism. The process is illustrated in Fig. 5.

These calculations have been extended to III-V semiconductor microclusters,[26] GaAs, GaP and AlAs, with stoichiometric composition. The

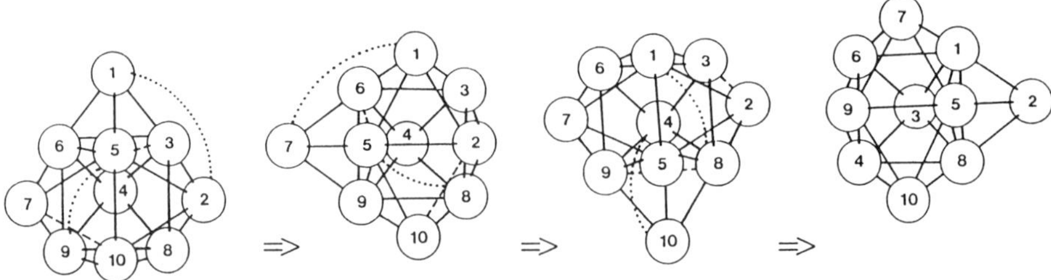

Figure 5. Si_{10}: Schematic representation of the pseudorotation. Dashed lines = "breaking" bonds. Dotted lines = "forming" bonds. From Ref. 15.

main results can be summarized as follows: i) In order to determine their relevant structures, simple alternate "decoration" of the monoatomic isoelectronic clusters (Si and Ge), although being a good first approximation, is not sufficient. This comes mainly from the presence of strong anion-anion bonds which involve nontrivial distortions and may not favor the alternate type of configuration. ii) In agreement with indications from reactivity experiments on GaAs,[27] alternating arrangements of cations and anions require a critical size of the cluster to establish itself. This is the 10-atom cluster for GaAs and the 8-atom cluster for GaP. In the case of AlAs, no critical size exists, which leads us to predict a different behavior in the reactivity to e.g. ammonia. iii) In agreement also with indications from photoabsorption experiments on InP,[28] we find that at least in the size range up to 10 atoms all the atoms are severely undercoordinated with respect to the bulk. This is in contrast to our findings on silicon clusters in the same size range, which are slightly overcoordinated with respect to the bulk. Our findings also support the conclusions drawn from the observation of chemical reactivity, i.e. of high reactivity of GaAs microclusters analogous to that of GaAs surfaces and of low reactivity of silicon microclusters in clear contrast with the behavior of silicon surfaces. iv) Simulations at different temperatures on the GaAs systems show that, even at temperatures of the order of the melting point of the bulk, the characteristic features of the "T = 0" isomers is observed. In particular the tendency of the As ions is to be on the outside of the cluster. More complex behavior is observed for GaP. The tendency of the anions to "segregate" is understood in terms of an electronic mechanism, which in the case of GaP is contrasted by the size effect which would drive the larger species (Ga) to the "surface" of the cluster.

4. CARBON MICROCLUSTERS AND FULLERENES

Calculations on carbon microclusters are only partial.[6] They were performed mostly to test the computational scheme against previous ab-initio calculations and experimental data when available. Additional nice and even unexpected results were obtained as soon as we started to assess the thermal behavior. The four-atom cluster at low temperatures is a planar rhombus transformed at high temperatures into a linear chain, in agreement with several predictions based on the fact that a linear chain is favored by entropy.[29] The ten-atom cluster, which is ring shaped at low temperatures, also showed a tendency to open at high temperatures. More interestingly, this aggregate showed a floppy nature already at rather low temperatures (200 K), oscillating between one D_{5h} structure and its enantiomorph.

C_{60} is the only cluster for which structural, electronic and vibrational properties are known experimentally. Indeed, structural parameters have been measured on the crystal phase[30,31] where, however, the intramolecular structure is expected to be only very slightly modified. C_{60} constitutes therefore an ideal test of our whole scheme, although it is computationally rather heavy. An unconstrained annealing of the structure starting from a random configuration is prohibitive for this system. Owing mainly to the stiffness of the bonds, it requires long simulation times. The only simulated annealing search which has so far been able to obtain the Buckminsterfullerene (Bf) structure (truncated icosahedron) is from Ballone and Milani.[32] This uses Monte Carlo sampling and a classical potential[33] to describe the interatomic interactions, and the atoms are constrained to a spherical shell.

Starting with a Bf structure where all bonds are equal, we have relaxed[34] the atomic positions with a combined electronic and ionic steepest descent method, and further checked the stability of the structure with MD runs. The bond lengths split correctly into two sets, one corresponding to the bonds at the frontier between two hexagons and the other to bonds at the frontier between two pentagons. The calculated values of 1.39 A and 1.45 A turned out to be in very good agreement with the nuclear magnetic resonance (NMR) data (1.40±0.015, 1.45±0.015) obtained at room temperature by measuring the magnetic dipolar coupling.[30] Our values are also in perfect agreement with LDA results obtained simultaneously by other groups.[35,36] We then made MD simulations at finite temperature. Simulations at 450 K showed that the effects of temperature were only minor both on the structure and on the electronic properties. Fig. 6 illustrates the calculated radial distribution function compared with the one obtained from neutron diffraction experiments.[31] We notice that thermal broadening masks the splitting of, for example, the nearest-neighbor distance. Fig. 7 contains the plot of (a) our time-averaged density of electron states and (b) the photoelectron spectrum measured in the gas phase.[37] The density of states is known to be only an approximation of the observed spectrum, and to be less and less valid for increasing ionization energy. The agreement is however very good in the range of a few electron volts below the highest occupied states. We notice in particular that the positions of the first two peaks are reproduced very well and that the third peak is already shifted by 0.4 eV with respect to experiment. From the trajectories at 450 K we have[34] also extracted the vibrational frequencies corresponding to the optically active modes. Both IR and Raman active modes were identified with accuracies better than 6%. We note that for high frequencies (1500 K) however this means an error of 90 cm^{-1}, and that these frequencies are systematically underestimated. This discrepancy must be primarily associated with the LDA inaccuracy. We note however that ours constitutes the only ab-initio determination of the vibrational spectrum of C_{60} and the agreement is not worse than for results obtained

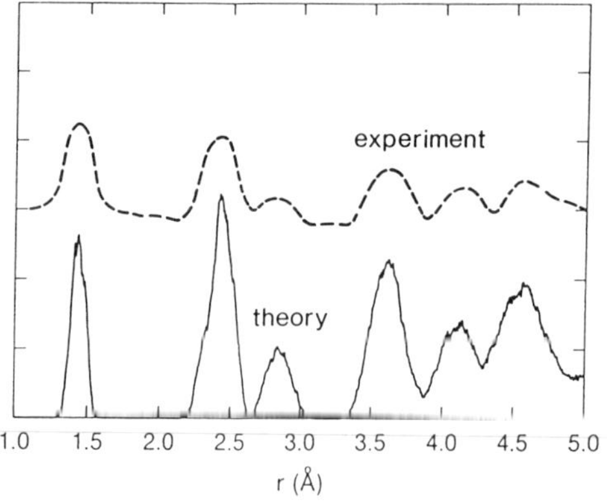

Figure 6. C_{60}: Distribution of interatomic distances calculated[34] at ~450 K (solid line) compared with the radial distribution function obtained from ND data (Ref. 31).

with empirical (or semi-empirical) methods.

C_{70} is known to assume a so-called rugby-ball structure with a D_{5h} symmetry, as demonstrated by NMR experiments[38] showing the presence of five non-equivalent sites. Structural parameters have not been measured yet. Our calculated values are intermediate between the Hartree-Fock ones[39] and those obtained with MNDO,[40] in perfect analogy with the results for C_{60}. The shortest bond lengths are only very slightly shorter than in C_{60}, and the largest bond length, split by about 0.08 Å, corresponds to the smallest interatomic distance between the 10 atoms on the additional ring at the equator of the structure. Comparison of the DOS with the photoemission spectrum measured on the solid[41] reveals the same type of agreement than for C_{60}. The photoelectron spectrum calculated in Ref. 36 also shows the same type of agreement, in spite of the fact that the assumed structure is somewhat different from our optimized one and that the transition matrix elements are explicitly calculated.

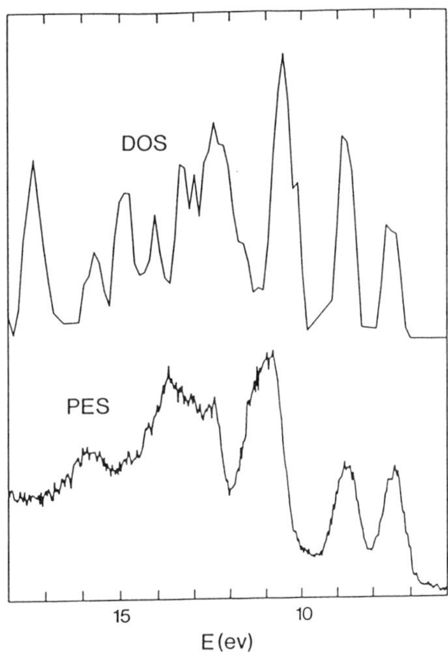

Figure 7. (a) Average density of states (DOS) calculated[34] at ~450 K; (b) photoelectron spectrum (PES) from Ref. 37. The zero of the energy in (a) is adjusted such that the position of the first peak (which corresponds to the HOMO) coincides with that in (b). From Ref. 34.

In spite of their large number of atoms, C_{60} and C_{70} can be studied with standard quantum-mechanical methods due to their high symmetry. In contrast, higher fullerenes and especially doped fullerenes have low symmetry so that the prediction of structural and electronic properties is certainly more complicated for standard ab-initio methods. The Car-Parrinello method does not rely on any symmetry constraint so it is an especially useful tool for the engineering of new materials at the computer. In the search for new fullerenes, the B- and N-doped C_{60} are the most natural ones. $C_{59}B$ has indeed be recently synthesized in beam.[42] Starting from the C_{60} atomic positions and relaxing them, we have found[43] that both in $C_{59}B$ and $C_{59}N$ the optimum arrangements involve significant distortions of the network, mainly localized around the impurity atom. These are more important in the case of $C_{59}B$, as can be expected on the basis of the larger core size mismatch between the impurity and the host, and also owing to the nature of the electron states mostly involved in the doping: the HOMO and the LUMO (lowest unoccupied molecular orbital) of C_{60} which are bonding and antibonding, respectively, on the double bond. In particular, B-C bonds are ~10% larger than C-C bonds while N-C bonds are only 2% larger, and the C-C bonds vary by at most 2% with respect to the undoped cluster. Especially interesting is the appearance of new features in the electron energy spectrum, i.e. impurity levels in the middle of the HOMO-LUMO gap of the undoped cluster. The picture we obtain is analogous to that of deep impurity levels in semiconductors. This is illustrated in Fig. 8. In both cases the doped fullerene is a radical, with one singly occupied state which can act as an acceptor in $C_{59}B$ and as a donor in $C_{59}N$. Indeed $C_{59}B$ has been found to act like a Lewis acid in reactions with ammonia.[42] The presence of the heteroatom also creates a permanent dipole moment which we have calculated to be ~1.4 D in $C_{59}B$ and ~2.2 D in $C_{59}N$.

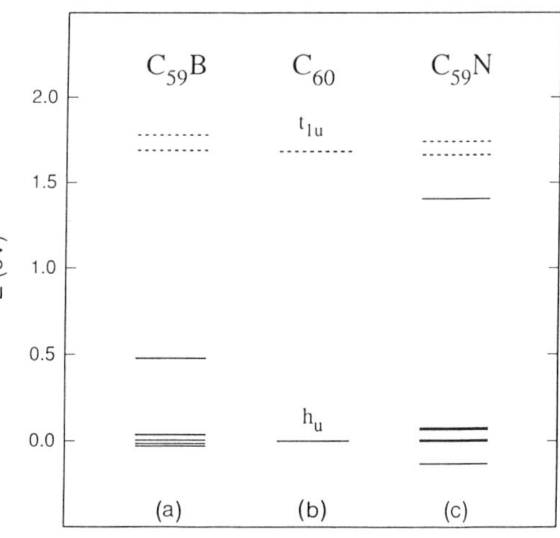

Figure 8. Energy level diagram of (a), (c) doped and (b) undoped C_{60}. Solid and dashed lines refer to occupied and unoccupied states, respectively. Note that the $C_{59}B$ LUMO is almost doubly degenerate. From Ref. 43.

The work on fullerenes has recently been extended to the solid phase and in particular to alkali-metal intercalated C_{60}.[44]

5. OTHER APPLICATIONS OF THE CP METHOD AND CONCLUSIONS

Several other applications of the CP method to microclusters have appeared in the literature, mainly devoted to the search for low-energy structures of selected aggregates of S,[45] Se,[46] P,[47] Al,[48,49] Be,[50] Mg[51,52] and more recently B.[53] Most interesting is the study of B_{12}[53] owing to its "revolutionary" results and the accuracy of the investigation. In fact, in contrast to the common belief that one aggregate of twelve atoms of boron would be stable in a hollow distorted icosahedron, the result of the dynamical SA[53] is a new structure, energetically favored by ~0.6 eV and having lower symmetry and lower coordination. The work on S, Se, P, Be and Mg is primarily aimed at determining a common "seed" in the growing sequence. While the presence of 8-atom rings in S, Se aggregates and of roof-shaped tetramer units in P clusters is clear and agrees also in some cases with experimental evidence, the results on Be and Mg are more intriguing. In general, as in the case of Si clusters, the limited size range investigated does not yet allow us to recognize a clear "motif", much less a key structural trend leading to the bulk structure. In addition, the limited number of structures investigated does not yet give a satisfactory and complete picture. In the case of Al, very special structural characteristics are found[48] for sizes up to 10 atoms which reflect the directional character of the chemical bonding and reveal the non-prototypical nature of Al clusters as "simple metal" clusters.

The work in Ref. 49, instead, concentrates on the issue of the stability of the icosahedral shape for Al_{13} and Al_{55}. While the 13-atom cluster turns out to be a slightly distorted icosahedron (as also confirmed in Ref. 25), for Al_{55} several isoenergetic isomers with "distorted" configurations are energetically favored. This result is interesting, although more work is needed to identify the structural pattern of such isomers and also to assess the relative stability of "ordered" structures (other than the icosahedron and the cubooctahedron). In fact, due to the necessarily short times of the simulated annealing, it is not surprising that amorphous structures are obtained for aggregates as large as ~50 atoms.

In conclusion I wish to emphasize that in spite of the considerable amount of applications of the CP method to the study of cluster physics, we are only at the beginning. In particular, the study of the dynamical aspect is in its infancy. The recent application to fullerenes, as yet limited to the study of T=0 properties, reveals another advantage of the CP method: as a tool for engineering new materials via computer.

REFERENCES

1. R. Car and M. Parrinello, Phys. Rev. Lett. **55**, 2471 (1985).
2. W. Kohn and P. Vashishta, in *Theory of the inhomogeneous electron gas,* S. Lundqvist and N. March, eds. (Plenum, New York, 1983), pp. 79-147.
3. A. Rahman, Phys. Rev. **A136**, 405 (1964).
4. W. Kohn and L.J. Sham, Phys. Rev. **A140**, 1133 (1964).
5. W. Andreoni, in *The chemical physics of atomic and molecular clusters*, G. Scoles and S. Stringari, eds. (North-Holland, Amsterdam, 1990), pp. 159-175.
6. W. Andreoni, D. Scharf and P. Giannozzi, Chem. Phys. Lett. **173**, 449 (1990).
7. U. Roethlisberger and W. Andreoni, J. Chem. Phys. **94**, 8129 (1991).
8. K. Garzon and J. Jellinek, Z. Phys. **D19**, 239 (1991).
9. J. Jellinek, Proc. 24th Jerusalem Symposium on Quantum Chemistry (Kluwer, Dordrecht, 1991) in press.
10. F. Ercolessi, W. Andreoni and E. Tosatti, Phys. Rev. Lett. **66**, 911 (1991).
11. W.D. Knight et al., Phys. Rev. Lett. **52**, 2141 (1984).
12. J.L. Martins, J. Buttet and R. Car, Phys. Rev. **B31**, 1804 (1985).
13. I. Moullet, J.L. Martins, F. Reuse and J. Buttet, Phys. Rev. Lett. **65**, 476 (1990).
14. P. Fantucci, V. Bonacic-Koutecky and J. Koutecky, Phys. Rev. **B37**, 4369 (1988).
15. U. Roethlisberger and W. Andreoni, Z. Phys. **D19**, 243 (1991).
16. W. Ekardt and Z. Penzar, Phys. Rev. B **38**, 4273 (1988).
17. K. Raghavachari, J. Chem. Phys. **84**, 5672 (1986).

18. G. Pacchioni and J. Koutecky, J. Chem. Phys. **84**, 3301 (1986).
19. D. Tomanek and M. Schlueter, Phys. Rev. **B36**, 1208 (1987).
20. P. Ballone, W. Andreoni, R. Car and M. Parrinello, Phys. Rev. Lett. **60**, 271 (1988).
21. W. Andreoni and G. Pastore, Phys. Rev. **B41**, 10243 (1990).
22. W. Andreoni, Z. Phys. **D19**, 31 (1991).
23. K. Raghavachari and C.M. Rohlfing, J. Chem. Phys. **89**, 2219 (1988).
24. M.F. Jarrold and V.A. Constant, preprint (1991).
25. U. Roethlisberger, W. Andreoni, P. Giannozzi, J. Chem. Phys., submitted.
26. W. Andreoni, submitted for publication.
27. L. Wang et al., Chem. Phys. Lett. **172**, 335 (1990).
28. K.D. Kolander and M.L. Mandich, Phys. Rev. Lett. **65**, 2169 (1990).
29. K.S. Pitzer and E. Clementi, J. Am. Chem. Soc. **81**, 4477 (1959).
30. C.S. Yannoni et al., J. Am. Chem Soc. **113**, 3190 (1991).
31. J.S. Lannin, Fang Li, D. Ramage, J. Conceicao, preprint.
32. P. Ballone and P. Milani, Phys. Rev. **B42**, 3201 (1990).
33. J. Tersoff, Phys. Rev. Lett. **61**, 2879 (1988).
34. B.P. Feuston, W. Andreoni, M. Parrinello and E. Clementi, Phys. Rev. B **44**, 4506 (1991).
35. Jae-Yel Yi, Q. Zhang and J. Bernholc, Phys. Rev. Lett. **66**, 2633 (1991).
36. J.W. Mintmire et al., Phys. Rev. **B43**, 14281 (1991).
37. D.L. Lichtenberger et al., Chem. Phys. Lett. **176**, 203 (1991).
38. A.K. Abdul-Sada, R. Taylor, J.P. Hare and H.W. Kroto, J. Chem. Soc. Chem. Commun., 1423 (1990); R.D. Johnson, G. Meijer, J.R. Salem and D.S. Bethune, J. Am. Chem. Soc. **113**, 3619 (1991).
39. G.E. Scuseria, Chem. Phys. Lett. **180**, 451 (1991).
40. K. Raghavachari and C.M. Rohlfing, J. Phys. Chem. **95**, 5768 (1991).
41. M.B. Jost et al., Chem. Phys. Lett. (1991) in press.
42. Guo Ting, Jin Changming, and R.E. Smalley, J. Phys. Chem., **95**, 4948 (1991).
43. W. Andreoni, F. Gygi, M. Parrinello, submitted for publication.
44. W. Andreoni, F. Gygi, M. Parrinello, submitted for publication.
45. D. Hohl, R.O. Jones, R. Car. M. Parrinello, J. Chem. Phys. **89**, 6823 (1988).
46. D. Hohl, R.O. Jones, R Car, M. Parrinello, Chem. Phys. Lett. **139**, 540 (1987).
47. R.O. Jones and D. Hohl, J. Chem. Phys. **92**, 6710 (1990).
48. R.O. Jones, preprint.
49. J.-Y. Yi, D.J. Oh, J. Bernholc, R. Car, Chem. Phys. Lett. **74**, 461 (1990); J.-Y. Yi, D.J. Oh, J. Bernholc, Phys. Rev. Lett. **67**, 1594 (1991).
50. R. Kawai and J.H. Wearie, Phys. Rev. Lett. **65**, 80 (1990).
51. V. Kumar and R. Car, Z. Phys. D **19**, 177 (1991) and Phys. Rev. B (1991) in press.
52. V. de Coulon et al. Z. Phys. D. **19**, 173 (1991).
53. R. Kawai and J.H. Wearie, preprint.

DISCUSSION

BIANCONI - You have shown that C_{60} exhibits a mode at 530 cm^{-1} that you can predict by your theoretical method. Can you discuss the symmetry of this mode and its changes induced by B or N doping ?

ANDREONI - The 530 cm^{-1} is IR active. It has T_{1u} symmetry. The experimental value if 527cm^{-1}. We have not studied the vibrational spectrum of the doped C_{60}.

DUJARDIN - I understood that in silicon cluster, the silicon atoms are undercoordinated. How is this compatible with a reduced reactivity of silicon clusters as compared to solid silicon surfaces ?

ANDREONI - No, what I said is that in Si microclusters, at least in the investigated range of size (from 6 to 13 atoms) the Si atoms are slightly overcoordinated with respect to the bulk. It is in the GaAs microclusters that the atoms are undercoordinated.

ELAND - Is there a reason why calculations using empirical potentials give poor results for clusters ?

ANDREONI - Yes, I think so. The semi-empirical potentials I have mentioned were obtained from fitting to a very limited data base taken mostly from the bulk. The chemical bonding in small clusters is very different from that in the bulk.

BESWICK - Could you comment on the advantages of the Carr-Parinello method as compared with other ab-initio approaches ?

ANDREONI - There are several advantages. In the study of structural properties, it does not rely on a specific symmetry and also allows the use of efficient optimization procedures such as simulated annealing. It also allows the study of finite-T properties and dynamics with the same accuracy as T=0 properties.

WONG - Do you expect any temperature-induced phase transformation in a strongly bonded system like C_{60} or C_{70} ?

ANDREONI - In C_{60} we observe very high stability of the molecular structure.

BIANCONI- Upon doping by B, $C_{59}B$ shows the formation of an impurity state. How is this electronic state localized close to the impurity B atom ? How does it depend on the C_{60} crystalline structure deformation ?

ANDREONI - The impurity state is localized on the carbon atoms in the near environment of the impurity. The "degree" of localization is not affected by the ionic relaxation. Also the "binding energy" of the impurity level turns out to be practically independent of the ionic relaxation.

SPECTROSCOPY OF MOLECULAR CLUSTERS WITH SYNCHROTRON RADIATON

B. Brutschy
Freie Universität Berlin, Institut für Physikalische Chemie und Quantenchemie, Takustr 3, 1000 Berlin 33

ABSTRACT

This paper focuses on various aspects of chemical reactivity of ionized molecular clusters. By using synchrotron radiation as a light source their energetics and dynamics may be studied both by classical photoionization mass spectrometry and by TPEPICO. In addition to thermochemical data on the ionic states of clusters, examples of intra-cluster ion-molecule and charge-transfer reactions are presented. This contribution tries to elaborate the specific features of chemical reactions in clusters as compared to those in the gas phase.

INTRODUCTION

In recent years molecular clusters have attracted considerable interest as model systems to study the physical and chemical properties of molecules in the transition from gas to condensed phase. By varying the number and composition of the molecular subunits in a complex and by controlling the amount of energy deposited in it the intra- and intermolecular factors governing chemical reactivity may be disentangled /1-4/. In the ionization of clusters two questions need to be unraveled, 1. the energies at which a size-selected ionic complex is produced and 2. the energies at which possible reaction channels open up. A cluster may thus be viewed as a microscopic test tube for observing complex reaction behavior.

Since neutral clusters, produced in supersonic beams, are only available in more or less broad distributions of size and composition, whose widths may be varied by changing the expansion parameters, the characterization and selection of a specific model cluster requires ionization with mass spectrometric detection. As an ionized cluster in general undergoes consecutive relaxation processes such as chemical reactions or fragmentation of molecular subunits /5/ the expected information on the size of the parent cluster is often lost or falsified by the detection process itself. The problem is hence to retrieve ionic pre-cursors by characteristic properties like appearance potentials (APs) or breakdown (BDs) curves.

Up to now, in spite of the development of several sophisticated techniques /5,6/, no universal method for characterizing an ionic (neutral) precursor is available. Although lasers are nowadays very powerful and popular light sources for photoionization of molecules and molecular clusters /4/, synchrotron radiation (SR) still offers some hitherto unequalled advantages over competing ionization sources. Its wide tunability, extending from the visible into the far vacuum ultraviolet (VUV), and the availability of high flux, high resolution monochromators allow one to study clusters with high energy resolution and over a wide energy range. Several groups are using this tool for cluster studies /6-8/.

In the following a short summary of the development of this field will be given. The selection focuses on studies in the valence shell region of molecular clusters and tries to elaborate the specific features of their ion-chemistry.

EXPERIMENTAL SETUP

A typical experimental setup /9/ for photoionization mass spectrometry of clusters is shown in Fig. 1.

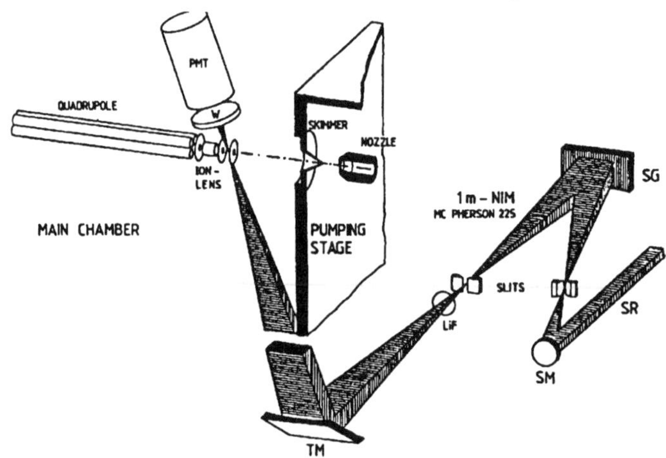

Fig. 1 Schematic view of the experiment. (SM,(TM) sperical (toroidal) mirror, PMT= photo multiplier). Typical beam parameters: p_0= 2 bar, nozzle diam.d=80 μm, T_0= 300 K

The clusters are synthesized in an adiabatic expansion of the compounds of interest, either pure or dissolved in a rare gas. The latter method, known as the "seeded beam technique" is important for cooling down larger molecules. Monochromatized SR, in our case from the electron storage ring BESSY in Berlin, is focused onto the skimmed, supersonic beam and the ions produced by photoionization (PI) are mass analyzed in a quadrupole mass spectrometer. The ion count rate is normalized to intensity variation of the light beam by measuring the spectral photon flux in the ionization volume, which is typically 10^8-10^9 photons/s Å. Second order and higher order light, a typical problem of SR sources, is suppressed in the spectral region up to 11.9 eV by a lithium fluoride cut-off filter (LiF) (transmission T=80%) and from 11.9 up to 18 eV by an indium filter (T= 10%).

Alternatively, several groups use a TOF mass spectrometer either of the Wiley/McLaren /10/ or of the reflectron design /11/. Since storage rings are operated for most of the time in multi bunch mode the SR is quasicontinuous. Therefore a pulsed time structure must be impressed onto the ion detection by pulsing the acceleration field. In our group we use a setup depicted schematically in Fig. 2.

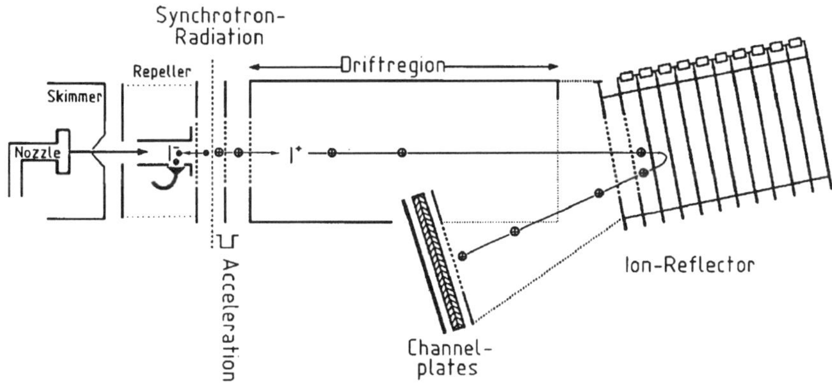

Fig. 2 Schematic view of the TPEPICO measurement unit.

Both photoelectrons and photoions are analyzed parallel to the molecular beam.With tunable SR one preferentially measures electrons of nearly zero kinetic energy /12/. These threshold photoelectrons (TPEs) are separated from the others by a modified steradiancy analyzer /12/ with an off-axis channeltron. To demonstrate the energy resolution Fig. 3 shows the $^2P_{3/2}$, $^2P_{1/2}$ doublet of Ar^+. While the collection efficiency of the analyzer is excellent and the energy resolution quite satisfactory, a certain percentage of fast electrons may also pass it. The ions are mass selected in a homemade reflectron TOF mass spectrometer. The advantage of this design over the classical one of Wiley/ McLaren is a better mass resolution ($m/\Delta m \approx 400$) and the possibility of distinguishing the metastable fragment ions from their unfragmented precursors. The masses of the metastables may easily and very accurately be determined by using the deceleration field in the ion mirror as a high-pass energy filter. The TPEs are accelerated by a field of 2 V/cm and the ions by a fast high voltage draw-out pulse of 250 V and a second static acceleration field of 500 V/cm. The pulser is triggered by the threshold electrons. The mass spectra are recorded by a fast multi-hit TOF analyzer (5 ns/channel) and averaged in a multichannel averager. In order to normalize the TOF mass spectra to intensity variations of the light, the accumulation time is controlled by a preset counter, which measures a fraction of the photon counts.

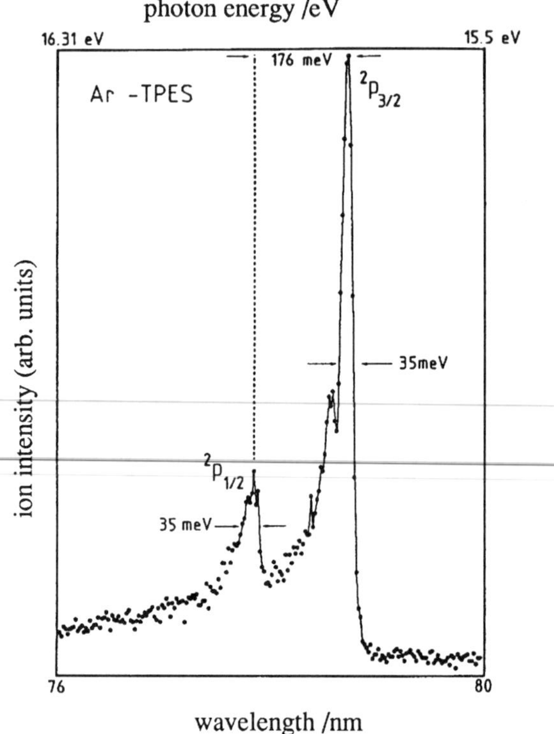

Fig.3 Threshold photoelectron spectrum of Ar($^2P_{3/2}$, $^2P_{1/2}$) measured with the steradiancy analyzer.

RESULTS

1) Energetic information on clusters and thermochemical data.

The first values to be determined in photoionization mass spectrometry of clusters with SR are the ionization (IPs) and fragmentation appearance potentials (FAPs). By varying the photon energy the photoionization efficiency curve (PIE) of each product ion may be determined and the thresholds for its formation derived /1,3/. From the APs and by simple Born-Haber cycles the values of the ionic and neutral dissociation energy D_n^{+0} and D_n^0 respectively, may be derived. They may be compared with solvation enthalpies as determined from ion-molecule equilibrium measurements /13/. While in the past there was some confusion about the adiabaticity of the measured appearance potentials, it is meanwhile generally accepted /14/ that the vertical potentials are most often considerably lower than the adiabatic values. Therefore the thermochemical data derived from photoionization work are lower bounds to the adiabatic values /3/ and are often

too small. The reason for this are most often poor Franck-Condon factors for an adiabatic transition from a weakly bound van der Waals (vdW) to the considerably more strongly bound ion-induced-dipole potential minimum. Hence the ions produced are often vibrationally hot.

A cluster may get rid of excess energy by its partial evaporation (vdW fragmentation) via predissociation. Fragmentation typically sets in 0.1 -0.5 eV above the ionization threshold. For a near adiabatic transition an additional amount of excess energy in the range of the neutral binding energy D_n^0 is necessary for the dissociation of one molecule. Due to this fragmentation, the PIE curves are often pressure dependent. Various examples have already been published in the literature /4,15-17/. Hence the PIE curves are at best free from fragmentation near the thresholds. Of course the cluster size distribution is a crucial parameter determining the absolute amount of the contributions dueto fragmentation.In vertical transitions far from the adiabatic limit fragmentation may be observed even at threshold, if the initial excess energy exceeds D_n^0. Therefore measurements must be made for different expansion conditions, i.e. for different cluster size distributions. Often the onset region of the PIE-curves is rather broad, reflecting the non-adiabatic character of the ionizing transition or the existence of isomers. Hence conservative error limits of the AP values are typically 0.05 - 0.1 eV.

In general the IPs of a cluster system decrease with the number of subunits, the change being pronounced for small clusters /9/. While in heterogeneous systems the additional stabilization of the ion is due to polarization forces, in homogeneous clusters, charge exchange forces may come into play. Recently Greer at al./18/ proposed a Hückel-type model for the size dependence of the IPs of the N_2O cluster system based on a hole hopping mechanism similar to the small polaron model.

Besides vdW fragmentation and direct ionization the clusters rearrange in the ionized state by chemical reactions. For polar molecules, particularly those with hydrogen bonds (H-bonds) a dominant process is proton transfer (PT) /1,2,3/. With cluster photoionization mass spectrometry PT was studied in a pioneering work by Y.T. Lee's group /19/. From this type of measurement lower bounds to the absolute proton affinities of clusters may be derived, allowing one to fix the rather uncertain absolute PA ladder, as constructed from relative PA values from PT equilibria. These vertical values represent the adiabatic proton affinities, as long as the proton transfer takes place without any additional barrier viz. by direct dissociation of a primarily formed (protonated ion• neutral radical) -complex. This being taken for granted for the dissociation of $NH_4^+ \bullet NH_2$, the PA value of ammonia (203 kcal/mol) as derived by Ceyer et al. /19/ and later reproduced by Kamke et al. /8/ for a long time was accepted as a genuinely adia-

compound	n	PA_{ver} ± 0.15/eV	$PA_{/20/}$	$PA_{/21/}$
$(CH_3NH_2)_n$	1	9.64	9.28	9.53
	2	9.95		
	3	10.00		
$(CH_3CH_2NH_2)_n$	1	9.74	9.41	9.68
	2	10.1		
	3	10.3		

Table 1 The absolute proton affinities of some alkyl amines clusters, determined by using PI mass spectro metry /19/

batic value. The absolute PA values for some alkyl amines (Table 1), determined by Bisling et al./9/ with the same method, exceeded however those of the ammonia based PA scale of Lias et al /20/.by about 6 kcal/mol. Since the values derived in PI are lower bounds to the adiabatic values, this meant that the Lias/Ceyer value of ammonia was several kcal/mol too low.In a recent MD study, Greer et al. /38/ could give a possible an explanation for this discrepancy. Meanwhile Meot-Ner et al. /21/ have reevaluated the upper range of the proton affinity scale from the enthalpies of proton-transfer equilibria, by using variable-temperature pulsed high-pressure mass spectrometry. These new values are in good agreement with that from the cluster photoionization work. It is not yet understood, why the proton affinity values of the alkyl amines are obviously very close to their adiabatic values, while that of ammonia is not.

Table 1 also illustrates the increase of the PA with cluster size. This may be explained by an increased stabilization of the additional proton by polarization forces. As was shown by laser induced resonant two photon ionization /22/, this increase plays a crucial role in the cluster size dependence of PT reactions in mixed clusters.

2) Ion-molecule reactions in clusters.

The study of reaction products from ionized vdW clusters opens the possibility to study reactions in an environment of solvent molecules. Since the latter may act as a heat bath, product ions, normally not found in a mass spectrum, may be stabilized. Hence clusters are hoped to be more realistic models for chemistry in the condensed phase. Another difference to reactive scattering in the gas phase is the greatly enhanced interaction time, as compared with that in more or less short lived collision complexes. This is due to the fact that the reacting species are bound. Therefore the reaction efficiency is generally considerably enhanced over that in bimolecular gas phase ion-molecule reactions (IMRs), as was proved recently in a comparative study of nucleophilic reactions of aromatic radical cations with ammonia /23/. From these differences one anticipates being able to observe reactions in clusters, which although exothermic are not observed in a collision complex. An example are the product ions formed in the methane cluster system as reported by Ding et al. /24/.

In the ion spectrum several mass series appear, corresponding to an ionic fragment X^+ solvated by neutral molecules S ($X^+ \cdot S_n$). The series observed are:

1) $CH_4^+ \bullet S_n$ 4) $C_2H_5^+ \bullet S_n$

2) $CH_5^+ \bullet S_n$ 5) $C_3H_9^+ \bullet S_n$

3) $C_2H_4^+ \bullet S_n$ 6) $C_2H_7^+ \bullet S_n$

Some of the solvated ions listed above are also produced in analogous IMRs in the gas-phase:

7) $CH_4^+ + CH_4 \longrightarrow CH_5^+ + CH_3$ - 6.2 kcal/mol

8) $CH_3^+ + CH_4 \longrightarrow C_2H_5^+ + H_2$ - 27.1 kcal/mol

9) $CH_4^+ + CH_4 \longrightarrow C_2H_4^+ + 2\,H_2$ - 2.1 kcal/mol

10) $CH_3^+ + CH_4 \longrightarrow C_2H_7^+$ - 36.9 kcal/mol

The reactions 9and 10 are energetically allowed in the energy range of the PI but not observed in the gas-phase presumably due to a more complex reaction dynamics caused by unfavorable reaction barriers or by symmetry selection rules. Their rates may dramatically increase upon the addition of solvent molecules. Reaction 10 for example is an exothermic condensation reaction which only requires a third body for the disposal of the condensation energy.

Similar observations were made for the methyl fluoride cluster system /25,13,4/. Here the ionic reaction dynamics is obviously dominated by a fast proton transfer in dimeric and trimeric reaction centers. The subsequent rearrangement takes place by ejection of HF, H_2 or other stable neutral fragments. Again reaction products appear which should be formed in exothermic IMRs but are never observed.

By measuring accurate APs of the product ions usung SR, their formation enthalpies may be derived and the number of possible structures of the product ions may be reduced /13/. Often one may also derive information on the transition states of IMRs.

Going from homogeneous to heterogeneous clusters opens up additional possibilities. Grover et al /7 / for example studied mixed clusters comprised of a hydrocarbon plus a simple di- or triatomic inorganic molecule, with the ionization potential of the organic being substantially lower than that of the inorganic. In one example very interesting reaction behavior was observed for a C_6H_6/HCl mixture. At a photon energy of 14.58 eV $C_6H_6Cl^+$ appeared in the mass spectrum and was assigned as a cationic Wheland or σ-complex,

belonging to an important class of transient intermediates in aromatic substitution reactions. From the pressure dependence of this ion and from the correspondence of the AP with the excitation of the B_{2u} state of benzene$^+$, these authors assigned the trimer, $(benzene)_2^+\cdot$ HCl, as precursor and supported this assignment by a number of ancillary measurements. The structure and formation energetics of the complex urged the assumption of the following sequential mechanism:

$$C_6H_6\cdot C_6H_6^+(B_{2u})\cdot HCl \longrightarrow C_6H_6\cdot C_6H_6^+\cdot Cl + H$$

$$\downarrow$$

$$[C_6H_6Cl^+\cdot C_6H_6]^* \longrightarrow C_6H_6Cl^+ + C_6H_6$$

The extra benzene molecule plays a crucial role, since the excited Wheland intermediate would dissociate promptly and thus be unobservable in the mass spectrum. The same group also studied reactions of clusters of 1,3 butadiene with sulfur dioxide /26/ and nucleophilic substitution in ($C_3H_4\cdot$HCl) clusters /27/.

Another example where clusters obviously act as heat bath or even as a cage is the system of ionized carbon tetrafluoride. The molecular ion CF_4^+ had not been observed until recently by means of photon or electron impact ionization mass spectrometry as it dissociates at the threshold to CF_3^+ + F /28/ . The first signs of its existence were found -although in negligible abundance - by Kime et al /29/ in the mass spectrum of CF_4 after ionization by electron impact. The cluster ion system of this molecule, studied recently by Hagenow et al /30/ with both electron impact and photon ionization showed quite large abundances for the molecular ions $(CF_4^+)_n$ n≥1 (Fig. 5).

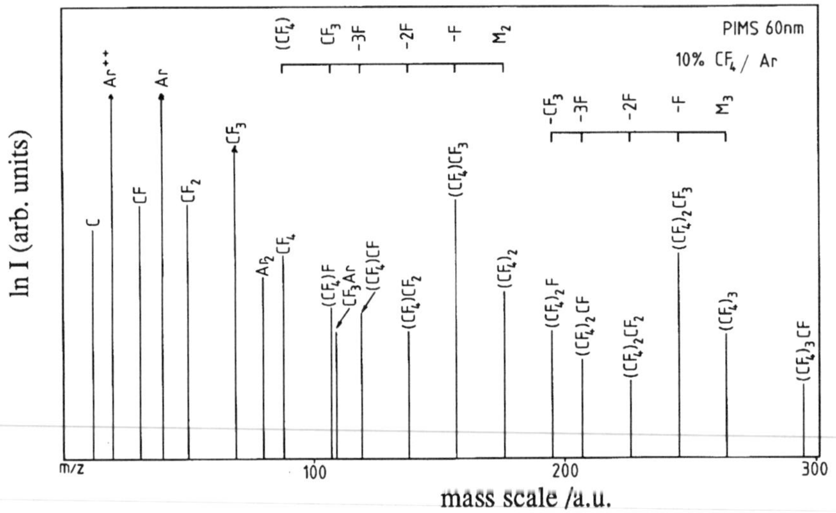

Fig. 4 Photoionization mass spectrum of an expansion of CF_4 in Ar.

The AP of CF_4^+, measured with PI, is 15.77 eV, i.e. very similar to the AP of CF_3^+ with a value of 15.5 eV. The dependence of its intensity on the stagnation pressure in the beam source was very similar to that of a dimer or of larger complexes.

The stability of molecular cations with higher symmetry is often strongly reduced by the Jahn-Teller (JT) effect. CF_4^+ has a threefold $^2T_1$ lower state, whose T_d symmetry may be reduced by JT distortions. For CF_4^+, these have been observed to be small, when compared with other tetrahedral ions /31/. Therefore the rapid dissociation was assigned to a repulsive potential in the Franck-Condon region of the ionization transition. Recent ab initio calculations by Garcia de la Vega and San Fabian /32 / support this assumption. They found for the most stable structure of CF_4^+ a C_{3v} symmetry with a very shallow potential minimum along the C-F bond corresponding to the dissociation coordinate of CF_4^+ into CF_3^+ + F . This minimum was unstable when the zero point vibrational energy was considered. According to these calculations in a vertical ionization transition there should be an excess energy of about 2 eV, which has to be dispersed in a cluster, presumably by evaporization of several molecular subunits. Hence the assumption of a dimer precursor would certainly not explain such efficient energy transfer.

While the molecular ion was observed easily after photoionization and mass analysis in a quadrupole, we found until now no CF_4^+ in the TOF mass spectrum measured with the reflectron. However the CF_3^+ ion showed a slightly delayed peak which could possibly be due to the fragmentation of a CF_4^+ ion in the acceleration region. If this assignment is correct -and presently ancillary measurements to examine this are under way - this would indicate that the CF_4^+ ion is metastable, dissociating on a microsecond time scale in accordance with the theoretical predictions.

3.Electron transfer in mixed molecular aggregates

After a molecular subunit is ionized it may be reneutralized by charge transfer from an associated neutral molecule if the reaction is exothermic.Since ET is a resonant process, the resonance condition is in general satisfied by the large amount of intra- and intermolecular ionic states. Since mixed clusters are bound after a vertical ionization transition - or at least are stabilized by vdW fragmentation- the reagents may interact for a much longer time than in a short lived ion-molecule collision complex. Hence the general remarks on ion-molecule reactions in clusters made in the previous section should also be applicable for electron transfer reactions.

In resonant two-photon ionization with lasers, where the primary formed ion may be selected by the resonant step in the ionization transition, ample evidence has been found for the great importance of electron transfer /4/. Since ET is energetically allowed if the ionic ground state of the acceptor lies in the ionization continuum of the donor, both ET and direct ionization may take place simultaneously and are therefore difficult to disentangle by classical one-photon ionization mass spectrometry. In that case the PIE curve of the ionic donor is always a superposition of the PIE curves of both channels. One way to avoid such ambiguities is to define the internal energy of the primary photoion produced by measuring the energy of the photoelectron. This is done in Photoelectron-Photoion-

Coincidence (PEPICO) measurements /12,28/.With freely tunable SR one preferentially measures the electrons with nearly zero kinetic energy (threshold electrons), known by the acronym TPEPICO (Threshold PEPICO). The ion efficiency curves normalized to the total ion yield are the breakdown curves /28/.They generally reflect the photoelectron spectra of the primary formed ion and may be viewed as the ionic fingerprint of the primary ion.

If in a cluster a reaction takes place the product ion should exhibit the breakdown curves of its ionic precursor, thus allowing a better characterization of the initial state of reacting complex.

As an example ET in the mixed cluster system Ar/CO_2 will be discussed in the following. Fig.5a shows the TPEs spectrum of CO_2, measured for an effusive beam at room temperature using a high resolution electrostatic electron energy analyzer/33/, while Fig.5b shows that of a supersonic beam of pure CO_2, measured with a steradiancy analyzer /34/.

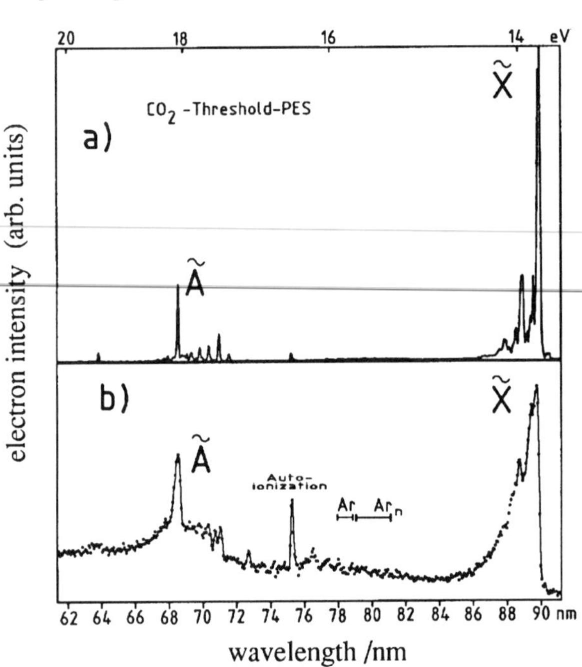

Fig. 5 Threshold photoelectron spectrum of CO_2,measured for
a an effusive beam (electr.analyzer)
b)a supersonic beam of pure CO_2, (steradiancy analyzer)

Since the latter spectrum represents the total electron count rate at various photon energies, part of the unstructured bands are due to electrons from clusters and part from hot electrons. The first two ionic states are the X 1 $^2\Pi_g$ at 13.78 eV, and the A 1 $^2\Pi_u$ at 17.59 eV /35/. In the energy region of the X state the cluster mass spectrum measured with TPEPICO shows a good intensity. Fig.6 depicts the dependence of the breakdown curves in the vicinity of the X state, measured for a pure expansion of CO_2. The maximum of the curve shifts with increasing number of molecules to lower energies, reflecting the stabilizing action of electrostatic forces. In the gap between the X and A state the ionization efficiency is very low as expected from the ionization cross section of the monomer.

If however CO_2 is expanded in Ar one observes in the region of the Ar cluster absorption a considerable intensity increase in the $(CO_2)_n^+$ cluster spectrum. The break-down curve of several ions, not normalized to the total ion intensity, demonstrates this(Fig.7).

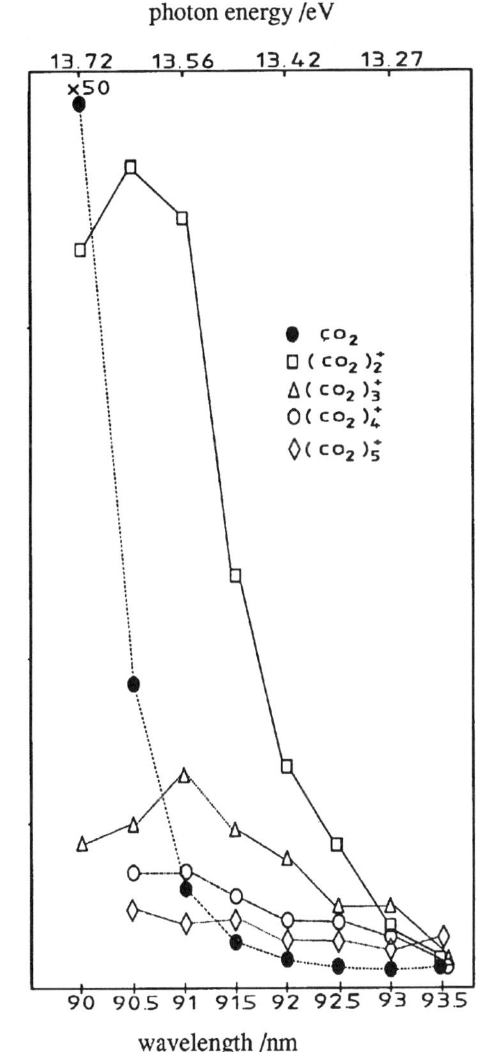

Fig.6 Breakdown curves for some ions from a pure expansion of CO_2 in the region of the first ionic state (X), (curves not normalized to the total ion intensity)

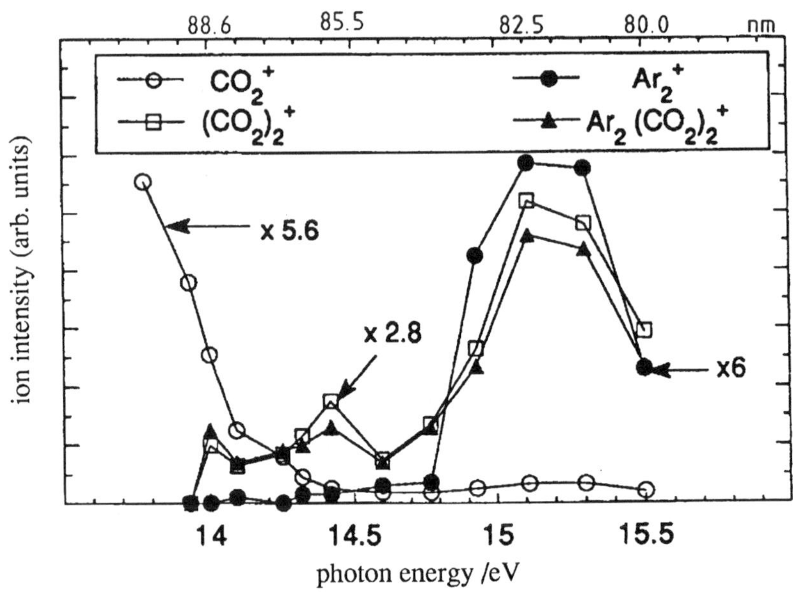

Fig. 7 Breakdown curves for some ions from a mixed expansion of 2% CO_2 in Ar, measured in the region between the X and A ionic states .

Although the scanning width is still rather broad, one clearly observes an enhanced ionization probability in the efficiency curve of $(CO_2)_2^+$ between 14.7 and 15.5. For comparison the breakdown curve of the same ion from a pure CO_2 beam shows no such increase. On the other hand, the argon clusters show a similar resonance in their breakdown curves. It corresponds to the unresolved cluster analogue of the monomer's ion doublet $^2P_{3/2}$ $^2P_{1/2}$ (Fig. 3) shifted considerably to the red. From this agreement we assume larger argon clusters as chromophores which after their ionization are neutralized by an exothermic ET from CO_2 clusters subunits. In some of the aggregates the excess energy is enough to allow boiling off of all the argon atoms. Although the collection of data is not yet complete the results are encouraging. The method may provide new insight into complex intracluster ion chemistry.

If in a cluster a molecule is electronically excited to a state, with an energy exceeding the IP of another associated molecule, autoionization may takes place. This process, traditionally called Penning ionization, was identified as an electron transfer from the donor into the orbital from which the excited electron was promoted /3,4/. In clusters a similar process is observed /36/. A nice example of the transition of the electronic properties of clusters to that of the corresponding solid phase is the observation of resonances in the PIE curves of molecules expanded in Ar. The dependence of these resonances on the expansion parameters,as measured by Kamke et al. /36/, revealed intracluster Penning ionization at different cluster size distributions. Under beam conditions, where only small mixed aggregates are formed, sharp resonances appear in the PIE curve of the heterogeneous aggregates, red-shifted relative to the resonances of the free atom. For expansion conditions favoring large Ar aggregates the shift turns to the blue. The PIE curve of the benzene cation, from a mixed benzene/argon expansion, shows for example resonances assignable to the surface excitons of larger argon clusters /37/. Hence they reflect an exciton induced intracluster dissociative Penning ionization of benzene molecules, attached to the surface of these Ar substrate clusters.

SUMMARY AND CONCLUSION

This survey mainly focused on the chemical aspects of ionized molecular clusters. By this selection very interesting work, e.g. on the fragmentation dynamics of clusters and on other more physical questions has been omitted. From the early mass spectrometric studies of the energetics of ion production more sophisticated methods are now under way, by which a better characterization of the intracluster processes should be possible.Many investigations on clusters suffer presently from low ion intensity due to low cluster concentrations and the still moderate intensity of the present SR sources. In the near future high brilliance wiggler/undulator sources will increase the sensitivity by up to two orders of magnitude allowing new experimental techniques like laser/SR coupling for pump/probe experiments or multiphoton spectroscopy.

FEFERENCES

/1/ C.Y. Ng; Adv. Chem. Phys. **52**, 263 (1983)

/2/ T.D. Märk and A.W. Castleman Jr.; Adv. At. Mol. Phys. **20**, 65 (1985)
/3/ B. Brutschy, P. Bisling, E. Rühl and H. Baumgärtel; Z. Phys. D **5**, 217 (1987)
/4/ B. Brutschy J. Phys. Chem. **94**, 8637 (1990) and references therein
/5/ U. Buck; J.Phys. Chem. **92**, 1023 (1988)
/6/ E. Holub-Krappe, G. Ganteför, G. Bröker and A.Ding; Z.Phys. D **10**, 319 (1988)
/7/ E. A. Walters, J.R. Grover, M.G. White and E.T. Hui; J.Phys. Chem. **91**, 2758 (1987)
/8/ W. Kamke, R. Herrmann, Z. Wang and I.V. Hertel; Z. Physik. D **10**, 491 (1988)
/9/ P.G.F. Bisling, E. Rühl, B. Brutschy and H. Baumgärtel; J. Phys. Chem. **91**, 4310 (1987)
/10/ W.C. Wiley and I.H. McLaren; Rev. Sci. Instrum. **26**, 1150 (1955)
/11/ U. Boesl, H.J. Neusser, R. Weinkauf and E.W. Schlag; J. Phys. Chem.**86**, 4857 (1982)
/12/ T. Baer; Adv. Chem. Phys. **64**, 111 (1986)
/13/ E. Rühl, P. Bisling, B. Brutschy and H. Baumgärtel; J. Electron. Spectrosc. Relat. Phenom **41**, 411 (1986)
/14/ W. Kamke, R. Herrmann, Z. Wang and I.V. Hertel; Z. Phys. D **10**, 491 (1988)
/15/ P.M. Dehmer and S.T. Pratt; Chem.Phys.**75**,5265 (1981)
/16/ E. Rühl, B. Brutschy and H. Baumgärtel; Chem. Phys. Letters **157**, 379 (1989)
/17/ J. Krauß, J. de Vries, H. Steger, E. Kaiser, B. Kamke and W. Kamke; Z. Phys. D, in press (1991)
/18/ J.C. Greer, W. Gotzeina, W. Kamke, H.Holland and I.V. Hertel; Chem. Phys. Letters **168**, 330 (1990)
/19/ S.T. Ceyer, P.W. Tiedemann, B.H. Mahan and Y.T. Lee; J. Chem. Phys. **70**, 14 (1979)
/20/ S.G. Lias, J.F. Liebmann, R.D. Levin; J. Phys. Chem. Ref. Data **13**, 685 (1984)
/21/ M. Meot-Ner and L.W. Sieck; J. Am. Chem. Soc. **113**, 4448 (1991)
/22/ B. Brutschy, C. Janes and J. Eggert; Ber. Bunsenges. Phys. Chemie **92**, 74 (1988)
/23/ Ch. Riehn, Ch. Lahmann and B. Brutschy; sumitted to J.Am. Chem.Soc.

/24/ A. Ding, R.A. Cassidy, J.H. Futrell and L. Cordis; J.Phys. Chem. **91**, 2562 (1987)

/25/ J.F. Garvey and R.B. Bernstein; J. Am. Chem. Soc. **109**, 1921 (1987)

/26/ J.R. Grover, E.A. Walters, J.K. Newmann and M.G. White; J. Am. Chem. Soc. **112**, 6499,(1990)

/27/ E.A. Walters, J.R. Grover, D.L. Arneberg, C.J.Santandrea and M.G. White; Z. Phys. D **16**, 283 (1990)

/28/ J.H.D. Eland;"Photoelectron Spectroscopy", Butterwoths, London

/29/ Y.J. Kime, D.C. Dricoll and P.A. Dowben; J. Chem. Soc. Faraday Trans. II **83**, 403 (1987)

/30/ G.Hagenow, W. Denzer, B. Brutschy and H. Baumgärtel; J. Phys. Chem. **92**, 6487 (1988)

/31/ C. A. Coulson and H.L. Strauss; Proc. Roy. Soc. A **269**, 443 (1962)

/32/ J.M. Garcia de la Vega and E. San Fabian; Chem. Phys. **151**, 335 (1991)

/33/ spectrum by K. Hottmann, Freie Universität Berlin

/34/ W. Denzer doctor thesis, Freie Universität Berlin (1991)

/35/ Kimura et al., "Handbook of the He I Photoelectron Spectra of Organic Molecules",Halsted Press,New York

/36/ B.Kamke, W. Kamke, Z. Wang, E. Rühl and B. Brutschy; J. Chem. Phys. **86**, 2525 (1987) and references cited

/37/ J. Wörmer, Gruzielski, J.Stapelfeldt, G. Zimmerer and T. Möller; Phys.Scr. **41**, 490 (1990)

/38/ J.C. Greer, R. Ahlrichs and I.V. Hertel; Z. Phys. D **18**, 413 (1991)

ACKNOWLEDGMENT

The experiments at BESSY were funded by the "Bundesministerium für Forschung und Technologie " (project #: 05313FX B), by the "Freie Universität Berlin" and by the "Fonds der Chemischen Industrie". Their support is gratefully acknowledged. I also would like to express my gratitude to Prof.Dr.H. Baumgärtel for fruitful discussions and permanent support and to my colleagues P. Bisling, W. Denzer, M. Fieber, H. Hofmann, G. Hagenow, E. Rühl and U. Rockland, for their engaged collaboration in the project.

QUESTION

BAER - It is interesting that you observe stable CH_4^+ when it is produced from the dimer or higher order clusters. What is known about the geometry and the binding energy of the CH_4^+ ?

BRUTSCHY - In an ab-initio calculation on the geometry of CH_4^+ by Garcia de la Vega et al. (Ref. 32)., the minimum energy geometry had a C_{3v} geometry with a very shallow minimum along the C-I axis, along which probably dissociation takes place. Obviously, the structure is $CF_3^+ + F^\bullet$ with weak bonding energy. In the calculation no bond state was found.

ELAND - What is the origin of the peak broadening that you reported in the reflectron mass spectrum of CF_4 ?

BRUTSCHY - The peak-broadening is due to fragmentation of CF_4^+ in the acceleration region into $CF_3^+ + F^\bullet$. Therefore in the reflectron the lifetime is in the submicrosecond region, but not in ns-region.

DUJARDIN - I could not understand very well how you can discriminate between surface and bulk excitons in clusters.

BRUTSCHY - Surface excitons in the bulk of Ar are blue-shifted relative to the atomic resonances by 90 meV while bulk excitons are shifted by 500 meV (Schwentner et al). The shift of the Penning structure in Benzene+ is 110 meV. Möller et al also found these less blue-shifted resonances for Ar-clusters and assigned them to surface excitons.

UKAI - About Penning ionization : is it possible to determine the auto-ionization width from your experimental results ? My continuing question about the Penning collision is what amount the auto-ionization width is.

BRUTSCHY - The width of the auto-ionization resonances is the experimental width. So we cannot deduce the lifetime, but only give an upper limit.

Electronic Excitations in Rare Gas Clusters: Decay and Relaxation Processes probed with Fluorescence Spectroscopic Methods

M. Joppien, F. Grotelüschen, M. Lengen, R. Müller, J. Wörmer and T. Möller

II. Institut für Experimental-Physik, Universität Hamburg

D-2000 Hamburg 50

ABSTRACT

The electronic structure and relaxation processes of rare gas clusters has been probed with time- and spectrally resolved fluorescence methods. Pure (Ar, Kr and Xe) and doped ($XeAr_N$) clusters ($N = 2-10^6$) exhibit strong absorption bands showing a close similarity to bulk- and surface excitons of rare gas solids. Selective excitation leads to several decay- and relaxation processes (exciton trapping, desorption of electronically excited species) depending both on the cluster size and the character of the primary excited state.

INTRODUCTION

The remarkable progress in the area of supersonic beam techniques [1] provides the basis for the exciting field of the physics and chemistry of clusters that bridge the gap between molecules and condensed matter. Experimental investigations are directed in part to the evolution of energy levels from finite to infinite systems. Van der Waals clusters, which are bound by dispersion forces, are particularly appealing because their elementary excitations delineate from atomic or molecular states. Rare gas clusters (RGC) which are the simplest van der Waals clusters, may therefore play the role of a prototype. Since the electronic excitations of rare gases are located in the VUV-spectral range, excitation with synchrotron radiation (SR) is the first choice for experimental investigations in this field.

So far, a series of investiagations devoted to the evolution of the ionisation potential [2-4] with the cluster size and the fragmentation dynamics [5] are reported, which benefit from the high intensity and the tunability of SR. Here we present a few examples of our recent work on the bound electronic excitations in rare gas clusters located energetically below the ionisation limit. Within the scope of the present paper, we only give a short description of the different excitation channels [6-10] which is followed by a more detailed discussion of various relaxation and decay processes. Time- and energy resolved fluoresence methods with selective photo-excitation are used for the experimental investigation of both the excitation and decay processes of pure and doped rare gas clusters containing up to 10^6 atoms.

EXPERIMENTAL

The experiments were performed at the experimental station Clulu at HASYLAB (Hamburg). Rare gas clusters are prepared in an adiabatic nozzle expansion of pure gas or a gas mixture and excited with tunable SR (energy range 7 eV-20 eV, bandpass between 3 meV and 50 meV). The fluorescence light is either detected directly with different photomultipliers or spectrally analysed with a monochromator which is equipped with a position sensitive detector. This detector is coated with a thin layer of CsI which considerably increases the sensitivity between 6 eV and 10 eV. Additionally, the time structure of SR allows time-resolved measurements in the regime 100 ps - 10 µs. The size distribution and the average cluster size is analyzed with a time-of-flight mass spectrometer in a different set of measurements. For small clusters a correction for the fragmentation process is taken into account[11]. The width (FWHM) of the size distribution roughly corresponds to the average number N of atoms per cluster. In a few cases the width of the size distribution has been reduced by cross jet deflection [12,13]. Two different types of experiments have been performed:

- fluorescence excitation measurements which probe the absorption profiles of clusters

- time- and energy resolved fluorescence measurements for the investigation of decay and relaxation processes.

RESULTS AND DISCUSSION

A main question in cluster physics is the evolution of energy levels from the atom to the bulk limit, e. g. the transition from atomic and molecular Rydberg states to excitons of the solid. Here the discussion focusses on decay and relaxation processes which take place after selective excitation into different excited states. However, we shall start with a short description of the different excitations of RGCs which are necessary for the understanding of the decay processes.

Fluorescence excitation spectra of Kr_N clusters containing between 3 and 10^4 atoms are presented in figure 1 in comparison with the imaginary part of the dielectric constant of liquid Kr [14] in the energy range of the first spin-orbit split atomic resonance line. The fluorescence yield approximately corresponds to the absorption coefficient because dark relaxation channels are of minor importance [6]. Large Kr_N clusters ($N = 10^4$) exhibit several absorption bands which are assigned to surface (s) and bulk (t: transverse; l: longitudinal) n =1 and n = 1' excitons of solid Kr. Since the radius of these tightly bound states is small compared to the nearest-neighbour distance the excitation is similar to an atomic excitation (Frenkel-exciton). With decreasing cluster size the contribution of the bulk excitons decreases. In particular, small clusters

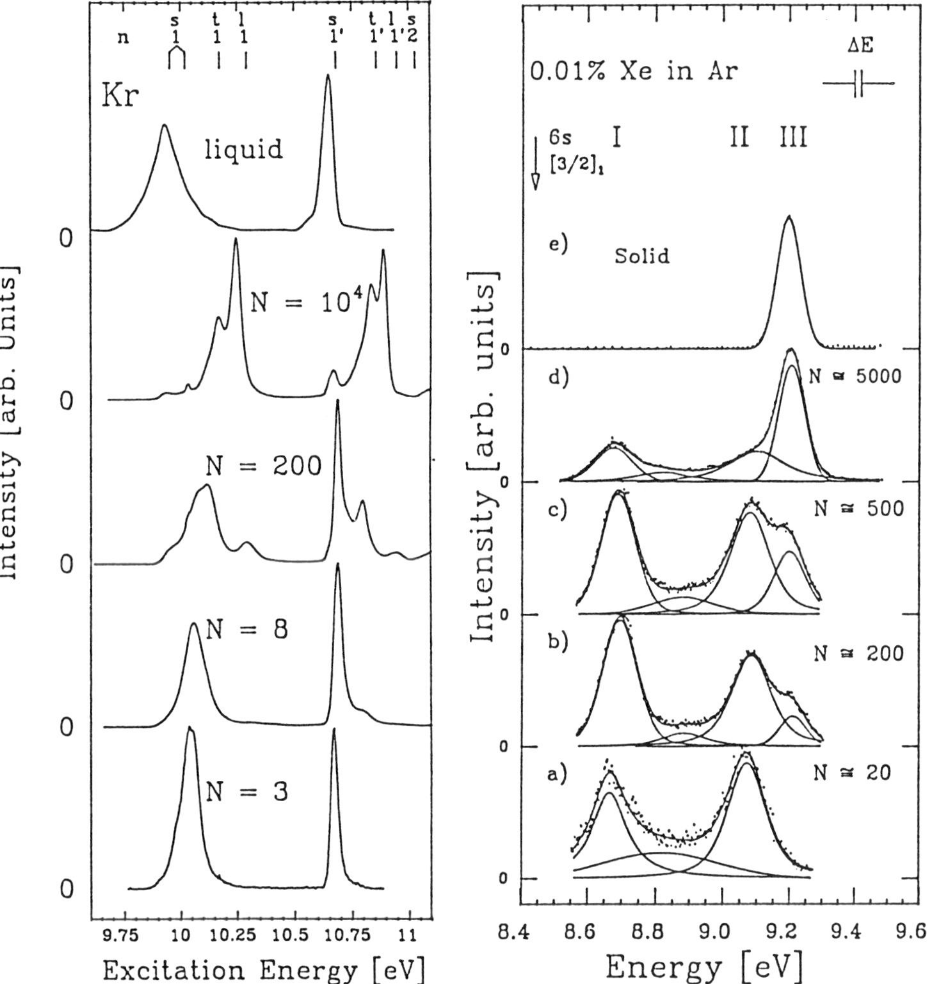

Fig. 1 Fluorescence excitation spectra of Kr_N clusters in comparison with the dielectric function of liquid Kr at 122 K (from ref.14). The energetic positions of excitonic absorption bands (s: surface, t, l: bulk) of solid Kr are indicated at the top of the figure[19].

Fig.2 Fluorescence excitation spectra of $XeAr_N$ clusters in comparison with an absorption spectrum of Xe doped Ar[16]. *The energetic position of the atomic resonance transition is indicated by an arrow.*

containing roughly ten atoms only show surface-type excitations. More quantitatively, atoms located at the surface of Kr_N contribute to surface excitations whereas atoms located in the interior of the cluster contribute to bulk excitations [6]. Similar results are obtained for Ar_N and Xe_N clusters [7,9].

In doped clusters ($XeAr_N$) the situation is somewhat different. Figure 2 presents fluorescence excitation spectra of $XeAr_N$ clusters (N = 20 - 5000) in comparison with an absorption spectrum of Xe atoms embedded in an Ar matrix [16] recorded in the vicinity of the first atomic resonance line $5p \rightarrow 6s[3/2]_1$. Again the fluorescence yield corresponds to the absorption coefficient. In this particular case the first excited state is split into three absorption bands all blue-shifted relative to the atomic resonance line. The band with the largest excitation energy (here labeled III) is identical with the band in Xe doped solid Ar. It is therefore assigned to an excitation of Xe atoms in a substitutional site (twelve nearest neighbours) in the interior of the cluster. Band I and band II are assigned to excitations of Xe atoms *inside* the cluster surface (nine nearest neighbours) and Xe atoms *on top* of the cluster surface (three nearest neighbours), respectively. This assignment is based on the evolution of the band intensity with the cluster size, the energetic shift relative to the atomic resonace line, which is assumed to be roughly proportional to the number of nearest neighbours. Furthermore, this assignment is supported by measurements on pure Ar_N clusters covered with Xe atoms from an atomic cross jet [15].

For the discussion of relaxation and decay processes it is important to keep in mind, that in both cases (pure and doped clusters) atoms located in different positions (bulk, surface) can be selectively excited by the choice of a suitable excitation energy. This is important because relaxation processes are expected to follow different path-ways depending on the character of the primary excited state [17].

Spectrally resolved fluorescence spectra of Kr_2 and Kr_3 recorded after primary excitation (band pass 0.25 nm) in the vicinity of the first atomic resonance line are presented in figure 3. In case of Kr_2, depending on the excitation wavelength, sharp continua are observed, which are assigned to the well-known 'first' continua of the Kr excimer, which result from transitions from vibrationally highly excited levels of the strongly bound excimer states ($A\,^3\Sigma_{1u}$; $B\,^1\Sigma_{0u}^+$) to the repulsive ground state [18]. The life-time of these states is determined to 1.8 ns (singlet state) and 60 ns (triplet state). For primary excitation of Kr_3 broad asymmetric continua are observed. Depending on the excitation wavelength, additionally a small sharp peak is observed, similar to the first continuum of the dimer. The assignment of this emission is not yet clear. The broad continua are interpreted as emissions of the fragment Kr_2 which emits from vibrationally excited levels of the $A\,^3\Sigma_{1u}$ state to the ground state. Time-correlated measurements (see figure 3 c2) support this assignment. The large width of the continua reflects the broad vibrational distribution, indicating that the fragmentation process of Kr_3 takes place on a ps time scale. Furthermore, the long life-time of the emitting state indicates that the triplet state is efficiently populated from the initially excited singlet state.

Fluorescence spectra of Kr_N clusters containing between 2 and 30 atoms

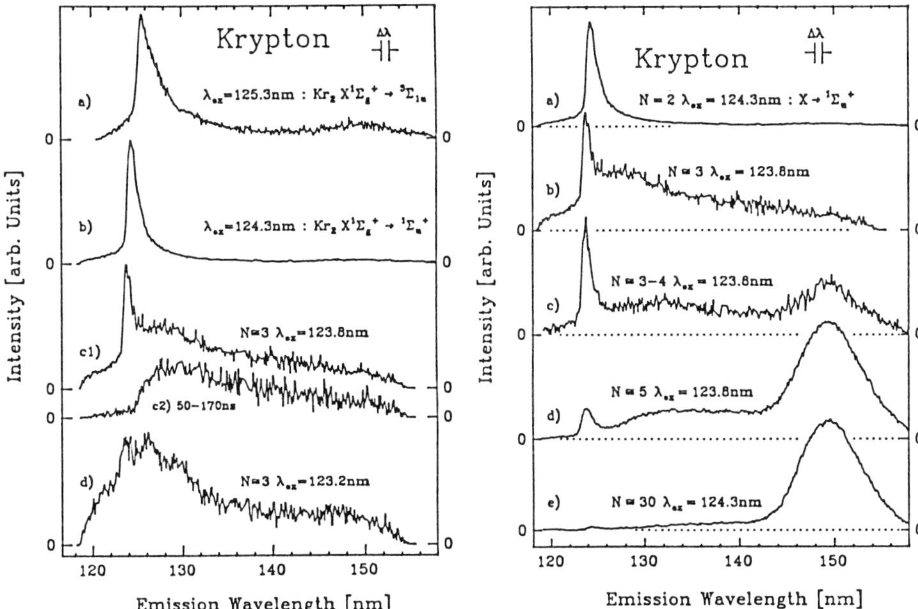

Fig. 3 Fluorescence spectra of Kr_2 (a, b) and Kr_3 (c,d) .The excitation wavelength and the assignement of the transitions is indicated in the figure.The spectrum of Kr_3 is recorded in a mass seperated beam by cross-jet deflection.

Fig. 4 Fluorescence spectra of Kr_N clusters containing between 2 and 30 atoms recorded after primary excitation into the energetically lowest state.

are presented in figure 4. Three emissions are observed, a sharp band at 124 nm, a broad continuum between 124 nm and 150 nm and additionally a continuum with a maximum at ~ 149 nm. The energetic position of the broad continuum between 125 nm and 150 nm slightly changes with the cluster size. As it is the case for Kr_3 it is assigned to emissions of Kr_2 in the $A\,^3\Sigma_{1u}$ state, that means fragmentation takes place and a strongly bound electronically excited dimer leaves the cluster. Consequently, the energetic shift of the maximum to larger wavelengths is interpreted as an increase in vibrational relaxation. It should be pointed out, that this emission is not observed in solid Kr. However, solid Ar exhibits a corresponding band which is interpreted accordingly. In contrast to Ar_N clusters [20],the evaporation of electronically excited dimers from Kr_N clusters is only observed after primary excitation of surface states. With increasing cluster size a broad continuum centered at 149 nm becomes the prominent emission band. This emission is assigned to the so called 'self-trapped' exciton emission which is the dominant decay channel in liquid and solid Kr[19] and large Kr clusters [21]. It is interpreted as the decay from the vibrationally relaxed lowest excimer state of a strongly bound molecular center in the cluster. In large Kr_N clusters both bulk and surface states contribute with nearly identical efficiency to the formation of this state.

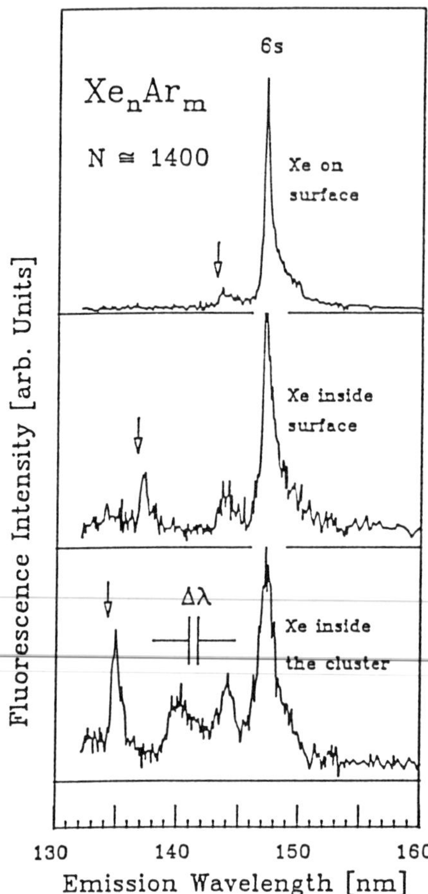

Fig. 5 Fluorescence spectra of $XeAr_N$ clusters recorded after primary excitation of Xe atoms in different sites. The excitation wavelength is indicated by an arrow. Resonant emission are due to scattered light.

The most important difference between pure and doped clusters is that in the latter case the electronic excitation is generally localized at the guest atom or molecule. Consequently, in this case the movement of the electronic excitation is always accompanied with a mass flow. It has been pointed out in the preceding section that it is possible to selectively excite Xe atoms in different sites in Ar_N clusters by the choice of a suitable excitation energy . Fluorescence spectra recorded after primary excitation of Xe atoms *on top* of the cluster surface, *inside* the cluster surface and *in the interior* of Ar_N clusters (N = 1400) are presented in figure 5. In case of excitation of Xe *on top* of the cluster surface, the resolution - limited emission line is assigned to the atomic resonance transition $6s[3/2]_1 \Rightarrow 5p$. It is therefore concluded, that the electronically excited Xe atom leaves the cluster. If Xe atoms located *in* the cluster surface are primarily excited, the emission line becomes slightly broader and shifts of longer wavelength. This indicates, that the electronically excited atom remains weakly bound at the cluster surface. It should be noted, that in this case an excess energy of approximately 0.6 eV must be transfered to the cluster. For the surface state molecular- dynamics calculations for $XeAr_{54}$ state that this relaxation takes place within 2 ps. The excited Xe atom remains at the cluster surface [17], in agreement with our results. Excitation of Xe atoms *in* the interior of the Ar_N cluster results in the emission of three bands. Two of them ($\lambda = 140.4$ nm and $\lambda = 147.2$ nm) are well known in Xe doped solid Ar[22]. In analogy the emission at 140.4 nm is assigned to the decay of Xe atoms in substitutional sites *inside* the cluster, that means the excited atom remains inside the cluster. The other emission bands are assigned to the decay of Xe atoms inside a micro-cavity or at the cluster surface. This comparison shows that the decay processes depend sensitively on the site of the primary excited Xe atom. This general trend is in good agreement with the molecular dynamic calculations [17].

CONCLUSION

Time- and energy resolved fluorescence spectroscopy with VUV-SR excitation is used to investigate the character of electronically excited states and decay processes in rare gas clusters. In pure (Ar_N, Kr_N and Xe_N) and doped ($XeAr_N$) bulk and surface states correlated with the lowest electronically excited state are energetically clearly separated. This offers the opportunity to investigate site - specific relaxation processes. Depending on the cluster size and the character (bulk, surface) of the primary excited state, reactive (fragmentation, predissocation) and non-reactive ('exciton-trapping') decay processes are observed. While in homogenous systems (e.g. Kr_N) the formation of strongly bound excimer centers leads to the evaporation of electronically excited dimers, relaxation processes in doped clusters ($XeAr_N$) are dominated by a strongly repulsive interaction between the excited electron and the cluster. In case of excitation of Xe atoms *on top* of the surface of Ar_N clusters, this repulsion leads to the evaporation of the electronically excited atom. The overall trend is in good agreement with molecular dynamic calculations [17,23].

We are grateful to W. Laasch and G. Zimmerer (Universität Hamburg) for fruitful discussions. Financial support from the Bundesministerium für Forschung und Technologie (BMFT) under grant 05 405 AXB TP 2 is kindly acknowledged.

REFERENCES

1 O.F. Hagena, W. Obert, J. Chem. Phys. 56, 1793 (1972)
2 L. Cordis, G. Ganteför, J. Heßlich, A. Ding, Z. Phys. D 3, 323 (1986)
3 W. Kamke, J. de Vries, J. Krauss, E. Kaiser, B. Kamke, I.V. Hertel,

Z. Phys. D 14, 339 (1989)

4 G. Ganteför, G. Bröker, E. Holub- Krappe, A. Ding, J. Chem. Phys. 91, 7072 (1989)

5 E. Holub - Krappe, G. Ganteför, G. Bröker, A. Ding, Z. Phys. D 10, 319 (1989)

6 J. Stapelfeldt, J. Wörmer, T. Möller, Phys. Rev. Lett. 62, 98 (1989)

7 J. Wörmer, V. Guzielski, J. Stapelfeldt, T. Möller, Chem. Phys. Lett. 159, 321 (1989)

8 J. Wörmer, T. Möller, Z. Phys. D 20, 39 (1989)

9 J. Wörmer, V. Guzielski, J. Stapelfeldt, G. Zimmerer, T. Möller, Phys. Scr. 41, 490 (1990)

10 T. Möller, Z. Phys. D 20, 1 (1991)

11 R. Müller, Diplom-thesis, Hamburg 1990

12 J. Gspann, Ber. Bunsenges. Phys. Chem. 88, 256 (1984)

13 U. Buck, H. Meyer, Phys. Rev. Lett. 52, 109 (1984)

14 P. Laporte, J.L. Subtil, R. Reininger, V. Saile, S. Bernstorff, I.T. Steinberger, Phys. Rev. B 35, 6270 (1987)

15 M. Lengen, Diplom-thesis, Hamburg 1990

16 R. Reininger, private communication

17 D. Scharf, J. Jortner, U. Landman, J. Chem. Phys. 88, 4273 (1988)

18 M. C. Castex, in Spectral Line Shapes, R.J. Exton, ed., p. 289, Deepek Publishing (Hampton 1987)

19 G. Zimmerer, in Excited State Spectroscopy in Solids, U. M. Grassano, N. Terzi eds. , (North-Holland, Amsterdam 1987), p. 37

20 M. Joppien, F. Groteluschen, T. Kloiber, M. Lengen. T. Möller, J. Wörmer, G. Zimmerer, J. Keto, M. Kytka, M.C. Castex, J. Luminesc. 49 & 49, 601 (1991)

21 E.T. Verkhovtseva, E.A. Bondarenko, Yu.S. Doronin, Chem. Phys. Lett. 140, 233 (1987)

22 U. Hahn, R. Haensel, N. Schwentner, Phys. Stat. Sol. b 109, 233 (1982)

23 D. Scharf, J. Jortner, U. Landman, Chem. Phys. Lett. 126, 495 (1986)

QUESTION

ELAND - How do you know how many atoms are in the clusters responsible for the observed fluorescence ?

MOLLER - The size distribution is measured with a time-of-flight mass-spectrometer. For small clusters (N≤100 atoms/cluster) the fragmentation of the clusters has to be considered. Therefore, the average size of neutral clusters is estimated by a calculation of the loss of atoms due to the fragmentation. Additionally, in case of micro clusters containing less than 10 atoms, we make use of mass separation by cross jet deflection according to the ideas of Ospann.

BAUMGARTEL - How is the cluster size measured ? What is known of the size distributions of the doping molecules on the surface of a cluster ?

MOLLER - The size distribution is measured with a time-of-flight mass-spectrometer. This method is sufficient for large clusters (N Larger than hundred). In case of Ar clusters doped with Xe atoms, the number of Ar atoms inside the clusters is additionally estimated by a comparision with equivalent absorption spectra of pure Ar clusters in the spectral range where Ar absorbs. This enables us to account for changes in the cluster size due to a melting of the cluster if Xe atoms are deposited on top of the cluster. The concentration of Xe atoms on top of the surface of Ar clusters can be varied from single, isolated atoms up to nearly closed films by changing the intensity at the Xe cross jet.

CORE LEVEL EXCITATION OF ATOMIC AND MOLECULAR CLUSTERS

E. Rühl, C. Schmale, H.W. Jochims, E. Biller, R. Locht*, A.P. Hitchcock**, and H. Baumgärtel
Institut für Physikalische und Theoretische Chemie, Freie Universität Berlin, Takustr. 3, D-1000 Berlin 33, F.R. Germany

ABSTRACT

Core level excitation of weakly bonded atomic and molecular species is reported. Excitations into unoccupied orbitals with valence or Rydberg character as well as into the corresponding K- and L-continua are compared with the isolated and condensed species. Core level excitation of clusters is also accompanied with extensive fragmentation of singly and doubly charged clusters. Coincidence techniques such as photoelectron-photoion-coincidences (PEPICO) and photoion-photoion-coincidences (PIPICO) are used to study the photochemical fragmentation pathways of core excited clusters. Kinetic energy releases related to charge separation of cluster dications (Coulomb explosion) are derived from PIPICO spectra. The fragmentation energetics is discussed in relation to cluster dication threshold energies.

INTRODUCTION

The field of core excitation processes of clusters is relatively unexplored. Most studies that have been performed in the energy regime of core level excitation (E>150 eV) have concentrated on isolated atoms or molecules in the gas phase and condensed species, i.e. solids or adsorbates.[1,2] Valuable information is obtained from these studies, such as energy positions and relative intensities of resonant processes above or below the corresponding ionization energies. It has been found that the shapes of inner shell absorption spectra of isolated atoms differ from those of the corresponding condensed species.[3] In contrast to this valence transitions of core excited molecular species ($1s\rightarrow\pi^*$) show typically the same energy positions in the gas and condensed phase.[4] However, intensity changes of discrete resonances relative to the corresponding ionization continua are observed.[4] Recently, we have reported progress in the field of core level excitation of isolated clusters in the gas phase.[5] It has been shown that in the case of argon clusters the cation yield spectra of mass selected clusters differ considerably from those of the isolated cation in the L_3/L_2 excitation regime (240-300 eV). Cluster specific resonances are observed, which are more similar to the absorption spectrum of the solid rather than to the atom. Broad features are observed in the corresponding L_3/L_2 continuum, which are assigned as single scattering (EXAFS) processes in clusters. More recent results in

* permanent address: Départment de Chimie Générale, Université de Liège, Sart-Tilman, B-4000 Liège, Belgium
** permanent address: Department of Chemistry, McMaster University, Hamilton, Ontario, L8S 4M1, Canada

this field show that also above the L_1 edge similar EXAFS structure is observed (320-500 eV), which gives further confidence in the given assignments.[6] We have also studied fragmentation pathways of core excited argon cluster dications.[7] With the photoion-photoion-coincidence (PIPICO) technique it was shown that doubly charged clusters undergo asymmetric fission, leading to series of products, such as monomer cations, dimer cations, and trimer cations in coincidence with cluster cations. The kinetic energy releases are significantly lower than expected from a simple electrostatic picture, which is in agreement with findings for collision induced fragmentation carbon dioxide cluster dications, where internal exitation has been assumed to be responsible for the low kinetic energy releases.[8]

In this paper we report recent progress in the field of core level excitation in atomic and molecular clusters in the energy regimes of element K- and L-edges: argon 2p (240-280 eV), carbon 1s (270-330 eV), and oxygen 1s (520-570 eV). Besides photoion yield curves of mass selected cluster cations, which provide spectroscopic information, we have also studied fragmentation reactions of core excited cluster monocations and dications. Threshold energies of unstable argon cluster dications are measured, which are used to interpret the energetics of core excited cluster dication decay.

EXPERIMENTAL

The experiments were performed at the electron storage ring BESSY. The high energy TGM-II beam line allows the dispersion of soft x rays between 150 eV and 600 eV. The energy resolution (ΔE) is obtained from the full width at half maximum signal of the argon (L_3^{-1}, 4s) ion signal at 244.390 eV. The experiments were performed typically with an energy resolving power ($E/\Delta E$) of 300-500.

The TGM-7 beam line is used for threshold experiments of direct cluster double ionization. This monochromator disperses synchrotron radiation between 7 eV and 150 eV. The experiments are performed with a 950 l/mm grating with an optimized dispersion range between 20-120 eV. Higher order light and stray light was filtered by an aluminum foil.

The experimental setup shown in Fig. 1 is similar to that reported earlier.[5] A time of flight mass spectrometer (TOF-MS) is used for cation detection. The design of the mass spectrometer involves conventional Wiley/McLaren space focussing. Under typical experimental conditions a mass resolution ($M/\Delta M$) of 125 is obtained for Ar^+ (m/z=40). The spectrometer is operated with moderate acceleration fields in the ionization region (100 to 200 V/cm).

Opposite to the TOF-MS a threshold photoelectron spectrometer is mounted. This device can be replaced by a total yield detector.

With this setup we have measured mass spectra at resonant excitation energies, total cation and electron yield curves, photoionization efficiency curves of mass selected cations, threshold photoelectron (TPES) spectra, photoelectron-photoion-coincidence (PEPICO) spectra, and PIPICO spectra. Mass spectra were recorded by the use of pulsed extraction of cations into the mass spectrometer, whereas PEPICO and PIPICO spectra were obtained with static voltages. Further experimental details are given in ref. 7.

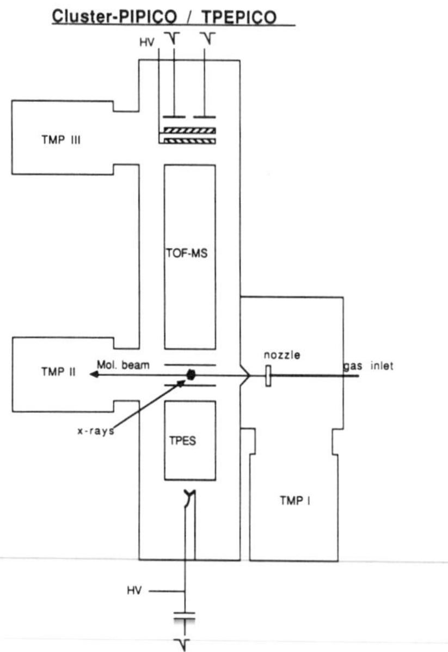

Fig. 1. Experimental setup

The stagnation conditions for the supersonic expansion are similar to those reported earlier.[5,7] We have extended the stagnation temperature range by cooling the gas inlet line and the nozzle down to 90 K with liquid nitrogen. From scaling laws and correlations to average cluster sizes we expect for T_0=300 K an average cluster size ($\overline{N}$) of 2 (p_0= 2 bar) to 10 (p_0=5 bar), depending on the expansion conditions.[5] Kinetic energy releases (KER) for charge separation in cluster dications are calculated according to the formulas given in ref. 9. The samples were of commercial quality and were used without further purification.

RESULTS AND DISCUSSION

I. ATOMIC CLUSTERS

Fig. 2 shows the photoion-photoion-coincidence (PIPICO) spectrum of argon clusters recorded at 248 eV. This energy corresponds to 3d (3/2) exciton state in clusters and the solid.[3,6,7] Three series of charge separation product channels are identifed: Ar^+/Ar_n^+, Ar_2^+/Ar_n^+ and Ar_3^+/Ar_n^+. These series of charge separation show that asymmetric fission of argon cluster dications is dominant. The signal at Δt=0 is shown to be not due to symmetric charge separation processes. However, forthcoming photoelectron-photoion-photoion-coincidence experiments will unequivocally show the fraction of symmetric versus asymmetric charge separation in core excited argon clusters.

The series of coincidences involving cation pairs of dimers and clusters is stronger than the other ones. This is consistent with the high stability of the argon dimer cation. Its formation is a dominant process in single ionization of argon clusters. Recent experimental results from mass resolved molecular beam scattering show that small neutral clusters form preferentially monomer and dimer cations after electron impact ionization.[10a] This is in agreement with photofragmentation studies of mass selected cluster cations, where for species smaller than Ar_{15}^+ the major fragments Ar_2^+ and Ar_3^+ are detected.[10b] From photoelectron spectra of argon cluster beams and DIM calculations the chromophors Ar_3^+ (for small clusters) and Ar_{13}^+ (for large clusters) are proposed.[11]

Asymmetric fission of molecular cluster dications has been studied with electron impact ionization techniques. It is reported that more than 90% of the mass is retained

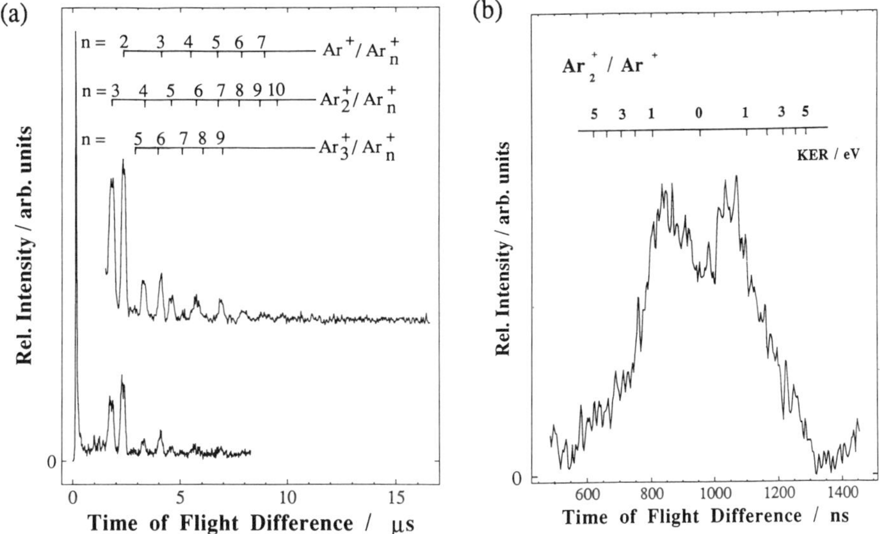

Fig. 2. (a) Photoion-photoion-coincidence (PIPICO) spectrum of argon clusters at 248 eV excitation energy. (b) shows a detailed view of the Ar^+/Ar_2^+ signal. The energy scale refers to the two body dissociation Ar_3^{++}->Ar^+/Ar_2^+. Experimental conditions: p_0=4.7 bar, T_0=215 K.

in the observed fagments,[8] whereas stable doubly charged argon clusters undergo a loss of one or two single atoms, where no Coulomb explosion is observed.[12]

Fig. 2 (b) shows a detailed view of the Ar^+/Ar_2^+-PIPICO signal. From the width of the coincidence signal one obtains the corresponding kinetic energy release (KER), which is indicated in Fig. 2b for a two body dissociation reaction (Ar_3^{++}->Ar^+/Ar_2^+). Various other charge separation reactions involving three body dissociations and evaporation of neutrals are discussed in ref. 7. In a purely electrostatic picture the KER corresponds to a charge separation distance. For the widest portion of the PIPICO signal a charge separation distance of 3.8(3) Å is calculated. This value is in agreement with the value found from EXAFS in argon clusters (3.9(1) Å),[5] the solid (3.75 Å), and matrices (3.7 Å).[13]

The time separation of both maxima of the PIPICO signal corresponds to a KER of 0.7(2) eV. It is remarkable that the shape of the PIPICO signal remains nearly unchanged if the average cluster size is decreased.

A likely explanation for the low KER's can be found by considering a complex charge separation mechanism in cluster dications: According to the double ionization model for solids a two step process leads to double ionization, where first step is the photoionization of one atom within the cluster followed by subsequent electron impact ionization of a neighboring atom. Immediate Coulomb exposion will result in a high KER, which accounts for the widest part of the PIPICO signal. As a competing process charge separation within the clusters prior to Coulomb explosion is likely. Some of the total energy in the cluster dication may be consumed by evaporation of neutrals or can be stored as internal energy of the singly charged fragments. Similar results are reported by Gotts and Stace for collision induced fragmentation of carbon dioxide cluster dications.[8] The difference between the predicted KER's for Coulomb repulsion and the experimental results are interpreted in terms of simultaneous internal excitation.

In an oversimplified picture of purely electrostatic charge separation the low KER can be understood in terms of medium sized cluster dications, such as Ar_7^{++} to Ar_8^{++}, where a charge separation distance of 20.5 Å appears to be in a reasonable range, if non-compact cluster structures are assumed to exist prior to charge separation.

Related to the KER's in clusters is the question of energy position of the dissociating dication state. The asymptote of the Ar^+/Ar_2^+ product channel is 30 eV as calculated from thermodynamical reference data.[14] With the experimental KER of 3.8 eV one obtains 33.8 eV as a lower limit for the dissociating dication state of Ar_3^{++}.

Experimental efforts have been made to measure thresholds of unstable cluster dications in order to correlate the above estimated energy of dissociative dication states with experimental data. Since small cluster dications are unstable we have used the PIPICO-technique to investigate both the thresholds of the Ar^+/Ar_2^+ and Ar_2^+/Ar_3^+ PIPICO signals as a function of average cluster size and photon energy. It is observed that the thresholds of both product channels depend on the average cluster size. The results are listed in Tab. I:

Table I: Threshold energies for double ionization in atomic, clustered and solid argon. The average cluster sizes are estimated to the references listed in ref. 5. The error limits for cluster threshold determination are 0.3 eV.

species		threshold energy / eV	reference
atom		43.4	15a
cluster	$\overline{N}=3$	36.0	this work
	$\overline{N}=4$	35.5	this work
	$\overline{N}=6$	35.0	this work
	$\overline{N}=10$	33.5	this work
	$\overline{N}=20$	32.8	this work
Ar_{101}		32	15b
solid		28.6	15c

The data in Tab. I show that the double ionization thresholds of the solid and the atom differ by nearly 15 eV. The value for solid argon (28.6 eV) is obtained from recent photoelectron-photoelectron-coincidence measurements, which is in good agreement with the double ionization energy of the solid.[15c] From electron impact mass spectrometry a threshold for the stable cluster dication Ar_{101}^{++} is reported.[15b] The thresholds of cluster dications undergoing Coulomb explosion show values that fall between the stable cluster dication (Ar_{101}^{++}) and the atomic dication. The estimated lower limit of the Ar_3^{++} dissociative state (33.8 eV) agrees well with the appearance energies of small cluster dications ($\overline{N}$<6).

II. MOLECULAR CLUSTERS

Fig. 3 shows photoionization efficieny curves of CO and its dimer in the energy regime of the carbon K-edge as well as the C^+ and dimer spectra at the oxygen K-edge. All spectra are dominated by the C (1s)->π^* transition (287.4 eV). and O (1s)->π^* transition (534.11 eV), respectively.

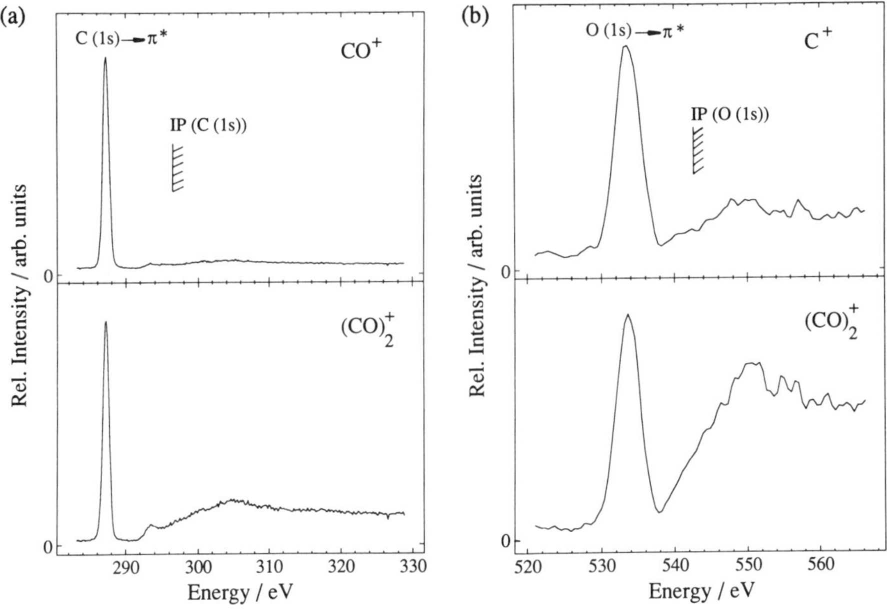

Fig. 3. Photoionization efficiency curves of (a) CO^+ and $(CO)_2{}^+$ in the regime of the carbon K-edge and (b) C^+ and $(CO)_2{}^+$ in the regime of the oxygen K-edge. Experimental conditions: p_0=4.7 bar, T_0=170 K.

In contrast to the cluster size dependent energy positions of Rydberg states below the argon 2p edges, we do not observe energy shifts for this valence transition in clusters. This is in agreement with photon stimulated desorption of solid carbon monoxide, where no shift relative to the gas phase spectrum is observed.[4] The most interesting difference in spectral shape is the decreased intensity of both 1s->π^* bands relative to the K-continuum intensities and changes in spectral shape in both Rydberg regimes. The x ray absorption spectrum of CO shows under high resolution conditions evidence for several Rydberg series below the carbon K-ionization continuum.[16] In the dimer spectrum we find an onset of intensity at 292 eV without any fine structure, a shoulder at 293 eV, and an increase in intensity up to 305 eV, which corresponds to the energy of the C(1s)->σ^* transition. The spectral shape of the dimer cation mimicks that of cations desorbing from the solid (photon stimulated desorption (PSD)).[4] The ion yield curve of CO^+ shows also weak contributions of cluster ion fragmentation above 292 eV. In gas phase CO one finds that there is no parent cation intensity above the carbon K-ionization energy,[17] since double and multiple ionization dominate this energy regime. This shows that cluster fragmentation leads to CO^+ as a final product of dissociative ionzation processes. This is also consistent with the PEPICO spectrum of CO clusters recorded at 305 eV, where a considerable CO^+ signal is observed (cf. Fig. 4a). Another interesting difference of the cluster cation yield is a comparison with coordinated CO in transition metal carbonyls.[18] It is found that quenching of Rydberg states is evident as observed for van der Waals clusters of CO. Due to metal 3d level backbonding the intensity of the C(1s)->π^* transtion is decreased in organometallics.

Similar differences in spectral shape are observed for oxygen K-electron excitation, where a considerably stronger K-continuum is observed relative to the O (1s)->π^* resonance. This is in good agreement with PSD of cations from the solid.[4] Relative changes in intensity of π^* resonances vs. K-continuum intensities have been rationalized for the solids in terms of a delayed onset behavior.[4] These phenomena are consistent with reneutralization or screening effects by the solid. It is interesting to note that the spectral shape of the C^+ fragment shows a considerably stronger O (1s)->π^* resonance compared to the dimer cation. This is consistent with the spectra shown in ref. 17, indicating that C^+ is a product of unclustered CO fragmentation.

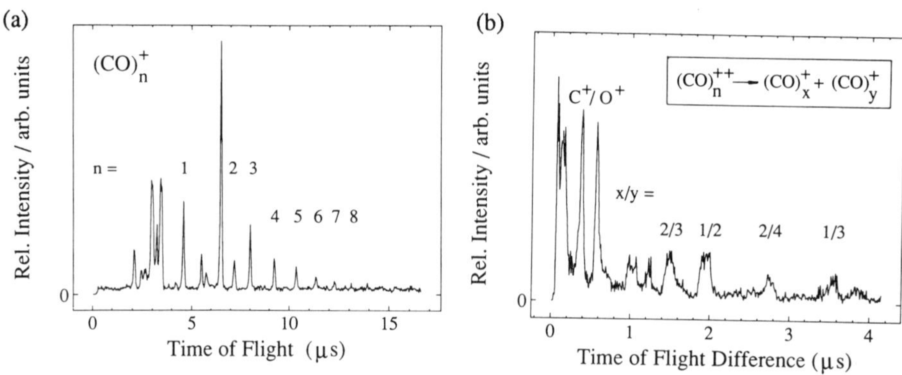

Fig. 4. (a) Photoelectron-photoion-coincidence (PEPICO) spectrum and (b) photoion-photoion-coincidence (PIPICO) spectrum of carbon monoxide clusters at 305 eV excitation energy. Experimental conditions: p_0=4.7 bar, T_0=170 K.

The photochemical behavior of carbon monoxide clusters are studied with the PEPICO and PIPICO techniques. The PEPICO spectrum of CO clusters is shown in Fig. 4a. The dominant peak in the spectrum is the dimer cation, which is one of the preferred fragmentation products of CO cluster dications, as deduced from the width of the PEPICO signal. Besides the dimer signal we find a relative strong monomer intensity as well as singly or multiply charged atomic fragments. The latter ones show unchanged intensity upon clustering, indicating that they are not involved in cluster dissociation pathways.

The PIPICO spectrum of CO clusters shows besides the well-known molecular charge separation channels ($C^{+}+O^{+}$) several coincidences corresponding to cluster dication decays. Most products involve asymmetric charge separation into entire molecular and cluster units. The widest portion of the $CO^{+}/(CO)_2^{+}$ PIPICO channel corresponds to a KER of 2.6(3) eV for a two body dissociation, whereas the peak maximum gives 0.9(2) eV. In terms of charge separation distances in an electrostatic (Coulomb explosion) picture this corresponds to 5.5 Å and 17 Å, respectively. The lower value is twice the molecular diameter,[19] which is maximum charge separation distance in a dimer. The higher value corresponds to a low KER which is probably due to internal excitation of the cluster fragments, as discussed for argon clusters.

ACKNOWLEDGEMENTS

Financial support by the Bundesminsterium für Forschung und Technologie is gratefully acknowledged. R.L. acknowlegdes financial support by the European Community. A.P.H. acknowledges NATO for a travel grant. We thank H. Schmelz for technical assistance.

REFERENCES

1 A.P. Hitchcock, J. El. Spectrosc. Relat. Phenom. 25, 245 (1982), and update 1990 (A.P. Hitchcock, private communication); A.P. Hitchcock Physica Scripta T31, 159 (1990)
2 J. Stöhr and D. A. Outka, J. Vac. Sci. Tech. A 5, 919 (1987); D. Outka and J. Stöhr, J. Chem. Phys. 88, 3539 (1989); J. Stöhr, NEXAFS Spectroscopy (Springer Series in Surface Science, Springer, Berlin, 1991)
3 R. Haensel, G. Keitel, N. Kosuch, U. Nielsen, and P. Schreiber, J. de Physique C 4, 236 (1971)
4 R.A. Rosenberg, P.J. Love, P.R. LaRoe, V. Rehn, and C.C. Parks, Phys. Rev. B 31, 2534 (1985)
5 E. Rühl, H.W. Jochims, C. Schmale, E. Biller, A.P. Hitchcock and H. Baumgärtel, Chem. Phys. Lett. 178, 558 (1991), and references therein
6 E. Rühl, C. Schmale, H.W. Jochims, E. Biller A.P. Hitchcock, and H. Baumgärtel, to be published
7 E. Rühl, C. Schmale, H.W. Jochims, E. Biller, M. Simon, and H. Baumgärtel, J. Chem. Phys., in press (1991)
8 N.G. Gotts and A.J. Stace, Phys. Rev. Lett. 66, 21 (1991)
9 D.M. Curtis and J.H.D. Eland, Int. J. Mass Spectrom. Ion Proc. 63, 241 (1985)

10 (a) U. Buck and H. Meyer, J. Chem. Phys. 84, 4854 (1986); (b) N.E. Levinger, D. Ray, M.L. Alexander, and W.C. Lineberger, J. Chem. Phys. 89, 5654 (1988)
11 F. Carnovale, J.B. Peel, R.G. Rothwell, J. Valldorf, and P.J. Kuntz, J. Chem. Phys. 90, 1452 (1989)
12 A.J. Stace, P.J. Lethbridge, and J.E. Upham, J. Chem. Phys. 93, 333 (1989)
13 W. Niemann, W. Malzfeld, P. Rabe, R. Haensel, and M. Lübcke, Phys. Rev. B 35, 1099 (1987)
14 H.M. Rosenstock, K. Draxl, B.W. Steiner, and J.T. Herron, J. Phys. Chem. Ref. Data 6 (Suppl. 1) (1977); J. Jortner, Ber. Bunsenges. Phys. Chem. 88, 188 (1984)
15 (a) D.M.P. Holland, K. Codling, J.B. West, and G.V. Marr, J. Phys. B, At. Mol. Phys. 12, 2465 (1979); (b) P. Scheier and T.D. Märk, Chem. Phys. Lett. 136, 423 (1987); (c) H.W. Biester, M.J. Besnard, G. Dujardin, L. Hellner, and E.E. Koch, Phys. Rev. Lett. 59, 1277 (1987)
16 M. Domke, C. Xue, A. Puschmann, T. Mandel, E. Hudson, D.A. Shirley, and G. Kaindl, Chem. Phys. Lett. 173, 122 (1990), Chem. Phys. Lett. 174, 668 (1990)
17 A.P. Hitchcock, P. Lablanquie, P. Morin, E. Lizon, E. Lugrin, M. Simon, P. Thiry, and I. Nenner, Phys. Rev. A 37, 2448 (1988)
18 E. Rühl and A.P. Hitchcock, J. Am. Chem. Soc. 111, 2614 (1989)
19 Gmelins Handbuch der Anorganischen Chemie, "Kohlenstoff Teil C - Lieferung 1", (Verlag Chemie, Weinheim, 1970), p. 123

QUESTION

BAER - I am surprised that you obtained similar bond distances for the Ar_2 clusters as for the Ar_n clusters from your NEXAFS data. What about multiple scattering ?

RUHL - The cluster EXAFS analysis shows that an Ar-Ar distance of 3.9 ± 0.1Å is obtained. This value is slightly higher than values obtained for the solid (3.75Å) or matrices (3.7Å). An obvious reason for the slightly increased Ar-Ar distance in small argon clusters (with $\bar{N}$ = 10) can be found in the spherical (icosahedral) structure of argon clusters. In contrast to this, the solid is known to exist in a f.c.c. lattice. Further EXAFS experiments in the energy regime of Ar(1s) excitation are needed to confirm the results obtained at the Ar(2p) - edge (250eV). The advantage of Ar(1s) - EXAFS will be, that the energy space is not limited as in the case of Ar(2p) - EXAFS, where the Ar(2s)- edge disturbs the 2p EXAFS above 330 eV. Another advantage of Ar(1s)-EXAFS will be that only one threshold energy has to be considered in the EXAFS analysis. Our EXAFS analysis has shown, that the oscillations that are observed in photoion yield curves of argon clusters above the 2p-and 2s-edges can be described with a single scattering mode, whereas the resonance centered at approximatively 256eV is probably due to multiple scattering.

DUJARDIN - Experimental question about the PIPICO curves you showed. They have a non zero background. So it seems that the false coincidences have not been substituted. Could this change some of the results ?

RUHL - The false coincidence background is extremely weak, so that is was not necessary to substract any background. Therefore, we do not think the results are affected by false (or random) coincidences.

MENZEL - Two questions related to the comparison of clusters and solids :

1. In solid Argon, surface and bulk 2p resonances can be distinguished by using the detection of different ion signals such as Ar^+ and Ar_2^+ (see e.g. D. Menzel, Appl. Phys., 1990 and work cited there) or by detecting decay electrons at different exit angles (G. Rocker et al., to be published in Phys. Rev. B). Can a similar distinction be made in the case of clusters ?
2. Again in solid Argon, evidence for a bi-exciton excitable by electrons (P. Feulner et al., PRL June 1991) or even by simple photons (P. Feulner et al., to be published) has been found. Is there any evidence that such an excitation also exists in Ar clusters ?

RUHL -

1. Different cation channels will not give a distinction between different processes. However if the beam expansion conditions are changed, the NEXAFS structure at the Ar(2p)-edges changes its shape independent on the cation channel. We have analyzed the different components of spectral features in the energy regime of the 4s Rydberg/exciton state (244-246eV) using a total electron detection method (E. Ruhl et al., to be published). When the average cluster size is changed we observe how the individual contributions of small clusters (surface excitons), medium sized clusters, and large clusters (bulk excitons) change their relative intensity.
2. We have found no evidence of these processes in clusters so far.

WONG - Was the Fourier transform of the Ar EXAFS phase corrected to give the Ar-Ar distance you quoted ? If so, which (whose) phase shift value did you use ?

RUHL - The EXAFS analysis was done by the use of Mc Kale parameters. Further details can be found in: E. Rühl et al., Chem. phys. lett. 178, 558 (1991).

STUDY OF SILICON COMPOUNDS PRODUCED BY RADIO FREQUENCY DISCHARGE OF SILANE

J.L. Maréchal, N. Herlin, C. Reynaud, I. Nenner
Service des Photons, Atomes et Molécules
C.E.N. Saclay, F- 91191 Gif/Yvette Cedex
and
Laboratoire pour l'Utilisation du Rayonnement Electromagnétique
CEA, CNRS et MEN, Université de Paris-Sud
F-91405 Orsay Cedex

ABSTRACT

We present a new experimental set-up designed for the study of free semi-conductor clusters. The cluster production is based on the technique of radio-frequency discharge decomposition of a gas. In a preliminary step, we study the gaseous particles (Si_nH_x) generated by pure silane (SiH_4) discharges at low pressures (< 10 millitorr) under continuous RF excitation conditions. Particles produced in the discharge cell are then driven in the ionization chamber of a time of flight mass spectrometer. The UV light used to ionise the particles is either a Helium lamp or synchrotron radiation. This set-up allows to study the mass distribution of particles and the evolution of electronic properties as a function of the size, using the technique of core hole photoionization. We have studied the neutral species present in the post discharge zone and the positive ions present in the discharge.

INTRODUCTION

The synthesis of small powders (Si, SiC, SiN...) with controlled properties (size, stoechiometry ...) is an increasing field of interest because of its technological implications. These particles can be obtained by infra-red pyrolysis[1] or radio-frequency (RF) discharge[2] in a gas or a mixture of gases. The RF discharge was first used for surface processing and more recently for nanometric powders synthesis. The chemistry that leads to particle nucleation and growth in discharges has been investigated (for example[3,4]), but the detailed mechanism is still badly known. The size of the particles is highly correlated to the plasma conditions[2]. During the powder formation, gaseous particles with a small number of silicon atoms are formed. Very few studies on these particles are existing. We present here results obtained with a new experimental set-up designed for the characterization of such particles using mass spectrometry. We observed positive ions present in the RF discharge of silane and hydrogenated particles present in the post discharge zone. The study of the electronic properties of these species is interesting because it corresponds to the growing of a semi-conductor material. A first attempt was done using the technique of core-valence monoelectronic transitions in the 100 eV photon energy region because it corresponds to the transition (Si 2p->σ*) in silane.

EXPERIMENTAL SET UP

A general view of the experimental set up is presented in Figure 1. It is composed of two parts : 1) a RF discharge chamber 2) an expansion chamber with a time of flight mass spectrometer[5]. The pressure in the discharge cell is measured by a capacitive barocell gauge. The pressure in the spectrometer is measured by a ionization gauge. Without gas flow, the pressure in the discharge cell is below 10^{-4} Torr and in the mass spectrometer 5 10^{-7} Torr.

The mass spectrometer is divided into two parts which are differentially pumped by turbomolecular pumps (330 l/s) : the ionization zone and the ions detection zone. Particles entering the mass spectrometer are ionised at 21.2 eV using a laboratory He source (VG UVL100) or around 100 eV using the synchrotron radiation delivered by Super ACO storage ring at LURE (SA 71 beam line). The positive ions are extracted by a (0-500 V) pulsed electric field (20 kHz frequency, 2 µs without field), they are then accelerated by a 4000 V continuous field and enter the free field drift tube (450 mm long) through a 3 mm diameter hole. The detectors are 50 mm diameter microchannel plates. Two modes of acquisition can be used. At a fixed wavelength, mass spectra are obtained by detecting the ions in coincidence with the extraction pulse. Tuning the wavelength, excitation spectra can be obtained either by collecting the total ions signal or by selecting a specific mass.

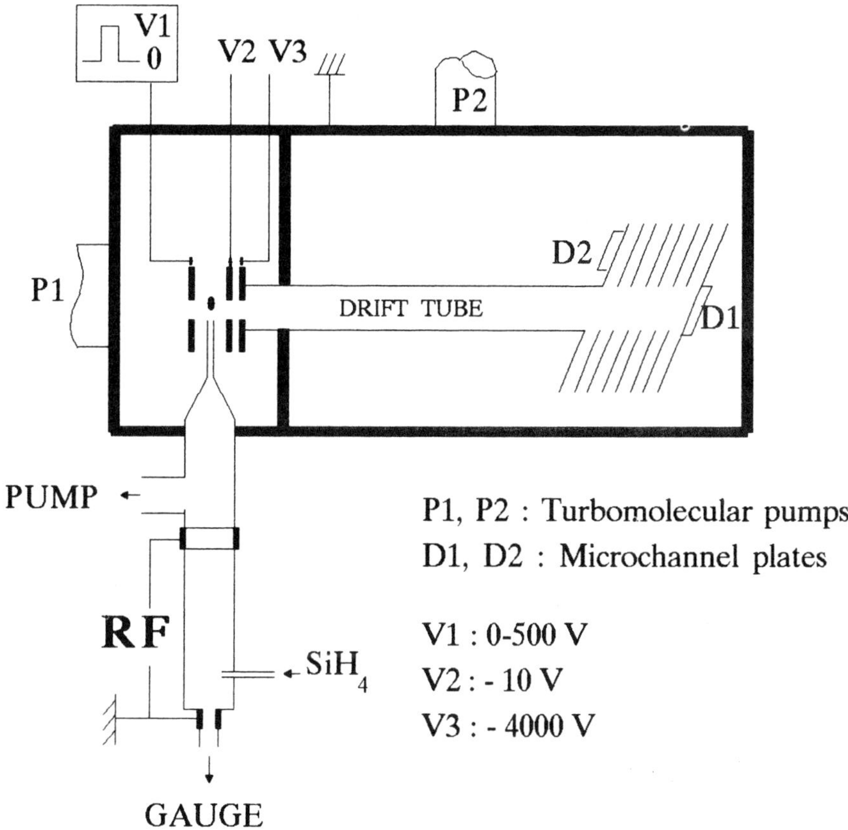

Figure 1 : Schematic drawing of the experiment

The gas used in the experiment is silane (SiH_4) obtained from Alphagaz with 99.99% purity level. The plasma chamber is a 100 mm high, 10 mm diameter cylindrical glass cell. In plasma conditions, the pressure in the cell is kept constant by adjusting the SiH_4 flow. The plasma is established using a 13.56 MHz RF generator which provides an output power level adjustable between 0 and 300 Watt. SiH_4 enters the cell through a microleak valve, the residual products are pumped out towards a primary pump (30 m^3/h) or are driven to the mass spectrometer ionization chamber through a 0.3 mm diameter nozzle. Under continuous RF excitation, brown powders appear immediately in the cell and within a few minutes, the walls are covered with a dark film.

RESULTS AND DISCUSSION

In a preliminary step, silane mass spectra were recorded without discharge, at 21.2 eV or around 100 eV. The main peaks at 21 eV can be attributed to SiH_2^+ and SiH_3^+, while at 100 eV the main peaks are Si^+ and SiH^+. At high photon energy, fragmentation processes due to core-hole relaxation phenomena occurs and explain the dehydrogenation effect observed in the mass spectra. In silane, these processes have already been reported[11] and analyzed.

As soon as the discharge is established, the mass spectrum is drastically modified : some features which are attributed to $Si_nH_x^+$ appear. At very low pressures, the discharge diffuses through the nozzle and positive ions due to the discharge are observed without any light source. Ions mass distribution presented in Figure 2 shows hydrogenated ions up to $Si_5H_x^+$. The most abundant species are always $Si_nH_{2n}^+$ and $Si_nH_{2n+2}^+$. These results are comparable to the observations of Haller[7] and of Perrin *et al*[6] in a multipole discharge. Hydrogenated ions are supposed to be important intermediate reaction products in the particle nucleation phenomena and in the deposition of hydrogenated silicon film[7,8,9,10].

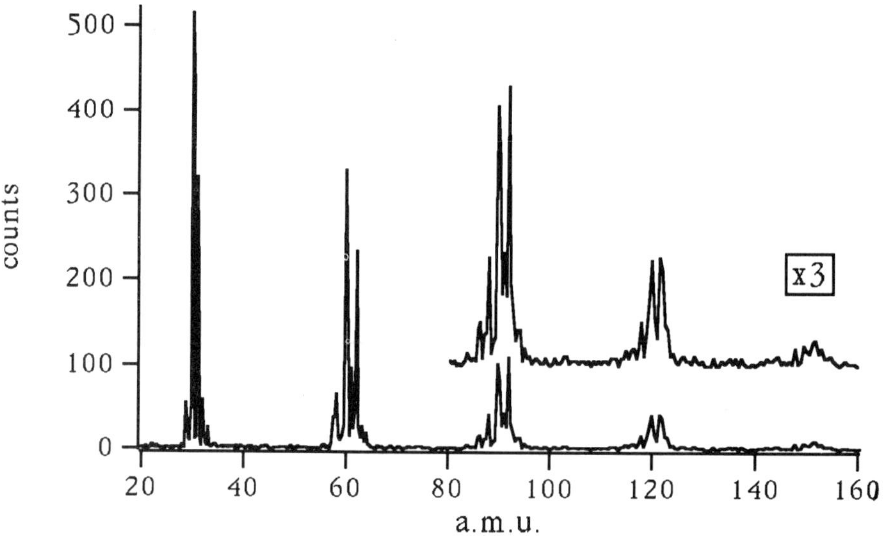

Figure 2 : Mass spectrum of the positive ions present in the discharge

At higher pressures (1-10 millitorr), the discharge remains confined in the glass cell, and no more ions can be directly observed. Mass spectra obtained from the photoionization of neutral species present in the post discharge zone are studied as a function of plasma conditions and ionization wavelength. Silane decomposition increases with increasing RF power, but there is no significant effect on the relative intensity of the different peaks (between 5 and 50 W).

A typical mass spectrum (1 millitorr pressure, 30 W RF power, 101.5 eV photon energy) is presented in Figure 3. Several groups of peaks corresponding to different values of n can be identified. The relative abundance of the ions decreases rapidly with n. The highest value we ever observed for n is 6. The mass spectrum corresponding to a specific value of n shows a wide distribution of hydrogenated ions ($x \geq 0$). The low hydrogenated ions observed in the spectra around 100 eV can be attributed, as in the case of silane without discharge, to dehydrogenation processes following photoexcitation. Between 100 and 106 eV photon energy, we have checked for n=1 that our mass distribution spectra are in agreement with previous work[11] *i.e.* the relative intensity of low hydrogenated ions increases with increasing photon energy. In order to have a better identification of the 'mother' species, the mass spectrum at 21 eV was recorded. It shows more highly hydrogenated ions than around 100 eV. For n=2, our mass distribution is quite similar to the one observed for Si_2H_6. It seems to indicate that the observed reaction products are essentially higher silanes.

By varying the photon energy between 100 and 106 eV, we. observe a resonance of the ions signal intensity close SiH_4 to resonance (103 eV). Moreover, the relative intensities for different n do not remain constant. To examine this last feature more precisely, we have deduced the branching ratio between different values of n from the mass spectra. Due to the low level signal this work was done only for n=1,2 and 3 (Figure 4).

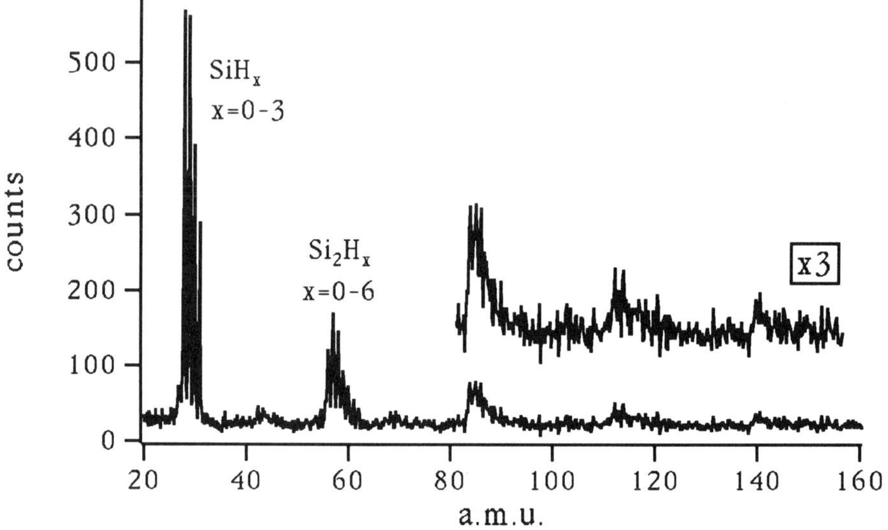

Figure 3 : Mass spectrum of the neutral particles in the post-discharge with a photon energy of 101.5 eV.

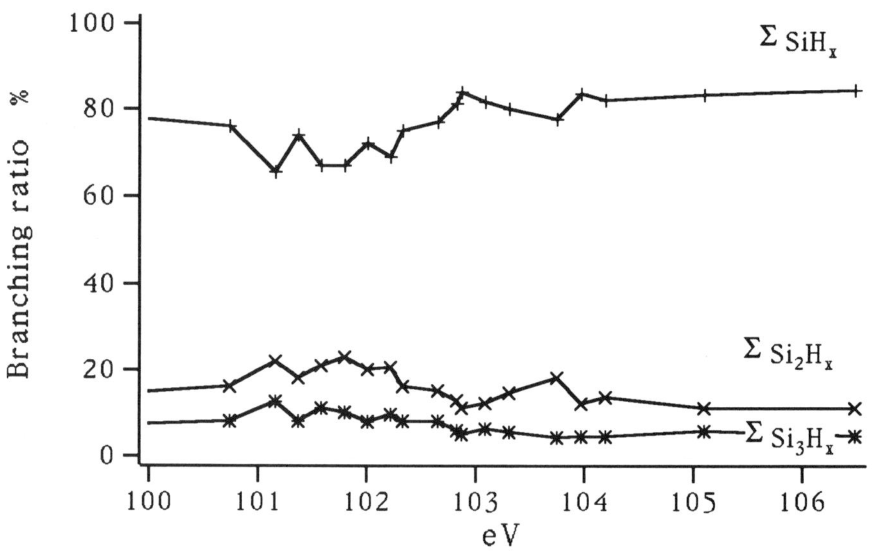

Figure 4 : Branching ratio for SiHx, Si_2H_x and Si_3H_x particles as a function of the wavelength.

The branching ratio were then used to obtain excitation spectra from the measured total ion excitation spectrum. Spectra were then normalized to the spectrum of pure SiH_4 (Figure 5).

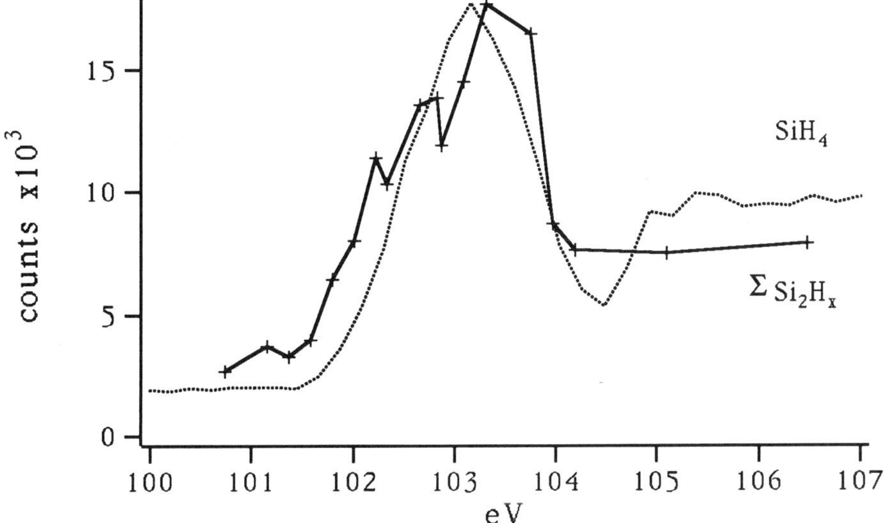

Figure 5 : Excitation spectrum of Si_2H_x particles compared to the spectrum of pure SiH_4.

The structure due to Si_2H_x seems to be wider on the low energy side than the one due to SiH_4. This is comparable to the shoulder observed in the spectrum of $Si_2(CH_3)_6$ compared to $Si(CH_3)_4$ (12, 13). It is tempting to attribute this broadening to the σ^* antibonding orbital of the Si-Si bond. A more detailed comparison, between SiH_4 and Si_2H_6 spectroscopy near Si 2p edge is in progress in order to test this hypothesis.

CONCLUSION

We have presented a new experimental set-up which has allowed core hole spectroscopy study of mass-selected particles. Modifications of the experiment are in progress in order to increase the signal to noise ratio.

We have analyzed the particles produced in a RF discharge of pure silane. They are always highly hydrogenated. To our knowledge, it is the first time that neutral molecules containing from 1 to 6 silicon atoms have been observed in the post-discharge zone. The powders which can be seen in the cell are never detected in the mass spectrometer due to the low signal to noise ratio and probably to the wide size dispersion of the powders obtained under continuous RF excitation. In order to increase the size selectivity, we are modifying the experiment to operate under pulsed RF excitation[2].

AKNOWLEDGEMENTS : The authors are very thankful to P. Morin,.M. Cauchetier, M. Luce, O. Croix, J. Perrin, P. Dubot and M.A. Gaveau.

REFERENCES

1. M. Cauchetier, O. Croix, M. Luce, Adv. Ceram. Materials, 3, 548 (1988)
2. A. Bouchoule, A. Plain, L. Bouffendi, J.P. Blondeau, C. Laure, J. Appl. Phys, to appear in 1991
3. K.G. Spears, T.J. Robinson, R.M. Roth, IEETr. Pl. Sc., PS14, 179 (1986)
4. K.G. Spears, R.P. Kampf, T.J. Robinson, J. Phys Chem, 92, 5297 (1988)
5. M.J. Ramage, P. Morin, T. Lebrun, I. Nenner, M.J. Hubin-Franskin, J. Delwiche, P. Lablanquie, J.H.D. Eland, Rev. Sci. Instr., 60, 7078 (1989)
6. J. Perrin, A. Lloret, G. de Rosny, J.P.M. Schmitt, Int. J. Mass Spectr. and Ion Process, 57, 249 (1984)
7. I. Haller, Appl. Phys. Lett., 37, 282 (1980)
8. I. Haller, J. Vac. Sci. Technol., A1, 1376 (1983)
9. D.A. Doughty, A. Gallagher, Phys. Rev. A, 42, 6166 (1990)
10. M.L. Mandich, W.D. Reents Jr, J. Chem. Phys., 90, 3121, (1989)
11. Y. Sato, K. Ueda, A. Yagishita, T. Sasaki, T. Nagata, T. Hayaishi, M. Yoshino, T.Koizumi, A. MacDowell, Phys. Scripta, 41, 55 (1990)
12. R. Sodhi, S. Daviel, C. Brion, G. de Souza, J. Electron. Spectrosc. and Relat. Phenom., 35, 45, (1985)
13. I. Nenner, Private communication

HIGH ENERGY CHEMISTRY OF CONDENSED SPECIES

G. Dujardin
Laboratoire pour l'Utilisation du Rayonnement Electromagnétique (LURE)
Bât. 209d, Université Paris-Sud
91405 Orsay Cédex, France.
and Laboratoire de Photophysique Moléculaire, CNRS
Bât. 213, Université Paris-Sud
91405 Orsay Cédex, France.

ABSTRACT

We discuss the high energy (15 - 200 eV) photon induced electronic processes producing positive and negative ion desorption from solid surfaces of condensed gases. We focus on simple systems which serve as prototypes : condensed argon and condensed oxygen.

I. INTRODUCTION

High energy photochemistry is concerned with reactions occuring in isolated atoms and molecules, small clusters or in condensed species when exciting these systems at energies higher than their ionization energy.

In the gas phase, the high energy valence photoexcitation (15 eV $\lesssim h\nu \lesssim$ 200 eV) is dominated by direct and indirect single ionization (ejection of <u>one</u> valence electron) which usually accounts for more than 90 % of the total photoabsorption[1]. It is known however that a number of "minority" processes, involving the excitation (ionization) of more than one electron, also exist at high energy. These are for example the formation of satellite states of ions[2], the formation and dynamics of multiply charged species[3], the formation and stability of doubly excited Rydberg states[4].

At solid surfaces, the high energy processes may be strongly modified as compared to the gas phase due to electronic coupling with the substrate and/or neighbour atoms or molecules. We will discuss in this paper how the above mentioned "minority" high energy processes are modified at solid surfaces and whether new processes, which are not observed in gas phase, may exist on surfaces.

The studied systems consist of rare gas atoms and simple diatomic molecules condensed as a few multilayers on a cold (T = 12 K) platinum substrate. Dynamic processes are investigated by means of positive and negative ion desorption. The absence of ion desorption is expected to be related to very rapid electronic relaxation of excited atoms and molecules at solid surfaces. On the opposite, the observation of ion desorption is expected to occur from quasi-stable excited states which are long lived enough to allow the dynamics to take place. Three examples of high energy photochemistry in condensed gases are studied in the following : (i) satellite states of ions in condenses argon, (ii) exciton pairs in condensed argon, (iii) positive energy excitons in condensed oxygen.

II. EXPERIMENTAL

The experimental setup has been described in reference 5. Briefly, the synchrotron radiation from Super-ACO at Orsay is used as a photon source of variable energy in the 15-200 eV range. Rare gas atoms or molecules are condensed at low temperature (T $\simeq$ 12 K) on a liquid helium flow cryostat. Positively or negatively charged desorbed ions are detected using a quadrupole mass spectrometer and are counted as a function of the photon energy. The residual gas pressure in the UHV chamber is kept below 10^{-10} Torr in order to ensure that no impurities are condensed at the surface of the sample.

III. POSITIVE ION PHOTODESORPTION FROM CONDENSED ARGON

Until very recently it was believed that positive ions could not be desorbed from condensed rare gases. Considering the case of argon, it is well known from the calculated repulsive potential energy curves of Ar^+ - Ar [5] that the ionization of an argon atom may lead to some repulsion from neighbour atoms. This effect could produce ion desorption. However, efficient resonant charge exchange with neighbour atoms may lead to a rapid delocalization of the excitation and then prevent any desorption to occur.

Nevertheless the Ar^+ ion desorption has been observed in the valence energy range[5,7]. The results are shown in figures 1 and 2.

The Ar^+ ion yield recorded as a function of the photon energy in the 32 - 46 eV range is shown in figure 1. The ion desorption threshold energy at about 33.4 eV is well above the 3 p and 3s ionization threshold energies of solid argon at 13.8 eV and 28.4 eV respectively[8]. The step structures shown in figure 1 are located just below the double ionization threshold energy at 40 eV[8]. It follows that the Ar^+ ion desorption observed in the 32 - 46 eV energy range may be ascribed to the formation of satellite states $(Ar^+)^*$, i.e. Rydberg states of the singly charged ion, which are known from gas phase data [5] to lie in this energy range.

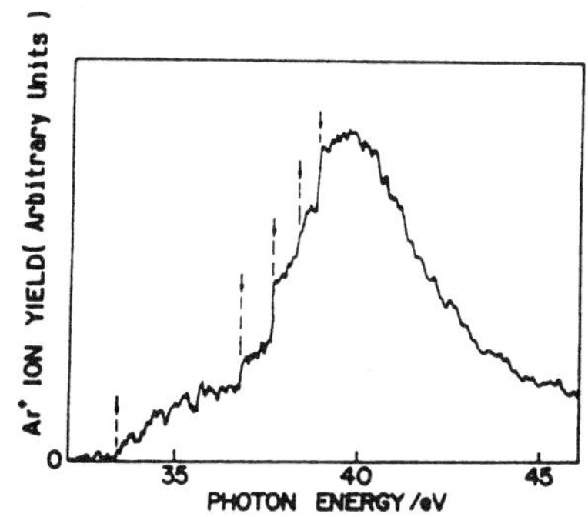

Fig. 1 : Ar^+ ion yield from condensed argon (≃ 100 ML). The arrows indicate step structures.

Ion desorption from multilayers of condensed argon is also observed at lower photon energies. We see in figure 2 two broad resonances at about 23.5 eV and 25 eV producing Ar^+, Ar_2^+ and Ar_3^+ ion desorption. These processes have been previously observed by electron impact [7] and photon impact[9] on condensed argon. They have been assigned to the formation of pairs of surface and bulk excitons since the energy of the resonance at 23.5 eV is twice that of surface 3p excitons in condensed argon[10] whereas the energy of the resonance at 25 eV is twice that of bulk 3p excitons[10]. As compared to electron impact experiments, the photon impact studies allow to investigate these exciton pair processes into much more details. Focussing our attention on the surface exciton pair resonance at 23.5 eV, we see in figure 2 that the resonance energy varies depending on whether Ar^+, Ar_2^+ or Ar_3^+ ions are detected. Furthermore the Ar^+ and Ar_2^+ resonances seem to be the superposition of two or more resonances. It follows

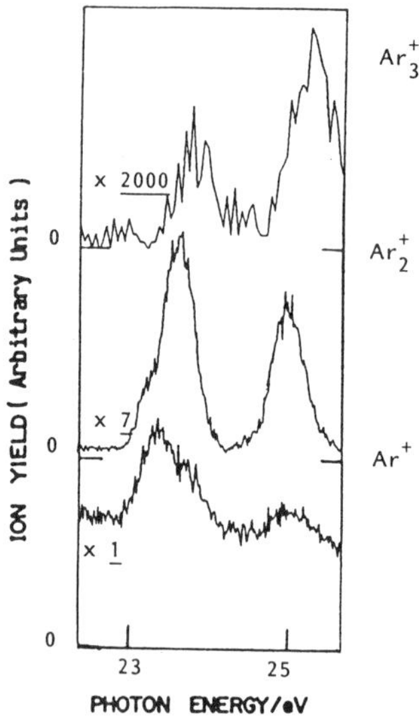

Fig. 2 : Ar_2^+, Ar_2^+ and Ar_3^+ ion yields from condensed argon (~ 3ML).

that the formation of surface exciton pairs gives rise to several seperate resonances close in energy and which are hardly resolved in the curves of figure 2. The multi-components of the surface exciton pairs may be related to the multi-components of the surface single exciton. It is known indeed [10] that the 3p surface single exciton in condensed argon has three $^3P_2$, $^3P_1$ and $^1P_1$ components at 11.71 eV, 11.81 eV and 11.93 eV respectively. Combining two of these single surface excitons may then result in several components of exciton pairs.

The formation of a pair of excitons following the absorption of a single photon is a quite surprising result. Exciton pairs are known to exist in semiconductors [11] . However in this latter case, excitons are produced separately and combine into pairs after their formation when their density is high enough. The one photon multiple electron excitation is well known to play an important role in the photoionization of gas phase atoms and molecules and of solids[12] . In solids, several multielectron processes such as Auger processes[13] or single photon two core electron ionization[14] have been extensively studied. Most of these processes involve the simultaneous excitation of two electrons from the same atom or molecule and have been shown to be mainly due to electron correlation effects in the initial state. There is very few evidence of one photon collective excitation of two electrons from separate atoms in solids. A major exception was the observation of two atom double ionization in condensed rare gases[8]. The exciton pair process is of the same nature as the two atom double ionization. Both processes involve the excitation (ionization) of two electrons which have no correlation in the initial state and interact only in the final state.

The two ion desorption processes that we discussed here, i.e. satellite state ion and exciton pair formations, illustrate the dominant role of multiple electron excitation with respect to the one electron excitation processes which produce no ion desorption.

IV. NEGATIVE ION DESORPTION FROM CONDENSED OXYGEN

Negative ion desorption may be a powerful probe of photon stimulated charge transfer (PSCT) processes at solid surfaces including substrate to adsorbate charge transfer, adsorbate to adsorbate charge transfer and intramolecular charge transfer.

PSCT processes have been studies by this method in multilayers of condensed O_2 [15] . We focus here on resonant processes involving positive energy excitons.

The O^- ion desorption yield from condensed O_2 as a function of the photon energy is shown in figure 3.

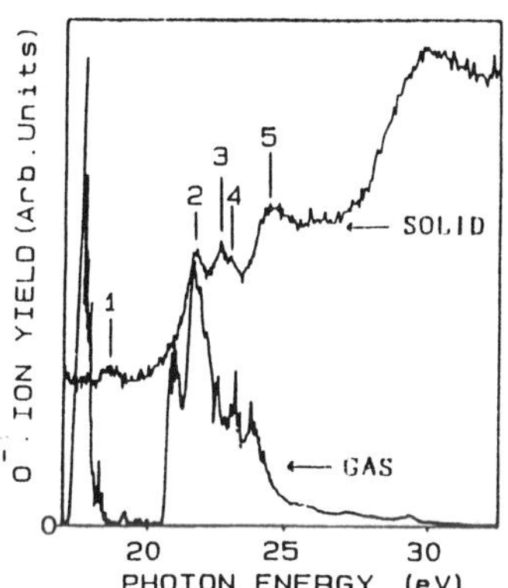

Fig. 3 : Comparison of the O^- yield curve from condensed O_2 (thickness ≃ 5 monolayers) with the O^- ion yield curve from gas phase O_2 [16] . Observed narrow resonances are numbered from 1 to 5 and their energies are 18.6, 21.7, 22.6, 22.9 and 24.4 eV respectively.

The observation of resonances in the O^- ion yield curve is a surprising result. Resonances due to the formation of super-excited Rydberg states are known to exist in the photoabsorption and photoionization spectra of isolated molecules. However these excited states are expected to be very rapidly quenched at solid surfaces before any desorption can occur.

phase resonances lying in the 17-19 eV energy range belong to Rydberg series converging to the $b^4\Sigma_g^-$ state of O_2 of which the adiabatic energy is 18.17 eV[18]. The $b^4\Sigma_g^-$ adiabatic ionization energy in solid O_2 is 16.5 eV [19]. The strong red gas-to-solid shift (1.7 eV) of this ionization energy leads to an intersting effect. It appears indeed that the Rydberg resonances in condensed O_2 at about 18.6 eV (peak number 1 in figure 3) are lying above the corresponding $b^4\Sigma_g^-$ ionization energy at 16.5 eV. This unusual situation may be explained by a strong cavity effect due to the repulsion between the Rydberg electron and the neighbour O_2 moleucles which could maintain the electron in a quasibound-state although the internal energy of the system is higher than the ionization energy. This positive energy Rydberg resonance effect is very similar in nature to the positive energy states observed above inner shell ionization of polyatomic molecules [20]. The gas phase resonances in the 20-25 eV energy range have been shown [16] to correspond to a shape resonance and to Rydberg and doubly excited Rydberg state series converging to higher ionic states of O_2 of which the energies are unknown in solid O_2. However a similar positive energy state effect is expected to occur if we assume that the red shift of these ionization energies is comparable to that of the $b^4\Sigma_g^-$. We note that resonances have been observed in the photoionization spectra of O_2 clusters in the same 17-25 eV energy range [21]. However in this latter case positive ions are mainly issued from direct or indirect single ionization like in gas phase O_2 whereas such ionization processes are believed to produce no ion desorption from condensed O_2 [22].

The resonant desorption of O^- and O^+ ions following the formation of Rydberg resonances O_2^* may be ascribed to the ion pair formation ($O_2^* \rightarrow O^+ + O^-$) occuring eitheir at the surface of the condensed gas or from gas phase O_2^* molecules expeled from the surface through the action of the repulsive forces from neighbour molecules.

V. CONCLUSIONS

Interaction of Synchrotron Radiation with surfaces is mostly used to probe the structure (electronic and geometric) of substrates or adsorbates by using photoemission, EXAFS, etc... methods. However the Synchrotron Radiation may also be used to induce molecular or chemical reactions at solid surfaces. The easiest probe of this high energy chemistry is to study positive and negative ion desorption.

The few examples that we discussed in this paper demonstrate that new electronic states at solid surfaces (exciton pairs in condensed argon, positive energy excitons in condensed oxygen) may be observed by positive and negative ion desorption. These electronic excitations have not been observed by photoabsorption, photoemission or electron energy loss spectroscopy since they are very minority processes as compared to single ionization (excitation) processes. The reason which makes these minority processes more "visible" in positive and negative ion desorption is that the dominant single ionization or excitation usually do not produce any ion desorption.

ACKNOWLEDGEMENT

I thank L. Hellner, M.J. Besnard-Ramage, L. Philippe and R. Azria for their continuous participation in this work.

REFERENCES

1. I. Nenner and J.A. Beswick, Handbook on Synchrotron Radiation, 2 (Elsevier Science Publishers B.V., 1987).

2. M.Y. Adam, P. Morin and G. Wendin, Phys. Rev. A31, 1426 (1985).

3. G. Dujardin, S. Leach, O. Dutuit, P.M. Guyon et M. Richard-Viard, Chem. Phys. 88, 339 (1984).

4. A. Dadouch, G. Dujardin, L. Hellner et M.J. Besnard-Ramage, Phys. Rev. A43, 6057 (1991).

5. G. Dujardin, L. Hellner, M.J. Besnard-Ramage and R. Azria, Phys. Rev. Lett. 64, 1289 (1990).

6. H.U. Böhmer et S.D. Peyerimhoff, Z. Phys. D 3, 195 (1986).

7. Y. Baba, G. Dujardin, P. Feulner and D. Menzel, Phys. Rev. Lett. 66, 3269 (1991).

8. H.W. Biester, M.J. Besnard, G. Dujardin, L. Hellner and E.E. Koch, Phys. Rev. Lett. 59, 1277 (1987).

9. T. Schwabenthan, R. Scheuerer, E. Hudel and P. Feulner, Solid State Com., in press.

10. Electronic excitations in condensed rare gases, N. Schwenter, E.E. Koch and J. Jortner, Springer-Verlag Berlin Heidelberg 1985.

11. H. Sumi, Submitted to Surf. Science.

12. Photoionization and other probes of many-electron interactions, ed. F.J. Wuillemier, Plenum Press, New York (1976).

13. P.A. Bennet, J.C. Fuggle, F.U. Hillebrecht, A. Lenselink and G.A. Sawatzky, Phys. Rev. B 27, 2194 (1983).

14. S.I. Salem, B. Dev and P.L. Lee, Phys. Rev. A22, 2679 (1980).

15. G. Dujardin, L. Hellner, L. Philippe, R. Azria and M.J. Besnard-Ramage, Phys. Rev. Lett., in press.

16. A. Dadouch, Thèse Université Paris-Sud (Jul. 1991).

17. M. Chergui, N. Schwentner and V. Chandrasekharan, J. Chem. Phys. 89, 1277 (1988).

18. H. Oertel, H. Schenk and H. Baumgartel, Chem. Phys. 46, 251 (1980).

19. F.J. Himpsel, N. Schwentner and E.E. Koch, Phys. Stat. Sol. (b) 71, 615 (1975).

20. J.L. Dehmer, J. Chem. Phys. 56, 4496 (1972).

21. M. Ukai, K. Kometa, K. Shinsaka, Y. Hatano, T. Hirayama, S. Nagaoka and K. Kimara, Chem. Phys. Lett. 167, 334 (1990).

22. L. Hellner, M.J. Besnard-Ramage, G. Dujardin and R. Azria, in DIET 4, ed. by G. Betz and P. Vargor (Springer-Verlag Berlin, 1990).

DISCUSSION

BESWICK - Do you think possible to observe the exciton pairs states in Ar clusters starting from Ar_2, Ar_3, etc. ?

DUJARDIN - I thought, a priori, that it would be difficult to observe exciton pair states in Ar clusters. The reason is that, at this photon energy (≅25eV), the dominant process is the 3p ionization which gives rise to intense signals of ions in Ar clusters. The case of condensed argon is more favourable for observing exciton pair states producing Ar^+ ion desorption since the 3p ionization does not produce any ion desorption from the solid. However, Dr. Hertel reported this morning the observation of resonances in the ionization curves of metastable Ar clusters around 26eV. I guess that these resonances are also due to the formation of exciton pair states in clusters.

RUHL - You have explained fine structure of O^- desorption from solid oxygen with an ion pair formation process. Is the same fine structure found in O^+ desorption from solid O_2 ?

DUJARDIN - Yes, but the resonances are not so visible in the O^+ in yield curve. The reason is that there is a large background of desorbed O^+ ions coming from ionization processes.

BROCKLEHURST - In aromatic crystals highly excited states can undergo fission into two triplet excitons (or one molecular triplet and one ion pair). Is such a process possible in argon ?

DUJARDIN - This process should be possible in condensed argon if there was some excited state lying at about two times the energy of the single exciton. However, contrary to the case of aromatic crystals, there is no electronic state of argon at this energy. Therefore the fission into two excitons is not possible in condensed argon.

HERTEL - 1. How can you be sure these are p-excitons.
2. There are "Rydberg states" as "Excitons" above threshold ; also seen clusters (Ar_n, Kr_n) $(N_2O)_n$.

DUJARDIN - 1. 3p and 3s exciton energies are very well known in condensed argon from the photoabsorption spectrum. The observed resonances are exactly at two times the energy of surface and bulk 3p excitons. The 3s exciton energy is much higher than these resonance energies.
2. It is time that Rydberg states above the lower threshold ionization limit are known in clusters. What is very unusual in condensed O_2 is that these "Rydberg states" or "Excitons" are above **their own** ionization energy limit.

NENNER - The interpretation of a high energy resonance in your oxygen desorption data, in terms of a Rydberg or a shape resonance decaying into $O^+ + O^-$ is not convincing. Is it possible that it is a doubly excited state ? Can the dissociation be a secundary process ?

DUJARDIN - The only ways to produce O^- ion desorption from condensed O_2 are either by secondary processes involving the photoelectrons produced in the solid or by $O^+ + O^-$ in pair formation from electronically excited states. Secondary processes can only give rise to continuous background or very broad resonances (with a width larger than 3eV) by dissociative attachment of electrons on O_2 molecules. The observed narrow resonances can then be assigned only to the formation of unknown excited states in condensed O_2. The comparison with gas phase data suggests that these "excitons" are derived from O_2 Rydberg states and that these excitons are above their ionization limit in condensed O_2. These excited states are not "Shape resonances" but their formation above their ionization limit has the same qualitative description as shape resonances produced by core excitation in polyatomic molecules, i.e. an electron trapped by a Coulomb barrier above the ionization energy?
Except the above mentioned Rydberg states, there is no doubly excited states of the isolated O_2 molecule in the studied energy range. We cannot completely exclude that pairs of excited states, as in condensed argon, also give rise to resonances in condensed O_2. However in this latter case it would not be obvious to explain how these resonances may relax to $O^+ + O^-$ ion pair states.

DYNAMICS OF VALENCE HOLE EXCITATIONS IN ADSORBATES

M. Ohno
Department of Physics
Uppsala University, Box 530, S 751 21 Uppsala, Sweden

W. von Niessen
Institut für Physikalische und Theoretische Chemie,
Technische Universität Braunschweig, W 3300 Braunschweig
Fed. Rep. of Germany

ABSTRACT

The valence photoemission spectra of NiCO, PdCO and PtCO are calculated by the ab-initio Green's function method using an extended basis set. The present ab initio many-body appoaches give a reasonably good description of the valence photoemission spectra of CO adsorbed on Ni, Pd and Pt metal surfaces. A detailed study of the dynamics of valence photoemission of CO adsorbed on different substrates is made. For the 1π level of NiCO and PtCO the quasi-particle picture (Q.P.P.) breaks down, however, for PdCO the Q.P.P. is still valid. For the 4σ and 5σ levels of NiCO, the lowest energy state is the main line single hole (1h) state not the CT (charge transfer) 2h1p (two hole one particle) shakedown state. However, for PdCO and PtCO, the σ (π) to π^* metal - ligand CT 2h1p shakedown state of a non-negligible intensity becomes the lowest energy state. The main line state is still 1h state and the Q.P.P. is valid. In PdCO and PtCO where the π bonding is weaker than in NiCO, the excitation from the σ metal orbital becomes substantial because of a weaker π bonding.

1. INTRODUCTION

The conventional picture of chemisorption of CO on transition metal surfaces in terms of the σ donation from the 5σ orbital of CO to the metal $d\sigma$ orbital and the π donation from the $d\pi$ metal orbital to the CO π^* orbital seems to be well accepted[1]. Even single metal atom clusters such as NiCO have beeen extensively and successfully employed to model on top chemisorption of small molecules. For NiCO, it has been shown that the ground state of the cluster reproduces well the qualitative aspects of the local electronic structure of the adsorption site and the behavior of ground state properties[2]. The ab initio molecular many-body calculations of the core and valence core hole spectra of NiCO and NiN_2 by the CI (configuration interaction) method and Green's function method using an extended basis set, show that the ab initio many-body approaches seem to be able to give a detailed description of the dynamics of photoionization in the core and valence shell of adsorbates which seems to depend mainly on the localized nature of the excitations and even a small model molecule such as NiCO appears suitable to describe the relevant states of the adsorbates when the many-electron effects are treated in a proper way (see references in ref.3 and refs.3-5)

In contrast to NiCO, much less theoretical efforts have been made so far for the CO chemisorption on Pd and Pt metal surfaces. Theoretical studies of PdCO and PtCO show that the π bonding in NiCO is stronger than in PdCO and PtCO [6]. It would be certainly interesting to see whether there is any significant difference among the valence photoemission spectra of the CO/Ni, the CO/Pd and the CO/Pt systems because of a different degree of the π bonding in the ground state. By referring to the recent work for the detailed analysis of the valence spectra of PdCO and PtCO calculated by the ab initio ADC(3) (the third order algebraic diagrammatric construction) Green's function method, using an extended basis set [7], in the present work we focus on the spectral features of NiCO, PdCO and PtCO in order to study how the spectral features change according to the different degree of the π bonding.

2. METHOD

We use the Green's function formalism to calculate the IP's (ionization potential) and their spectral intensities directly. Using the spectral representation of the Green's function, the (vertical-electronic) ionization energies are given by the negative-pole positions in the Green's function. The residue provides the pole strength. This is, in the sudden approximation, a measure for the relative intensities of the states which derive their intensity from the same orbital. The Green's function can be calculated by using a well-established diagrammatic perturbation expansion in terms of the self-energy. The essence of the perturbation expansion lies in the renormalization of the self-energy. In the present work we calculate the self-energy by the ADC(3) scheme so that all diagrams appearing in the third order are summed to infinity and the 1p1h (one particle one hole) and 1h1h interactions in the 2h1p configurations are approximated within the framework of the random-phase approximation (RPA) and the ladder approximation, respectively. We refer refs. 3 and 8 for details of the method. The M (metal) - C distance optimised for the linear NiCO, PdCO and PtCO molecule is 3.2, 3.55 and 3.75 a.u., respectively [6,9]. The C - O distance is 2.2 bohrs [6,9]. The experimental M - C distance for the CO/Ni(100) and CO/Pd(100) systems is 3.5 and 3.65 a.u., respectively [10,11]. The spectra of NiCO, PdCO and PtCO presented in the present work were calculated at 3.2, 3.65 and 3.75, respectively. The spectra calculated at different M - C distances are reported elsewhere [3,7]. The ground state of NiCO, PdCO and PtCO is the $^1\Sigma^+$ state [6,9]. We use basis sets of Cartesian

Gaussian functions on the atoms to expand the MOs. We refer to refs. 3 and 7 for a detailed description of the basis sets used. The integral and SCF calculations have been performed with the program system MOLECULE [12]. In the ADC(3) Green's function calculations the eigenvalues and eigenvectors were extracted with a block Davidson method [13].

3. RESULTS AND DISCUSSION

In table 1 we list the HF(Hartree-Fock) eigenvalues of free CO, NiCO, PdCO and PtCO. In table 2 we list the ADC(3) results and experimental ionization energies. The present results refer to an isolated molecule and solid state effects associated with CO chemisorbed on a metal surface are totally neglected. However, it is interesting to compare the present results with the experimental data from the adsorbate CO on a metal surface to see whether the linear model molecule can be a suitable model for the adsorbate. Here a comparison with the theoretical results for the molecule is made simply by adding the workfunction to the binding energy of the adsorbed molecule measured relative to the Fermi level of the substrate, to bring them to a common reference level.

Table 1. Theoretical valence IP by Koopmans' approximation (eV)

level	CO	NiCO	PdCO	PtCO
1π	17.49	16.41	16.83	16.93
5σ	15.11	16.81	17.25	17.52
4σ	21.89	21.10	21.58	21.69

The HF eigenvalues have no physical relevance because of the neglect of ground state and final state correlation and relaxation in the final state. However, the eigenvalue shift due to a change of enviroment should indicate the initial state (chemical) shift (neglecting the ground state correlation energy shifts). For all coordinated molecules of interest now, the ordering of the 5σ and 1π levels (eigenvalues) is reversed compared to that of a free CO molecule. In comparison with free CO, in the coordinated molecule the 5σ orbital which is directed towards the metal atom is shifted to larger binding energy (B.E.), whereas the 4σ and 1π which are predominantly localized on the oxygen are shifted to smaller B.E.. For the 1π and 4σ levels, the chemical shifts at a fixed metal - carbon bond length are fairly independent of the substrate, whereas for the 5σ level it increases by as much as 1 eV in the sequence of Ni, Pd and Pt, thus reflecting the differences in the σ bonding in the ground state. When the C

- O distance is kept constant, the energy separation of the 1π and 4σ levels is fairly independent not only of the M - C bond length but also of the substrate. Furthermore it is only 0.4 eV larger than that of free CO. This indicates that the influence of the bonding on the 4σ and 1π levels is not significant. With an increase of the M - C bond length, the changes in the chemical shifts for the 1π and 4σ levels are negative, but that of the 5σ level is positive. This indicates also the presence of a specific initial state interaction between the CO 5σ orbital and the metal atom through the σ bonding. With an increase of the bond length the eigenvalues come closer to those of the free CO molecule.

We make an analysis of the spectra (spectral functions) obtained by the ADC(3) calculations. In the present work we did not take into account the photoionization cross section of each ligand level, so it is not possible to make a direct comparison of the results with the experimental photoemission spectrum, except for the ionization energies. The agreement with experiment is good.

(i) 1π level

The final state shift of the largest intensity state for NiCO, PdCO and PtCO are 2.1, 1.9 and 2.0 eV, respectively. It is fairly independent of substrates but much larger than that for free CO (0.5 eV). In NiCO the quasi-particle picture (Q.P.P.) of the 1π level breaks down due to the strong $1\pi^{-1} \leftrightarrow 1\pi^{-1}\pi_M^{-1}2(n)\pi$ metal - ligand CT (charge transfer) non hole-hopping static relaxation (SR). The spectral strength is distributed over several peaks, the intensities of which are comparable. In PdCO and PtCO the spectral lines are found almost at the same energies. This is because of the involvement of the same CT 2h1p configurations. The first two spectral lines are mainly due to the $\sigma_M^{-2}2(3)\pi$ CT 2h1p configuration induced by the hole-hopping dynamical relaxation (DR). The remaining spectral lines are mainly due to the $1\pi^{-1}\pi_M^{-1}2(3)\pi$ metal - ligand CT 2h1p configuration induced by the non hole-hopping SR. In PdCO the lowest energy state takes about 60 percent of the total spectral intensity and is the main line 1h state. The Q.P.P. is still valid. However, in PtCO the aforementioned configuration which involves the two holes in the same metal orbital interacts so strongly with the $1\pi^{-1}$ configuration that the lowest energy state looses a large part of its intensity to the first shakeup satellite and the Q.P.P. breaks down, e.g. the sum of the intensities of the first two spectral lines in PtCO and PdCO is very close. This is analogous to the case of $Cr(CO)_6$ and $Fe(CO)_5$ for which the CT 2h1p configurations which involve two holes in the same metal orbital leads to the breakdown of the Q.P.P. [14]. In free CO the 1π spectrum consists of one prominent peak. The dominant configuration is the $1\pi^{-1}$ configuration and the rest is the $1\pi^{-2}$ 2π 2h1p configuration. We thus see that the many-body effects are completely different in the free CO molecule and when it is bound to a transition metal atom.

Table 2. Theoretical valence IP of free CO, NiCO, PdCO and PtCO and experimental data from free CO and CO on Ni, Pd and Pt metal surfaces (eV)

level	theory CO	exp (a)	theory NiCO	exp. (b)	theory PdCO	exp. (c)	theory PtCO	exp. (d)
1π	16.98(0.9)	16.9	14.29(0.33)	13.8	14.94(0.62)	13.9	14.98(0.40)	14.3
			14.94(0.14)		15.59(0.05)		15.37(0.28)	
			15.53(0.14)		16.49(0.06)		16.71(0.05)	
			16.06(0.07)		17.38(0.03)		17.64(0.02)	
			18.34(0.06)		20.24(0.11)		20.30(0.09)	
			19.51(0.11)				20.55(0.03)	
5σ	13.91(0.89)	14.0	15.01(0.75)	13.8	14.87(0.14)	13.9	14.90(0.03)	14.3
			17.40(0.08)		15.21(0.67)		15.48(0.74)	
			20.23(0.03)		18.57(0.04)		15.70(0.03)	
					20.39(0.03)		18.46(0.04)	
							19.99(0.02)	
							20.08(0.01)	
4σ	20.08(0.79)	19.7	18.6(0.65)	17.0	18.57(0.09)	16.8	18.46(0.06)	17.4
	23.0(0.11)		21.04(0.06)		18.97(0.65)		19.13(0.65)	
			23.92(0.05)		20.39(0.02)		20.08(0.03)	

(a) ref.15 (b) ref.16 (Ni(100) surface with 6.3 eV workfunction)
(c) ref. 11 (Pd(100) surface with 6.0 eV workfunction)
(d) ref. 17 (Pt(111) surface with 5.7 eV workfunction)

(ii) 5σ level

The final state shift for NiCO, PdCO and PtCO are 1.8, 2.0 and 2.0 eV, respectively. It is fairly independent of substrates but much larger than that for free CO (1.2 eV). In free CO the 5σ spectrum consists of one prominent peak. The main line is the 1h state. In NiCO the 5σ main line state is the 1h state where the screening charge resides on the bonding orbital, the character of which does not change much from that of the ground state. The satellite in NiCO is dominated by the $5\sigma^{-1}\pi_M^{-1}$ $2(n)\pi$ CT 2h1p configurations induced by the non hole-hopping SR. In contrast to NiCO, in both PdCO and PtCO the $1\pi^{-1}\sigma_M^{-1}2(3)\pi$ CT 2h1p shakedown satellite of the non-negligible intensity induced by the hole-hopping DR, becomes the lowest energy state.

The intensity of this satellite is considerably smaller in PtCO than in PdCO mainly because of the extra satellite due to the $\sigma_M^{-2}6(\mathrm{n})\sigma$ 2h1p configuration where two holes are created in the same metal orbital as in the case of 1π ionization. The rest of the satellite lines are mainly due to the $\sigma_M^{-1}\pi_M^{-1}2(3)\pi$, $5\sigma^{-1}\pi_M^{-1}$ $2(3)\pi$ and $\mathrm{M}^{-2}6(\mathrm{n})\sigma$ (metal - ligand CT) 2h1p shakeup configurations.

(iii) 4σ level

The final state shift for NiCO, PdCO and PtCO is 2.5 eV. It is independent of substrates, but much large than that for free CO (1.8 eV). In free CO the 4σ main line is the 1h state. The satellite is dominated by the $5\sigma^{-1}1\pi^{-1}$ $2(\mathrm{n})\pi$ 2h1p configuration which is induced by the intra ligand DR. In NiCO the 4σ main line state is the 1h state. The satellite in NiCO is dominated by the $4\sigma^{-1}\pi_M^{-1}$ $2(\mathrm{n})\pi$ CT 2h1p configuration induced by the metal - ligand CT SR process. In PdCO and PtCO as for the 5σ ionization, the lowest energy state is not the main line 1h state but the shakedown state of a substantial intensity. The second spectral line is the main line 1h state. The shakedown satellite line is mainly due to the σ_M^{-1} $\pi_M^{-1}2(3)\pi$ metal - ligand CT 2h1p shakedown configuration. The shakeup satellite is mainly due to the same CT 2h1p configurations.

For the 5σ and 4σ levels of NiCO the lowest energy state is the 1h main line state, however for PdCO and PtCO it is the metal - ligand CT 2h1p shakedown state. This dramatic difference in the spectral feature of the 5σ and 4σ levels among NiCO, PdCO and PtCO is most likely due to the different degree of the π CT. The π bonding is considered to be stronger in NiCO than in PdCO and PtCO [6]. Then for PdCO and PtCO, in the ground state the π bonding orbital is more metal like than in NiCO. In order to screen the valence hole, the $2\pi^*$ empty level has to be pulled down below the Fermi level so that the bonding orbital becomes much more CO like. For the 5σ and 4σ levels, the 2h1p shakedown state induced by the hole-hopping σ (π) to π^* metal - ligand CT DR becomes more stable than the 1h state. Here the CT occurs from the σ metal orbital rather than the π metal orbital because of a weaker π bonding. For NiCO, the bonding orbital has already some CO like character in the ground state. Then the metal - ligand CT 2h1p shakeup leads to an unstable excited state which eventually relaxes to the more stable main line 1h state.

The theoretical total spectral intensity of the lower shakeup energy region where the metal - ligand CT excitations dominate is almost same as the theoretical main line intensity of free CO. For a coordinated molecule the main line intensity of free CO is distributed over several spectral lines by the metal - ligand CT excitations and the intra ligand exitations appear in the higher energy region, although the excitations mix very much with the metal-ligand CT excitations and much less pronounced than in a free molecule. This is also the case with metal carbonyls [14] and the core hole spectra of the adsorbates [5].

By using an angular resolved resonant photoemission spectroscopy, one can boost the intra ligand shakeup satellite intensity of a particular symmetry. However, the metal-ligand shakeup(down) satellite intensity is less pronounced because of a much smaller spectator Auger decay matirx element, although the metal-ligand shakeup(down) satellites can be separated from the intra ligand ones because of their smaller excitation energies.

4. CONCLUSION

The valence photoemission spectra of NiCO, PdCO and PtCO were calculated by the ab-initio Green's function method, using an extended basis set. The overall good agreement with the experimental spectra of the adsorbates shows that the ab-initio molecular many-body approach such as the present Green's function method seems to be able to give a detailed description of the dynamics of the valence photoemission from the adsorbates. Even small molecules such as NiCO, PdCO and PtCO seem to be suitable for the description of the relevant states of the adsorbates when the many-body effects are treated in a proper manner. For the 5σ and 4σ levels, in NiCO the lowest energy state is the main line 1h state and not the CT 2h1p shakedown state. On the other hand, in PdCO and PtCO the metal - ligand CT 2h1p shakedown satellite state of a non-negligible intensity induced by the hole-hopping σ (π) to π^* metal-ligand CT DR becomes the lowest energy state. The Q.P.P. is still valid and the main line state is still the 1h state. This dramatic spectral difference arises because of a weaker π bonding in PdCO and PtCO than NiCO. For the systems where the π CT is weak (the CO/Pd, CO/Pt and N_2/Ni systems), the CT from the σ metal orbital becomes substantial. For the 1π ionization, the Q.P.P. is valid for PdCO. In PtCO the lowest energy state which corresponds to the main line in PdCO looses a large part of the intensity to the first satellite and the intensity of both spectral lines becomes comparable. The Q.P.P. begins to break down. This intensity transfer occurs due to the involment of the $\sigma_M^{-2}2(3)\pi$ 2h1p CT configuration induced by the hole - hopping σ to π^* CT DR. In NiCO, the Q.P.P. breaks down due to the π metal - ligand CT SR. The calculation also shows that the main line intensity of free CO is distributed over several spectral lines in the coordinated molecule because of the metal - ligand CT excitations. This shows that the basic spectral feature difference between free molecules and adsorbates.

ACKNOWLEDGEMENT

One of the authors (M.O.) would like to thank DFG (Germany) and NFR (Sweden) for financial support.

REFERENCES

1. G. Blyholder, J. Chem. Phys. 68, 2772, (1964), J. Vac. Sci. Technol. 11, 865 (1974)

2. C. W. Bauschlicher Jr., S. R. Langhoff, and L. A. Barnes, Chem. Phys. 129, 431 (1989) and references therein

3. M. Ohno and W. von Niessen, Phys. Rev. B 42, 7370, (1990) and references therein

4. M. Ohno and W. von Niessen, Phys. Rev. B in press

5. M. Ohno and P. Decleva, in this proceedings, Sur. Sci. in press and Chem. Phys. in press.

6. P. J. Hay and C. M. Rohlfing in Quantum Chemistry : The challenge of transition metals and coordination chemistry, edited by A. Veillard (Reidel, Dordrecht 1986) p.135 and references therein

7. M. Ohno and W. von Niessen, Phys. Rev. B submitted

8. J. Schirmer, Phys. Rev. A 26, 2395, (1982), J. Schirmer, L.S. Cederbaum , and O. Walter, Phys. Rev. A 28, 1237, (1983)

9. C.W. Bauschlicher, Jr. , Chem. Phys. lett. 115 ,387, (1985) and references therein.

10. S. Anderson and J. B. Pendry, Phys. Rev. Lett. 43, 363, (1979)

11. R. J. Behm, K. Christmann, G. Ertl and M. A. Van Hove, J. Chem. Phys. 73, 2984 (1980)

12. J. Almlöf, A vectorized MOLECULE code, Final report on the project supported by NASA Ames Grant No. NCC 2 - 158 (Dec. 1984)

13. E.R. Davidson, J. Comput. Phys. 17 , 87, (75).B. Liu, NRCC Workshop on Numerical Algorithms in Chemistry, Algebraic Methods 1978, p.49; F. Tarantelli, unpublished, we used a program by this author.

14. M. Ohno and W. von Niessen, Phys. Rev. B , J. Chem. Phys. in press.

15. D. W. Turner, C. Barker, A. D. Barker and C. R. Brundle, Molecular Photoelectron Spectroscopy (Wiley - Interscience, London 1970)

16. A. Nilsson and N. Mårtensson, Phys. Rev. B40, 10249 (1989)

17. R. Murphy, E. W. Plummer, C. T. Chen, W. Eberhardt and R. Carr, Phys. Rev. B39, 7517 (1989)

PHOTOCHEMICAL ETCHING OF Cu AND GaAs WITH Cl_2 IN THE VUV

B. Li, I. Twesten, M. Chergui and N. Schwentner
Inst. für Experimentalphysik, FU Berlin, Arnimallee 14, D–1000 Berlin 33

ABSTRACT

Photochemical reactions and etching efficiencies of polycristalline Cu and GaAs (100) with Cl_2 have been studied in the spectral range of 105 – 300 nm by using synchrotron radiation and for Cl_2 pressures between 10^{-6} and 10 mbar. Dissociation of Cl_2 leads in the case of Cu to the formation of $CuCl_x$ films (typical thickness 10^4 nm) with an efficiency of about 10^7 $CuCl_x$ molecules per generated Cl atom at high Cl_2 exposures. For low Cl_2 exposures, anisotropic $CuCl_x$ growth is observed in the irradiated area (typical thickness 50 nm) and the efficiency decreases with increasing photon energy. Anisotropic etching shows up for GaAs with typical etch depth of 100 nm in the irradiated area. The efficiency increases in general with photon energy, but exhibits in addition strong variations within small photon energy ranges for example between 118 and 122 nm. The efficiency increases strongly with pressure in the range from 10^{-1} to 10 mbar.

1. INTRODUCTION

Chemically assisted etching processes like plasma and reactive ion etching are widely applied in present–day microelectronic circuit fabrication[1]. Improved techniques should reduce damage on the active area of devices and increase selectivity as well as anisotropy of etching[2]. Photon–assisted etching of semiconductors and metals with gases has been demonstrated in recent years, employing lasers[3-12] and synchrotron radiation[13,14]. An extension of these studies to shorter wavelengths into the vacuum ultraviolet and soft X–ray range is an obvious research direction. Spatial resolution can be improved in principle due to less stringent diffraction limitations. In addition many gases possess high absorption cross sections and efficient dissociation and reaction channels in the VUV range. Synchrotron radiation (SR) provides several advantages for example a continuum of wavelengths and a small divergence. It is a well suited tool to determine efficiencies of etching processes at short wavelengths, an information required to choose optimal wavelengths and etching conditions.

Our experiments are designed to investigate etching effiencies in the spectral range from 105 to 300 nm by using SR from BESSY in Berlin. Measurements have been performed on combinations of Cu, GaAs and Si with Cl_2, HCl, CF_4 and XeF_2. In this paper we present results on the photon induced reaction of poly–Cu and GaAs (100) with Cl_2. SR investigations on etching of Si and GaAs with Cl_2 and of SiO_2 with SF_6 have been reported[13,14,15]. The dark reaction of Cu with Cl_2 has been studied in detail[16] and laser assisted etching has been performed[4,5,17].

2. EXPERIMENT

A cell of 2 x 2 x 1.5 cm^3 volume covered with a LiF window of 2 mm thickness is placed close to the exit slit of a 3m normal–incidence monochromator (beamline 3mNIM2[18]) between the exit slit and the grating of the monochromator. The sample is mounted at the back side of the cell just in the focal area of the monochromator. Two different modes of irradiation are possible. Operating with the grating in first order yields an illuminated 2 cm long area on the sample and wavelengths in a range of about 10 nm are dispersed along this 2 cm. This 10 nm wavelength range can be tuned from 100 up to 300 nm by turning the grating. The efficiency variation within the 10 nm range is obtained from one illumination by measuring the change of the etch depth along the illuminated part. In the second mode of irradiation, the grating is turned to zeroth order yielding a spot of white light of 1 x 2 mm at the sample. The spectral bandwidth is limited to wavelengths λ longer than 150 nm, 125 nm and 105 nm respectively by introducing a quartz or a CaF_2 filter and by the LiF window.

The cell is evacuated with a turbomolecular pump to a background pressure of 10^{-6} mbar. A continuous flow of gas in a pressure range from 10 to 10^{-3} mbar is delivered by a stainless steel UHV gas handling system. Cl_2 with a purity of 99.998 %, polycristalline Cu samples and silicon–doped GaAs (100) wafers have been used. The distance between the sample and the LiF window, which is filled with gas corresponds to 1 cm. The samples are covered with a Ni mesh of 64% transmission consisting of 10 μ thick wires with a separation of the wire centers of 50 μ. The samples are removed after typical irradiation times of 1 to 100 minutes. The mesh structures which have been transferred to the sample inside and outside the focal spot are fotographed in a microscope. Depth profiles of the surface are measured by a mechanical stylus (Dectac 3030).

3. RESULTS FOR THE REACTION OF Cu WITH Cl_2

3.1 PHOTOCHEMICAL EFFICIENCIES

Cu reacts with Cl_2 already in the dark and a white $CuCl_x$ layer is formed with a composition x depending on depth and exposure conditions and varying between $0 < x < 2$ [19]. We have measured the growth of the $CuCl_x$ films in the dark for our typical exposure conditions (Cl_2: 3 x 10^{-2} mbar, 10 min.) and a result is shown in fig. 1a. The left hand part of the sample has been protected from Cl_2 and the stage between protected and unprotected part is displayed. The density of $CuCl_x$ is about 3.1 times smaller than that of Cu and the sample swells in vertical direction due to the growth of the film[19]. The step height of 0.5 μ therefore corresponds to reaction of a pure Cu film with a thickness of 0.25 μ and the total $CuCl_x$ height corresponds to about 0.75 μ (fig. 1a).

A similar Cu sample has been covered with a Ni mesh, has been exposed for 15 min. to a Cl_2 pressure of 1.5 x 10^{-2} mbar and to the Synchrotron radiation with a white spectrum of wavelengths with $\lambda > 150$ nm by insertion of a quartz filter, with $\lambda > 125$ nm by a CaF_2 filter and $\lambda > 105$

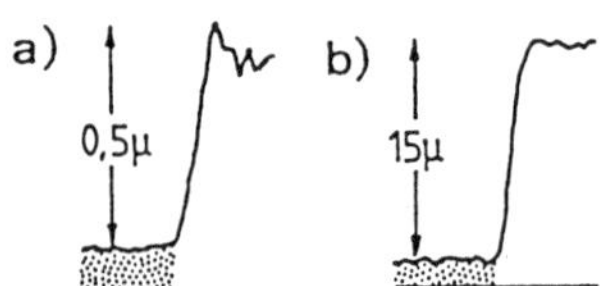

Fig. 1: Step height from Cu surface to top of the $CuCl_x$ film after exposure to 1.5 x 10^{-2} mbar Cl_2 for 15 min. a) dark reaction, b) after irradiation with $\lambda > 105$ nm.

nm by the LiF window, respectively.The intensity decreases in all cases slowly for wavelengths longer than 300 nm[18]. The three wavelength regions yield similar; results for all parts outside of the focal area.
A $CuCl_x$ film is formed on the whole plate of 2.3 x 2.5 cm^2 which is substantially thicker than without irradiation (fig. 1). Part of the film has been removed from the Cu substrate with a tape and the step height between the substrate and the top of the film is shown in fig. 1b. The $CuCl_x$ step height of 15 μ observed with irradiation is more than a factor of 10 larger than without irradiation despite the fact that the exposure times and exposure pressures are comparable. It has to be pointed out that films of such a large thickness are not observed in the dark reaction even if the exposure pressure would be increased by an order of magnitude[19]. Therefore this increase has to be attributed to the influence of the irradiation. It is remarkable that the whole Cu plate is covered with $CuCl_x$ of comparable thickness and not only the focal spot on the sample. This excludes a light induced reaction at the surface and the excitation of the Cl_2 gas has to be responsible for the increased reactivity. The mean free path at the emploied pressures is large and Cl fragments can reach any point of the Cu plate without appreaciable losses by recombination.

The absorption cross sections of Cl_2 [20,21,22] show two major absorption bands, one centered around 330 nm and one around 170 nm and both lead to dissociation of Cl_2. A calculation based on flux and absorption cross sections and a test experiment with a Hg lamp indicate, that the Cl_2 dissociation in the 330 nm range is mainly responsible for the reaction. A comparison of the number of CuCl molecules in the film with the number of generated atoms shows, that one Cl atom from the photodissociation process stimulates 10^7 Cl_2 molecules to form $CuCl_x$ [23]. This huge amplification factor asks for further investigations to clarify the microscopic process.

3.2 WAVELENGTH DEPENDENT EFFECTS

Cu samples covered with the Ni mesh have been exposed for 10 minutes to a Cl_2/Ar (1.6:100) mixture with a pressure of 4 x 10^{-3} mbar (Cl_2 partial pressure of 6.4 x 10^{-5} mbar) and to white light in the wavelength regions $\lambda >$ 150 nm, $\lambda >$ 125 nm and $\lambda >$ 105 nm. Fig. 2 shows a picture of the $CuCl_x$ surface in the optical microscope after etching with $\lambda >$ 105 nm (LiF window). Similar pictures were obtained for the quartz and CaF_2 filters. The structured centre represents the focal spot. This picture demonstrates that the etching inside the focal spot is different from that outside at these low Cl_2 exposures. The contrast of the structures inside the focal point is caused by a different composition x of the $CuCl_x$ compounds because CuCl and $CuCl_2$ crystals have different colours. The thickness of the $CuCl_x$ film outside the focal spot is smaller than 100 Å and the mesh structure is absent in this part. A clear reproduction of the structure can be seen within the focal spot with a modulation depth of several hundred Å. The structures at the sharp edge on the right hand side of the focal spot (fig. 2) measured by the stylus are shown in fig. 3a. The focal area is at a higher level than the surrounding due to the swelling of the $CuCl_x$ film. It has a pronounced mesh–structure and the $CuCl_x$ thickness decreases in leaving the focal spot. Obviously an anisotropic reaction is obtained in the illuminated area at low Cl_2 pressures (6.4 x 10^{-5} mbar) and in addition the film thickness can be controlled by the photon energy. The step height between the wires (fig. 3b) varies for otherwise identical

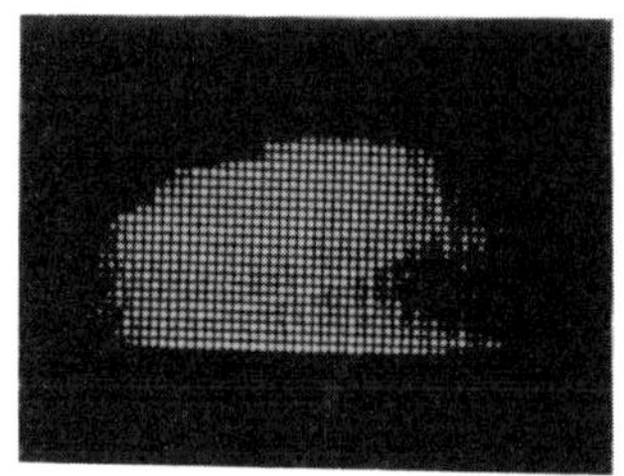

Fig. 2: Mesh structure (grid size 50 μ) on Cu surface in focal area due to irradiation by $\lambda > 105$ nm for 10 min with Cl_2 partial pressure of 6.4 x 10^{-5} mbar.

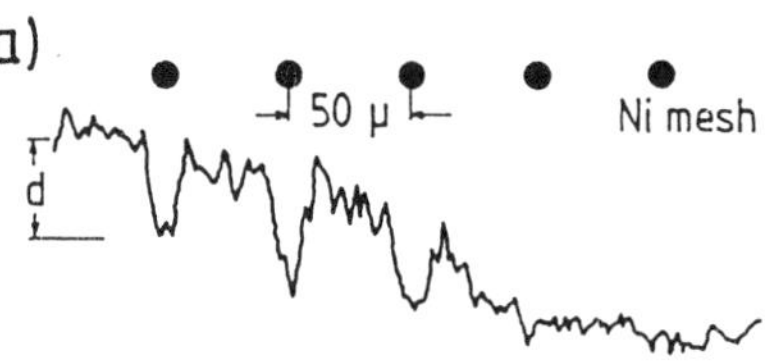

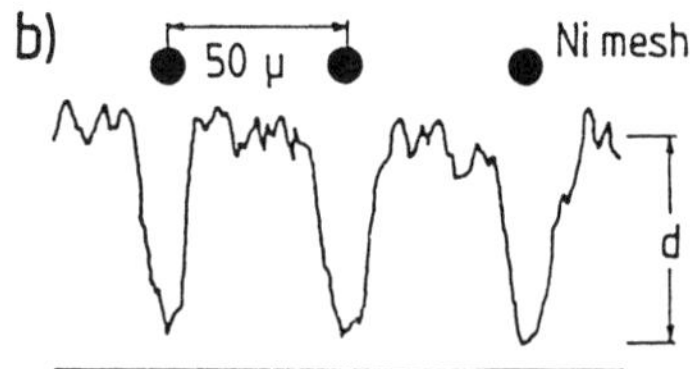

Fig. 3: Depth profile along surface in the focal spot of fig. 2. a) close to border of focal area, b) center of focal spot. Depth d corresponds to 460 Å, 310 Å and 220 Å for quartz, CaF_2 and LiF filters respectively.

illumination conditions with the filters. It decreases from 460 Å with quartz (6.5 Å/mA h) to 310 Å with CaF_2 (4.4 Å/mA h) and to 220 Å with LiF (3.1 Å/mA h). The decrease of efficiency by inclusion of shorter wavelengths indicates that the growth initiated by the long wavelengths is counteracted by the short wavelengths. Photon induced desorption of Cl or Cl_2 resulting in a smaller effective sticking coefficient and/or dissociation of $CuCl_x$ molecules in the film and thus a change in the dynamical equilibrium of $CuCl_x$ formation can be responsible for the decrease of efficiency with shorter wavelenghts.

4. RESULTS FOR THE ETCHING OF GaAs WITH Cl_2

4.1 ETCHING IN TWO DIFFERENT WAVELENGTH REGIONS (LiF, QUARTZ).

GaAs (100) wafers were etched with Cl_2 (pressure 1.5 mbar) and by exposure to light in zeroth order of the grating with a quartz filter or a LiF window. A dose of 314 mA h was accumulated during an irradiation time of 50 min. A clear pattern of the etched mesh structures appears in the focal spot (fig. 4) for irradiation with the LiF window ($\lambda > 105$ nm) and the structures are absent outside of the illuminated part.

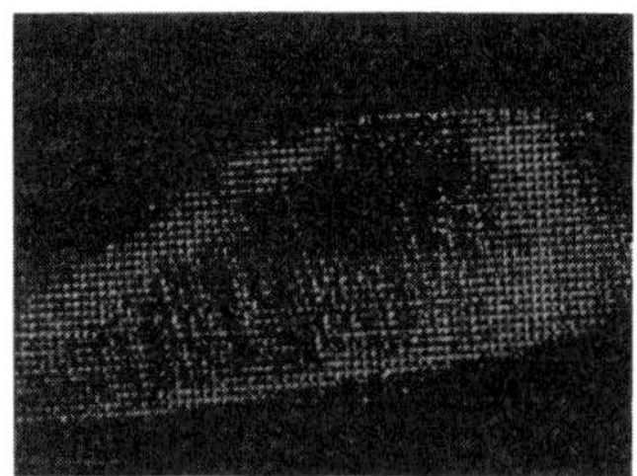

Fig. 4: Microscopic picture of etched mesh structure (grid size 50 μ) on GaAs (100) for irradiation with $\lambda > 105$ nm at a dose of 314 mA h and a Cl_2 pressure of 1.5 mbar.

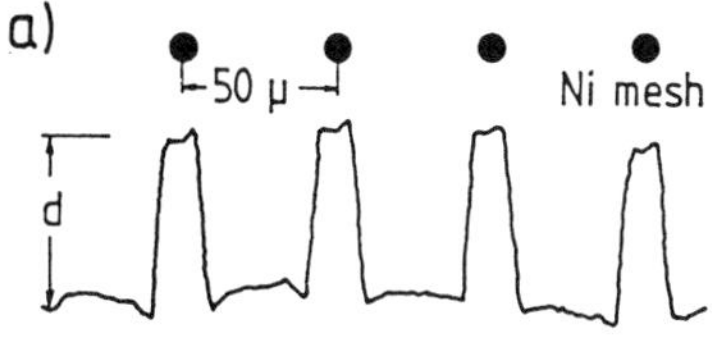

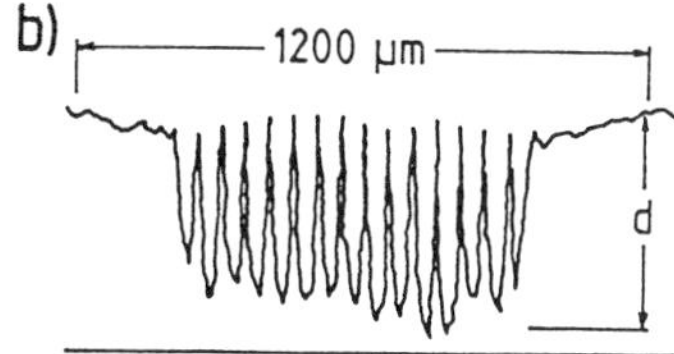

Fig. 5: Etching depth profile of GaAs with Cl_2 and $\lambda > 105$ nm a) in the center of focal spot, b) survey of whole focal area. d = 1300 Å for $\lambda > 105$ nm (d = 670 Å for $\lambda > 150$ nm).

Fig. 5 shows depth profiles measured with the stylus. The etching depth between the wires reached 1300 Å in the focal spot (fig. 5a). A survey of the profile across the whole diameter of the focal spot is reproduced in fig. 5b. It demonstrates that there is no deposition or etching outside the focal spot and that the etching goes into the depth of the GaAs sample. Comparable etching profiles have been obtained for irradiation with the quartz filter ($\lambda >$ 150 nm), but the etching depth was reduced to one half of that with the LiF window for similar exposure conditions. The etching rates amount to 4.5 Å/mA h for $\lambda > 105$ nm and to 2.2 Å/mA h for $\lambda > 150$ nm. Obviously the spectral composition plays an important role for the etching efficiency of GaAs with Cl_2. In GaAs the efficiencies increase for inclusion of the shorter wavelength ranges contrary to the case of Cu.

4.2 ENERGY RESOLVED ETCHING

One attempt has been made to etch with the wavelength dispersed radiation from the monochromator. A GaAs (100) sample covered with the Ni mesh was exposed to Cl_2 (1.5 mbar pressure) for an irradiation dose of 438 mA h. The grating was tuned in a position that the wavelength region from 117 nm to 123 nm was dispered along a 2 cm long region of the sample. The etched structures along the irradiated part have been fotographed by successive microscope pictures and the comparison of these pictures with the corresponding wavelength scale is shown in fig. 6. The wavelengths are

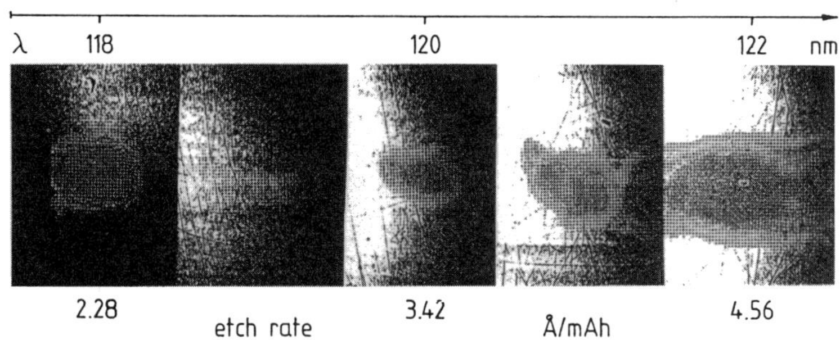

Fig. 6: Composed microscope picture of an etched GaAs surface for a wavelength region extending from 117 nm (left hand side) to 123 nm (right hand side) with the etching rates at 118, 120 and 122 nm given in Å/mA h.

dispersed horizontally and the vertical extension of the structures represents the size of the focus. Some parts are etched strongly and others not, or only weakly. This indicates that the etch rates vary significantly with wavelengths even within the very small displayed range of photon energies. The depths for some strongly etching wavelengths, for example 118 nm, 120 nm and 122 nm correspond to 1000 Å, 1500 Å and 2000 Å, respectively. It is interesting to note that the efficiency at λ = 180 nm is too weak to be determined in this way. The wavelength dependence of the etch rates in a large range of photon energies will be investigated in future experiments.

4.3 DEPENDENCE OF ETCH RATE ON Cl_2 PRESSURE

The variations of the etch rate with Cl_2 pressure have been determined for white light exposure (LiF window, 50 minutes) in a pressure range from 1.5 x 10^{-2} mbar up to 10 mbar. The etching rates increase progressively with Cl_2 pressure (fig. 7). An extremely strong growth of the etch rate was observed for a Cl_2 pressure of 3 mbar and a prolonged irradiation time of 100 min (dose 510 mA h). In this case not only the focal area but also the outside parts were etched with a mean depth of about 2 μ. The largest depth of 6.5 μm was observed in the irradiated area. This result at very high Cl_2 pressures is similar to those of Terakado et al.[14]. The etching reaction in the nonirradiated region can be suppressed according to these experiments by reducing the substrate temperature and selective etching in the irradiated region is dominant.

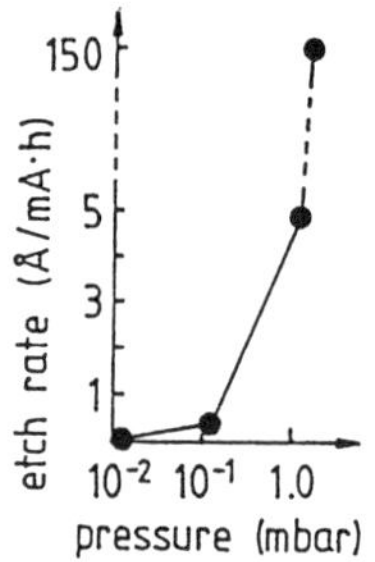

Fig. 7: Etching rate of GaAs versus Cl_2 pressure for $\lambda > 105$ nm at room temperature.

Our observations and the results of Terakado et al. suggest that the photon–assisted etching of GaAs with Cl_2 represents a complex superposition of several processes including direct and indirect dissociation of Cl_2 [23]. The balance between the processes can be controlled by three parameters: photon energy, gas pressure and substrate temperature. Variation of one of the parameters strengthens some of these processes and supresses others. A gas pressure of 1 mbar Cl_2 represents the optimal condition for a highly selective etching of GaAs at room temperature in the wavelength range of 105 – 300 nm according to our results. Further investigations of the dependence of the etch rate on wavelength, pressure and temperature will be performed to resolve the complicated interplay of the different processes.

ACKNOWLEDGEMENTS

The authors would like to thank Dr. J. Janes (IMT) for helpful discussions, Dr. M.–P. Macht (HMI) for providing the mechanical stylus and Wacker Chemitronic GmbH for delivering GaAs samples. We also want to acknowledge the invaluable experimental backing of R. Schriever and specially H. Kunz.
This work has been supported by the Bundesministerium fuer Forschung und Technologie under contract No. 05 413 AXI7TP5.

REFERENCES

1. N.G. Einspruch, D.M. Brown: Plasma processing for VLSI; VLSI Electronics Microstructure Sci., Vol. 8, Academic, New York (1984)
2. M. Hirose: Photon, Beam and Plasma Assisted Processing, ed. by I.W. Boyd and E.F. Krimmel (1989); Euro. Mat. Res. Soc. Sym. Proceeding **2**, 1988
3. W. Sesselmann and T.J. Chuang: J. Vac. Sci. Technol. **B3** (5), 1507 (1985)
4. T.S. Baller, G.N.A. van Veen, J. Dieleman: J. Vac. Sci. Technol. **A6** (3), 1409 (1988)
5. J.H. Brannon, K.W. Brannon: J. Vac. Sci. Technol. **B7** (5), 1275 (1989)
6. P. Brewer, S. Halle, R.M. Osgood Jr.: Appl. Phys. Lett. **45** (4), 475 (1984)
7. Q.Z. Qin, Y.L. Li, Z.K. Jin, Z.J. Zhang, Y.Y. Yang, W.J. Jia, Q.K. Zheng: Surf. Sci. **207**, 142–158 (1988)
8. V. Liberman, G. Haase, R.M. Osgood Jr.: Chem. Phys. Lett. **176** (3,4), 379 (1991)
9. V. Svorcik, V. Rybka: Jpn. J. Appl. Phys. **28** (9), L1669 (1989)
10. D. Bäuerle: Chemical Processing with Lasers, Springer Series Mat. Sci. **1**, (Springer, Berlin, Heidelberg 1986)
11. R. Kullmer, D. Bäuerle: Appl. Phys., **A47**, 377 (1988)
12. S. Affrossman, R.T. Bailey, C.H. Cramer, F.R. Cruicksbank, J.M.R. MacAllister, J. Alderman: Appl. Phys. **A49**, 533 (1989)
13. N. Hayasaka, A. Hiraya and K. Shobatake: Jpn. J. Appl. Phys. **26** (7), L1110 (1987)
14. S. Terakado, J. Nishino, M. Morigami, M. Harada, S. Suzuki, K. Tanaka, J. Chikawa: Jpn. J. Appl. Phys. **29** (5), L709 (1990)
15. K. Shobatake, H. Ohashi, K. Fukui, A. Hiraya, N. Hayasaka, H. Okano, A. Yoshida, H. Kume: Appl. Phys. Lett. **56** (22), 2189 (1990)
16. W. Sesselmann and T.J. Chuang: Surf. Sci. **176**, 32 and 67 (1986)
17. W. Sesselmann, E.E. Marinero and T.J. Chuang: Surf. Sci. **178**, 787 (1986)
18. J. Bahrdt: Dissertation (1987); J. Bahrdt, P. Gürtler, N. Schwentner: J. Chem. Phys. **86**, 6108 (1987)
19. W. Sesselmann and T.J. Chuang: Surf. Sci. **176**, 32 and 67 (1986)
20. H. Okabe: Photochemistry of Small Molecules (John Wiley & Sons, New York, 1978)
21. C. Roxlo, A. Mandl: J. Appl. Phys., **51** (6), 2969 (1980)
22. L.C. Lee, M. Suto: J. Chem. Phys., **84** (10), 5277 (1986)
23. B. Li, I. Twesten, N. Schwentner: to be published

II. CORE-HOLE EXCITATION AND RELAXATION

PHOTOIONIZATION OF LASER INDUCED DISSOCIATED MOLECULES

P.Morin[1,2], L.Nahon[1,3] and I.Nenner[1,2]

[1]LURE, bat 209d, Univ. Paris-Sud, 91405 ORSAY Cedex, FRANCE

[2]Commissariat à l'Energie Atomique, CEN Saclay, DRECAM, SPAM, 91191 GIF sur YVETTE Cedex, FRANCE

[3]X-RS, Parc Club, 28 rue Jean Rostand, 91893 ORSAY Cedex, FRANCE

Abstract:

Experiments combining laser and synchrotron radiation to respectively photodissociate molecules and ionize the resulting fragments are shown to be suitable either to study photodissociation reactions ("pump and probe" experiment) as well as photoionization dynamics of open shell systems.

1) dissociation products issued from the laser induced fragmentation of a polyatomic molecule are probed by photoelectron spectroscopy. Synchrotron radiation is used as a probe, able to photoionize all the fragments. It is thus possible, on one hand to determine without any ambiguity the products of the reaction, and on the other hand to measure the internal (vibrational) energy carried out by the fragments. As an example we have been able to establish that in the case of s-tetrazine ($C_2N_4H_2$), the fragments (N_2 and HCN) are rather cold in their streching modes, and rather hot in the bending mode of HCN. We have shown that the s-tetrazine dissociation is well described by the so called "equilibrium geometry of the transition state" model.

2) radicals produced by the laser induced dissociation of a suitable precursor molecule, can be photoionized with synchrotron radiation. Photoelectron spectroscopy is used to study the resonances located below and above edges and to look at their relaxation. Photoionization of atomic iodine in the 4d subshell reveals specific behaviour due to the open shell character of the species.

Introduction

Molecular fragmentation induced by absorption of light plays a prominent role in photochemistry. Indeed it can be viewed as an elementary process of a chemical reaction and it is a basic process in various related fields like atmospheric pollution, interstellar chemistry or combustion dynamics which recently received much attention.

Dissociation dynamics has intrigued chemists for a long time; especially in the case where three or more fragments are emitted, the question emerges to know if the chemical bonds are simultaneously or sequentially broken and how fast those processes take place; we thus have to distinguish between concerted and not concerted reactions. Various models try to describe simplified situations (like the "statistical model" (RRKM theory), impulsive model, equilibrium geometry model) and give predictions on the energy partitioning in the photodissociation[1]. Useful comparison can be made with theory only if the experiment is able to give enough informations like, ideally:

- branching ratio between the various dissociation channels
- internal and translational energy of the fragments
- correlations between the various kinetic momenta of the particles

It is the reason why " two color" or "pump and probe" experiments have successfully been developed. In this kind of experiment, a first laser beam induces a transition in the parent molecule into a selected (pre)dissociative state. Interaction with a second laser beam allows a precise determination of the state of selected fragments (by mean of Laser Induced Fluorescence). This state to state photochemistry is nevertheless limited because of the low energy photons available with usual lasers. The advantages of using a laser as a probe are the high intensity available which makes this technique very sensitive, the very narrow spectral width which allows very precise spectroscopy (determination of vibrational, rotational and translational energy) and finally the possibility to use synchronised pulsed laser allowing in some cases a direct determination of time evolution of the system.

The main advantages of using the synchrotron radiation as a probe are its high energy which allows to ionize (i.e. to probe) all the nascent fragments, and its broad tuning range which allows to chose the more adequate photon wavelength to ionize a selected fragment.

The use of either a laser or synchrotron radiation as a probe appears as two complementary approaches: the good resolution obtained with laser allows determination of precise internal energy like rotation of fragment. But on the other hand this precision is so good that it can be applied only to systems which spectroscopy is well known. Also the informations obtained are generally focussed on one particular fragment.

Synchrotron radiation however, combined with electron spectroscopy is able to give an overview of the dissociation process by giving immediately branching ratios between the various products of fragmentation[2]. It can also provide informations on the vibrational internal energy of some fragments.

Ionization with synchrotron radiation of laser dissociated molecules may also be used for an other purpose: the study of photoionization of short lived or very reactive radicals. In that case the interest is not the laser induced dissociation of the parent molecule, but the ionization dynamics of a particular fragment. When the precursor molecule is correctly chosen (high dissociation rate, production of few type of fragments), this method provide a "clean" *in situ* source of transient species, that can further undergo ionization with synchrotron radiation. The absence of any electric or magnetic field allows proper operation of electron analyzer or mass spectrometer.

Of special interest is the study of halogen radicals for which photoionization studies are very scarce, and which are expected to show specific behaviour due to their open shell character.

In the present paper, we present some applications of the combination of laser and synchrotron radiation illustrating the physical as well as the chemical point of view of laser induced molecular photodissociation. The ionization of atomic iodine is first described as it also shows the efficiency of the method. Then we present the case of s-tetrazine photodissociation, for which the equilibrium geometry model is shown to describe correctly the fragmentation process. Both works have been subject of detailed publications[3-6].

PHOTOIONIZATION OF RADICALS: ATOMIC IODINE

In this experiment iodine molecules are photodissociated in two ground state iodine atoms with a cw laser and then photoionized by synchrotron radiation in the 4d shell excitation range. The basic interest of such an experiment was to emphasize the role played by the unfilled 5p orbital, when the 4d shell was excited or ionized.

A schematic of the experimental set up is shown in figure 1. Atomic iodine is produced by the laser dissociation of I_2, by inducing mainly a transition into the ($^1\Pi_{1u}$) repulsive state of I_2, correlated to two atoms in the ($^2P_{3/2}$) ground state. A high power (8 Watts) laser beam was obtained from a cw Argon-ion laser used in a visible mode (mostly 514 and 488 nm).

Collinearly to the laser beam, the S.R. emitted from SuperACO storage ring, monochromatized by a 2.5 toroidal grating monochromator, crosses the interaction zone and photoionizes the iodine atoms. The ejected electrons are energy analyzed with a 127° electrostatic cylindrical analyzer. We have shown[3] that a laser power of 8 Watts was enough to achieve a quasicomplete photodissociation of iodine. Indeed, in figure 2 we show a set of photoelectron spectra recorded without laser (2a), intermediate (2b) and with a 8 Watts (2c) power. We readily see in figure 2c that the iodine molecular dissociation is almost fully completed. We measured a dissociation rate close to 95%.

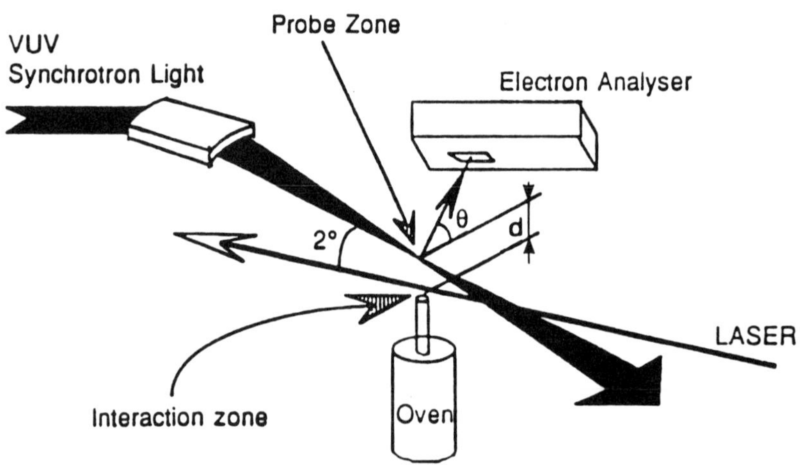

Figure 1 Schematic

of the experimental set up

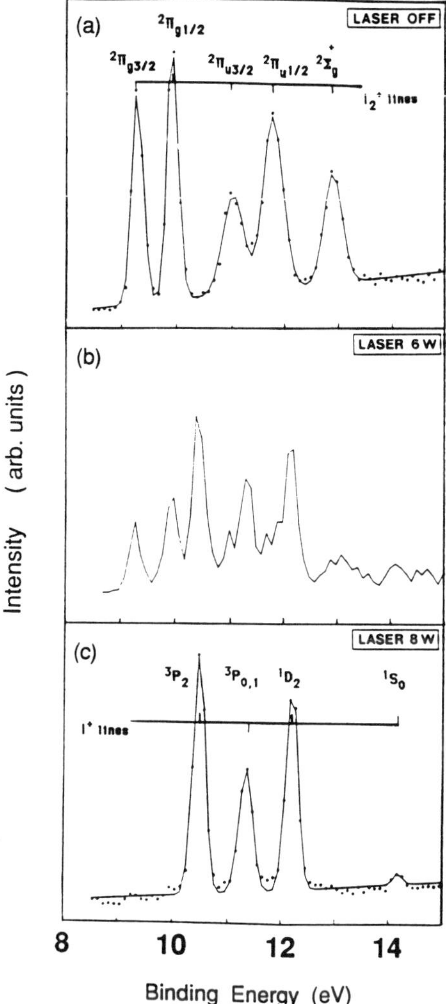

Figure 2 Photoelectron spectra of iodine recorded at 21.21 eV

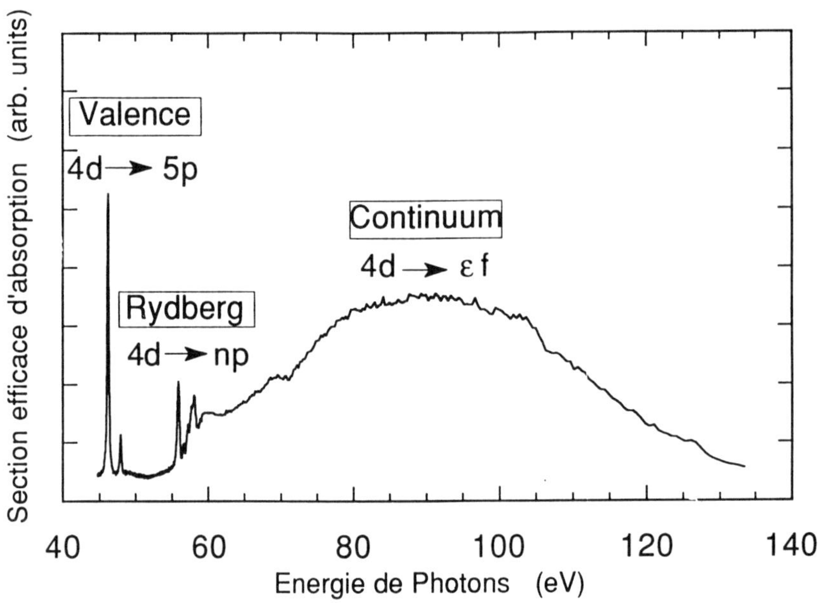

Figure 3. Total ion yield spectrum of atomic iodine. $4d^{-1}$ thresholds are spread between 57.42 and 60.93 eV.

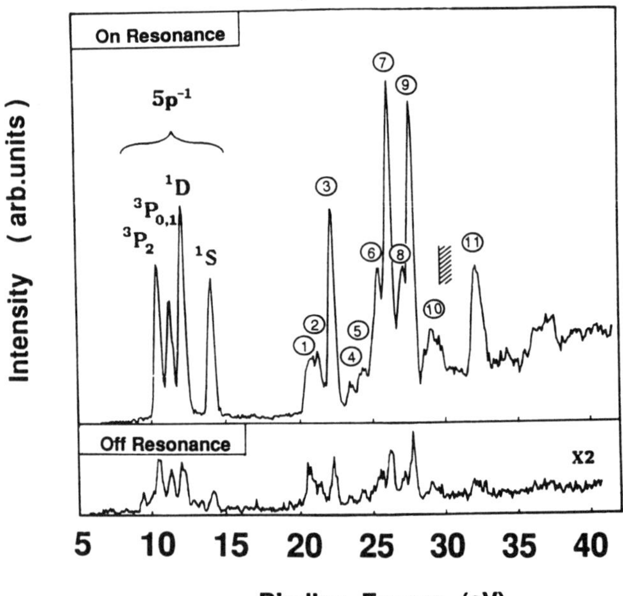

Figure 4. Photoelectron spectrum of atomic iodine recorded on (46.2 eV) and off (45.5 eV) the 4d-->5p ($^2D_{5/2}$) resonance.

We show in figure 3 a total ion yield recorded in the region of the 4d excitation. This total ion yield spectrum has been obtained by inverting polarities in the electron analyzer. The spectrum shows clearly three spectral regions of interest: the first one located around 46 eV exhibits two strong resonances attributed to the transitions $4d^{10}5s^25p^5$ ($^2P_{3/2}$) ----> $4d^95s^25p^6$ ($^2D_{5/2}$, $^2D_{3/2}$); the second one corresponds to transitions into the Rydberg np (n > 5) orbitals ($4d^{10}5s^25p^5$ ($^2P_{3/2}$) ----> $4d^95s^25p^5$ np); and the third one is assigned to a giant shape resonance in the 4d-->εf channel.

We discuss here only the relaxation of the first resonance 4d-->5p ($^2D_{5/2}$) . PES recorded on (45.5 eV) and off (46.2 eV) this resonance are presented in figure 4. The dramatic change in intensity between the two spectra reflects the high amplitude of the resonance; it can be noticed that all the electronic states populated out the resonance are resonating. This indicates that this core to valence transition decays into outer shell channels via a strong aotoionization process. Nevertheless some lines are more amplified than other; they correspond to singlet states. The very interesting region of the resonant spectrum extends between 20-30 eV (peaks labelled 1-10). In a simplified picture one would expect only two lines corresponding to the $5s^{-1}$ (^3P,^1P) state. In fact one observes numerous lines spread over more than 8 eV, due to strong electronic correlations. Ab initio multi-configurational calculations[7] gave evidence for a very strong interaction between the parent configuration $5s5p^5(^1P)$ and the satellites configurations $5s^25p^3(^2P)5d$ (^1P) and $5s^25p^3(^2D)5d$ (^1P).

Finally peak labelled 11, assigned to the $5s^{-2}5p^6$ state, is the only one produced by a resonant Auger process with a 5p spectator electron. This peak, located above the double-ionization limit relaxes by a two-step autoionization process into the double-ionization continuum.

In conclusion of this first part, we have shown that laser induced dissociation appears thus to be a very powerful method of production of radicals, whose electronic properties, different from the one observed in closed-shell species, can be investigated, in a very efficient way, by photoionization probed by electron spectroscopy. In the following section we show how this technique can be applied to disentangle dissociation mechanisms in case of laser irradiation of a complex molecule.

PHOTODISSOCIATION DYNAMICS: s-TETRAZINE

The photodissociation of ground state s-tetrazine ($C_2N_4H_2$), near the S^1 state using visible light raised considerable attention as it results in the production of three fragments: N_2 and 2HCN. Until recently, the interpretation of this mechanism as a concerted three body dissociation was questioned; stepwise process was discussed for instance in reference [8]. In figure 5 are reported the relevant energy levels in the region of the S_0 and S_1 states of the molecule.

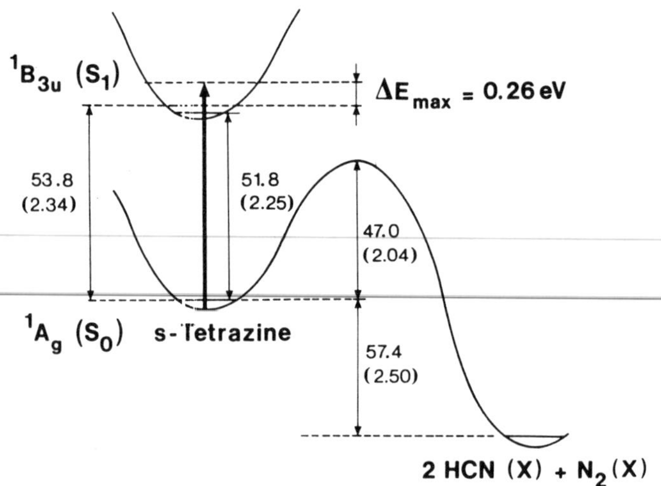

Figure5 Energy diagram of s-tetrazine photodissociation (energies in

Excitation from the ground state into S_1 state can be achieved with a cw Argon-ion laser operated in the visible. From the large exothermicity of the reaction (4.5 eV), one would expect, according to statistical theory, the fragments to carry some appreciable internal (vibrational) energy.

Using the technique described in the previous section, we have investigated the nature of the fragments as well as their internal (vibrational) energy, by mean of electron spectroscopy. Figure 6 shows photoelectron spectra of s-tetrazine recorded at 23 eV photon energy without (a) and with laser (b). In spectrum 6b, lines due to the parent molecule have desappeared and simultaneously one can clearly see the contribution of N_2 and HCN molecules whose photoelectron spectra are well known. No other contributions can be seen, which shows that on the experiment time scale (100 ns) we deal with a three body fragmentation.

Of special interest is the vibrational energy of the products. In figure 7 is reported a photoelectron spectrum restricted to the nitrogen lines. It has been recorded with a better resolution (.13 eV total resolution) than spectrum 6b. The observed peak at 16.44 eV is due to ionization of N_2 (X) (v=1) into N_2^+ (A) (v=0). The other "hot bands", namely 1->1, 1->2 etc cannot be directly seen because they are quasi degenerate with the other vibrational components of the A state. It's also the reason why the X state of N_2^+ is not a good candidate to observe hot bands: Franck-Condon factors allow only the 0->0, 1->1, etc... transitions to be observed. Of course they are also quasi degenerate. We have deduced that 5.4%+- .5% of the fragment N_2 was found in v=1.

We made a similar analysis for the HCN fragment, not reported here. Unfortunately, we could not resolve individual bands, because in addition to streching mode, we have to account for bending motion which gives very close lying lines. Nevertheless it has been possible to estimate the nascent HCN to have some vibrational excitation with a distribution in the bending mode (ν_2) extending to ν_2=6 and peaking at ν_2=3 (negligible C-N stretching excitation). The comparison between our findings and the prediction of a simple statistical distribution of energy is presented in figure 8. Obviously we observe that the nascent fragments are colder than expected in a pure statistical model. These results provide more accurate and additional informations compared to previous work of Coulter et al[9] who found 1% excitation of C-N stretching in HCN and estimated some 10% excitation of v=1 in N_2.

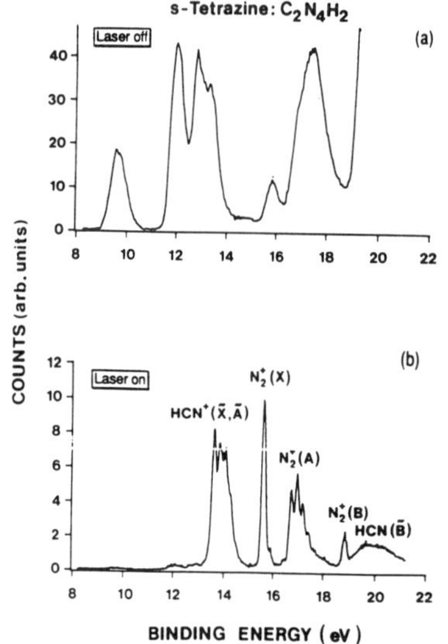

Figure 6 Photoelectron spectra of s-tetrazine a) laser off b) laser on (3Watts) recorded at 23eV

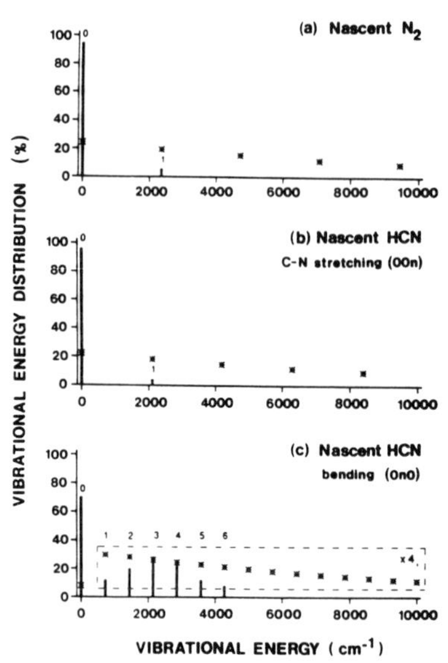

Figure 8 Vibrational energy distributions for N2 (a) and HCN (b and c) fragments as deduced from this work (bars) or in a pure statistical model (stars).

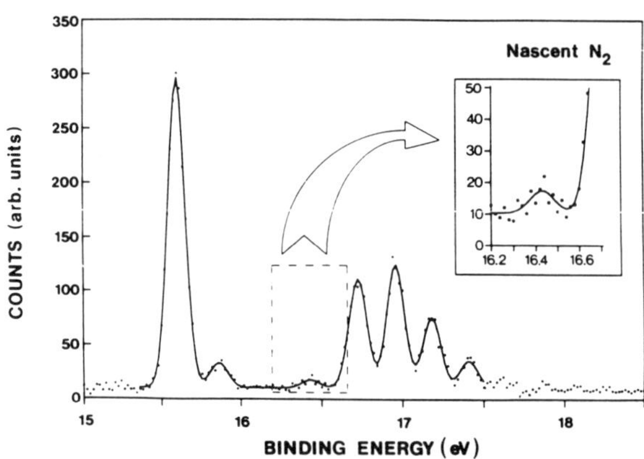

Figure 7 Photoelectron spectrum of nascent N2 fragment recorded at 23eV.

Those findings can be rationalized if we consider the equilibium geometry model. In such a model, it is assumed that after photon absorption, the system evoluates to form a transition complex, defined at the top of the potential barrier, before undergoing fast dissociation along a purely repulsive surface. In this second step chemical bonds are broken in a concerted way and the internal energy of the fragments is just given by the change in geometry between the equilibrium state and the equilibrium geometry of each fragment. In the case of tetrazine, Scheiner et al[10] calculated the geometry of the transition state; it has been reported in figure 9. The reaction coordinates are represented as dashed lines, the "perpendicular" coordinates are the one of interest which we consider to test the model. In figure 9, we have reported the potential energy curves of N_2 and HCN (stretching and bending motion) in order to relate the perpendicular coordinates to the maximum internal energy expected in the nascent fragments. The 1.141A equilibrium N-N distance,
when reported into the N_2 potential curve, gives a maximum of one quantum number of vibrational excitation. Similarly a weak excitation along C-N stretching is available. On the other hand a large change in the HCN angle (133° vs 180°) explains the rather large excitation of HCN bending. These results are consistent with those obtained by translational laser spectroscopy[11]; indeed most of the exothermicity is found in translation.

In conclusion we confirm the fragmentation of s-tetrazine is consistent with a concerted three body dissociation. Furthermore the internal energy is governded by the geometry of the transition state, in agreement with recent dynamical calculations of Strauss and Houston[12].

CONCLUSION

We have shown the combination of laser and synchrotron radiation is a powerful tool to study both photoionization dynamics of transient species and molecular photodissociation process. The laser can be used also to excite some electronic states and synchrotron radiation to ionize

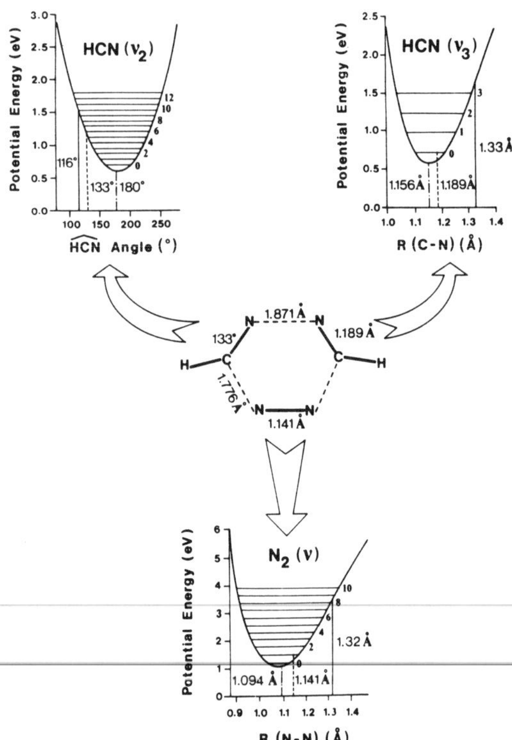

Figure 9 Transition state for the unimolecular dissociation of s-tetrazine as calculated by Scheiner et al with the potential energy curves of N_2 and HCN in the relevant coodinates.

the excited system. This has already been done in several experiments for atoms (Na,Li,Ba,Ca,K etc...[13,14]). The purpose in that case is different: one try to emphasize the influence of the excited electron on the other electrons.

In the future, it will be of high interest to use pulsed instead of cw laser. It would be then possible to probe the dynamics of molecular dissociation on a subnanosecond time scale, through a pump and probe experiment.

References

[1] Okabe H. , *Photochemistry of small molecules*, (Wiley, New York, 1978).

[2] Nahon L. Thèse d'université, Orsay, 1991

[3] Nahon L. , J. Tremblay, M. Larzillière, L. Duffy et P. Morin, Nucl. Instrum. Methods B **47**, 72 (1990).

[4] Nahon L. , L. Duffy, P. Morin, F. Combet-Farnoux, J. Tremblay and M. Larzillière, Phys. Rev. A **41**, 4879 (1990).

[5] Nahon L. , A. Svensson and P. Morin, Phys. Rev. A **43**, 2328 (1991).

[6] Nahon L. , P. Morin, M. Larzillière et I. Nenner, J. Chem. Phys. (1991.bis) (in press).

[7] Combet-Farnoux F. , J. Phys. (Paris) **C1**, 73 (1991).

[8] Glownia J. H. et S. J. Riley, Chem. Phys. Lett. **71**, 429 (1980).

[9] Coulter D. , D. Dows, H. Reisler et C. Wittig, Chem. Phys. **32**, 429 (1978).

[10] Scheiner A. C. , G. E. Scuseria et H. F. Scheafer III, J. Am. Chem. Soc. **108**, 8160 (1986).

[11] Zhao X. , W. B. Miller, E. J. Hintsa, Y. T. Lee, J. Chem. Phys. **90**, 5527 (1989).

[12] Strauss C.E. and P.L.Houston, J.Phys.Chem. **94**,8751(1990)

[13] Cubaynes D. , J. M. Bizau, F. J. Wuilleumier, B. Carré et F. Gounand, Phys. Rev. lett. **63**, 2460 (1989).

[14] Meyer M. , B. Müller, A. Nunnemann, Th. Prescher, E. v. Raven, M. Richter, M. Schmidt, B. Sonntag et P. Zimmermann, Phys. Rev. Lett. **59**, 2963 (1987).

DISCUSSION

BESWICK - In the case of tetrazine photofragmentation, can you comment on the possible rotational excitation of the HCN fragments ?

MORIN - From our experiment we are able to give the maximum vibrational energy carried out by N_2 and HCN (ie 1.88eV) as well as the average value (.188 eV). Those numbers are significantly smaller than those deduced from vibrational energy distribution of Zhao and co-workers. We deduced from this previous work a 1.26 eV available internal energy. This energy gap is certainly due to rotation. Indeed, if we assume the equilibrium geometry model to be valid, we expect HCN fragments to be rotationally excited, as we known from Zhao et al., that repulsion between both HCN is stronger than between N_2 and HCN. In other words, when N_2 is produced, two identical bonds (C-N) are broken. In a concerted dissociation process, this cannot produce rotational excitation of N_2. On the other hand, when HCN is produced, two identical bonds are broken (C-N and N-N) ; in a concerted dissociation process, there is no reason to assume the same repulsion energy to be released in both case. This can lead to rotational excitations of HCN.

DE LANGE - To achieve photodissociation one would preferably employ photons of about 200nm. Using a CW laser is of course compatible with the pulse structure of the synchrotron radiation but is not sufficiently intense for general purposes. What pulsed lasers that could operate synchronously with the synchrotron radiation pulses would you envisage to employ in the future ?

MORIN - The use of a pulsed laser synchronized with synchrotron radiation could make possible several type of experiment. The more interesting are probably experiments where short lived species are created (< a few ns). If we can ionize with S.R. those species, we would be able to achieve a real time decay determination. Of course this necessitates a perfect synchronization between S.R. and the laser. This is possible if we use a mode-locked laser (let say Ar ion laser) with a resonating frequency equal to a multiple of the frequency corresponding to one bunch operation of the S.R. The point is to match the phase shift between the two independent sources. Another kind of experiment is also possible with pulsed laser, without any synchronization: using for example a pulsed laser in the UV to obtain efficient dissociation of various molecules. In that case, as the S.R. appears as a continuous source, the repetition rate of the laser has to be as high as possible. Cooper vapor lasers offer a good repetition rate (10kHz) which make feasible this kind of experiment.

PIANCASTELLI - How critical is the choice of the angle between the laser beam and the synchrotron beam ?

<u>MORIN</u> - In our experiment, this angle is not critical because we are not interested in angular effects at the present time. In other experiments, taking advantage of the polarization of both synchrotron and laser radiation, one would have to check this point carefully. For instance it would be of interest to use a linearly polarized laser beam to photodissociate I_2 molecules. In that case, it will produce aligned iodine atoms: this aligment will change the angular distribution of electrons after ionization with S.R., and will include higher terms than $\cos^2\theta$ in the distribution.

<u>CHEREPKOV</u> - You presented absolute photoabsorption cross section for the 4d-subshell of atomic iodine, which appears to be much lower than the analogous cross section for Xe. It is well known that the dipole sum rule is approximately applied for each subshell separately. Have you checked that your cross section for iodine satisfies the dipole sum rule ?

<u>MORIN</u> - I totally agree with you that dipole sum rule has to apply to the 4d ionization of iodine. Our conclusion concerning the missing oscillator strength relative to the 4d cross section, is that there are only two possible explanations:

1. a second maximum in the 4d cross section may be present after the Cooper minimum, with an appreciable oscillator strength. This would be surprising in view of similar measurements on Xenon which show only a very weak enhancement after the 4d Cooper minimum.
2. Our absolute cross section measurement is not direct: it involves the knowledge of the absolute absorption cross section of I_2 in the region of the 4d shell. Clearly the sum rule cannot apply correctly to I_2 measurements also. Thus it is more likely to conclude that the absolute cross section measurements on I_2 are questionable.

FRAGMENTATION OF CORE-EXCITED MOLECULES WITH SYNCHROTRON RADIATION

J. Delwiche(°)

Laboratoire de Spectroscopie de photoélectrons
Université de Liège, Institut de Chimie B-6
Sart Tilman par B-4000 Liège 1, Belgium

ABSTRACT

The fragmentation pathways of singly and doubly charged ions formed by autoionization following excitation of a core electron are reviewed and illustrated by results on the fluorescence of SiX_4 (with X = F or Cl) and photoelectron-photoion-(photoion)-coincidence spectroscopy of C_6F_6.

INTRODUCTION

A great deal of efforts has been put recently on the understanding of the production and dissociation mechanisms of doubly charged ions[1,2] by means of photoionization.
The availability of the synchrotron radiation is for a great part in the development of this new field, as well as the introduction of new sophisticated coincidence detection methods[3-5].
In this paper, we wish to concentrate on the fate of ions that are produced by means of a so-called resonant Auger process which is better refered to as an autoionization process. We shall be dealing with medium-sized polyatomic molecules for which available data are very scarce[6], contrarily to the situation for small molecules[7,8].
Thanks to the tunability of the synchrotron radiation over a wide range of photon energies it is now possible to selectively excite an inner-shell electron[6] of a molecule into an unoccupied orbital either of the valence or of the Rydberg type.
In Fig. 1, we schematically summarize the two different

(°) Maître de Recherches of the F.N.R.S. of Belgium

types of the so-called resonant Auger processes.
We are dealing with a two-step autoionization mechanism.
In the first step, an inner-shell electron in promoted into an empty orbital of the molecule by a resonant absorption.

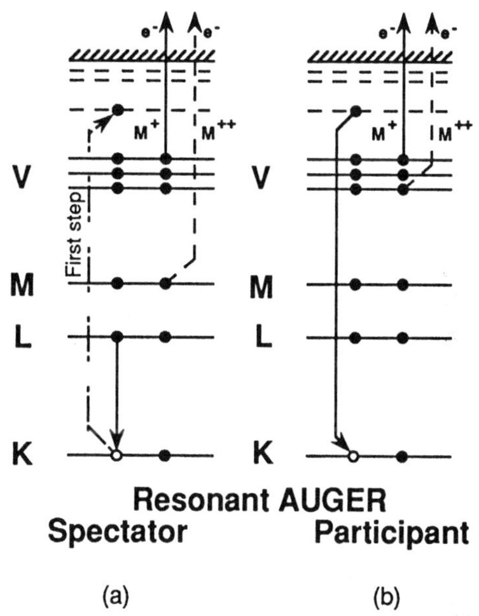

Fig. 1 : The resonant Auger process

Two different decay pathways are then open for the relaxation of the excess energy :

- The spectator decay process in which the core-electron remains on its valence or Rydberg orbital and a valence shell electron fills the inner-shell gap (Fig. 1-a). The energy released in this transition causes one, two or more valence shell electron(s) to be ejected.

- The participant decay process (Fig. 1-b) in which it is the core-electron that jumps back into it initial position. The energy released also leads to the removal of one, two or more valence shell electron(s).

In both case, one ends with a singly or a doubly charged ion. At this point, two important questions arise :First, what are the decay channels of these molecular-ions formed by autoionization, and secondly, are their decay channels different from those of the ions formed by direct photoionization into the continuum.
We shall now illustrate what kind of answer to these two questions can be given for medium-sized molecules with one exemple taken from the litterature and one from our Laboratory work.

FORMATION OF NEUTRAL FRAGMENTS

While it is obvious that the highly excited ions produced by autoionization will decay with production of excited charged fragments, the question of the production of neutral species and of their state of excitation is rather poorly documented for medium-sized molecules.
The fluorescence of the neutral fragments formed after the photoionization of SiX_4 (where X = F and Cl) has been studied by Rosenberg et al.[9,10] at the Synchrotron Radiation Center, University of Wisconsin, Madison, USA.
In particular, they have observed the fluorescence from neutral Si^* atoms (from $SiCl_4$) and of Si^*, SiF^*, and F^* fragments (from SiF_4) formed in the decay of SiF_4^+ ions created by the following spectator process :

$$SiX_4\,(2p)^{-1}R \rightarrow SiX_4^+\,(V_a)^{-1}\,(V_b)^{-1}R$$

where V_a and V_b represent occupied valence levels and R a Rydberg level.
They also showed that the fluorescence yield of the neutral fragments is greatly enhanced when the 2p core electron is ecxited on a Rygberg orbital, as opposed to the behaviour of the molecules where the excited electron is located on a valence orbital.
This can be rather easily rationalized if one accepts the model of a Rydberg orbital rather large and diffuse,

having little overlap with the molecular-ion core. Therefore, there will be very little interaction between the excited electron and the molecular-ion core during the dissociation process.

FORMATION OF SINGLY AND DOUBLY CHARGED FRAGMENTS ION PAIR PROCESSES

To illustrate the production of singly and doubly charged fragment ions from resonant Auger formed ions, we whish to report some new results that have been obtained at LURE (Orsay) by Ibrahim[11,12], using the synchrotron radiation from the SuperACO storage ring, on C_6F_6.
In these experiments, PEPICO[13] (PhotoElectron-PhotoIon Coincidence) and PEPIPICO (PhotoElectron-Photoion-Photoion Coincidence) spectra have been recorded as a function of the photon wavelength in the vicinity of the C1s and of the F1s edges : at 288 eV and 689 eV, respectively. These energies correspond to the transition of a C1s or of a F1s electron to the first $\pi^*(2e_u)$ orbital of C_6F_6, which is C-C antibonding in character (Fig. 2).
In the PEPICO spectra, which correspond to the production of only one charged species together with neutral fragments, it is observed, in addition to singly charged fragment ions, several doubly charged fragments : F^{2+}, C_2F^{2+}, CF_3^{2+}, C_5F^{2+} (not at 288 eV), and $C_6F_5^{2+}$ at 288 eV and at 689 eV, with total relative intensities of 3.06 % and 2.76 % respectively. Triply charged species are also observed at both photon energies : $C_2F_2^{3+}$ (not at 288 eV), C_4F^{3+}, CF_3^{3+}, and $C_5F_4^{3+}$ with total relative intensities of 2.81 % and 2.42 %, respectively. It is interesting also to note the existence of rearrangment ions (CF_3^+) which means that some fragmentation processes are slow enough to allow the migration of F atoms.
In the PEPIPICO spectra, which correspond to fragment ions formed by pairs, ion pairs involving a doubly charged fragment and a singly charged fragment are observed meaning that triple ionization also plays a role in the production of the fragment ions.

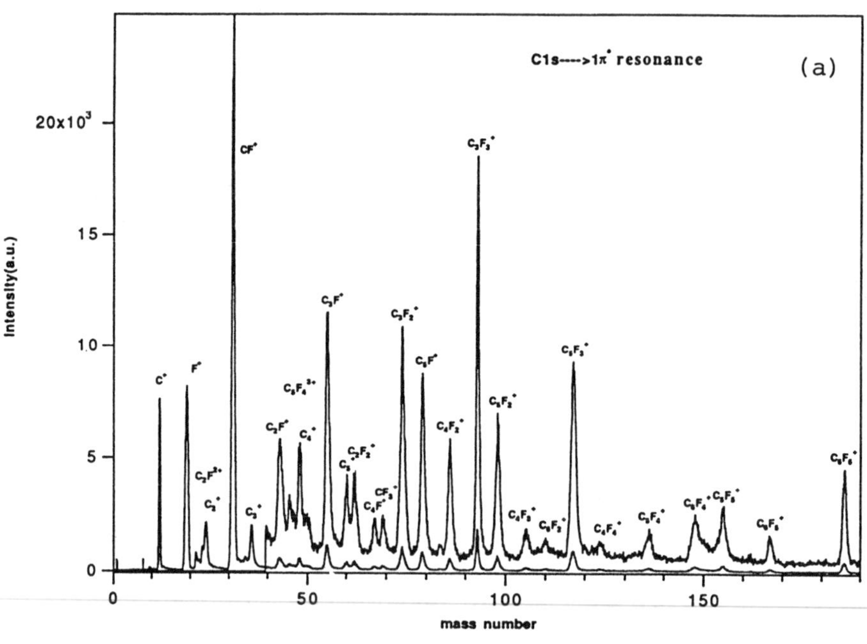

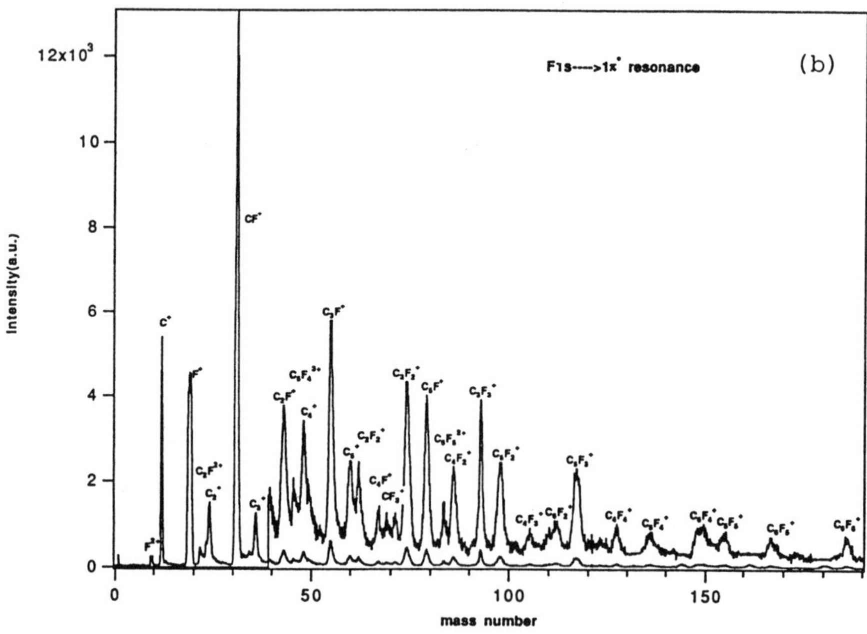

Fig. 2 : The PEPICO spectra of C_6F_6 at 288 eV (a) and 689 eV (b).

It is also interesting to investigate if the nature of the primary core hole (C1s or F1s) has an influence on the fragmentation pattern of the molecule, in other terms, if there exists some kind of selectivity in the breakage probability of some bonds within the molecule.

Table I : Relative intensities of selected ionic groups and of selected ion pair groups observed in the PEPICO and PEPIPICO spectra of C_6F_6.

Transition	F^+, F^{2+} (ions)	C_n^+ (n≥2) (ions)	CF^+ (ions)	F^+ + (others)$^+$ (ion pairs)	CF^+ + (others)$^+$ (ion pairs)
$1s_C \rightarrow 1\pi^*$ (288 eV)	19.08	10.09	38.32	36.11	25.42
$1s_F \rightarrow 1\pi^*$ (689 eV)	22.12	11.02	36.07	38.05	21.72

In Table I, the relatives intensities of several groups of fragment ions and of ions pairs are reported.
It is clear that the fragmentation pattern is somewhat different according to the nature of the primary exitation. Excitation of the C1s electron induces preferentially the breakage of the C-C bonds, while the breakage of the C-F bonds is enhanced when it is a F1s electron that is excited into the first π^* orbital. The same tendancy is observed for the production of singly charged ions and for the production ot ion pairs

CONCLUSIONS

These two exemples show that the fate of the molecular-ions formed by autoionization of a highly excited neutral species can lead to a great variety of fragments either neutral or charged and more or less excited.
To get a deeper insight of the different decay pathways of these molecules, coincidence measurements with energy analyzed electrons are needed.

ACKNOWLEGMENTS

I am very grateful to the staff of LURE at Orsay for their technical support. I also aknowledge my colleagues of the University of Liège (Dr. M.- J. Hubin-Franskin and Mr K. Ibrahim) for their collaboration, the treatment of the data on C_6F_6, and very helpful discussions, and my colleagues of Orsay (Drs. P. Morin, M. Lavollée, I. Nenner, and Mr M. Simon) for their assistance in obtaining the data on C_6F_6. The work on C_6F_6 was supported by the Fonds National de la Recherche, the Fonds de la Recherche fondamentale collective, and the Ministère des Finances (Loterie Nationale of Belgium, as well as by the NATO (contract n° 484/87), who are gratefully acknowledged.

REFERENCES

1. I. Nenner, M.-J. Hubin-Franskin, J. Delwiche, P. Morin, and S. Bodeur, J. Mol. Struct. **173**, 269 (1988).
2. I. Nenner, J.H.D. Eland, P. Lablanquie, J. Delwiche, M.-J. Hubin-Franskin, P. Roy, P. morin, and A. Hitchcock in "Electron-Molecule Scattering and Photoionization", P.G. Burke and .B. West ed., (Plenum, N.Y., 1988), p 15.
3. M.-J. Besnard-Ramage, P. Morin, T. Lebrun, I. Nenner, M.-J. Hubin-Franskin, J. Delwiche, and J.H.D. Eland, Rev. Sci. Instrum. **60**, 2128 (1988).
4. J.H.D. Eland, Acc. Chem Res. **22**, 381 (1989).
5. J.H.D. Eland and D.A. Hagan, Int. J. Mass Sperctrom. Ion Proc. **100**, 489 (1990).
6. M.-J. Hubin-Franskin, M. Furlan, J. Delwiche, Physicalia Mag. **12**, 49 (1990)
7. F.P. Larkins, W. Eberhardt, I.W. Lyo, R. Murphy, and E.W. Plumer, J. Chem. Phys. **88**, 2948 (1988).
8. R. Murphy, I.N. Lyo, and W. Eberhardt, J. Chem. Phys. **88**, 6078 (1988).
9. R.A. Rosenberg, C.R.Wen, K. Tan, and J.-M. Chen in DIET IV, G. Betz and P. Varga ed., (Springer, Berlin, 1990), p 97.

10. R.A. Rosenberg, C.R. Wen, K. Tan, and J.-M. Chen, Phys. Scripta **41**, 475 (1990).
11. K. Ibrahim, PhD Thesis, Université de Liège, 1991.
12. K. Ibrahim, M.-J. Hubin-Franskin, J. Delwiche, M. Simon, M. Lavollée, I. Nenner, to be published.
13. P. Morin, M.Y. Adam, I. Nenner, J. Delwiche, M.-J. Hubin-Franskin, and P. Lablanquie, Nucl. Instr. and Meth. **208**, 761 (1983).

DOUBLE CORE VACANCY STATES IN X-RAY PHOTOABSORPTION SPECTRA OF GAS PHASE SULFUR MOLECULES

S. Bodeur*
Laboratoire de Chimie Physique, 11, rue P. et M. Curie, 75231 Paris cedex 05, France
C. Reynaud, K. Bisson, P. Millié and I. Nenner*
CEA-DSM/DRECAM/ Service des Photons, Atomes et Molécules, Bâtiment 522, Centre d'Etudes de Saclay 91191 Gif sur Yvette cedex, France
U. Rockland and H. Baumgärtel
Institute für Physikalische Chemie, Freie Universität Berlin, D-1000 Berlin 33, Germany
* also LURE, laboratoire mixte CNRS, CEA et MENJS, Centre Universitaire de Paris sud, 91495 Orsay cedex, France

ABSTRACT

Double-core-vacancy excited states are observed in the photoabsorption spectra of gas-phase sulfur compounds such as SF_6 and SF_5Cl, near the sulfur K edge, i.e. in the 2400-2800 eV photon energy range, where one observes EXAFS oscillations. We interpret the structure as originating from the 1s2p electron excitation into unoccupied valence orbitals. In SF_6, one observes a doublet with an energy separation of 8.7 eV. In SF_5Cl, only one resonance is detectable above the SK edge. We also report new theoretical calculations on the H_2S molecule, using a MCSCF formalism, which allows a better evaluation of the relaxation energy compared to previous results and especially of the singlet-triplet separation in cases where the two vacancies have the same symmetry.

INTRODUCTION

Photoabsorption spectra of molecules, in the X-ray region, are generally interpreted in terms of a one-electron scheme. Near the ionization edge, resonances called XANES are due to either core-valence or core -Rydberg simple excitations (discrete spectrum) or to multiple scattering resonances (continuum). Far from the edge, one observes wide oscillations due to simple scattering phenomena, also named EXAFS. Many-body effects are nevertheless present because of simultaneous excitation of a core and a valence electron, and this may account to a few percent of the total cross section in the XANES region. In the EXAFS domain, multiple excitation can be observed but two core electrons are involved. According to our previous related work on silicon compound molecules[1], the structures never exceed 1 % of the total intensity.

In the present paper, we report new observations of double-core-vacancy excited states in the photoabsorption cross section in sulfur compounds : SF_6 and SF_5Cl, near the sulfur K edge in the 2400-2800 eV photon energy range available with the DCI storage ring at LURE. This work follows a detailed work on the one electron part of the spectrum of these two molecules[2], performed at both the SK and the ClK edges. The purpose of this work is to verify the part of pure sulfur atomic effect against the effect of chemical bonding and symmetry of the molecule. The interpretation is based on a simple model described previously [1] , but in estimating the relaxation energies using a formula proposed by Ohrendorf et al[3]. Because it is evidently important to know the corrections of this formula, we present also here an original calculation method, in which the relaxation energy of the double-core-ion, is calculated directly with a MCSCF formalism. When applied to the H_2S molecule, the mean uncertainty never exceeds 10%.

EXPERIMENTAL AND RESULTS

Photoabsorption spectra have been obtained at LURE, the French synchrotron facility, on the EXAFS IV station implanted on the DCI storage ring and equipped with a double crystal monochromator and a set of two refecting mirrors for high harmonics rejection. The experimental set up consists of a gas cell described in details elsewhere [4]. Briefly it is a 205 mm long cell limited by propylene windows. The incident and transmitted intensities are measured by two ionization chambers. Special care is taken to eliminate water traces fom the cell and the inlet system. Spectra are obtained by scanning the photon energy automatically, in choosing small steps of 0.3 eV which allows to detect small structures within the resolving power (0.5 eV). The photon wavelength is calibrated by using the main XANES resonance of SF_6 located at 2486.0 eV. We estimate that the accuracy of the wavelength scale amounts to 0.5 eV and the accuracy for the relative energies of sharp features within a single spectrum to be better than 0.3 eV.

SF_6 sample is commercially available from Prodair compagny with a purity of 99 %. SF_5Cl is a gas at room temperature and has been synthetized by Professor Willner, Hannover, following the method of Schack et al[5]. The purity amounts to 99 %.

We present in Figures 1 and 2 the full photoabsorption spectra of SF_6 and SF_5Cl respectively. We have magnified the region of interest above 2650 eV in each case. The energies of the resonances of interest here are reported in Table I. The most intense resonances seen in figure 1 and 2 are found below the sulfur 1s ionization edge. They are due to $S1s \rightarrow \sigma^*$ transitions as seen in Table I. The most important observations of this work are seen as peaks in magnified spectra reported in the insert of Figures 1 and 2. In SF_6, there is a doublet whereas in SF_5Cl, there is only evidence of a single peak. Notice that we have explored the region above the chlorine edge without detecting any feature of this sort.

Table I. Experimental energies of double core vacancy states in SF_6 and SF_5Cl, near the SK edge. The single $1s \rightarrow \sigma^*$ excitation energy is given for comparison.

Molecule	Main single-core Energy ($1s \rightarrow \sigma^*$)	Double-core-excitation Energy (eV)	Energy splitting (eV)	assignement
SF_6	2486.0 (t_{1u}) (ref6)	2693.5	8.7	$1s^{-1}\,2p^{-1}\,(T)\sigma^{*2}$
		2702.8		$1s^{-1}\,2p^{-1}\,(S)\sigma^{*2}$
SF_5Cl	2484.8 (e) (ref2)	2700	-	$1s^{-1}\,2p^{-1}\,(S\text{ or }T)\sigma^{*2}$

Figure 1 : Photoaborption spectrum of SF_6 and SF_5Cl near the sulfur K edge.

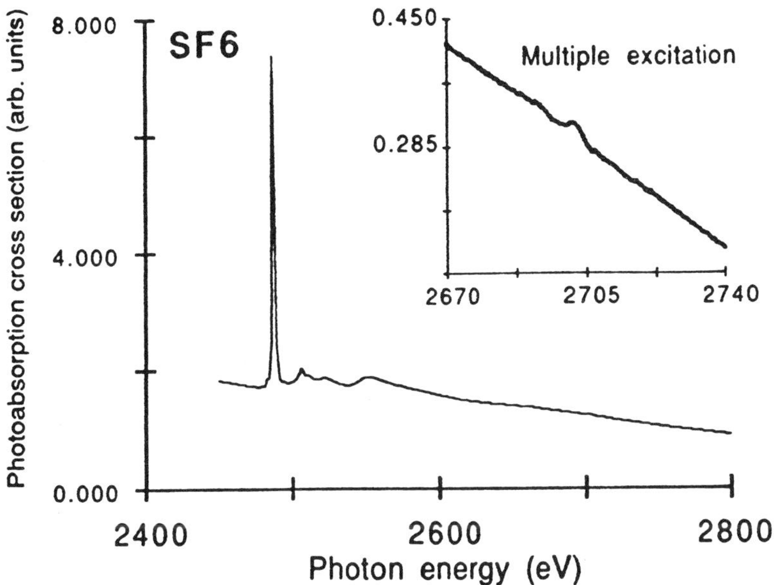

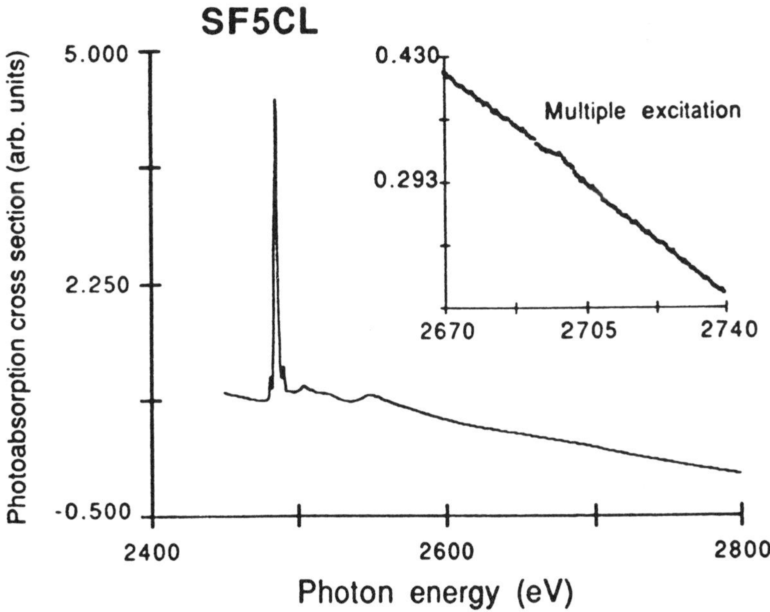

THEORETICAL AND RESULTS

I - Simple approach

In our first work on double-core-vacancy excited states in silicon compound molecules [1], we have used a simple Hartree-Fock model, in which the double-core-vacancy excitation energy, $E_{ij\to kl}$ is reduced to

$$E_{ij\to kl} = E_{i\to k} + E_{j\to l} + \mathcal{J} + \mathcal{K}$$

where i and j represent the core vacancies, k, l the unoccupied orbitals, $E_{i\to k}$ and $E_{j\to l}$ the single-core excitation energies given by experiment, $\mathcal{J}$ and $\mathcal{K}$ the total Coulomb and total exchange terms respectively. In this simple model, only part of the relaxation energy $E_R(ij)$ is taken into account in the single-core excitation energies but the cross term of the relaxation energy (see below equation (1)) and correlation effects are neglected. In the present work, we consider this cross term in some details, because it has a very large contribution, whereas we still neglect correlation effects. Ohrendorf et al [4] have proposed an approximated formula derived from a second order perturbation approach, to evaluate the relaxation energy for two vacancies i,j and for a given molecule.

$$E_R(ij) \simeq E_R(i) + E_R(j) + 2\sqrt{E_R(i)E_R(j)} \qquad (1)$$

We see that the relaxation energies are not following an additive rule and the cross term amounts to some 50% of the total. We have evaluated in Table II, the different terms which are necessary to calculate the double-core-vacancy states. We find a reasonable agreement with the experimental values of Table I (the mean value of the doublet in SF_6 is 2697.8 eV), showing the importance of the Coulomb term and of the cross term. The exchange term ($2\mathcal{K}_{1s2p}$) is compatible with the energy splitting of the doublet. Considering our previous study on silicon compounds [1], we suggest that the same interpretation is valid for the SF_6 molecule. For SF_5Cl, we cannot draw any conclusion since only a single peak is observed. We conclude that this first approach provides a first simple method for experimentalists to interpret double-core-vacancy states.

Table II. Energy terms for the evaluation of double-core-vacancy excited states in SF_6 and SF_5Cl, near the sulfur 1s edge.

	$E_{exp}(1s \to \sigma^*)$	$E_{exp}(2p \to \sigma^*)$ a)	$\mathcal{J}$	$-2\sqrt{E_R^{1s}E_R^{2p}}$	$2\mathcal{K}_{1s2p}$	Total $\binom{\text{Triplet}}{\text{Singlet}}$
SF_6	2482 (a_{1g})	172.6(a_{1g})	62	-38	7.5	2678.6 ∓ 3.5
	24866 (t_{1u})	176.4(t_{1u})	62	-38	7.5	2686.4 ∓ 3.5
SF_5Cl	2480.2[2] (a_1)	171 (a_1)	62	-38	7.5	2675.2 ∓ 3.5
	2484.8[2] (e)	176 (e)	62	-38	7.5	2684.8 ∓ 3.5

a) ref 6, for SF_6, 7 for SF_5 Cl

II - Direct evaluation of the relaxation energy

In order to go beyond the approximate formula of Ohrendorf et al[3], we have tried to determine the relaxation energy by the direct method, i.e.

$$E_R = E_{ij \to kl}(\text{frozen}) - E_{ij \to kl}(\text{relaxed})$$

where $E_{ij \to kl}(\text{frozen})$ is given using SCF orbitals of the neutral molecule and $E_{ij \to kl}(\text{relaxed})$ is given by a SCF calculation of the double-core-vacancy excited molecule. However, the convergence of a SCF calculation for such species is in general very difficult, or even impossible in the case of a double vacancy from the same orbital. This is being due to the presence of many other ionic states, which are either more stable than the core hole state, or have the same electronic symmetry[3,8] (see for example in reference 3, the illuminating discussion on the singlet-triplet separation in the $1s^{-1}2s^{-1}$ case). We present a simple, reliable methodological tool to simulate the SCF calculations of most of the double core hole states, using a MCSCF formalism, with an application to a simple sulfur molecule H_2S, for which single core ionization energies are known[9,10].

In the present method, if the vacancy leads to N-1 ionic configurations of lower energy, we optimize the energy of the Nth state of the configuration interaction by a MCSCF program[8]. In this

Table III. Calculated ionization and relaxation energies for H_2S

M^+ or M^{2+}	Ionization energies (eV) This work	Experiment	Relaxation energies (eV) direct calculation	Formula (1)	relative difference
$2p^{-1}$					
$1b_2^{-1}$	170.3	170.5[9]	10.7		
$3a_1^{-1}$	170.2		10.8		
$1b_1^{-1}$	170.3		10.7		
$2s^{-1}$ $2a_1^{-1}$	234.4	234.5[9]	9.8		
$1s^{-1}$					
$1a_1^{-1}$	2471.9	2478.5[10]	30.3		
$2p^{-2}$					
$1b_1^{-1}1b_2^{-1}$(T)	372.3		41.1	42.8	4%
$1b_1^{-1}1b_2^{-1}$(S)	379.1		40.7	42.8	5%
$\alpha 1b_2^{-2}$	379.3		40.4	42.8	5%
$+\beta 3a_1^{-2}$	379.4		40.3	42.8	5%
$+\gamma 1b_1^{-2}$	389.4		39.9		
$2s^{-1}2p^{-1}$					
$2a_1^{-1}1b_2^{-1}$(T)	429.0		39.3	41.1	5%
$2a_1^{-1}1b_2^{-1}$(S)	448.4		41.1	41.1	0%
$2s^{-2}$					
$2a_1^{-2}$	503.9		36.2	39.2	8%
$1s^{-1}2p^{-1}$					
$1a_1^{-1}1b_2^{-1}$(T)	2689.9		68.8	77.1	12%
$1a_1^{-1}1b_2^{-1}$(S)	2699.5		66.7	77.1	16%
$1s^{-1}2s^{-1}$					
$1a_1^{-1}2a_1^{-1}$(T)	2745.6		67.9	74.5	10%
$1a_1^{-1}2a_1^{-1}$(S)	2759.1		66.6	74.5	12%
$1s^{-2}$					
$1a_1^{-2}$	5156.3		113.1	121.1	7%

NB. T and S refer to triplet and singlet coupling of the core vacancies

MCSCF frame, the convergence problem can be solved quite easily by

i) freezing the orbitals higher than the hole(s) one(s) and of the same symmetry, i.e. constraining them to remain identical to the neutral ones during the optimization, ii) adding a new set of orbitals to relax the frozen ones via the corresponding single and di excitation configurations [8]. The diexcitation configuration excitations are not indispensable in the single core calculations, but they have been found to play an important role in the calculations of the diions. We present in Table III our results on the IE's in the H_2S case. Note that the $1a_1^{-2}$ ion is the 15th state of the $^1A_1$ symmetry in the calculation. In the special case of a double hole from one 2p shell orbital, the interaction of the three almost degenerate configurations $1b_2^{-2}$, $3a_1^{-2}$ and $1b_2^{-2}$ has to be considered, and this does not raise any special difficulty with our method. A good test of the fiability of our method is the singlet-triplet separation in the case of $1a_1^{-1}2a_1^{-1}$ $(1s^{-1}2s^{-1})$. Its value is found with the same order of magnitude (slightly larger) than the one calculated with the neutral orbitals, as observed by Ohrendorf et al[3].

We now have some trust in the relaxation energy values for the H_2S molecule, as reported in Table III. Comparing them with the results obtained with formula (1), given by Ohrendorf et al[3], we obtain systematically smaller values (~10%). In conclusion, this method is a great improvement compared to the simple approach of reference 1. It shows that the cross term of the relaxation energy cannot be neglected. It allows a clear identification of the nature of the core holes, but does not give any information on the nature of the excited orbitals.

ACKNOWLEDGEMENTS

We are very grateful to J.L. Maréchal (CEA-SPAM, Saclay, France) and D. Bazin (LURE, Orsay France) for their efficient help in the measurements. The support of the LURE staff is acknowledged. This work has been suported by the PROCOPE French-German exchange program.

REFERENCES

1. S. Bodeur, P. Millié, E. Lizon à Lugrin, I. Nenner, A. Filipponi, F. Boscherini and S. Mobilio, Phys. Rev. A. 39, 5075 (1989) ; J.L. Ferrer, S. Bodeur, I. Nenner, J. Elec. Spectrosc. 52, 711 (1990)

2. C. Reynaud, S. Bodeur, J.L. Maréchal, D. Bazin, P. Millié, I. Nenner, U. Rockland, H. Baumgärtel (to be published)

3. E.M.-L. Ohrendorf and L.S. Cederbaum, F. Tarantelli, Phys. Rev. A 44, 205 (1991)

4. S. Bodeur, J.L. Maréchal, C. Reynaud, D. Bazin, I. Nenner, Z. Phys. D. - Atoms, Molecules and Clusters, 17, 291 (1990)

5. C.J. Schack, R.D. Wilson and M.G. Warner, Chem. Commun. 1110 (1969)

6. A.P. Hitchcock, C.E. Brion, Chem. Phys. 33, 55 (1978)

7. B.M. Addison, K.H. Tan, G.M. Bancroft and F. Cerrina, Chem. Phys. Letters 129, 468 (1986)

8. K. Bisson, Thèse de Doctorat (1991), Université de Paris sud (Unpublished) ; P. Archirel, K. Bisson and B. Levy (to be published)

9. K. Siegbahn, C. Nordling, G. Johansson, J. Hedman, P.F. Heden, K. Hamrin, U. Gellus, T. Bergmark, L.O. Werme, R. Manne and Y. Baer, ESCA applied to free molecules (North-Holland, Amsterdam, 1969)

10. O. Keski-Rahkonen, M.O. Krause, J. Elec. Spectrosc. 9, 371 (1976)

DYNAMICS OF CORE HOLE EXCITATIONS IN ADSORBATES

M. Ohno
Department of Physics
Uppsala University, Box 530, S-751 21 Uppsala, Sweden

P. Decleva
Dipartimento di Scienze Chimiche
Universita di Trieste, Via Valerio 22
I 34127 Trieste, Italy

ABSTRACT

We performed the ab-initio 2h-2p/3h2p CI calculations of core hole spectra of NiCO and NiN_2 using an extended basis set. The main line satellite line energy separations and intensity ratios of the new high resolution XPS core hole spectra of the CO/Ni(100) and N_2/Ni(100) systems are well reproduced by the present calculations. We give a new interpretation of the 2.1,5.5 and 8.5 eV satellite lines. For the N_2/Ni(100) system, the present calculations confirm the interpretation that the N core peak splitting is due to two inequivalent N atoms and the binding energy of the outer N atom is smaller than the inner N atom. For the weakly coupled N_2/Ni system where the π bonding is much weaker than for the strongly coupled CO/Ni system, the σ to σ^* excitation associated with the local metal configurational changes becomes much more significant. The importance of this excitation is pointed out.

1. INTRODUCTION

Recent high-resolution XPS core hole spectra of the CO/Ni(100) and N_2/Ni(100) systems show a number of newly resolved satellite lines which were not theoretically predicted before [1]. The most spectacular aspect of these core hole spectra is that a large part of the spectral intensity (more than 80 percent for the N_2/Ni(100) system) goes to the shakeup/off satellite lines extending over more than 75 eV. This shows that inclusion of the most prominent sattelite peaks are not sufficient for the determination of the spectral intensity[1]. Indeed the main line satellite line intensity ratios of these new high resolution XPS spectra(see table 1) differ considerably from some of the previous measurements (0.74/0.26 and 0.83/0.17 for the C (carbon) and O (oxygen), respectively [2], 0.5/0.5 and 0.5/0.5 [2] and 0.25/0.75 and 0.15/0.85 [3] for N_a (inner N) and N_b (outer N), respectively). Consequently some of the core hole spectral intensities calculated for NiCO and NiN_2 using molecular approaches also differ from the new experimental data (see ref. 4 for the results of NiCO by different theoretical approaches. For NiN_2 see refs.2,3,5-8). However, this does not imply that the small cluster model of the adsorption site is inadequate. On the contrary even single metal atom clusters have been extensively and successfully employed to model on top chemisorption of small molecules and reproduce well qualitative aspects of the both local electronic structure of the adsorption site and the behavior of the ground state properties [9]. For NiCO and NiN_2 the ground state is $^1\Sigma^+$ state and is well described by a single closed shell configuration, although configuration interaction (CI) turns out very important for a correct representation of the bonding energy [9]. Moreover, the core hole spectra of closed shell molecular carbonyls show a close resemblance to the core hole spectra of the adsorbates [1,8]. So there appears to be significant evidence for the strongly localized nature of the excitations in the adsorbate molecule and the adequacy of a small molecular cluster approach for a description of their basic features [4,5,7,8,10,11].

For even a qualitative understanding of the core hole spectra of adsorbates where the satellite lines take more than 80 percent of the spectral intensity, it is essential to take into account the multiple excitations in a consistent manner. For incorporation of such many-body effects, it is an advantage of the many-body technique such as Green's function method and CI method beyond the simple SCF molecular orbital approach, e.g. ΔSCF method, that one can deal simultaneously with local metal, local ligand and metal-ligand charge transfer (CT) excitations. Indeed recent ab-initio Green's function calculations of the valence photoemissiobn spectra of NiCO, NiN_2, PdCO, PtCO, metal carbonyls and metal carbonyl nitrosyls using an extended basis set, show good agreement with experiment [4,10]. In order to study these new core hole spectral features of the CO/Ni and N_2/Ni systems, we performed the ab-initio 3h2p (three hole two particle) (with the 2h-2p ground state correlation) CI calculations of the core hole spectra of NiCO and NiN_2 using an extended basis set. The results by 2h1p CI calculations will be reported elsewhere [11].

2. THEORY

We refer to ref. 12 the details of the present scheme which has been proved to give a reliable semiquantitative description for the core hole spectra of small molecules. The present 3h2p CI scheme includes the metal-ligand CT excitations as additional excitations to the 2h1p configurations, e.g. 3h2p configurations and the many-electron effects such as the 2h-2p double excitations in the presence of a core hole, the relaxation of two holes and one particle and the screening of the 1h1h and 1h1p interactions in the 2h1p configuration. The contracted Gaussian basis sets similar to those descirbed in refs. 4 and 10 are employed. The calculation for NiCO (NiN_2) was performed, keeping the CO (N_2) distance at 2.156 (2.069) a.u. and the Ni - C (N) distance at 3.5(3.3) a.u.. The 2h1p and 3h2p CI results of these moleccules at different distances will be reported elsewhere [11].

3. RESULTS AND DISCUSSION

In table 1 we list the present 2h-2p/3h2p CI results for NiCO and NiN_2. For the configurations we denote $1s^{-1}i^{-1}j$ by i →j.

Table 1. Theoretical core hole spectra of NiCO and NiN_2 by the 2h-2p/3h2p CI method and experimental spectra of CO and N_2 adsorbed on a Ni metal surface

	Exp[1]		Theory (3h2p CI)			
level	E(eV)	I	E(eV)	I	I_{rel}	configuration
1σ	0.	.36	0.	.370	100.	$1s^{-1}$
O_{1s}	5.5	} .64	1.59		1.2	π-π^*
	8.5		5.83	.103	28.0	σ-σ^*
	15.0		8.94	.090	24.3	π-π^*
	26.0		9.71		7.3	π-π^*
	36.0		10.23		3.9	π-$\pi*$
	45.0		10.87		0.1	σ-σ^*
	55.0		11.45		4.4	same
			12.63		0.3	δ-δ^*
			12.71		0.8	δ-δ^*
			12.86		0.3	π-π^*
2σ	0.	.29	0.	.392	100.	$1s^{-1}$
C_{1s}	2.1	} .71	2.03		4.8	π-π^*
	5.5		7.15		1.1	σ-σ^*
	9.5		7.34	.147	37.6	σ-σ^*
	33.0		8.69		1.3	π-π^*
			9.32		0.3	1π-$\pi*$ + π - π^*
			10.23		14.5	π-π^*
			10.45		12.7	same
			11.51		1.5	1π-π^*
			11.98		0.1	π-π^*
			12.29		1.2	1π-π^* + π-π^*
			12.95		0.6	σ-σ^*

Table 1.Cont.

level	Exp[1] E(eV)	Exp[1] I	Theory (3h2p CI) E(eV)	I	I_{rel}	configuration
	0.	.16	0.	.1557	42.95	$1s^{-1} + \pi - \pi^* + \pi^2 - \pi^{*2}$
	5.3		1.88		0.6	π-$\pi^* + \pi^2 - \pi^{*2} + 1s^{-1}$
			5.85		6.98	σ-σ^* ($+ \sigma\pi - \sigma^*\pi^*$)
			5.93	.3625	100.	σ-$\sigma^* + \pi - \pi^*$ ($+ \sigma\pi - \sigma^*\pi^*$)
inner		.84	9.15		0.14	π-π^*
N_{1s}			10.09		0.68	σ-$\sigma* + \pi - \pi^*$
			11.75		0.30	σ-σ^* + double exc.
	15.0		12.92		0.33	same
			13.28		0.07	σ-σ^*
			13.30		0.04	σ-σ^*
	0.	.11	0.	.1354	36.74	$1s^{-1} + \pi - \pi^* + \pi^2 - \pi^{*2}$
	2.1		3.08		0.89	π-$\pi^* + \pi^2 - \pi^{*2} + 1s^{-1}$
	5.8		6.92	.3685	100.	σ-$\sigma^* + \sigma\pi - \sigma^*\pi^*$
outer	8.5	.89	9.80		3.88	σ-$\sigma^* + \pi - \pi^*$
N_{1s}			11.63		1.16	σ-$\sigma* + \pi - \pi^*$
			12.99		0.52	σ-$\sigma^* + 1\pi$-π^*
	15.0		13.37		0.53	π-$\pi^* + \pi^2 - \pi^{*2}$
			13.88		0.21	σ-$\sigma^* + 5\sigma - \sigma^*$
			14.46		0.18	π-π^*

The calculated C and O main line satellite line intensity ratios are 0.39/0.61 and 0.37/0.63, respectively. They agree well with experiment[1]. The results by the ab-initio GVB CI method are 0.56/0.44 and 0.53/0.47, respectively [7]. For both core levels, the lowest energy state is the 1h main line state closer to the Koopmans state, not the 2h1p CT shakedown state.

The calculated N_a and N_b main line satellite line intensity ratios are 0.16/0.84 and 0.14/0.86, respectively. So the agreement with experiment is very good. The ratios by the GVB CI method are 0.28/0.72 and 0.34/0.66 [7], respectively and 0.18/0.82 and 0.25/0.75 [8], respectively. The calculated 1s peak splitting by the present scheme and the GVB CI scheme [8] is 1.6 and 0.7 eV, respectively, whereas the experimental one is 1.3(1.5) eV [1,3]. This splitting is indeed due to the two inequivalent N atoms in accord with the interpretation by others [1,5,8] and not due to the CT screening from the different substrate bands [3]. The Koopmans energy splitting is very small (about

0.1 eV). However, the relaxation energy shift for N_b is larger than for N_a because of the different degree of the π bonding. This results in the smaller binding energy for N_b than N_a, in accord with others [1,5,8]. The lowest energy state is the relaxed 1h state rather than the Koopmans state because of large σ to σ^* excitation associated with the local metal configurational changes.

The C and O satellite of a small intensity is obtained at 2.03 and 1.59 eV, respectively. The corresponding N_a and N_b satellite of a small intensity is obtained at 1.9 and 3.1 eV, respectively. These satellites for CO and N_2 are the π(metal) to π^* ligand CT 2h1p shakeup state (with respect to the lowest energy state). Nilsson and Mårtesson(NM) [1] interpreted as the π CT shakeup from the CT 2h1p shakedown state. However, in the case of N_2 Freund et al [8] interpreted this as the ionic state resulting from different spin couplings of the CT screened lowest energy state. They obtained this state at 1.12 and 2.33 eV, respectively. The differences between the XPS lowest energy and XAS energy for the C and O are 1.5 and 1.1 eV, respectively [14]. For NiCO this satellite energy is very close to the 1s to $2\pi^*$ resonant excitation energy. Then the O satellite should be around 1.7 eV (=2.1– 0.4) eV. Indeed the present calculation predicts it at 1.6 eV and its intensity is very small, in accord with experiment. With an increase of the Ni - C (N) bond length (weaker coupling), the main line satellite line energy separation becomes smaller. This is mainly because of the changes of the magnitude of the 1h1h and 1h1p interaction in the 2h1p configurations (see ref. 4. for a discussion on the spectral behavior with a change of the bond length). The C core hole spectrum of the coadsorption CO/H/Ni(100) system shows indeed the spectral features of the weakly coupled NiCO such as a smaller lowest energy state intensity (0.23) and a smaller satellite energy separation in comparison to the CO/Ni(100) system [15]. The N_a satellite energy separation is much smaller than the N_b. This indicates that N_a hole state is much more weakly coupled with the metal atom than N_b one. This is also in accord with others [1,5,8]. The bonding strength of NiN_2 is about half of that of NiCO and it is the π bonding which is much weaker in NiN_2 than in NiCO [9,13]. Then the π bonding for the N_a hole state is much weaker than for the N_b hole state. The larger polarization of the π bonding charge in the presence of the N_b core hole explains a larger relaxation energy shift and satellite intensity for the N_b core hole ionization. For the CO/Ni system the 1s to $2\pi^*$ resonant excitation energy is very close to the satellite energy because the 1s to $2\pi^*$ resonantly excited 1h1p state is similar to the π CT 2h1p shakeup satellite state except for the presence of a hole in the substrate band in the latter state. For the N_2/Ni system, however, the N_a resonant excitation energy is closer (0.3 eV) to the corresponding XPS lowest energy than for the N_b (1.0 eV). The weaker the coupling of the hole state is, the closer the resonant excitation energy is to the XPS lowest energy. More screening charge has to be provided for the N_a core hole than for the N_b one either by the σ relaxation which becomes more significant than for the CO/Ni system because of a much less π CT capability for the N_2/Ni system or by the resonantly excited 2π electron. Two differently screened states become very similar in the energy. For the N_b and C (O) core hole excitaion, the resonantly excited 2π electron induces too large polarization in the core hole region. Then the resonantly excited state as well as the XPS π CT shakeup satellite state may relax to the XPS lowest energy state when the 2π charge flows toward the substrate. If this occurs before the participant and spectator Auger decay occurs or the presence of the 2π electron is negligible, then the resonant photoemission spectrum becomes identical to the normal AES spectrum. This problem is discussed in details elsewhere [16].

The 5.5 eV C and O satellite is obtained at 7.3 and 5.8 eV. respectively. The relative intensity ratio of the C and O satellite line to the main line by the present scheme is 0.38 and 0.28, respectively. The ratio by the ΔSCF [2,6] is 0.25 and 0.14, and by the GVB CI scheme [7] is 0.34 and 0.15, respectively. The previous experimental intensity ratios are 0.35 and 0.20 for the CO/Ni system [2] and 0.4 and 0.29 for solid $Ni(CO)_4$ [17], respectively. So the agreement with experiment is good. The N_a and N_b 5.5 eV satellite is obtained at 5.9 and 6.9 eV, respectively. The relative intensity ratio of the N_a and N_b satellite line to the main line by the present scheme is 2.49 and 2.72, respectively. The ratio by the ΔSCF scheme [2,6] is 1.0 and 1.0, and by the GVB CI scheme [7,8] is 1.57 (3.17 by ref. 8) and 1.14 (2.28), respectively. The previous experimental intensity ratios are 1.0 and 1.0 [2] and 2.4 and 4.0 [3], respectively. The new experimental N_a and N_b broad satellite intensity (including the 8.5 eV satellite on the shoulder) is 0.26 and 0.35, respectively [1]. The present theoretical intensity is 0.37 and 0.39, respectively. The latter intensity is very close to the corresponding C satellite intensity(0.37). Theoretical intensity is about 0.28. If one transfers the overestimated main line intensity (by 0.1) to this satellite, we obtain 0.38 in good agreement with experiment. The 5.5 eV satellite was interpreted by others as the π CT 2h1p shakeup [4,5] or the Koopmans state [2,6-8] (or for NiCO, local metal excitation according to the anaylsis of the results of ref. 6 by the authors of ref. 5) or the 2π screening bonding orbital to Rydberg like level ($n\pi$) shakeup [1]. However, the present calculation gives a quite different interpretation, namely the 5.5 eV satellite is the σ to $n\sigma^*$ shakeup excitations associated with the local metal configurational changes. For NiN_2 the lowest energy state looses much intensity. This satellite state takes the largest intensity and it is closer to the Koopmans state.

The newly resolved 9.5 eV C satellite is interpreted as the 1π to $2\pi^*$ intraligand shakeup by NM. The 9.5 eV C π CT shakeup (to $2\pi^*$ and $3\pi^*$) satellite is obtained around 10 eV. The newly resolved 8.5 eV O satellite is interpreted as the 2π to Rydberg state excitations by NM. The 8.5 eV O π CT shakeup satellite is obtained at 8.9 eV.

The N_b π to $2(3)\pi^*$ CT shakeup satellite of a very small intensity obtained around 10 eV corresponds to the 8.5 eV N_b satellite.

In free CO and N_2 the dynamics of the screening of the core hole is governed by the σ to σ^* and π to π^* intraligand excitations [11,12], whereas for NiCO and NiN_2 the lower shakeup energy region is dominated by the π metal-ligand CT excitations and the σ excitation instead of the intraligand excitations of the same symmetry. The intraligand excitations show up in the higher energy region, although the excitations mix very much with the metal-ligand CT excitations and local metal excitations and much less pronounced than in free molecule. For the coordinated molecule the main line intensity of the free molecule is distributed to the main line of a much smaller intensity and several satellite lines by the metal-ligand CT and local metal excitations. Thus the main line intensity of the adsorbate is much smaller than in the free molecule, in accord with experiment (also true for the valence photoemission spectra [4,10]).

4. CONCLUSION

The main line satellite line energy separations, intensity ratios and the two core hole peak energy splitting of the core hole spectra of the CO/Ni and N_2/Ni systems are well reproduced by the 2h-2p/3h2p CI calculations. For NiCO the lowest energy state is the 1h main line state closer to the Koopmans state, whereas for NiN_2 it is the relaxed 1h state due to the strong σ CT relaxation because of a weaker π bonding in NiN_2 than NiCO. The 2.1 eV C and N_b satellite is the π CT 2h1p shakeup satellite rather than the ionic state due to the different spin couplings. For the CO/Ni system the interpretation and the existence of this satellite is supported by the closeness of the 1s to $2\pi^*$ resonance energy to the satellite energy. The 5.5 eV satellites are the σ to σ^* shakeup state. For NiN_2 this state takes the largest intensity and is closer to the Koopmans state. The importance of the σ to σ^* excitation associated with the local metal configurational change is emphasized. The 9 eV C, 8.5 eV O and N_b satellites are the π CT shakeup (to $2\pi^*$ and to the higher Rydberg levels) states. The ab initio molecular many-body approach seems to be able to give a detailed and consistent description of the dynamics of photoionization from adsorbates, and even small molecules such as NiCO and NiN_2 appear to be suitable to describe both ground and ionized states of the adsorbates when the many-body effects are taken into account in a proper manner.

ACKNOWLEDGEMENT

One of the authors (MO) would like to thank the surface physics group in Uppsala for delightful discussions and to Uppsala University and NFR for financial support.

REFERENCES

1. N. Mårtensson and A. Nilsson, J. Electro. Spectr. and Rel. Phenom. 52, 1, (1990), Phys. Rev. B 40, 10249, (1989) and references therein.

2. C.R. Brundle, P.S. Bagus, D. Menzel, and K. Hermann, Phys. Rev. B 24, 7041, (1981) and references therein.

3. E. Umbach, Sur. Scie. 117, 482 (1982) and references therein.

4. M.Ohno and W. von Niessen, Phys. Rev. B 42 , 7370 (1990) and references therein. In table 7 of this ref. the intensites of ref.42 are incorrectly quoted. The correct values are 0.56/0.19 and 0.53/ 0.18.

5. D. Saddei, H.-J. Freund and G. Hohlneicher, Sur. Sci. 102, 359 (1981) and references therein

6. P. S. Bagus and K. Hermann, Sur. Sci. 89, 588 (1979) and Solid State Commun. 38, 1257 (1981)

7. C. M. Kao and P. P. Messmer, Phys. Rev. B31, 4835, (1985)

8. H.-J. Freund, R. P. Messmer, C. M. Kao and E. W. Plummer, Phys. Rev. B31, 4848 (1985) and references therein.

9. C.W. Bauschlicher, Jr., S. R. Langhoff and A. Barnes Chem. Phys. Letters 129, 431, (1989) and references therein.

10. M. Ohno and W. von Niessen, Phys. Rev. B in press, J. Chem. Phys. in press and proceeding of this conference

11. M. Ohno and P. Decleva, Chem. Phys. in press and Sur. Sci. in press

12. A. Lisini, G. Fronzoni and P. Decleva, J. Phys. B21, 3653, (1988)

13. C.W. Bauschlicher, Jr., Chem. Phys. Lett. 115, 389, (1985)

14. A. Nilsson, E. Zdansky, H. Antonsson, O. Björneholm, N. Mårtensson, J. N. Andersen and R. Nyholm, to be published.

15. H. Antonsson, A. Nilsson, N. Mårtensson, I. Pana and P. E. M. Siegbahn, J. Electron Spectro. and rel. phenom. 54, 601, (1990)

16. M. Ohno, submitted

17. N. Barber, J. A. Connor and I. H. Hillier, Chem. Phys. Lett. 9, 570 (1971)

RADIATIVE CORE-VALENCE TRANSITIONS IN $KMgF_3$ AND KF CRYSTALS

P. A. Rodnyi

Leningrad State Technical University, Leningrad (St. Petersburg) 195251, USSR

M. A. Terekhin

I. V. Kurchatov Institute of Atomic Energy, Moscow 123182, USSR

ABSTRACT

Using the VUV synchrotron radiation, the emission properties of $KMgF_3$ and KF are inverstigated. It is shown that the short-time (~1ns) radiation of the crystals belongs to core-valence transitions.

INTRODUCTION

Lately, a new form of luminescence involving the hole transitions between upper core and valence bands has been actively investigated in ionic crystals [1-3]. The core-valence transitions (CVT) require the incident excitation quantum energy $h\nu_i$ being in excess of the crystal upper core - conduction bands gap energy ΔE_{cc}. The validity of the $h\nu_i > \Delta E_{cc}$ condition is indeed the main criterion for relating detected radiation to CVT. Besides that, the main features of the new luminescence are short decay times (~1ns) and the high thermal stability of all of its parameters [3]. CVT crystals have already found applications in scintillator technique [4].

The physical mechanism involved in core-valence bands transitions have been fairly well clarified in barium fluoride [1,3] and in some alkali halides crystals [2,3]. It seems interesting to investigate CVT parameters in double-cation halogenides [3]. In the present paper we have studied $KMgF_3$ crystals and also KF crystals for the sake of comparison.

EXPERIMENTAL TECHNIQUE

Experiments were performed using synchrotron radiation from 450MeV electron storage ring Siberia-1. The measuments of reflection and luminescence exsitation spectra were taken at the VUV spectroscopy installation comprising normal incidence monochromator, cryostat, automatic control system. Luminescence (fast component) emitted from the sample was viewed at a right angle to the exciting. For further details on the experimental lay-out see ref.[5].

Nominally pure $KMgF_3$ and KF single crystals were used in measurements. Freshly cleaved sample (about $2*12*12\ mm^3$) was used for the investigation.

EXPERIMENTAL DATA AND DISCUSSION

The solid curve in Fig.1 represents our $KMgF_3$ crystal reflection spectrum, exhibiting a fairly good agreement with the previously published one [6,7]. The lower most peak at E_{ex}=11.8eV is due to anion Γ-excitons formation. The 21.3eV peak in the $KMgF_3$ reflection spectrum is due to core (cation) Γ-excitons. The fast decay (~1ns) $KMgF_3$ luminescence bands is located in the 5.2 9.0eV interval (curve 1 Fig.2) with quasiconstant intensity in the 77-550K temperature range. The excitation threshold of this luminescence was found to be at 21,8eV (dotted line in Fig.1).

The short decay time, thermal stability and disposition of excitation threshold suggest that the observed radiation of $KMgF_3$ involve CVT. In KF we hav detected not wide band of short-time luminescence with peak at 7,9eV (curve 2 in Fig.2). Upon heating KF from liquid nitrogen up to room temperature luminescence intensity is reduced by the factor of 10^2. However the luminescence band 7.9eV KF also belongs to CVT. Let us show it.

Fig.2 presents the KF and $KMgF_3$ energy diagram as deduced from our present data and from experimental photoelectron and optical spectra [6-8] and calculation results [9]. Accoding to the CVT model [3] the hole created due to a high energy excitation in the core band comes up to the band upper boudery in $\sim 10^{-13}$sec and then recombines during $\sim 10^{-9}$sec with a valence band electron. Therefore the radiative CVT occur between the core band top and various valence band levels (Fig.2). The condition of radiative CVT without reabsorption is

$$(E_{g2} + \Delta E_v) < E^a_{ex} \qquad (1)$$

with E_{g2} representing the core-valence bands gap (i.e. the second energy gap of the crystal),ΔE_v-valence band width and E^{a}_{ex} - the anion exciton production energy. The latter determines the crystal fundamental absorption edge. In $KMgF_3$ E_{g2}=5.2eV, ΔE_v=4.5eV , E^{a}_{ex}=11.8eV. Consequently, the condition (1) is valid and the radiative CVT band lies within the crystal transperence region.

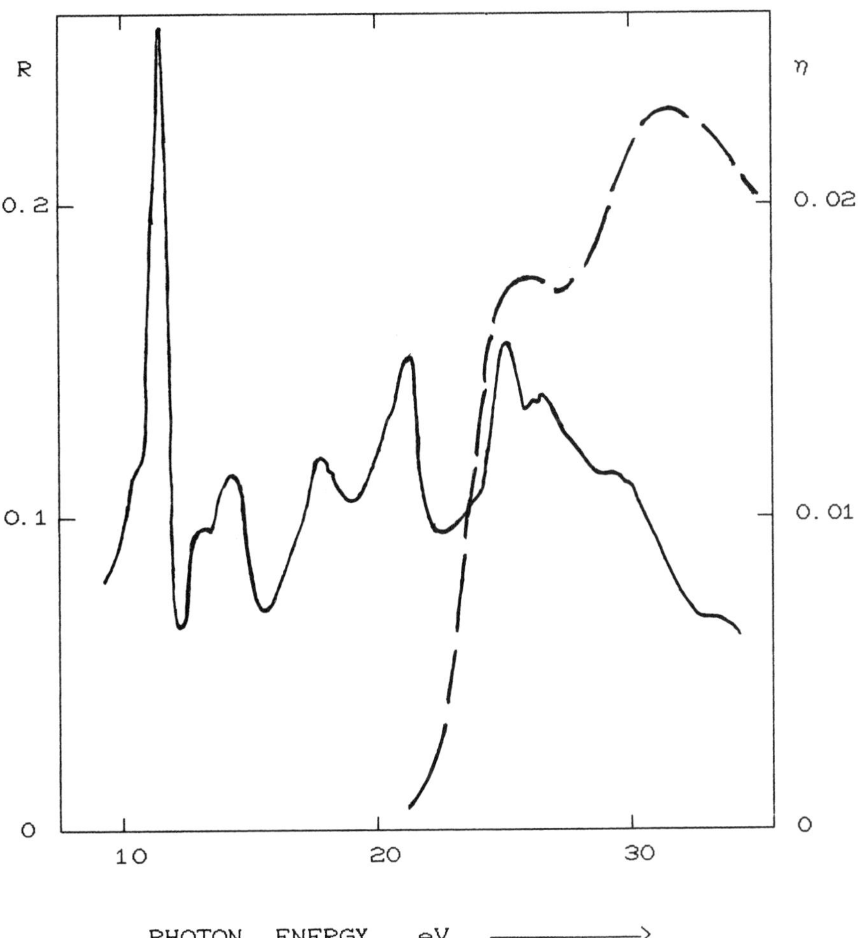

Fig.1 Reflectance (solid curve) and short-time luminescence excitation (dotted curve) of $KMgF_3$ at 295K.

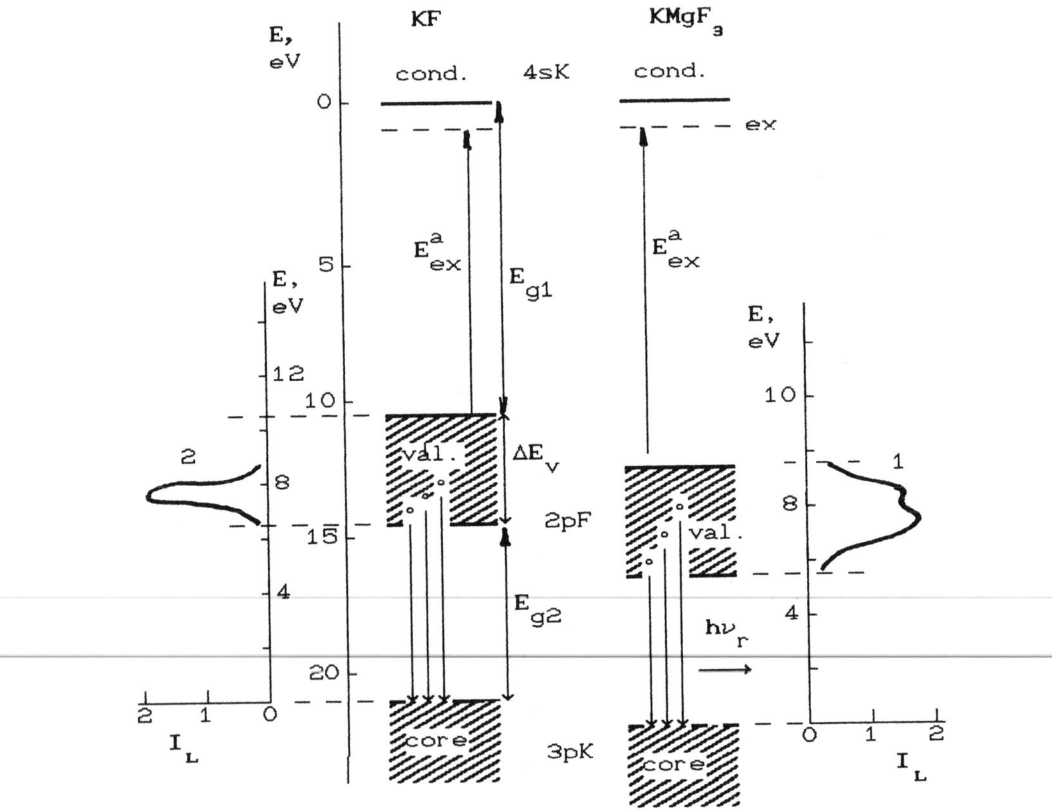

Fig 2. Scheme of energy bands of KF and $KMgF_3$ crystals and possible transitions. Short-time luminescence spectra of $KMgF_3$ (1) and KF(2) crystals at 80K.

In case the non-reabsorbtion condition is only partially valid, e.g.

$$E_{g2} < E^{a}_{ex} < (E_{g2} + \Delta E_v) \qquad (2)$$

the high-energy edge of the CVT band is reabsorbed within the crystal. *Eq.* 2 is valid for the *KF* crystal: E_{g2}=7.3eV, ΔE_v=3.0eV [8], E^{a}_{ex}=9.6eV. Due to this fact, only the lower-energy part of the CVT band is detectable (curve 1 Fig.2). So the substitution of KF to $KMgF_3$ means the change of eq.2 to eq.1. Upon heating KF the luminescence intensity is reduced due to the shift of the fundamental absorption edge and the increasing of the reabsorbtion. The *CVT*-bound luminescence intensity in $KMgF_3$ is substantially higher than in *KF*.

Fig.2 shows that in $KMgF_3$ the upper energy gap E_{g1} and valence band width ΔEv are larger then in the KF crystal. Taking into account the previosly published data on *CsCl*, $CsCaCl_3$ and BaF_2, $BaLiF_3$ [10] one can make the conclusion that the shift and broadering of valence band due to substituting of single-cation to double-cation halogenides is typical to ionic crystals. The core band position is less sensetive to addition of another cation in the crystals.

SUMMARY

The short-time luminescence excitation threshold is 21.8*eV* for $KMgF_3$, it lies amidst the core exciton formation range. The crystal luminescence parameters varied very little over a broad temperature range. Our experimental data make it possible to indentify the observed $KMgF_3$ and *KF* radiation to core-valence transitions.

The comparison of single-cation and double-cation halogenides luminescence parameters suggests that the latter are promising fast phosphors.

ACKNOWLEDGEMENTS

We would like to thank Prof. S.D.Fanchenko from Kurchatov Institute of Atomic Energy and Dr. S.V.Petrov from Institute for Physical Problems for their close cooperation during the course of this research..

REFERENS

1. S. Kubota, N. Kanai, J. Ruan (Gen). Phys. Status Solidi (b). **139** 635 (1987).

2. P.A. Rodnyi, M.A. Terekhin. Phys. Status Solidi (b). **166** 283 (1991).

3. A.V. Golovin, N.G. Zakharov and P.A. Rodnyi. Opt.i Spektrosk. **65**, 176 (1988) [Opt. Spektrosk. (USSR) **65** 102 (1988)].

4. P. Schotanus, C.V.E. van Eijk, R.W. Hollander and J. Pijpelink. IEEE Trans. on Nuclear Science, **NS-34** 272 (1987).

5. V.V. Mikhaylin, M.A. Terekhin. Rev. Sci. Instrum. **60** 2545 (1989).

6. J.H. Beamont, A.J. Bourdillon and J. Bordas. J. Phys. C. **10** 333 (1977).

7. H. Takahashi and R. Onaka. J. Phys. Soc. Jap. **43** 2021 (1977).

8. R.T. Poole, J. Liesegang, R.C.G. Leckey and J.G. Jenkin. Phys. Rev. **B11** 5179 (1975).

9. R.A. Heaton and C.C. Lin. Phys. Rev. **B25** 3538 (1982).

10. P.A. Rodnyi, M.A. Terekhin, E.N. Mel'chakov. J. of Lumines. **47** 281 (1991).

Dissociation dynamics of Core Excited Iron Carbonyl-Nitrosyl

M.Simon[1,2], M.Lavollée[1], T. Lebrun[1,2],J. Delwiche[3], M. J. Hubin[3] and P. Morin[1,2]

(1) LURE, Université Paris-sud, bat 209d, 91405 Orsay Cedex

(2) CEA/CENSaclay/DRECAM/SPAM 91191 Gif sur Yvette Cédex

(3) LSE, Sart Tilman par B-4000 Liège1, Belgique

Abstract

Core excitation of a complex molecule like Iron carbonyl nitrosyl allows to show that site selective excitation is not necessarily followed by selective fragmentation, probably due to a fast electronic to vibrational energy conversion. The use of multicoincidence electron/ion/ion technique shows however that fragmentation takes place on a much slower time scale allowing stewise processes to be observed.

Introduction

Special attention has been paid recently to core excitation of polyatomic molecule in the quest for selective photochemistry [1,3]. Indeed taking advantage of core hole localization as well as localization of antibonding orbitals along a specific bond, the simple idea merged up to expect a selective fragmentation following inner-shell excitation using monochromatized synchrotron radiation. Recent studies with "prototype" molecules (ex HBr [4] or N_2O [5]) revealed that dissociation dynamics of simple molecules (diatomic or triatomic) answered to the chemical selectivity criterion: the observed fragmentation crucially depends on the very nature of the excited resonance. The extension of such findings to more complex systems for which internal energy conversion mechanisms occur very quickly remains however an open question. The Iron Carbonyl Nitrosyl molecule $Fe(CO)_2(NO)_2$ is interesting from that point of view because there are two kinds of ligands (CO and NO) for which the transitions C_{1s} and N_{1s} to π^* or σ^* orbitals can be selectively excited .

The core excitation mostly products unstable multiply charged ions and requires detection of ionic fragments in coincidence. Thanks to the multicoincidence technique (electron-ion = PEPICO, electron-ion-ion = PEPIPICO) initiated by J.Eland[6], we are able to distinguish most of the dissociation channels which gives us information about electronic dissociation dynamics. Kinetic momenta correlation of ionized fragments are also determined, giving deep insight into fragmentation dynamics.

I Experimental section

The experimental set-up as well as the related data treatment has been described in detail in a previous publication [7]. Briefly, the monochromatized synchrotron radiation crosses at right angle an effusive jet of gas. In the ionization chamber, a high DC electrostatic field (2000V/cm) extracts all the ions on one side and the electrons in the opposite side. Electrons and ions are detected by microchannel plates, supplying respectively the start and stop of a multihit TDC. The main advantage of such a TDC is that several consecutive stops can be registred within an adjusted time window. Time of flight of ions arising from various coincidence type (single, double, triple coincidence) are simultaneously stored and treated in real time through a MacII microcomputer.

II Relaxation dynamics

Figure 1 shows the ion yield spectrum of Iron Carbonyl-nitrosyl around the nitrogen K edge. The ionization threshold is estimated at 405 eV. Below threshold (400.3 eV), an intense resonance appears we assign to the N_{1s}---> π^* transition. Above threshold, a broad resonance is observed with its maximum at 412.2 eV, which corresponds to a molecular shape resonance. Indeed the core electron may be temporarily trapped behind a potential barrier originating from the anisotropy of the molecular field. We obtained the same kind of spectrum (not shown) when exciting the carbon K shell (C_{1s}--->π^* at 287.8 eV and C_{1s}--->σ^* at 305 eV). We measured coincidence spectra at those 4 photon energies.

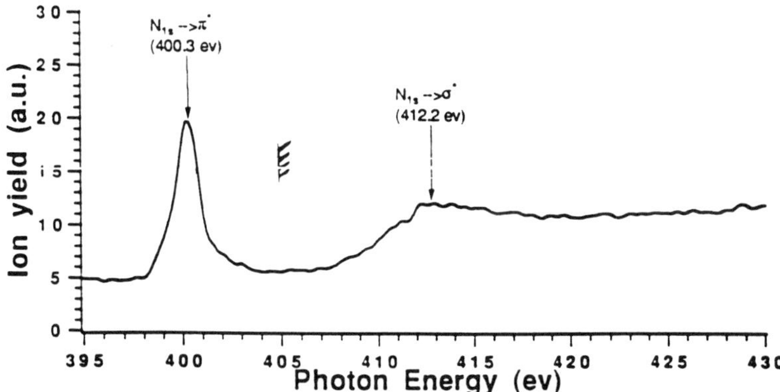

figure 1 : Ion yield of $Fe(CO)_2(NO)_2$ in the nitrogen K shell region.

We establish fragmentation spectra corresponding to different ionization degree of the molecule (simple, double, triple), producing one, two or three ions. As a result of the measurement we are able to obtain, after appropriate data reduction, the fragmentation pattern related to the production of one two or three ionic fragments. Those spectra correspond to single, double or triple ionization as far as no doubly charged fragments are produced. The only exception is the observation of $FeCO^{2+}$. This ion of course is detected as a one fragmentation process corresponding to the reaction:

$Fe(CO)_2(NO)_2^{2+}$ ---> $Fe(CO)^{2+}$ + 2(NO) + CO , which obviously belong to a double ionization process.

Figure 2 shows spectra of simple, double and triple fragmentation obtained after excitation at 400.3 eV.

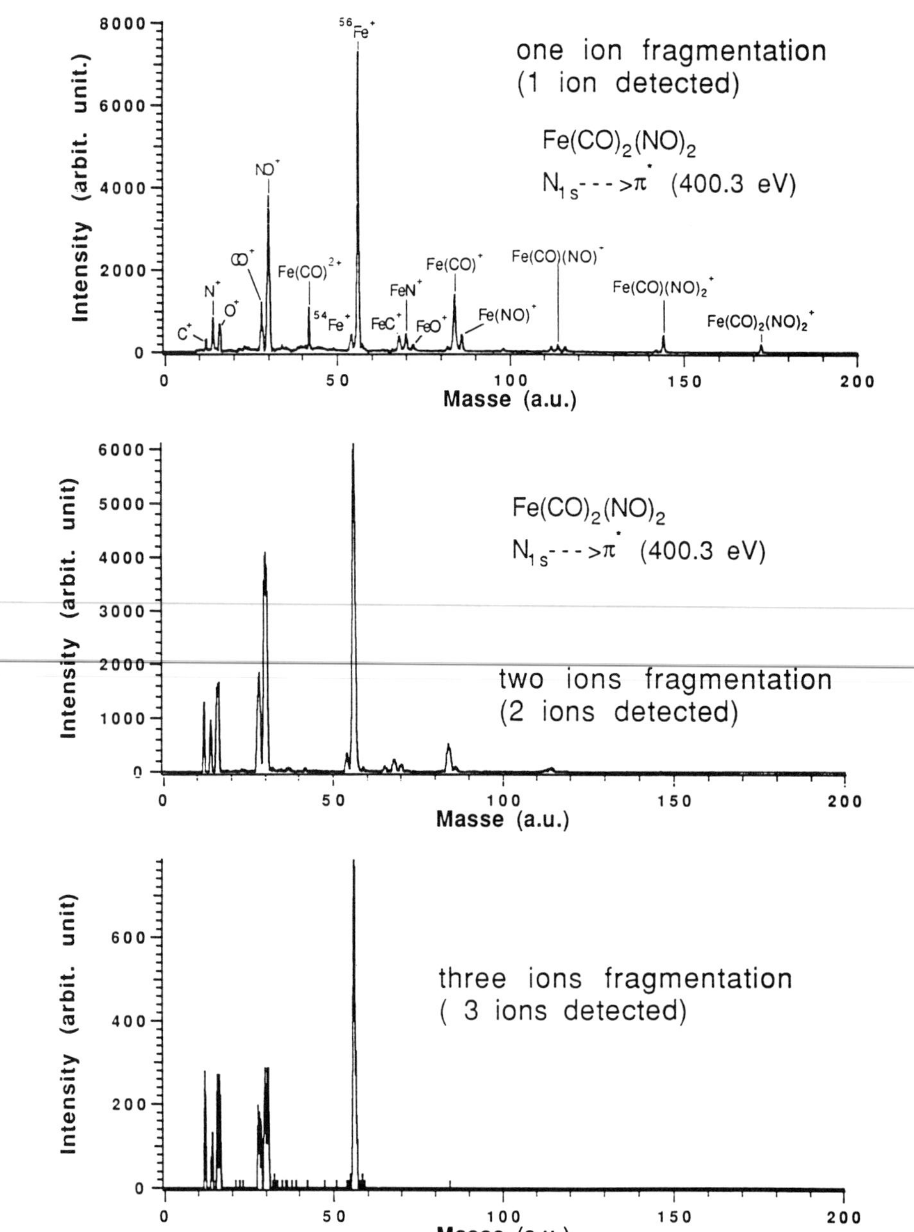

figure 2 : simple, double and triple fragmentation spectra obtained at 400.3 eV corresponding to the transition N_{1s} --->Π^*

The assignment of the different peaks is directly reported on the figure. We observed than the more the ionization degree is, the more atomization of the molecule is observed. The first observation we can draw is that the spectra obtained after ionization in the carbon K shell or in the nitrogen K shell are perfectly identical. This means that the initial localization of the excitation is lost in the fragmentation process: no "memory effects are observed.

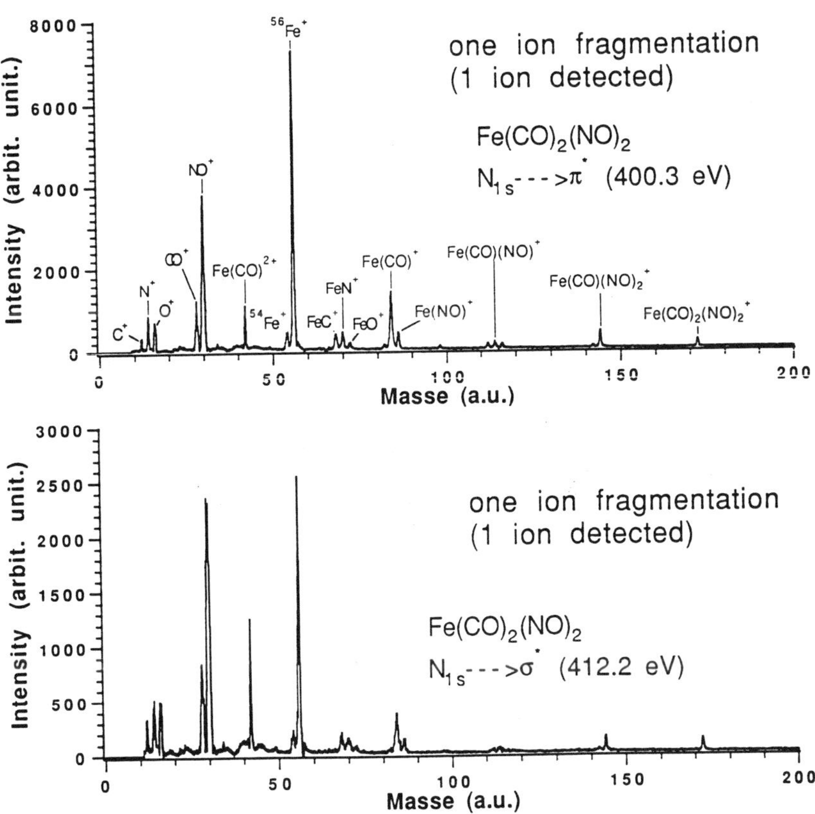

figure 3 : comparison of simple fragmentation spectra at 400.3 ev and 412.2 ev

However there are slight differences between simple fragmentation spectra arising from π^* and σ^* excitation (similar results are obtained at carbon and nitrigen K shell). The comparison of the two spectra is shown in figure 3. Note that branching ratios of the C^+, N^+, O^+, CO^+ and NO^+ ions are more important above threshold than below (on resonance). The $Fe(CO)^{2+}$ ion is also more abundant above threshold as it can be easily explained by the increase of double ionization due to Auger decay.

We interpret those results in terms of stepwise relaxation processes. After photon absorption, an electronic relaxation of Auger type occurs. Below threshold this process leads, according to resonant Auger model, to electronically excited singly charged ion and above threshold, according to normal Auger model, to doubly charged ion. The core hole life time is in the femtosecond time scale and the fragmentation process takes place in a picosecond time scale. Consequently, between the electronic relaxation and the fragmentation, the molecule has enough time to redistribute electronical energy into vibrational energy along all its freedom degrees according to the various density of states. This result is drastically different from what is observed on small molecules. There are two main reasons for a more efficient energy redistribution within a polyatomic molecule than within a simple one : a polyatomic molecule requires a longer time to dissociate because several chemical bonds are to be broken and the vibrational density of states is much more important for a polyatomic molecule than for a small molecule. Such energy redistribution has already been observed for neutral Iron carbonyl [8] and singly charged Iron carbonyl [9] but has never been pointed out in the case of multiply charged molecule.

Those results desagree with the Coulomb explosion model[10] which predicts a fast multiply charged ion dissociation governed by Coulombic charges repulsion. The difference observed between simple fragmentation spectra taken below and above threshold could be explained by the fact that resonant Auger process products excited singly charged ion given therefore more energy to redistribute before fragmentation. This excess energy may induce a different molecule fragmentation.

The crucial point of our experiment is the fragmentation channels separation depending on the molecule ionization degree. For instance, chemical selectivity has been observed with acetone [2,3] but the difference in the observed spectra may be only due to the increase of double ionization above threshold.

III Fragmentation dynamics

In this section, we focus on the fragmentation dynamics of doubly charged Iron carbonyl-nitrosyl, with special emphasis on sequential processes. Indeed, if we consider a doubly charged triatomic molecule ABC^{2+} dissociating into $A^+ + B^+ + C$, two chemical bonds are broken and the question is to determine the sequentiality or the simultaneity of those two breakings. J.Eland [11] showed that kinetic momenta correlation study of the detected ions gives access to this information, as revealed by the correlation slope measurement of selected coincidence peaks. A slope of -1 is observed if neutral ejection of C occurs before charge separation of A^+ and B^+ and is equal to $-m_B/m_{BC}$ if the first reaction has been charge separation into A^+ and BC^+ followed by neutral ejection of C. We apply this model on Iron carbonyl nitrosyl. We selected among the different pairs three demonstrative exemples. The results concerning those three pairs are reported on figure 4.

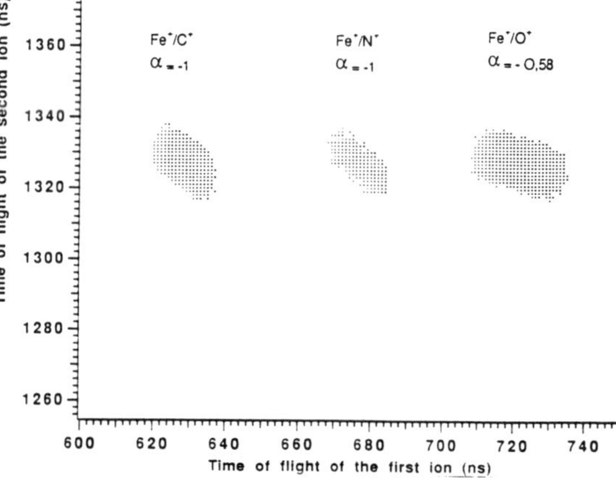

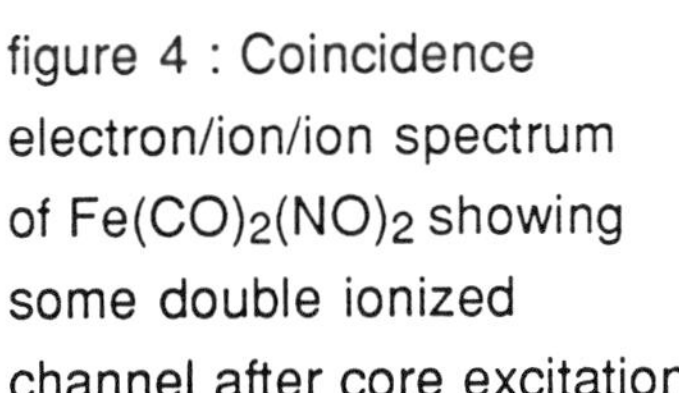
figure 4 : Coincidence electron/ion/ion spectrum of $Fe(CO)_2(NO)_2$ showing some double ionized channel after core excitation

The peaks appear as "cigars" of various orientation. The slope -1 of the Fe^+/C^+ (and Fe^+/N^+) peak indicates that neutral evaporation occured before charge separation, according to the following mechanism:

$Fe(CO)_2(NO)_2{}^{2+}$ ---> FeC^{2+} + 2 NO + CO + O

FeC^{2+} ---> Fe^+ + C^+

Of course we don't know much about the dynamics of the first step, but we do know that charge separation occurs as a final step.

The slope -0.58 measured for the Fe^+/O^+ peak suggests the following mechanism :

$Fe(CO)_2(NO)_2{}^{2+}$ ---> $Fe(CO)_2{}^{2+}$ + 2 NO

$Fe(CO)_2{}^{2+}$ ---> $FeC(CO)^+$ + O^+

$FeC(CO)^+$ ---> Fe^+ + C + CO

which is the only mechanism leading to a -.58 slope. This mechanism shows, in particular, that the ligand evaporation preceed the charge separation. The other coincidence peaks analysis provides the same kind of results. This successive ligand evaporation has been established for neutral Iron carbonyl dissociation [8] but has never been measured for this double ionized molecule. This mechanism also shows that charge separation is not always the faster process. We give evidence here that Coulomb explosion model is not well adapted to describe the fragmentation dynamics of this molecule.

Our data also reveal that an end chain atom is ejected with more kinetic energy than a middle chain atom if we compare for instance the ejection of C^+ and N^+ as compared to O^+. A possible explanation is the following. The ejection of O+ results in the CO bond breaking, which releases a significant energy. On the other hand, C^+ and N^+ originate from FeC^{2+} and FeN^{2+} fragmentation releasing less energy. Further theoretical investigation of the involved dissociation energies is required at that point to confirm this hypothesis.

Conclusion

We have shown that core excited Iron carbonyl nitrosyl relaxation is a multistep process. After photon absorption, the electronic relaxation occurs very quickly. Before dissociation, energy is converted along all vibrational modes inducing a statistical dissociation. The dissociation is itself a sequential process.

We are particularly endebted to F. Villain for the synthesis of Iron carbonyl nitrosyl.

References

[1] I. NENNER, P. MORIN, M. SIMON, N. LEVASSEUR and P. MILLIE, J. Elec. Spec. and Rel. Phenom., 52 (1990) 623-648

[2] W. EBERHARDT, T. SHAM, R. CARR, S. KRUMMACHER, M. STRONGIN, S. WENG, D. WESNER, Phys. Rev. Let., 50(1984)1038

[3] M. C. NELSON, J. MURAKANI, S. ANDERSON and D. HANSON, J. Chem. Phys., 8(1986) 4442

[4] P. MORIN and I. NENNER, Phys. Rev. Lett., 56(1986)1913

[5] T. LEBRUN, M. LAVOLLEE, M. SIMON and P. MORIN, to be published

[6]J.H.D.ELAND, F.S.WORT AND R.N.ROYDS, J.Electr.Spectr.Relat.Phen. 41(1986) 297

[7] M. SIMON, T. LEBRUN, M. LAVOLLEE and P. MORIN, in press Nucl.Instr. and Meth (1991).

[8] I. WALLER et J. HEPBURN, J. Chem. Phys., 10(1988)6658

[9] K. NORWOOD, A. ALI, G. FLESH and C. NG, in press J. A. C. S (1991)

[10] T. A. CARLSON, proceedings of DIET 1

[11] J. ELAND, Accounts for Chemical Research, 22(1989)381

ION DESORPTION FROM Si(100)-H_2O/D_2O BY CORE-ELECTRON EXCITATION

K. Tanaka and H. Ikeura
National Laboratory for High Energy Physics, Tsukuba, Ibaraki 305
and
University of Tokyo, Bunkyo, Tokyo 113

N. Ueno and Y. Kobayashi
Chiba University, Chiba 260

K. Obi and T. Sekiguchi
Tokyo Institute of Technology, Meguro, Tokyo 152

K. Honma
Himeji Institute of Technology, Akoh, Hyogo 678-12

ABSTRACT

The photon stimulated ion desorption (PSID) from the adsorption system of H_2O/D_2O on the Si(100) surface has been studied using pulsed synchrotron radiation in the range 100-800 eV. Ions were detected and analyzed by a simple time of flight spectrometer. Not only H^+, D^+ and F^+ but also O^+ ion was observed. The relative ion yield curves of these ions indicate characteristic behavior near and above the O K edge (539.7 eV); H^+ and D^+ ions exhibit sharp rises at ca. 530 eV and two broad peaks below (ca. 535 eV) and above (ca. 555 eV) the O K edge, O^+ exhibits a delayed threshold at ca. 570 eV and gradual increase up to 700 eV, and F^+ does not show any significant structure in this region. The results are discussed in terms of the primary excitation followed by the Auger decay and delocalization of the excitation.

INTRODUCTION

Recently photochemical reactions on the solid surface have received much attention concerning their important role in the fabrication of microelectronic devices.[1] Among them, photon stimulated desorption (PSD) from surfaces has been studied intensively during the last decade. The photon stimulate ion desorption (PSID) is very efficiently observed by core excitations both in ionic and covalent systems. The role of the Auger process in highly ionic system is well known in the context of the Knotek-Feibelman (KF) model.[2] According to the KF model, interatomic Auger decay of the core hole creates a two hole positive ion at an initially negative ion site and the expulsion of the positive ion results from the reversal of the Madelung potential.

In covalent systems, however, the Auger process has also been demonstrated to be of importance in PSID, but its role is not always clear.[3] The Auger stimulated desorption (ASD) model[4] assumes an intra-atomic Auger decay of a core hole state followed by desorption via a localized two holes state where the hole-hole repulsion U is greater than some appropriate covalent interaction V. In simple covalent systems such as CO and NO on transition metal surfaces, however, delocalization of the excitation, which can strongly quench the desorption process, is easy to take place by screening effect on metal surfaces. This effect has been found to be important and to depend sensitively on the primary excitation.[5,6] Multielectron excitations with energies much above the pure core level have also been found to play an important role for the ion desorption. In particular the multielectron excitation can counteract the screening effect on the metal surfaces.

In this paper, we present the results of the PSID study of H_2O/D_2O on the Si(100) surface and interpret our results in terms of the primary excitation followed by the Auger decay and delocalization of the excitation.

EXPERIMENTAL

The experiment was carried out at the Photon Factory of the National Laboratory for High Energy Physics on the grasshopper monochromator beamline 11A. The ion spectrometer used was a simple time of flight (TOF) one consisting of an accelerating plate, a drift tube of 9 cm in length and a multichannel plate (MCP) detector. The TOF spectrometer was mounted at an angle of 75 degrees with respect to the incident photon beam in a UHV chamber with a base pressure of typically 3 x 10^{-10} Torr. The chamber was also equipped with a double-pass cylindrical mirror analyzer (CMA) for photoelectron and Auger electron analysis, a quadrupole mass spectrometer and an ion gun for sputter cleaning the sample.

A schematic diagram of the experimental arrangement is shown in Fig. 1. The PF ring was operating at 2.5 GeV with a maximum current of ca. 35 mA in a single bunch mode. As shown in Fig. 1, the soft X-ray pulse with a turn time of 624 ns and a width of 100 ps was incident on the sample at an angle of 15 degrees from the surface

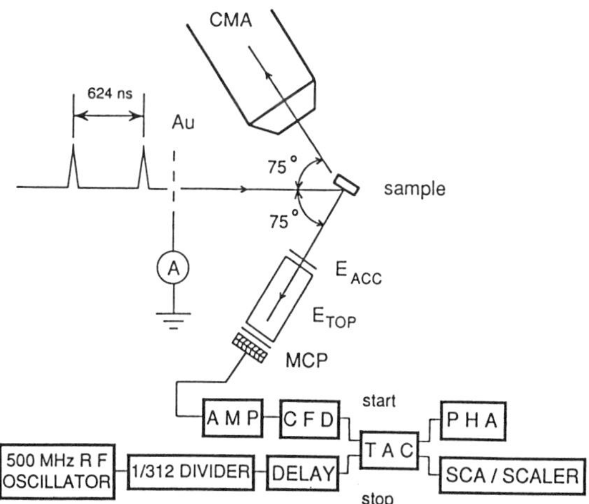

Fig. 1. Schematic diagram of the experimental arrangement.

through an in situ Au coated W grid for beam flux monitoring. The PSD ions were detected by the TOF spectrometer in the normal direction to the surface. The ion TOF spectrum was obtained by using the usual system of TAC and PHA where the ion signal was fed into the start input and RF timing signal into the stop input of the TAC. A relative PSID yield curve for each ion was obtained by acquiring the output signals of the TAC within a particular SCA window as a function of photon energy after flux normalization.

The sample was cleaned following standard procedures to form the Si(100)2x1 surface. A 10 L saturation exposure to D_2O was carried out with the sample at room temperature. It is well known that the adsorption of H_2O on the Si(100) surface at room temperature results in dissociation to H_a and OH_a.[7]

RESULTS AND DISCUSSION

A typical ion TOF spectrum from Si(100)-10 L D_2O obtained at photon energy of 720 eV is shown in Fig. 2. This spectrum was obtained by using the TAC in reverse mode where the ion signal was fed into the stop and RF timing signal into the start input. Two sharp peaks observed at 18 and 631 channels are the prompt peaks caused by the scattering of the incident photon at the sample. The interval of these peaks is exactly known as 624 ns and may be used for the calibration of time scale on the abscissa. H^+ ion is the dominant one and its TOF is typically 255 ns. D^+, O^+, F^+, and possibly O^{2+} ions can easily be identified from their TOF. Despite the 10 L exposure to D_2O, the intensity of H^+ ion is much larger than D^+. The reason for this can be assumed that the residual H_2O (ca. 3 x 10^{-10} Torr) preferentially adsorbs to the clean surface prior to the exposure to D_2O.

Relative PSID yield curves for these ions in the photon energy region of 100-800 eV obtained are shown in Fig. 3 together with a total electron yield (TEY) from the same sample, where the relative

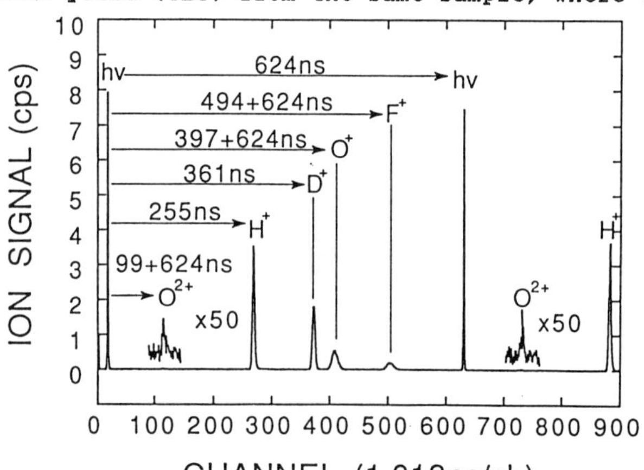

Fig. 2. Typical ion TOF spectrum from Si(100)-10 L D_2O obtained at photon energy of 720 eV.

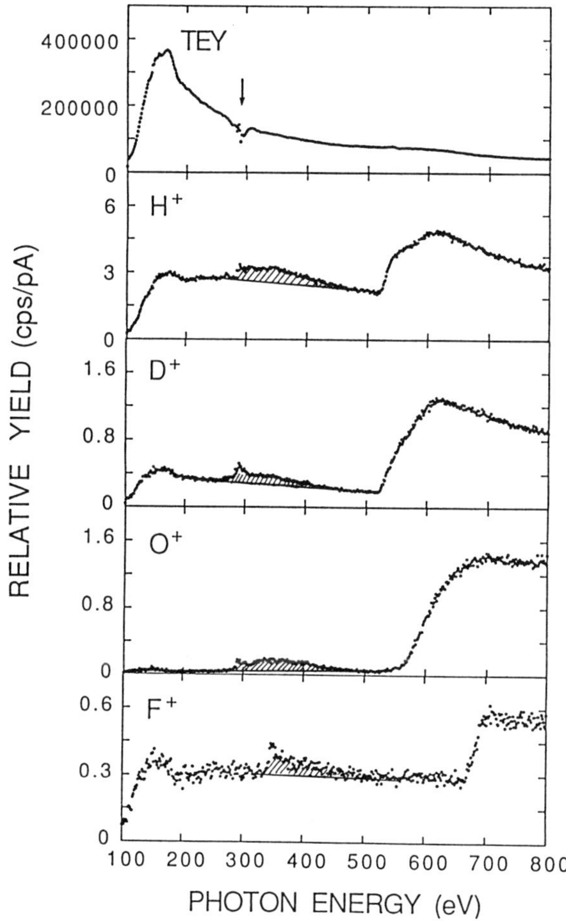

Fig. 3. Relative PSID yield of H^+, D^+, O^+ and F^+ and total electron yield from Si(100)-10L D_2O.

yield was obtained simply by acquiring the individual signal (cps) and dividing by the photoelectric current I_0 (pA) of the beam flux monitor. A slit width of 200 μm and a 1200 line/mm grating of the monochromator were used in this measurement which resulted in a resolution of ca. 13 eV at 500 eV and 1.2 eV at 150 eV. The actual intensity of H^+, D^+, O^+, F^+ ions, TE and I_0 obtained are typically 60, 40, 20, 9, $9x10^5$ cps and 15 pA, respectively, at 700 eV.

Several interesting features are found in PSID yield curves in Fig. 3. In 100-200 eV region, the profile of PSID yield of H^+, D^+ and F^+ resembles that of TEY, which increase rapidly at ca. 100 eV (Si L edge) and has a broad maximum at ca. 150 eV. Thus, it can be assumed that the core electron excitation of Si results in these ions desorption. It should be added that O^+ yield is exceptionally small in this region. Then, the question has been raised whether ions are desorbed by the excitation of Si atom to which the adsorbates are directly bound or desorbed by a secondary electron excitation following the excitation of nearby Si atoms in the substrate. The latter mechanism is called x-ray-excited electron stimulated desorption (XESD) mechanism proposed by Jaeger and coworkers in the study of PSID from NH_3/Ni system.[8] It is difficult to answer the question from the present study, mainly because of lack of energy resolution. Recently, Rosenberg and coworkers[9] studied PSID from H_2O/Si(111) in this region with higher resolution, and they concluded that XESD cannot be a major factor in the H^+ ion desorption from apparent discrepancy between TEY and H^+ ion yield curves.

In 200-500 eV region, irregular structures near 290eV (C K edge) and hatched inflation of yields are seen in Fig. 3. It can be assumed, however, that these irregular structures and hatched inflation of yields arise from the difficulty in the normalization to the beam flux which is strongly influenced by the carbon contamination of the optical elements and from the contribution in ion yield by the second order light, respectively.

Near the O K edge (539.7 eV), not only H^+ and D^+ but also O^+ ion was observed and their PSID yield curves indicate interesting threshold behavior; H^+ and D^+ ions exhibit sharp rises at ca. 530 eV and O^+ ion exhibits a delayed threshold at ca. 570 eV. F^+ ion does not show any significant structure in this region, and exhibits a sharp rise at ca. 685 eV (F K edge). In order to examine the threshold behavior in detail, PSID yield curves of H^+, D^+ and O^+ in the photon energy region of 520-610 eV were obtained with energy resolution of 2 eV (slit width of 30 μm). The results obtained are summarized in Fig. 4. As shown in the figure, H^+ and D^+ ions exhibit sharp rises at 530 eV and two broad peaks below (ca. 535 eV) and above (ca. 558 eV) the O K edge, whereas O^+ ion exhibits a delayed threshold at ca. 570 eV. These results indicate that the

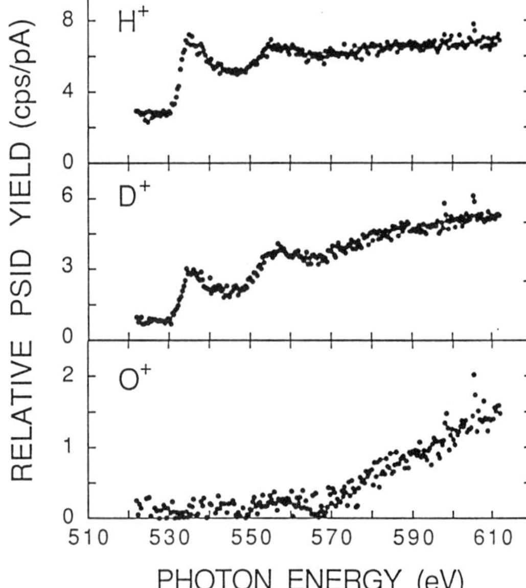

Fig. 4. Relative PSID yield of H^+, D^+ and O^+ from Si (100)-10L D_2O.

major source of ion desorption under the experimental conditions is the surface OH radical. PSID yields from Si(111)-100L D_2O[9] and Ru(001)-monolayer H_2O/D_2O[10] have recently been obtained in this energy region. In both cases, only H^+ ion was observed and its PSID yield exhibits a relatively narrow peak at 534 eV and a broad peak at ca. 560 eV which are in good agreement with our results within the experimental error. In both experiments, special attention was paid to the photon energy region of 520-580 eV. This may be one of the reason why O^+ ion was not observed.

PSID yield curves should be compared to the absorption spectrum of the surface OH radical. In Fig. 5, oxygen Auger electron yield (AEY) spectrum obtained for this purpose are shown. Oxygen Auger electrons of 503 eV are created by the decay of the O 1s core hole, and thus AEY represents the primary absorption cross section of the adsorbed OH. Several important features are found in the AEY spectrum. AEY exhibits sharp rise at 530 eV, a large broad peak (A) at ca. 538 eV and broad shoulder peak (B) at ca. 558 ev, which exactly coincides in position with those of PSID yield of H^+ and D^+ shown in Fig. 4. The AEY spectrum shows little structure above

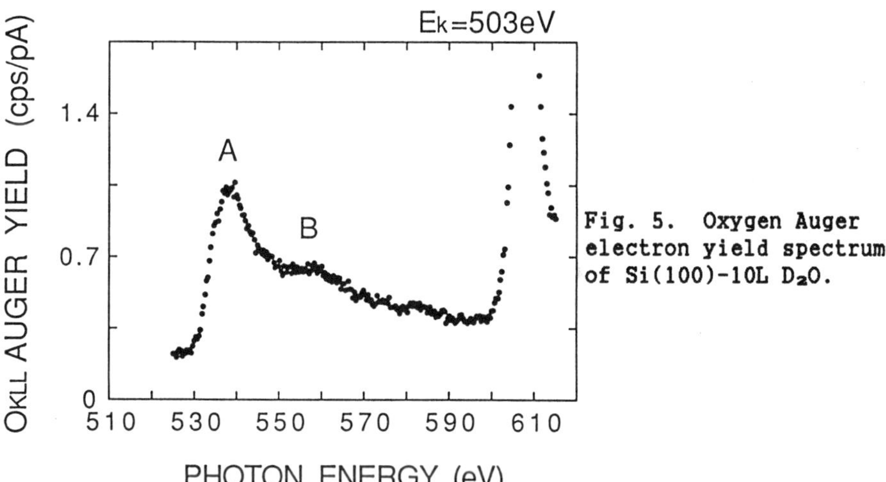

Fig. 5. Oxygen Auger electron yield spectrum of Si(100)-10L D_2O.

570 eV where O^+ yield starts to increase. Extremely large peak seen at 606 eV can easily be attributed to the Si 2p photoelectrons from the surface. Because the absorption spectrum of OH radical in this energy region has not been reported, the AEY spectrum is compared with the gas phase EELS spectrum[11] and the recent CI calculations[12] of excited states of H_2O. Although the AEY spectrum is not well resolved, its overall structure resembles that of EELS, and, by comparison with the CI calculation, peak A at 538 eV is tentatively assigned to a bound state transition 1s → 4a1 (534.6 eV) and/or 1s → 2b2 (536.5 eV) into the two lowest unoccupied molecular orbitals of H_2O and peak B at 558 eV to the shake up transition 1s,3a1 → c,4a1 (556.3 eV) and/or 1s,1b2 → c,2b2 (562.2 eV). Shake off transition 1s,3a1 → c,c (571.7 eV) and doubly excitation 1s,2a1 → 4a1,4a1 or 4a1,2b2 or 2b2,2b2 (ca. 570 eV) are possible transitions which

suggestively coincide with the threshold energy of O^+ ion.

As stated above, two broad peaks in PSID curves of H^+ and D^+ both coincide in position with those of the AEY spectrum, but the relative intensity of these two peaks show opposite tendency in the PSID and AEY spectra. This finding may suggest that H^+ and D^+ desorptions are enhanced by the excess energy above the pure core level and the lower excitation such as bound state transition (peak A) is easily delocalized by strong interaction with surface. In order to examine the role of the primary excitation, the probability of ion desorption is obtained by dividing the PSID yield by the Auger electron yield as a function of photon energy. This probability for D^+ ion increases monotonously with increasing photon energy in the region of 530-590 eV. This effect is particularly remarkable in O^+ desorption; PSID yield of O^+ is negligibly small at these excitation peaks (A and B) and increases by higher excitation above 570 eV.

As OH is bound with the O-atom pointing to the surface the probability of neutralization of O^+ is assumed to be much higher than that of H^+. Then, even if O^+ ion is formed by these excitations, it may immediately neutralize and desorb as neutral O atom. Thus, the appearance of O^+ ion above 570 eV has to be explained by the formation of the highly charged ion such as O^{2+} and O^{3+}. As mentioned above, shake off transition and/or doubly excitation can take place in this region and form OH^{m+} ($m \geq 3$) through the following Auger decay. OH^{m+} may decompose into O^{2+} + H^+, O^{3+} + H^+, and so on, which can survive the neutralization and desorb as H^+, O^+ and O^{2+} ions. It is considered one of possible evidence for this explanation that O^{2+} is surely observed in the ion TOF spectrum obtained at 720 eV (Fig. 2). Similar explanation has been applied by Treichler and coworkers[6] to the PSID from Ru(001)-CO system. They observed the delayed onset for the fragment ions (O^+, C^+ and O^{2+}) and interpreted this due to the doubly excitation.

In conclusion, comparison of PSID yields of H^+, D^+ and O^+ with Auger electron Yield shows that ion desorption yield strongly depends on the primary excitation and is much enhanced by excess energy. Shake off transition and/or doubly excitation are considered to play an important role in the PSID above 570 eV.

Further work including theoretical analysis and detailed comparison of gas phase data with chemisorbed data is needed to understand the whole dynamics of the PSID by core-electron excitation. Electron-ion and ion-ion coincidence experiments for both gas phase and chemisorbed H_2O are in progress.

ACKNOWLEDGEMENT

The authors would like to thank the staff of the Photon Factory for their support particularly in the single bunch operation. This study was performed under the Photon Factory Program Advisory Committee (Proposal No. 90-135). Part of this study was financially supported by the Grant-In-Aid from the Japanese Ministry of Education, Science and Culture.

REFERENCES

1. T. J. Chuang, Surf. Sci. Rept., 3, 1 (1983).
2. M. L. Knotek and P. J. Feibelman, Phys. Rev. Lett., **40**, 964 (1978); Sur. Sci., 90, 78 (1979); Phys. Rev., B18, 6531 (1978); P. J. Feibelman, Surf. Sci., **102**, L51 (1981).
3. R. Jaeger, J. Stöhr, R. Treichler and K. Baberscke, Phys. Rev. Lett., **47**, 1300 (1981); R. Jaeger, R. Treichler and J. Stöhr, Surf. Sci., **117**, 533 (1982).
4. D. E. Ramaker, C. T. White and J. S. Murday, J. Vac. Sci. Technol., 18, 748 (1981); Phys. Lett. **A89**, 211 (1982).
5. D. Menzel, P. Feulner, R. Treichler, E. Umbach and W. Wurth, Phys. Scr. **T17**, 166 (1987); W. Wurth, C. Schneider, R. Treichler, E. Umbach and D. Menzel, Phys. Rev. **B35**, 7741 (1987).
6. R. Treichler, W. Riedl, W. Wurth, P. Feulner and D. Menzel, Phys. Rev. Lett., **54**, 462 (1985).
7. C. U. S. Larsson, A. L. Johnson, A. Flodström and T. E. Madey, J. Vac. Sci. Technol., **A5**, 842 (1987).
8. R. Jaeger, J. Stöhr and T. Kendelewicz, Surf. Sci., **134**, 547 (1983).
9. R. A. Rosenberg, P. J. Love, V. Rehn, I. Owen and G. Thornton, J. Vac. Sci. Technol., **A4**, 1451 (1986); R. A. Rosenberg, C. -R. Wen and D. C. Mancini, in Desorption Induced by Electronic Transitions, DIET III, edited by R. H. Stulen and M. L. Knotek (Springer, Heidelberg, 1988), p.220.
10. D. Coulman, A. Puschmann, W. Wurth, H. -P. Steinrück and D. Menzel, Che. Phys. Lett., **148**, 371 (1988); D. Coulman, A. Puschmann, U. Höfer, H. -P. Steinrück, W. Wurth, P. Feulner and D. Menzel, J. Chem. Phys., **93**, 58 (1990).
11. G. R. Wight and C. E. Brion, J. Electron Spectrosc. Relat. Phenom., **4**, 25 (1974).
12. N. Kosugi, private communication.

DISCUSSION

DUJARDIN - 1. Have you never observed the desorption of Si^+ ions or Si neutrals by multiple excitation in Silicium ?

2. From the PIPICO experiment, could you measure the kinetic energy of the desorbed ions ?

3. How do you know that the H and OH fragments issued from the dissociative chemisorption of H_2O stay attached to neighbour silicon atoms ?

TANAKA - 1. In the present study, we observed only ionic species, and we could not observe Si^+ ions in the excitation range 100-800 eV. However, I should like to point out that this fact does not necessarily rule out the possibility of the neutral Si desorption.

2. It is interesting to know the kinetic energy of the desorbed ion. We have, however, observed the PIPICO spectrum for the O^+ and H^+ ions mainly in order to make sure that the O^+ and H^+ ions are produced simultaneoustly from the surface OH molecule. I would say that in principle it is possible to get information on the kinetic energy of the ion by detailed analysis of the PIPICO spectrum, but somewhat higher TOF resolution seems to be necessary for this purpose.

3. We have not performed any structural analyses of the adsorption system. Thus, our assumption about the dissociative chemisorption of H_2O is based on the results of an electron stimulated desorption ion angular distributions (ESDIAD) study of H_2O on Si(100) by Larsson and co-workers (Ref. 7 in the paper).

FRAGMENTATION OF CORE EXCITED METHYLAMINE

R. Thissen, J. Delwiche, M.-J. Hubin-Franskin, M. Furlan
Laboratoires de Spectroscopie de Photoélectrons et d'Electrons Diffusés, Université de Liège, Sart Tilman B6, 4000 Liège 1, Belgium

P. Morin, M. Lavollée, I. Nenner
L.U.R.E., Université de Paris Sud, Bât 209 D, 91405 Orsay, France

ABSTRACT

The dynamics of CH_3NH_2 after Nitrogen 1s electron excitation has been studied using the synchrotron radiation from SuperACO, L.U.R.E. storage ring, and by the photoionization method coupled to time-of-flight mass spectrometry and coincidence spectroscopy.

The dissociation patterns of the singly and doubly charged molecular ions formed after relaxation of the Nitrogen 1s core hole have been shown to be dependent of the initial electronic transition (Rydberg state or ionization continuum) and of the one or two hole(s) final configuration. We have also shown the existence of stable states of the dication and important rearrangement processes inside of the singly or doubly charged ions

INTRODUCTION

The study of the relaxation channels of core excited molecules in the gas phase is a subject of recent interest. First of all, because there is not such a long time that scientists have the possibility to use monochromatized synchrotron light[1.2], that is, by now, the most efficient mean to study the effects induced by an absorption of photon in the region of the electromagnetic field between 50 and 1000 eV. This region of energy includes the core ionization thresholds of most of the elements forming molecules of chemical and biological interest. Moreover, these thresholds are very well separated for the different atoms (1s C: 284 eV,1s N: 410 eV)[3]. Therefore, by use of the monochromatized synchrotron light, it is possible to excite selectively one type of atom inside of a molecule[4]. Hence, a lot of questions are opened. Is there a relation between the excited atom and the localization of the fragmentation? Is there an effect of the details of the electronic relaxation leading either to mono or multiply charged cations on the fragmentation of these ions? Is there an effect of the initial excitation process, inducing either a resonant excitation of the core electron to an unoccupied molecular orbital or the direct ionization? But to answer all these questions, it was necessary to solve another problem. It is the fact that doubly charged ions, those are the typical result of the electronic relaxation process of a core hole[5], are very scarce in classical mass spectra of molecules. This fact is due to an intermediate fragmentation of the dication, That can be explained by a completely repulsive shape of the potential surface of the dication states, or by a great vibrational excitation of the binding states[6]. However, to extract an important information in these studies, it has been necessary to develop very powerful coincidence electronics, that gave to scientists the possibility to study selectively and in detail the dynamics of doubly charged ions[7.8], and then to try to answer part of the here above questions.

This paper presents results that have been obtained with the methylamine (CH_3NH_2) molecule. Nitrogen 1s core electrons have been excited with monochromatized synchrotron light in the region of 400 eV. We have recorded time-of-flight mass spectra, and also coincidences spectra to study the fragmentation channels of the doubly charged ions. We have studied the fragmentation of the ions, its dependence on the initial excitation process,and also on the nature of the ionization continuum (simple or double) to which the core hole states are coupled.

EXPERIMENTAL SET-UP

The experimental set-up makes use of a synchrotron radiation light source, a monochromator, a time-of-flight mass spectrometer, and a coincidence electronic.

This set-up has already been described in detail [9]. Briefly, we used light produced by the Super ACO storage ring at L.U.R.E, and mounted our time-of-flight mass spectrometer behind a toroidal grating monochromator (beam line SU7) providing photons in the 200-800 eV energy range, with an ultimate resolution of 0.25 eV at 250 eV. The time-of-flight spectrometer is 15 cm long. The ionization region is continuously submitted at an electric field of 2000 V/cm, and the second acceleration field for the ions is of 2500 V/cm. These characteris-tics follow the MacLaren conditions [10] to ensure a good mass resolution. During the experiments, the spectrometer has been maintained at a working pressure of about $4*10^{-6}$ torr. This pressure being sufficiently low to avoid collision effects or a saturation of the acquisition devices. The partly home-made electronic system [9] contain a 4208 Multistop Lecroy TDC, and is controlled by a microcomputer. With this acquisition system, it is possible to record PEPICO and PEPIPICO spectra [8].

RESULTS AND DISCUSSION

Figure 1 presents the total ionization cross section obtained for the methylamine in the 399-413 eV range. We can observe two resonances at 400.78 and 402.3 eV and a shoulder at 403 eV corresponding to the excitation of the core electrons to 3s, 3p and 3d Rydberg orbitals, respectively. A shape resonance, associated to the C-N bond, is centered around 404.8 eV. The ionization threshold is situated at 405.17 eV[11]. We recorded PEPICO and PEPIPICO spectra at several photon energies corresponding either to the excitation of the core electron to the Rydberg orbitals, or to the shape resonance. These energies are situated on the spectrum by continuous lines for PEPICO and dotted lines for PEPIPICO spectra.

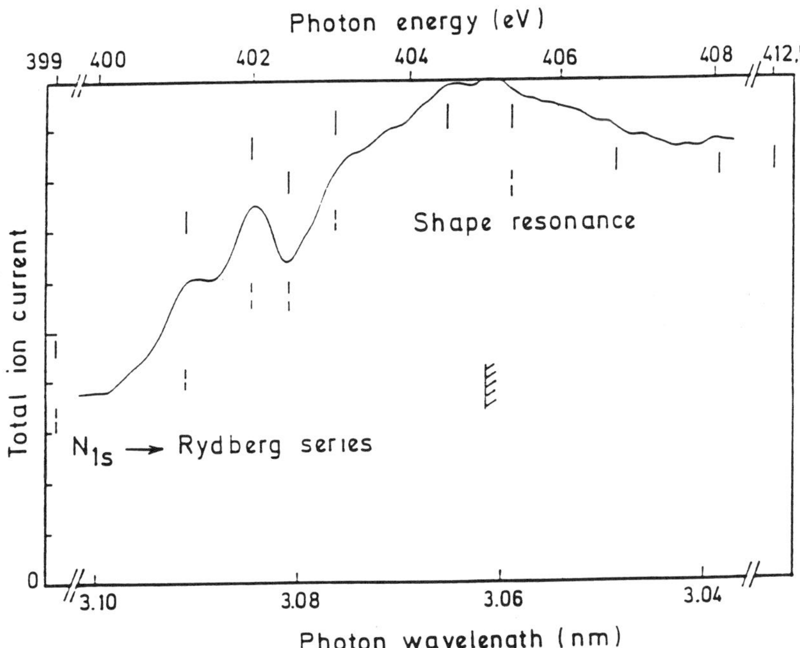

Fig. 1. Total ionization cross section of the methylamine around the nitrogen 1s threshold. Dotted and continuous lines indicate the energies at which PEPIPICO and PEPICO spectra have been recorded, respectively.

We present in figure 2 a typical PEPICO spectrum recorded at 403 eV. In this spectrum we have truncated the H^+ peak by a factor of 10 to highlight the other ions. The H^+ ion is the principal one and constitutes 50% of the detected ions. There are nearly no doubly charged ions in this spectrum. This fact confirms that dications are generally unstable and have to be studied by use of coincidence measurements. Nevertheless, there is a trace of a dication at a flight time of 630 ns and corresponding to a mass of 29. This unexpected ion had already been observed at lower excitation energies [12], and is the proof that at least one state of this dication is stable from the point of view of the fragmentation in ion pair, even for excitation energies as high as 400 eV.

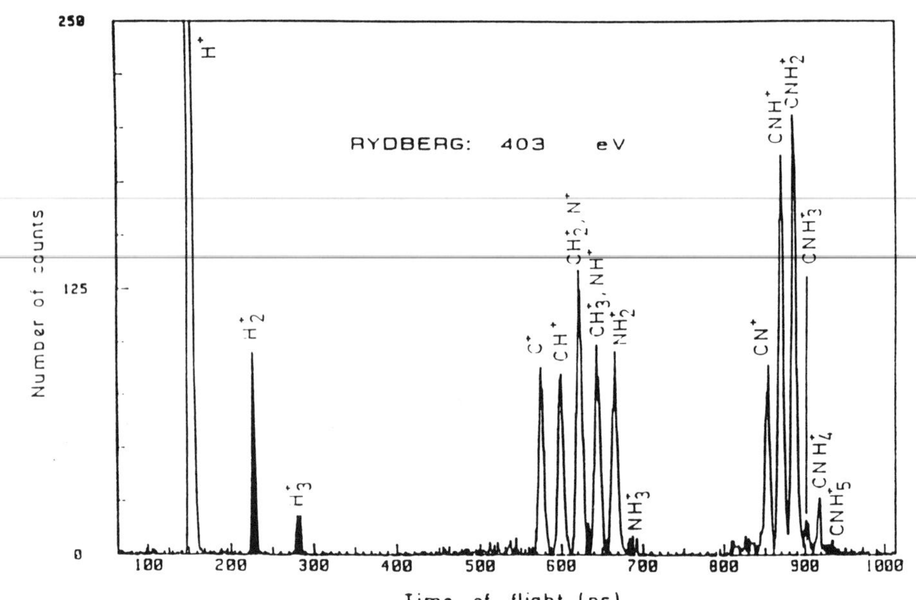

Fig. 2. Typical PEPICO spectrum of the methylamine recorded on the $N_{1s} \Rightarrow$ Rydberg 3d resonance (403 eV).

Another point is the presence, in this PEPICO spectrum, of a non-negligible quantity (4% of the total ion yield) of rearrangement ions (H_2^+, H_3^+, NH_3^+). These had also been observed previously at lower excitation enrgies[13], but were completely unexpected at such energies. They are the proof of very strong and rapid changes in the geometry of the molecule.

The effects of the photon energy on the fragmentation of the molecule have been studied. We have measured the relative intensities of each ion for several different energies. The result is presented in table I, for nitrogen fragments corresponding to the breakage of the C-N bond.

table I: relative intensities of different ions in the PEPICO spectrum with respect to the excitation process

excitation process	NH^+ % of the total ion yield	NH_2^+ % of the total ion yield
Rydberg	4.8	4.4
shape resonance	7.6	6.9

For these ions, there is a neat enhancement at photon energies corresponding to the shape resonance. This is a proof of a relation between the initial excitation process on the final fragmentation, and it is the sign of a certain localization of the relaxation process around the Nitrogen atom after an excitation into the shape resonance.

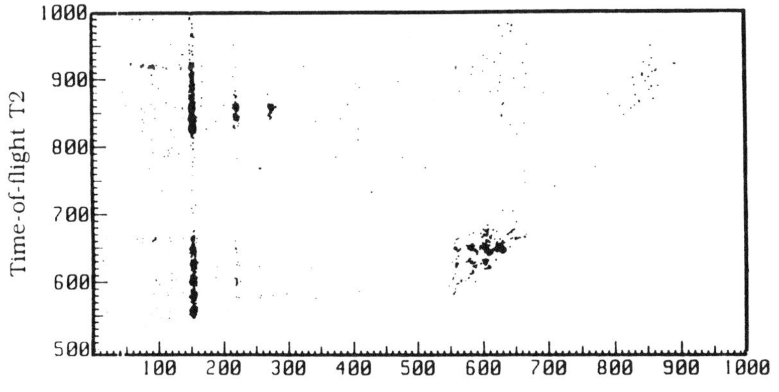

Fig. 3. Typical PEPIPICO spectrum of the methylamine recorded on the $N_{1s} \Rightarrow$ Rydberg 3d resonance (403 eV). The flight time of the first ion detected is plotted on the horizontal axis, and the flight time of the second on the vertical one.

To study selectively the behaviour of the dications, we have also recorded PEPIPICO spectra, and one of these is presented in figure 3. This spectrum contain three principal groups of ion pairs. There are pairs corresponding to the fragmentation of the C-N bond with a charge localized on C and N fragments (600 ns/ 600 ns), pairs corresponding to the same fragmentation but with one charge localized on the H atom (150 ns/ 600 ns), and finally the pairs involving only a fragmentation of C-H or N-H bond (150 ns/ 850 ns). Remark also the presence of ion pairs involving rearrangement ions (H_2^+ : 220 ns; H_3^+ : 290 ns), constituing 7% of the total ion pairs detected. The spectrum is dominated by pairs involving H^+ ion (87% of the total ion pairs).

As for PEPICO spectra, we recorded PEPIPICO spectra at different excitation energies corresponding either to the population of Rydberg orbitals or of the shape resonance. The principal result of this study is presented in table II. We can see that the behaviour is completely different of the one observed in the PEPICO spectrum. Here, there is a neat decrease of the formation of nitrogen fragments after excitation to the shape resonance. We can then conclude that the behaviour of the cation and the one of the dication are not sytematically similar after one particular excitation process.

table II: relative intensities of different ion pairs in the PEPIPICO spectrum with respect to the excitation Process

excitation process	H^+/NH^+ % of the total ion pair yield	H^+/N^+ % of the total ion pair yield
Rydberg	5.8	10.8
shape resonance	3.81	8.82

CONCLUSION

In this paper, we have studied the fragmentation of singly or doubly charged cations by means of PEPICO and PEPIPICO acquisitions. The treatment of the results obtained has permit to show that the fragmentation of the methylamine molecule after Nitrogen 1s electron excitation is dependent on several factors. There is the initial excitation process, that induce different track of relaxation, and finally induce different fragmentations. And also, there is the nature of the ionization continuum (simple or double) to which the core hole states are coupled.

Likewise, this study realized at high excitation energy has shown that the cations and dications states formed are sometimes sufficiently stable to permit the detection of doubly charged ions, and also are subject to very important changes in geometries those induce the formation of non-negligible quantities of rearrangement ions.

REFERENCES

1. D.M. Hanson, R. Stochbauer, T.F. Madey, J. Chem. Phys. 68, 377 (1982)
2. R.A. Rosenberg, V. Rehn, V.D. Jones, A.K. Green, C.C. Parks, G. Loubriel, R.H. Stulen, Chem. Phys. Lett., 80, 488 (1981)
3. L. Ley, M. Cardona, Eds., Photoemission In Solids II, (1979) (Springer-Verlag, Berlin)
4. W. Eberhardt, T.K. Sham, SPIE Proc, 447, 143 (1984)
5. I. Nenner, Giant Resonances in Atoms, Molecules and Solids, J.P. Connerade, J.M. Esteva, R.C. Karnatek, Eds (Plenum, New-York) (1987)
6. I. Nenner, P. Morin, P. Lablanquie, M. Simon, N. Levasseur, P. Millie, J. Electron Spectrsc. Relat. Phenom. 52, 623 (1990)
7. G. Dujardin, S. Leach, O. Dutuit, P.M. Guyon, M. Richard-Viard, Chem. Phys., 88, 339 (1984)
8. J.H.D. Eland, F.S. Wort, R.N. Royds, J. Electron Spectrsc. Relat. Phenom., 41, 297 (1986)
9. M. Simon, T. Lebrun, P. Morin, M. Lavollée, J.L. Marechal, to be published
10. W.C. Wiley, I.H. McLaren; Rev. Sci. Instr. , 6, 1150, (1955)
11. R.N.S. Sodhi, C.E. Brion, J Electron Spectrsc. Relat. Phenom., 36, 187 (1985)
12. J.E. Collin, M.J. Hubin-Franskin, Bull. Soc. Royale Sciences Liege, 3-4, 267 (1966)
13. E. Ruhl, S.D. Price, S. Leach and J.H.D. Eland, Int. J. Mass Spectrom. Ion Processes, 97, 175 (1990)

RELAXATION PROCESSES FOLLOWING EXCITATION AND IONIZATION OF THE IODINE 4d AND BROMINE 3d CORE ELECTRONS IN C_2H_5I AND C_2H_4IBr

F. Motte-Tollet, M.-J. Hubin-Franskin,* and J. Delwiche*
Laboratoire de Spectroscopie d'Electrons diffusés and Laboratoire de Spectroscopie de Photoélectrons, Université de Liège, Institut de Chimie-Bât.B6, Sart Tilman par 4000 Liège 1, Belgium

P. Morin
Département de Recherche sur l'Etat condensé, les Atomes et les Molécules, CEA, CEN Saclay, 91191 Gif sur Yvette Cedex, France and LURE Université de Paris Sud, 91405 Orsay Cedex, France

ABSTRACT

The photoelectron and Auger spectra of C_2H_5I (I *4d)* and C_2H_4IBr (I *4d*, Br *3d*) have been obtained using the synchrotron radiation on and off resonances above the I *4d* ionization threshold up to 118 eV. Two Auger bands related to the relaxation of the I *4d* core hole have been observed at 14 and 28.4 eV electron energies. For C_2H_4IBr, an additional Auger band has been observed at 45.7 eV and is related to the decay of the Br *3d*-vacancy states.

In addition resonant excitation of the Br *3d* electrons into the lowest empty σ^* has led to a selective decay into the iodine *4d* continuum and the iodine *4d* Auger processes rather than into the outer-valence-shell ionization continuum.

INTRODUCTION

The relaxation channels available after core electron excitation have gained recently a large interest due to the availability of the tunable continuous photon source of the synchrotron radiation. However, these studies have been devoted mainly to small polyatomic systems.[1,2] They have shown that the Auger relaxation of the core hole excited state can lead to the formation of doubly charged ions.[3] These molecular ions can then decay through dissociation into a pair of correlated fragment ions and/or into a neutral and a doubly charged fragment ion that generally undergoes a subsequent charge separation.[4-6] The dissociation and fluorescence decay channels can also sometimes compete with the Auger relaxation process.[3,7]

The C_2H_4IBr molecule is an interesting molecule as it has eight atoms with two different atomic sites I and Br on each carbon atom which both present shallow core electrons. In addition, the excitation of the I *4d* core electrons into the lowest lying σ^* orbital is followed at least partly by the relaxation into singly and doubly charged molecular ions with a selective dissociation of the C-I bond.[6,8] The absence of any doubly charged or multiply charged molecular or fragment ions after the excitation of the I *4d*

---

* Maître de recherches FNRS.

electrons is indicative of a quite low degree of ionization. This is not the case for HI where multicharged fragment ions I^{n+} have been observed after the excitation of the *L* shell of iodine.[3]

Complementary to these results, we have centered in this paper our interest on the Auger process itself which can lead to a molecule with two holes in the valence shell after the ionization of a shallow core electron.[4] In this study, we have investigated the Auger relaxation processes relating to the ionization of the iodine 4*d* and bromine 3*d* shallow core electrons in the 1-bromo-2-iodoethane. Using the synchrotron radiation, we have obtained the photoelectron and Auger spectrum at a photon energy above the iodine 4*d* edge (67 eV). Moreover, the spectra have also been measured at 97 and 118 eV, i.e. above the Br 3*d* ionization threshold on the I 4*d* -> εf and the Br 3*d* -> εf giant shape resonances respectively.[1,9,10] Finally, the photoelectron and Auger spectra have been recorded when the bromine 3*d* electrons are resonantly excited into the unoccupied valence molecular orbital σ^* at 71 eV. A comparison between the behaviors of the spectra on and outside resonances has been made.

In order to help the analysis of the spectra and of the results, the same kind of measurements has been made for iodoethane.

EXPERIMENTAL

The experimental set up was described in detail elsewhere.[11] Briefly, the VUV synchrotron radiation emitted by the SuperAco storage ring at Orsay was focused on the entrance of a toroidal grating monochromator (2400 g/mm). The monochromatic light was refocused at the center of the interaction chamber where it crosses an effusive jet of gas at right angles. After ionization, the ejected electrons were detected by a 127° cylindrical electrostatic analyzer.

The apparatus was operated in two different modes; a, the photoelectron and Auger spectrum at fixed photon energy; b, the threshold electron spectrum obtained by measuring the ejected electrons of near zero kinetic energy as a function of the photon energy.[11] The spectra were corrected for the incident light intensity (monitored by a gold mesh photoemission) and for the electron transmission of the analyzer retardation system. All spectra were recorded at the magic angle which eliminates angular effects and the analyzer was operated at a constant pass energy of 5 eV. The overall resolution of the experiment was about 0.6 eV at 70 eV.

The calibration of the binding energy scale as well as of the kinetic energy scale was performed using the well known 4*d* ionization energy values and the $N_{4,5}OO$ Auger lines in xenon, respectively.[12-14]

The commercial sample of C_2H_5I was used without further purification other than outgassing the liquid. The 1-bromo-2-iodoethane was prepared in our laboratory according to Simpson's procedure.[15] By comparison with known geometries of similar compounds[16] and as discussed in the HeI photoelectron spectrum,[17] the molecule exhibits very likely at room temperature a trans structure with severely hindered rotation.

RESULTS

A. Iodoethane

The I 4*d* photoelectron and Auger spectra recorded at 70 and 74 eV and restricted to the binding energy range of 35-65 eV are displayed in Figs. 1a and 1b, respectively. The spectra are dominated by two bands located at 56.5 and 58.2 eV binding energies which are associated with the ionization of the iodine $4d_{5/2}$ and $4d_{3/2}$ core electrons, respectively. The splitting $4d_{5/2}$-$4d_{3/2}$ is 1.7 eV, as it is expected in iodine compounds.[1,18-20] Two broad bands of weaker relative intensity are seen to be shifted with the photon energy on the binding energy scale but always appear at the same kinetic electron energies. They are related to Auger electrons of 14 and 28.4 eV here labelled I(4*d*)VV1 and I(4*d*)VV2 as the electronic relaxation of the I 4*d* hole involves two electrons of the valence shells.[19-21] According to these results, electronic states of the doubly charged molecular ion are located around 29 and 43 eV (difference between the I 4*d* ionization energy and the Auger electron kinetic energy). The Auger bands at these energies have been reported in iodomethane[21] where they are also related to the I 4*d* core hole decay.

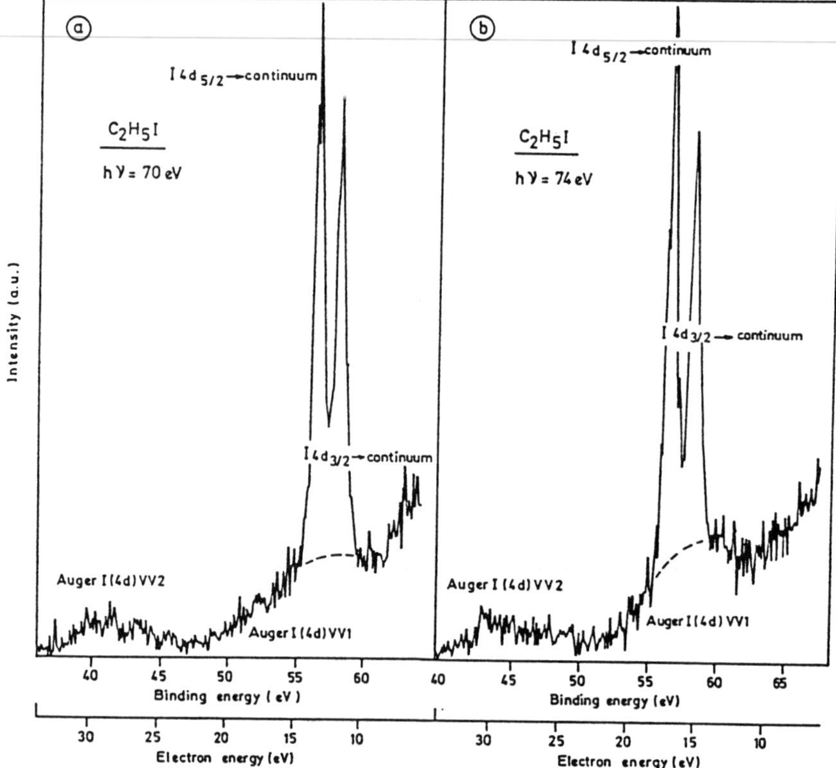

Fig. 1. Photoelectron and Auger spectra of C_2H_5I recorded at (a) 70 and (b) 74 eV photon energies.

B. 1-Bromo-2-Iodoethane

1. The threshold electron spectrum

The threshold electron spectrum of C_2H_4IBr in the 50 to 90 eV region is shown in Fig. 2. The peaks seen below the I 4*d* edge and located in the 51-53 eV region are due to the I 4*d* electron excitation into the unoccupied valence molecular orbital σ^*. The higher energy bands located just below the I 4*d* threshold are related to the I 4*d* electron excitation into Rydberg orbitals. Finally, the band at 71 eV evidenced below the Br 3*d* edge is associated with the bromine 3*d* electron resonant excitation into the lowest energy orbital σ^*. The cross section is strongly enhanced above 65 eV due to the I 4*d* -> εf giant shape resonance, already observed in other iodine compounds[1,18,20] and which has its maximum at 97 eV.[10]

2. The I 4*d* and Br 3*d* photoelectron and Auger spectra off resonance and on the giant shape resonance

The photoelectron spectrum recorded at 67 eV, i.e. off resonance and restricted to the binding energy range of 30-60 eV is presented in Fig. 3. The bands due to the iodine 4*d* electron ionization are present at 56.6 and 58.3 eV. The splitting $4d_{5/2}$-$4d_{3/2}$ is 1.7 eV, the same value as that observed in C_2H_5I. The two broad bands centered around 14 and 28.4 eV electron energies do not shift with the photon energy on the electron energy scale and thus are of Auger type. On the basis that the corresponding energies are very similar to those observed in C_2H_5I

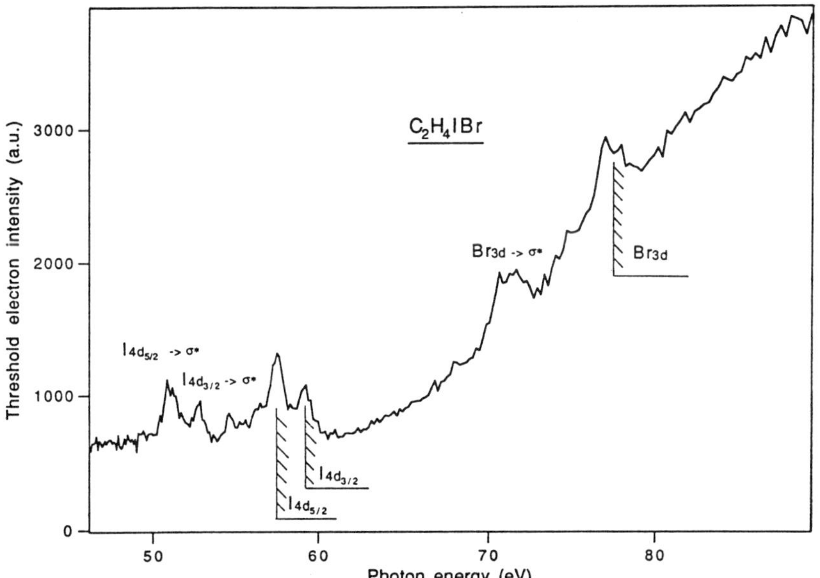

Fig. 2. Threshold electron spectrum of C_2H_4IBr recorded in the 50-90 eV photon energy range.

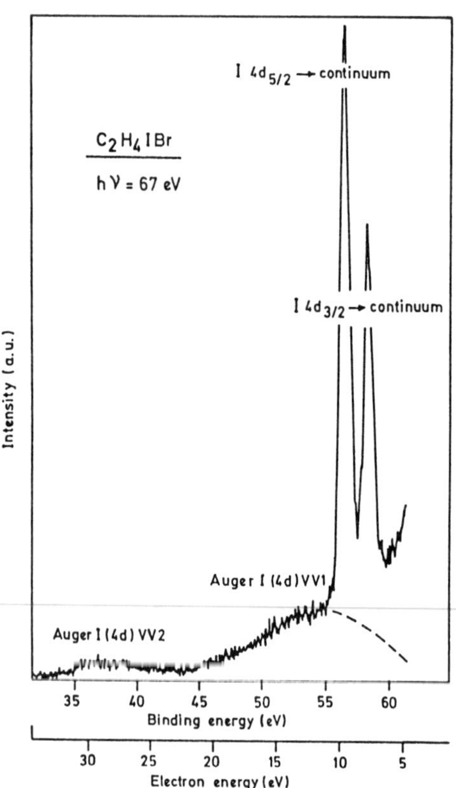

Fig. 3. Photoelectron and Auger spectrum of C_2H_4IBr recorded at 67 eV photon energy.

(Figs. 1a and 1b) and in CH_3I,[21] we suggest that they are related to the I 4*d* core hole electronic decay.

The photoelectron spectra in the 50 to 85 eV binding energy range on the I 4*d* -> εf and the Br 3*d* -> εf giant shape resonances at 97 and 118 eV also exhibit the bands due to the I 4*d* electron ionization and those due to the Auger processes related to the I 4*d* hole decay at 14 and 28.4 eV electron energies (Figs. 4 and 5). It should be noted that in the 118 eV spectrum the photon band width is insufficient to resolve the $4d_{5/2}$ and $4d_{3/2}$ components separately. The Br 3*d* electron ionization manifests itself by a single band with a maximum at 76.7 eV binding energy, the experimental resolution being too poor to separate the $3d_{5/2}$ and $3d_{3/2}$ components. An additional Auger band associated with the relaxation of the Br 3*d* hole is evidenced in the 118 eV spectrum at 45.7 eV electron energy and allows us to localize electronic states of the doubly charged molecular ion around 31 eV. It should be noted that on the Br 3*d* -> εf giant shape resonance, the ratio of the Br 3*d* ionization band relative to the I 4*d* one (0.4) is enhanced compared to its ratio on the I 4*d* -> εf giant shape resonance (0.15).

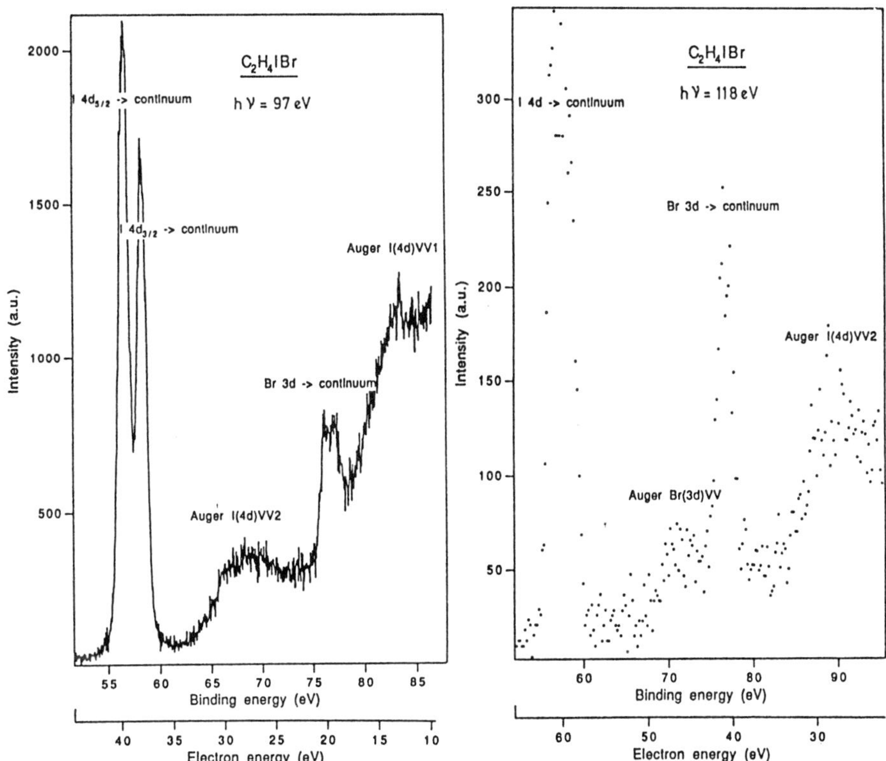

Figs. 4 and 5. Photoelectron and Auger spectra of C_2H_4IBr recorded on the I4*d* -> εf giant shape resonance at 97 eV and on the Br3*d* -> εf giant shape resonance at 118 eV.

3. The photoelectron and Auger spectrum on the Br 3*d* -> σ^* resonance

The photoelectron spectra recorded at 69.8 eV and on the maximum of the Br 3*d* -> σ^* resonance at 71 eV from the outer-valence-shell ionization region up to the I 4*d* electron ionization bands are displayed in Figs. 6a and 6b. The I 4*d* ionization peaks as well as the I(4*d*)VV1 and the I(4*d*)VV2 Auger bands are enhanced at the Br 3*d* -> σ^* resonance compared to the valence bands. So, the decay of the resonantly excited molecule (Br 3*d* -> σ^*) takes place preferentially to the doubly charged ion states centered at 29 and 43 eV and to the iodine 4*d* ionization continuum rather than to the outer-valence-shell ionization. It should also be noted that this enhancement is much more important than that which could result from the presence of the I 4*d* -> εf giant shape resonance (Fig. 2).

DISCUSSION

A. Auger decay after core electron ionization

1. The iodine 4*d* Auger decay

The relaxation of the molecule after the I 4*d* core electron ionization in C_2H_5I and in C_2H_4IBr leads to the formation of doubly charged molecular ions around 29 and 43 eV (Figs. 1a and b, Fig. 3). The corresponding Auger bands are quite broad. This means that many electronic states of the dication are populated. The iodomethane has three ionic states located in the 29 eV region. In addition, as estimated from the rule of the thumb,[24] the double ionization threshold in C_2H_5I and in C_2H_4IBr for two electrons ionized from the iodine "lone pair" are of 27 and 27.7 eV. So the 29 eV band is due to the excitation of the ground and the lowest energy excited states of the dication.

The 43 eV band corresponds very likely to numerous excited states of the dication of which the density is expected to be quite high for polyatomic molecules.

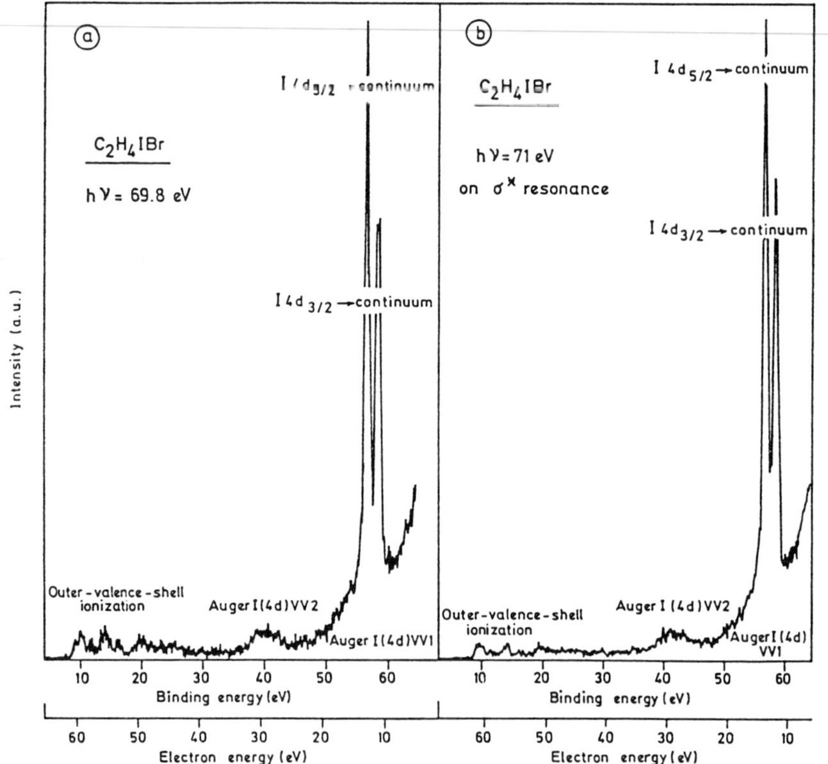

Fig. 6. Photoelectron and Auger spectra of C_2H_4IBr recorded at (a) 69.8 eV photon energy and at (b) 71 eV on the Br3*d* -> σ^* resonance.

2. The bromine 3*d* Auger decay

The Auger band related to the decay of the Br 3*d* core hole shows evidence of the presence of doubly charged $C_2H_4IBr^{2+}$ ions at about 31 eV (Fig. 5). Electronic states of the CH_3Br^{2+} ion have also been evidenced at 28.8 and 31.5 eV.[25] Notice that the bromine lone pair double ionization threshold as calculated on the basis of the rule of Tsai and Eland[24] and of the single ionization value[17] is about 29.9 eV. This means that the bromine core ionization leads to electronic decay channels involving valence electron emission of the bromine lone pair and population of the lowest energy electronic states of the dication.

B. Behavior on the Br 3*d* resonances

1. The bromine 3*d* -> εf giant shape resonance

The comparison of the 97 and 118 eV photoelectron spectra

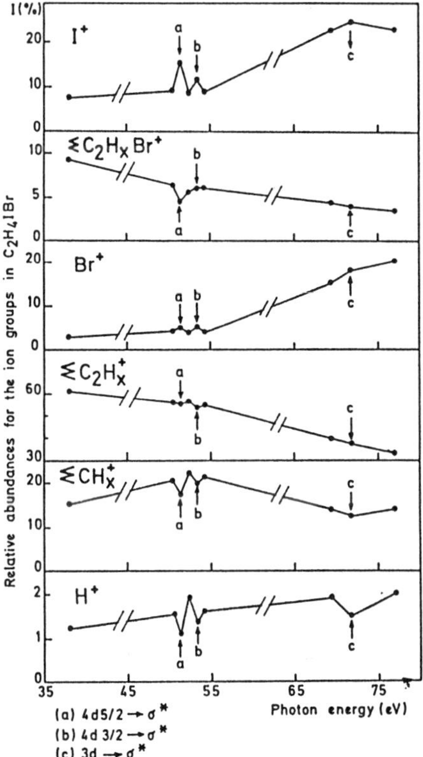

Fig. 7. Relative abundances for the various ion groups evidenced in the mass spectra of C_2H_4IBr versus the photon energy. They have been obtained by measuring the ratio of the area of the ion group feature and of the sum of the various ion group features.

(Figs. 4 and 5) shows that the Br 3*d* -> εf giant shape resonance decays selectively into the Br 3*d* continuum rather than to the I 4*d* one. This suggests that the excitation remains localized on the bromine atom.

2. The Br 3*d* -> σ^* resonance

The Br 3*d* -> σ^* resonance at 71 eV decays partly into the iodine 4*d* continuum and through the I(4*d*)VV1 and I(4*d*)VV2 states respectively at 29 and 43 eV. The fragmentation of the resulting ionic states leads to a selective enhancement of the C-I and C-Br bond dissociations as shown on Fig. 7 where the relative intensities of the fragment ions have been reported at selected photon energies ("on" and "off" resonances) in the 50-55 eV and the 70-77 eV photon energy regions.[10]

ACKNOWLEDGMENTS

We are very grateful to the staff of the L.U.R.E. for their technical support. We acknowledge the Fonds National de la Recherche Scientifique, the Fonds de la Recherche Fondamentale Collective and the Ministère des Finances (Loterie Nationale) of Belgium for financial support. This work has also benefited from a NATO contract (no. 484/87). The authors also gratefully acknowledge Dr J.-L. Piette from the Laboratoire de Synthèses Organiques (University of Liège) for his highly valued advice on the trans-1-bromo-2-iodoethane preparation, B. Kempgens for the measurements of the C_2H_4IBr mass spectra and I. Nenner (L.U.R.E., CEA) for her suggestions.

REFERENCES

1. P. Morin and I. Nenner, Phys. Scr. T17, 171 (1987).
2. I. Nenner, J. H. D. Eland, P. Lablanquie, J. Delwiche, M.-J. Hubin-Franskin, P. Roy, P. Morin, and A. Hitchcock, Electron-Molecule Scattering and Photoionization (Plenum Publishing Corporation, 1988), p. 15.
3. D. M. Hanson, Adv. Chem. Phys. 77, 1 (1990).
4. I. Nenner, P. Morin, and P. Lablanquie, Comments At. Mol. Phys. 22, 51 (1988).
5. W. Eberhardt, W. Plummer, I. W. Lyo, R. Reiniger, R. Carr, W. K. Ford, and D. Sondericker, Austr. J. Phys. 39, 633 (1986).
6. R. Thissen, M. Furlan, J. Delwiche, M.-J. Hubin-Franskin, J.-L. Piette, P. Morin, and I. Nenner, to be published.
7. R. A. Rosenberg, C.-R. Wen, K. Tan and J.-M. Chen, Phys. Scr. 41, 475 (1990).
8. M.-J. Hubin-Franskin, J. Delwiche, M. Furlan, K. Ibrahim, R. Thissen and J. E. Collin, Int. J. Mass Spectrom. Ion Proc. 101, 273 (1990).
9. I. Nenner, P. Morin, M. Simon, P. Lablanquie, and G. G. B. de Souza, Desorption Induced by Electronic Transitions,DIET III (Springer-Verlag, Berlin, 1988), p. 10.
10. B. Kempgens, J. Delwiche, M.-J. Hubin-Franskin, M. Furlan, unpublished results.

11. P. Morin, M. Y. Adam, I. Nenner, J. Delwiche, M.-J. Hubin-Franskin, and P. Lablanquie, Nucl. Instr. and Meth. 208, 761 (1983).
12. K. Colding and R. P. Madden, Phys. Rev. Lett. 12, 106 (1964).
13. G. C. King, M. Tronc, F. H. Read, and R. C. Bradford, J. Phys. B: Atom. Molec. Phys. 10, 2479 (1977).
14. H. Aksela, S. Aksela, G. M. Bancroft, K. H. Tan, and H. Pulkkinen, Phys. Rev. A 33, 3867 (1986).
15. M. Simpson, Ber. 7, 130 (1874).
16. T. H. Gan, J. B. Peel, and G. D. Willet, J. Mol. Struct. 44, 211 (1978).
17. A. Bouguerne, J. Delwiche, M.-J. Hubin-Franskin, and J. E. Collin, J. Electron Spectrosc. Relat. Phenom. 56, 303 (1991).
18. G. O'Sullivan, J. Phys. B: At. Mol. Phys. 15, L327 (1982).
19. L. Karlsson, S. Svensson, P. Baltzer, M. Carlsson-Göthe, M. P. Keane, A. Naves de Brito, N. Correia and B. Wannberg, J. Phys. B: At. Mol. Opt. Phys. 22, 3001 (1989).
20. I. Nenner, P. Morin, P. Lablanquie, M. Simon, N. Levasseur et P. Millié, J. Electron Spectrosc. Relat. Phenom. 52, 623 (1990).
21. D. W. Lindle, P. H. Kobrin, C. M. Truesdale, T. A. Ferret, P. A. Heimann, H. G. Kerkhoff, U. Becker, and D. A. Shirley, Phys. Rev. A 30, 239 (1984).
22. G. Dujardin, S. Leach, O. Dutuit, P-M. Guyon, and M. Richard-Viard, Chem. Phys. 88, 339 (1984).
23. W. J. Griffiths and F. M. Harris, J. Chem. Soc. Faraday Trans. 86, 2801 (1990).
24. B. P. Tsai and J. H. D. Eland, Int. J. Mass Spectrom. Ion Phys. 36, 143 (1980).
25. W. J. Griffiths and F. M. Harris, Org. Mass Spectrom. 25, 375 (1990).

DICATIONS. SPECTROSCOPY AND PHOTON INDUCED DYNAMICS.

M.-J. Hubin-Franskin*
Institut de Chimie, B6
Université de Liège,
Sart Tilman par 4000 LIEGE 1 (Belgium)

ABSTRACT

Some recent progress in the spectroscopy, fragmentation and photoionisation cross sections of dications are brievly reviewed and illustrated for dications formed by single photon valence shell ionisation.

INTRODUCTION

Interest to dications is related to quite fundamental aspects of physical chemistry such as electronic correlations which are of importance for the characterisation of the condensed matter and also in inner shell photochemistry and its selectivity.

Two electrons excitation or emission is a forbidden process. It is induced by electronic correlations. Thus its study helps to test the theoretical models and their validity according to the photoelectron energy, to the target..., the configuration mixing and the two hole localisation in case of molecules.

In inner shell photochemistry excitation of a core electron into the LUMO or into the continuum is generally followed by electronic relaxation involving photoemission of one or two electrons and formation of a singly or multiply charged molecular cation. The detailed mechanism of relaxation requires a knowledge of the spectroscopy and of the dynamics of these ionised species.

---

* Maître de Recherche FNRS.

For singly charged cations photoelectron spectroscopy at fixed wavelengths or with the synchrotron radiation has provided quite numerous data for the spectroscopy of the ions related to the valence and inner valence vacancy. The dynamics is less well known despite some effort has been made by mass spectrometry and photoionisation with the synchrotron radiation sometimes coupled to coincidence methods achieving then a state-selected dynamics generally restricted to small molecules[1-3]. Previously doubly charged ions were identified in electron impact mass spectra. However recently their spectroscopy has benefited of improved experimental methods such as Time-Of-Flight (TOF) mass spectrometry coupled to (é,2e) Dipolar Spectroscopy[4] or to photoionisation[5-9] , Double Charge Transfer (DCT)[9-11] and Auger (photo)electron spectroscopy[10]. Theoretical calculations have also contributed mainly for small polyatomic molecules by semi-empirical[12] and ab-initio methods and also using the Valence-Bond model[13]. Accurate quantum chemical investigations have been extended to larger systems due to the advances in vector and parallel computer technology and architecture together with development of suitable computational strategies[14]. The outer valence shell double ionisation transitions of polyatomic molecules have been computed using the two particles Green's function method.
Generally cations are not stable as such. Insight into their dynamics has required the development of specific experimental methods in order to observe their ionic fragments and thus the dissociative double ionisation independently of the single ionisation one, especially TOF mass spectrometry coupled to detection in coincidence of two or more charged particles formed at the same initial event[8,15]. In case of polyatomic ions numerous dissociation channels are available. In addition isomerisation and rearrangement may occur.

In this paper a briev review is given of some progress made in the knowledge of the spectroscopy, the dynamics and the double photoionisation cross sections of dications. It will be illustrated with two extreme cases:

carbon monoxide and hexafluorobenzene studied at least partly with the synchrotron radiation.

RESULTS AND DISCUSSION

A. The Spectroscopy

The density of electronic states is quite high even

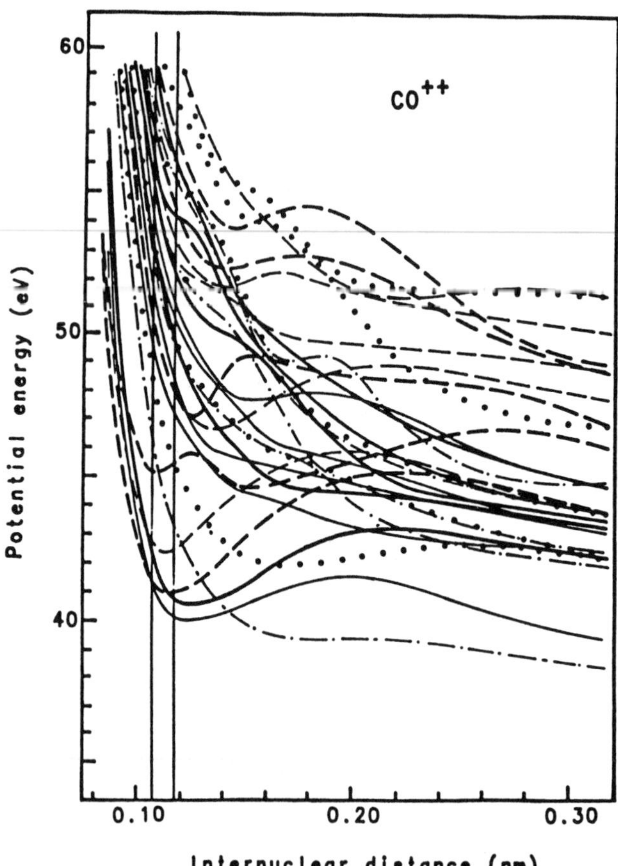

Fig.1 Calculated potential-energy curves of the electronic states of CO^{2+} lying between 35 and 60 eV. ($^1\Pi$; $^3\Pi$; $^1\Sigma^+$; $^3\Sigma^+$; $^1\Delta$; $^3\Sigma^-$) (from ref.16).

in the double ionisation threshold region partly due to the larger number of configurations accessible when two electrons are ionised for a given set of orbitals. This is illustrated for carbon monoxide[16], for which ab-initio calculations predicted numerous states in the 35 to 60 eV region i.e. from the double ionisation threshold (35.9 eV) up to 25 eV above as shown in Fig.1.
The six lowest energy ones displayed on Fig.2 have been observed in various complementary experiments such as Double Charge Transfer (DCT), Mass Spectrometry (MS) and PhotoIon-PhotoIon Coincidence (PIPICO) photoionisation with the synchrotron radiation.

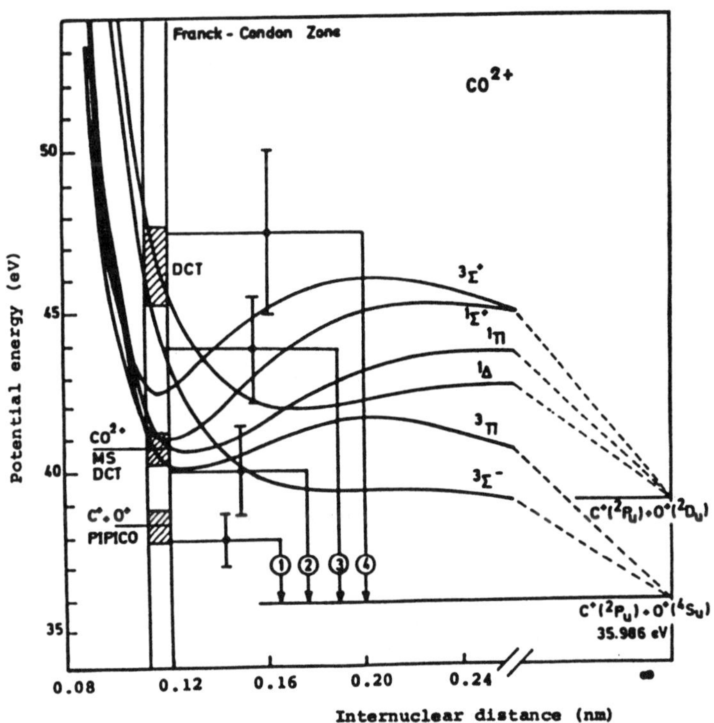

Fig.2 Details of the potential-energy curves for the six lowest-energy electronic states of CO^{2+} (from ref.16).

The lowest energy states are triplets $^3\Sigma^-$ and $^3\Pi$ both being at large internuclear distances dissociative into the pair of correlated ionic fragments in their ground states $C^+(^2P_u)+O^+(^4S_u)$. For the triatomics nitrous oxide[9], carbon dioxide, carbon disulfide and carbonyl suphide[8,11] a triplet is also the lowest energy one observed.
For larger polyatomic molecules the density of states is even higher. In case of hexafluorobenzene the lowest energy electronic state has been calculated by the INDO method to be a triplet, the singlet being very close in energy[12]. Experimentally[17] this dication has been observed with an ionisation energy of 27.3 eV matching qualitatively the computed value. For benzene a quite high density of electronic states (226) has been computed for the dication in the double ionisation threshold region up to 40 eV using the two particle Green's function method[14]. Interestlingly seventy of them have "main" states character with 2h contribution over 50%. Even at low double ionisation energy the familiar concept of a "main" state as derived from a single 2h configuration is inapplicable and the configuration mixing is very substantial[18]. In the case of BF_3 the dications computed in the double ionisation threshold region have been analysed in terms of an atomic localisation of the positive charges and a two-hole state population on the fluorine atom(s) giving rise to two-site dicationic states characterised by two holes on distinct fluorine site and one-site states where the two holes are located on the same fluorine atom[19].

B. Stability and Fragmentation

A fundamental question is related to the stability of dications with respect to the Coulomb repulsion between the positive charges, and to the mechanisms of fragmentation i.e. two-body charge separation, multistep dissociation with preloss of neutral(s).

The dication of CO has a lifetime on the microsecond timescale and is observable as such in mass spectra[16].

Dications of polyatomic molecules may have lifetimes

sufficiently long to support rearrangement with migration

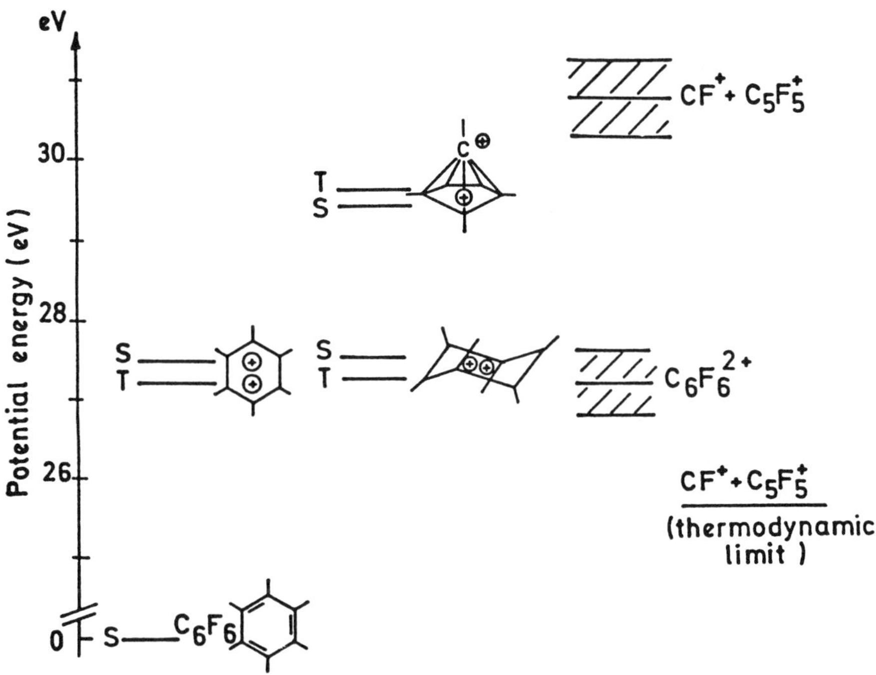

Fig.3 Schematic of the potential energy of the lowest hexafluorobenzene dication isomers, scaled on the experimental value of 27.3 eV for the double ionisation energy (from ref.17).

of hydrogen[21] or fluorine atoms[17] and isomerisation. As an example the hexafluorobenzene dication has three isomers[12]. The two lowest energy ones corresponding to a pseudohexagonal ring and a chair form respectively are quite close in energy and the third one , a C_{5v} pyramidal isomer is located 2 eV above them (Fig.3). It is characterized by a broad distribution of lifetimes[17]. This distribution has been interpreted in terms of a potential energy curve being partly attractive and of the population of numerous vibrational levels of which lifetimes diminishe according to their location with respect to the potential barrier to isomerisation.

Generally in the double ionisation region the direct

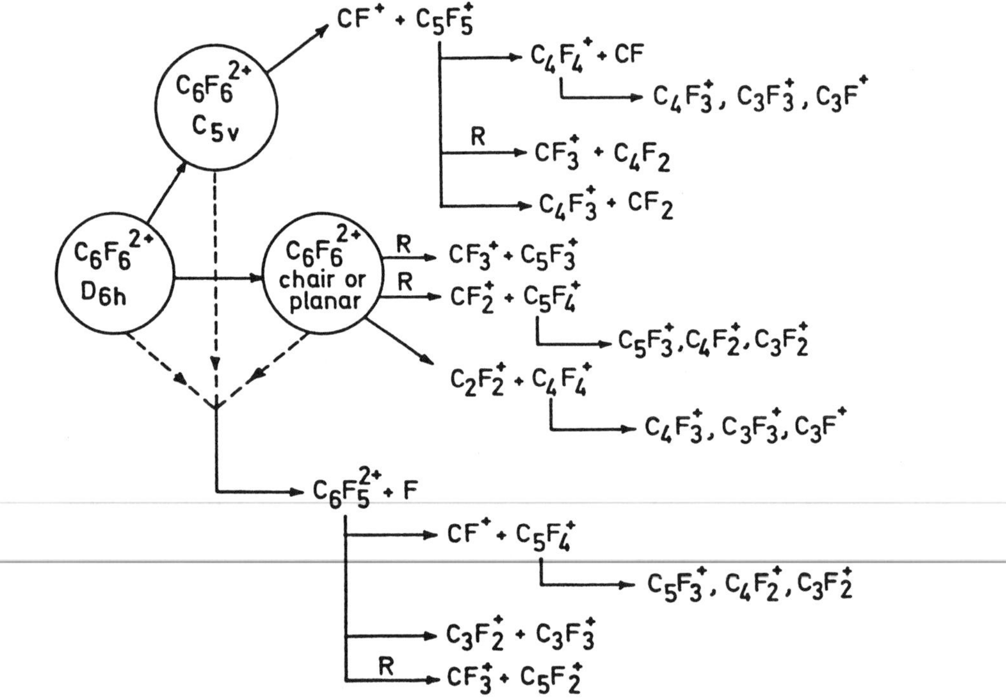

Fig.4 Fragmentation pattern of the hexafluorobenzene dication at 40 eV derived from the PEPIPICO measurements. R=rarrangement (from ref.17).

two-body charge separation due to Coulomb repulsion holds. The dissociation into a doubly charged fragment and a neutral requires higher energy; this reaction may be followed by charge separation[8,17,20] (deferred charge separation). For large polyatomic systems such as methylorganic compounds[21], hexafluorobenzene[17] and polyaromatic polyclic compounds[23] the fragmentation proceeds via two-and three-body pathways as reported by triple PhotoElectron PhotoIon PhotoIon COincidence spectroscopy (PEPIPICO)[20]. In the case of hexafluorobenzene[17] illustrated in Fig.3,4 the lowest dissociation limit of the dication is the charge separation into $CF^+ + C_5F_5^+$ which involves very likely the C_{5v} pyramidal isomer due to the charge distribution in the isomer. At higher energy opens the channel with the loss of a fluorine atom and the formation of $C_6F_5^{2+}$ followed by its charge separation into $CF^+ + C_5F_4^+$. Remarkably the dominating channel in the whole double ionisation region up to 120 eV is CF^+ + another$^+$.

C. Photoionisation Cross Section for dications

The double photoionisation cross section of argon is well described by the Wannier-Rau theory of direct ionisation forming a pair of correlated electrons in the final state, at energies up to at least 12.5 eV above threshold, even though it contains substantial resonant states of the neutral atom[23].
At a broader energy scale for Ne, N and O the ratios of the oscillator strength for double photoionisation to the total photoionisation oscillator strength is proportional to N-1, N being the total number of electrons in the atom. In addition a simple model for double photoionisation involving an internal electron impact process has been proposed[24,25].

In the double ionisation of molecules the understanding of the dynamics of two-electron ejection near threshold from a neutral molecule is much more difficult than for an atom mainly due to the well higher density of electronic states of the dication and to the "width" due to the Franck-Condon envelopes, part of these states being repulsive.
However at higher energy the double photoionisation cross section may reflect indirect processes such as autoionisation of an excited state of the singly charged ion or of a doubly excited state of the neutral molecule[8,17]. As shown in Fig.5 the cross section for the $CF^+ + C_5F_5^+$ ion pair of hexafluorobenzene[17] is enhanced by two resonances centered at 36 and 45 eV respectively.

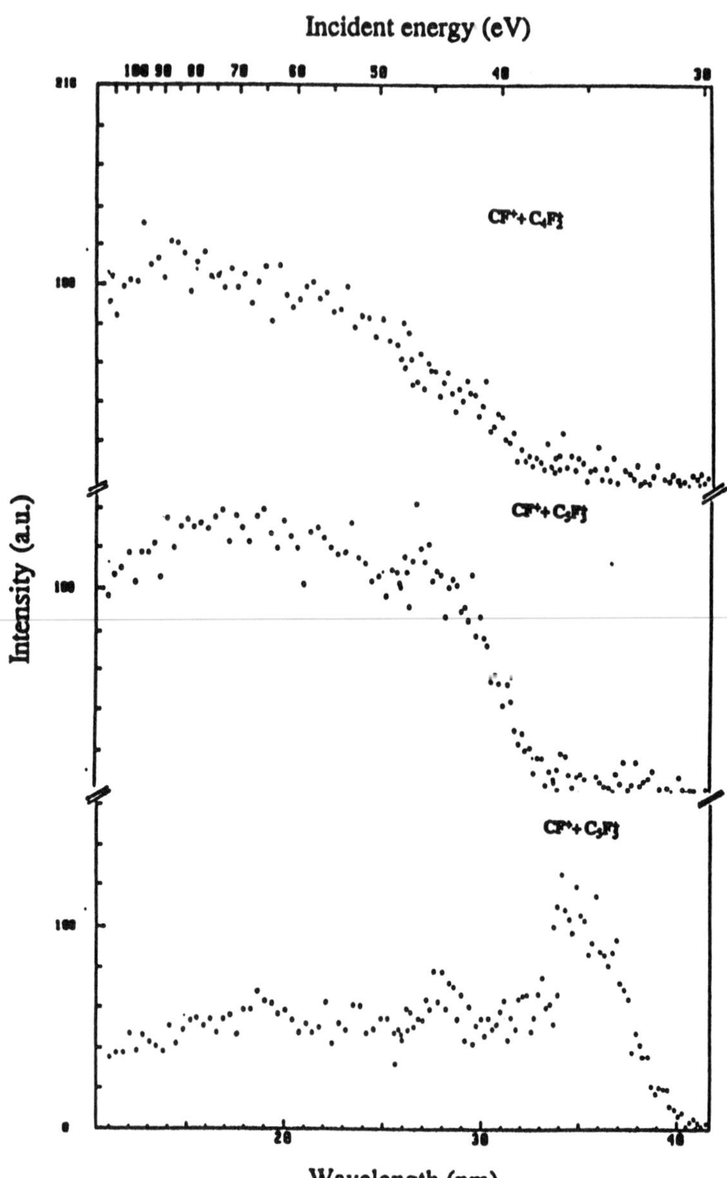

Fig.5 Relative photoionisation cross sections for some correlated ion pairs (from ref.17).

This latter contributes also to the $CF^{+}+C_5F_3^{+}$ cross section. Thus the same broad resonance enhances two different final ionic channels.

ACKNOWLEDGEMENTS

I am pleased to thank the staff of LURE for their expert operation of the machines, the "Fonds National de la Recherche Scientifique", the "Fonds de la Recherche Fondamentale Collective" and the "Service de la Programmation de la Politique Scientifique" and the "Patrimoine de l'Université de Liège" for their financial support. I also gratefully acknowledge the support from NATO (Contract 484/87).

REFERENCES

1. J.H.D. Eland, Int. J. Mass Spectrom. Ion Phys. 12, 389 (1973).
2. M. Richard-Viard, O. Atabek, O. Dutuit and P.M. Guyon, J. Chem. Phys. 93, 3213 (1990).
3. M.-J. Hubin-Franskin, J. Delwiche, P.-M. Guyon, O. Dutuit, M. Richard-Viard and M. Lavollée, to be published.
4. A.P. Hitchcock, C.E. Brion and M.J. Van der Wiel, Chem. Phys. Lett. 66, 213 (1979).
5. G. Dujardin and S. Winkoun, J. Chem. Phys. 83, 6222 (1985).
6. B.P. Tsai and J.H.D. Eland, Int. J. Mass Spectrom. Ion Phys. 36, 143 (1980).
7. T. Masuoka and J.A.R. Samson, J. Chimie Phys. 77, 623 (1980).
8. P. Lablanquie, I. Nenner, P. Millié, P. Morin, J.H.D. Eland, M.-J. Hubin-Franskin and J. Delwiche, J. Chem. Phys. 82, 2951 (1985).
9. S.D. Price, J.H.D. Eland, P.G. Fournier, J. Fournier and P. Millié, J. Chem. Phys. 88, 1511 (1988).
10. A. Cesar, H. Agren, A. Naves de Brito, S. Svensson, L. Karlsson, M.P. Keane, B. Wannberg,

P. Baltzer, P.G. Fournier and J. Fournier, J. Chem. Phys. 93, 918 (1990).
11. M.L. Langford, F.M. Harris, C.J. Reid, J.A. Ballantine and D.E. Parry, Chem. Phys. 149, 445 (1991).
12. M.J.S. Dewar and M.K. Holloway, J. Am. Chem. Soc. 106, 6619 (1984).
13. N. Levasseur and P. Millié, J. Chem. Phys. 92, 2974 (1990).
14. E. Ohrendorf, H. Köppel, L.S. Cederbaum, F. Tarantelli and A. Sgamellotti, J. Chem. Phys. 91, 1734 (1989).
15. G. Dujardin, S. Leach, O. Dutuit, P.-M. Guyon and M. Richard-Viard, Chem. Phys. 88, 339 (1984).
16. P. Lablanquie, J. Delwiche, M.-J. Hubin-Franskin, I. Nenner, P. Morin, K. Ito, J.H.D. Eland, J.-M. Robbe, G. Gandara, J. Fournier and P.G. Fournier, Phys. Rev. A40, 5673 (1989).
17. K. Ibrahim, P. Lablanquie, M.-J. Hubin-Franskin, J. Delwiche, M. Furlan, I. Nenner, D. Hagan and J.H.D. Eland, submitted to J.Chem. Phys.
18. F. Tarantelli, A. Sgamellotti, L.S. Cederbaum and J. Schirmer, J. Chem. Phys. 86, 2201 (1987).
19. F. Tarantelli, A. Sgamellotti and L.S. Cederbaum, J. Chem. Phys. 94, 523 (1991).
20. J.H.D. Eland, Acc. Chem. Res. 22, 381 (1989).
21. E. Ruhl, S.D. Price, S. Leach and J.H.D. Eland, Int. J. Mass Spectrom. Ion Phys. 97, 175 (1990).
22. S. Leach, J.H.D. Eland and S.D. Price, J. Phys. Chem. 93, 7583 (1989).
23. P. Lablanquie, J.H.D. Eland, I. Nenner, P. Morin, J. Delwiche, M.-J. Hubin-Franskin, Phys. Rev. Lett. 58, 992 (1987).
24. J.A.R. Samson and G.C. Angel, Phys. Rev. A42, 5328 (1990).
25. J.A.R. Samson, Phys. Rev. Lett. 65, 2861 (1990).

ADSORPTION AND REACTION OF ORGANIC AND ORGANOMETALLIC SPECIES ON SEMICONDUCTORS: DYNAMIC ASPECTS

M.N.Piancastelli
Dipartimento di Scienze e Tecnologie Chimiche,
II Universita' di Roma "Tor Vergata", 00173 Rome, ITALY

ABSTRACT

The adsorption properties and chemical reactivity of semiconductor surfaces are of great interest from both a fundamental and an applied point of view. We are carrying out since many years a systematic investigation on the adsorption of various organic and organometallic species on semiconductors, in particular the silicon (111) cleaved surface. The UHV technique we primarily use is synchrotron radiation photoemission. The main results of this research line are: the complete characterization of the adsorption properties of various classes of organic systems (alkanes, alkenes, alkynes, aromatics, alcohols); the discovery of a surprisingly high reactivity of the Si surface with respect to aromatic molecules; the characterization for the first time of the adsorption properties of organometallic systems on Si (carbonyls and cyclopentadienides of various transition metals); the possibility of using unmonochromatized synchrotron radiation ("white light") to induce chemical reactions on the Si surface saturated with molecular adsorbates, in particular the capability of depositing metallic films using organometallic species as precursors.

INTRODUCTION

The adsorption bond and geometry of organic and organometallic simple molecules on metal substrates have been widely investigated in the last ten years with UHV techniques, while no such attention has been paid to the adsorption properties of the same systems on semiconductor surfaces. Studies of this kind can shed light on several topics, as e.g. the reactivity of semiconductor substrates, and consequently their cleaning procedures and industrial treatment; the capability of these substrates to form localized bonds, and therefore a possible different behavior with respect to metal surfaces in general; the possibility to find alternate ways to prepare substrates of high technological interest, such as thin metallic films on semiconductors, with methods based on the use of unmonochromatized synchrotron radiation ("white light") to induce photodecomposition reactions.

A series of experiments we performed in the last few years on the adsorption properties of simple organic and organometallic molecules on semiconductor substrates has revealed many intriguing aspects of this interaction. We review the main results we obtained, which are part of an extensive investigation on the topic. The technique we used is primarily synchrotron radiation photoemission. Other complementary techniques as HREELS (High-Resolution Electron Energy Loss Spectroscopy), LEED (Low-Energy Electron Diffraction), Auger, XAS (X-ray Absorption Spectroscopy) were also used. The measurements were carried out at the Synchrotron Radiation Center, University of Wisconsin, USA, and at LURE, Orsay, France.

The semiconductor surface we investigated more extensively is the single-crystal silicon (111) cleaved surface, reconstructed 2X1. The organic molecules we studied on the above mentioned substrate include several aromatic systems (benzene, pyridine, thiophene, furan), molecules representative of systems with single, double and triple carbon-carbon bonds (neopentane, ethylene, acetylene), and simple alcohols (methanol and ethanol). The organometallic species we examined include Mo, W and Fe carbonyls and Ni and Fe cyclopentadienides. In this paper we will review several selected topics among the whole bulk of results, and we will provide a complete list of literature references[1-26].

The topics we will deal with are: the surprisingly high reactivity shown by aromatic systems ; the high sticking coefficient and the different reactivity exhibited as a function of temperature by alcohols; the adsorption properties of organometallic species and the capability of depositing metallic films on Si by using metal carbonyls as precursors.

AROMATIC MOLECULE REACTIVITY ON SILICON

The most striking result we obtained during the study of the interaction between aromatic molecules and semiconductor surfaces is the discovery of an unusually high reactivity of the Si(111)2X1 surface with respect to benzene. The exposure of this surface to few langmuirs ($1L=10^{-6}$ torr.sec) of benzene at room temperature (RT) is enough to obtain a stable chemisorbed state. Furthermore, this state is completely different from that one obtained in the same conditions for C_6H_6 on most metal surfaces. On metals, benzene forms a π bond with the surface, and the corresponding molecular geometry implies the ring plane parallel or almost parallel to the surface. On Si, the chemisorption bond is a strong σ bond: in the adsorption process, a C-H σ molecular bond is broken and a C-Si σ bond is formed between the molecule and a Si surface atom. The corresponding chemisorption geometry implies the ring plane at a large angle with respect to the surface. The situation is different at low temperature (60 K). In the latter case, benzene is physisorbed on the surface, and no chemical reaction takes place between the molecules and the substrate. The adsorption behavior of benzene on silicon has been investigated by synchrotron radiation photoemission of both core and valence levels[2,11]. In Figure 1 we report valence photoemission spectra for benzene on Si(111)2X1 taken at 60 K after a 5 L exposure (bottom curve), at 60 K after a 100 L exposure (middle curve) and at RT after a 10^3 L exposure. The photoemission peaks related to benzene molecular orbitals are labelled C through G. In the two spectra taken at 60 K, the number and relative position of the spectral features corresponds to those reported in the literature for free benzene, thus indicating a physisorbed state. The situation is significantly different at RT (top curve in Figure 1): the peak labelled C in the low-temperature spectra (bottom and middle curve in Figure 1) is split in two well resolved components. We interpret this experimental finding in the following way: peak C corresponds to the first occupied molecular orbital, which in free benzene is the doubly degenerate $1e_{1g}$ state, with π symmetry. The

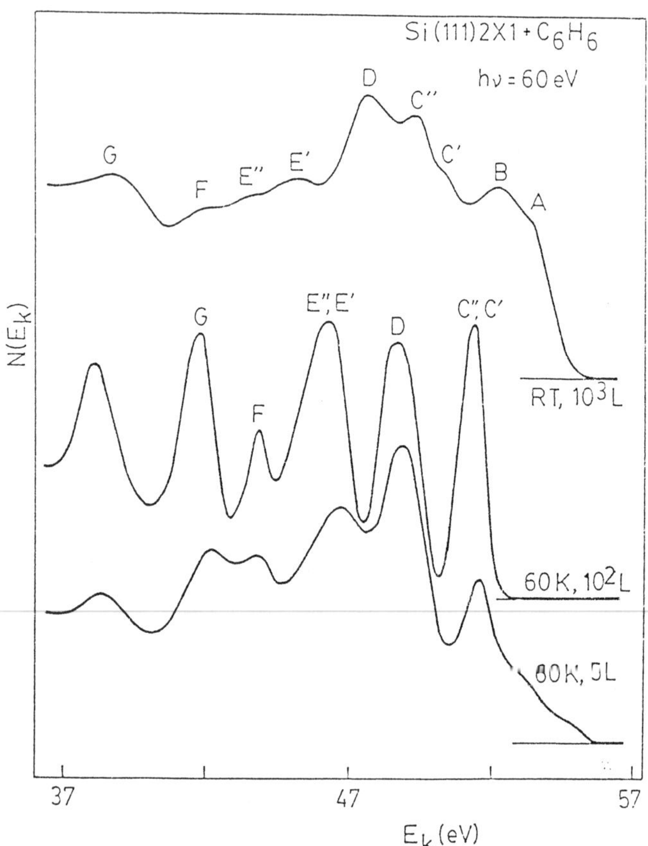

Figure 1. Valence photoemission spectra taken (from bottom to top) for cleaved Si exposed to 5 and 100 L of benzene while at 60 K, and exposed to 1000 L while at RT. The spectra were taken at 60 eV photon energy.

degeneracy is conserved in the physisorbed state (low-temperature spectra) and removed in the RT spectrum. The two π orbitals are not equivalent with respect to monosubstitution of hydrogen on the aromatic ring. Therefore, we can assign the degeneracy removal as due to a process in which one molecular carbon-hydrogen bond is broken and one carbon-silicon bond is formed between the molecule and the Si substrate. This interpretation of photoemission data has been independently supported by HREELS results [3,5]. We obtained analogous results for a series of aromatic molecules on Si(111)2X1 (pyridine, thiophene, furan), which leads us to the general and quite unexpected conclusion that the reactivity of the cleaved Si(111) surface with respect to aromatic systems is rather high [1-8,10,11,14,15,19].

ALCOHOL REACTIVITY ON SILICON

To clarify the adsorption properties of various classes of organic molecules on Si, we investigated simple alcohols, methanol and ethanol, on Si(111)2X1 as a function of temperature in the range 80-300 K. We used synchrotron radiation photoemission of both core and valence levels[23]. A first interesting experimental finding is that at RT the sticking coefficient is rather high: exposures of the Si surface to few L of alcohol vapors are sufficient to produce intense adsorbate-induced spectral features. A practical consequence of such a high sticking coefficient is that cleaning procedures for Si crystals or related vacuum components with alcohols should be performed with special care, since the risk of accidental contamination is present. As for the adsorption process, we were able to identify two different states: a chemisorbed state at RT and a physisorbed state at 80 K. The investigation with photoemission is particularly important for these systems, since this technique is able to provide a direct evidence for the presence of the two adsorption states.

In Figure 2 we show a photoemission spectrum taken at a photon energy of 40 V on a 80 K Si(111)2X1 substrate exposed to 10 L of C_2H_5OH. Six spectral features labelled A through F are evident, which can be related to ethanol molecular orbitals. Since there is a one-to-one correspondence in peak number and relative position between free ethanol and ethanol adsorbed on Si, the adsorption state at 80 K is likely to be physisorbed ethanol. A very interesting point results from the comparison of curves a) and b) in Figure 2. The two spectra are quite similar, apart from the disappearance of the spectral feature labelled D on switching from a) to b). This feature is related to a molecular orbital

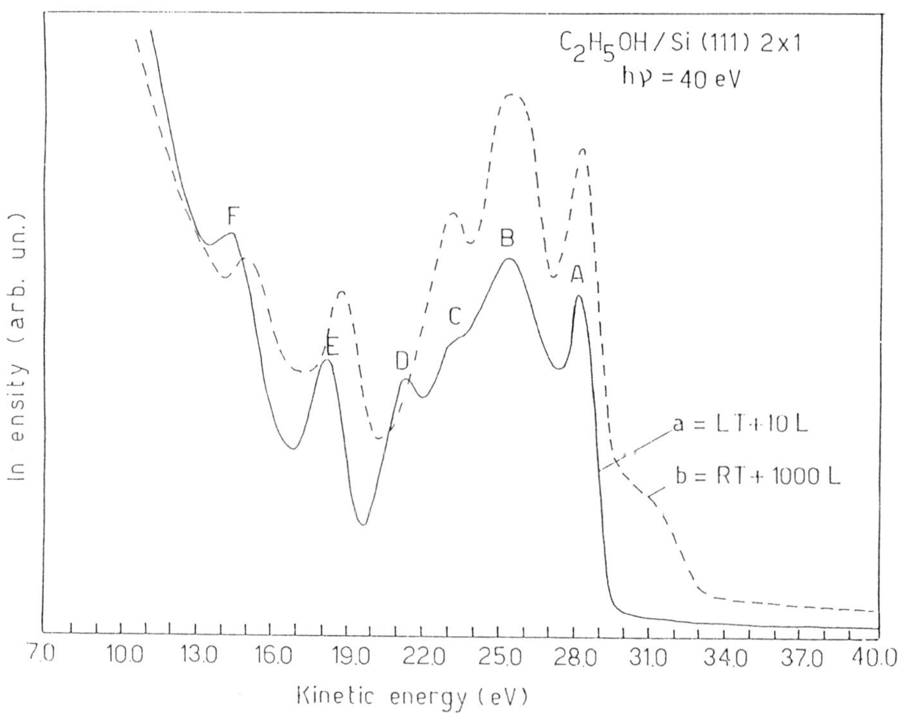

Figure 2. Valence photoemission spectra taken on Si(111) 2X1 a) at 80 K after a 10 L exposure to ethanol; b)at RT after a 1000 L exposure to ethanol. For both spectra the photon energy is 40 eV.

with mainly σ_{O-H} character. Its absence from the RT spectrum suggests a chemical reaction involving the breaking of the molecular σ_{O-H} bond and the formation of a σ_{O-Si} bond between the molecule and the substrate. Therefore the adsorbed state at RT is an ethoxy species. We obtained completely analogous results for methanol on Si. We can conclude that simple alcohols on Si show a rather high reactivity, as it was the case for aromatic molecules.

SILICON METALLIZATION VIA ORGANOMETALLIC PRECURSORS

The study of the adsorption properties of organometallic compounds on silicon is a rather new field, very interesting from both a fundamental and an applied viewpoint. The fundamental interest resides in the possibility of identifying the adsorption state(s) as a function of temperature and coverage, and to characterize the possible chemical reactions which take place between the molecule and the surface. The practical applications are based on the possibility of inducing a decomposition of the organometallic species on the surface, which can lead to the formation of a metallic film on silicon. Since no previous data on this topic were reported in the literature, we recently began a systematic study of the interaction of various organometallic species with the Si(111)2X1 surface, aiming at both characterizing this interaction in general and developing a method to prepare thin metallic films on Si which is alternative to the currently used procedures. The systems we investigated include $Mo(CO)_6$[17,18], $W(CO)_6$[21,22], $Fe(C_5H_5)_2$[24], $Ni(C_5H_5)_2$[24] and $Fe(CO)_5$[25,26]. In all cases the adsorption was studied as a function of exposure and in the temperature range 60-300 K. For the metalocenes, the adsorption is molecular in the whole temperature range, as demonstrated by the fact that there is a one-to-one correspondence between spectral features of free molecules and molecules adsorbed on Si. For carbonyls, the situation is different: a molecular adsorption is generally characterized at low temperature, while as the temperature is increased (for Mo and W hexacarbonyls) or as a function of time (for Fe pentacarbonyl) a partial decomposition reaction is detected, leading to subcarbonyl species on the Si surface. This different behavior for the two classes of compounds is on line with the higher chemical stability of the free metalocenes with respect to the corresponding carbonyls. As an example of this behavior, in Figure 3 we show spectra of $Fe(CO)_5$ on Si(111)2X1 taken on a 80 K substrate immediately after the exposure (curve a) and few minutes after the exposure (curve b)[26]. There is an evident difference in the low-binding energy part of the spectra: the first two well resolved peaks of curve a merge into a single spectral feature as a function of time. This experimental finding has been interpreted as due to the fact that as a function of time one or more CO groups are released from the adsorbed iron pentacarbonyl molecule, and therefore the symmetry of the adsorption site is lowered, leading to a single broad peak in place of two resolved features.

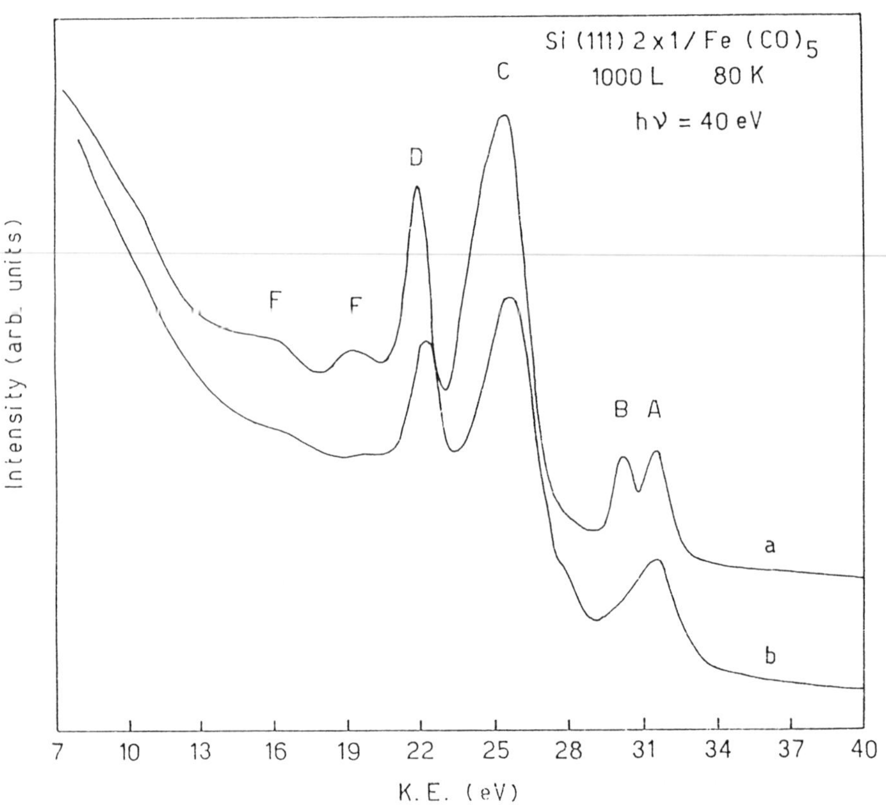

Figure 3. Valence photoemission spectra of a Si(111)2X1 surface exposed to 1000 L of $Fe(CO)_5$ at 80 K: a) immediately after exposure; b) after 20 min. The photon energy is 40 eV.

The photoassisted decomposition of organometallic species offers many applications in the field of electronic devices. This general method for depositing metal films presents several advantages over conventional techniques, in particular the limited number of process steps (typically one or two) and the mild conditions under which the substrate is processed. As a light source for assisting metal deposition, synchrotron radiation has unique characteristics, being highly intense in a wide photon energy range. Our approach consists in exposing the semiconductor surface to unmonochromatized synchrotron radiation ("white light") after it has reacted <u>in situ</u> with an organometallic species. We have successfully deposited films of several metals (Mo[17,18], W[21,22], Fe[25,26]) on Si(111)2X1 using the corresponding carbonyls as precursors. We report in Figure 4, bottom curve, the valence photoemission spectrum taken at a photon energy of 48 eV on cleaved Si, exposed to 100 L of $Mo(CO)_6$, at 50 K. Several features are evident which can be related to the adsorption process. The peak at lower binding energy corresponds to Mo 4d orbitals, while the features at higher binding energy are related to molecular orbitals mainly localized on the CO ligands. The number and relative position of the photoemission peaks are consistent with the adsorption state being molecular. As the adsorbate-saturated Si surface is exposed to white light for increasing amounts of time (Figure 4, second curve from bottom to top curve), a dramatic change is evident in the spectra. The deepest-in-energy features, related to shakeup satellites of CO-localized states, gradually disappear, while a clear Fermi edge appears, demonstrating the occurrence of a photoinduced metallization process. The peak immediately below the Fermi edge is caused by Mo valence electrons, while the other features at higher binding energy can be related to residual carbon on the surface.

We consider different possible mechanisms to explain the observed phenomena. We exclude a pyrolitic metallization, since sample regions close to those illuminated by the white light still exhibit spectral features of adsorbed $Mo(CO)_6$. Our results are consistent with the decomposition of adsorbed carbonyl due to photon adsorption. We also found that high-energy photons are necessary to induce the process, since exposures to white light filtered through a sapphire window (which cuts off photons of energy higher than 8 eV) do not induce metallization.

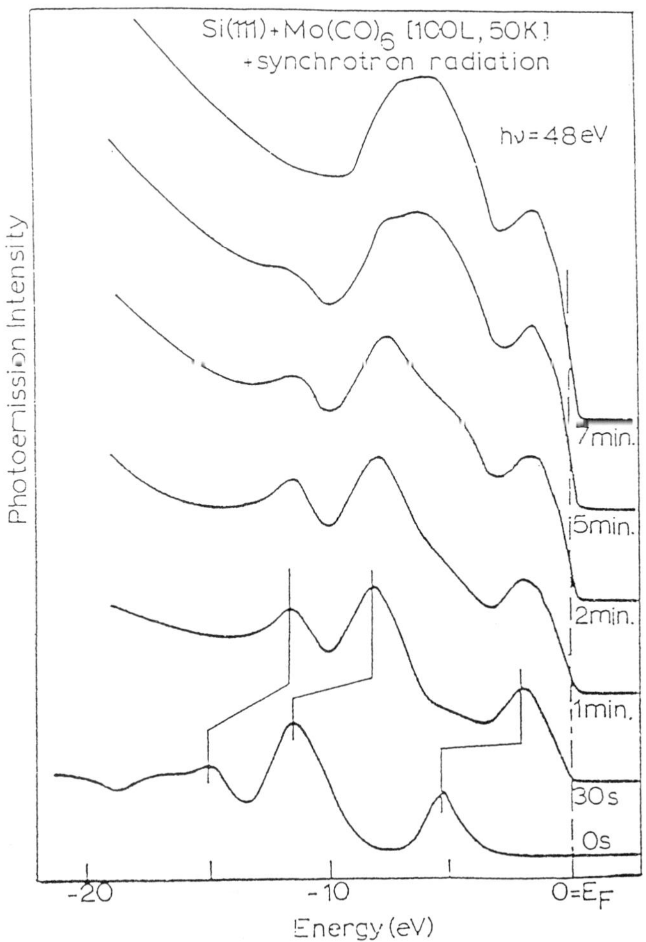

Figure 4. Bottom curve: valence photoemission spectra taken at 48 eV photon energy on a 50 K Si(111)2X1 substrate exposed to 100 L of $Mo(CO)_6$. The other curves are spectra of the same surface taken after irradiation with white light for increasing lengths of time.

Strictly analogous results were obtained for the photoassisted decomposition of $W(CO)_6$ on Si, which leads to the formation of a tungsten film on the surface[21,22].

The deposition of iron films on Si is a process of high technological interest, due to the possibility of preparing substrates with different magnetic and electric properties. We were able to obtain Fe films on Si(111)2X1 using the previously described photodecomposition method. We tried both $Fe(C_5H_5)_2$[24] and $Fe(CO)_5$[25,26] as precursors. It turned out that only the carbonyl molecules undergo the photodecomposition. We studied the white light-induced photolysis of $Fe(CO)_5$ on Si with both synchrotron radiation photoemission and XAS techniques. The results obtained by photoemission are analogous to those previously described for Mo and W hexacarbonyls on Si: as a function of irradiation, a Fermi edge becomes clearly visible and the molecular orbital-related spectral features disappear from the valence spectra.

In Figure 5 we report the results of XAS measurements performed at the Fe $L_{2,3}$ edge. To obtain a better insight into the photodecomposition process, we compared the data from the Fe film prepared with our method with a thick Fe film grown by the usual evaporation procedure. Figure 5a shows the Fe $L_{2,3}$ XAS spectrum taken on a 80 K Si substrate exposed to 1000 L of $Fe(CO)_5$. While from the photoemission data a partially decomposed state (iron subcarbonyl) was detected (see Figure 4 and related discussion), the XAS measurements suggest that the nature of the adsorption state is even more complex. The spectrum reported in Figure 5a reveals the presence of at least two components. We propose to assign the component at lower energy to metallic Fe, on the ground of a comparison between curves a and b of Figure 5. Figure 5b is the XAS spectrum of a thick Fe layer, deposited by e-beam evaporation. The energy position of metallic Fe is coincident with the shallower of the two components in Figure 5a, which hints for a common assignment. The second component in Figure 5a is assigned to the adsorbed iron subcarbonyl.

The Fe $L_{2,3}$ XAS spectrum of Figure 5c is obtained after white-light irradiation of the surface of Figure

5a. It is very similar to that one related to the evaporated Fe layer (Figure 5b). Therefore the occurrence of the photodecomposition process leading to a Fe metallic film on the Si surface, already suggested by photoemission data, is clearly confirmed by XAS measurements.

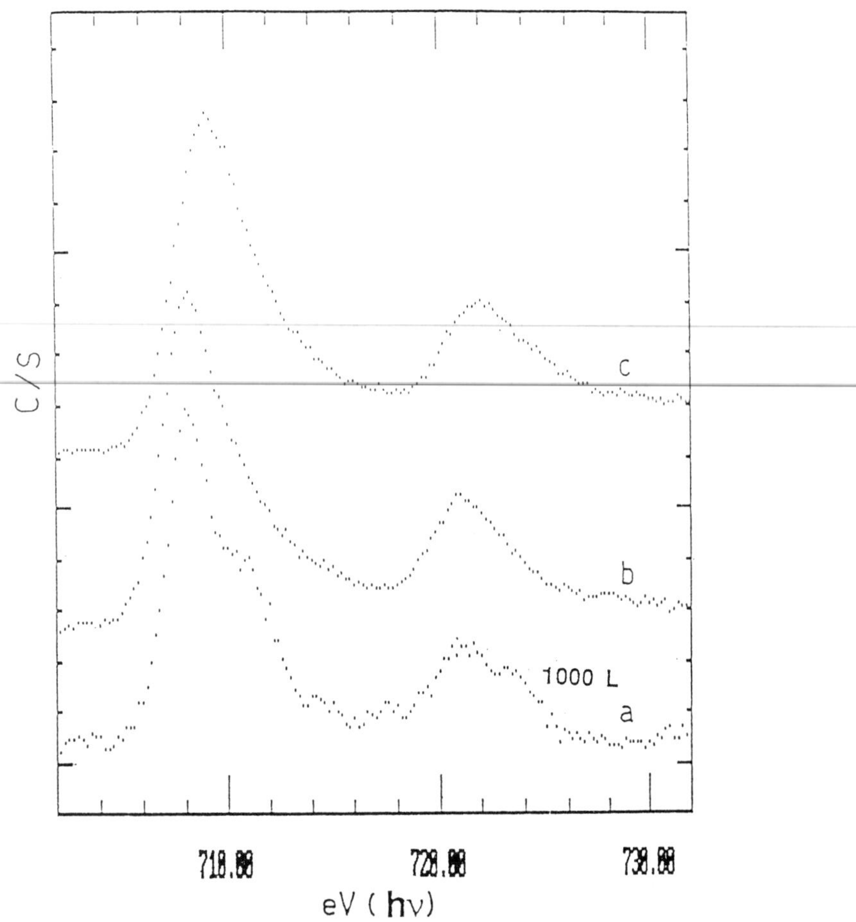

Figure 5. XAS spectra of Fe $L_{2,3}$ thresholds taken at 80 K on: a) a Si(111)2X1 surface exposed to 1000 L of $Fe(CO)_5$; b) a thick evaporated Fe layer; c) same surface as a), after 30 min. irradiation with white light.

LITERATURE REFERENCES

1) M. N. Piancastelli, G. Margaritondo and J. E. Rowe "Pyridine Chemisorption on Silicon and Germanium (111) Surfaces". Solid State Commun. 45, (1983) 219

2) M. N. Piancastelli, F. Cerrina, G. Margaritondo, A. Franciosi and J. H. Weaver "A Strongly Bound Chemisorption State for Benzene on Si(111)" Appl. Phys. Letters 42, (1983) 990

3) M. N. Piancastelli, G. Margaritondo, J. Anderson, D. J. Frankel and G. J. Lapeyre "Strongly Bound State of Benzene on Cleaved Si(111): Vibrational Modes and Chemisorption Bonds". Phys. Rev. B 30, (1984) 1945

4) M. N. Piancastelli, M. K. Kelly, G. Margaritondo, J. Anderson, D. J. Frankel and G. J. Lapeyre "Pyridine on Cleaved Si(111): Analysis of the Vibrational Modes and Nature of the Chemisorption Process". Phys. Rev. B 32, (1985) 2351

5) M. N. Piancastelli, M. K. Kelly, G. Margaritondo, D. J. Frankel and G. J. Lapeyre "Temperature-Dependent Adsorption of Aromatic Molecules on Silicon". Phys. Rev. B 34, (1986) 2511; Phys. Rev. B 35, (1987) 873

6) M. N. Piancastelli, M. K. Kelly, G. Margaritondo, D. J. Frankel and G. J. Lapeyre "Surface Vibrational Spectroscopy Studies of Aromatic Molecule Fragmentation on Silicon Surfaces". Phys. Rev. B 34, (1986) 3988

7) M. N. Piancastelli, M. K. Kelly, G. Margaritondo, J. Anderson, D. J. Frankel and G. J. Lapeyre "Synchrotron Radiation Photoemission and Surface Vibrational Studies of Organic Molecule Chemisorption on Silicon". Appl. Surf. Sci. 26, (1986) 498

8) N. Tache, Y. Chang, M. K. Kelly, G. Margaritondo, C. Quaresima, M. Capozi, P. Perfetti and M. N. Piancastelli "Benzene Chemisorption on Amorphous Silicon" Appl.Phys.Lett. 50, (1987) 531

9) M. N. Piancastelli, M. K. Kelly, D. G. Kilday, G. Margaritondo, D. J. Frankel and G. J. Lapeyre "Ethylene on Cleaved Silicon:High-Resolution Electron-Energy- Loss Study of an Unusual Adsorption State in the Temperature Range 85-300 K" Phys.Rev. B 35, (1987) 1461

10) M. N. Piancastelli, R. Zanoni, M. K. Kelly, D. G. Kilday, Y. Chiang, J. T. McKinley, G. Margaritondo, P. Perfetti, C. Quaresima and M. Capozi "Thiophene on Si(111)2X1: Synchrotron Radiation Study of a Desulfurization Process" Solid State Commun. 63, (1987) 85

11) M. N. Piancastelli, M. K. Kelly, Y. Chang, J. T. McKinley and G. Margaritondo "Benzene Adsorption on Low-Temperature Silicon: Synchrotron Radiation Photoemission Study of Valence and Core States" Phys.Rev. B 35, (1987) 9218

12) M. N. Piancastelli, R. Zanoni, M. K. Kelly, J. T. McKinley, G. Margaritondo, C. Quaresima, M. Capozi and P. Perfetti "Reactivity of Organic Molecules on Amorphous Si and Ge Films" J.Vac.Sci.Technol.A6, (1988) 762

13) M. N. Piancastelli, M. K. Kelly, G. Margaritondo, D. J. Frankel and G. J. Lapeyre
"Surface Vibrational Study of Acetylene Adsorption on Cleaved Silicon"
Solid State Commun.65, (1988) 1295

14) M. N. Piancastelli, M. K. Kelly, G. Margaritondo, D. J. Frankel and G.J.Lapeyre
"Neopentane on Cleaved Si: Vibrational Study of the Adsorption of a Highly Symmetric Hydrocarbon"
Solid State Commun. 68, (1988) 985

15) M. N. Piancastelli, M. K. Kelly, G. Margaritondo, D. J. Frankel and G. J. Lapeyre
"A Deoxygenation Process on Cleaved Si"
Surf.Sci. 211/212, (1989) 1018

16) M. N. Piancastelli, R. Zanoni, D. W. Niles and G. Margaritondo
"Ethylene and Acetylene Adsorption on Cleaved Si: a Photoemission Study with Synchrotron Radiation"
Solid State Commun. 72, (1989) 635

17) R.Zanoni, M.N.Piancastelli, J.T.McKinley and G.Margaritondo
"Synchrotron-Radiation-Induced Metal Deposition on Semiconductors: $Mo(CO)_6$ on Si(111)"
Appl.Phys.Lett. 55, (1989) 1020

18) R.Zanoni, M.N.Piancastelli, J.T.McKinley and G.Margaritondo
"$Mo(CO)_6$ on Si(111)2X1: a Synchrotron Radiation-Excited Photoemission Study"
Phys.Scripta, 41, (1990) 636

19) M.N.Piancastelli, R.Zanoni, J.T.McKinley and G.Margaritondo
"Low-Temperature Pyridine Adsorption on Cleaved Silicon: a Synchrotron Radiation Photoemission Study"
Solid State Commun. 75, (1990) 285

20) M.N.Piancastelli, R.Zanoni and G.Margaritondo "Reactivity of Amorphous Silicon with respect to Organic Molecules: a Synchrotron Radiation Photoemission Study" Rend.Fis.Acc.Lincei serie 9 1, (1990) 291

21) R.Zanoni, M.N.Piancastelli, M.Marsi and G.Margaritondo "Silicon Metallization by Synchrotron-Radiation-Induced $W(CO)_6$ Surface Reaction" Solid State Commun. 76, (1990) 1239

22) R.Zanoni, M.N.Piancastelli, M.Marsi and G.Margaritondo "Synchrotron-Radiation-Stimulated Tungsten Deposition on Silicon from $W(CO)_6$" J.Vac.Sci.Technol. A 9, (1991) 931

23) M.N.Piancastelli, R.Zanoni, M.Marsi and G.Margaritondo "Chemisorption and Physisorption of Simple Alcohols on Cleaved Silicon" Solid State Commun., in press

24) R.Zanoni, M.N.Piancastelli, M.Marsi and G.Margaritondo "Organometallic Adsorption on Semiconductors: a Synchrotron Radiation Photoemission Study of Ferrocene and Nickelocene on Si(111)2X1" J.Electron Spectrosc.Relat.Phenom, in press

25) R.Zanoni, M.N.Piancastelli, F.Sirotti and G.Rossi "A Soft-X-Ray Photoemission Study of Iron Deposition on Si(111)2X1 by Synchrotron Radiation Excited Photodecomposition of Adsorbed $Fe(CO)_5$" Appl.Phys.Lett., in press

26) R.Zanoni, M.N.Piancastelli and G.Margaritondo "Organometallic Adsorption on Semiconductors: a Synchrotron Radiation Photoemission Study of Iron Pentacarbonyl on Si(111)2X1" J.Electron Spectrosc.Relat.Phenom., submitted

DISCUSSION

DE LANGE - In the case of organic molecules absorbed on a silicon substrate, can you tell which of the carbon atoms get involved in chemical bonding to the susbstrate ?

PIANCASTELLI - Yes, for aromatic molecules containing heteroatoms the carbon atom involved is that one adjacent to the heteroatom (in the α position).

DUJARDIN - I am surprised that there are so many molecular states at room temperature on the Si(111)2x1 cleaved surface. Would it be the same on the Si(111)7x7 reconstructed surface covered for example with C_6H_6 on $Fe(CO)_5$? What is the reason for choosing the cleaved Si(111)2x1 surface instead of other Si(111) reconstructed surfaces ?

PIANCASTELLI - There are not very strong differences between the 7x7 and the 2x1 reconstructed surfaces as far as reactivity is concerned. We chose the cleaved surface under the hypothesis that the higher density of defects (i.e. cleavage steps) would make it more reactive with respect to the annealed surface, but for many molecular adsorbates it is not the case.

LEACH - 1. Is there reconstruction of the Si(111) surface on deposition of organic species ? For example, in the case of benzene deposition, reconstruction could lead to the formation of some very reactive Si sites giving rise to the reaction with benzene that you observe.

2. In the benzene reaction case, do you also see splitting of degenerate orbitals other than the e_{1g} M.O. ?

3. What diagnostic tool did you use in the case of surface reactions of organic species other than benzene (e.g. pyridine, ...) for which there are no degenerate orbitals ?

PIANCASTELLI - 1. No, there is no reconstruction in my opinion leading to more reactive sites.

2. The answer is no, but the peaks at higher ionization energy are broader, so the effect can go unnoticed.

3. HREELS (High-Resolution Electron Energy Loss Spectroscopy).

ELAND - Do you know what coverages were involved in your experiments ?

PIANCASTELLI - We did not measure for coverage in most cases.

WONG - 1. What are the geometries of these aromatic molecules on the Si(111) surface ?

2. Do you see similar non-parallel arrangements of these molecules on other Si surfaces, say Si(100) ?

PIANCASTELLI - 1. We know that at Room Temperature (RT) the molecular plane sits at some angle with respect to the surface which is neither O° nor 90°, but we did not measure yet the exact value.

2. We concentrated on the Si(111)2x1 surface. So I have no data for the Si(100) surface.

CORE-INDUCED PHOTODISSOCIATION OF SURFACE MOLECULES: SELECTIVITY, LOCALIZATION, AND EVIDENCE FOR ULTRAFAST PROCESSES

Dietrich Menzel
Physik-Department E 20, Technische Universität München
W-8046 Garching b. München, Germany

ABSTRACT

Photodissociation of molecules at surfaces, be they chemisorbed species on metal or semiconductor surfaces or the topmost layer of a molecular condensate, shows some interesting specificities. The well documented, strong changes observed in photodissociation of chemisorbed molecules on metal surfaces, as compared to their free counterparts, can be understood by the influences of coupling to the substrate and to the neighbouring adsorbates, and the consequent transfer of energy and charge. This can lead to selective quenching of many of the excitations before dissociation takes place; only strongly localized excitations survive. The evidence for such effects from the prominence of core shake-up excitations in photodissociation of chemisorbates is briefly summarized and the derived models are discussed.

While these observations could be interpreted within the Franck-Condon picture (all electronic processes fast compared to motion of atoms), time scale and other considerations suggest that a non-Franck-Condon picture is more appropriate in which atomic motion can be competitive with hole decay. Even clearer evidence for the existence of such ultrafast processes has been obtained from the selective dissociative action of certain core-to-bound excitations involving strongly antibonding MO's at or close to the core absorption threshold, in condensed or adsorbed layers of small hydrogen-containing molecules (water, ammonia, benzene). It is proposed that no separation of electronic and nuclear motions is possible in these cases. These effects should exist in the free molecules as well.

Thus the influence of coupling at surfaces leads to selectivity and localization of the primary excitation by quenching of all delocalizable excitations which become dissociative only after time-consuming redistribution. Thereby, primary excitations are projected out which are either dissociative on an ultrafast timescale and therefore do not need localization, or are localized by interactions and thus preserved for longer times against quenching than others.

INTRODUCTION

Photodissociation is usually discussed in either of two limiting models, depending on the size of the molecule concerned. For small molecules consisting of two to four atoms, a direct model is usually favoured which argues in terms of potential energy curves of the quasi-diatomic system representing the bond to be broken; while these curves intersect which leads to the possibility of redistribution of energy, the main dissociation processes are considered to be direct. On the other extreme, larger molecules are usually treated in a statistical model in which the excitation energy is redistributed in the molecule for considerable times, and dissociation is determined by the statistics of energy build-up at the bond concerned. In the latter model, the excitation and dissociation steps should be essentially decoupled, while considerable selectivity can exist in the first case.

Many interesting aspects and possibilities arise when the primary excitations involved are core excitations[1]. In these cases, the primary locus of excitation can be well defined, so that a decision should become possible about the directness of a dissociation event, i.e. about its time scale. Nevertheless, evidence for selectivity and localization[2] in molecular photodissociation is scarce and often controversial.

A very interesting special case is the photodissociation of surface molecules[3]. Molecules coupled to a substrate, be it a very different solid (e.g., chemisorbed molecules in submonolayers on metal or semiconductor surfaces) or the surface molecules of a condensate of the same species, would be expected to belong to the second class, that of large systems, so that the statistical approach might appear to be more appropriate at first glance. The evidence accumulated in the past decade has shown that this not at all the case; in fact surface systems may be those types of systems in which strong localization and direct, ultrafast reactions become most clearly obvious. This is due to the fact that essentially any transfer of excitation away from the primary locus in a surface molecule will prevent dissociation of this or another surface molecule. This short survey summarizes the evidence from which such a conclusion can be derived.

We shall discuss observations of core-induced photodissociation in two very different classes of surface systems which have been mentioned as limiting cases in the last paragraph. Small chemisorbed molecules at metal or semiconductor surfaces which exist there in submonolayer to monolayer quantities, constitute the limiting case of very strong coupling to a semiinfinite bath. Such systems have the advantage that they can be very accurately defined in all geometrical, electronic and dynamic properties by the arsenal of methods developed by modern surface science. In the examples given, it should be understood that such a good definition has always been accomplished prior to the photodissociation experiments. The other extreme, the surface layer of a molecular condensate, constitutes usually a much more weakly coupled system. In all cases, the experiment consists in the measurement of photodesorption, i.e. the recording of the particles (ions or neutrals, constituting fragments or intact species) which leave the surface under irradiation, as a function of the primary excitation. While early work has been done using electron impact, most recent work has used photons from synchrotron radiation (SR) sources. Apart from the much more specific excitation induced by photons as compared to electrons due to the well-known characteristic threshold behavior of excitation cross sections in both cases, SR photons also allow the use of their well-defined polarization to exploit dipole selection rules for an even better definition of primary excitation and for other symmetry arguments, as will be shown below. Since it has been shown that core excitations lead mainly to positively charged fragments[5], these signals are the ones of concern here. As the argument often consists in the demonstration of some kind of specificity for a certain primary excitation (see below), a simultaneous measurement of the absorption cross section of the same system is important. This is usually accomplished by measuring the decay electrons leaving the system - either their sum, at all energies ("total electron yield, TY or TEY") or a part of them in a certain energy range. The latter can either be a broad band (often used: the "partial yield, PY or PEY" of all electrons above a cut-off energy so that electrons originating deep inside the substrate are eliminated and the surface contribution is accentuated) or a narrow band (important: the "Auger electron yield, AY or AEY" at a certain kinetic energy corresponding to a characteristic decay of the molecules in the top layer - the signal of choice for a submonolayer on a metal substrate, for obvious reasons).

CHEMISORBED SMALL MOLECULES ON METAL SINGLE CRYSTAL SURFACES

An extensive literature, starting back in the early sixties, has shown[3,4] that electronically induced dissociation of chemisorbed molecules on metal and semiconductor surfaces possess some unique properties setting them off from the corresponding free molecules. For our present purposes, the most important aspects are:

1) Generally the dissociation probabilities are strongly reduced on the surface (by several orders of magnitude in many cases).

2) There are huge differences of this reduction for different adsorbate states of the same molecules on the same (and, of course, on different) surfaces, and for different primary excitations. In particular, core excitations are much more effective than valence excitations[5,6], and for both a general rule is that the decrease of dissociation probability induced by coupling to the substrate is the smaller the more complex the excitation is.

As an example, we show in fig. 1 the behavior of the yield of the fragment ion O^+ from chemisorbed NO on a Ru(001) surface[7], in the region above the O1s threshold. Below it, the dissociation probability is much smaller, even though valence excitation cross sections are much larger than core cross sections. Furthermore, as obvious from the figure comparing ion yield and AY (i.e. absorption of the chemisorbed species), higher core excitations (core shake-ups, multielectron excitations) are much more efficient for dissociation than single core excitations, be they neutral or ionic. At and above the N1s threshold, the O^+ signal is lower by more than two orders of magnitude and does not show enhancement at the satellite energies; it is still the biggest ion signal, though. Furthermore, it has been

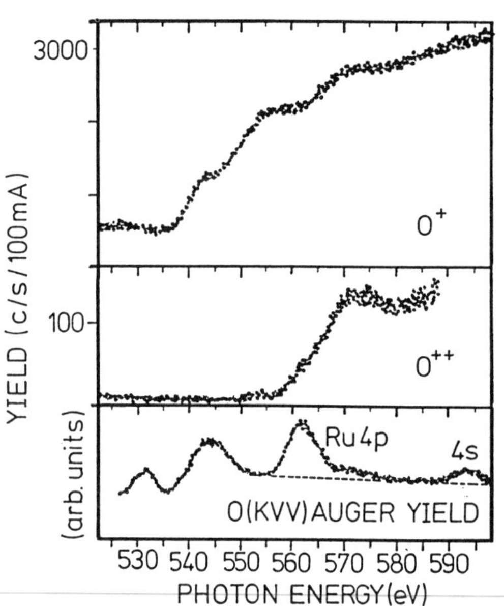

Fig. 1: Yields of the fragment ions O^+ and O^{2+} desorbed from chemisorbed NO on Ru(001) by photons in the O1s region, compared to the absorption cross section of the adsorbate as given by the Auger yield (the Ru photoelectron peaks should be disregarded). After ref. 7.

shown[7] that of the two adsorbed NO species[8] present on this surface, a linear one with a positive surface dipole and a bridged negative one, essentially only the linear one contributes to the ion signal. Similar results have been obtained[6,9] for adsorbed CO on various surfaces, and to some degree for adsorbed N_2O[10]. The general explanation that has been given and that follows from the generalized so-called MGR mechanism[11], is that all these specificities in the decrease of dissociation are due to specific quenching of the various primary excitations into the substrate. Indeed, any effect which is expected to increase the coupling of the adsorbate to the substrate (increased adsorption energy; closer geometry and proximity of the primary excitation to the substrate; increased coverage and two-dimensional order which lead to lateral delocalization of the O^{2+} excitation; coadsorption of alkali atoms; simple excitations) have been found to decrease the dissociation yield, while any effect which decreases the coupling (geometry with smaller overlap; vanishing lateral interactions at low coverage; inert spacer layers; strongly correlated and thus strongly Coulomb-localized[12] excitations) increases the dissociation yield.

The mentioned recapture/delocalization mechanism assumes that the overall dissociation probability is the product of the primary excitation probability and a factor defining the survival of the excitation at its primary location up to the bond-breaking event. The latter is due to the fact that the metal substrate with its huge charge and excitation density acts as a perfect excitation sponge: anything transferred to it (excitation or charge) never comes back to a surface molecule, but is efficiently thermalized. Thus, the competition of bond breaking with quenching/delocalization will enhance the strongly localized excitations in the actual fragmentation yield observed. For the core ionization and even more the core shake-up excitations, the increased localization becomes understandable from the complexness of the products of their core decay which are m*h* n*e* states (where *h* are valence holes and *e* are excited electrons, and m and n are their numbers), with m ≥ n and m up to 4 or more[6,10].

Scrutinizing these arguments further shows that survival of the primary excitation is possible not only by increasing the localization time of the excitation, but also by decreasing the need for it, i.e. shortening the time scale of dissociation. The latter will be the case if more repulsive electronic states are excited. In fact, the specificity of core shake-up states and their decay states can also be understood in this way, namely by their increased repulsiveness which in the extreme corresponds to the Coulomb explosion of an m-hole system. However, the latter may even be unnecessary for long-living core holes (such as C1s, N1s, O1s with their life times of up to 10 fs[13]) and very repulsive primary core states, for which dissociation can become competitive with core decay in terms of time scale. Such processes have been observed by spectroscopic means in some gas phase molecules[14]. Thus the efficient quenching of longer living states dissociating more slowly may in fact project out ultrafast photodissociation processes in surface species.

SURFACE MOLECULES OF CONDENSATES OF HYDROGENIC MOLECULES

The considerations concerning chemisorbates in the preceding chapter have shown the importance of coupling-induced quenching for the modifications of the dissociation behavior at a metal surface. The arguments leading to the assumption of the existence of ultrafast processes in these systems are appealing but not conclusive. In a search for more direct evidence, we have examined condensed layers of hydrogenic molecules (H_2O/D_2O, NH_3, C_6D_6) in their core regions. Their selection was made since any ultrafast dissociations should be most obvious for the fast-moving very light H^+/D^+ ions. The following facts have been found:

1) In all these molecules, the fragment ion yield is strongly enhanced (as seen from comparison with the absorption cross section as measured by TY, PY or AY) in specific neutral core-to-bound excitations very close to threshold. For water[15,16] (fig.2) and ammonia[16], this is in fact the lowest possible core excitation (to the strongly antibonding $4a_1$ state); for benzene[17], the lowest core excitations are of π type which do not involve the C-H bond, so that it is understandable that the enhanced state with lowest energy is of σ^*(C-H) character (fig.3).

2) A comparison made for the water case[15] between the surface molecules of the condensate and submonolayer chemisorbed molecules on a metal surface has shown that this effect is independent of coupling strength. We suspect, in fact, that it exists even in the free molecules. (See also footnote 20).

3) For water, a detailed investigation of the direction of polarization relative to that of ion emission has been made[15] and has shown

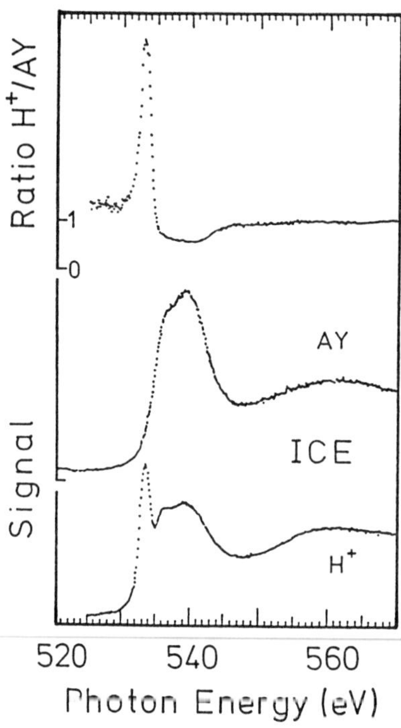

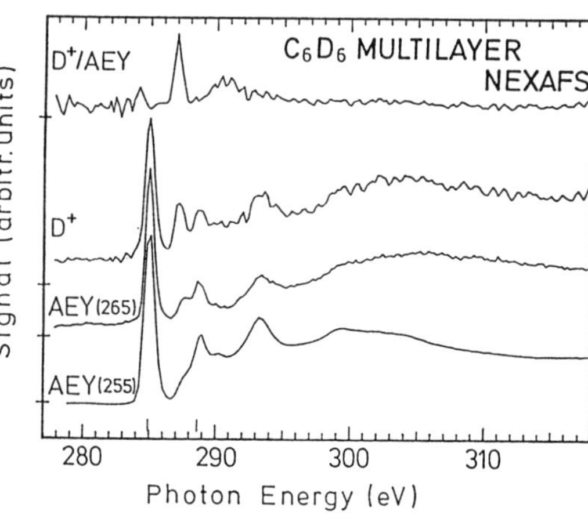

Fig. 3: Similar data (D^+ fragment yield, Auger yield at the two kinetic energies indicated, and their ratio) from a condensed d_6-benzene layer under irradiation with photons in the C1s region. After ref. 17.

Fig. 2: Yields of Auger electrons (representative of absorption) and of H^+ fragment ions from a condensed layer of H_2O, induced by irradiation with photons in the O1s excitation region. The ratio of the two signals (top curve) indicates the selectiveness of the lowest core resonance. After ref. 15.

that the maximum of H^+ signal is <u>not</u> found for the E-vector in the molecular axis, but in the direction of the O-H bond to be broken.

4) Studying decay electron spectra for the various core resonances for the benzene case it has been found[17] that the resonance showing enhanced D^+ production possesses a decay spectrum which <u>cannot</u> be understood as related to an intact benzene core state. This corroborates a change of the molecular structure before the core hole is filled.

All these finding are most easily understood if the existence of ultrafast processes is accepted, i.e. if dissociation <u>starts</u> during core hole life time. The fact that hydrogen <u>ions</u> are detected shows that core decay must occur before the bond has <u>fully</u> separated, otherwise neutral H atoms would result as is the case for HBr and CH_3Br in the gas phase after Br core excitation[14]. Such a signal may well exist as well but is much more difficult to detect. We admit that findings 1) and 2) would still be understandable in a normal Franck-Condon mechanism in which core hole decay takes place fast compared to dissociation, and the electronic states relevant for dissociation are the products of spectator decay of the core hole (see the discussion in ref.15). Findings 3) and 4) are not understandable in this way, however, and the general time scale arguments given are quite suggestive as well[20]. We believe, therefore, that our results make the existence of ultrafast, direct channels of photodissociation at least very probable, with localization of the dissociative event to the locus of primary excitation and concomitant strong selectivity.

OUTLOOK

While our results have been obtained with surface systems and surface specific methods, we believe that their importance is not restricted to surfaces. By the selective quenching occurring at surfaces, the events which we interpret as ultrafast channels become much easier to observe than in free molecules, where they should also exist but may be swamped (and in fact appear to be so, at least for CO[18]) by the other "normal" processes. For the hydrogenic molecules the situation should be much more favorable, and work on free molecules should definitely be done, preferably with a facility allowing to measure the correlation of the directions of polarization and ion detection, to examine a possible symmetry-breaking. On our part, we are in the process to extend the measurements to systems with more short-lived core holes, and to vibrationally resolved excitation. In view of the additional information which can be derived from recording of the decay spectra[19] of the relevant core excitations (see the benzene case), such measurements will be carried out where this has not yet been the case. Theoretical work on the very interesting coupled processes becoming obvious here would be highly desirable.

ACKOWLEDGEMENTS

I thank the many coworkers who participated in the described work, in particular D. Coulman, P. Feulner, G. Rocker, H.-P. Steinrück, R. Treichler, and W. Wurth, and who have made important contributions to these developments, both in the actual work carried out and in many discussions which shaped the ideas set out here and in the cited work.

This work has been supported by the German Federal Ministry of Research and Technology (under grant No. 05 466 CAB) and by the Deutsche Forschungsgemeinschaft (through SFB 338).

REFERENCES

1. See for instance D. M. Hanson, Advances Chem. Physics vol.77 (Academic press, New York 1990), p.1; and references given therein.
2. See for instance W. Habenicht, H. Baiter, K. Müller-Dethlefs and

E.W. Schlag, Physica Scripta 41, 814 (1990).
3. See for instance D. Menzel, Nucl. Instrum. Methods Phys. Res. B13, 507 (1986); D. Menzel, Laser Chem., in press (1991).
4. See T.E. Madey, D.E. Ramaker and R. Stockbauer, Ann. Rev. Phys. Chem. 35, 215 (1984);
Ph. Avouris and R.E. Walkup, Ann. Rev. Phys. Chem. 40, 173 (1989);
and the monographs "Desorption Induced by Electronic Transitions" I to IV, Springer, Berlin, 1983 to 1990.
5. R. Franchy and D. Menzel, Phys. Rev. Lett. 43, 865 (1979);
P. Feulner, R. Treichler and D. Menzel, Phys. Rev. B 26, 554 (1981).
6. R. Treichler, W. Riedl, W. Wurth, P. Feulner and D. Menzel, Phys. Rev. Lett. 54, 462 (1985).
7. R. Treichler, W. Riedl, P. Feulner and D. Menzel, Surface Sci. 243, 239 (1991).
8. E. Umbach, S. Kulkarni, P. Feulner and D. Menzel, Surface Sci. 88, 65 (1979);
P. Feulner, S. Kulkarni, E. Umbach and D. Menzel, Surface Sci. 99, 489 (1980);
H. Conrad, R. Scala, W. Stenzel and R. Unwin, Surface Sci. 145, 1 (1984).
9. R. Treichler, W. Wurth, W. Riedl, P. Feulner and D. Menzel, Chem. Physics 153, 259 (1991).
10. R. Treichler and D. Menzel, unpublished.
11. D. Menzel and R. Gomer, J. Chem. Phys. 41, 3311 (1964);
D. Menzel, in: Desorption Induced by Electronic Transitions, DIET-I, Eds. N.H. Tolk et al. (Springer, Berlin 1983), p. 68.
12. M. Cini, Solid State Commun. 24, 681 (1977;
G.A. Sawatzki, Phys. Rev. Lett. 39, 504 (1977).
13. J. McGuire, Phys. Rev. 185, 1 (1969);
F.C. Brown, Solid State Phys. 29, 1 (1974).
14. P. Morin and I. Nenner, Phys. Rev. Letters 56, 1913 (1986);
P. Morin and I. Nenner, Physica Scripta T17 171 (1987).
15. D. Coulman, A. Puschmann, W. Wurth, H.-P. Steinrück and D. Menzel, Chem. Phys. Lett. 148, 371 (1988);
D. Coulman, A. Puschmann, U. Höfer, H.-P. Steinrück, W. Wurth, P. Feulner and D. Menzel, J. Chem. Phys. 93, 58 (1990).
16. D. Menzel, G. Rocker, D. Coulman, P. Feulner and W. Wurth, Physica Scripta 41, 588 (1990).
17. D. Menzel, G. Rocker, H.-P. Steinrück, D. Coulman, P.A. Heimann, W. Huber, P. Zebisch and D.R. Lloyd, J. Chem. Phys. (in press).
18. A.P. Hitchcock, P. Lablanquie, P. Morin, E. Lizon A Lugrin, M. Simon, P. Thiry and I. Nenner, Phys. Rev. A 37, 2448 (1988).
19. W. Wurth, P. Feulner and D. Menzel, paper presented at the Auger Workshop IWASES-II in Lund, Sweden, Sept. 1991 (to be published in Physica Scripta).
20. Indirect proof of our conclusion of ultrafast dissociation in H_2O and NH_3 can be taken from the highly resolved absorption spectra of the isolated molecules reported by K.J. Randall, J. Feldhaus, A.M. Bradshaw, J. Schirmer, Y. Ma, F. Sette, and C.T. Chen at this conference which show that the $1s \rightarrow 4a_1$ excitations do not show vibrational structure and thus must be dissociative.

DISCUSSION

DUJARDIN - Following core excitation of surface molecules (and/or of the bulk) a lot of Auger electrons are produced and may induce reactions. May this explain in some cases the high reactivity observed by core excitation ? May this also lead to a loss of selectivity ?

MENZEL - The effect you mention is well known and is called XESD (X-ray induced Electron Stimulated Desorption). Only for condensed layers it can interfere with direct photoeffects, since for monolayers on a different substrate the excitation energies are usually sufficiently different for adsorbate and substrate, so that separation is easy. But even for condensed layers of identical molecules, these processes —which indeed lead to loss of selectivity— can be distinguished from direct photoeffects. The strongly selective processes in the hydrogenic molecules reported here are clearly direct photoeffects (see the discussion in the cited literature, especially in refs. 15 and 17).

NENNER - 1. Have you tried to investigate fluorescence signals which are seen in core excited molecules, having in mind that those radiative channels originate from excited fragments produced after dissociation of ions after the core hole relaxation ?

2. Since the desorption phenomena are especially sensitive to shake up, have you tried to investigate double core vacancy shake up effects in chemisorbed molecules ?

For example in H_2S 1s + 2p excitation
2p + 2p
etc...

MENZEL - 1. So far we have not yet tried to record fluorescence signals. It would be interesting to do so for radiation originating from desorbed particles in the gas phase, since one could learn about desorption of metastables. This has been done by other authors in the valence region of, for instance, rare gases, but not for core excitations. Looking directly onto the surface, i.e. for fluorescence of molecules there, would be less informative since the signal would be dominated by bulk processes.

2. For the light atoms mainly investigated so far (C, N, O) there are no double core excitations since even the 2s levels are essentially valence levels. In H_2S we have not yet looked for such excitations, but it would certainly be interesting to do so.

PIANCASTELLI - Would you consider the partial ion yield spectra as a general criterion to distinguish between various kinds of excitations (doubly excited states, shape resonances, etc.) ?

MENZEL - Yes, indeed, for the case of (sub)monolayers on metal surfaces we think that the fragment yield spectra are a good means to selectively investigate multiple excitations and distinguish them from single particle excitations such as Rydberg and shape resonances. Due to the limited time it was not possible to elaborate on this, but I refer to references 6, 7 and 9 for the use of symmetry as well as energy which is possible by applying dipole selection rules to the surface molecules with their known orientation. Even for the surface molecules of condensed layers, we believe that the fragment ion yields constitute a means to distinguish between repulsive molecular states and Rydberg excitations (see the discussion of the decomposition of mixed valence/Rydberg states in refs. 15-17).

LEACH - The $4a_1$ state in water should have a tendency to be linear. This implies excitation of bending vibrations which might be observable if dissociation was slow enough. Was your energy resolution sufficient to pick up such bending vibration features ?

MENZEL - There are valence excitations which in their ground state are linear (or more strongly bent than the ground state), so that vibrational excitations will be connected with the electronic excitation. However, the $1a_1^{-1}\ 4a_1^{+1}$ states of H_2O and NH_3 appear, according to our findings as well as to the high resolution molecular core absorption spectra reported at this conference (see my ref. 20), to be fully dissociative, so that they do not possess a stable geometry. Nevertheless, angle changes can be expected on the same time scale as dissociation, in analogy to the valence excitations.

BAUMGARTEL - Can you say something about anisotropy effects in photon stimulated desorption process ?

MENZEL - For (sub)monolayers on metal surfaces, there are very strong angular anisotropies (FWHM angles down to 15°) which in fact are the basis of a method for the determination of molecular orientations of adsorbates called ESDIAD (Electron Stimulated Desorption Ion Angular Distribution). It rests on the assumption that the direction of ion emission always reflects the bond direction in the ground state (apart from a correctable bending by the image force) —which may be questionable for ultrafast processes (see answer to Leach). For ions from the surfaces of molecular condensates, maximum ion emission is invariably found in the surface normal. The correlation between polarization and ion emission angles has been used in the case of H_2O to conclude to symmetry breaking for the selectively fragmenting excitations (see refs. 15).

DYNAMICS OF CO RECOMBINATION IN SPERM WHALE MYOGLOBIN BY DISPERSIVE XANES

S. Della Longa, A. Bianconi,
Dipartimento di Medicina Sperimentale, Università dell'Aquila,
v. S. Sisto 20, 67100, L' Aquila, Italy

I. Ascone, A. Fontaine,
LURE (Lab. CNRS, CEA, MEN) Bat. 209D, 91405 Orsay, France

A. Congiu-Castellano, G. Borghini
Dipartimento di Fisica, Università di Roma "La Sapienza", P.le A. Moro 5, 00185, Roma, Italy

ABSTRACT

The structural changes of the Fe active site following the low temperature photodissociation of sperm whale carboxymyoglobin (MbCO) has been studied by X-ray Absorption Near Edge Structure (XANES) with the dispersive X-ray absorption method. This experimental method allows to study time dependent biological processes in diluted samples in the time scale of 10-100 seconds. From the Fe K-edge XANES difference spectra between the MbCO and its photoproduct (Mb*), a model of the intermediate state Mb* at low temperature (15K) can be inferred. The structure of the photoproduct Mb* is non-ligated, with the iron out of the heme plane, but distinct from the deoxy state Mb. A structural model built taking into account EXAFS and Raman data on Mb* explains some but not all the features observed by XANES spectroscopy. A smaller displacement of the iron from the pyrrolic nitrogen plane, coupled to a more domed structure of the porphyrin plane could explain better the spectral differences between the Mb* and Mb states.

INTRODUCTION

Ligand binding to hemoproteins is a process on which the scientific interest has been focussed on in order to understand the dynamics of the protein. Myoglobin (Mb) provides the simplest protein in order to study this process. In the flash photolysis experiments on MbCO[1,2], the light pulse strikes the MbCO sample and breaks the bond between the CO and the heme iron, and the rebinding process is studied. This process has been described as a three-well model, with time-dependent barriers between three main states, each of one can have many substates:

$$MbCO \underset{<}{\rightsquigarrow} Mb^* \underset{<}{\longrightarrow} Mb$$

where Mb* indicates the protein conformations when CO is dissociated but confined in the heme pocket, and Mb indicates the deoxy-Mb conformations when CO is in the solvent.

The CO rebinding to hemoproteins after photolysis depends on several factors, for example temperature, pH, protein species and viscosity of the solvent. Temperature dependence is the most constraining factor for dynamical studies. Below about 180K, in a 2:1 glycerol-water solvent, a single recombination process, Mb*-> MbCO is reported from kinetic data[2]. Therefore at such a low temperature, Mb* represent the photoproduct, because the ligand cannot exit the protein and the binding occurs directly from the heme pocket. The existence of conformational substates explains that this 'geminate' process is non-exponential in time.

Fourier transform infrared spectroscopy studies[3] of MbCO have identified three CO stretching frequency bands, A_0, A_1 and A_3 in sperm whale. Using the photoselection technique[4] combined with psec resolved infrared spectroscopy, Moore et al.[5] have related the C-O tilting angle α from the heme normal to the A_n bands, being $\alpha(A_0)=15^o\pm10^o$, $\alpha(A_1)=28^o\pm2^o$, $\alpha(A_3)=33^o\pm4^o$. Each band after photolysis exhibits a different non-exponential rebinding. So the A_n bands are taken as probes of three substates of tier 0 of the ground state. No linear dichroism is found after photolysis, so that the Fe-C bond is broken and CO has moved away and randomly oriented, but no detailed information on the structure(s) of the photoproduct Mb* have been obtained.

EXAFS (Extended X-ray Absorption Fine Structure) studies at low temperatures gave the variation of the averaged distance Fe-N_p between the Fe and four pyrrolic nitrogens of the iron in sperm whale carboxymyoglobin after photolysis. The EXAFS studies report Fe-N_p = 2.01±0.02 Å for MbCO, 2.03±0.02 Å for Mb*, and 2.06±0.02 Å for Mb in frozen solutions[6,7,8,9] and the values 1.96, 1.96 and 2.06±0.02 Å for MbCO, Mb* and Mb respectively in dry film samples[10]. These values have to be compared with the Fe site structure determined by X-ray crystallography (XRD) giving an average Fe-N_p =2.06 Å in Mb[11] and 2.01 Å in MbCO[12]. The EXAFS experiments indicate that the structure of the Fe heme complex in the low temperature photoproduct is not completely relaxed in the deoxy-Mb configuration.

On the other side the optical, infrared, and magnetic susceptibility studies[13,14] have indicated that the configuration of the heme in the Mb* photoproduct is very close to the fivefold Fe site configuration of deoxy-Mb.

Raman investigation[15] of the photoproduct structure compared with that of deoxy -Mb have pointed out a small expansion of the heme core size (the distance from the heme center and the pyrrolic nitrogens, C_t-N_p) in the photoproduct.

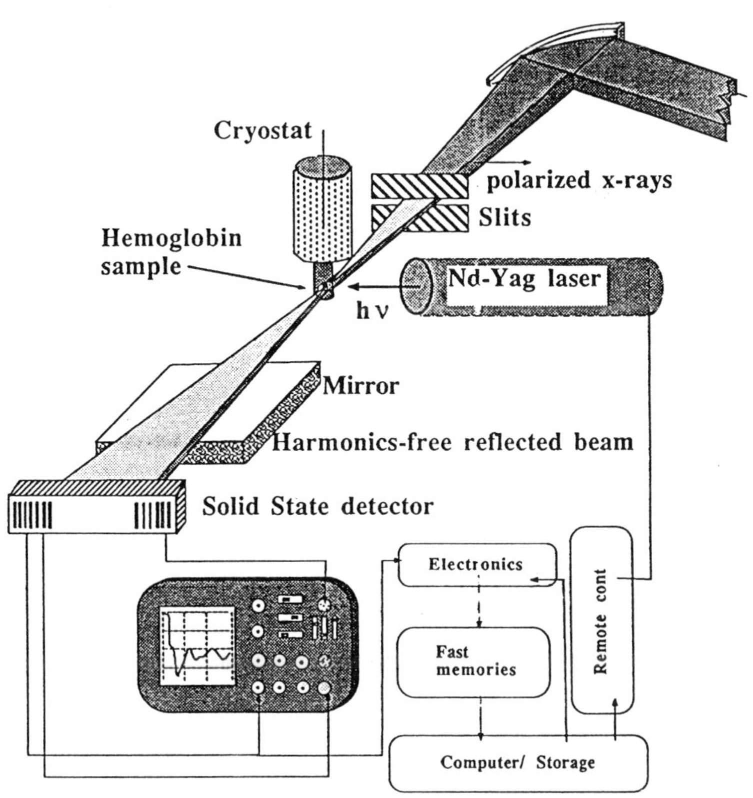

Fig. 1 Experimental apparatus for time resolved XANES study of CO rebinding in hemoproteins combining the X-ray dispersive method and flash photolysis techniques

The XANES (X-ray Absorption Near Edge Structure) spectroscopy is very sensitive, via multiple scattering resonances, to the local structure (bonding angles and geometry) around the metal binding site. In the case of hemoproteins, a cluster of atoms including all the heme plane, the proximal histidine, and the ligand, is known to give detectable contributions to the spectrum. We have investigated the structural differences between the photoproduct Mb* and MbCO in sperm whale myoglobin by XANES, and we have compared them with the differences between deoxy and MbCO form. Previous time resolved XANES studies have been reported with very low energy resolution in which the fine details of the XANES spectra are smeared out [6,10,16,17].

We have performed a XANES study on sperm whale myoglobin with high energy resolution in the temperature range within 10 and 100 K, in order to elucidate the structural parameters of the photoproduct. We have used the dispersive method to collect spectra in the time scale of hundreds of seconds for proteins in solution at millimolar concentrations[18].

MATERIALS AND METHODS

Carboxymyoglobin has been prepared as 7 mM samples in a 2:1 glycerol-H_2O solution at pH 6.7 from sperm-whale met-myoglobin from Sigma, by standard methods described elsewhere[19]. A slight excess of concentrated solution of sodium dithionite was added and finally CO has been passed over the protein solution for at least 30 min just before the experiment to obtain MbCO.

The spectrometer of the dispersive line simply combines a focusing dispersive X-ray optical system including a triangular bent crystal and a position sensitive linear detector able to work under high flux conditions[20]. This system gives an high stability in the energy scale, because there is no mechanical movement during the data collection, so an high precision about the energy shift of the absorption edge can be obtained. The zero of the energy scale has been fixed at the Fe metal K-edge defined as the first maximum of the derivative spectrum. The use of the dispersive technique have allowed us to collect spectra of frozen solutions of hemoproteins with good resolution in 4 min (presented here) or less. An x-ray flat fused quartz mirror at grazing incident is introduced in the optical system between the sample and the detector, as a low pass filter to get rid of harmonics contamination. The MbCO sample in a 1 mm thick cell has been cooled at 15K by a close circuit cryostat and it has been photolyzed by using a

Nd-Yag laser or with continuous illuminations using optical fiber optics. The experimental set-up is shown in fig.1.

RESULTS AND DISCUSSION

In fig. 2 the Fe K-edge normalized absorption spectra at 300K of sperm whale MbCO, deoxy-Mb, detected by scanning mode, and their difference spectrum are shown as reference spectra, In the following discussion we will refer the features named A, B, C1, D and C2 at about 9, 14, 16, 22 and 36 eV respectively.

Going from the CO-form to the deoxy-form, an energy-shift of about 2 eV of the absorption threshold is seen. In a previous XANES work on oxymyoglobin[21], a red-shift going from the oxy-form to the deoxy-form was reproduced by theoretical calculations of the XANES spectra for deoxy- and oxy- structures obtained from crystallography. It was shown that the energy of the absorption threshold is highly sensitive to the Fe-N_p distance and to the effective charge at the iron site.

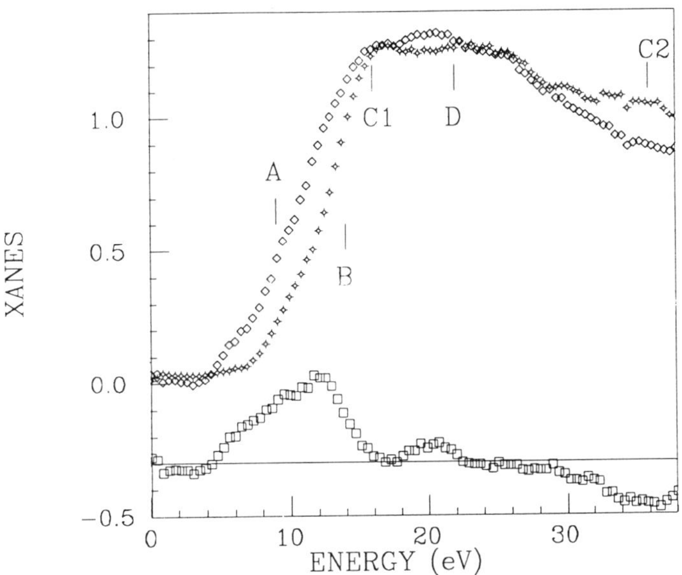

Fig. 2 Fe K-edge absorption spectra, detected by scanning mode, of sperm whale myoglobin in 2:1 Gly-H_2O at pH 6.7: MbCO (plotted) at 300 K, Mb (solid) at 300 K, and the difference spectrum Mb - MbCO (lower curve).

The sensitivity of XANES features C1 and C2 to the Fe-C-O angle in MbCO and the sensitivity of the XANES feature D to the distortion of the heme plane have been discussed in previous papers[22,23,24].

We have reported in the same figure also the difference spectrum Mb/MbCO, as an easy reference for the difference spectra Mb*/MbCO directly measured 'in situ' by the dispersive technique, presented here without any data manipulation (only a straight line, fitting the pre-edge, is subtracted from the data). In fig. 3 we report, from bottom to top, the measured difference spectra Mb*/MbCO at 15K, 30K, 40K, 60K and 80K under continuous illumination. As it is clear from the reference Mb/MbCO, a red-shift of the Fe K-edge results as the main positive peak in the difference spectrum. The peaks C1 and D present little variations, while the feature C2 strongly decreases. The zeroes of the difference spectrum (crossing points of the XANES spectra) are at about 4, 23 and 48 eV.

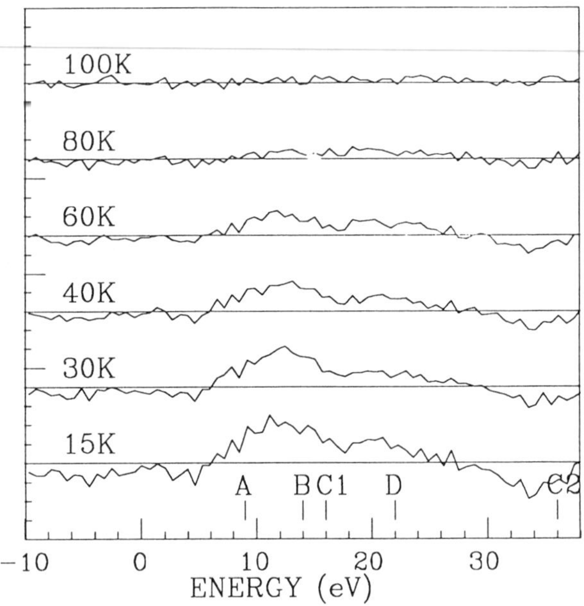

Fig. 3 Temperature dependence of Fe K-edge difference absorption spectra of sperm whale myoglobin. From bottom to top: difference spectra Mb*/MbCO at 15 K, 30 K, 40 K, 60 K, 80 K, 100K.

The zeroes of the difference spectra don't depend on temperature, even if they are slightly different from those of the difference spectrum between deoxy-Mb and

MbCO. It is worth to remember that in the difference spectra the intensities, but not the line shapes, depend on the extent of photolysis. It indicates that, in our experimental conditions, we observe only the two main conformational states MbCO and Mb*, separated by an enthalpic barrier whose variations with temperature are less than the sensitivity of the technique. In 4 min of acquisition time, at 80K, the recombination rate is too fast and no difference is seen, as well known from kinetic data of optical spectroscopy.

Fig. 4 shows the difference spectrum between Mb e MbCO (bottom curve) and the difference spectrum between Mb* and MbCO at 15K (upper curve). From this figure it is clear that the energy of the absorption threshold changes in the photoproduct in the same way as in deoxy-Mb. The photolysis induces a red-shift of the Fe K-edge of about 1.5 eV, slightly less than in the deoxy-Mb. Moreover, the feature C2 decreases as strongly as in deoxy-Mb. As regard the features C1 and D, we measure a not negligible increase, different between the two curves. So, the photoproduct seems to be non ligated, with an increased distance Fe-N_p, but distinct from the deoxy state of the protein.

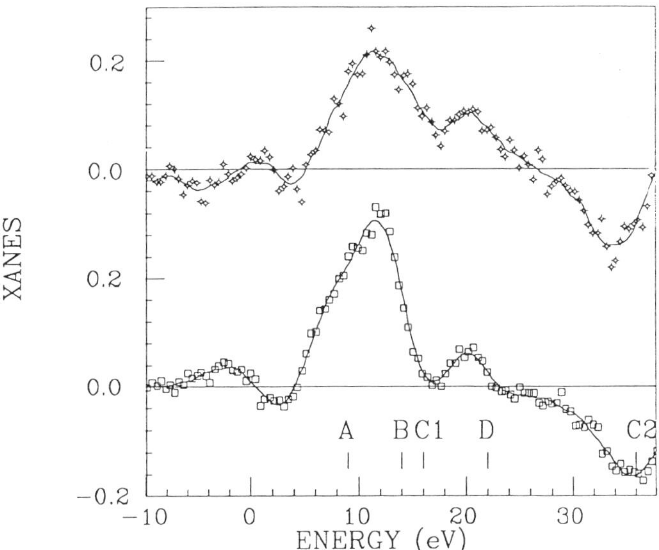

Fig. 4 Difference spectrum between deoxy- e CO-myoglobin at 300K (bottom curve); difference spectrum between photoproduct Mb* and MbCO at 15K (upper curve). Row data are interpolated with a smoothing curve.

We have attempted a series of simulations of the XANES spectra[25] of Mb, MbCO and Mb* via the multiple scattering approach, on the basis of the crystallographic and EXAFS data, as regard to Mb and MbCO, and from a structural model that could take into account some of the experimental observations, as regard to Mb*. We have considered a molecular cluster that include the Fe, a perfectly planar 4-fold symmetric tetrapyrrolic macrocycle, the himidazole ring of the proximal histidine, and the CO ligand.

At first, we have reproduced by calculation the XANES spectra of MbCO and Mb. By this way we observed the 2.0 eV red-shift of the absorption threshold as mainly due to the lengthening of 0.05 Å of the Fe-N_p distance going from MbCO to Mb, while the disappearing of the C2 peak was consequent to moving away the CO molecule from the cluster. Then, in order to simulate the Mb* structure, we have considered a non ligated (deoxy-like) structure with Fe-N_p=2.045 Å (from EXAFS), N_p-C_t=2.027 Å, (from Raman), Fe-C_t=0.27 Å (calculated by triangulation from N_p-C_t and Fe-N_p) and Fe-N_ε=2.12 Å, as in deoxy-Mb. The results are very similar to that of the Mb model, and the right value of the energy shift is found (1.5 eV). Alternatively, we have added to this structure a CO molecule parallel to the heme plane, near but non ligated to the iron (at a distance of about 2.4 Å, as it be in Van der Wall contact with both the heme core and the N_ε of the distal histidine), simulating a sort of physisorption event. This model was suggested, besides the increasing of C1 and D in the photoproduct Mb*, by some EXAFS data that don't report decreasing of the coordination number of the first shell around the iron after photolysis. This model gave a worse agreement to the experimental data, moreover don't produced the desired effect of increase C1 and D, and so it was rejected. Further calculations are in course in order to reproduce the experimental data. In any case, the C1 and D features, in non ligated hemoproteins, show a noticeable dependence on distortions of the porphyrin plane, that play an important role in the transfer mechanism of the information of the ligand binding between the protein and the iron site. A possibility is that the strain determined by the movement of the iron out of plane after photolysis would be stronger to the heme, in a rigid protein matrix that in the CO form contacts more strictly the heme than in the deoxy form (although the distal pocket in MbCO is larger than in Mb, 90 Van der Wall contacts in MbCO against 74 in Mb are reported by crystallography[11]). By experiment, the relations between the XANES features and the distortions of the porphyrin plane have been studied[24] on the so called "basket-handle" Fe(II) porphyrins, in which distortions of the tetrapyrrolic macrocycle are induced by tuning the length of the basket-handle. In all the compounds studied, the Fe-C-O angle is linear, and a positive correlation between the intensity of the peak D and the degree of the distortion of the heme plane has been found.

In summary, a structural model can be inferred for the photoproduct Mb* at low temperature: when the light strikes the protein the CO moves away, the iron move out of plane, going to an average distance of 2.045 Å from N_p, consistently with EXAFS and Raman data, but the protein matrix could remain in the same configuration of MbCO. It could determine a different strain and a distortion of the heme plane explaining the different behavior of the C1 and D features.

REFERENCES

1 P.J. Steinbach, A. Ansari, J. Berendsen, D. Braunstein, K. Chu, B.R. Cowen, D. Ehrenstein, H. Frauenfelder, J.B. Johnson, D.C. Lamb, S.Luck, J.R. Mourant, G.U. Nienhaus, P. Ormos, R. Philipp, R. Scholl, A. Xie, and R.D. Young, Biochemistry, (1990); M.K.Hong, D. Braunstein, B.R. Cowen, H. Frauenfelder, I.E.T. Iben, J.R. Mourant, P. Ormos, R. Scholl, A. Schulte, P.J. Steinbach, A. Xie, and R.D. Young, Bioph. J., 58, 429-436 (1990).

2 R.H. Austin, K.W. Benson, L. Eisenstein, H. Frauenfelder, I.C. Gunsalus, Biochemistry, 14, 5355-5373 (1975).

3 P. Ormos, D. Braunstein, H. Frauenfelder, M.K. Hong, S. Lin, T.B. Sauke, and R.D.Young, Proc.Nat.Acad.Sc , 85, 8492-8496 (1988).

4 A.C. Albrecht, J. Mol. Spectr. , 6, 84-108 (1961).

5 J.N. Moore, P.A. Hansen, and R.M. Hochstrasser, Proc.Natl. Acad. Sci. USA, 85, 5062-5066 (1988).

6 B. Chance, R. Fischetti, L. Powers, Biochemistry, 22,3820 (1983)

7 L. Powers, J.L. Sessler, G.L. Woolery and B. Chance, Biochemistry, 23, 5519-5523, (1984).

8 L. Powers, B. Chance, M. Chance, B. Campbell, J. Khalid, C. Kumar, A. Naqui, K.S. Reddy and Y. Zhou, Biochemistry, 26, 4785-4796, (1987).

9 K. Zhang, B. Chance, K.S. Reddy, Physica B, 158, 121-122, (1989).

10 Tsu-Yi Tieng, Huey W. Huang, G.A. Olah, "5K EXAFS and 40K 10-s resolved EXAFS studies of photolyzed MbCO",Biochemistry , 26, 8066-8072 (1987).

11 T. Takano, J. Mol. Biol. , 110, 569 (1977); S.E.V. Phillips " X-ray Structure of Deoxy -Mb (pH8.5) at 1.4 Å resolution, Brookhaven Protein Data Bank (1981) G. Fermi, M.F. Perutz, B. Shaanan, R. Fourme, J. Mol. Biol., 175, 159 (1984).

12 J. Kuriyan, S. Wilz, M. Karplus, G.A. Petsko, J. Mol. Biol., 192, 133 (1986).

13 F.G. Fiamingo, and J.O. Alben, Biochemistry, 24, 7964-7970 (1985).

14 M. Sassaroli, and D.L. Rousseau, J. Biological Chemistry, 261, 16292-16294 (1986).

15 D.L. Rousseau , and P.V. Argade Proc. Natl. Acad. Sci. USA, 83, 1310-1314 (1986).

16 D.M. Mills, A. Lewis, A. Harootunian, J. Huang, and B. Smith, Science, 223, 811-813, (1984).

17 A. Bianconi in EXAFS and Near Edge Structure edited by A. Bianconi, L. Incoccia, and S. Stipcich, Springer Verlag, Berlin, 1983 pag. 118-129

18 I. Ascone, A. Fontaine, A. Bianconi, A. Congiu-Castellano, A. Giovannelli, S. Della Longa, M. Momentau, in "Biophisics and Synchrotron radiation", eds. A. Bianconi and A. Congiu Castellano (Springer series in Biophysics, 2, 202,1987).

19 E. Antonini, M. Brunori, in "Hemoglobin and myoglobin in their reactions with ligand" (North-Holland Publ. Amsterdam-London).

20 A. Fontaine, E. Dartyge, J.P. Itie, A. Jucha, A. Polian, H. Tolentino, G. Tourillon, Topics in Current Chemistry, vol. 151, pp. 179-203 (Springer-Verlag Berlin Heidelberg, 1989).

21 A. Bianconi, A. Congiu-Castellano, M. Dell' Ariccia, A. Giovannelli, E. Burattini, P.J.Durham, Bioch. Bioph. Res. Comm.,131,98-102,(1985).

22 A. Bianconi, A. Congiu-Castellano,P.J. Durham, S.S. Hasnain, S. Phillips, Nature, 318, 685-687 (1985)

23 A. Bianconi, A. Congiu-Castellano, M. Dell' Ariccia, A. Giovannelli, P.J. Durham, E. Burattini, M. Barteri, FEBS 2039, 178, 165-170 (1984).

24 C. Cartier, M. Momenteau, E. Dartyge, A. Fontaine, G. Tourillon, A. Bianconi, M. Verdaguer, Bioch. Bioph. Acta, submitted.

25 S. Della Longa, A. Bianconi, I. Ascone, A. Fontaine, A. Congiu Castellano, G. Borghini, in preparation (1991).

III. TIME-DEPENDENT FLUORESCENCE DECAYS

TIME-RESOLVED SPECTROSCOPY OF NUCLEIC ACID SYSTEMS USING SYNCHROTRON RADIATION FROM 230 NM TO 354 NM

Malcolm Daniels*
Chemistry Department and Radiation Center
Oregon State University, Corvallis, OR 97331-5903, USA

Jean-Pierre Ballini* and Paul Vigny*
Laboratoire de Physique et Chimie Biomoleculaire-CNRS (UA198)
Institut Curie, Section de Physique et Chimie
11 rue Pierre et Marie Curie
75231 Paris Cedex 05, France

*Laboratoire d'Utilisation du Rayonnement Electromagnetic (LURE)
CNRS and Université Paris-Sud, Bat. 209C
91404 Orsay Cedex, France

ABSTRACT

The excited states of nucleic acids are complex, both at the individual chromophore level and because of the effect of stacking interactions on the electronic states. Considerable progress has been made recently by studying the lifetimes of the stacked states and by utilizing the technique of time-resolved spectroscopy. Experimental results obtained using the ACO synchrotron at LURE, Orsay, will be presented. Resolution of the decay data gives a model-based estimate of the number of emitting species and their lifetimes, and this information is then used to deconvolate experimental time-windowed spectra (time-delayed spectra) to give true time-resolved spectra. It is a unique feature of the synchrotron, compared with the laser, that the combination of delayed detection (photon counting) with the continuous wavelength distribution of the synchrotron allows the acquisition of excitation spectra by uninterrupted repetitive scanning over a wide range of UV exciting wavelengths, in the present work from 230 nm to 354 nm. Such time-delayed excitation spectra can also be deconvoluted into components corresponding to the various time-resolved emission spectra. In this way we are able to demonstrate for the first time that ground state stacking interactions are directly responsible for excimer-like emissions. Time-resolved emission spectra and time-resolved excitation spectra will be presented for the dinucleoside phosphate d(CG) and the synthetic alternating polynucleotide poly d(GC), a 'B-type' DNA structure.

INTRODUCTION

The study of the photophysics of nucleic acids (NAs) falls naturally into two parts. First there is the behavior of the chromophores of the constituent major purine and pyrimidine nucleosides (adenosine A, guanosine G, thymidine T and cytidine C) which can be regarded as monomers. Then there are the interactions, both in the ground state and excited state, which can occur when the chromophores are stacked in polynucleotides and the well-known A, B and Z helical structures of DNA.

Continuing efforts over the past three decades have established major features of the monomer parameters-corrected fluorescence spectra, accurate quantum yields and transition moment directions. However, a major area of continuing concern is the monomer lifetimes, important because of their relevance to the nature of the emitting state. It was early recognized[1] that the low fluorescence quantum yields (10^{-5}-10^{-4}) imply fluorescence lifetimes in the low picosecond (ps) range. Values from 0.2 p to 1.4 ps were estimated from the relation $\tau_f = \phi_f\tau_o$, calculating the intrinsic radiative lifetimes by the Strickler-Berg procedure (and hence implicitly assuming the absorption and emission to be directly correlated). Experimentally these lifetimes have been very difficult to determine and only one direct determination has been reported, τ_f=6 ps for 9-methyladenine[2]. Coupled with the quantum yield of $5x10^{-5}$ for the adenyl group this gives an intrinsic radiative lifetime of 150 ns, indicating clearly that the transition in emission is quite forbidden. This is also the conclusion from work on adenosine at 77 K, where quantum yields are higher and the fluorescence lifetimes longer[4]. In this case we were also able to estimate the total radiative oscillator strength ~$1.7x10^{-2}$. Comparison with the total absorption oscillator strength of 0.29 shows the degree of forbiddenness of the lowest-lying transition and indicates that in absorption this transition is hidden beneath the strong first absorption band.

A similar conclusion has been reached for the uracil chromophore from studies of the resonance raman excitation spectrum[5]. Here a fluorescence emission spectrum which is in agreement with the observed (77 K) spectrum has been calculated from the raman data. In turn the emission parameters allow the calculation of the corresponding absorption spectrum. However this does not agree with the observed absorption. As in the case of the adenyl chromophore the emission and the observed strong absorption are not corresponding transitions. This situation may well hold for the other chromophores, G&C.

Study of low or sub-picosecond lifetimes of weak emitters such as these requires an appropriate picosecond excitation source. Synchrotron radiation is not useful in the respect as the single-bunch pulse widths of current and future machines are not less than ~150 ps. However, using the ACO synchrotron at LURE, Orsay we observed red-shifted emissions from ApA and poly rA[6] and from DNA and poly d(AT)[7] having lifetimes in the low nanosecond (ns) range at room temperature.

These emissions, which are observed from pH7 aqueous solution of oligomers, have been assigned as originating from excimer-like states developed in stacked structures. Synchrotron radiation at ACO has proved to be a most useful excitation source for investigating the photophysics of these structures. Indeed in one respect, for the determination of time-resolved excitation spectra, it is so far a unique source. In this work we have combined analysis of decay profiles at discrete emission wavelengths with time-windowed emission spectra and time-windowed excitation spectra to give us true time-resolved emission spectra and their corresponding excitation spectra. Of the many possible combinations of nucleotides we have investigated systematically the sequence isomers TpdA and dApT, and the dinucleoside phosphate dCpdG(dCG) and its polymer poly d(GC).

The results on the A,T systems have been reported recently in detail.[8] Here we present some results for d(CG) and poly d(GC) and compare them with the (AT) system.

EXPERIMENTAL AND DATA ANALYSIS

As full descriptions of the ACO synchrotron and detection electronics at the UV/vis port have been given elsewhere[6-8] we outline only the salient features to provide a framework for subsequent data presentation and discussion. The 13.6 MHZ synchrotron source provided excitation pulses 1.6 ns wide through a sapphire window. In practice we work with dilute transmitting solutions below the vacuum UV and water cut-off. Excitation wavelengths utilized were between 220 nm and 348 nm, selected by an SLM double monochromator, and emissions from 280 nm to 536 nm were collected in right-angle geometry through an SLM single monochromator. Photomultiplier detection was used in the time-correlated single photon counting mode in which the time-interval between excitation pulse and emission pulses ('stop' and 'start' inputs) is converted by a time→amplitude converter (TAC in Fig. 1) into a voltage pulse. These pulses are then accumulated in a pulse-height analyzer (PHA) to give a histogram of the emission intensity for the particular excitation and emission wavelengths in use.

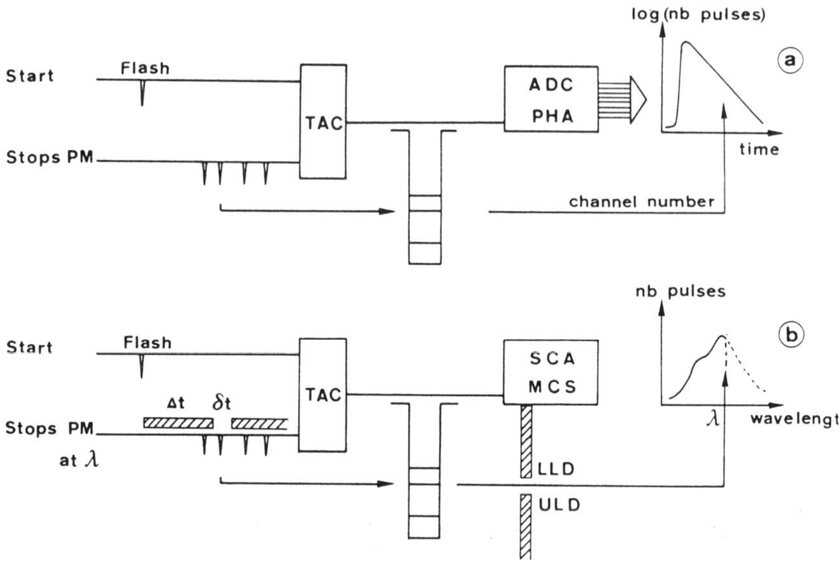

Fig. 1

This is the well-known lifetime mode illustrated in Fig. 1a. The range of the time-axis is determined by the upper and lower levels of the voltage discriminator, usually set to encompass the full range of decay into the noise-level. However, by changing the discriminator width, signals will be selected for a particular time-interval, a 'time-window'. If now the emission monochromator is scanned, while the analyzer is used in multichannel-scaling, then a 'time-windowed emission spectrum' is obtained, Fig. 1b. Several of these may be collected for different windows subsequently processed (taking account of excitation pulse-width and decay of the excited ensemble during the time-window) to give the true time-resolved emission spectra which result from excitation at a given wavelength. In a complementary procedure, the emission monochromator can be fixed at some interesting wavelength while the excitation monochromator is scanned, thus producing 'time-windowed excitation spectra', which can similarly be processed to give an excitation spectrum corresponding to each time-resolved emission spectrum. Somewhat awkwardly and inexactly we have termed these 'time-resolved excitation spectra'. Such spectra are quite rare, indeed at the time of writing we know of no examples other than our own work. The reason for this is that the ability to determine these spectra depends on a combination of synchrotron radiation characteristics, of which we may mention rapid tuning of the continuous spectral distribution, with photon output increasing towards the VUV, and of course high brightness of the pulsed radiation, together with the time resolution and sensitivity of time-correlated single photon detection. At present we know of no other spectroscopic system with this combination of characteristics.

To summarize, three classes of data are collected: 1) fluorescence decay profiles at three or sometimes four widely spaced wavelengths such as 320 nm, 350 nm, 390 nm and 430 nm. These are then analyzed both globally and independently using a model of discrete exponential lifetimes (one to three) to allow us to decide on the number of independent emitters and their lifetimes. 2) time-resolved emission spectra; three windows are used viz/ 0-2.35 ns, 2.35 ns-4.70 ns and 4.70 ns-36.0 ns, and results are resolved into time-independent spectra based on lifetimes obtained from decay-profile analysis. 3) time-windowed excitation spectra, resolved on the same basis. Overall, all classes of data are analyzed in terms of the extended impulse response function for the fluorescence intensity

$$I_f(\lambda_{ex}, \lambda_{em}, t) = \sum_{t=1}^{t=3} \sigma_i(\lambda_{ex})\, \alpha_i(\lambda_{em})\, \exp\text{-}t/\tau$$

RESULTS AND DISCUSSION

As in the case of the (A,T) systems reported earlier we find the decay profile is best fit by three exponential components, one in the lower ps range with the nominal value ≈100 ps-140 ps, and two in the ns range. It must be emphasized that little weight or molecular significance can be attributed to the ps lifetime because of system limitations of 1.6 ns exciting pulse width and timing resolution of 83 ps/channel. We regard it simply as a fitting parameter. In the direct application of these lifetimes to the extraction of time-resolved spectra we obtain three spectra (one corresponding to each lifetime). However, the time-windowed spectra are weak and noisy (a consequence of the low quantum yield and subdivision of the total emission), resulting in quite noisy time-resolved spectra. To overcome this our strategy has been to treat the system as two-component, using the ps lifetime and a mean ns lifetime derived from two-component global decay analysis. Because of the great difference in 'lifetimes' the two classes of spectra which result are quite insensitive to the precise lifetime values which are used.

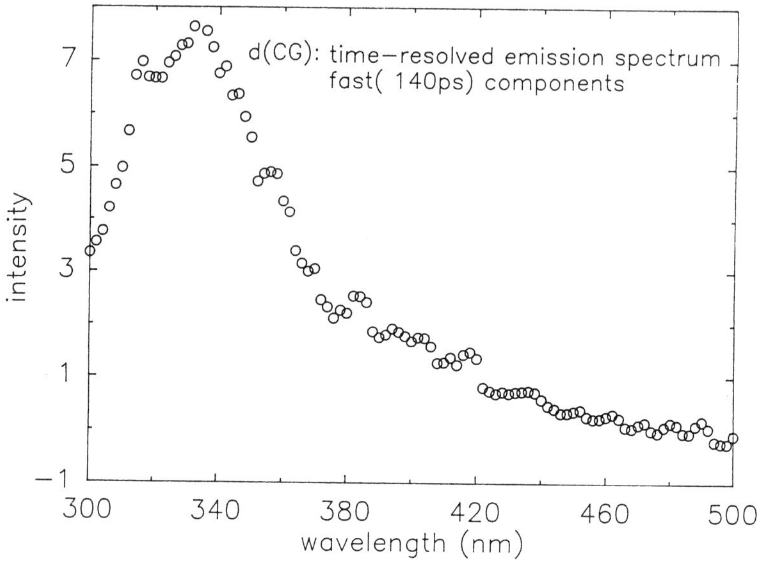
d(CG): time-resolved emission spectrum
fast(140ps) components
intensity
wavelength (nm)
7
5
3
1
−1
300
340
380
420
460
500

Fig. 2a

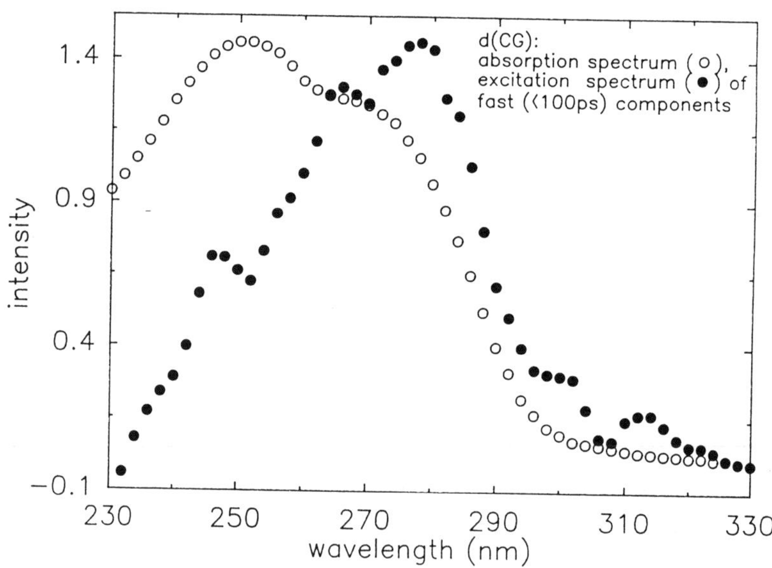
d(CG):
absorption spectrum (○),
excitation spectrum (●) of
fast (<100ps) components
intensity
wavelength (nm)
1.4
0.9
0.4
−0.1
230
250
270
290
310
330

Fig. 2b

The 'fast' spectra which result from applying this procedure to the d(CG) data are shown in Fig. 2a & b. The emission spectrum (Fig. 2a) is consistent with expectations for an equimolar mixture of (non-interacting) C+G based on earlier work on the monomers and we assign it to emissions from the unstacked fraction (d(CG) exists as an equilibrium mixture of stacked and unstacked conformers). The excitation spectrum for these fast emissions (Fig. 2b) is quite unusual in that it differs considerably from the absorption spectrum (Fig. 2b). This absorption range encompasses two absorption transitions in G and three in C and the behavior of Fig. 2b can be understood if internal conversion from the higher levels to the emitting levels of C or G is incomplete. After considering all possible combinations we can account for the unusual result only if excitation of the higher states S_2 and S_3 of C does not lead to fluorescence, either through internal conversion directly to ground or perhaps by rapidly quenching due to hydrate formation. The satisfactory fit obtained by this mechanism is shown in Fig. 3.

Fig. 3

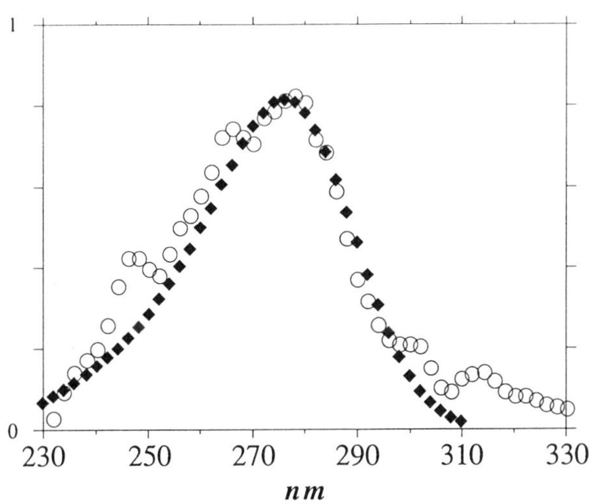

O, Expt. pts.; ♦, calculated curve

The long lifetime (ns) spectra for d(CG) are shown in Fig. 4.

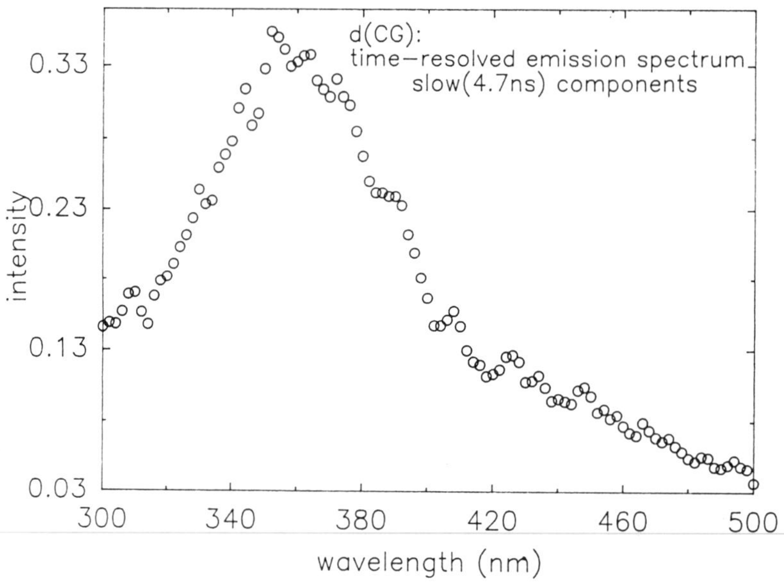

Fig. 4a

The emission spectrum is red-shifted from the monomers and peaks between 350 nm and 360 nm. Like the ns (AT) spectra and consistent with the lifetime analyses, it consists of two components, but in contrast to the (A,T) systems the lower energy components I is much weaker (integrated intensity of I is ~36% of total emission intensity compared with 67% for d(TA) and 77% for d(AT). Based on laser studies[9] with Z-form crystalline $d(CG)_3$ the assignment of this component is for it to originate from bases stacked in the geometry of the Z-conformation. The greater fraction of the bases in d(CG) are considered to have B-conformation stacking in low ionic-strength aqueous solution and the emission peaking ~355 nm may originate from this stacking.

Comparing excitation spectra in Figs. 4b and 3b, two striking features are apparent. First the decrease in emission probability in the S_2 region of C absorption i.e. from about 280 nm to 250 nm is again prominent and presumably finds the same explanation. Second, at wavelengths below 250 nm a steep, almost linear, increase in emission probability is found. If this were due to the incidence of S_3 transition of C (which it roughly parallels) then it should be observed in the ps

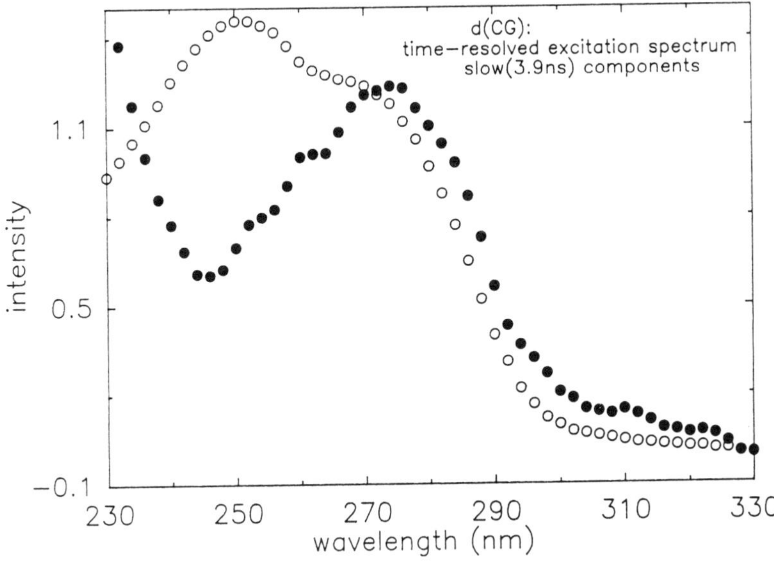

Fig. 4b

spectrum. As it is not, some other cause must be found. Similar behavior (linear increase in the ns spectrum which is not found in the ps spectrum) has also been observed for the (A,T) system cf. Fig. 5 of ref. 8. There it could be assigned to a perpendicular transition between the planes of stacked bases and we suggest the same behavior may be occurring here.

poly d(GC)

Space limitation precludes a full presentation and discussion of the results for poly d(GC) which is considered to exist as a double-strand B-form helix under the conditions of our experiments. Emission spectra, both ps and ns, and the ps excitation spectrum are very similar to those of d(CG), the most noticeable feature being that the polymer emissions are considerably quenched. The ns excitation spectrum though similar to d(CG), with the same marked linear increase at low wavelengths, differs in the region of λ_{max} in a way which indicates that either the base-planes are tilted or twisted, or we are seeing the result of absorption by a weak out-of-plane component of the strong in-plane transitions, as suggested by Ho et. al.[10] However, the major difference between d(CG) and the polymer lies in the absorption spectra which are currently under investigation (Fu and Daniels).

ACKNOWLEDGEMENTS

We thank Dr. J.-C. Brochon for facilitating our access to the UV-visible fluorescence lifetimes facility at LURE in the limited time available, also for the emission spectra correction data. Ding-Guo Hu (Beijing University and OSU) implemented the global analysis program for the lifetime data and wrote the programs for the deconvolution of the time-windowed emission and excitation spectra. Much data treatment was carried out by Ying-Jian Fu. In Paris we thank Mme Aliè Brunissen for HPLC work and much other assistance. This work would not have been possible without the support of PHS Grant 30474.

REFERENCES

1. M. Daniels and W. Hauswirth, Science 171 675 (1971).

2. M. Yamashita, S. Kobayashi, K. Toriznka and T. Sato, Chem. Phys. Letts. 137 578 (1987).

3. J.P. Morgan and M. Daniels, Photochem. Photobiol. 31 101 (1980).

4. L.P. Hart and M. Daniels, Biochem. Biophys. Res. Commun. 162 781 (1989).

5. P.-Y. Turpin and W.L. Peticolas, J. Phys. Chem. 89 5156 (1985).

6. J.-P. Ballini, M. Daniels and P. Vigny, J. Lumines 27 389 (1982).

7. J.-P. Ballini, M. Daniels and P. Vigny, Biophys. Chem. 18 61 (1983).

8. J.-P. Ballini, M. Daniels and P. Vigny, Biophys. Chem. 39 253 (1991).

9. M. Daniels, L.P. Hart, P. Shing Ho, J.-P. Ballini and P. Vigny, SPIE 1204 304 (1990).

10. P.S. Ho, G. Zhou and L.B. Clark, Biopolymers 30 151 (1990).

X-RAY EXCITED FLUORESCENCE CORRELATION SPECTROSCOPY: A NEW TOOL FOR MOLECULAR DYNAMICS STUDIES

J. Goulon, C. Goulon*, C. Gauthier, H. Emerich
European Synchrotron Radiation Laboratory, B.P. 220, F-38043 Grenoble Cedex

ABSTRACT

We explore the feasability of X-ray *Fluorescence Correlation Spectroscopy* (FCS), a technique analyzing the time-dependent fluctuations of the local concentrations of slowly diffusing macromolecules carrying n fluorescent labels. Small probe volumes ($\leq 10^{-10} cm^3$) and ultra low concentrations are a prerequisite. The small volume requirement can be met by exploiting the short penetration depth of X-rays at very grazing incidence. Two strategies are envisaged : (i) the *X-ray microprobe* option is practically hopeless due to excessive sample heating; (ii) in the *micro-collimated detection of fluorescence*, the sample contributes to a large number of fluorescent pixels, each of which has an extremely small volume. Eventhough the microcollimator causes heavy losses of intensity, it can act as an efficient polarization filter to eliminate the scattering background. The main difficulty is in the design of a CCD-type camera resolving the fluorescence signal from the background. This novel technique could be a valuable complement to FCS experiments at optical wavelengths and may have an attractive potentiality in both biophysics and chemical physics.

INTRODUCTION TO FLUORESCENCE CORRELATION SPECTROSCOPY

We wish to start with the basic concept of this technique used for more than 15 years at optical wavelengths [1-4]. The intensity of the fluorescent signal of a steadily illuminated volume V depends on the fluctuating number density of excited species at a given time τ : the fluorescence intensity $I_f(\tau)$ is thus a random process and its correlation function contains information on the local fluctuations of the concentration of species j, *i.e.* $C_j(\mathbf{r}, \tau)$, because [1,3] :

$$\langle \delta I_f(0).\delta I_f(\tau) \rangle = (gQ_j\,\varepsilon_j)^2 \iint I(\mathbf{r})\, I(\mathbf{r}')\, \langle \delta C_j(\mathbf{r}, 0).\, \delta C_j(\mathbf{r}', \tau) \rangle\, d\mathbf{r}\, .d\mathbf{r}' = K_f^{(1,1)}(\tau) \qquad (1)$$

where Q_j and ε_j denote respectively the fluorescence quantum efficiency and the extinction coefficient of the emitting species, g being a geometrical factor accounting for the solid angle in which the fluorescence photons are collected. A *normalized* autocorrelation function may be defined as[1,3] :

$$G_f^{(1,1)}(\tau) = \frac{\langle \delta I_f(0).\delta I_f(\tau) \rangle}{\langle \delta I_f(0).\delta I_f(0) \rangle} = \frac{\int dq\, |\, I(q)\, |^2\, F(q,\tau)}{\int dq\, |\, I(q)\, |^2\, F(q,0)} \qquad (2)$$

$I(\mathbf{q})$ being the spatial FT of $I(\mathbf{r})$. Assuming a fully Gaussian spatial profile of the beam at the focal spot where the sample is placed [1,3] :

$$I(\mathbf{q}) = (2\pi)^{3/2}.\,(\sigma_1^2\sigma_2)\, I_0\, \exp\{-1/2\,(q_x^2 + q_z^2)\sigma_1^2\}.\, \exp\{-1/2\, q_y^2\sigma_2^2\} \qquad (3)$$

For isotropic diffusion into and out of the illuminated volume :

* Present affiliation : University of Nancy-1, UFR des Sciences Pharmaceutiques et Biologiques, Nancy, France

$$F(q,t) = \langle | \delta C_j (q, 0)|^2 \rangle . \exp \{- q^2 . D^{(j)} . t\} \quad (4)$$

and therefore [3] :

$$G_f^{(1,1)}(t) = \left[1+\frac{t}{\tau_1}\right]^{-1} . \left[1+\frac{t}{\tau_2}\right]^{-1/2} \qquad \text{with : } \tau_j = \sigma_j^2 / D^{(j)} \quad (5)$$

the correlation time τ_j varying with the geometry of the beam. With : $\sigma_1 \approx 10^{-3}$ cm and $D \approx 10^{-6}$ $cm^2.s^{-1}$, the correlation time τ_1 is of the order of 1.0 s : this gives access to slow diffusion motions.

FCS at optical wavelengths is also sensitive to chemical relaxation processes. Let us consider for example a chemical equilibrium : $A \rightleftharpoons B$ with transformation rate constants $\{k_a ; k_b\}$ and where only B fluoresces when the system is steadily irradiated. Chemical fluctuations around the equilibrium state induce local variations of the concentration of species B so-that $G_f^{(1,1)}$ is now multiplied by another factor $G_R(t)$ which is independent of the geometry of the incident beam but characterizes the chemical relaxation process. As long as A and B have nearly the same diffusion coefficient [3]:

$$G_R(t) = \frac{1}{[C_A + C_B]} \left[C_B + C_A \exp \{ -\frac{t}{\tau_R} \} \right] \qquad \text{with : } \frac{1}{\tau_R} = k_a + k_b \quad (6)$$

Bimolecular reactions of the type $A + B \rightleftharpoons AB$ describing the binding of a small ligand to a macromolecule result in even more complicated expressions. Access to the chemical or the diffusion rates parameters is usually difficult unless one process dominates over the other one. This is where X-ray FCS and optical FCS may complement each other : X-ray fluorescence is an atomic spectroscopy and is inherently insensitive to chemical relaxation processes affecting optical spectroscopy. Alternatively, fluctuations in the X-ray excited optical Luminescence (XEOL) yield [5] could complement the information content of X-ray FCS.

A key parameter in fluorescence correlation analyses is the relative mean square fluctuation [1] :

$$\frac{[\langle [\delta I_f]^2 \rangle]^{1/2}}{[\langle I_f \rangle^2]^{1/2}} = \frac{[\langle [\delta N_j]^2 \rangle]^{1/2}}{\langle N_j \rangle} = [\langle N_j \rangle]^{-1/2} = [V C_j]^{-1/2} \quad (7)$$

Thus, the amplitude of the fluctuations of the fluorescent intensity will exceed 10^{-3} (rms) only if : (i) the solutions of diffusing macromolecules are very dilute; (ii) the excited volume is very small ($\leq 10^{-10}$ cm^3). There is a "trick" that has been exploited in optical FCS experiments [6] and that could be transposed into X-ray FCS : it consists in enhancing the sensitivity of the detection by increasing the number n of fluorescent labels bound to the diffusing macromolecule. In practice, up to n = 200 molecules of Ethidium bromide can be bound to DNA helices[6] !..

Suppose that we want to resolve $G_f^{(1,1)}(\tau)$ with a time-scale $\Delta\tau$. For stationary processes, the data acquisition time Δt can be larger than $\Delta\tau$ and is limited only by the sample photodegradation. From the $\langle N_{ev} \rangle . \Delta\tau$ detected events, we wish to extract the number of fluorescent species $\langle N_j \rangle N_a$ with sufficient precision so as to discern fluctuations of magnitude $[\langle N_j \rangle N_a]^{1/2}$ where N_a denotes the Avogadro's number, $\langle N_j \rangle$ being expressed in moles. It can be shown that [1,4] :

(i) at low count rates : $\langle N_{ev} \rangle . \Delta\tau < \langle N_j \rangle N_a \Rightarrow S/N \approx \langle N_{ev} \rangle [\Delta t \,.\, \Delta\tau]^{1/2}$

(ii) at high count rates : $\langle N_{ev} \rangle . \Delta\tau >> \langle N_j \rangle N_a \Rightarrow S/N \approx [\Delta t \,/\, \tau_c]^{1/2}$

(where τ_c is the characteristic relaxation time of $G_f^{(1,1)}(\tau)$)

Thus the S/N ratio can be improved either (i) by sacrifying the resolution $\Delta\tau$, or (ii) by cycling many times the full relaxation process. The empirical condition [4] : $\langle N_{ev}\rangle.\Delta\tau \approx \langle N_j\rangle N_a = [VC_j]\, N_a$ gives the *optimal* time scale $\Delta\tau$ to carry out *one single* measurement of the autocorrelation function for the incident flux N_{ph}. Under X-ray excitation at grazing incidences the X-ray fluorescence count rate is :

$$\langle N_{ev}\rangle.\Delta\tau \approx g_\Omega\, [g_Q]_j\, [\Delta\mu^0(C_j) / \mu_{total}]\, G_\phi \,.\, [DQE] \,.\, N_{ph} \,.\, \Delta\tau \qquad (8)$$

where $g_\Omega = \Omega/4\pi \approx 0.5$ is the solid angle collection efficiency, $[g_Q]_j$ is the quantum yield of fluorescence of the absorbing element in species j, DQE is the detector quantum efficiency ; $\Delta\mu^0(C_j)$ refers also to the absorption edge jump, μ_{total} being the absorption coefficient of the whole solution at the energy of excitation; finally G_ϕ is a factor which depends on the angle of incidence ϕ and on the amplitude of the evanescent waves[1,8] . Thus :

$$N_{ph}\,\Delta\tau \approx \frac{N_a\, V\, \sigma_i\, A_i\, \rho_{Solv}}{g_\Omega [g_Q]_j\, G_\phi\, [n\Delta\sigma_j^0]\, A_j\, M_{Solv}} \qquad (9)$$

where, for more clarity, we have explicited μ_{total} and $[\Delta\mu_j^0]$: σ_i denotes the atomic cross sections (with edge jump $\Delta\sigma_i^0$) for an absorbing element of atomic weight A_j; ρ_{Solv} and M_{Solv} are the density and the molecular weight of the solvent respectively. For water, $\sigma_i\, A_i$ refer to oxygen only.

For example [1] : $V \approx 10^{-10}cm^3$; $[M_{Solv}/\rho_{Solv}] = 18\ cm^3$

$g_\Omega[g_Q]_j \approx 0.3$; $G_\phi \approx 0.05$; $DQE \approx 1$ $\Leftrightarrow$ $N_{ph}\,\Delta\tau \approx 5.10^{12}$

$[A_j\, \Delta\sigma_j^0] / [A_{Ox}\, \sigma_{Ox}] \approx 54.4$; $n = 1$

Thus, even with the most intense Synchrotron Radiation sources, the temporal resolution $\Delta\tau$ of X-ray FCS experiments will remain rather coarse when compared to FCS experiments currently performed in the visible range ($\Delta\tau$ =30 μs). This is a consequence of the lower absorption cross section of X-rays and of the reflection losses decreasing G_ϕ. Changing the nature of the probe element does not help much. The only way to reduce $\Delta\tau$ is by decreasing V down to *ca.* 10^{-11} cm^3. This is practically hopeless unless the short penetration depth of X-rays under T.I.R. is exploited

F.C.S. EXPERIMENTS IN THE X-RAY T.I.R. REGIME

Starting with the complex refractive index : $n^* = 1 - \delta - i\beta$, Snell's law predicts that total reflection of X-rays occurs only at grazing incidences [7,8] : $\phi \le \phi_c = [2\delta]^{1/2}$. For aqueous solutions, ϕ_c is 2.64 mrad at 8.10 keV [9] or 3.94 mrad at 5.41 keV [10]. The penetration depth zz" is defined as resulting in a reduction of the *intensity* by the factor an 1/e [8] :

$$zz'' = \frac{\lambda\sqrt{2}}{4\pi\,\phi_c(\lambda)}\,\{w^2 - u^2 + 1\}^{-1/2} \quad \text{where : } u = \phi/\phi_c\ ;\ v = \beta/\delta\ ;\ w^2 = [(u^2-1)^2 + v^2]^{1/2} \qquad (10)$$

At very glancing angles ($\phi \to 0$), $zz'' \to zz_0'' = [2\pi N_a r_0\, \Sigma\rho n_j Z_j / M_w]^{-1/2}$ where M_w and ρ are the molecular weight and density, Z_j is the atomic number of atoms j present in proportion x_j, r_0 is the classical electron radius. For water : $zz_0'' \approx 46$ Å. Thus, by tuning ϕ very slightly below the critical angle ϕ_c, the penetration depth zz" could be kept below 150 Å. Typically, with a well collimated beam

(20 µm vertical x 100 µm horizontal), the illuminated volume is still 0.8 10^{-8} cm^3 if zz" = 105 Å : this suggests that a microfocus is required to reach the ultimate limit of $10^{-10}cm^3$. We will discuss an alternative solution below.

Below $\phi = \phi_c$, a standing wave field builds up *via* the interference between the incident and reflected plane waves [1,8] and this effect is included in the factor G_ϕ introduced in Eq.(8) :

$$G_\phi = |\rho_A|^2 . \frac{\xi}{u} . L_\phi . zz''^{-1} \quad \text{with :} \quad |\rho_A|^2 = \frac{4u^2}{1 + 2u\xi + 2\xi^2} \; ; \tag{11}$$

$$\text{and : } L_\phi = \int_0^\infty dz.\rho_f(z).\, e^{-[1/zz'' + \mu_f].z} \qquad \xi = \frac{1}{\sqrt{2}}\left[w^2 + u^2 - 1\right]^{1/2}$$

For a uniform distribution of the dilute probe element $\rho_f(z)$, $L_\phi . zz''^{-1} \approx 1$ and G_ϕ is given by the product : $G_\phi \approx |\rho_A|^2.\xi.u^{-1}$, where ρ_A is the normalized amplitude of the evanescent wave. Numerical simulations show that $G_\phi(u)$ will decay rapidly below the critical angle ϕ_c and the requirement to have *simultaneously* a short penetration depth and a large fluorescence intensity imposes to critically optimize the glancing angle $(0.95 \le u \le 1.01)$, with $0.02 \le G_\phi \le 0.5$.

For liquids, capillary waves driven by thermal fluctuations and propagating across the surface may be another source of problems. The surface profile $\zeta(y,t)$ can be written as [11] :

$$\zeta(y,t) = \sum_{q_r} \zeta_q(q_r,t).\, e^{-iq_r y} \qquad \text{with : } q_r = 2\pi / \lambda_r \tag{12}$$

where $\zeta_q(q_r,t)$ is the time-dependent amplitude of the viscoelastic capillary waves of characteristic ripple wavelength λ_r. Another correlation function $G_q(\tau)$ needs then to be introduced[11] :

$$G_q(\tau) = \langle\, \zeta_q(q_r,t).\, \zeta_q(q_r,t+\tau)\, \rangle = \frac{k_B T}{A(\gamma\, q_r^2 + \rho g)} . e^{-\Gamma_q \tau} . \cos \omega_q \tau \tag{13}$$

where k_B is the Boltzman constant, γ the interfacial tension, ρ the density of the liquid phase and g the gravity constant. The time-characteristic damping coefficient Γ_q and the frequency ω_q are explicit functions of the wavenumber q_r which can be calculated from the interfacial tension γ, from the density ρ and shear viscosity μ_v of the liquid phase. Schematically, the surface tension acts as a restoring force and the volume shear viscosity as a damping mechanism for capillary waves which are at the origin of the "surface roughness" that limits the reflectivity of liquids at X-ray wavelengths [9]. Hydrodynamic surface fluctuations of the solvent (with amplitudes [11] of the order of 10Å) can well interfere with FCS experiments : u is locally *a time-dependent random variable* inducing very large fluctuations of zz" and $G_\phi(u)$ and fluorescent macromolecules are randomly swept into and out of the X-ray beam *without* diffusing. Note, however, that : (i) the fluctuations of zz" should have no effect on FCS experiments if $\rho_f(z)$ is characteristic of a *monolayer* of proteins diffusing in a membrane; (ii) capillary waves being induced by the *solvent* , their characteristic times ($\le 10^{-3}$ s) are expected to be much shorter than the τ_j of the slowly diffusing macromolecules. Alternatively, the sensitivity of X-ray FCS experiments to the dynamics of capillary waves could, in itself, open new experimental perspectives in Molecular Dynamics. Note that in the latter case, no very high dilution is necessary nor even desirable because the fluctuations of the active volume should dominate over the fluctuations of the number of fluorescent species. This should result in a considerable experimental simplification.

THE X-RAY MICROPROBE OPTION

X-ray FCS clearly requires very intense sources, especially if a fine temporal resolution $\Delta\tau$ is desired. The spectral brilliance is also essential because the X-ray beam will have to be refocused at the sample. *A priori*, the excitation beam does not need to be monochromatic but it is desirable to filter out all photons of energy lower than the absorption edge because the latter contribute only to the photodegradation of the sample and to the scattering background. Thus, a (low K) X-ray undulator installed on a high β section of a high energy machine such as the ESRF appears as most suitable for FCS experiments, especially when operated with a reasonably small pinhole.

An obvious strategy is to concentrate as many X-ray photons as we can onto the smallest volume that can be optically defined. In the T.I.R. regime, the penetration depth can be reduced down to a few hundred Å but we still need to limit the beam footprint at the surface of the sample. One may think of adapting a kind of "microprobe optics", *e.g.* using cylindrically bent SiC mirrors possibly coated with multilayers and arranged in a Kirkpatrick-Baez configuration (see for instance Underwood *et al.*[12]). Submicron focusing performances could eventually be achieved with a second stage of Bragg-Fresnel Multilayer Lenses (BFML) of the type developed by Erko *et al.*[13]. Unfortunately, numerical simulations[1] have confirmed that the sample will be exposed to a very damageable heating, unless a narrow bandpass 2-crystal monochromator is inserted.

Another serious difficulty arises from the level of scattered light, which is expected to be much higher than for FCS experiments at optical wavelengths : at very high dilution, the scattering intensity will exceed the fluorescence signal by several orders of magnitude [1]. Fortunately, a scattering background (B_{sc}) that is truly uncorrelated at the time scale of fluorescence fluctuations, does not spoil the time dependence of the autocorrelation function : it merely introduces a scaling factor $1/A = \langle I_f \rangle / \langle B_{sc} \rangle$ which, nevertheless, has to be maximized[1]. In this respect, the optical configuration of an X-ray microprobe is remarkably propitious to the design of *Energy dispersive optical filters* [13-15] : this is still the most efficient solution[14] to discriminate weak fluorescence signals in situations where $A \approx 10^4$. Unfortunately, such filters are known to suffer from a poor efficiency (in terms of photon collection) and their energy tunability is most often limited with the consequence that a design needs to be optimized principally for a selected probe element. Solid-state detectors offer far more flexibility as regards tunability but they cannot respond to high fluxes : even with a (still futuristic!) 100 channel Silicon Drift Detector[17] operated at a few μs shaping time, the ultimate count rate (*i.e.* 10^6- 10^7 events) would be largely overpassed with only scattering. Note that at low energy ($E \leq 4.0$ keV), the *Bremsstrahlung* of the secondary photoelectrons will contribute anyway to a broadband background[18] that cannot be energy discriminated : at the ultimate dilutions required for X-ray FCS experiments, this background may have nearly the same intensity as the fluorescence peak[17,18] and noise correlation analyses may be the only way to discriminate the fluorescence signal.

MICROCOLLIMATED DETECTION OF FLUORESCENCE

We have proposed another strategy [1] that makes use of a Micro Channel Collimator (MCC)[19] (See fig.1) and takes advantage of the high linear polarization rate of the ESRF BL#6 source. For a plate of thickness e_z set rigourously parallel to the surface of the sample located at a distance Δz, the field of view delimited by one cylindrical channel inclined at an angle α with respect to the normal of the surface will have an elliptical section with main axes :

$$l_x = 2a = \frac{d}{\cos\alpha} . \left[1 + \frac{2\Delta z}{e_z} \right] \qquad l_y = 2b = d. \left[1 + \frac{2\Delta z}{e_z} \right] \tag{14}$$

Clearly, the field of view of each individual channel makes it possible to delimit on the sample a large number of *fluorescent pixels* all associated with an extremely small active volume provided that the sample is still excited under very grazing incidence and T.I.R. regime. The radiometrical efficiency of the plate is independent of Δz and the number of fluorescence photons transmitted *per* channel is still given by Eq. (9), but with g_Ω replaced by g_{MCC} defined as following [1]:

$$g_{MCC} \approx \frac{d^2}{16\, e_z^2} \cdot \frac{\delta S}{S} = \frac{d^2}{16\, e_z^2} \cdot \frac{\pi\, d^2}{4 \cos\alpha\, .\, S} = g_{\delta\Omega} \cdot g_{\delta S} \tag{15}$$

Here S and δS refer to the total illuminated surface at glancing angle and to the channel entrance cross section respectively. For a plate of standard thickness e_z = 0.5mm and a typical pore diameter d=20µm, $g_{\delta\Omega} = (d/4e_z)^2 \approx 10^{-4}$. One may like to reduce $\{ l_x , l_y \}$ by diminishing both d and e_z but $g_{\delta\Omega}$ should not be decreased further. Indeed, the poor efficiency of MCCs (due to the low values of $g_{\delta\Omega}$) is a dramatic penalty but which is still a reasonable price for real advantages :

(i) The MCC will act as a *polarization filter* transmitting very little scattered radiation if the exciting X-ray beam is linearly polarized with its polarization vector oriented along the direction of the channel axes. The filter efficiency is given by [1] : $g_{Filter} \approx 1/2\, (d/e_z)^2$ and is implicitely proportional to the polarization rate of the source. This is where the new helical undulator[20] developed for the ESRF BL#6 will offer remarkable advantages : (i) it will produce very intense beams of linearly polarized X-ray photons; (ii) the polarization vector can be oriented at 45° with respect to the horizontal plane; (iii) the polarization rates should be $\geq$ 99%. It is most essential for the present application that the polarization vector *is not in the horizontal plane.* The MCC should be combined to a pixel detector featuring a very good spatial resolution and, ideally, energy discrimination capability.

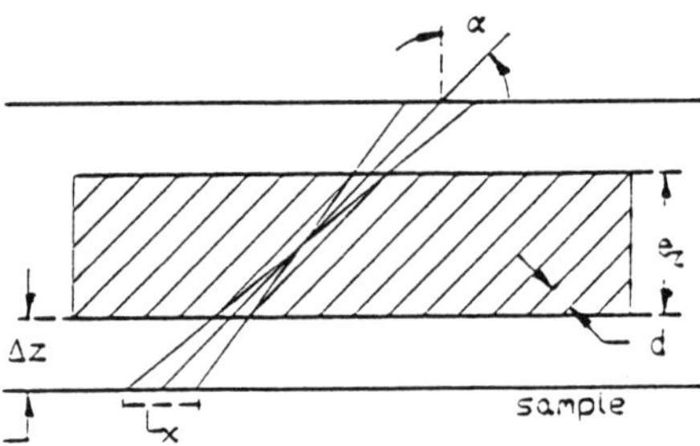

Fig. 1 : MCC with the channel pores inclined at 45° and oriented along the polarization vector

(ii) Given the small volume ($\approx 10^{-11}cm^3$) of each fluorescent pixel excited in the T.I.R. regime, Eq. (7) predicts "large" fluctuations in the number density of the diffusing macromolecules. It may be preferable, whenever possible, to raise the concentration (up to $C_j \approx 10^{-4}$ mole.l^{-1}) in order to improve the contrast over the residual scattering background. Assuming that the sample *and* the source are spatially homogeneous, the *second moment* of the fluctuations, *i.e.* $[\delta I_f(0).\delta I_f(t)]$, can be averaged over all illuminated pixels (≥ 5120) and the factor $g_{\delta S}$ get more or less eliminated. Note, however, that the temporal resolution $\Delta\tau$ still suffers from replacing g_Ω by $g_{\delta\Omega}$ in Eq. [9] : with $N_{ph} = 10^{15}$ (polychromatic) photons.s^{-1} available at the output of the undulator, $\Delta\tau \approx 25$ s. At this stage, the only practical way to improve the time resolution is by having $n \geq 1$ in Eq. [9]. With a monochromatic beam ($N_{ph} = 10^{12}$ photons.s^{-1}), the temporal resolution becomes exceedingly poor and one should reasonably forget about X-ray FCS experiments unless $n \geq 100$.

(iii) In the case of non-homogeneous systems, one could try to map the dynamics of the local fluctuations : the price to be paid is again a dramatic degradation of the temporal resolution (and the data acquisition time) which both increase $\propto g_{\delta S}^{-1}$. As far as several fluorescent pixels ($p \geq 50$) can be grouped in "local area of interest", then the experiment appears still feasible for $n \geq 10$. This is, however, out of question with a monochromatic beam.

(iv) The larger the beam footprint area S will be at the sample, the smaller the heating effects and radiation damages. With $N_{ph} = 10^{15}$ photons.s^{-1} at energy $E_{ph} = 4.0$ keV illuminating a total active area $S = 20x2\ mm^2 = 4.10^{-5}m^{-2}$, the power absorbed *per* unit area would be ≈ 0.8 kW.m^{-2} for a surface reflectivity $R_\phi = 0.95$ and, with a common sample holder, the temperature may rise by less than 4°C. This would certainly not hold true for a microprobe, due to the extremely small value of S. There should not be any significant heating effect with a monochromatic beam.

From the experimental point of view, the microcollimated detection of fluorescence is not demanding any additional exotic X-ray optics : a vertical compression of the incident beam is obviously desirable and could be obtained with a cylindrically bent SiC mirror resulting in a source demagnification of *ca.* 10. A two-crystal monochromator would reduce the energy bandpass but also enhance the linear polarization rate [21] if the plane of incidence is correctly oriented and if the Bragg angle is close to 45°. The key difficulty is indeed in the design of a pixel detector. Given the rather low count rates expected from each fluorescent pixel (*i.e.* $\leq 10^3$ events.s^{-1}), one could try to use a back illuminated CCD camera to detect directly the emitted X-rays. In a CCD, each sensing element has a very small capacitance : thus, at cryogenic temperatures, the low readout noise (4 e^- rms) makes energy discrimination possible [22,23], at least at very low counting rates. Unfortunately, the requirement to mount the CCD sensor extremely close (spacing : 200μm) to the MCC is hardly compatible with cryogenic constraints. Radiation damages in MOS structures could be another limitation [23]. Clearly, this experiment could benefit of the development of radiation hard, fully depleted pixel detectors [24] combining a high energy resolution, a good spatial resolution and a good DQE. Standard CCDs suffer from a long readout time. Recently, new devices with parallel readout channels became available. These devices could accommodate the strategy illustrated by fig.2 and which consists in : (i) detecting the fluorescence pixels in a dedicated zone (*ca.* 10 rows out of 512); (ii) using the non-illuminated rows for fast signal transfer and temporary storage.

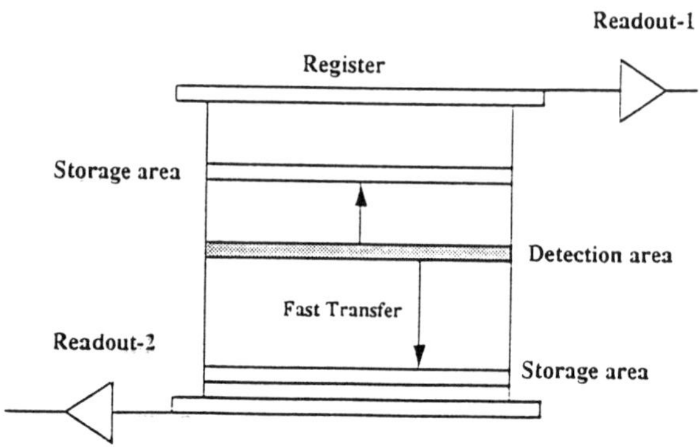

Fig. 2 : Sensing and storage area in a CCD optimized for (parallel) delayed readout.

At present, the typical transfer time requested to shift a "pixel row" being *ca.* 10μs, one could move the whole image of 10x512 pixels in less than 100μs. Transfer time even shorter than 10μs may well become realistic in the near future. This is important for reducing the dead time of the detector.

PERSPECTIVES

We have already investigated the concept of microcollimated detection of fluorescence for mapping with EXAFS or XANES the environment of a given absorber in heterogeneous samples [25]. X-ray FCS may be a valuable extension of this earlier project : we hope to extract from these experiments quantitative information on the very slow diffusion of metallic clusters on a surface, on lateral diffusion of large size proteins in oriented lipid membranes, such small diffusion constants being hardly measurable by NMR. In favourable cases, the dynamical mapping capability discussed in the previous section could be exploited to display completely immobilized patches contrasting with slowly diffusing macromolecules. Practical limitation arises unfortunately from the poor efficiency ($g_{\delta\Omega}$) of MCCs and from the reflectivity losses decreasing G_ϕ: this is why extremely intense fluxes ($N_{ph} = 10^{15}$photons.s^{-1}) are needed. XEOL-Correlation Spectroscopy certainly deserves further attention : its sensitivity to chemical relaxation processes, the very low penetration depth of the exciting beam, the high quantum efficiency of XEOL of selected phosphors, the easy discrimination between optical luminescence and X-ray scattering together with the resources of optics in the visible range could be decisive advantages.

As far as time-correlated fluctuations of the intensity of X-ray fluorescence can be detected within realistic data acquisition times under monochromatic X-ray excitation, then some additional selectivity could be transferred into fluorescence excitation spectroscopy (Fl-EXAFS) : (i) a temporal discrimination of any uncorrelated background should be possible; (ii) the characteristic correlation times of slowly diffusing macromolecules with bound fluorescent labels would be very long whereas unbound/solvated species with extremely short correlation cannot be discriminated from shot noise.

Finally, we want to lay emphasis on possible correlations between time-dependent X-ray fluorescence fluctuations and hydrodynamic surface capillary waves : such correlations could perhaps stimulate specific experiments at liquid interfaces.

REFERENCES

[1] J. Goulon, C. Goulon, C. Gauthier, H. Emerich, ESRF Int. Report, JG/XAS/02-91, (1991)

[2] D. Magde, E. Elson and W.W. Webb, Phys. Rev. Lett., **29**, 705-8, (1972),

[3] B.J. Berne, R. Pecora, *"Dynamic Light Scattering, Applications to Chemistry, Biology and Physics "* Wiley-Interscience Pub., (1976)

[4] D. Magde, in *" Chemical Relaxation in molecular Biology "* ed. by I. Pecht and R. Rigler, Springer Ser. in Mol. Biology, Biochemistry and Biophysics, **24**, 43-83, (1977)

[5] A.P. D'Silva, V.M. Fassel, Anal. Chem., **45**, 542-47, (1972)

[6] B.A. Scalettar, J.E. Hearst, M.P. Klein, Macromol. , **22**, 4550-59, (1989)

[7] L.G. Paratt, Phys. Rev., **95**, 359-69, (1954)

[8] B. Lengeler, IFF - Ferienkurs / KFA (Jülich), Synchrotronstrahlung in der Festkörperforschung Chapt. **18**, (1988)

[9] A. Braslau, M. Deutsch, P.S. Pershan, A.H. Weiss, J. Als-Nielsen, J. Bohr, Phys. Rev. Lett. **54**, 114-17, (1985)

[10] L. Bosio, R. Cortès, G. Folcher, M. Oumezine, Rev. Phys. Appl. **20**, 437-43, (1985)

[11] L.B. Shih, Rev. Sci. Instrum., **55**, 716-26, (1984)

[12] J.H. Underwood, A.C. Thomson, Y. Wu, R.D. Giauque, N.I.M. , **A266**, 296-302, (1988)

[13] A. Erko, E. Khzmalian, L. Panchenko, S. Redkin, V. Zinenko, P. Chevallier, P. Dhez, C. Khan Malek, A. Freund, B. Vidal, to be published in N.I.M. (1991)

[14] J.B. Hastings, P. Eisenberger, B. Lengeler, M.L. Perlman, Phys. Rev. Lett., **43**, 1807-10, (1979)

[15] M. Marcus, L.S. Powers, A.R. Storm and B.M. Kincaid, Rev. Sci. Instrum., **51**, 1023-29, (1980)

[16] I.G. Grigoryeva, A.A. Antonov, V.B. Baryshev, Synchr. Rad. News, **3**, 15-17, (1990)

[17] C. Gauthier, *Ph. D. thesis* , (1991)

[18] F.S. Goulding, J.M. Jaklevic, N.I.M., **142**, 323-32, (1977)

[19] N. Yamagushi, S. Aoki, S. Miyoshi, Rev. Sci. Instrum., **58**, 43-44, (1987)

[20] P. Elleaume, ESRF Internal Report SR/ID 88-23 (1989); N.I.M, **A291**, 371-77, (1990)

[21] C. Malgrange, C. Carvalho, L. Braicovich, J. Goulon, N.I.M., in press (1991)

[22] J. Janesik, T. Elliott, R. Bredthauser, C. Chandler, B. Burke, SPIE, **982**, 70-95, (1988)

[23] D.H. Lumb, A.D. Holland, SPIE, **982**, 116-122, (1988)

[24] L.D. Strüder, H. Bräuninger, M. Meier, P. Predehl, C. Reppin, M. Sterzik and J. Trümper P. Cattaneo, D. Hauff, G. Lutz, K.F. Schuster, A. Schwarz, E. Kenziorra, A. Staubert, E. Gatti, A. Longoni, M. Sampietro, V.Radeka, P. Rehak, S. Rescia, P.F. Manfredi, W. Buttler, P. Holl, J. Kemmer, U. Prechtel, T. Zieman., N.I.M. , **A288**, 227-35, (1990)

[25] J. Goulon, Physica, **B158**, 5-13, (1989)

CROSSLUMINESCENCE FOR DETECTION

OF STRUCTURAL PHASE TRANSITION IN $RbCaF_3$

I.A.Kamenskikh
Moscow State University, 117234, Moscow, USSR

M.A.MacDonald, I.H.Munro
Daresbury Laboratory, Warrington, WA4 4AD, U.K.

M.A.Terekhin
Kurchatov Institute of Atomic Energy
123182, Moscow, USSR

ABSTRACT

Crossluminescence excitation and decay for fluorperovskite $RbCaF_3$ has been studied in the VUV region. It has been shown that decay curves were not single exponential. The effect of structural phase transition on the crossluminescence decay time of $RbCaF_3$ has been discovered.

INTRODUCTION

Optical intrinsic fluorescence of some wide band gap crystals (BaF_2, CsCl, RbF, etc.[1-5]) originating from radiative transitions between the valence band and the upper core-band has been discovered quite recently. Crystals demonstrating this type of luminescence, called crossluminescence (or Auger-free luminescence), are promissing materials for scintillation technique[6], hence detailed investigation of their properties seems to be a vital task. It has been shown[7] that the fluorperovskite $RbCaF_3$ demonstrates a higher yield of crossluminescence compared to binary crystal RbF. According to a conventional physical model of crossluminescence, a single exponential decay is expected, determined by the lifetime of a core hole.

Many perovskite crystals ABX_3 are known to have succesive structural phase transitions on cooling. In the case of $RbCaF_3$ the first one occurs at T_c=193 K, and it is a first-order phase transition $O_h^1(Pm3m) \rightarrow D_{4h}^{18}(I4/mcm)$[8], relating to the condensation of the R_{25}^z-type soft phonon, as in $SrTiO_3$ at 106 K.

Here we report on the investigation of excitation spectrum and decay for the 250 nm band[7] of crossluminescence $RbCaF_3$. It seem us interesting to see how these structural changes effect on decay time. Measurements of crossluminescence of $CsCaCl_3$ and $CsSrCl_3$ did not reveal any changes due to structural phase transitions[9].

EXPERIMENTAL TECHNIQUES

The measurements have been performed on the Seya-Namioka monochromator at the beamline 3.1 of SRS, Daresbury Laboratory, U.K. Interference filters 250nm have been used for separation of luminescence band. Both multi-bunch and single-bunch modes of machine operation have been used: in the first case we measured luminescence excitation spectra, in the second one - luminescence decay. Time structure of synchrotron radiation from SRS, characterised by ~200 ps bunches separated by 320 ns together with single photon counting technique using XP2020Q photomultiplier allowed us to measure lifetimes less than hundred of picoseconds[10].

RESULTS AND DISCUSSION

Fig.1 shows crossluminescence excitation spectra. First forbidden energy gap of $RbCaF_3$ E_g=10.8eV has been evaluated from impurity luminescence excitation spectrum[7]. The onset of crossluminescence excitation is observed when the energy of exciting photons becomes large enough to excite electrons from the upper core band (fig.2,a).

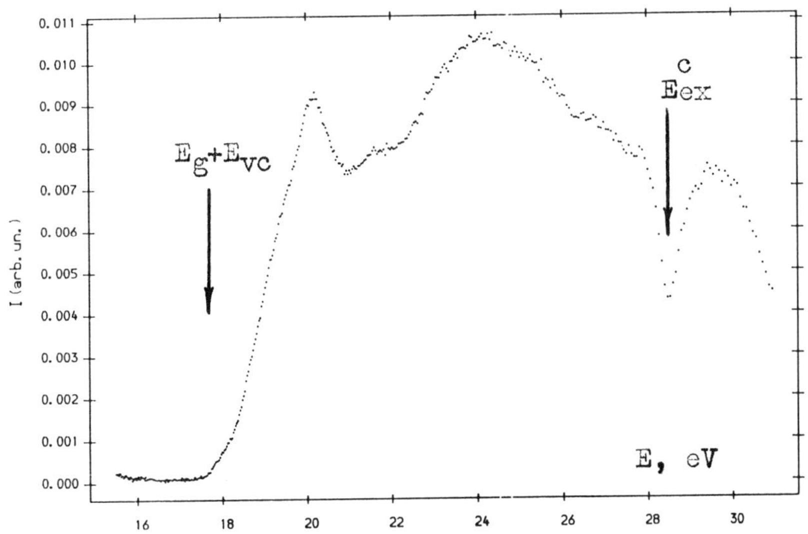

Fig.1. Luminescence excitation spectrum of $RbCaF_3$ at 300 K, interference filter with λ_{max}= 250nm.

It starts at $E_g + E_{VC} = 17.6$ eV, where E_{VC} - is the difference between of the tops of valence and upper core band.

A remarkable feature of $RbCaF_3$ excitation is a sharp dip with a minimum at 28.52 eV. Its energy location is close to the position of a $(3p,Ca^{2+})$-core exciton in the reflectivity spectrum of CaF_2 crystal[11] but it is too deep to be caused by reflection or surface losses. It is more natural to attribute it to the existance of another relaxation channel for calcium core-excitons, not leading to the creation of holes at the $(4p,Rb^{+})$-core band. On cooling the depth and width of the dip remain almost the same in a good accord with the results[12] where it has been shown for CaF_2 that the main mechanism of decay for $(3p,Ca^{2+})$-core excitons (27.7 eV) should be Auger-process. For $RbCaF_3$ energy separation between $E^c_{ex} = 28.52$ eV and the onset of crossluminescence excitation at $(E_g + E_{CV})$ provides information about the energy difference between the tops of $(4p,Rb^{+})$- and $(3p,Ca^{2+})$-core bands. Its value ~ 11 eV is comparable with $E_g = 10.8$ eV, so the condition for the Auger-decay of a hole from $(3p,Ca^{2+})$-core band with the creation of a hole at $(4p,Rb^{+})$-core band (required for the existance of crossluminescence) and another one in the valence band is satisfied at its limit, Auger relaxation with the creation of two holes in the valence band will dominate (fig.2,b).

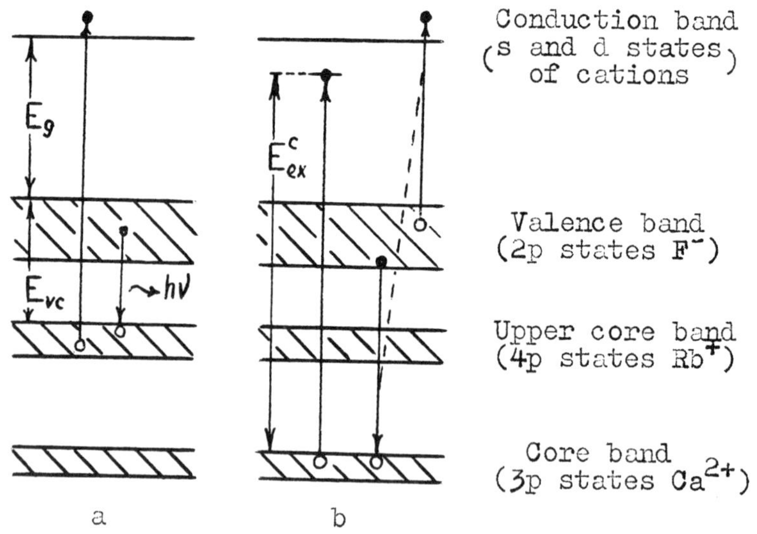

Fig.2. Schematic energy diagrams showing the relaxation processes of $(4p,Rb^{+})$-core hole (a) and $(3p,Ca^{2+})$-core exciton (b) in $RbCaF_3$.

In the single-bunch mode decay times of crossluminescence of $RbCaF_3$ have been measured using single photon counting technique. The time structure of the exciting radiation was measured as zero-order scattered visible light signal, providing the promt spectrum P(t). For non of the mearesurements a single-exponential decay curve has been observed. Much better fitting of the experimental data was obtained using two exponents:

$$L(t)=\int_0^t P(x)[A_1\exp(-(t-x)/\tau_1) + A_2\exp(-(t-x)/\tau_2)]dx. \quad (1)$$

Typical shape of the decay curve together with the two-exponential best fitting and promt spectrum is presented in fig.3. Table 1 summarizes fitting parameters for three measurements at various temperatures.

Table 1. Parameters of two-exponential fitting of crossluminescence decay for $RbCaF_3$ under 24 eV photon excitation.

T (K)	τ_1(ns)	τ_2(ns)	A_2/A_1
300	0.84	2.76	2.50
225	0.92	2.75	2.43
165	0.64	2.54	1.83
87	0.64	2.26	1.28

Results obtained are not consistent with the simple physical model for "pure" crossluminescence, not quenched by other processes. Unfortunately in the papers dedicated to crossluminecsence the authors usually present only the values of decay times without comments on the exact decay law, so we can not state that the observed discrepency is characteristic only for this compound. However, non-exponential decay have been observed also for BaF_2 where fast luminescene was excited with X-rays from a wiggler of Novosibirsk storage-ring VEPP-3[13]. Decay law is explained by the authors by the interaction of the core holes with fast defects or free electrons produced by the radition. Both reasons can scarcely be applied in our case due to low energy and flux of exciting radiation. From our point of view the processes of lattice relaxation in the presence of a core hole should be taken into account. The decay time of crossluminescence seems to be depend on the energy of excitation or, saying with other words, on the energy of created electronic excitations. As this difference is not large to be sure in effect one requied more accurate and systematic measurements.

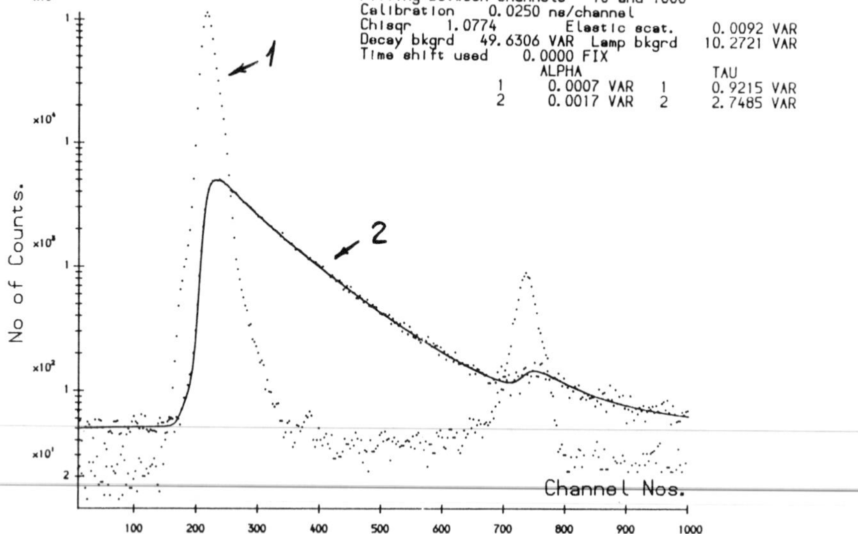

Fig.3. Fitting of $RbCaF_3$ crossluminescence decay curve, excitation energy 24eV, T= 225 K: curve 1- promt spectrum, 2(doted)- decay, 2(solid)- calculated fit.

For temperatures lower than T_C= 193 K a steep decrease of decay time is observed (fig.4). Further decrease of for $T < T_C$ is in a good accord with the structural changes : during the phase transition octahedra CaF_6 in unit cell of $RbCaF_3$ rotate by the angle φ^o which depends on the temperature[8]. The rotation of the octahedra results in the changes in Rb^+- F^- distance leading to the modification of the overlap integral for wave functions (4p,Rb^-) and (2p,F^+). Using their asymptotic expression valid for $\gamma r >> 1$:

$$\Psi_{nl}(r) = B(r)\exp(-\gamma r) \quad , \qquad (2)$$

where B(r) is a function slowly changing with distance, we can roughly evaluate this modification. Each Rb^- ion has 12 nearest-neighbour fluorine ions, after the phase transition the distance to four of them remains the same-r = 3.15 Å, for other four F^- ions it becomes shorter - r = 2.93 Å, and for the rest four is longer: r = 3.40 Å. It can be shown that the probability W of the transition of an electron from F^- to Rb^+ is :

$$W \sim \exp(-2r \cdot \min(\gamma_1, \gamma_2)). \quad (3)$$

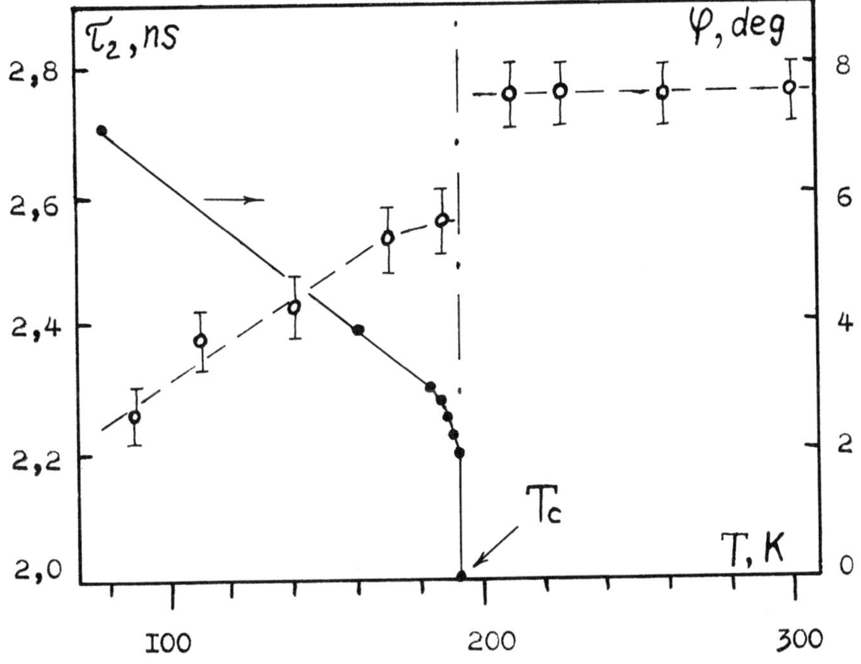

Fig.4. ⌶ - Temperature dependence of crossluminescence decay time for $RbCaF_3$.
● - Temperature dependence[8] of the rotation angle φ of the octahedra CaF_6 in $RbCaF_3$.

According to ref.14 for F^- is equal to 0.5, for Rb^+ - to 1.416. Averaging over all nearest-neighbor Rb^+ - F^- pairs we obtain that after the phase transition the probability increases by 5%. This is in reasonable agreement with experimentally obtained 7% of lifetime decrease at 190 K (fig.4).

CONCLUSION

In this report we showed for the first time that the measurements of crossluminescence decay time seem to be more sensitive to structural transitions than luminescnece spectra and may be used for detection of structural phase transitions.

ACKNOWLEDGMENTS

The authors are very grateful to Dr.A.N.Vasil'ev and Prof.P.A.Rodnyi for valuable discussions. We also thank Dr.D.Shaw for his assistance in time-resolved measurements. Data analysis has been made using the programs modified by M.Hayes.

REFERENCIES

1. Yu.M.Aleksandrov, V.N.Makhov, P.A.Rodnyi et al.,- Solid State Phys. 26 (1984) 1734 (In Russ.).
2. J.L.Jansons, V.J.Krumins, Z.A.Rachko, J.A.Valbis,- Phys.Stat.Sol.(b) 144 (1987) 835.
3. S.Kubota, M.Itoh, J. Ruan et.al.,- Phys.Rev.Lett. 60 (1988) 2319.
4. A.V.Golovin, P.A.Rodnyi, M.A.Terekhin,- ZhTF Pis'ma 15 (1989) 29 (In Russ.).
5. P.A.Rodnyi, M.A.Terekhin,- Phys.Stat.Sol.(b) 166 (1991) 283.
6. P.Schotanus, C.W.E. van Eijk, R.W.Hollander, J.Pijpelink,- IEEE Trans. Nucl.Sci. NS-34 (1987) 272.
7. P.A.Rodnyi, M.A.Terekhin and S.V.Petrov,- Solid State Phys. 32 (1990) 3171 (In Russ.).
8. M.Hidaka, S.Maeda, J.S.Story,- Phase Transitions 5 (1985) 219.
9. Ya.A.Valbis, Z.A.Rachko, Ya.L.Yansons et al.,- In: Radiation-Stimulated Processes in wide band-gap materials (in Russ.), Latvian State University, Riga 1987, p.72.
10. I.H.Munro, D.Shaw, G.R.Jones and M.M.Martin,- Analytical Instr. 14(3&4) (1985) 465.
11. G.W.Rubloff,- Phys. Rev. B5 (1972) 662.
12. H.Wiesner, B.Hoenerlage,- Z.Physik 256 (1972) 43.
13. A.N.Belskiy, V.V.Mikhailin, A.L.Rogalev.,et al.,- In: Conference Proceedings Vol.25 "2nd European Conference on Progress in X-ray Synchrotron Radiation Research" SIF, Bologna 1990, p.797.
14. E.Clementi,- IBM J. Res. Develop. Suppl. 9 (1965).

TIME-RESOLVED FLUORESCENCE STUDY OF DYNAMICS PARAMETERS IN BIOSYSTEM

J.C. Brochon, F. Merola
L.U.R.E., CNRS-CEA-MENJS, Centre Universitaire Paris-Sud, Orsay F91405

A.K. Livesey*
MRC and DAMTP, University of Cambrige, U.K

ABSTRACT

The time-resolved fluorometry can be used to probe the rather complex biological macromolecule structure in solution. The interrelationship between structure variability, structure dynamics in the nanosecond time scale, and biological function are studied in measuring spectroscopic and rotational dynamics parameters of the fluorescent moiety in biosystems. The time-correlated single photon counting technique, using synchrotron radiation as an excitation pulse, has been used to measure the decays of polarised components of fluorescence. This technique has a very high sensitivity that so extremely low quantum yield or very dilute material can be measured. Furthermore, the statistical errors on the data obeys Poisson statistics. A description of the successful analysis of polarised pulse-fluorescence data by maximum entropy method, MEM, is given. The determination of fluorescence lifetime distributions provide insights into the local environment of fluorophores and the conformational heterogeneity and intermolecular interactions in biosystem. We also demonstrate that MEM is able to extract full and correct description of conformational dynamics behaviour, in term of rotational correlation time distributions of an heterogeneous fluorescent emission. However, it is shown that the data still leave some ambiguity in the allowable distributions and that the MEM results reflect this ambiguity. Several applications on single tryptophan proteins are presented.

INTRODUCTION

Time-resolved fluorescence spectroscopy is now a classical method for studying macromolecules in solution and in particular biological systems such as proteins, membranes and nucleic acids. The diversity of physical and chemical environments of fluorophores in biological molecules often leads to an heterogeneous emission pattern. Recently the use of the maximum entropy method (MEM) of analysis made a significant advance in the understanding of the photophysics of such molecules[1,2,3,4,5] . Fluorescence measurements also play an important role in studying **the rotational dynamics** of biological molecules which can be related to their activity[6] .By using a linearly polarized pulsed excitation light and measuring the parallel and perpendicular components decays of the emission one can monitor

* present address: Thornton SHELL research centre, CHESTER, U.K.

the orientational dynamics in solution of the entire macromolecule and local segmental motions of the emiting species. This technique can measure rotational correlation times ranging from a few picoseconds to a several times the lifetime of emitting species which itself ranges from picoseconds to hundreds of nanoseconds. The single-photon correlated counting technique has a very high sensitivity so that extremely low quantum yield or very dilute material can be measured. Furthermore, the statistical errors on the data are accurately determined because the noise obeys Poisson statistics.

Most current analyses of these data assume that all fluorescent species have identical rotational dynamics (correlation times). We introduce a new method of analysis for polarised pulse-fluorimetry based on the Maximum Entropy Method (MEM). It is an extension of the total fluorescence analysis we have already reported [1,2,3]. MEM is able to analyse polarised fluorescence data without setting any limits on the number, shape and positions of fluorescence lifetime and rotational correlation time distributions. We show however that the data leave some ambiguity in the allowable distributions and that the MEM results reflect this ambiguity. The analysis of polarised fluorescence in terms of distributions is now able to extract full and correct description of conformational dynamics behaviour of heterogeneous fluorescent species.

THE INTERPRETATION OF PULSE-FLUORIMETRY DATA

The measurement technique is, in principle, deceptively simple. An infinitely sharp flash of vertically polarised light is passed through a solution of macromolecules which has one or more natural fluorescing centres (e.g. tryptophan) or has had then chemicaly attached. The fluorescing centres are excited and after a characteristic time, depending on the local chemical and physical environment, decay and produce a photon of a longer wavelength. In the meantime, the fluorescing centre may have rotated or flexed from its original position so that the relative polarisation of the emitted signal is now a function both of the angle between the absorption and emission dipoles and the rates of rotation and flexion. We measure the parallel I_{vv} and perpendicular I_{vh} components of this emitted signal. In practice, the exciting flash has a finite width compared to the time scale of the phenomenon under study and so the sum of exponentials has to be convoluted with the known (by measurement) excitation temporal profile $E_\lambda(t)$:

$$I_{vv}(t) = \frac{1}{3} E(t) *[\int_0^\infty \int_0^\infty \int_{-0.2}^{0.4} \gamma(\tau,\theta,A)\, e^{-t/\tau} (1 + 2A\, e^{-t/\theta})\, d\tau\, d\theta\, dA] \quad (1)$$

$$I_{vh}(t) = \frac{1}{3} E(t) *[\int_0^\infty \int_0^\infty \int_{-0.2}^{0.4} \gamma(\tau,\theta,A)\, e^{-t/\tau} (1 - A\, e^{-t/\theta})\, d\tau\, d\theta\, dA] \quad (2)$$

where $\gamma(\tau,\theta,A)$ are the number of fluorophores with fluorescence decay τ, rotation correlation time θ, and initial amplitude of anisotropy A (related to angle between absorption and emission dipoles). Symbol * denotes the convolution with time.

Although we fundamentally wish to analyse the distribution in terms of a 3-dimensional distribution of density $\gamma(\tau,\theta,A)$ we only get limited information about the initial anisotropy A. We can re-write equation 1,2 as:

$$I_{vv}(t) = \frac{1}{3} E_{\lambda}(t) * \{ \int_0^{\infty}\int_0^{\infty}\int_{-0.2}^{0.4} \gamma(\tau,\theta,A)\, e^{-t/\tau} d\tau d\theta dA$$
$$+ \frac{2}{3} E_{\lambda}(t) * \{ \int_0^{\infty}\int_0^{\infty}\int_{-0.2}^{0.4} A\,\gamma(\tau,\theta,A)\, e^{-t/\tau} e^{-t/\theta} d\tau d\theta dA \} \quad (3)$$

$$I_{vh}(t) = \frac{1}{3} E_{\lambda}(t) *\{ \int_0^{\infty}\int_0^{\infty}\int_{-0.2}^{0.4} \gamma(\tau,\theta,A)\, e^{-t/\tau} d\tau d\theta dA$$
$$- \frac{1}{3} E_{\lambda}(t) * \{ \int_0^{\infty}\int_0^{\infty}\int_{-0.2}^{0.4} A\,\gamma(\tau,\theta,A) e^{-t/\tau} e^{-t/\theta} d\tau d\theta dA \} \quad (4)$$

The first term of the equations tell us nothing about the structure of γ along the anisotropy axis A. In both of the second term A only appears as the product $A\gamma(\tau,\theta,A)$. Thus we only have information on

$$\int_{-0.2}^{0.4} A\,\gamma(\tau,\theta,A)\, dA$$

Formally this reduces our problem to the determination of two lower 2-dimensional projected distribution β of $\gamma(\tau,\theta,A)$ on the (τ,θ) plane:

$$\beta(\tau,\theta) = \int_{-0.2}^{0.4} \gamma(\tau,\theta,A)\, dA \quad (5)$$

The second term cannot distinguish between points which lie on a curve, (iso-κ),in the (τ,θ) plane whose τ and θ values are such that $1/\tau + 1/\theta = 1/\kappa$ (κ is a constant determined by the data).

Total fluorescence

If we are only interested in the fluorescence decay constants τ we can considerably simplify the data by summing the parallel and (twice) the perpendicular components.

$$T(t) = I_{vv}(t) + 2\, I_{vh}(t) = E_{\lambda}(t) * \int_0^{\infty} \alpha(\tau)\, e^{-t/\tau} d\tau \quad (6)$$

where $\alpha(\tau)$ is the distribution of fluorescence decays given by :

$$\alpha(t)=\int_{-0.2}^{0.4} A\ \gamma(\tau,\theta,A)\ dA \qquad (7)$$

T(t) can also be measured in one experiment by setting the polariser at the "magic" angle of 54.75°.

COMPARISON WITH CURRENT METHOD OF ANALYSIS

At the moment many fluorometry experimenters use the sum T(t) and ratio Q(t) functions to analyse their data.

$$Q(t) = \frac{I_{vv}(t) - I_{vh}(t)}{I_{vv}(t) + 2I_{vh}(t)} \qquad (8)$$

First the allowed τ values are extracted from the sum and then assuming that all the θ values have the same structure, this spectrum is used in equations 1,2 to derive the allowed θ values. This gives a biased representation of the fluorescence anisotropy decay.

$$Q(t) = \frac{E_\lambda(t) * \int_0^\infty \int_0^\infty \int_{-0.2}^{0.4} A\ \gamma(\tau,\theta,A)\ e^{-t/\tau}\ e^{-t/\theta}\ d\tau\ d\theta\ dA}{E_\lambda(t) * \int_0^\infty \alpha(\tau)\ e^{-t/\tau}\ d\tau} \qquad (9)$$

Implicitly they assume that the structure in the (τ, θ) plane is fully correlated i.e that the probability for a given chromophore having a given correlation time θ is independant of its fluorescence lifetime τ. In addition, they assume that all chromophores have the same A value. In this model, the population distribution $\gamma(\tau,\theta,A)$ can be written as the product of two independant functions:

$$\beta(\tau,\theta) = \alpha(\tau)\ \rho(\theta) \qquad (10)$$

where $\alpha(\tau)$ is the previously determined distribution of fluorescence lifetimes and $\rho(\theta)$ is the distribution function of rotational correlation times. Then Q(t) can be rewritten as

$$Q(t) = \int_0^\infty \rho(\theta)\ e^{-t/\theta}\ d\theta \qquad (11)$$

From (9) and (10) it follows that $\rho(\theta)$ is normalized:

$$\int_0^\infty \rho(\theta)\ d\theta = A \qquad (12)$$

and the complete expressions (1) and (2) of the polarized components can be re-written as:

$$I_{vv}(t) = \frac{1}{3} E_\lambda(t) * \{ \int_0^\infty \alpha(\tau)\, e^{-t/\tau} d\tau \; (1 + 2 \int_0^\infty \rho(\theta)\, e^{-t/\theta} d\theta) \} \quad (13)$$

$$I_{vh}(t) = \frac{1}{3} E_\lambda(t) * \{ \int_0^\infty \alpha(\tau)\, e^{-t/\tau} d\tau \; (1 - \int_0^\infty \rho(\theta)\, e^{-t/\theta} d\theta) \} \quad (14)$$

THE MAXIMUM ENTROPY METHOD

We wish to determine the relative properties of species, $\gamma(\tau,\theta,A)$ with decay constants (τ_i) rotational time constants (θ_j) and initial anisotropies (A_k) having measured, (inevitably), sampled, noisy, and incomplete representations of the temporal polarised emitted light and flash profiles. At the kernel of the data analysis we need the inverse Laplace transform of the measured light deconvoluted by the flash E(t). Inverting the Laplace transform is a very ill-conditioned problem. As a result, small errors in the measurement of the fluorescence curves and the flash profile can lead to very large errors in the reconstruction of $\gamma(\tau,\theta,A)$.

We can view this ill conditioning, which leads to a multiplicity of allowable solutions, in a different way. Consider the set A of all possible shapes of the curves $\gamma(\tau,\theta,A)$ displayed as a rectangle in Figure 1. We can calculate "mock" data sets form each $\gamma(\tau,\theta,A)$ in turn and test where they agree with the noisy data set. All those $\gamma(\tau,\theta,A)$ that agree with the data within the experimental precision are bounded by a dot-dashed line. Some of these curves however are unphysical (e.g. contain negative numbers), and can be rejected (by the dashed line).

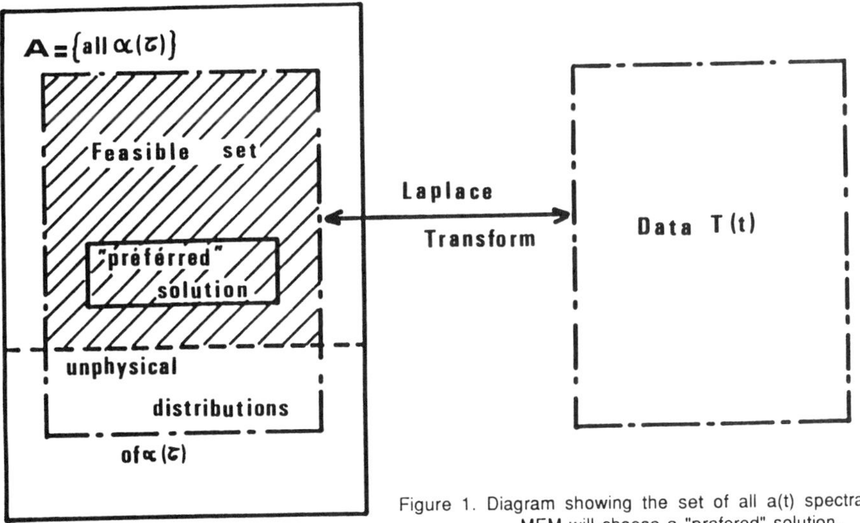

Figure 1. Diagram showing the set of all a(t) spectra. MEM will choose a "prefered" solution.

The remaining subset of spectra (shown shaded) we call the feasible set. Every member of this set agrees with the data and is physically allowable.Since the feasible set is infinite, we are forced to choose one member and we do so directly by maximising some function $M[\gamma(\tau,\theta,A)]$ of the spectrum. This function is chosen so that it introduces the fewest artifacts into our resulting distribution. It has been proved by Livesey and Skilling[7] that only the Skilling-Jaynes[8,9] entropy function, S, will give the least correlated solution in $\gamma(\tau,\theta,A)$. The function is defined as :

$$S = \int_0^{\infty}\int_0^{\infty}\int_{-0.2}^{0.4} \gamma(\tau,\theta,A) - m(\tau,\theta,A) - \gamma(\tau,\theta,A)\log\frac{\gamma(\tau,\theta,A)}{m(\tau,\theta,A)}\, d\tau\, d\theta\, dA \quad (15)$$

where $m(\tau,\theta,A)$ is the prior model for the distribution. For the total fluorescence, the entropy is also independent of θ and A and is given by:

$$S = \int_0^{\infty} \alpha(\tau) - m(\tau) - \alpha(\tau)\log\frac{\alpha(\tau)}{m(\tau)}\, d\tau \quad (16)$$

and for the fully correlated anisotropy analysis, an entropy function independant of τ and A is defined:

$$S = \int_0^{\infty} \rho(\theta) - m(\theta) - \rho(\theta)\log\frac{\rho(\theta)}{m(\theta)}\, d\theta \quad (17)$$

where m is the model which encodes our prior knowledge about the system before the experiment. The user may sometimes have some prior model[1,2] for the distribution but if not, it can be shown that the correct model is flat in $\log\tau$, $\log\theta$ and $\cos\delta$ space where $A = (3\cos^2\delta - 1)/5$
We chose to bound the feasible set by a chi-squared statistic.

$$\sum_{k=1}^{M}\frac{(I^{calc}_{k,vv} - I^{obs}_{k,vv})^2}{\sigma^2_{k,vv}} + \sum_{k=1}^{M}\frac{(I^{calc}_{k,vh} - I^{obs}_{k,vh})^2}{\sigma^2_{k,vh}} \leq 2M \quad (18)$$

where I^{calc}, I^{obs} are the kth calculated and observed intensities, σ^2_k is the variance of the kth point, and M is the total number of points in each components.

MEASUREMENTS

Polarised fluorescence decays were measured by the time correlated single photon counting technique[10,11,12].

γ-Cardiotoxin and Erabutoxin proteins were prepared from the total venom of Naja nigricollis[13]. An additional HPLC step was necessary to completely eliminate agregated forms of the cardiotoxin protein. The samples were sieved before

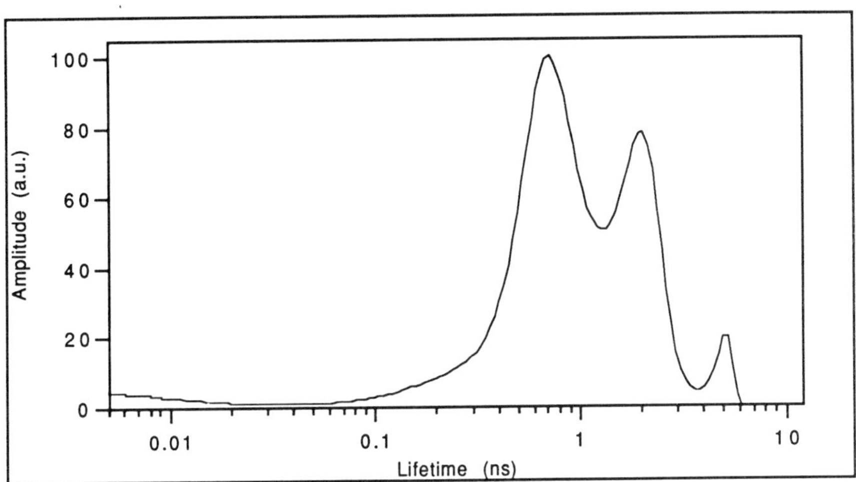

Figure 2: Lifetime distribution of Cardiotoxine total fluorescence decay.

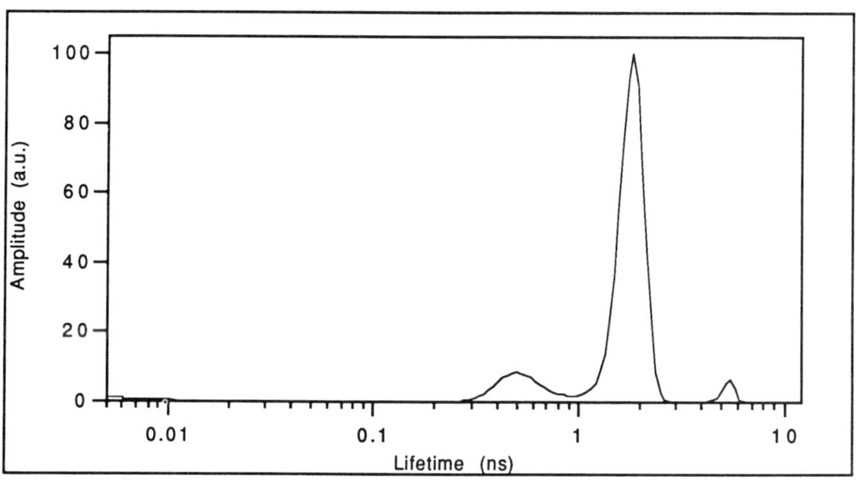

Figure 3: Lifetime distribution of Erabutoxine total fluorescence decay.

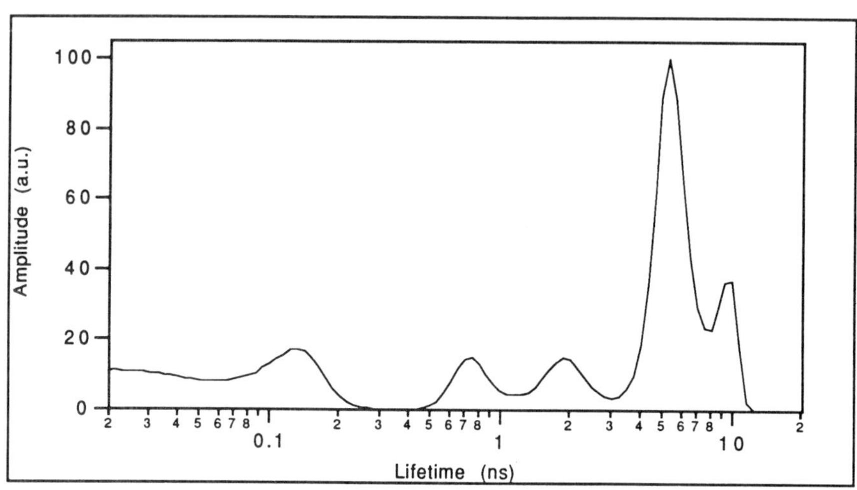

Figure 4: Lifetime distribution of HSA total fluorescence decay at 20°C

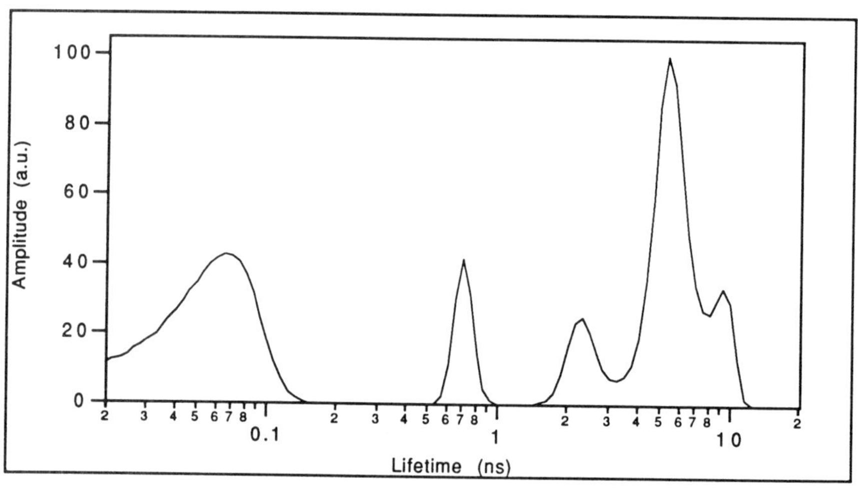

Figure 5: Lifetime distribution of HSA total fluorescence decay at 52°C

experiments on a Sephadex G25 column equilibrated with a 10mM cacodylate buffer pH 7, 10 mM NaCl, 0.1 mM EDTA. The data were recorded at 20°C with an excitation at 300nm ($\Delta\lambda$=6nm) and fluorescence was detected through a monochromator at 347nm ($\Delta\lambda$=6nm). The relative levels of scattered light and buffer fluorescence were detected to be negligible in these conditions. About 3.10^7 total counts were stored for each experiment.

Human serum albumin from Sigma was disolved in sodium phosphate 0.05M, NaCl 0.1M, pH 7.4 and purified by HPLC to remove oligomers. THe fluorescence was filtered by a 1M CuSO4 (path 1cm) to eliminate any scattered excitation light. Fluorescence decays were measured, at 20° and 52°C, using the synchrotron radiation, emitted by the positron storage ring SUPER-ACO at Orsay, as the excitation light pulse. The repetition rate was 8.3 Mhz corresponding to the double bunch mode. The measured FWMH of the excitation was routinely 500 picoseconds for a stored current of 50mA. The detector was also a Hamatsu microchannel plate photomutiplier.

Ethidium bromide (2,6-diamino-10-ethyl-9-phenyl phenanthridinium bromide) and glycerol were used without any further purification .The oligonucleotide $pd(CG)_3$ was from P-L Biochemicals. The measurements were done in a mixture of Tris Buffer-Glycerol 55-45% w/w at 0.2°C. The buffer was Tris-Hcl 10mM NaCl 0.1M pH7.2. Under these conditions all the ethidium bromide is specifically bound to the double-stranded part of the oligonucleotide. The ratio of molecular concentration was determined by absorbance. There is no more than one molecule of dye bound to each molecule of oligonucleotide. For the corresponding experiments, the synchrotron radiation from the electron storage ring A.C.O. at Orsay was the excitation light pulse. The repetition rate was 13.6 Mhz (single bunch mode). The photons from both free and oligonucleotide-bound ethidium bromide were collected from each compartement of a double compartment quartz cell, all faces having been optically polished. The excitation was 520nm ($\Delta\lambda$=4 nm) the fluorescence was selected through a cut-off filter (603nm) which eliminated any scattered light. The detector was a Philips-RTC xp2020Q photomutiplier. Temperature was 0.2 °C regulated to +/- 0.1°C.

RESULTS

Lifetime distribution in proteins:

γ-Cardiotoxin and Erabutoxin each contain only one tryptophan residue at position 11 and 29 respectively and their structures are very similar. In cardiotoxin protein, the total fluorescence decay was multiexponential as shown in figure 2. However, the lifetime distribution for Erabutoxin is more homogeneous since the main peak accounts for 84% of the intensity (figure 3). In addition, the cardiotoxin displays fast local rotational dynamics allowing the tryptophane residue to "probe" various local environments. Not surprisingly the Erabutoxin appears as a rather rigid body. The fluorescence heterogeneity of cardiotoxin may be related to its ability to flex 3-dimensional structure in the subnanosecond time range.

Human serum albumin contains only one tryptophan residue. The total fluorescence intensity decays were mutiexponential. MEM analysis of T(t) showed the existence of five well resolved excited state lifetimes at 20°C (figure 4) with a major contribution (52%) of the peak centered at 5.26ns . At 52°C the lifetime distribution is rather broads, four families of lifetime are observed (figure 5) and

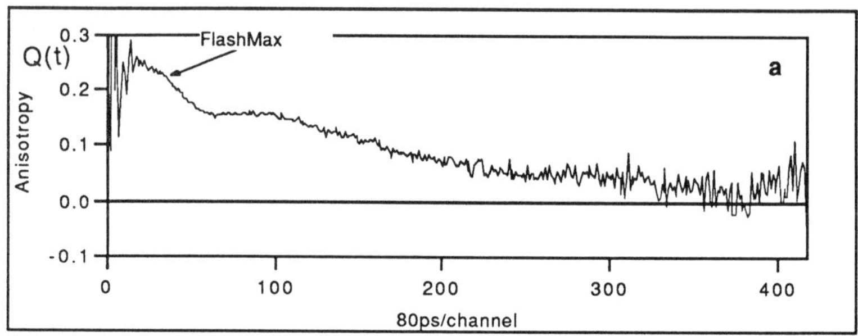

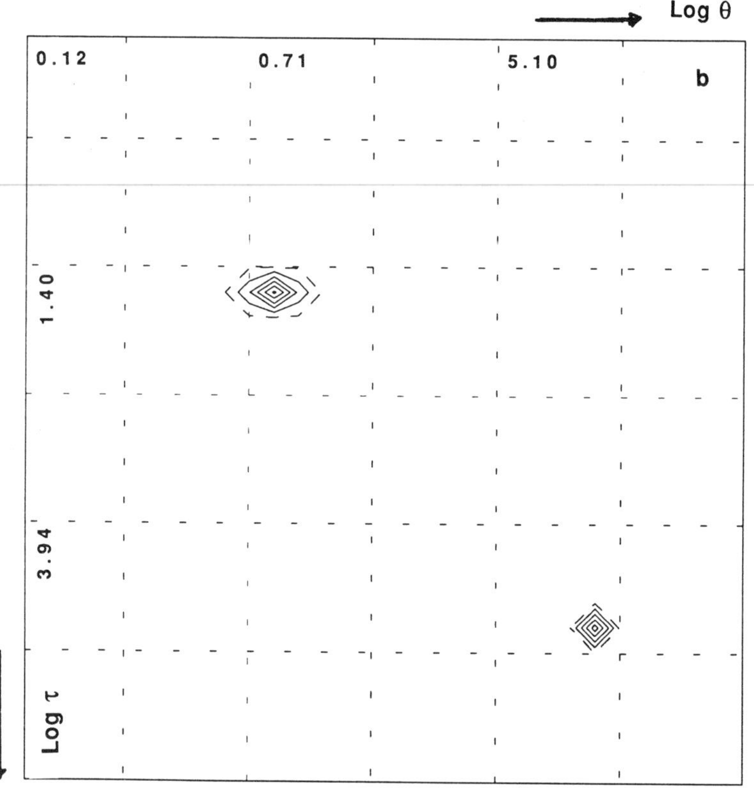

Figure 6: (**a**) Computer generated "data" to test MEM algorithm with τ_1=1.40ns,θ_1=0.71ns and τ_2=5.38ns,θ_2=9.22ns (**b**) Contour plot of 2-dimensional (τ,θ) MEM analysis of the corresponding simulated "data".

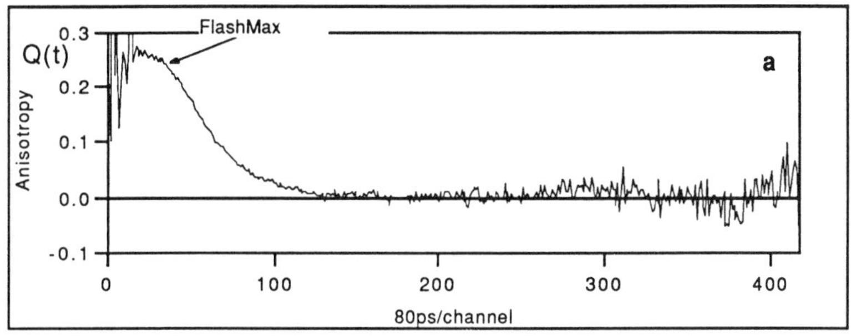

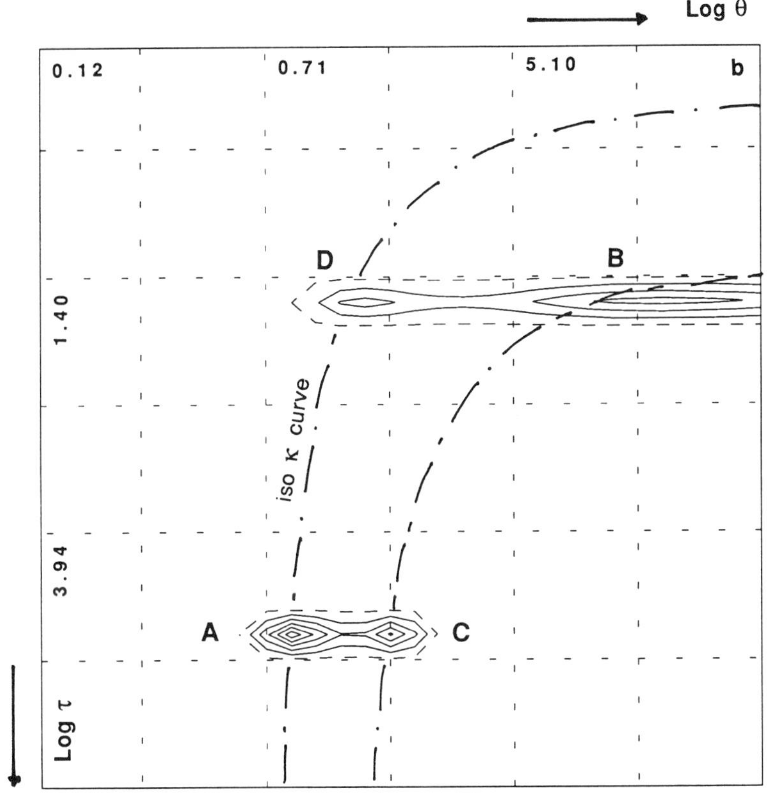

Figure 7: (**a**) Computer generated "data" to test MEM algorithm with τ_1=5.38ns,θ_1=0.71ns and τ_2=1.40ns,θ_2=9.22ns (**b**) Contour plot of 2-dimensional (τ,θ) MEM analysis of the corresponding simulated "data".

their relative contributions are changed. The short lifetime (centered at 0.06ns) displays a major contribution (42%) since the estimated contribution of the peak centerd at 5.13ns falls to 28%. The lifetime distributions can be attributed to the structural heterogeneity of HSA.which is modulated by temperature. As previously observed[14] ,the protein becomes more flexible at 52°C. The broadening of the peak is explained by an equilibrium between different conformational substates with a time constant of the order of magnitude of the average lifetime.

Rotational dynamics:

Computer simulations to test the MEM of analysing rotational dynamics

We calculated the theoretical $I_{vv}(t)$ and $I_{vh}(t)$ from a known spectrum of decay times and rotational times. The simulated $I_{vv}(t)$ and $I_{vh}(t)$ were convoluted with the experimental flash profile of ACO (FWMH =1.6ns) The ¨data¨ were produced at 400 points each separated by 0.080ns. Synthetic noise drawn from a Gaussian probability distribution of zero mean and a variance equal to the number of ¨fluorescence¨counts was added to each point .This ¨noise¨ was calculated from a random number generator[15] . Each trial consists of a "mixture" of two fluorophores characterised by their (τ_i,θ_j,A_k) with τ_1=1.40ns, τ_2=5.38ns and θ_1=0.71ns ,θ_2=9.22ns; all initial anisotropies are A=0.3. The corresponding apparent anisotropy decays Q(t) are shown on figures 6a and 7a. In the first test the two fluorophores have a short decay τ_1 together with a short rotational time constant θ_1 and a long decay τ_2 with a long rotational time constant θ_2. In the second set the rotational time constants are swapped so that the two fluorophores have a short decay with long rotational time and long decay with short rotational time. In figure 6a we can see that Q(t) cannot be correctly described by the sum of two rotational decays because the anisotropy actually rises between 60th and 90th channels. In figure 7a only a single short rotational decay is apparent. Although the longer rotational decay is present, the total fluorescence decays away quickly so its presence is only weakly determinable. In assuming that Q(t) can be analysed as a sum of exponentials, the present examples give a pictorial representation of the quality of data available to determine the rotational time constants.

In figures 6b and 7b, we display the maximum entropy analysis of these data sets. In Figure 6b we note that the two entities are correctly determined as sharp peaks in their true positions. In figure 7b the MEM pattern clearly displays the two lifetime components centered at their true positions. Along the θ axis we can see four separated peaks. From lifetime τ_2=5.4ns there are two sharp peaks centered at θ_B=0.7ns and θ_D=1.57ns. From lifetimes τ_1=1.4 ns two peaks are recovered but with a lower accuracy. In particular we are very uncertain of the long correlation time because both Ivv and Ivh measurements have decayed rapidly (due to the short decay τ_1) before we could get an accurate estimation of the long rotational decay. These four peaks belong to two different iso-k curves (see page 3) and the three combinason A+B, C+D, or A+B+C+D fit accurately with the data, only the relative proportion of these peaks being changed. Additional information is needed to resolve the ambiguity[16].

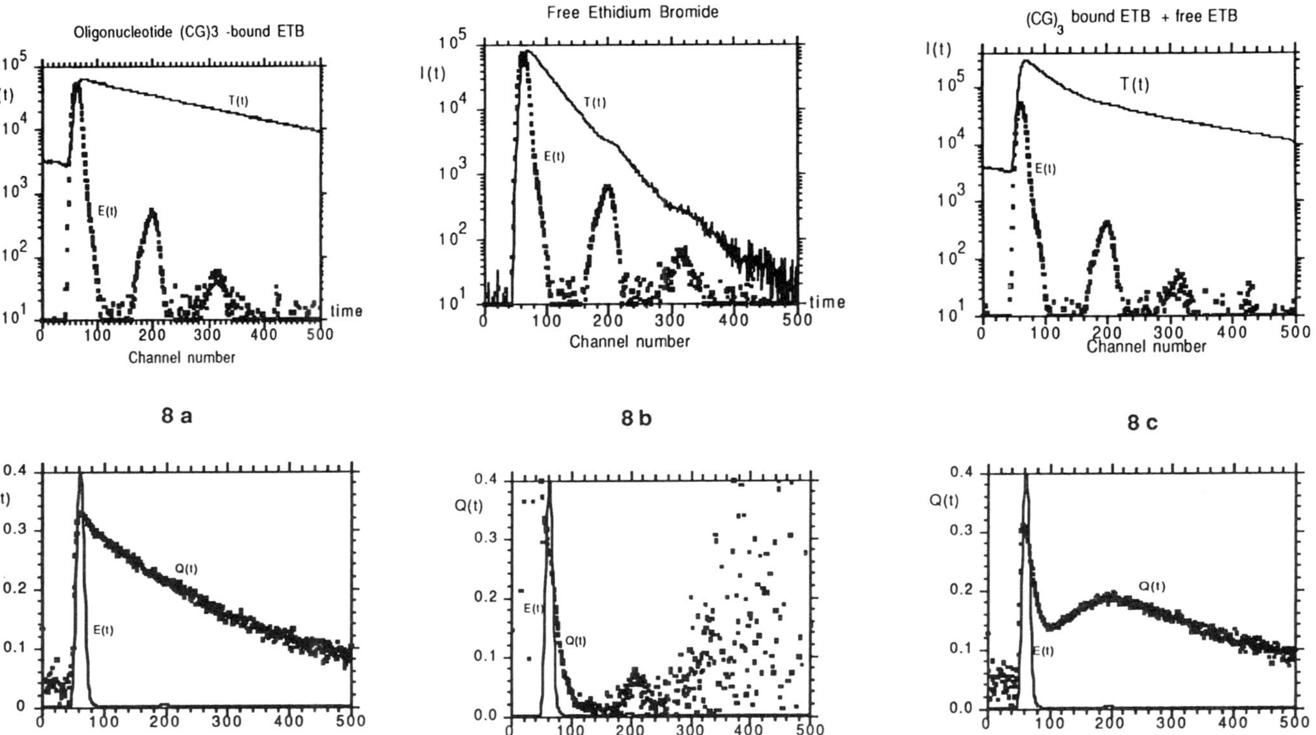

Figure8: Data on ETB free and bound to oligonucleotide. In upper case T(t) is the total fluorescence decay. E(t) is the measured excitation profile, both in semi-logaritmic scale. In lower cases the apparent fluorescence anisotropy decays Q(t) and E(t) in linear scaling.

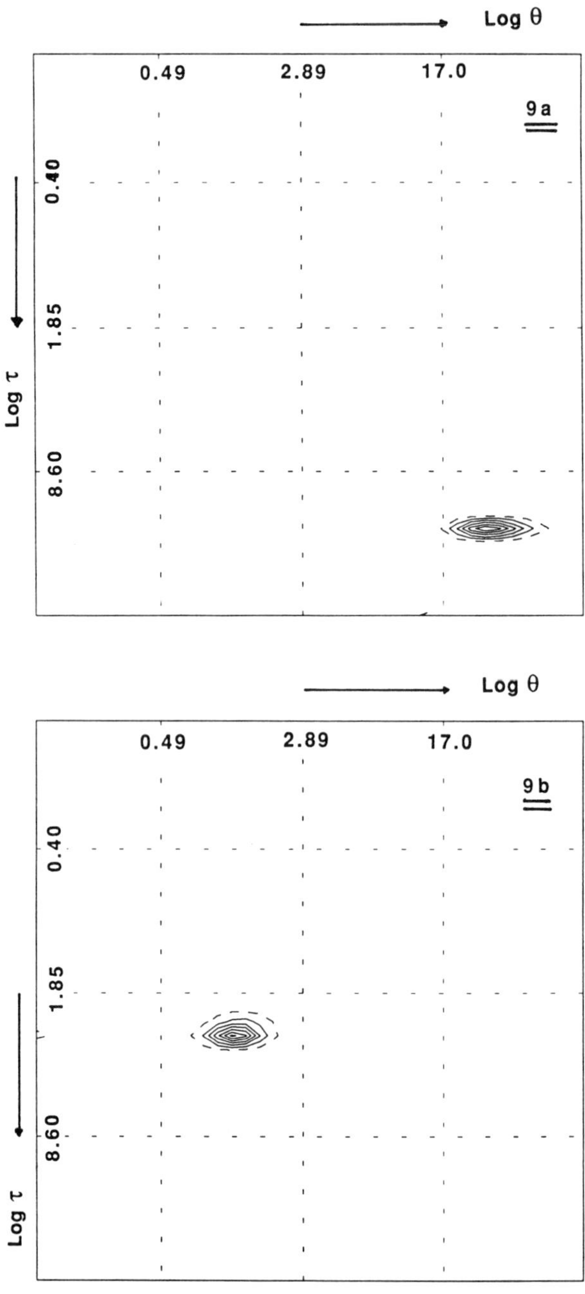

Figure 9: Contour plot (τ,θ) at A=0.36 for Ethidium bromide
(a) ETB-oligonucleotide complex alone
(b) Free dye

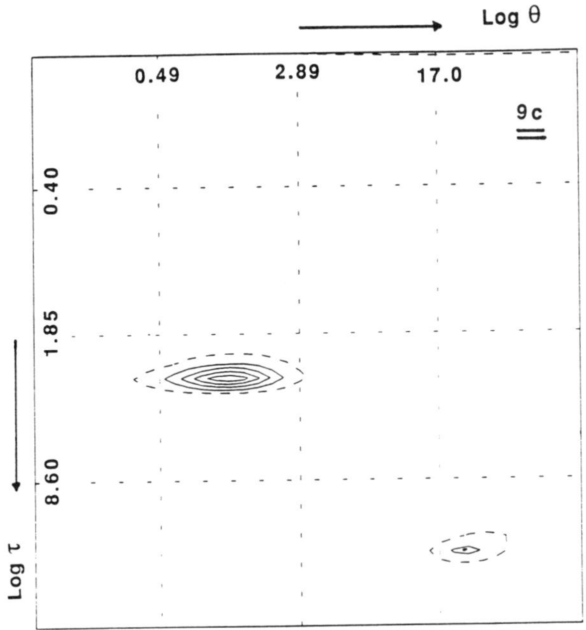

Figure 9: Contour plot (τ,θ) at A=0.36 for Ethidium bromide (c) Mixture of free and bound ETB

Ethidium Bromide-Oligonucleotide complex

The measured total fluorescence decay, flash profile and apparent fluorescence anisotropy decay Q(t) for oligo$(CG)_3$ bound ETB and free ETB are presented in figure 8a and 8b respectively. Both samples in each compartment of the double cell give measured decays presented in figure 8c. The MEM reconstruction of contours $\beta(\tau,\theta)$ was performed on 40 points equally spaced in $\log\tau$ between 0.1 and 40 ns and on 40 points equally spaced in $\log\theta$ between 0.1 and 100ns, the initial anisotropy was 0.36 Not surprisingly, the contour $\beta(\tau,\theta)$ for the free ETB (figure 9b) displays a strong single peak centered at τ=2.94ns and θ=1.20ns and for the bound BET the contour (fig.9a) is also a single peak at τ=15.9ns and θ=28.9ns. The analysis of the "mutiplexed" data gives the contour plot presented in figure 9c. There are two peaks centered respectively at τ_1=2.94ns, θ_1=1.40ns and τ_2=15.5ns, θ_2=26.8ns and a small amount of scattered light which can be attributed to internal optical reflection in the double compartment cell.

CONCLUSION

MEM is able to analyse polarised fluorescence data without setting any limitation on the number, shape and positions of fluorescence lifetime and rotational correlation time distributions. We show however that the data leave some ambiguity in the allowable distributions and that the MEM results reflect this ambiguity. Analysing polarised fluorescence in terms of distributions is now able to extract full description of conformational dynamic behaviour of heterogeneous fluorescent species.

REFERENCES

1. A.K. Livesey and J.C. Brochon, Biophys. J. 52,693 (1987)
2. A.K. Livesey, M. Delaye, P. Licinio and J.C. Brochon, Faraday Discuss.Chem. Soc. 83,247 (1987)
3. J.C. Brochon and A.K. Livesey in Fluorescent Biomolecules, D. Jameson and G.D Reinhart Eds, Pleum Press p.351
4. M. Vincent, J.C. Brochon, F. Merola, W. Jordi and J. Gallay,Biochemistry 27, 8752 (1988)
5. F. Merola, R. Rigler, A. Holmgren and J.C. Brochon,Biochemistry 28, 3383 (1989)
6. M. Karplus and J.M. Mac Cammon, Ann. Rev. Biochem. 53,263 (1983)
7. A.K. Livesey and J. Skilling, Acta Crystallogr. Sect. B Struct. Crystallogr. Cryst. Chem. A41, 113 (1985)
8. E.T. Jaynes, I.E.E.E. (Inst. Electr. Electron. Eng.) Trans. Biomed. 4, 227 (1968)
9. E.T. Jaynes, in Papers on Probability, statistics and statistical physics. (R.D. Rosenkrantz Eds.) D. Reidel,Dordrecht (1983)
10. J. Yguerabide, Methods Enzymol. 26, 498 (1972)11.Ph. Wahl in New Techniques in Biophysics and Cell Biology (M. Pain and B. Smith Eds) Vol. 2 ,233 John Wiley &Sons London (1975)
12. D.V. O'Connor and D. PHilips in Time-Correlated Single Photon counting, Academic Press ,London (1984)
13. J.M. Grognet Third cycle Thesis , Université R. Descartes, Paris-V (1988)
14. I. Munro, I. pecht, L. Stryer Proc. Natl. Acad. Sci. USA 76, 56 (1979)
15. W.H. Press, B.P. Flannery, S.A. Teukolsky, W.T. Vetterling in "Numerical Recepes" Cambridge University Press, Cambridge (1986) pp. 191

16 J..C. Brochon in preparation

DISCUSSION

BROCKLEHURST - In the maximum entropy method, you use a model Σm_i : how do you choose the model values, m_i?

BROCHON - $m_i(\tau)$ is the model which encodes our prior knowledge about the system under study. If we have no prior knowledge any τ has the same probability to exist in the final pattern, so $m_i(\tau)$ should be flat in τ.

GRATTON - How will the maximum entropy method handle cases in which the decay times are not associated with a particular molecular species, while the rotational correlation times are always associated with molecular species.

BROCHON - In measuring separately total fluorescence decay (i.e. at the magic angle) we can only deal with photophysics of the fluorescence without any assumption on rotational correlation times (molecular species). In addition we can handle cases where lifetimes (or other kinetics parameters) are fully uncorrelated with molecular species, assuming that all "lifetimes" contribute to all rotational processes.

MILLS - 1. What is the pulse width of Super ACO ?

2. You mentioned in your presentation that the pulse width changes during the fill. How much does it change and is the change just in width or in shape too ?

3. With the pulse width of Super ACO, what is the shortest lifetimes that you can measure ?

BROCHON - 1. The measured FWMH of Super-ACO is routinely 500ps at 60mA stored in the machine.

2. In 1991, the lifetime of the beam at 60mA ranges between 5 to 8 hours. In measuring alternatively excitation profile and fluorescence decays every two minutes we minimize the artifact due to slow change in measured width and shape of the excitation pulse. Typically an experiment takes place in 30-40 minutes depending on quantum yield of sample and chosen spectroscopic conditions.

3. The shortest lifetime measured for a strongly quenched trytophan fluorescence in protein was 40 picoseconds.

BAER - What is the wavelength range that you normally use ? What is the advantage of a synchrotron over a picoseoond laser system ?

BROCHON - We can use wavelengths from 200nm to 700nm determined by the transparency of the quartz window on the machine and the upper sensitivity limit of our MCP detector. The wavelength most commonly used for time-resolved fluorescence of tryptophan protein study is 300nm. The synchrotron offers two mains advantages among others:

- there is a wide-ranging tunability of excitation, particularly in the UV, for example very useful for energy transfer process studies. That allows a large flexibility for various scientific projects in a facility laboratory for biophysicists among others ,
- the laser is a quite complementary tool if it is dedicated to a fixed wavelength for specific study.

LEACH - You have described very nice experiments and techniques for data extraction. The determination of physico-chemical properties from fluorescence lifetime distributions of complex systems must be based on adequate models. What is the present situation with respect to these models (i.e. what sort of models are used) and to their validity ?

BROCHON - The data analysis provides numerical values for parameters of the chosen kinetics model. A precise model to fit should be a priori selected. The model is supposed to be selected from previous validation in studying simpler systems. When the systems become more complex or the time resolution more sharp (picosecond time domain) more sophiscated models should be taken into account in data analysis.

BAUMGARTEL - Can this method be used to distinguish different binding sites of the fluorescent dye is nucleotides ? Can the exchange of fluorescence labels between different binding sites be studied ?

BROCHON - For dye intercalation the time-resolved fluorescence provides insight in binding heterogeneity in measuring lifetime distribution. By increasing the dye concentration we can populate different binding sites having different binding constants and fluorescence properties. The exchange of fluorescent labels between different binding sites can be studied by determining the changes in lifetime distribution.

FREQUENCY DOMAIN FLUORESCENCE LIFETIME MEASUREMENTS USING SYNCHROTRON RADIATION

Enrico Gratton
Laboratory for Fluorescence Dynamics,
University of Illinois at Urbana-Champaign
Urbana, IL 61801

ABSTRACT

The study of ultrafast phenomena is important in many areas of physics, chemistry and biology. Current picosecond methods generally involve the use of short pulse, high intensity lasers. Even with today's striding advances in laser technology, a convenient, tunable and stable light source has not been found. Synchrotron radiation can provide an interesting alternative. Given the intensity generally available in the UV spectral region, where most biomolecules absorb, only direct detection can be achieved. In this respect frequency-domain techniques provide an alternative to correlated single photon counting techniques. We have built a frequency-domain fluorometer at the ADONE Synchrotron at Frascati. Several applications to the study of proteins and membranes have been obtained during the past five years. We will discuss the advantages of using frequency-domain methods and synchrotron radiation in these studies.

INTRODUCTION

Fluorescence spectroscopy has been established as one of the major physical methods to investigate biological molecules, providing detailed information of structural and dynamic properties of macromolecules. A variety of fluorescence techniques have been used to investigate all the major components of cells: proteins, nucleic acids and membranes. In particular, the study of the fluorescence decay and the decay of the emission anisotropy provides information on the chemico-physical environment of macromolecules and their dynamics. Fluorescence methods are now also being used in biomedical areas where sensitivity and specificity are crucial requirements. These important applications are necessarily dependent upon the study of the basic fluorescence processes.

The measurement of the fluorescence decay can be obtained by two alternative methods: 1) in the time domain, the decay is recorded using the technique of correlated single photon counting; 2) in the frequency domain, the harmonic response is measured. Both techniques have comparable sensitivity and time resolution when used in conjunction with a picosecond mode-locked laser as a light source. Few biomedical laboratories can afford to invest in such specialized instrumentation. After the initial laser acquisition, the maintenance costs, in terms of specialized personnel, are considerable. Moreover, even for those facilities, it is prohibitively expensive to maintain a battery of dye lasers and optics to provide the kind of broad wavelength range required for most of the experiments on biological samples. Synchrotron radiation (SR) provides all the advantages of continuously variable excitation and facile wavelength selection, while maintaining the time structure and intensity requisite for time resolved fluorescence spectroscopy of biological materials.

The instrumentation for fluorescence spectroscopy and data analysis has undergone rapid development in the last 10 years. New optical components, microelectronics and affordable computers have spurred the technological development. However, the basic instrumentation for time resolved fluorescence is still expensive and, in general, requires specific technical expertise. In the past years several centralized facilities for this purpose have been created in the United States (Illinois, Maryland, Texas,Pennsylvania, Ohio) with the specific goal of serving as user facilities and to foster new technical developments. The advantages such facilities offer to biomedical research include the combination of state-of-the-art instrumentation with the expertise to carry out successful experiments. Synchrotron radiation sources are a viable alternative to laser based facilities because of the broad wavelength range, high intensity obtainable with most of the new low energy synchrotron radiation sources and the excellent time characteristic.

Soon after SR became readily available, several synchrotron sources were equipped with instrumentation for fluorescence decay measurements. In the United States all sources currently utilize time domain techniques (Stanford, Brookhaven). Also, in England, France, and Germany, time domain techniques are used. Instead, in Italy frequency domain methods are employed. By far the major problem with the time domain method is the requirement for the storage ring to operate in single batch mode. In contrast, frequency domain methods can operate in multi-batch mode. Among the most successful sources used in biomedical applications is the Frascati Synchrotron, where we installed our first frequency domain fluorometer in 1983 [1]. A new beam line ("Plastique") has recently been activated at Frascati to be used exclusively for studying the fluorescence of biological materials using the frequency domain method [2].

METHODS

The conceptual basis for exploiting the time structure of synchrotron radiation (SR) in time-resolved fluorescence is based on the use of the harmonic content of the SR pulse [6]. The SR pulse is quasi-gaussian with a width dependent upon the radio-frequency used in the storage ring. For the second generation synchrotrons the pulse width is about 150 ps, which provides useful harmonic content to about 4 GHz. For to new synchrotrons the pulse width is expected to be on the order of 30 to 50 ps providing useful harmonic content in the 10 GHz region. An important property is that the signals at low and high frequencies have approximately the same intensity. We have fully demonstrated these properties during our development of time-resolved fluorescence instrumentation in Frascati [1]. In the frequency domain, the time structure of SR can be represented as in Figure 1. The basic frequency separation of the harmonics depends on the repetition rate of the SR. In particular, for operation in single bunch mode, the minimum separation is typically about 3 MHz (300 ns); whereas when the maximum number of bunches is injected the minimum separation between pulses can be less than 2 ns depending on the synchrotron. This pulse separation is too short for most fluorescence time-resolved experiments. Since for most of the other experiments performed around the storage ring, the maximum bunch mode of operation is usually required, then the use of SR for time.resolved fluorescence spectroscopy would be severely limited using pulse techniques. This high frequency operation is not a problem using frequency domain methods. To produce lower frequencies, we simply amplitude-modulate the pulse train using an acousto-optic modulator (AOM). The AOM medium is

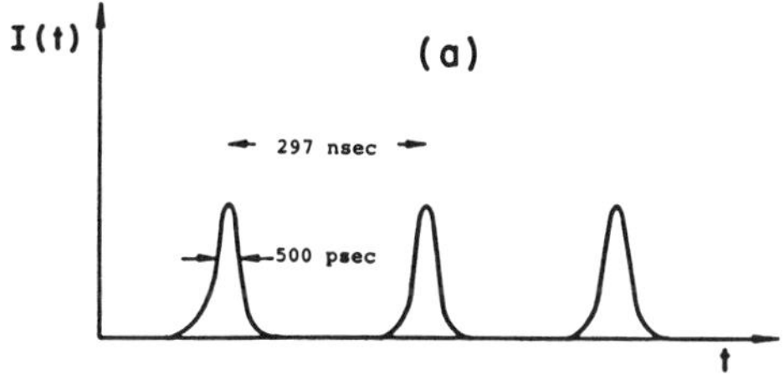

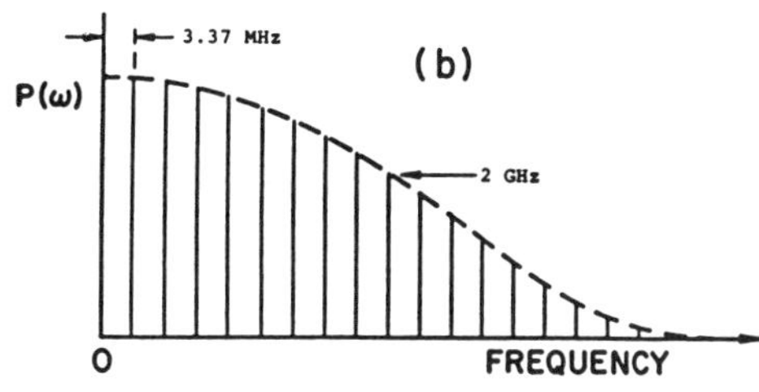

FIGURE 1. Time-domain and frequency-domain representation of the synchrotron radiation pulse structure.

fused silica with optimal UV transmission characteristics. This technique has been successfully used to insert additional frequencies using a high repetition mode-locked laser based instrument [7].

A different aspect of combining frequency methods in conjunction with SR relates to obtaining a signal which is phase coherent with the storage ring radio-frequency. However, it is possible to locally generate this radio-frequency using the superheterodyning method. This novel approach will facilitate the use of frequency domain methods with SR [5].

DESCRIPTION OF SUPERHETERODYNING METHOD

All modern frequency domain instruments use the cross correlation techniques first introduced by Spencer and Weber [4]. These instruments thus necessitate a sophisticated light source, containing harmonic content in the frequency range of the instrument's operation. Using the new synchrotron radiation sources such as ALS and ELETTRA, we anticipate a bandwidth of about 10 GHz [8]. The light detector generally limits the bandwidth of a given instrument. Obtaining the maximum bandwidth from a given detector is a crucial point. With photomultiplier tube (PMT) detectors, the maximum bandwidth can be obtained using only one or two dynodes of the PMT amplification chain. This result is achieved by injecting a radio frequency at the second dynode in order to frequency convert the signal inside the detector to a low frequency of operation [7]. In this scheme, it is possible to effectively increase the photomultiplier tube's anode bandwidth to a value similar to the cathode bandwidth. This operation mode can be standard mode for all those experiments which do not require ultra high frequency. For microchannel plate (MCP) detectors, the anode bandwidth is essentially the same as the cathode bandwidth and the signal can be handled directly at the MCP output. The output of the MCP, after DC/AC splitting, is amplified and connected to an ultra-wide-band mixer, providing frequency conversion as the first step in the detection. Te frequency translation process is performed in two steps. First, the high frequency signal is heterodyne down to an intermediate frequency of approximately 100 KHz. This is a very convenient frequency for intermediate amplification. Ceramic high-Q filters are available to filter the signal in this frequency region. The microwave local oscillator needs to be accurate only to a few KHz, the bandwidth of those filters. A second conversion process takes place by heterodyning the 100 KHz down to 40 Hz, a very convenient frequency for digital signal processing and averaging. The superheterodyning method substantially decreases the noise, thereby increasing the overall S/N of the MCP. Since most instruments always operate with two channels, and one channel, the reference, has high intensity, it is possible to provided a self-tracking filter based on a phase-locked-loop (PLL) configuration. The PLL automatically tracks the frequency variation difference between the laser repetition rate and the synthesizer, providing a stable output at exactly 40 Hz. By independently controlling the AC and DC signal, the AC signal does not need to match the DC level.

The electronic circuit has a flat frequency response to about 4 GHz, is highly sensitive, and is immune from synthesizer phase noise. We have performed phase measurements accurate to about 0.2 degrees with a 100 s integration time up to 4 GHz using a microchannel plate detector in a laser-based instrument [5]. The ultimate time resolution using phase measurements of this configuration with the microchannel plate is about 100 fs, which corresponds to a phase delay of 0.2 degrees at 4 GHz. Of course, other factors may limit the maximum resolution.

DETAILED DESCRIPTION OF THE ELECTRONIC DETECTION SYSTEM

A block diagram of a fluorometer for synchrotron radiation use is shown in Figure 2. The light source is the synchrotron. The light beam is delivered to the optical module using a mirror steering device. The beam is intensity modulated by an acousto-optic modulator to provide a continuum of modulation frequencies ranging from 0 to the first harmonic frequency of the SR. The very low frequency limit of this configuration can be useful for

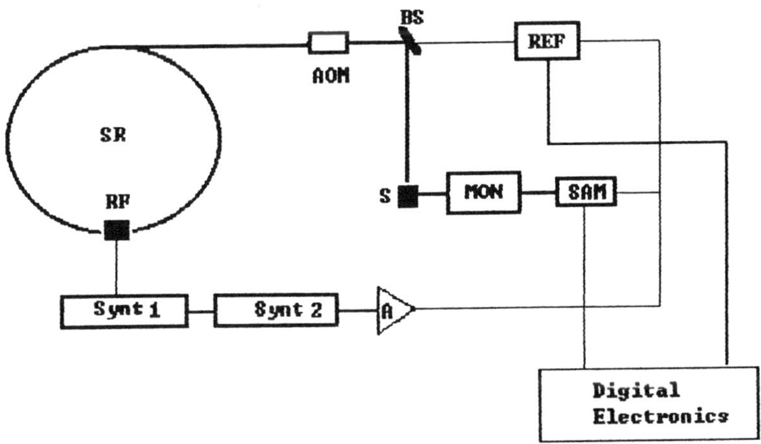

FIGURE 2. Schematic of a synchrotron radiation frequency-domain fluorometer; AOM = Acousto-Optic Modulator, quartz (Intra-Action Corp., model SWM-804Q); SAM, REF = Sample and Reference Photomultipliers (Hamamatsu, model R928); S=Sample compartment, with thermostated sample holder, magnetic stirrer and controlled atmosphere (ISS, Inc.); MON = Double dispersive grating monochromator; A1,= Radio-frequency amplifier (ENI, model 603L 3W, 0.8-1000 MHz); SYNT 1, 2 = Frequency synthesizers 10 KHz-1 GHz (Marconi, model 2022A); DIGITAL ELECTRONICS with internal or external cross-correlation.

the measurements of phosphorescence lifetimes. A master oscillator from the SR console can be used as a reference for the frequency synthesizer that drives the data acquisition electronics. However, a local tracking filter makes this mode of operation unnecessary. All optical parts of the instrument are mounted on an optical table. A personal computer controls all instrument operations, which include shutters, wavelength selection, high voltage ramping, rotation of the sample/reference turret and polarizer motion functions. Also the synthesizer frequency is selected using the computer interface.

The scheme of the electronic superheterodyning detection system is shown in Figure 3 and is based on recent research at the LFD [5]. The output of the MCP (Hamamatsu model R2566U-01) is separated into its AC and DC components (Figure 4). Special care is used in this part of the circuit. Since this is the only microwave frequency part of all the signal processing, all components (capacitor, cables, connectors) must be rated for

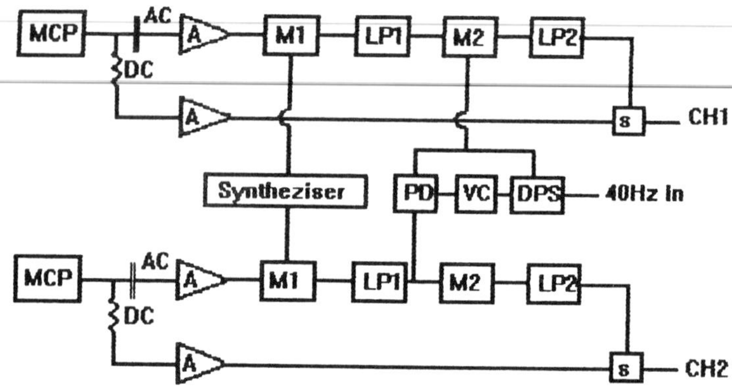

FIGURE 3. Block diagram of the superheterodyning detection system. MCP=Detectors: 2 stage, 6um Hamamatsu microchannel plate photomultiplier, AC,DC=radio frequency signals split in AC and DC components; Syntheziser=microwave signal generators; M1=Double balanced mixers; LP1=narrow bandpass electronic filters for intermediate frequency of 100 KHz; LP2=narrow bandpass electronic filters for cross-correlation frequency of 40 Hz; A=variable amplifier for AC signal; and DC signal amplifier; Tracking filter consisting of PD (phase detector); VC (voltage controlled oscillator); and DPS (digital phase shifter).

up to 4 GHz operation. After the microwave mixer, M1 (Anzac, model MDC123), a frequency component of the signal is at low frequency (100 KHz). It can then be handled using standard electronic components. It should be noted that the average DC current from our MCP must not exceed 200 nA, i.e., approximately 2 V after the DC current-to-voltage converter (gain 107 V/A). About the same signal intensity with respect to the DC signal is available at each one of the harmonic frequencies over a spectrum which extends up to the limit imposed by the detector. However, the AC part of the signal has a 50 Ohms impedance due to the mixer input characteristics, which typically results in a 10 uV signal. These signal levels are very low, and a direct conversion to the 40 Hz region is difficult, since the ubiquitous 60 Hz or 50 Hz line signal dominates. This difficulty is one of the reasons for using the intermediate conversion frequency, where amplification can be obtained in a low noise frequency region. The mixer has a conversion loss of about 7 dB (a factor of ~5), which reduces the signal out from the mixer to about 2 uV. The voltage level at the RF input of the mixer was fixed at about +10 dBm. This value was found to be optimal for the Anzac mixer. Some harmonics are produced at this high RF signal level, but we found that filtering of the second harmonic of the microwave frequency carrier is excellent and the linearity of the mixer output is also good. The mixer output is filtered using a 42IF301 ceramic IF filter transformer (Mouser Electronics, Mansfield, TX). This filter performs an impedance conversion from about 500 ohms to about 50 Kohms, providing a voltage gain of about a factor of 100. The output of the filter is then amplified by about a factor of 1000 and fed to the RF input of a second mixer (M2). The LO input of the second mixer is connected to the output of the tracking synthesizer, which provides a frequency exactly 40 Hz below this intermediate frequency. After bandpass filtering and amplification, the output of M2 contains all the information of the original high frequency signal. The reference channel operates in a similar fashion.

The amplification of the signal at 100 KHz, second stage mixing, and further amplification and low pass filtering is obtained using a modified PAR, model 5204, lock-in amplifier, equipped with a tunable input filter set at 100 KHz. The low frequency filtering at 40 Hz and amplification is performed using the standard digital ross-correlation electronics which is part of commercial fluorometers. At this point, four signals, corresponding to the AC and DC of the sample and of the reference, respectively, have been generated. These signals are applied to the four inputs of the analog converter interface card for an IBM-PC computer, where the usual acquisition and processing is performed under computer control.

LIFETIME MEASUREMENTS USING THE FREQUENCY-DOMAIN METHOD

The only synchrotron source equipped with a frequency-domain fluorometer is the synchrotron at Frascati. The frequency domain fluorometer at Frascati uses a single heterodyning step for frequency conversion. Also the relatively broad pulse at Frascati limits the useful Frequency range to about 150-200 MHz. Typical frequency domain data are shown in Figure 4 for a molecule with a relatively short lifetime. The phase delay and the modulation ratio are measured over the range 8 MHz to 100 MHz. The intensity decay for p-terphenyl in ethanol is best fit using a single exponential decay of 996 ps. This value is in good agreement with values reported in the literature.

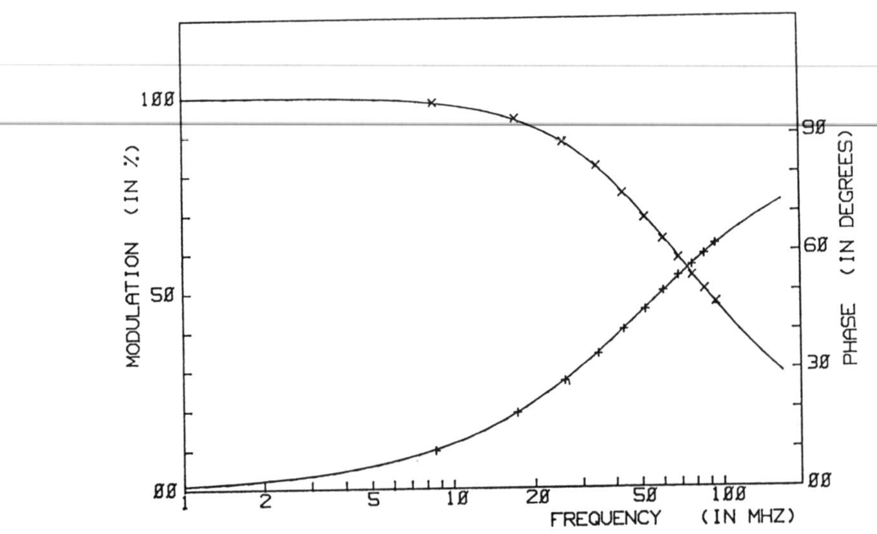

FIGURE 4. Phase and modulation measurements of p-terphenyl in ethanol. The solid curves correspond to a decay time of 996 ps. The excitation wavelength was 290nm and the emission was observed using a wg335 longpass filter. These data were acquired at the Frascati Synchrotron.

To illustrate the possibilities offered by the new superheterodyning method we report in figure 5 typical lifetime measurements performed using a mode-locked laser as the excitation source and a microchannel plate detector. In this case the frequency limit was about 4 GHz, due to the limited bandwidth of the mixers used in the first heterodyning step. We have measured the frequency response of rose bengal in ethanol and in buffer and of pnacyanol in ethanol. The fastest decay time measured using this apparatus was for the picocyanol dye that has a lifetime of approximately 11 ps.

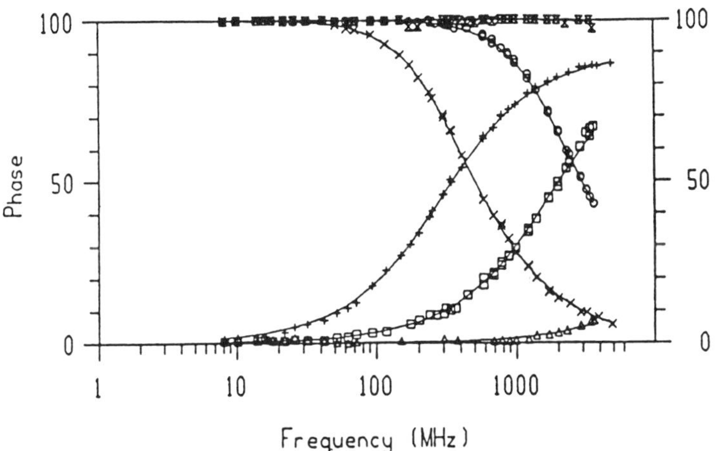

FIGURE 5. Phase and modulation measurements of different dyes using the 4-GHZ laser based instrument. Frequency response of the intensity decay of rose bengal, phase +, modulation x; excitation 565 nm. ; tau=543ps. Rose bengal in TRIS buffer pH9.5. Phase ;modulation 0; excitation 560 nm; emission, Hoya r60 filter; tau=93 ps. Pinacyanol in ethanol, phase , modulation , excitation 590 nm; emission filter Hoya 62; tau=11 ps.

The power of the frequency-domain method is best demonstrated by the measurement of the rotational rate of small molecules in low viscosity solvents. Figure 6 shows measurements of the differential phase and modulation ratio for indole in ethanol and p-terphenyl in ethanol. The frequency domain data were fitted using a rotational correlation time of 35s for indole and 65ps for p-terphenyl.

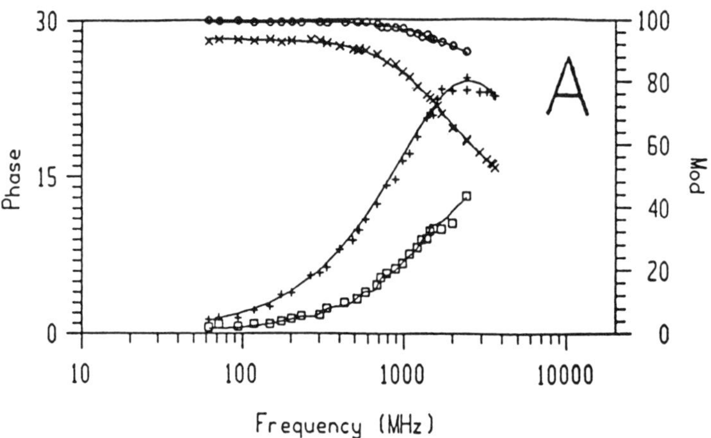

FIGURE 6. Polarized differential phase and demodulation ratio of indole and p-terphenyl in ethanol The solid curves correspond to rotational correlation times of 35ps and 65 ps, respectively.

The above example illustrate the ultra high time resolution that the frequency domain technique can provide. The superheterodyning method in conjunction with the radiation from the new synchrotrons will provide an unsurpassed capability for spectroscopy.

TABLE I. Unique features of SR for fluorescence spectroscopy

Spectral range in the UV not accessible with high repetition pulsed lasers
Simultaneous generation of all spectral components
Time structure with harmonic content in the GHz range
High intensity
Low cost access to SR
Polarized light
Pulse to pulse amplitude stability

REFERENCES

1. Gratton, E., D. M. Jameson, N. Rosato and G. Weber. Rev. Sci. Inst. 55, 486-494 (1984).
2. De Stasio, G., F. Antonangeli, N. Zema, T. Parasassi and A. Savoia. J. Vac. Sci. and Tech. In press (1989).
3. Gratton, E. and M. Limkeman. Biophys. J. 44, 315-324 (1983).
4. Spencer, R. D. and G. Weber. Ann. NY Acad. Sci. 158, 361-370 (1969).
5. Gratton, E. and M. vandeVen. Rev. Sci. Inst. Submitted (1989).
6. Gratton, E. and R. Lopez-Delgado. Rev. Sci. Inst. 50, 789-790 (1979).
7. Alcala, R., E. Gratton and D. M. Jameson. Anal. Inst. 14(3&4), 225-250 (1985).
8. Gratton, E. and B. Barbieri. Spectroscopy 1 (6), 28-36 (1986).
9. Bucci, E., H. Malak, C. Fronticelli, I. Gryczynski and J. R. Lakowicz. J. Biol. Chem. 263, 6972-6977 (1988).

DISCUSSION

DANIELS - What sort of life-time resolution can be obtained for multiple fluorescence emissions ? is this resolution obtained by curve-fitting to phase and modulation curves ?

GRATTON - Generally the lifetime resolution obtained in the frequency domain is better as compared to time domains due to:

1. less error in the determination of the values of phase + modulation compared to the error in determining the intensity of each time interval
2. Logarithmic frequency scale which allow a wider spanning of the "time" range, generally a run covers 3-4 decades in frequency.
3. The absence of deconvolution due to the instrument response allows to better determine very short time events ?

BROCHON - How do you manage the low level intensity of background signal in an analogic electronic system.

GRATTON - Background subtraction is performed by first recording the signal from a "blank" and then from the "sample + blank". The subtraction is performed by simple vector algebra using the phase and modulation of the sample and blank.

BESWICK - Could you comment on the possible interesting biological studies which can be achieved with your technique in the domain of a few ps?

GRATTON - Examples on :

1. Distribution of distances in proteins
2. Fast interconnexions between substates.
3. Fast rotational motions.
4. Early photochemical events

TIME-RESOLVED LUMINESCENCE OF VUV-EXCITED LIQUIDS AND SOLIDS.

Brian Brocklehurst
Chemistry Department, Sheffield University, Sheffield, S3 7HF, U.K.

Ian H. Munro
SERC, Daresbury Laboratory, Warrington, WA4 4AD, U. K.

SUMMARY

Vacuum ultraviolet radiation is used to study ´spurs´ - the groups of ion pairs and excited states that are produced along the tracks of fast particles. Luminescence decays, magnetic field effects and time-resolved excitation spectra are used to elucidate processes occurring between 1 and 320 ns - the spacing of pulses from the SRS at Daresbury operated in single-bunch mode. These techniques are illustrated by data for sodium salicylate with photon energies ranging from 4 to 10^4 eV.

INTRODUCTION

The general aim of this work is to increase understanding of the fundamental processes of radiation chemistry and radiation biology. Fast electrons (from accelerators, beta-sources, absorption of X- and gamma-rays) dissipate their energy slowly along sparse tracks with events separated by up to 200 nm in liquids and solids. These events vary greatly: they are arbitrarily classified as spurs (up to 100 eV, groups of excited states and ion pairs up to 5 or 6 in number), blobs (100-500 eV) and short tracks (500-5000 eV)[1]. The distribution of these events cannot be measured directly: theoretical estimates[1-3] for liquid water give distributions with 76% of the energy producing spurs, 10% blobs and 14% short tracks; the most probable energy loss for a fast electron is about 22 eV; experimental data is often treated in terms of single ion pairs, but in fact these account for only some 10-30% of the energy loss, depending on the assumptions made about distributions between ionisation and excitation.

Synchrotron radiation enables us to study these various species individually; in this paper we report results using photons ranging from 4 to 10^4 eV and compare them with beta-particles ($\sim$ 1 MeV); both valence shell and inner shell electrons can be excited. The role of carbon K-edge absorption is of particular interest in organic and biological systems.

Vacuum ultraviolet radiation is absorbed very strongly by all materials especially in the important region around 20 eV. This means that windows cannot be used and substrates must have very low vapour pressures. Also, penetration depths are very small - a few tens of nanometres: to obviate second-order effects of radiation damage in the surface layer it is essential to use a very sensitive technique, such as luminescence, to monitor primary processes. The time structure of synchrotron radiation is extremely valuable: when operated in single-bunch mode, the SRS at Daresbury provides pulses 160 ps wide separated by 320 ns. This enables us to monitor luminescence decays while the effects of magnetic field[4,5] help us to identify the processes involved.

We studied first a simple model hydrocarbon system (p-terphenyl in squalane, $C_{30}H_{62}$)[6]: the magnetic field effect was positive (ion pairs initially singlet) decreasing as expected with increasing photon energy in sharp contrast to the behaviour of crystalline aromatic hydrocarbons[7,8]. This led to work on an aromatic liquid (Santovac oil, a poly-phenoxy compound)[9], which showed intermediate behaviour; DNA has also given very interesting results[10]. In this paper we describe the techniques used and illustrate them with data for crystalline sodium salicylate; this material is widely used to monitor vacuum U.V. radiation[11] because of its stability, non-volatility and high fluorescence yield. Preliminary results have been published[12] but we have now greatly extended the range of our measurements and have made detailed measurements of the field effects in the low energy region (6-30 eV)[13].

EXPERIMENTAL

The apparatus and procedure used for work with the Seya monochromator (station 3.1 of the SRS) have been described previously[14]. In brief, before falling on the sample, light from the monochromator passes through a glass capillary, which facilitates differential pumping. The sample lies between the poles of a small electromagnet. Fluorescence is focussed onto a Mullard XP2020 photomultiplier which provides start signals to an Ortec 567 time-to-pulse-height converter; the SRS orbit clock provides stop signals. The TPHC output is sent to a multi-channel analyser which builds up a histogram containing precise information about the shape of the decay. Time resolution is limited by the synchrotron pulse-width and the transit time jitter of the photomultiplier which give a response fumction less than 1 ns fwhm; this can be deconvolved from the data using standard non-linear least squares analysis[15] or the maximum entropy method[15,16]. A computer programme running on a PDP-11 controls the experiment and provides automatic switching between memory sections of the MCA as the magnet is switched on and off. Various modifications of this equipment have been used on other stations. The continuous spectral output of the SRS makes it easy to measure time-resolved luminescence excitation spectra - a further diagnostic for the processes involved[14]. The TPHC has a provision for single-channel analyser output which can be set to a specific time-region.

The earlier results obtained in the vacuum ultraviolet[12] were complicated by the presence of higher order light in the toroidal grating monochromator (station 3.3). For studies with the Seya (3.1) we use a retractable lithium fluoride window to remove higher orders when working between 6 and 11 eV. At higher energies, the gratings are working close to their blaze angles where higher orders are negligible.

The fluorescence decays (Fig. 1) span a wide range - up to a thousand-fold between peak and tail. To reduce statistical fluctuations, it is necessary to group channels together in the tail region, before taking the ratios shown in the other figures, hence the change in spacing of points in Fig. 4.

RESULTS

Typical luminescence decay curves for sodium salicylate are shown in Fig. 1, which demonstrates the effect of excitation energy. At the lowest energy (320 nm), a single exponential represents the decay over some three orders of magnitude; a mean lifetime of 7.91 ns is obtained by deconvolution of the ´prompt´. At higher energies, the semi-log plots are not linear: the initial decay is faster and there is a long tail. Fig. 2 shows this over a wide energy range - results obtained on five different SRS stations. The apparent anomaly in the beta-particle results is probably an artefact: in that case the luminescence is viewed from the back of a thick sample so that the apparent decay rate is increased by absorpion and re-emission of fluorescence. Note the ´wraparound´ of the tail of the preceding pulse (Fig. 1 at left).

Measurements with the Seya on sodium salicylate have been made on a number of occasions. The results are reproducible for times up to about 100 ns after the luminescence peak; at longer times considerable differences have been found. This appears to be an effect of incident intensity: it is not easy to vary this in a uniform way across the sample but some data is available, as a function of beam current: Fig. 3 shows results obtained at zero field during field variation studies. This second order process complicates the measurement of time-resolved spectra; scans with the Seya show a maximum in the tail to peak ratio which corresponds to the blaze angle (maximum output) of the grating.

Fig. 4 shows some typical magnetic field data. One can distinguish two types of effect in three different time regions: at long times there is a positive effect (fluorescence enhanced by the field) requiring fields of only a few milli-Tesla; this effect has a threshold at ~ 8 eV. At shorter times there is a similar negative effect. There is also a positive effect on the peak requiring much larger fields (~ 100 mT). These two last have a higher energy threshold, their importance increasing gradually between 15 and 20 eV. A more detailed account of these effects in sodium salicylate is given elsewhere[13]. Qualitatively similar field effects have been found in Santovac oil[9] and in DNA[10]. These very different systems have in common that they are wholly or largely made up of aromatic units. Dilute solutions in squalane show much simpler behaviour[6] - a positive effect developing slowly and requiring low field, with no detectable effect on the luminescence peak.

Angel et al.[17] observed a sharp 75% drop in the total luminescence of sodium salicylate at the carbon K-edge. Calculation from the atomic absorption coefficients[18] gives ~75% excitation of the inner shell electrons, 25% of the valence electrons, suggesting that K-edge absorption produces very little luminescence; quenching

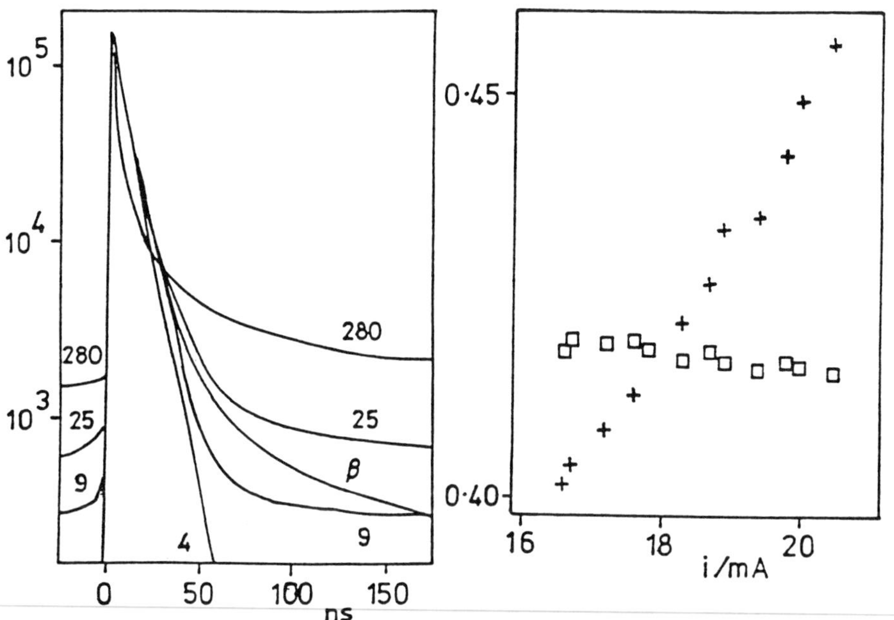

Fig. 1. Decay curves for sodium salicylate excited at various photon energies (eV - marked).

Fig. 3. Intensities relative to the peak of the 30-50 ns, □ , and 120-200 ns, + , regions, plotted against beam current.

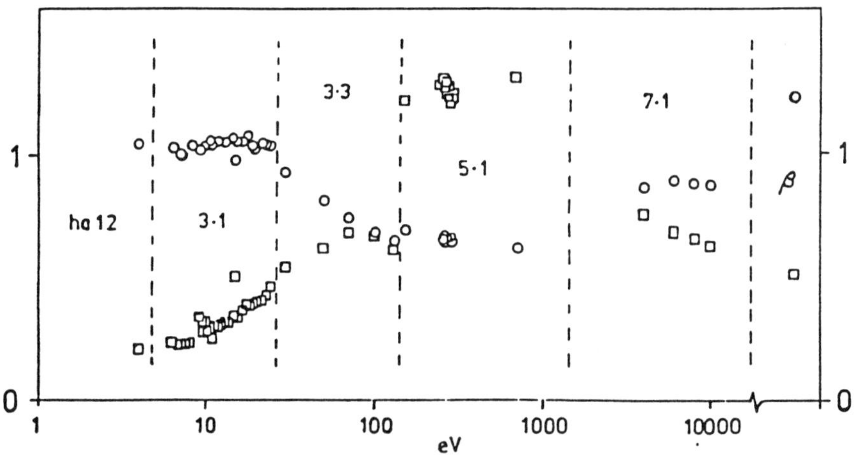

Fig. 2. Intensities (arbitrary units) of the 2.5-10 ns ,○, and 30-50 ns, □, regions relative to the peak intensity (-0.5-2 ns) for various photon energies and beta particles.

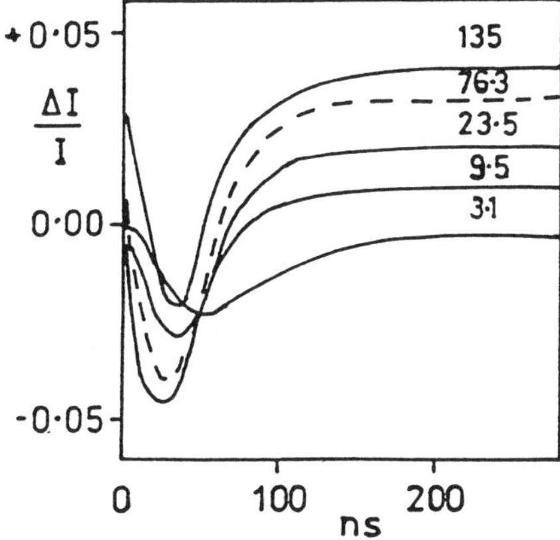

Fig. 4. Time-dependence of the effect of various fields (mT) on the pulse: excited at 24 eV.

of excited states by radicals, unpaired electrons and triplet molecules is very probable in the region of dense excitation produced by the Auger processes. Our time-resolved measurements show only a small change in pulse shape and no change in field effect across the edge, supporting this view. We observe a smaller effect on total intensity than Angel et al. (in comparison with a phosphor containing no carbon) but the presence of some stray light of other wavelengths, cannot be ruled out. More work in this region would be worthwhile.

DISCUSSION

Fig. 1 shows that excitation at low energies leads to a simple first order decay. We have not studied the emission spectrum as a function of energy but it is reasonable to suppose the emitter does not change and that the long-lived emission observed at higher energies reflects processes preceding its formation - recombination of ′ions′ or molecular triplets. (Electron loss in this case gives neutral salicylate radicals: the nature of the excess electron species is not known.) At times up to 100 ns these are entirely geminate processes - involving reaction between species produced by the same incident photon. Only at longer times do random reactions between products of absorption of two different photons become significant (Fig. 3).

The results can be understood qualitatively in terms of a reaction scheme adapted from that proposed by Martin, Klein and Voltz[8] to interpret their data for crystalline anthracene: it appears to be applicable to all the aromatic compounds we have studied[9,10,12]. At low energies (up to 8 eV here), direct excitation of singlet states takes place. Between 8 and 15 eV, ionisation occurs, followed by geminate recombination - mainly into singlet states since the spin correlation is retained. In the 15-20 eV region, a new process (fission) appears: one of the ions carries sufficient excess energy to excite a molecular triplet state leaving the ion pair also triplet so that the overall singlet multiplicity is retained. At higher energies still, groups of two or more ion pairs are formed, in which direct (singlet) or cross (random spin) recombination can occur.

There are two different types of field effect[4]. The first involves two doublets - radicals or radical ions: for example, ionisation of a singlet produces a singlet ion pair, but this is not a stationary state: the wave-function evolves into a mixture of singlet and triplet as a result of the electron-nuclear interaction (or, less commonly, of differences in g-values). Both the rate and extent of this evolution can be altered by an external field: small fields (~ 10-100 mT) suffice to saturate the effect and correspondingly, the effect develops slowly - over some 10s or 100s of nanoseconds. The sign of the effect depends on the initial spin state: initially, singlet pairs give a positive effect, i.e. fluorescence enhancement, triplet pairs a negative one.

The second effect involves triplet states: the principles are similar but the spin-spin interaction between two electrons is much larger. This is the ´fine structure´ rather than hyperfine structure of electron spin resonance spectra: it gives rise to the zero-field splitting of molecular triplets. Stronger fields are required (> 100 mT) and the time-scale is shorter (nanoseconds). Field effects are also found in triplet-doublet processes; they are dominated by the properties of the triplet.

At low energies, there is a positive field effect apparently of the first type: this persists at higher energies at long times (Fig. 4) but this result is suprising since, as Fig. 3 shows, this region is dominated by processes requiring two incident photons - recombination of ions or of molecular triplets with no initial spin correlation. At higher energies (Fig. 4), there is evidence of a high-field fission process at very short times: in agreement with the proposed mechanism, this has the same energy threshold[13] but different field dependence to the negative effect at intermediate times, due to recombination of initially triplet ion pairs.

The pulse shape changes in the 2-10 ns region (Fig. 2) can be understood in terms of a lifetime shortening due to quenching of the emitters by radicals or triplet molecules in the same spur: this is most marked in the dense ´blobs´ produced in 1-1000 eV region and decreases again when the primary photo-electron has sufficient energy to produce an extended track. The species involved at > 100 ns may include triplet states, but where detailed field effect studies have been made (8-24 eV, beta-particles) they appear to show that ion pairs are responsible. (It is puzzling that there is no apparent change in pulse-shape around 8 eV, the field effect threshold - Fig. 1). A similar but inverse energy dependence is observed at longer times (Fig. 2). Detailed interpretation of the field effect[13] at all energies is not yet possible: only the positive hyperfine effect is observed in the undulator data: the other effects reappear in the keV region. However, it is clear that the beta-particle results closely resemble those of photons in the 20 eV region.

CONCLUSION

Our results for sodium salicylate show clearly the relevance of this work to the study of mechanisms of radiation effects. The sparse tracks of fast betas approximate to the low energy VUV photons, but it is possible to study the less common events, ´blobs´ and short tracks, too. Other systems have been studied to date only with beta-particles[5] and in the 5-40 eV range. Qualitatively similar results have been obtained when aromatic residues predominate - solid anthracene[7,8] and DNA[10], and Santovac oil,[9]. However, the quantitative differences are considerable, both in pulse shape behaviour and in the field effects; they can be very complicated as in DNA[10] - not surprisingly. The model aliphatic system - a dilute solution of para-terphenyl in squalane - is simpler: it shows no fission effects - molecular triplets decompose immediately - and only the positive hyperfine effect[6]. Sodium salicylate is unusual in that the extent of this last is very small. To date, we have only reported qualitative results but data analysis using both non-linear least squares[15,19] and the maximum entropy method[16,19] is in progress, and computer models incorporating both spur dynamics and spin correlation are being developed to interpret the data[20].

The authors thank Mr. D. Fance, Mr. A. Hall, Drs. G. J. Baker, M. Hayes, I. R. Holton, A. Hopkirk, G. R. Jones, C. Mythen, D. A. Shaw and R. W. Sparrow for their assistance with the experimental work and the Science and Engineering Research Council for the provision of facilities.

REFERENCES

1. A. Mozumder and J. L. Magee, Radiat. Res., 1966, 28, 215.
2. S. M. Pimblott, N. J. B. Green, and M. J. Pilling, Radiat. Phys. Chem., 1991, 37, 377.
3. S. M. Pimblott, private communication.
4. K. Salikhov, Yu. M. Molin, R. Z. Sagdeev, and A. L. Buchachenko, "Spin Polarisation and Magnetic Effects in Radical Reactions", Elsevier, Amsterdam, 1984; U. E. Steiner and T. Ulrich, Chem. Rev, 1989, 89, 51.
5. B. Brocklehurst, Intern. Rev. Phys. Chem., 1985, 4, 279.
6. G. J. Baker, B. Brocklehurst and I. R. Holton, Chem. Phys. Lett., 1987, 134, 83; G. J. Baker, B. Brocklehurst, M. Hayes, D. M. P. Holland, A. Hopkirk, I. H. Munro, and D. J. Shaw, Chem. Phys. Lett., 1989, 161, 327.
7. J. Klein, J. chim. Phys., 1983, 80, 627; P. Martin, J. Klein and U. Hahn, in "Photophysics and photochemistry above 6 eV", ed. F. Lahmani, Elsevier, Amsterdam, 1985, p. 245.
8. P. Martin, J. Klein and R. Voltz, Physica Scripta, 1987, 35, 575.
9. B. Brocklehurst, A. Hopkirk, I. H. Munro, and R. W. Sparrow, J. phys. Chem., 1991, 95, 2662.
10. B. Brocklehurst, A. Hopkirk and I. H. Munro, Chem. Phys. Lett., 1990, 173, 129.
11. J. A. R. Samson, "Techniques of Vacuum Ultraviolet Spectroscopy", Wiley, New York, 1967; J. A. R. Samson and G. N. Haddad, J. Opt. Soc. Amer., 1974, 64, 1346.
12. G. J. Baker, B. Brocklehurst and I. R. Holton, J. Phys. B: At. Mol. Phys., 1987, 20, L305.
13. B. Brocklehurst, submitted to Chemical Physics.
14. B. Brocklehurst, A. Hopkirk and I. H. Munro, Radiat. Phys. Chem., 1991, 37, 487.

15. B. Brocklehurst, Chemistry in Britain, 1987, 23, 853.
16. A. K. Livesey and J. C. Brochon, Biophys. J., 1987, 52, 693.
17. G. C. Angel, J. A. R. Samson and G. Williams, Appl. Opt., 1986, 25, 3312.
18. B. L. Henke, P. Lee, T. J. Tanaka, R. L. Shimabukuro and B. J. Fujikawa, At. Data Nucl. Data Tables, 1982, 27, 1.
19. R. N. Young, D. A. Shaw and B. Brocklehurst, paper in this volume.
20. B. Brocklehurst, submitted to J. Chem. Soc. Faraday Trans.

DISCUSSION

LEACH - Sodium salicylate has often been used to detect UV and VUV photons because its fluorescence yield (ϕ_F) is supposed to be constant (about unity) over a very large wavelength range. However, there are some reports (L. Herman in the late 1940's) of ϕ_F variability, dependent in part on O_2 effects, size of Sodium (Na) Salicylate "flakes", radiation damage, etc. This leads me to ask how reproducible are your luminescence results on Na Salicylate and how they depend on sample preparation and treatment.

BROCKLEHURST - Samson's work shows that Sodium salicylate is very stable to irradiation in the VUV. Runs on station 3.1 (6-30eV) gave very reproducible results with the exception of the 'tail' intensity. This is shown in our paper to be an intensity effect. We did not investigate different methods of sample preparation.

WONG - Why were 5 beam lines used in your experiments ?

BROCKLEHURST - The different stations at the SRS are equipped with various monochromators suitable for specific energy ranges.

DANIELS - Does the work with sodium salicylate in solid state show any effects attributable to stacking of the molecules, either in change of emission spectrum or change in lifetimes ?

BROCKLEHURST - We have not looked at the emission spectrum as a function of excitation energy. The non-exponential nature of the changes and the general similarity of behaviour of very different systems suggest that slow processes (in recombination, triplet state diffusion) lead to a common emitter in each case; however, emission wavelength discrimination is planned for future work, especially on DNA.

TIME-RESOLVED LUMINESCENCE OF DIPHENYLALLYL ANIONS AND RELATED COMPOUNDS.

Ronald N. Young and Brian Brocklehurst
Chemistry Department, Sheffield University, Sheffield, S3 7HF, U.K.

David A. Shaw
SERC, Daresbury Laboratory, Warrington, WA4 4AD, U.K.

ABSTRACT

The Synchrotron Radiation Source at Daresbury is used to study luminescence decays of substituted allyl anions on a time-scale of 100 ps to 5 ns. Fluorescence competes with a twisting process about one of the central bonds which leads to cis-trans isomerisation. Time-resolved excitation spectra help in distinguishing between isomers and between tight and loose ion-pairs. Fluorescence depolarisation gives information about solvent behaviour.

INTRODUCTION

The study of cis-trans isomerisation is one of the most active areas of photochemistry at the present time[1,2]. It is one of the simplest excited state processes lending itself to theoretical modelling and it is of great practical importance in the context of the visual process. Experimental work on reaction dynamics, solvent effects etc., tends to concentrate on stilbene (1,2-diphenyl-ethene) which has readily accessible spectra in the near ultra-violet. We have studied the closely-related 1,3-diphenyl-allyl anions (DPA)[3,4]; they add an extra dimension, that of ion pairing, to the problem. DPA and its derivatives exist as trans,trans, cis,trans and cis,cis conformers (Fig. 1) and, depending on solvent and counter-ion, as loose (solvent-separated) and tight ion pairs. All these species can be distinguished by their absorption (excitation) spectra which peak in the visible region between 500 and 600 nm.

As in the case of stilbene, isomerisation takes place from the first excited singlet state and so competes with fluorescence. Lifetime measurements can therefore be used to study the effect of temperature, solvent and counter-ion on the process. The fluorescence transitions in these molecules are strongly allowed: lifetimes range from a maximum of 3-5 ns at low temperatures to less than 100 ps at room temperatures. The small pulse-width of the SRS at Daresbury combined with its high repetition rate (3 MHz) gives it advantages over both conventional flash-lamps and lasers; in simple cases, lifetimes can be measured in a few minutes and data for an Arrhenius plot can be obtained in 2-3 hours. Further, the ready tunability of the source makes possible the use of time-resolved excitation spectra to distinguish the species present. Fluorescence anisotropy measurements can be used to compare overall molecular rotation with internal rotation.

EXPERIMENTAL

As a result of the installation of the high brightness lattice at Daresbury the pulse-width has been reduced to 165 ps. Injection of a single bunch is not easy, but currents of 30 mA can be obtained with less than 1 part in 10000 in other radio-frequency buckets. Because of the very good machine vacuum, beam lifetime is long, typically 20-30 hours, so that beam currents may still exceed 15 mA at the end of a 24 hour run. Port HA12 is used for visible and the near ultraviolet; for a current of 20 mA, the number of photons available is 4.10^{12} /sec/1% bandwidth.

A Spex 1500 SP vacuum monochromator is used to disperse the incident radiation; filters are used to separate fluorescence from scattered incident light. Short wavelength cut-off filters suffice when only one emitting species is present: interference filters are used when more than one is involved.

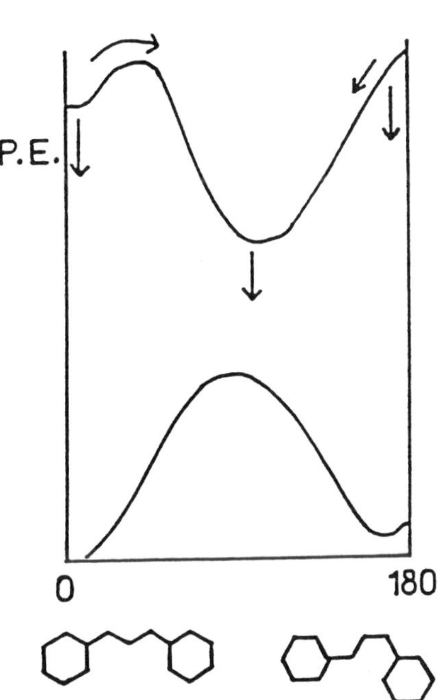

Fig. 1. Schematic potential energy diagram: ground & first excited singlets of DPA anions.

Routinely, fluorescence is detected with a Mullard XP2020Q or PM2254B red-sensitive photomultiplier. Time resolution has been further improved by using a Hamamatsu R1564U microchannel plate photomultiplier. Conventional single-photon counting is used to measure luminescence pulse shapes[5-8]. In brief, photomultiplier pulses are fed via an Ortec 583 or Tennelec TC 455 constant fraction discriminator to a time-to-pulse-height converter (TPHC); stop signals are provided by a pulse obtained from a timing button electrical pickup situated in the synchrotron vacuum vessel. The TPHC pulses are stored in a multichannel analyser: typically 10000-50000 counts are accumulated in the peak of the decay profile. The output pulses of the discriminator are counted while the monochromator is scanned to give a total luminescence excitation spectrum. The single channel output of an Ortec 567 TPHC can be recorded simultaneously to give the excitation spectrum of emission in a pre-set time region.

Fluorescence anisotropy decay measurements[6] are made by recording fluorescence decay profiles, with a polariser set alternately parallel and perpendicular to the plane of polarisation of the incident radiation.

DPA reacts with oxygen, carbon dioxide and proton donors such as water and alcohols: solutions are prepared under vacuum; to date, studies have been limited to solutions in amines and ethers. The cryostat and the preparation of samples have been described elsewhere[4].

DATA ANALYSIS

Synchrotron radiation has a major advantage over other sources in that the pulse shape is independent of wavelength: wavelength effects in the photomultiplier can be avoided. The response function or 'prompt' is recorded with a light scatterer at the fluorescence wavelength. Suspensions of 'Ludox' in water are used for this purpose. Typical prompts are shown in Fig. 2: note the log scale! Ideally the prompt should be recorded after each measurement, to obviate drifts in the electronics. However, Ludox solutions have the disadvantage that they cannot be cooled below 0 C. For data analysis we therefore use initially a non-linear least squares fitting programme which incorporates a time-shift as a variable[5]. This procedure is open to objection: short lifetimes and shifts do not separate easily so that systematic errors in one may affect the other. In some cases, the fitting parameter, 'chi-squared', varies erratically with shift, but in others, we find a monotonic change in the required shift with time, of some 20-50 ps over 1-2 hours which we regard as acceptable.

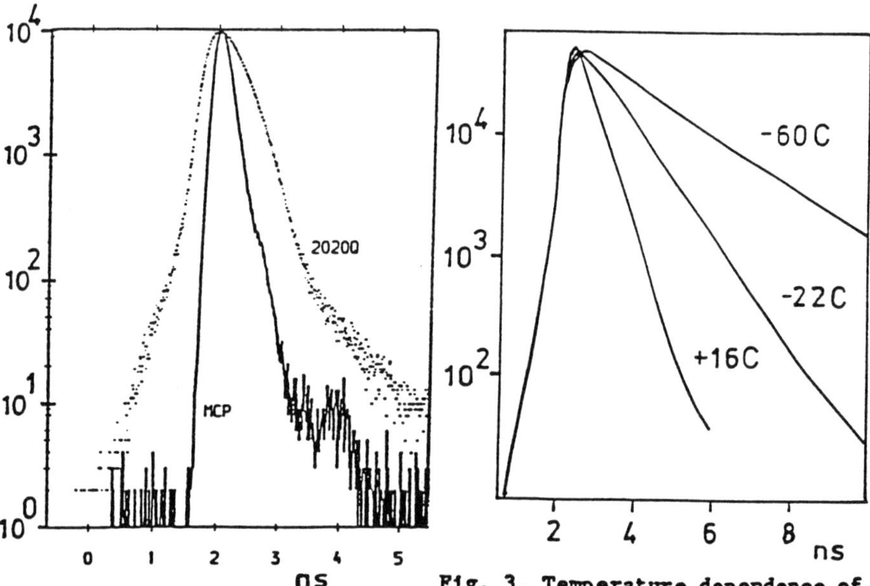

Fig. 2. Response functions.

Fig. 3. Temperature dependence of luminescence decays: Li-DPA in N,N-dimethyl-ethylene diamine.

Attempts to fit decay data to a number of exponentials can be misleading[9]. There is a danger of reading too much into a given set of numbers. An antidote to this is the use of the maximum entropy method (MEM)[10,11], fitting to a logarithmic series of exponentials - a exp($-t/\tau$) where the τ values are fixed, the a's variable: out of the range of possible results consistent with the data, MEM chooses that containing the minimum information consistent with fitting the data within pre-set limits.

RESULTS AND DISCUSSION

Typical decay curves are shown in Fig. 3. Non-linear least squares fitting requires only one exponential at ambient temperature but a second emitter - believed to be an aggregate of two or more ion pairs in this case - appears as the temperature is reduced. This is confirmed by the MEM analysis - Fig. 4 (note the log scale): the breadth of the distributions reflects the statistical uncertainties. One has to balance the shortage of beam time against the statistical limitations.

Synthetic data-sets are generated by convoluting one or two exponentials with the same prompt, adding noise appropriate to the number of counts in each channel: application of the MEM method then gives essentially the same results. In other cases, good fits cannot be obtained in this way, suggesting that a range of exponential decays is present.

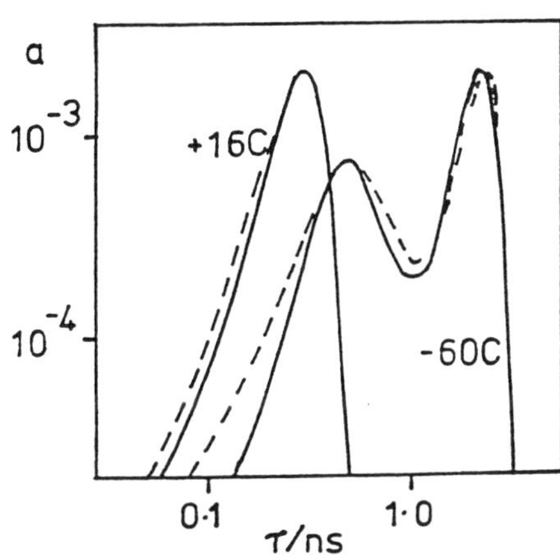

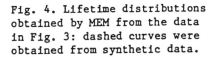

Fig. 4. Lifetime distributions obtained by MEM from the data in Fig. 3: dashed curves were obtained from synthetic data.

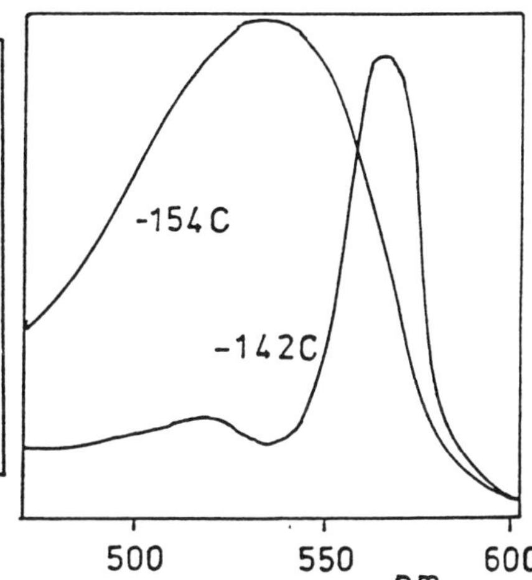

Fig. 5. Time-resolved excitation spectra (> 5 ns after the peak) for Li-2-methyl-DPA in MTHF.

Typically for these systems, lifetimes are in the range 10-100 ps at room temperature, increasing to limiting values around 3-5 ns on cooling. The results are interpreted in terms of the potential energy diagram shown in Fig. 1[1]. Fluorescence competes with temperature-dependent and temperature-independent non-radiative processes. The latter may involve both internal conversion to the ground state and inter- system crossing to the triplet manifold: internal conversion is likely to predominate because the energy gap is relatively small (~2 eV) in these anions. There is no evidence of the involvement of triplet states in the excited state behaviour, but it is not possible to test for this using triplet sensitisers or quenchers because of the likelihood of chemical reaction.

Fluorescence quantum yield measurements have been made as a function of temperature for the parent molecule[3]; these show that the temperature independent deactivation is a minor process - the yield rises to 0.82 at low temperatures. The major process is temperature dependent: by analogy with stilbene, this can be identified as a twisting process about one of the allyl bonds in the first excited singlet (Fig. 1); deactivation of the perpendicular form gives the cis,trans or the trans,trans isomer. This proposal is supported by a great deal of photochemical evidence[4,12].

The reaction scheme leads to equation (1), in which τ is the measured lifetime, τ_0 the low temperature limiting lifetime; A & E are the Arrhenius parameters for the activated decay process.

$$\ln(1/\tau - 1/\tau_0) = \ln A - E/RT \qquad (1)$$

A considerable body of data has been obtained for derivatives and vinylogues of DPA with lithium, sodium, potassium and caesium as counter-ions in a number of ether and amine solvents. Accurate determination of all three parameters is not easy: in many cases, the useful temperature range is limited by crystallisation of the solvent. However, τ_0 values all lie in the range 2-5 ns; the activation energy, E, for trans,trans compounds is usually of the order of 15-25 kJ mol^{-1}, much smaller than the 60-90 kJ mol^{-1} required for ground state interconversion between cis,trans and trans,trans forms; A is usually between 10^{12} and 10^{13} but occasionally larger - > 10^{15} s^{-1}. In particular, such large values have been found for tight ion pairs of DPA and have been ascribed to a large positive entropy of activation[4]. Studies of other systems show that tight ion pairs generally have much shorter lifetimes than the corresponding loose ion pairs, but determination of the Arrhenius parameters is usually difficult because tight pairs convert to the loose form on cooling.

In these complex systems, time-resolved excitation spectra help in identifying the emitters. They would also demonstrate the occurrence of excited state interconversion e.g. between ion pair types, but no evidence of such processes has been found. Unlike the trans form, cis-stilbene is non-planar in the ground state because of steric interaction between hydrogen atoms: as a result, its fluorescence is only observed from solids or very viscous liquids[1], showing that there is no intrinsic barrier to twisting. The

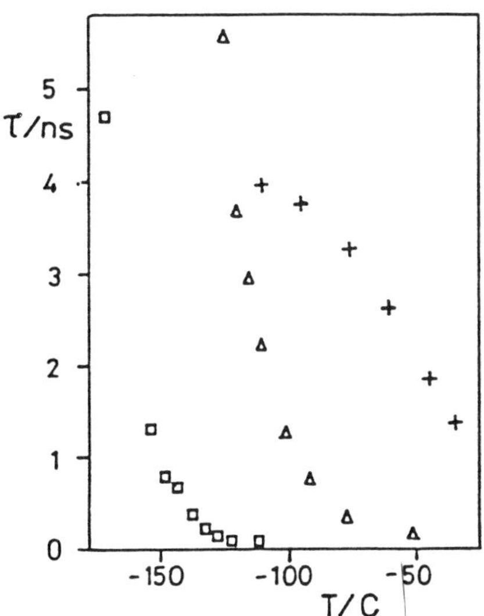

Fig. 6. Fluorescence lifetimes of cis,trans, □ , and trans,trans, +, Li-2-methyl-DPA and rotational correlation time, △, for Li-diphenyl-pentadienyl in MTHF.

same behaviour is observed in the case of DPA etc. No emission from the cis,trans form of the parent molecule has been observed, but substution, e.g. by methyl, on the central carbon - the 2 position - destabilises the trans,trans form relative to cis,trans in the ground state. Interestingly, the fluorescence of 2-methyl-DPA is dominated by emission from the minority ($\sim$ 7%) trans,trans form over most of the temperature range, showing that there is still a barrier to twisting despite the steric effects of the substituent. However, at low temperatures the main emission is that of the cis,trans form: the time-resolved excitation spectra in Fig. 5 show a dramatic change over a small temperature range as the cis,trans lifetime lengthens. The temperature dependence of the lifetimes of the two isomers in 2-methyl-tetrahydrofuran (MTHF) is shown in Fig. 6.

If there is no intrinsic barrier to internal rotation, or it is small, solvent viscosity will play a role in internal rotation[1]. Also shown in Fig. 6 is the rotational correlation time for lithium-pentadienyl in MTHF derived from fluorescence polarisation measurements. Isomerisation of the cis,trans form, which requires only a small 'free volume' is a faster process than the end over end rotation required for fluorescence depolarisation. Arrhenius plots show that the effective barrier is smaller for internal rotation.

In conclusion, the results presented illustrate the usefulness of synchrotron radiation for studies of photochemical processes occurring on the nanosecond time-scale. Interpretation of the results requires theoretical calculations of the potential energy surfaces involved and the effects of ion pairing thereon: this work is in progress[12].

The authors thank the Science and Engineering Research Council for providing facilities, Dr. M. Behan-Martin, Mrs. C. Gregory, Dr. M. Hayes, Dr. G. R. Jones, Dr. M. M. Martin, Professor I. H. Munro, Miss C. E. Oliver, and the workshop staff at Sheffield for their help with the experiments.

REFERENCES

1. D. H. Waldeck, Chem. Rev., 91, 415 (1991).
2. R. S. Becker, Photochem. Photobiol., 48, 369 (1988).
3. S. S. Parmar, B. Brocklehurst and R. N. Young, J. Photochem., 40, 121 (1987).
4. B. Brocklehurst, R. N. Young and S. S. Parmar, J. Photochem. and Photobiol., A: Chem., 41, 167 (1988).
5. D. V. O´Connor and D. Phillips, Time Correlated Single Photon Counting (Academic Press, London, 1984).
6. R. B. Cundall and R. E. Dale ed., Time-Resolved Fluorescence Spectroscopy in Biochemistry and Biology (NATO ASI Series A, Vol. 69, Plenum, New York and London, 1983).
7. I. H. Munro, D. Shaw, G. R. Jones and M. M. Martin, Anal. Instr., 14, 465 (1985)
8. B. Brocklehurst, Chemistry in Britain, 23, 853 (1987).
9. D. R. James and W. R. Ware, Chem. Phys. Lett., 120, 450, 455, (1985).
10. A. K. Livesey and J. C. Brochon, Biophys. J., 52, 693 (1987).
11. A. Siemiarczuk, B. D. Wagner and W. R. Ware, J. phys. Chem., 94, 1661 (1990).
12. C. E. Oliver, B. T. Pickup, R. N. Young and B. Brocklehurst, unpublished work.

IV. TIME-RESOLVED X-RAY SPECTROSCOPY AND SCATTERING

Time-Resolved X-ray Studies Using Third Generation Synchrotron Radiation Sources

Dennis M. Mills
Advanced Photon Source
Argonne National Laboratory
9700 S. Cass Avenue
Argonne, IL USA 60439

Abstract

The third generation, high-brilliance, hard x-ray, synchrotron radiation (SR) sources currently under construction (ESRF at Grenoble, France; APS at Argonne, Illinois; and SPring-8 at Harima, Japan) will usher in a new era of x-ray experimentation for both physical and biological sciences. One of the most exciting areas of experimentation will be the extension of x-ray scattering and diffraction techniques to the study of transient or time-evolving systems. The high repetition rate, short-pulse duration, high brilliance, and variable spectral bandwidth of these sources make them ideal for x-ray time-resolved studies.

The temporal properties (bunch length, interpulse period, etc.) of these new sources will be summarized. Finally, the scientific potential and the technological challenges of time-resolved x-ray scattering from these new sources will be described.

Introduction

The unique properties of synchrotron radiation, particularly, the high flux and temporal modulation, naturally lend it for use in time-resolved x-ray studies. Over the last several years, the kinetics of a variety of interesting transient physical systems with time scales spanning from hours to nanoseconds have been explored using radiation emitted from storage ring sources. Both the scientific and technical aspects of many of these time-resolved experiments have been chronicled in several recent review articles and/or workshop proceedings [1-4], and the reader is referred to these existing papers for a more detailed account of research in this expanding field. Hence, the content of this presentation will focus on the advances in the field of time-resolved studies that might arise with the advent of the next (third) generation of storage ring sources currently being constructed in Europe, Japan, and the United States. The discussion here will be limited to sources designed to optimize radiation output in the hard x-ray region (i.e., $E \geq 10$ keV).

Third Generation Sources

Third generation synchrotron radiation sources can be characterized as dedicated, low-emittance storage rings optimized for high brilliance x-ray beams (see Fig. 1). Put another way, these storage rings have been carefully designed to provide numerous, long, straight sections for insertion devices (wigglers and undulators) and a particle beam with a small source size and divergence. Three such storage ring sources are currently in various stages of completion: the European

Synchrotron Radiation Facility (ESRF) in Grenoble, France: the Super Photon Ring (SPring-8) in Harima, Japan; and the Advanced Photon Source (APS) in Argonne, Illinois, USA. Although the details of each of these storage rings are different, all have similar design goals and specifications. I will use the specifications for the APS as archetypical of a third generation synchrotron source because I am most familiar with them: I will also attempt to point out major differences between the sources if relevant to time-resolved studies. (Detailed specifications of the radiation properties of these new sources can be found in references [5], [6] and [7].) The storage ring is specifically designed to accommodate a large number (>30) of long (up to 5 meter) insertion devices. Particle beam emittances are approximately 1 nanometer-radian in the vertical and 10 nanometer-radian in the horizontal; the vertical emittance being determined by the need to be approximately equal to the natural vertical opening angle of the first harmonic of the undulator, typically 10-15 microradians. The third generation sources have particle beam energies in the range from 6 to 8 GeV. This beam energy is required to produce x-rays in the 4 to 40 keV range from the first and third harmonics of planned undulators. A by-product of this high particle energy is a high flux of very hard x-rays (E>50 KeV) produced from high field wigglers installed in straight sections. Because of the beam energy required, the circumference of these rings is large, 1104 meters in the case of the APS, corresponding to relatively long orbital periods (3.68 microseconds at the APS). However, that will not be the repetition rate of the radiation under normal operating conditions. In order to achieve the design current of 100 milliamperes while maintaining the low emittance, multi-bunch operation will be required. At this time, the APS plans to achieve this current level with 20 equally spaced bunches (184 nanosecond spacing), while the ESRF will attempt to accomplish this design goal with 992 bunches (2.8 nanosecond spacing). The typical bunch length for these sources will be 50-100 picoseconds (the APS calculated bunch length is 72 picoseconds FWHM).

Time-Resolved Experiments on Third Generation Sources

Flux increases of at least one order of magnitude over that at existing sources are expected from insertion devices on third generation sources through the 10 keV to 100 keV energy range (see Fig. 2). Clearly, an increase of this magnitude will make an impact on flux hungry experiments such as time-resolved x-ray diffraction or dispersive EXAFS and XANES. Increased flux will allow smaller apertures to be used with those experiments where small beam sizes are required, for example, to examine traveling interfaces due to physical processes such as pressure waves, combustion fronts, thermal waves, etc.

The large number of hard x-rays generated from these sources may facilitate time-resolved experiments under extreme conditions such as high pressure and high/low temperatures. High energy photons are particularly useful for these types of experiments in which the x-rays must pass through the walls of environmental cells (diamond anvil cells, furnaces, cryostats, etc.) to get to and from the sample. In addition, highly collimated beams may be required in order to illuminate the desired portion of the sample so that the parameters of interest (pressure, temperature, etc.) are relatively uniform throughout the volume of sample being probed.

It is clear, however, that these new sources will have the greatest impact on those experiments that depend on beam brilliance. Increases in beam brilliance four orders of magnitude greater than that available at existing synchrotron facilities are expected at the APS and equivalent sources. This improvement in beam brilliance should allow one to consider brilliance-limited experiments, which are currently difficult to perform in a static mode (such as high-resolution scattering experiments, surface scattering, ultra-small angle scattering, and microprobe experiments), in a time-resolved mode. In addition, high brilliance may also play an important role in flux-limited experiments since the well-collimated beams should reduce off-axis aberrations from optical components thereby facilitating better focusing performance for obtaining small beam sizes.

Although one generally associates high brilliance with the concept of more highly collimated photons on a sample, third generation sources will provide other interesting and useful radiation properties such as coherence. The applications of coherent x-ray beams to time-resolved studies may be prove to be very useful indeed. One such application is an extension of what is called in the visible region of the spectrum dynamic light scattering or intensity fluctuation spectroscopy [8]. In the visible portion of the electromagnetic spectrum, one can observe this phenomena via the so-called speckle patterns that result when coherent light is scattered. Using this effect, time-resolved studies of critical fluctuations near phase transitions, for instance, are investigated with visible light. The length scales that can be studied is limited by the wavelength of the probing light, which is several thousand Ångstroms for visible light. Using x-rays, one could probe the dynamics of processes at atomic-length scales, opening up a host of new possibilities for time-resolved, equilibrium and non-equilibrium thermodynamic studies. The key to the success of this type of experiment is the production of an x-ray beam that is both transversely and longitudially coherent over the scattering volume of the sample. The third generation of high-brilliance x-ray sources will have a considerable amount of coherent x-ray flux (see Fig. 3). Although only a fraction of the total number of photons emitted at that wavelength are coherent, there is ample flux for many types of time-resolved diffraction experiments. For instance, on an APS undulator nearly 10^{10} photons/sec-0.1%bw of transversely coherent photons will be emitted near 8keV. [The longitudinal coherence length, L_c, of the radiation is determined by the monochromaticity of the beam: $L_c = (\lambda/2)(\lambda/\Delta\lambda)$.]

None of the experiments mentioned above make direct use of the modulated time structure of the radiation. One of the more interesting proposed applications of the pulsed nature of the radiation is its use to capture transient phenomena in large macromolecular crystals. In particular, if it were possible to obtain a usable diffraction pattern from a crystal with the x-rays from one bunch of particles, then the temporal resolution of such a technique could be less than 100 picoseconds. Since it is clear that a crystal could not be rotated over reasonable angles on this timescale, monochromatic techniques that use angle integrating methods are not appropriate. An alternative to the monochromatic technique is a polychromatic Laue method in which a wavelength integration is performed by the diffracting planes. Such an approach is ideal for time-resolved experiments. It has been calculated that between 10^{10} and 10^{13} photons are required to produce a interpretable Laue photograph [9]. If a broad-energy bandwidth input beam is used, say 5-15 keV, then, in this bandwidth from one burst of x-rays from a wiggler on the APS running with 5 ma per bunch, approximately 4 x 10^{10} photons/horizontal milliradian will be emitted. By collecting over the full horizontal extent of the wiggler beam and slightly increasing the bandwidth, this number could be increased to over 10^{11} photons per burst. Several years ago, the feasibility of such experiments were explored using an APS-prototype undulator installed in the Cornell Electron/Position Storage Ring (CESR) [10]. Single-pulse diffraction patterns were collected, although not in a time-resolved mode nor with the necessary quality to permit new structural information to be extracted.

Summary

Although the third generation of synchrotron radiation sources will provide unparalleled opportunities for researchers interested in time-resolves studies, there are still many technological and experimental challenges that need attention. At the risk of being too parochial, I will mention the work the APS staff has been doing to address some of the important instrumentation issues. Obviously, one critical area is detector development. There are classes of time-resolved experiments that are already constrained by detector limitations with currently operating sources. New sources will only exacerbate the situation. Development of high-count-rate and fast-framing one-and two-dimensional detectors will be crucial to the success of many proposed time-resolved experiments. A programmable Charge Coupled Device (CCD) detector has been developed by the staff of the APS that allows operation in a "streak camera" mode (Fig. 4) permitting 1-D frame transfers to be completed in approximately 2 microseconds [11]. This is an important milestone since it means that, when the storage ring is run in single-bunch mode (as may be required for some timing experiments), a frame transfer can be completed before the next burst of x-rays arrives. Thus, the temporal resolution jumps from 2 microseconds to the duration of the burst itself, approximately 100 picoseconds. We have just begun to investigate two-dimensional arrays that have similar framing rates.

As mentioned above, there may be the need to run in a special timing mode (i.e., one-bunch mode) for certain types of experiments. Because it is expected that only 5 milliamperes of current will be in a single bunch, timing modes are usually not appreciated by the other users because the current in the storage ring and, hence, the flux and brilliance are reduced to a small fraction of the usual value: 5% in the example cited here. To lessen the effect of running in a timing mode on other users not interested in timing experiments, we are looking into possible techniques of altering the natural time structure of the ring in a local way (i.e., on selected beamlines) by using high speed mechanical choppers and modulating mirrors and monochromators [12]. Single pulses have been successfully isolated with mechanical choppers on large storage rings such as CESR [13] where the interpulse period is long (2.56 microseconds). We hope to pursue this approach further to see if it is applicable to the APS when more than one bunch of stored particles are in the ring.

Perhaps the most daunting hurdle, however, is uniform (spatial and temporal) excitation of the sample. This problem, however, is not unique to time-resolved studies involving synchrotron radiation, and, fortunately, one can borrow techniques from other disciplines (such as time-resolved laser techniques) to overcome this challenge. Certainly the next decade, as the ESRF, SPring-8, and APS all become fully operational, will be an exciting and stimulating time for those interested in the study of transient phenomena using x-rays.

Acknowledgements

The author would like to thank Prof. Mark Sutton, McGill University, and Dr. Brian Stephenson, IBM, for helpful discussions about their work on x-ray intensity fluctuation spectroscopy and Drs. Brian Rodricks, Dean Haeffner, and Wen Bing Yun, APS, who supplied the figures in this text. This work is supported by the U.S. Department of Energy, BES-Materials Science, under Contract No. W-31-109-ENG-38.

References

[1] S. Gruner, Science **238**, p 305, 1987.

[2] "Time-Resolved Studies and Ultra-FastDetectors: Workshop Report," R. Clarke, P. Sigler, and D. M. Mills, ANL/APS-TM-2, Argonne National Laboratory, Argonne, Illinois, USA.

[3] B. C. Larson and J. Z. Tischler, *Advanced X-ray/EUV Radiation Sources and Applications*, San Diego, CA, SPIE Proceedings **1345**, p 90,1990.

[4] "Time-Resolved Studies," D. M. Mills, *Handbook on Synchrotron Radiation, Vol. 3*, G. Brown and D. E. Moncton, eds., Elsevier Science Publishers, B. V., p 291, 1991.

[5] "The Red Book," European Synchrotron Radiation Facility, B.P. 200 38043 Grenoble Cedex, France.

[6] "Characteristics of the 7-GeV Advanced Photon Source: A Guide for Users," G. K. Shenoy, P.J. Viccaro, and D. M. Mills, ANL-88-9, Argonne National Laboratory, Argonne, Illinois, USA.

[7] "SPring-8 Project," The JAERI-RIKEN SPRing-8 Project Team, 2-28-8 Honkomagome, Bunkyo-ku, Tokyo, Japan.

[8] M. Sutton, S. G. J. Mochrie, T. Greytak, S. E. Nagler, L. E. Berman. G. A. Held, and G. B. Stephenson, Nature **352**, p 608, 1991.

[9] I. J. Clifton, V. Fulop, A. Hadfield, P. Norlund, I. Andersson, and J. Hajdu, Nucl. Instrum. and Meth. **A303**, p 476, 1991.

[10] D. M. E. Szebenyi, D. Bilderback, A. LeGrand, K. Moffat, W. Schildkamp, and T.-Y. Teng, Trans. Am. Cryst. Soc. **24**, p 167, 1988.

[11] B. Rodricks and C. Brizard, Nucl. Instrums. and Meth., in press.

[12] D. M. Mills, Rev. Sci. Instrum. **60**, p 2338, 1989.

[13] A. LeGrand, W. Schildkamp, and B. Blank, Nucl. Instrum. and Meth. **A275**, p 442, 1989.

Figure Captions

Fig.1 Plot of x-ray brilliance as a function of x-ray energy for the Advanced Photon Source and several existing sources currently operating in the United States.

Fig.2 A plot of the x-ray flux as a function of x-ray energy for one of the wigglers planned for the Advanced Photon Source and several existing sources currently operating in the United States.

Fig.3 A plot of the number of transversely (spatially) coherent photons as a function of energy for one of the proposed undulators to be placed on the Advanced Photon Source.

Fig.4 Schematic of the experimental arrangement for operating a CCD detector in a one dimensional fast timing mode. An x-ray opaque mask covering all pixels except the top row is placed between the detector and x-rays. The top, exposed, row of pixels is illuminated by the x-rays for a prescribed amount of time and then parallel transferred to the adjacent row behind the mask. This transfer time can occur in several microseconds. The pixels rows behind the mask are used as storage until the entire CCD array is filled, at which time it is read out serially.

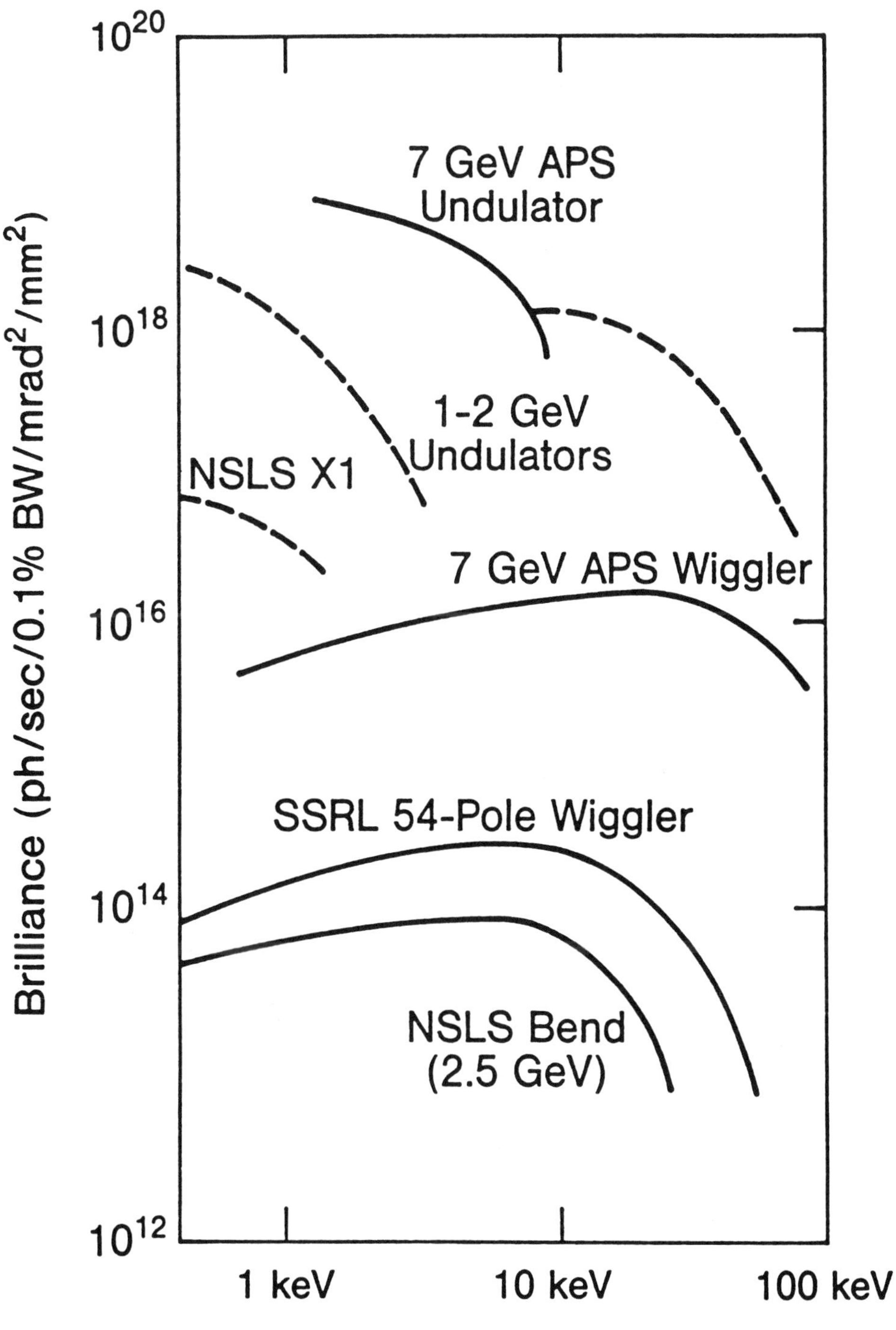

10^{20}
10^{18}
10^{16}
10^{14}
10^{12}
Brilliance (ph/sec/0.1% BW/mrad^2/mm^2)
7 GeV APS Undulator
1-2 GeV Undulators
NSLS X1
7 GeV APS Wiggler
SSRL 54-Pole Wiggler
NSLS Bend (2.5 GeV)
1 keV
10 keV
100 keV

Figure 1

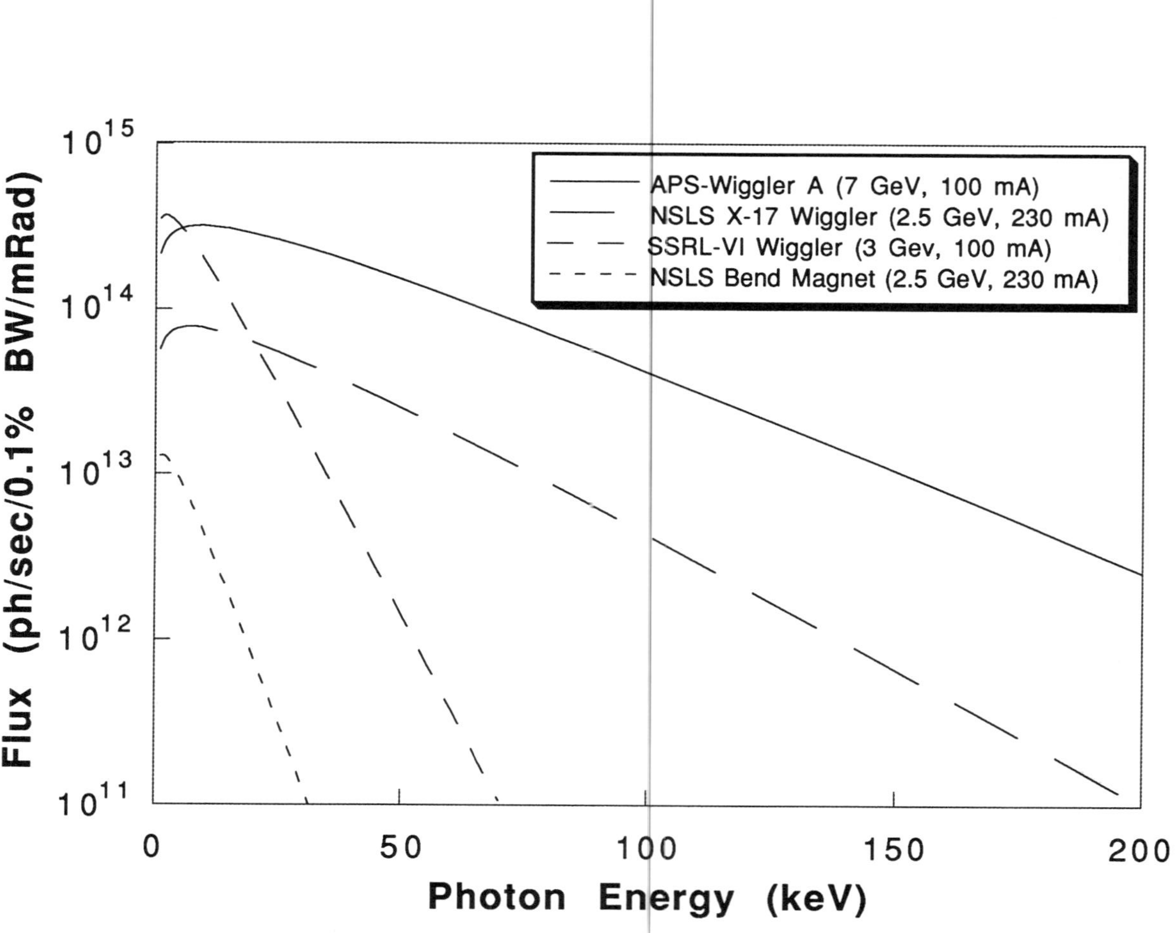

Flux (ph/sec/0.1% BW/mRad)
10^{15}
10^{14}
10^{13}
10^{12}
10^{11}
0
50
100
150
200
Photon Energy (keV)
APS-Wiggler A (7 GeV, 100 mA)
NSLS X-17 Wiggler (2.5 GeV, 230 mA)
SSRL-VI Wiggler (3 Gev, 100 mA)
NSLS Bend Magnet (2.5 GeV, 230 mA)

Figure 2

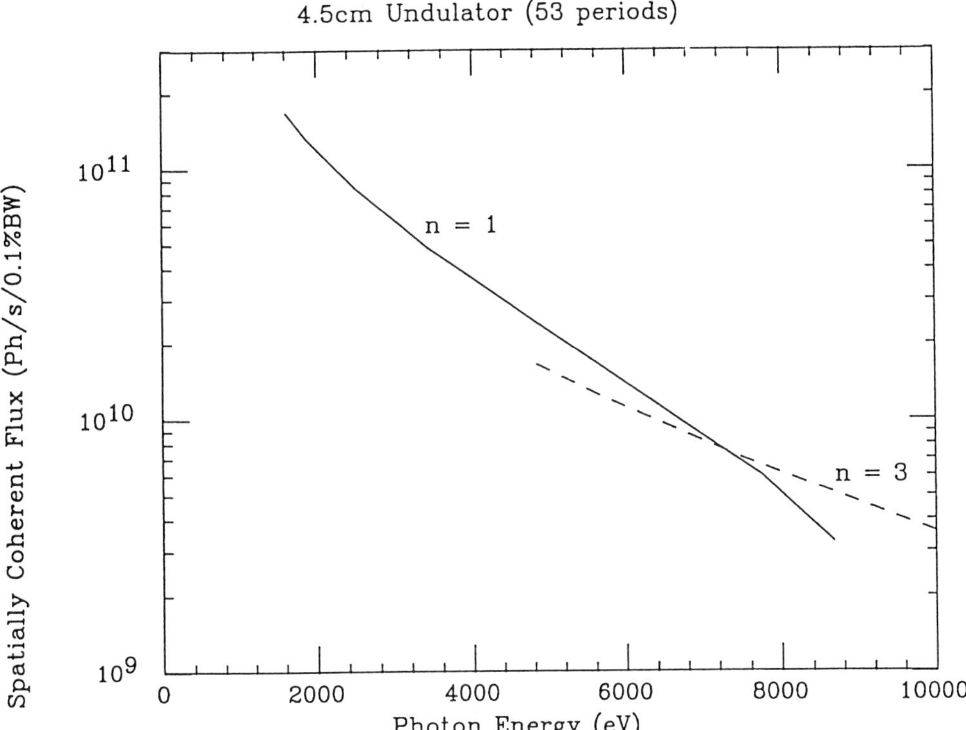
4.5cm Undulator (53 periods)
Spatially Coherent Flux (Ph/s/0.1%BW)
10^{11}
10^{10}
10^{9}
n = 1
n = 3
0
2000
4000
6000
8000
10000
Photon Energy (eV)

Figure 3

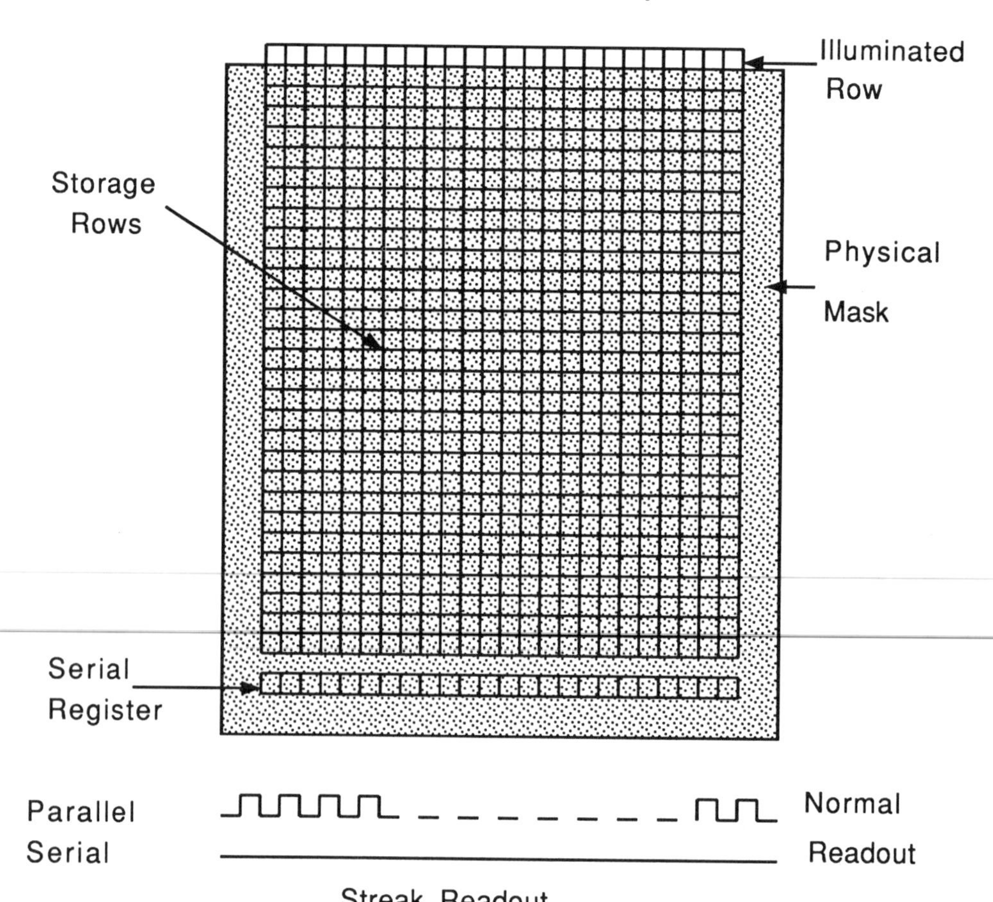
384x590 CCD Array
Illuminated Row
Storage Rows
Physical Mask
Serial Register
Parallel
Serial
Normal Readout
Streak Readout

Figure 4

DISCUSSION

BAER- In what way are your experiments dependent of brilliance rather than simply flux ?

MILLS - The experiment I spoke of concerning the use of coherent X-rays clearly depends on the beam brilliance since the ratio of the number of transversally coherent photons to the total number of photons emitted (at a given wavelength) increases as the phase space volume of the source decreases until the diffraction limit is reached (for that wavelength) where all the emitted photons will be coherent. Therefore in order to get a reasonable number of coherent X-rays, sources with small spatial and angular dimensions are required. Although the other experiments are probably flux limited, a brilliant beam can be useful in an indirect way. For many of these experiments small beam sizes are often required, due to either small samples and for small regions of the sample that is being excited. The focusing optics for brilliant beams are often much more effective since the aberrations are less when the angular divergences and source size are small.

NENNER - I wish to make a comment and inform you that the first time that a laser has been combined with synchrotron radiation comes from 1981 at Orsay-LURE ACO (Wuillermiet et al., 1981) and since then many groups have successfully used this 2 photon technique throughout the world.

MILLS - I totally agree with you. The first publication of the laser annealing experiment appeared in 1982. I perhaps should have stated that is was one of the first experiments to combine a laser with a synchrotron radiation X-ray diffraction experiment.

FONTAINE - Si-annealed experiment : what is the geometry of your 2 beams experiments and what is the quenching rate of the cooling you got and how is it depth dependent ?

MILLS - The laser beam is situated such that it comes at near-normal incidence to the sample, while the X-ray beam must satisfy Braggs law. For Si(III), with an X-ray wavelength of 1.5 Å, the Braggs angle is about 14.3° and hence this is the angle of incidence for the probing beam. As far as the quenching rate, we do not have a detailed analysis on the rates of cooling. From the time resolved "temperature" vs. depth into the crystal curves shown in the manuscript order of magnitude estimates can be made.

RUFFER - 1. Pico- and nano second resolved measurements are very valuable for many applications. Which kind of detectors in the X- and γ-ray regime are available ?

2. What about the time scale of fast shutters and modulated monochromators ?

MILLS - 1. For the experiments, I discussed pico-second (or even nano-second) detectors are not required since the source is pulsed. All we need is a detector that can discriminate on a time scale less than the inter pulse period, which is typically several hundred nanoseconds for large storage rings such as CHESS and new sources (APS, ESRF) when running in "several" bunch mode (1-20 bunches for instance).

2. The time scales for fast shutters and modulating monochromators that we are looking into are currently in the microsecond regime. At the APS, we have an active research program to look into possibilities of reducing this time-scale.

LEACH - A comment on the pulsed laser annealing of Si. Although your results support well the thermal model it is reasonable to suppose that at least in the initial stages when the laser pulse is applied, an electron gas or plasma is created. The rise and decay of this plasma could be studied by reflectivity measurements at specific wavelengths over the whole electromagnetic spectrum from the visible through the VUV. The density of the plasma could be followed in this way and in temporal fashion.

MILLS - Yes, I believe you are correct. The experiments I spoke of were in a nanosecond time scale, a time probably too slow to observe this phenomenon. However using pico or femto second probes such studies may be feasible.

BIANCONI - Can you describe the characteristics of the pumping laser ?

MILLS - The pumping laser was a pulsed UV excimer laser with a wavelength of 248nm. The width of the pulse was ~25nsec and had an output energy of approximately 1 Joule. The repetition rate was about 1Hz, and could be fired at 10Hz but with a reduced energy per pulse.

GRATTON - In relation to the possibility of nanosecond time resolution in X-ray protein crystallography, after the same process in initiated by same extend factor, for example a laser pulse, all protein molecules will relax independently, giving no diffraction. How do you plan to overcome this problem ?

MILLS - I dont know enough about protein crystallography, but I suspect it is a big problem ! See H. Bartunik's comment (following).

BARTUNIK - Excitation on ns and shorter time scales of protein structures will most probably involve local excitation within the conformation, e.g., the heme group in heme protein. Propagation of energy along covalent bonds towards remote parts of the molecule will occur on a time scale of tenths of a ps per interatomic distance. Within in total about 50-100 ps -, i.e. within the time width of a single SR pulse, the excitation will have propagated through a small protein molecule of ~ 50Å diameter. As a consequence, a perturbation of the protein conformation is to be expected which will relax to a new equilibrium state on a time scale of 10-100 ns which is characteristic for collective motions of extended path of the polypeptide chain. Hence, one may indeed have to expect that diffraction exposures on time scales shorter than 10-100 ns will correspond to significant incoherence at high resolution. Essentially the same arguments hold for simultaneous excitation of large parts of the molecule, e.g. in the absorption band region of peptides ! However, the temporary incoherence might not affect the diffraction pattern at low or medium resolution, i.e. on spatial scales of 5-10 Å. Thus, it might be possible to follow large amplitude domain motions, but those will occur on time scales longer than about 100 ns.

TIME-RESOLVED X-RAY ABSORPTION SPECTROSCOPY

A.FONTAINE, F.BAUDELET, E.DARTYGE, D.GUAY, H.TOLENTINO°, G.TOURILLON

LURE (CNRS-CEA-MENJS), Bât 209d F91405 Orsay -Cedex
° LNLS Rua Lauro Vanucci, 1020, Jardim Snata Cândida-13085 Campinas-Brasil

Abstract: Time-resolved experiments based on X-ray absorption spectroscopies involve generally mass transportation which means relatively slow processes. Selected experiments in elctrochemistry are presented and different options to improve the time-resolution are reviewed.

I - INTRODUCTION

X-ray Absorption Spectroscopy has suffered a great theoretical and experimental development in the last years. This technique has proved to be a powerful tool to elucidate number of questions in materials science. Great interest exists in time-resolved experiments achieved with extreme energy resolution and energy scale stability to take a full benefit of the strong correlation between the stereochemical environment of the absorbing atom and the exact shape and position of the absorption edge of the core levels.

Fast energy dispersive X-ray spectroscopy allows in-situ observations with data collected in a short time. Thence structural modifications are easily found. This scheme provides high energy resolution. However quantitative analysis for dilute systems are photon-limited. A great benefit is expected from the forthcoming storage ring (ESRF) which should be able to give flux larger by at least 3 orders of magnitude.

Nowadays the main limitation concerns very diluted samples since it is no longer possible to use the dispersive geometry. The fluorescent detection has proven to be efficient in photosynthesis study at concentrations of about 100 µmol. But the use of decay channels is merely irrelevant with the dispersive scheme. They requires a scanning monochromator.

We have to apologize for the uncompleteness of this report which is not an extensive review. In particular we do not report the important experiments, including stopped-flow measurements developed by the Japanese scientists (lead by T. Matsushita[2]) who have been the very first to associate synchrotron radiation and dispersive optics to perform X-ray absorption spectroscopy. In addition we would like to quote other studies carried out with the new born ports at Hamburg[3] and Daresbury[4].The readers interested can easily overcome this deficiency, looking through the proceedings of the four last EXAFS conferences hold at Stanford (1984) and Fontevraud (1986), Seattle (1988), and York (1990) as mentioned in the reference list.

Relevant of the fluorescent scheme and related to this field are the quick-Exafs spectrometers developed initially at Hamburg under R.Frahm[5] (see Joe Wong's paper, this volume) and at LURE under P.Lagarde,C.Prietto, H.Dexpert et M.Verdaguer[6]- which have been able in these three last years to open new routes in a wide variety of science. But the very first experiment to look for any kind of time-resolved X-ray Absorption Spectroscopy dealt with a stroboscopic approach. Denis Mills reports in this volume another scheme to take benefit of the time-structure of the emission from a storage ring.

II- TIME-RESOLVED EXPERIMENTS with a scanning monochromator

II-A- STROBOSCOPIC APPROACH.

Measuring a 100μs-resolved extended x-ray spectrum was made possible using the pulsed structure of a laser excitation as soon as 1983. The SR frequency at SPEAR (Standford) is about 1 MHz[7]. The laser flash used for the photolysis of carboxymyoglobin at 85K has a half-width of 60μs. X-ray absorption is measured for a deviation time τ, starting at a time t_1 after the flash trigger and measured again at t_2, etc,....The measuring interval τ and the times of measurement t_1, $t_2$are controlled by a gate circuit. The key component of the spectrometer were a pair of low-noise charge integrators. Each data point is the sum of 600 measurements, for XANES profiles and over 10^4 for EXAFS. Even if the specific experiment itself suffered from failure in photolysis efficiency, this first attempt proved the reliability of their spectrometer.

II-B- THE FREE STANDING FLOW APPROACH FOR THE MICROSECOND TIME-RESOLUTION.

Very recently, Livin and Stern et al.[8] (Seattle) demonstrated the possibility of the microsecond time-resolution using a flow technique. Preliminary results of photolysis of a di-platinum complex $Pt_2(P_2O_2H_2)_4^{4-}$ in a dilute solution. The free vertical standing flow is produced with technology borrowed from dye-laser and cytoanalysis, which includes a fine nozzle and a pumping system.

The nozzle produces a vertical jet of the solution with a vertical flatness. The flow can be adjusted to be 100-200μm-thick, and 4,5mm-wide.The laminar flow is achieved by adjusting the viscosity of the water solution by including viscous liquid such as admixtures of glycerol and ethylene glycol. This can be a serious limitation since this additive has to be neutral with respect to the studied molecule. A funnel catches and returns the solution to a reservoir which is then pumped back to the nozzle.

Time-resolution is obtained by line focusing the laser at some point on the flow below the nozzle and then probing X-rays crosses the solution at some distance d downstream on the flow, corresponding to a time t=vd, where v is the flow velocity.This time delay is tunable with the adjustment of d. Typical flow velocities are 20 m/s. Therefore an X-ray spot, 20μm-high gives a time-resolution of a microsecond.

To enhance the count rate, Livin and Stern were using a 1.5m-long tapered twin mirrors which acts as a waveguide. The bending was parabola-like to transform a 500 μm-high incoming beam into a 15 μm-high one at the exit. The overall efficiency was 30% and results in an improved flux reaching 10. EXAFS results show changes of distance of the Pt-Pt bond length within the microsecond. The limitations come from the need of a short excitation such as forphotolysis to initiate time-dependent process and from the need to make samples compatible with the form of a laminar steady flow. But it remains a very appealing scheme for photochemistry or any photo-induced phenomena.

II-C- QUICK EXAFS FOR MATERIALS SCIENCE.

This scheme relies on a fast rotation and encoding of a two-crystal generator to collect continuously the incident and the transmitted beams (or one of the decay channels) avoiding dead (wasted) time as generated with the scanning mode step by step. As a premium one gets a smoother rotation which eliminates a huge source of noise. Each data point results from an integration for typically 0.01 or 0.05 second. A full EXAFS spectrum (1300 eV) can be collected in 30 seconds approximately, instead of the 10-20 minutes required with the step by step scan. Taking into account the need to come back at the initial position to start the following acquisition, and the inherent delay between the first data point and the last data point of one single spectrum, one can speak of a time-resolution of a minute.

Since the two-crystal monochromator is decay-channel compatible the thin films investigation can be carried out using the QuicK-EXAFS approach.R.Frahm[5] collected data of high quality with only 5 monolayers of Copper deposited on wolfram.

M.Verdaguer is reporting in this volume, on a kinetical reaction study[6] where the Ce^{IV} ion in sulphate medium becomes reduced to Ce^{III} by adding a little quantity of alcohol. The redox reaction results in a change of the white line of the L_{III} edge and the EXAFS data as well, reflecting a change in the coordination..

III- THE TIME-RESOLVED SPECTROMETER BASED ON A DISPERSIVE OPTICS

Using the combination of X-ray energy dispersive optics and a position-sensitive detector able to work under high flux conditions we are able to proceed with both in-situ and time-dependent investigations[9,10].The cooled photodiode array is the unique tool which gives a good spatial resolution and a large dynamics[11]. Several papers have evidenced already what are the major benefits one can expect for chemistry[12], physics[13], material science[14], and biophyics[15,16,17]. The present paper is to discuss how valuable the energy dispersive scheme is for time-resolved experiments.

The theoretical time scale (2.7ms) is given by the faster frequency of the readout of the photodiode array used as the position sensitive detector to handle the high flux. Since focus have been put on electrochemistry which involves mass transportation along macroscopic distances, the true time scale results from the combination of the dilution of the sample under interest and the flux available at the national facility.

III-A - THE ENERGY-DISPERSIVE SCHEME

2.7 ms only are needed to collect a full EXAFS spectrum spread over 500 eV when a triangle-shaped Si bent crystal is used as a focusing dispersive optics and a photodiode array as a position sensitive detector. Each of the 1024 sensing elements of the photodiode array transforms an 6 10^4 8 keV-Xray photons into 8.8 x 10^7 electron-hole pairs. In total, each frame is made of 6x10^7 photons. It is worthwhile to summarize advantages and limitations related to this scheme:

a) The copper XANES spectra shown on fig. 1 have been recorded using a Si 311crystal to give an account of the **energy resolution which is Darwin width-limited.** The detector has to be put at twice the crystal-to-focus distance to achieve the optimized energy resolution[18]. For small sources as those of ESRF this restriction does no exist provided the detector is not close to the focus point.

b) In addition, because of the lack of mechanical movements, once the optics is tuned for a given absorption edge, **the energy scale is very stable**. This unique property yields an extreme sensitivity to detect accurately very minute energy shift of the absorption threshold induced by chemical change or magnetic dichroism.

Most of the current optics using Synchrotron Radiation diffracts in the vertical plane and thus is sensitive to vertical bouncing of the beam. The horizontal optical plane of the dispersive scheme adds this extra-advantage which helps to keep superior energy resolution since the orbit seems to show a better stability in the horizontal direction. Owe to the horizontal polarization of S.R. one must consider the $|\cos(2\theta)|$ attenuation factor which reduces the Darwin width of the crystal. This results in a lower reflectivity but in an improved energy resolution as well.

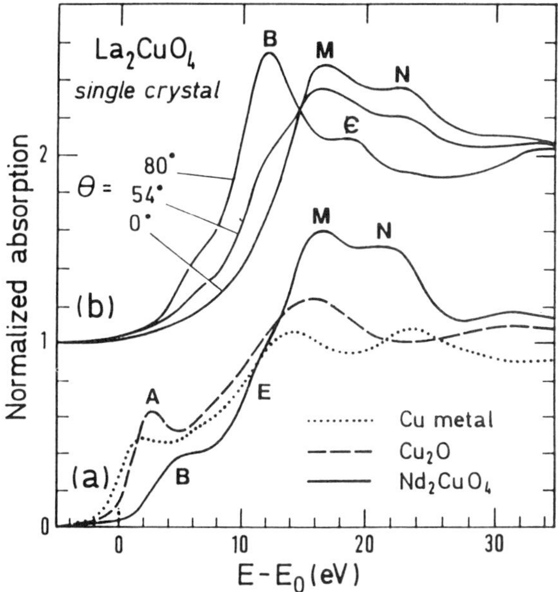

Fig.1 Copper K-edge spectra of metallic copper and cuprates measured with a Si(311) as a dispersive Bragg optics. M,N and B,E point out antibonding resonances in La_2CuO_4, B,E non bonding resonances for Nd_2CuO_4 and A for Cu_2O.

c) The linearity and the dynamics of the photodiode array give very good signal/noise ratio. The signal is essentially statistically limited.

Since this detector is made of silicon, the bad story is the decrease of the quantum efficiency for high energy X-ray photons. The good story is the relative insensitivity for harmonics which allows to work under high absorbance ($\mu t \sim 2 \rightarrow 6$). Nevertheless our spectrometer includes a mirror put just behind the sample to reject harmonics. It works also as horizontal slits of high definition which is required as soon as the experiment is polarization-specific.

d) Combined to time-resolved capability (60 spectra each of them being made of one or more - up to 32 - frames can be collected in one single shot) the in-situ observation permits a systematic investigation of the time-dependent system and an a-posteriori decision about stages of interest.

e) One has to stress that the dispersive optics does not reflect more photons than a flat crystal. The main feature which differentiates a monochromator from the dispersive optics is that the local rocking curve of the bent crystal - almost constant in width - scans across a broad energy range when one goes from one side of the crystal to the other one along the horizontal footprint of the beam span.

f) Since there is no mechanical movement during data collection the origins of possible shifts of the energy scale come from the radiation-produced thermal load on the crystal and from horizontal instabilities of the orbit the positron bunch (averaged over a few thousands turns in the storage ring). The stability of the dispersive scheme has been measured at the copper K-edge crystal (≈8980 eV) using a Si 111 monochromator and a reliability better than 30 meV was achieved after temperature stabilization[18].

This 30 meV can be converted into an angular fluctuation $d\theta=2x10^{-7}$ which, again is appreciable in terms of horizontal displacement of the source. This yields an upper magnitude of 3 µm, in the present situation where the source is at 15m from the crystal.

The understanding of the noise sources and the subsequent improvements permit to push down our limit of dilution in biophysics. Successful time-dependent and temperature-dependent photolysis of horse myoglobin have been carried out on May 1990 [This volume: S. Della-Longa]. The myoglobin dilution was 7.3 mM and maximum changes of $\Delta\mu \approx 2.4\ 10^{-3}$ (for a total jump at the edge of $1.\ 10^{-2}$) was observed with a noise kept lower than $2.0\ 10^{-4}$ and a time resolution of 80 seconds.

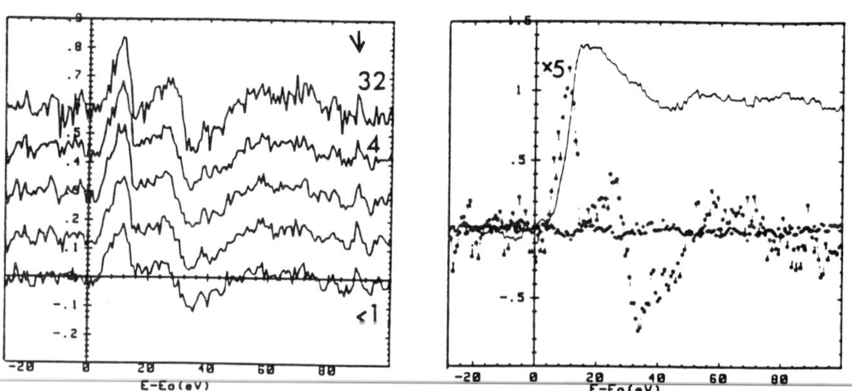

Figure 2. The time-dependent (<1, $1^{1/3}$, $2^{2/3}$, 4, 32 minutes, bottom to top) photolysis-induced signal showed here as differences from the initial spectrum shows an increasing noise and reveals the imperfection of the optics (arrows) which can be scratched of the surfaces with subsequent local misorientation. On right is the Fe-K edge myoglobin XANES spectrum and the photolysis-induced signal magnified by a factor 5 (square). The cross is given by the difference (x1) between spectra after the saturation of the photolysis effect. It is a kind of noise evaluation.

Energy resolution

The energy resolution δE of an X-ray reflector comes from the derivative of the Bragg's law,

$$\delta E = E \cot \theta\ \delta\theta$$

where **E** is the nominal energy and δθ is the overall angular resolution.
This includes:

i) the spatial resolution of the position sensitive detector **r**≈50µm, which proceeds via the angle of view of this effective pixel from the polychromatic focus point.
ii) the intrinsic Darwin width of the rocking curve of the perfect crystal;
iii) the penetration depth of the incident X-ray into the crystal,
iv) the size of the X-ray source whose contribution can be negligible under the conditions discussed below.

Besides the polychromatic imaging at the focus point, which is the optimized position for the sample, there exists another type of focusing, called herein monochromatic focusing, which strongly controls the achievable energy resolution.

The extended source does not contributes if the detector is put at the monochromatic focus. It is the interplay between the other parameters that determines the resolution. Owing to the good spatial resolution of our position sensitive detector (better than 50 μm), the dominant term is usually the Darwin width. Nevertheless, the Darwin width can be reduced by going to higher order and/or asymmetric reflections and thence the spatial resolution of the detector can become as well important.

"Aberration-free" optics

The small size of the polychromatic image allows spatially resolved experiments. This advantage is exploited, for instance, when samples are very small and when high pressure experiments are performed[19,20].

The cylindrical optics, which is obtained by bending a triangle-shaped crystal, gives rise to an aberration proportional to the square of the length of the crystal ($\approx l^2$)[21]. This parabolic energy-position correlation is known as the U-like aberration.

The ideal optics is *elliptical*, where the source and the image are located at the focus of an ellipse and the bent crystal profile fits an arc of this ellipse. This optics is aberration-free with the center of distribution of each energy being located at the same point (fig.5-b). To tackle the aberrations of the cylindrically bent crystal, corrections to the linear variation of the triangular shape were calculated and a set of Silicon crystals were tailored.

It was verified that the size of the image reduced drastically down to 400 μm, limited mostly by the demagnification factor (≈1:20) and the source size (S≈6 mm). This result is also essential for high pressure experiments since the beam must go through a tiny hole drilled in the gasket (≤ 500μm) squeezed between the diamond anvils.

Clearly new sources such as ESRF based on low emittance should allow better focusing. A gain of a factor 10 in each dimension of the source point is likely to be expected along with a better demagnification factor (1/60 instead of 1/15). Even if one cannot be sure to eradicate all the aberrations, one can envisage to go down to 50 μm as a typical focus size. It should lead to a dramatic increase of pressure making available the megabar range found in the center of the earth.

IV-TIME-DEPENDENT EXPERIMENTS WITH THE ENERGY-DISPERSIVE SPECTROMETER AT LURE

The $Ni(OH)_2$ electrode In-Situ investigation[22] is chosen to be included in this review since it combines Xanes and Exafs to trace electronic and structural modifications of the electrode at work and the great interest to work in an International ambience. Thanks to J.McBreen, W.E.O'Grady and G.Tourillon's expertise in both electrochemistry and XAFS, the data collected were of the uppermost quality.

The time resolution of such a class of experiments is flux-limited and a gain of three orders of magnitudes would be easily handled by the present detection scheme, assuming that the thermal load-related problems are solved.

Worthwhile to mention, even not detailed herein, is the investigation of the new high Tc superconductors. $YBa_2Cu_3O_{7-\delta}$ which get doped by oxygen uptake or undoped reversibly under removal have been measured in-situ, time being used to ramp up or down temperature and oxygen partial pressure independently[23]. From the quantitative evaluation of the amount of Cu^{I} ($3d^{10}$ in a linear coordination) H.Tolentino et al. were able to derive the possible sequence of the 2p hole injection in the oxygen bands and to correlate this symmetry-dependent holes to disappearance of the antiferromagnetic order (δ=0.60) and to the jump of T_c (60⇒92K) close to δ=0.25.

The last words of these comments are to stress that examples given in this brief presentation have been selected for they are well-conditioned to this spectrometer, but it is quite obvious that a lot of other measurements are better carried out, using the step by step scan or continuous scan with a two-crystal optics, taking advantage of the decay processes to be for example surface sensitive or high dilution-performant.

IV-A-IN-SITU TIME-RESOLVED XAFS STUDY OF THE NICKEL OXIDE ELECTRODE.

Structural determinations of the reactants and products of the nickel oxide electrode are difficult because of the inherently poor crystallinity of these materials. In-situ time-resolved X-ray absorption spectroscopy was coupled with cyclic voltammetry to investigate XANES[22] and EXAFS[24] during the anodic (oxidation) and the cathodic (reduction) sweeps. Both α-$Ni(OH)_2$ and β-$Ni(OH)_2$, with or without $Co(OH)_2$ additions, in KOH electrolytes were studied.

The electrode consisted of a composite of nickel hydroxide, graphite powder, vitreous carbon fibers and a plastic binder. The working electrode and the graphite counter electrode were separated by three layers of filter papers. The cell body consisted of two acrylic plastic blocks with thin (0.75 mm) acrylic plastic windows at the center of each block. A zinc wire was used as a reference, and all potentials quoted are with respect to it. The open circuit potential was ≈1.3V.

This technique and the electrochemical cell developed for regular XAS spectrometer (X11 at NSLS[24]) just worked fine for the focusing optics.

Cyclic voltammetry experiments were then done at 1mv/s between potentials of 0.6V and 2.1V for the β-$Ni(OH)_2$ electrode but up to 2.2, 2.4 V for the α-$Ni(OH)_2$ electrode more difficult to be fully oxidized.

The X-ray optics consisted of a bent Si-311 crystal and a mirror to reject the harmonics. Each XANES spectrum comes from the addition of 32x400ms-long frames which mean 13 seconds in total. The 60 spectra were spaced with one minute which leads to cover more than one complete cyclic voltammogram (50 spectra) started at 2.1V sweeping first the reduction which exhibits a double bump at 1.6 eV and 1.1 eV. The absorption edge step height was $\Delta\mu \approx 0.5$ for β-$Ni(OH)_2$ whereas $\Delta\mu \approx 0.35$ for α-$Ni(OH)_2$.

The cyclic voltammograms (fig.3) of α-$Ni(OH)_2$ and β-$Ni(OH)_2$ do not resolve the nickel oxidation process and oxygen evolution. The occurrence of two reduction steps in fig.4 has been seen in battery electrodes and is more common in electrodes with graphite as the conductive diluent. The second step is thought to be caused by the passive layer of reduced β-$Ni(OH)_2$ which is an insulating barrier. This assumption is well supported by the EXAFS analysis[25] of all the spectra collected during the reduction. There is no new intrinsically less active compound. The Ni-Ni distance found in the brucite-like reduced β-$Ni(OH)_2$ compound is 3.12Å whereas in the oxidized Ni-OOH form this distance is reduced down to 2.80Å. This 0.32Å difference generates a beat in the k-space whose location reflects the respective weight of the two end components. Because of this beating great accuracy has been achieved in the composition of the intermediate spectra which is nothing else than an admixture of the reduced hydroxide formed on surface which isolates the still oxidized core.

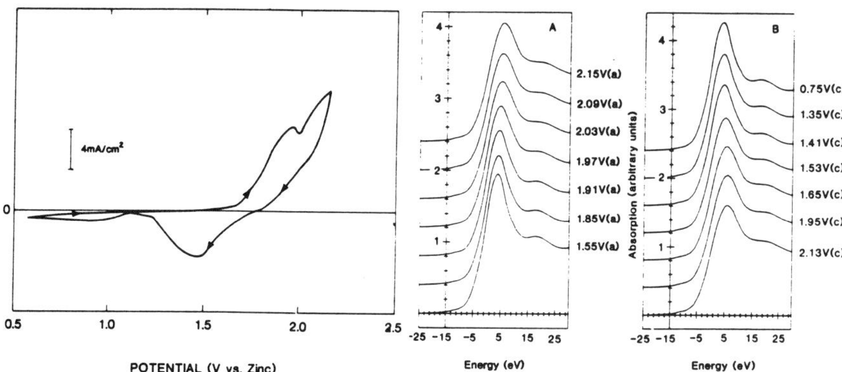

Figure 3.a- Cyclic Voltammogram of α-$Ni(OH)_2$ in 1.0 MKOH (sweep rate 1mV/s),

b- XANES spectra of α-$Ni(OH)_2$ during anodic sweep(A) and cathodic sweep(B).

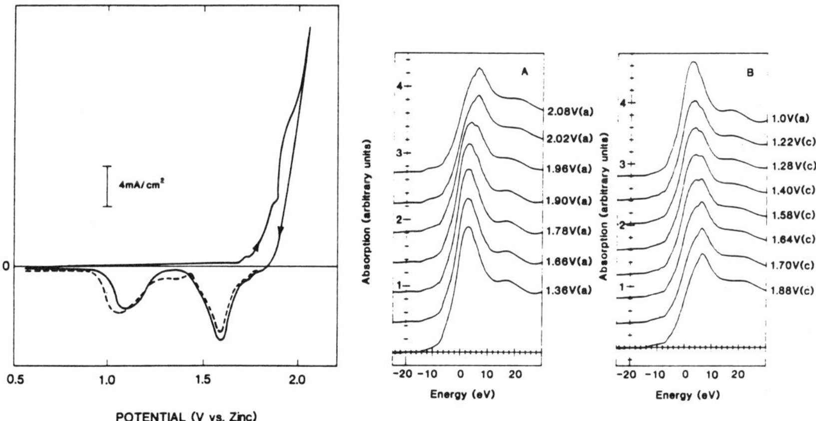

Figure 4.a- Cyclic Voltammogram of β-$Ni(OH)_2$ with $Co(OH)_2$ addition in 8.4 MKOH (sweep rate 1mV/s). Broken line is a second cathodic sweep b- XANES spectra of β-$Ni(OH)_2$ during anodic sweep(A) and cathodic sweep(B).

XANES spectra of the end compounds of the β-$Ni(OH)_2$ hydroxide is relevant of a Ni^{II} cation hexacoordinated by 6 oxygen atoms which build a regular octahedron in the reduced form. Previous EXAFS measurements carried out at X11-NSLS[24] have given 2.05Å as the Ni-O distance. The oxidized form departs from the regular octahedron. The shoulder found on the high energy side of the white line along with the overall reduction of the intensity of this white line evidence a splitting of the NiO distances with a set of short distances. The quoted EXAFS analysis concluded to a contraction of 4 NiO distances within a plane down to 1.88Å in conjunction with a slight increase of the apical distances up to 2.07Å. Theses distances come smaller than those found in the nickelates of the La_2NiO_4 family (4 at 1.95Å and 2 at 2.24Å[26]) or in $LiNiO_2$.

The correlated shift of the inflection point is rather small, 8343.8 to 8345eV. In the paper given in reference we correlated this edge shift as an evidence of the transformation of Ni^{2+} into Ni^{3+}. Due to the wide investigation of unusual Ni and Cu valencies induced by the superconductor burst and carried out by the authors and many other spectroscopists we are more reluctant now to support this edge shift as an evidence of the existence of a Ni^{3+}cation i.e. a $3d^7$ configuration.

With the "old words" this assumption was termed for a long time the “high spin—low spin” transition of the $3d^8$ (Ni^{2+}) cation which should end up as $3d^7$ (Ni^{3+}) with zero d electron in the d_{x2-y2} orbital allowing the in-plane movement of the oxygen anions towards Ni^{III}. Following the arguments detailed by P.Kuiper[26] in his thesis and several papers, we don’t believe that any crystal field energy can overcome the U_{dd} Coulomb repulsion. NiO and the Nickelates are insulators whose gap is controlled by the charge transfer energy Δ and the bandwidths w, and W of the narrow d band and the ligand 2p band, respectively .This is only true for NiO and CuO and related compounds. MnO and FeO are conversely insulators where U_{dd} determines the band gap. CoO seems to be still of the Hubbard type but U_{dd} is very close to Δ according to Zaanen’ thesis (1986)[27].

Therefore the oxidation of the Ni hydroxide proceeds very likely through holes created in the oxygen 2p band with a spin opposite to the Ni one. To enter into a more detailed discussion needs to look at the transport properties of the brucite-like structure if results are available. It is out of the scope of this paper.

To briefly summarize this section let’s recall that three results have been derived from this time-dependent XAFS study. First the two steps of the reduction comes without intermediate specy. Point 2, the large difference between the Ni-Ni distances of the two end products allows a quantitative evaluation of the progress of the reduction which correlates well to the voltammogram. Point 3, the oxidation does not seem to proceed via new Ni^{3+} specy as previously assumed. These data and more generally macroscopic data in NiO-related systems can be equally well explained by Ni^{2+} (S=1) with an antiferromagnetically coupled O 2p hole spin.

IV-B-ELECTROCHEMICAL INCLUSION OF COPPER AND IRON SPECIES IN A CONDUCTIVE POLYMER OBSERVED IN-SITU

A comparative in situ time resolved X ray absorption study of the electrochemical inclusion of copper and iron species in poly(3-methylthiophene) (PMeT) is reported. For both treatments, incorporation of metallic ions species and complexation with the sulfur atoms of the polymer backbone lead to significant increase of the *ex situ* macroscopic conductivity. However, the detailed mechanisms and kinetics of the various processes are specific to the metallic ion incorporated in the polymer.

IV-B-a BACKGROUNDS

Conductivity of organic conducting polymers can be varied from an insulating-semiconducting state to a conducting state when they are doped. From a technical point of view, polythiophene and their derivatives[28] appear to be the most promising organic conducting polymers, due to their high stability against oxygen and moisture in both their doped and undoped forms. Therefore, practical applications in the fields of organic batteries, display devices and photovoltaic systems have been tested.

In organic conducting polymers, the macroscopic conductivity is generally limited by both intra (presence of structural defects) and interchain (lack of ordering) defects. Intrachain conduction can be controlled and modified by varying the structure of the monomer units[28,29] and the nature of dopant[30], to produce a more regular polymer with a metallic-like behavior.

An elegant way of increasing the contact between the polymeric chains is to bridge them with metallic ions. In that respect, copper species were included in doped PMeT and an increase of the macroscopic conductivity (from 50 to 150 $\Omega^{-1}cm^{-1}$) was observed after cathodic polarization[3].

Energy dispersive X ray Absorption Spectroscopy is a particularly well suited method to investigate the electrochemical inclusion of metallic ions in conducting polymer[13] because

i) it provides informations on the oxidation state, the coordination geometry and the bonding angles and

ii) time resolved in situ investigation can be achieved. This spectroscopy is therefore a powerful technique to get informations on growth mechanisms and kinetics of various processes.

In this paper, we report a comparative in situ time resolved X ray absorption spectroscopy study of electrochemical inclusions of copper and iron metallic species in doped PMeT under continuous cathodic polarization.

IV-B-b EXPERIMENTS

Grafted polymer on a platinum wire (500 µm diameter) is obtained by the oxidation, at +1.35 V versus SCE (saturated calomel electrode), of 3-methyl-thiophene (5 x 10^{-1} M) in CH_3 CN with $N(Bu)_4SO_3CF_3$ (5 10^{-1} M) as the electrolytic salt[31]. The polymer is directly obtained in its doped conducting state with a typical 25-30% doping level and its thickness is controlled by the electrolysis time.

For the X ray absorption spectroscopy experiments, a 3 mm diameter thick PMeT is grafted on the Pt wire. The polymer is rinced with acetone, dried and immersed in an aqueous solution containing the metallic ions ($CuCl_2$ and $FeCl_3$, both solution concentration in the range 10^{-2} - 10^{-3} M).

Measurements of the time-dependent electrolysis current curve and polarization time-dependent macroscopic conductivity were obtained with more concentrated solutions (0.1-1.0 M).

The electrolytic cell used for the X ray absorption study has a thickness of 3 mm. It is composed of a teflon ring covered by two kapton windows. This cell is inserted in the X ray beam with the grafted polymer at the focusing spot of the optical system. A two-electrode configuration (the grafted electrode and a Pt wire) is used to reduce the metallic species (-5 V). The volume occupied by the polymer fibers (40% of the investigated volume) is estimated from the decrease in the $CuCl_2$ solution absorption coefficient observed when the grafted polymer is introduced in the electrolytic cell.

XANES spectrum were recorded with Si (311) which yields a resolution of 0.9 eV at the copper K edge, while the Si (111) was used to obtain the EXAFS data (resolution ~ 1.9 eV at the copper K edge). The whole spectrum of a thin iron or copper reference foil is recorded in 28 ms with a good signal to noise ratio. The size of the focused beam is 0.5 x 0.5 mm^2 and so a very small volume of the sample is analyzed. All the measurements have been made to probe a region of the polymer close to the Pt wire.

A number of reference compounds were used in order to get the backscattering amplitude and phase shift functions of the various pairs of atoms, namely Cu_2O (two oxygen atoms at 1.85Å) , CuO (four oxygen atoms at 1.96Å), metallic Cu, $FePO_4$ (four oxygens neighbor at 1.88Å), ferrocene (ten carbons atoms at 2.03Å)[32] and FeS_2 (pyrite, six sulfur atoms at 2.259Å). In the case of the Fe-O pair, phase shift and amplitude functions extracted from $FePO_4$ were checked by fitting the first coordination shell of α-Fe_2O_3 (6 oxygen atoms: 3 at 1.945Å and 3 at 2.116Å).

IV-B-c- RESULTS

The time-dependent evolution of the electrolysis current when cathodic polarization is applied to the grafted PMeT immersed in the $CuCl_2$ aqueous solutions is characterized by a sharp initial drop and a subsequent increase of the electrolysis current. For longer polarization time (more than six minutes) the current is decreasing slowly. This behavior is different from that observed when the polymer is soaked in a $FeCl_3$ aqueous solution: no increase in the electrolysis current is observed after the initial sharp drop. Instead, a slow decreasing current is measured over more than two hours.

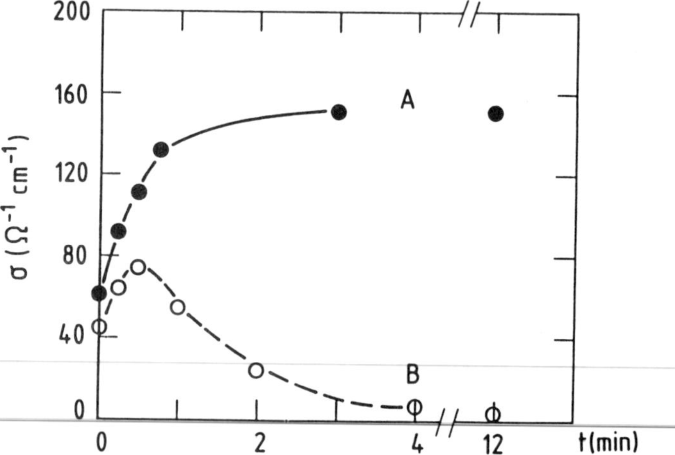

Figure 5, curves A and B, shows the ex situ macroscopic conductivity of the grafted PMeT measured at various time after cathodic polarization was applied to the polymer soaked in a $CuCl_2$ and $FeCl_3$ aqueous solution, respectively.

Curve A is characterized by a steady increase of the macroscopic conductivity from 50 to 150 $\Omega^{-1}cm^{-1}$. Longer polarization time does not induce any more change in the conductivity.

Curve B, shows first an increase of the macroscopic conductivity from 40 to 75 $\Omega^{-1}cm^{-1}$. For longer polarization time, σ values decrease slowly to 0.5 - 1.0 $\Omega^{-1}cm^{-1}$. At the end of the electrochemical process, the value of the macroscopic conductivity is less than that characteristic of the untreated polymer.

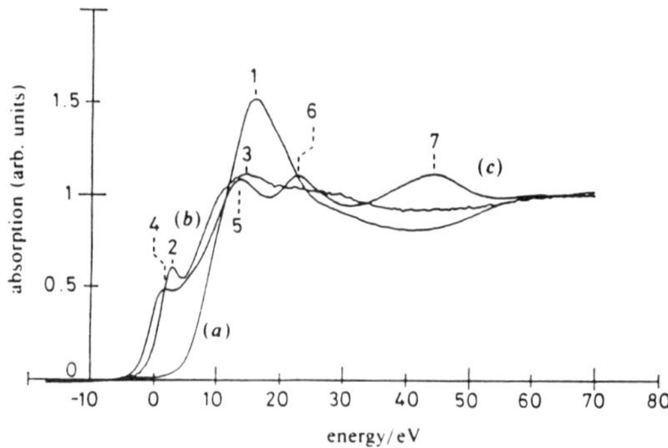

Fig. 6

Figure 6 shows the Cu K edge absorption spectrum of three copper species: curve A, 50 mM $CuCl_2$ aqueous solution; curve B, Cu^{I}-bipyridine complex in acetonitrile; curve C, metallic copper. Both the energy position and the shape of the absorption edge conveys chemical information on the absorbing atom. Compared to Cu^{0}, the shift in the energy position of the absorption edge is +1.2 eV for the Cu^{I}-bipyridine complex and +8.5 eV for the Cu^{II} ions in solution.

The white line observed at 15 eV in the absorption spectrum of Cu^{II} (peak 1) is due to a transition to an antibonding resonance, reflecting the presence of six oxygen atoms in a regular octahedral configuration. Interpretation of the edge features of the Cu^{I}-bipyridine complex is similar to that of Cu_2O, where Cu^{I} has a linear configuration with two oxygen atoms at 1.85 Å. If the z axis is chosen along the two Cu-N bonds, the $4p_x$, $4p_y$ and $4p_z$ metallic orbitals are no longer degenerated (as they are in the isolated atom), due to the lowering of the site symmetry. Peaks 2 of curve B corresponds therefore to the $1s - 4p_{x,y}$ non bonding Cu orbitals.

Multiple-scattering calculations were also conducted on Cu metallic clusters of varying dimensions and the shoulder in the middle of the rise of the absorption cross-section (peak 4) and the first three oscillations centered at 13.4, 22.4 and 43.8 eV (peak 5, 6 and 7, respectively) has been reproduced when the fourth shell surrounding a given copper atom is included[33].

The edge and pre-edge features of the absorption spectrum of compounds of figure 6 can be used as fingerprints of the chemical state of copper.

V-d- TIME-RESOLVED IN-SITU X-RAY ABSORPTION SPECTROSCOPY OF ELECTROCHEMICAL INCLUSION OF COPPER SPECIES IN PMeT

Electronic structure: XANES

In a first set of experiments, X ray absorption spectra are continuously recorded while cathodic polarization is applied to a PMeT grafted polymer soaked in a $CuCl_2$ aqueous solution, using a Si (311) crystal. Figure 7 shows a set of selected Cu K edge spectrum measured during the reduction process. The time elapsed from the beginning of the experiment is indicated on the right-hand scale and t=0s is the absorption spectrum of the Cu^{II} ions in the polymer before cathodic polarization is applied.

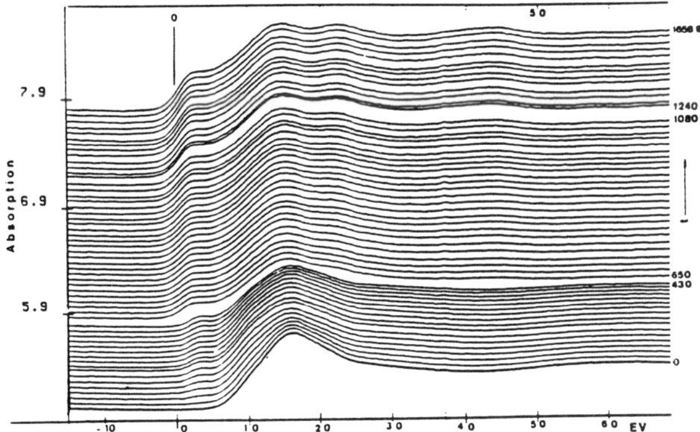

Fig. 7 Time dependent evolution of the Cu K edge absorption spectrum of cathodically polarized PMeT soaked in a $CuCl_2$ solution. The zero reference energy corresponds to 8979.12 eV and the polarization time is indicated on the right-hand scale.

There is a rapid shift of the absorption edge towards lower energy and the appearance of a small bump in the rise of the absorption cross-section, along with a decrease of the white line intensity after the onset of polarization (curve t=650s). Cu^{I} is therefore formed during the first step of the reduction process (fig.6). Their presence is also confirmed by the red coloration appearing when the polymer is treated with a bipyridine solution (Cu^{II} ions give a blue coloration).

Further modifications are observed in the absorption spectra upon prolonged cathodic polarization. The bump in the rise of the absorption edge is displaced a little more towards the low energy and the oscillations characteristic of the metallic state are observed (curve t=1656s). All these changes in the absorption spectrum are consistent with the evolution of the Cu^{I} species to Cu^{0} metallic copper. The overall electrochemical inclusion of copper species in PMeT proceeds therefore via a serie of subsequent reduction steps:

$$Cu^{II} \longrightarrow Cu^{I} \longrightarrow Cu^{0}$$

The inclusion process is fully reversible.

Figure 8, curve A, B and C, represents the time evolution of the Cu^{II}, Cu^{I}, and Cu^{0} concentration under continuous cathodic polarization, respectively, as determined by curve fitting of the absorption spectra of figure 7 with a linear combination of the absorption spectrum of the three various copper species (fig. 6).

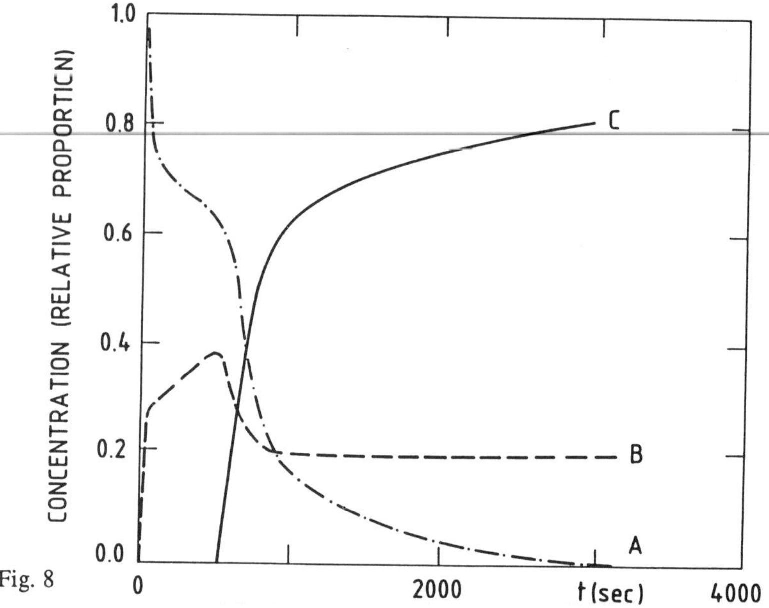

Fig. 8

The electrochemical time-dependent evolution of the various concentrations exhibits three kinetic domains. At the beginning, there is a sharp drop of the Cu^{II} concentration with a concomitant increase in the Cu^{I} concentration. This prompt change is followed by a slow increase of the Cu^{I} concentration (time constant of about 600s) . In the last kinetic domain the Cu^{0} concentration increases rapidly. Changes in the Cu^{II}, Cu^{I} and Cu^{0} concentrations are rather slow at the end of the electrochemical process.

Structural characteristics: EXAFS

A second set of experiments (same electrochemical procedure) was redone with a newly grafted polymer, using this time a Si (111) crystal in order to obtain a full energy domain to collect EXAFS data. Small differences may be observed in the time constant of the various processes, due to small variations in the size of the probed region of the polymer and in its location with respect to the Pt wire. Figure 9 shows a set of k^3-weighted Fourier transforms (FT) at different polarization time. Curve A is for polarization time t=0s and is typical of the Cu^{II} ions after their inclusion in the polymer. The first peak of the FT is symmetric. No peak is observed at larger distances, due to the poor atomic organization, characteristic of a solution. Curve fitting of the filtered inverse Fourier transform (FIFT) of the first peak shows that the first neighbors shell is composed of six oxygen atoms at 1.95 Å. The first neighbors shell makes therefore a regular octahedron around the Cu^{II} ions, as expected from the appearance of a strong white line resonance in the near edge absorption spectrum.

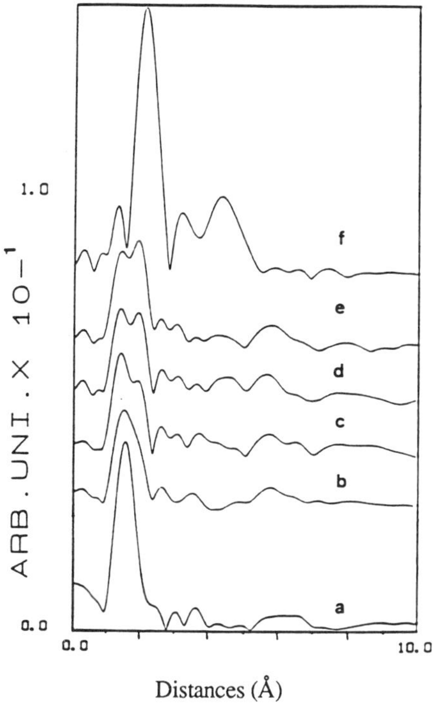

Fig. 9 Fourier transforms (k^3 - weighted) of Cu K edge spectra: curves A to F are for the cathodically polarized PMeT with polarization time t=0, 160, 168, 184, 200 and 460s, respectively.

At the beginning of the electrolysis (polarization time between 0 and 150s), curve fitting of the filtered inverse Fourier transform of that peak shows that the first coordination shell of the copper species is only constituted of oxygen atoms.

For t = 160s (curve B of fig.9) a shoulder is observed on the large-distance side of the peak.

For longer polarization time (curve C, D and E; t=168s, 184s and 200s, respectively), the first peak is now composed of two well resolved maxima, the intensity of the second one increasing at the expense of the first peak. The FT obtained for still longer polarization time (curve F, t=460s) shows the very presence of peaks at large distances (associated with second and third coordination shell): copper atoms are in a well organized crystalline structure.

The behavior of the first shell FIFT at various polarization time demonstrated that both oxygen and later on sulfur atoms are included in the first neighbor shell of the Cu^I ions. (fig.10)

The polarization time dependent evolution of the first peak of the FT curves reflects the increasing proportion of sulfur atoms in the first coordination shell of the Cu^I specy and prolonged cathodic polarization induces the formation of Cu^0 metallic clusters, which are responsible for the second and third coordination shells.

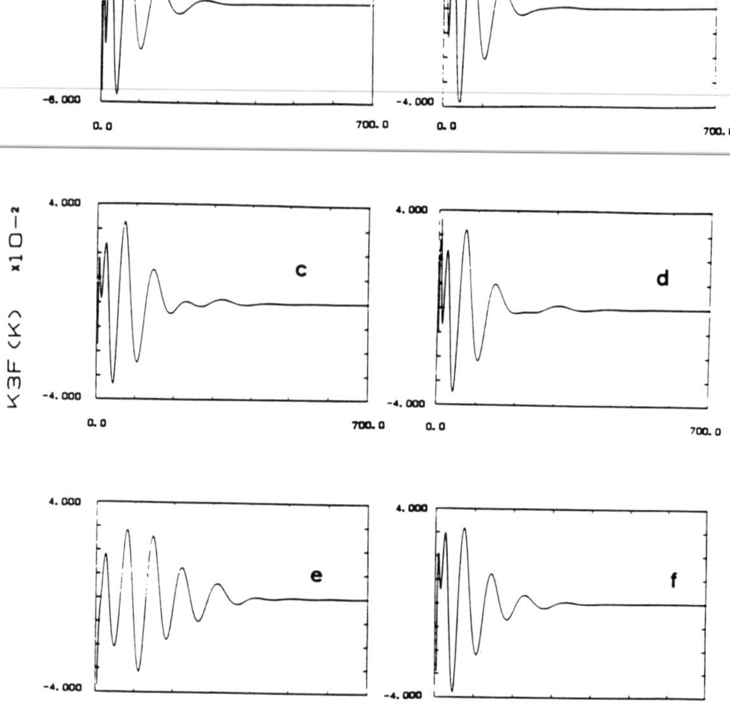

Fig. 10 First peak filtered inverse Fourier transform: curves A to E are for the cathodically polarized PMeT with polarization time t=0, 160, 168, 200 and 460s, respectively. Curve F is for Cu_2S reference compound.

IV-B-e ELECTROCHEMICAL INCLUSION OF IRON SPECIES IN PMeT

Electronic and structural characteristics: XANES and EXAFS[34]

Fe^{III} ions in solution are surrounded by 6.0 oxygen atoms; four of them are at 2.04 Å and two at 2.14 Å. The first coordination shell around the Fe^{III} ions is a slightly distorted octahedron.

The incorporation of Fe^{III} ions in the polymer causes major changes in the FT curve.

The first coordination shell of the Fe^{III} ions in PMeT evolves rapidly to four oxygen atoms at 2.04 Å and two sulfur atoms at 2.37 Å. This Fe-S distance corresponds to what is observed in a number of heme-iron enzymes[35].

Due to the slow kinetic, the cathodic potential applied to the grafted polymer was gradually increased during the course of the experiment to achieve completeness of the reduction process. (More than 140 mim elapsed between the start and the end of the experiment). On the basis of the observed spectra, several observations can be made.

There is a gradual shift of the absorption edge toward smaller energy under cathodic polarization. After completion of the reduction process, the absorption edge is displaced by 4.1 eV from its initial position, as observed exactly between $FeCl_3.6H_2O$ and $FeCl_2.4H_2O$ powders and should therefore be attributed to the reduction of Fe^{III} to Fe^{II}.

After about 1395s of electrolysis, the first coordination shell of iron species is composed of 3.7 oxygen atoms at 2.10 Å and 2.3 sulfur atoms at 2.43 Å. The number and type of the first coordination shell atoms are identical but the Fe-O and Fe-S distances are increased (compare to the t=0s coordination shell).

At the end of the reduction period, the first coordination shell surrounding the reduced Fe^{II} ions is only composed of 6 oxygen atoms at 2.19 Å and makes a regular octahedron around the ions. This geometry is consistent with the strong resonance observed in the near edge spectrum.

Complete reduction of the Fe^{III} ions requires more than 2 hours.

The presence of Fe^{II} ions at the end of the reduction process is also confirmed by the red coloration observed after a bipyridine treatment.

IV-B-f DISCUSSION

In situ X ray absorption spectroscopy investigation of the electrochemical treatment of PMeT with aqueous iron and copper ions solutions reveal that major differences occur between both treatments. The present discussion points out the roles played by both ions in promoting an enhancement of the macroscopic electrical conductivity.

Before electrochemical reduction, Cu^{II} ions are surrounded by oxygen atoms, coming from the surrounding water molecules. At the very beginning of Cu^I formation (first kinetic domain), the rate of Cu^{II} reduction is large. Since Cu^I ions are not stable in water, they interact with the oxygen atoms of the $SO_3CF_3^-$ doping salt: $Cu^ISO_3CF_3^-$ is already known to be stabilized by benzene, giving rise to a crystallized compound (34).

Within the second kinetic domain, stabilization of the Cu^I ions by the S atom of the polymer occurs because the oxygen complexing sites of the doping salt are already occupied.

Two sources of sulfur atom are available in the doped PMeT: the doping salt $SO_3\,CF_3^-$ and the heteroatom of the monomer unit of the polymer. Because of the steric hinderance due to the oxygen and carbon atoms, the sulfur atom of the salt can not be part of the first coordination shell of any other element. Therefore, Cu^{I} ions interact with the sulfur atom of the thiophene unit.

The 40% Cu^{I} concentration limit of the second kinetic domain should correspond to the easily accessible S sites of the polymer.

Finally, metallic copper is formed with a well platelet-organized cristalline structure. The Cu^{O} concentration increases rapidly with a concomitant increase in the rate of Cu^{II} reduction. These changes can be understood if the formation of Cu^{O} metallic clusters proceed from Cu^{I} ions through the following disproportionation mechanism:

Cu^{I} + Cu^{I} ----> Cu^{II} + Cu^{O}

The time dependent derivative of the Cu^{II} concentration curve is found therefore to be similar to the time dependent electrolysis current curve 1A (if the sign of the slope is neglected and with the possible exception of a difference in the time scale): the increase of the electrolysis current curve corresponding to the increase of the Cu^{II} reduction rate observed in the third kinetic domain.

The presence of Cu^{I} ions at the end of the electrochemical reduction treatment is consistent with the fact that Cu^{O} formation begins after all the stabilizing sites are already occupied.

Comparison of the ex-situ macroscopic conductivity of the cathodically polarized PMeT in $CuCl_2$ solution (fig. 5) with the time dependent concentration curve of the various copper species (fig. 6) shows that the increase of conductivity is not correlated to the formation of Cu^{O} metallic clusters. The Cu^{O} concentration increases at the end of the electrochemical reduction process while the change in the conductivity occurs at the beginning of the electrochemical treatment.

In fact, the conductivity increase is correlated to the appearance of Cu^{I}-S complex (second kinetic domain). The slight difference in the time scale is not a key issue. It should be due to difference in the $CuCl_2$ solution concentration and position dependent electric field vector intensity in a cylindrical geometry. The $\sigma = 150\ \Omega^{-1}\ cm^{-1}$ conductivity observed for long polarization time corresponds to the constant Cu^{I} ions concentration.

Inclusion of Cu^{I}-S bounded ions is therefore seen to be an effective way to increase the conductivity. According to a previously proposed scheme[35], the Cu^{I} interaction with two sulfur atoms on different polymeric chains bridges them together and reduces the interchain hopping time. In this model, the copper atom is linked to two sulfur atoms which should be in a linear configuration. T

The situation of the PMeT-$FeCl_3$ system is somewhat different .

There is a spontaneous association of Fe^{III} to the sulfur atoms of the polymer upon incorporation of the ions in the polymer.

As Cu^{I}-S, this bridging causes a reduction of the interchain hopping time and an increase of the macroscopic conductivity. This is what is observed at the very beginning of the ex situ macroscopic conductivity curve of a cathodically polarized PMeT in $FeCl_3$ aqueous solution.

However, for long polarization time, two phenomena act to reduce the conductivity. First of all, as it was demonstrated by EXAFS and XANES measurements, Fe^{II} ions are surrounded by six oxygen atoms and they do not interact with the polymer backbone. Upon reduction, the bridging of the polymeric chains by Fe^{III} ions is loosen and the macroscopic conductivity should decrease to its initial value. However, σ decreases to value lower than that of the untreated polymer. That is to say that, for long polarization time, Fe^{II} ions induces a dedoping effect. Iron clusters does not grow into PMeT , probably because the reduction potential of the Fe^{II} +2e-->Fe^{O} reaction is too negative compared to the reduction potential of the polymer.

V CONCLUSIONS

Time-dependent structural investigations can be correlated to electrical measurements to derive fundamental explanations of the increase of macroscopic conductivity of conducting polymer upon electrochemical inclusion of metallic species. Inclusion of metallic ions in polymers is an efficient way of increasing their conductivity by bridging together different chains and reducing the interchain hopping time. However, choice of the bridging ions is not a straightforward mater since solubility, oxido-reduction properties as well as complexation processes should be considered.

The major interest of most of the studies is to show the new possibilities of in-situ measurements since the dispersive scheme allows data to be collected at once. This is a major advantage over the step by step scheme.

The time scale achievable of the order of the tenth or the second restricts this time-resolved spectroscopy to materials science where mass transportation is involved. ESRF can open the millisecond time scale giving access to more dilute samples and perhaps visualisation of changes of conformation of large molecules as found in biophysical dedicated cases.

REFERENCES

Proceedings of the Int. Conf.(II), Frascati, Italy, 13-17 September 1982,
edited by A. Bianconi, L. Incocia and S. Stipicich
Springer Series in Chemical Physics Vol. 27

Proceedings of the Int. Conf. (III), Stanford, USA, 16-20 July 1984,
ed. by K.O.Hodgson, B.Hedman and J.E.Penner-Hahn
Springer Series in Physics Vol. 2

Proceedings of the IV Int. Conf. Fontevraud July 1986,
Journal de Physique,C8, supp n°12,42,Déc.86,
edited by J. Petiau, P. Lagarde, D. Raoux,

Proceedings of XAFS V Conference, Seattle August 1988,
edited by E.Stern, J.Rehr, J. of Less Common Metals

Proceedings XAFS VI York 1990,
Ellis Horwood Publishers, Ed by S.Hasnain, 1991

[1]-A.Fontaine, E.Dartyge, J.P.Itié, A.Polian, H.Tolentino, G.Tourillon, Topics in Curr. Chem. Vol. **151**, Spinger (1989)

[2]-T. Matsushita, H. Oyanagi, S. Saigo, H. Kihara, U. Kaminga, EXAFS and Near Edge Structure III, .

[3]-M.Hagelstein, S.Cunis, R.Frahm, W.Niemann, P.Rabe Phys.**B 158** (1989) 324 id + R.Piffer CB18 XAFS VI York 1990

[4]-G.Baker, C.Richard, A.Catlow, J.Couves, A.J.Dent, G.Derbyshire, G.N.Greaves, J.M.Thomas, XAFS VI York 1990

[5]-R.Frahm NIM **A270** 578 (1988), and Proceedings XAFS VI York 1990 Ellis Horwood Publishers, Ed by S.Hasnain, 1991

[6]-P.Lagarde, M.Lemonnier, H.Dexpert Physica **B 158** (1989) 337 C.Prietto, E.Prouzet et al XAFS VI York 1990

[7]-H.W.Huang, W.H.Liu, T.Y.Teng, X.F.Wang Rev. Ins.Sc. 54,11, 1983 , p1488

[8]-P.Livins, E.A.Stern, M.Newville, D.Thiel, A.Lewis, Proceedings XAFS VI York 1990
[9]-G.Tourillon, E.Dartyge, A.Fontaine, A.Jucha, PRL 1986, **57**, 603
[10]-D.Guay, G.Tourillon, A.Fontaine, Far.Disc Cem, Soc., 1990, **89**,
[11]-E. Dartyge, C. Depautex, J.M. Dubuisson, A. Fontaine, A. Jucha, P. Leboucher and G. Tourillon, NIM A246 (1986) p. 452-60
[12]-G. Tourillon, E. Dartyge, A. Fontaine, A. Jucha, PRL (1986) 57, 5, 506
[13]_E. Dartyge, A. Fontaine, G. Tourillon, R. Cortes, A. Jucha, Phys.Letters A, 113A, 7, p. 384
[14]-G. Maire, F. Garin, P. Bernhardt, P.. Girard, J.L. Schmitt, E. Dartyge, H. Dexpert, A. Fontaine, A. Jucha, P. Lagarde, Applied Catalysis 1986,26,305
[15]-I.Ascone, A.Fontaine, A.Bianconi, A.Congiu-Castellano, A.Giovanelli, S.Della-Longa, M.Momenteau. Springer series in Biophysics Vol2,1987 Biophysics and Synchrotron Radiation Ed. By A.Bianconi and A.Congiu-Castellano.
[16]-I. Ascone, A.Bianconi, E.Dartyge, S.Della-Longa, A.Fontaine, M.Momenteau. Biochimica & Biophisica Acta915 (1987),168.
[17] -J.S.Rohrer, M-S.Joo, E.Dartyge, D.E.Sayers, A.Fontaine, E.C.Theil. J. of Bio. Chem. 262,13385,(1987)
E.C. Theil et al. Proc. 3rd Int. Conf. on Bioorg. Chem. Noordwijkerhout July 1987
[18]-H.Tolentino, E.Dartyge, A.Fontaine, G.Tourillon, J.App.Crist.(1988) **21**,p.15H.Tolentino, F.Baudelet, E.Dartyge, A.Fontaine, G.Tourillon, NIM A289 (1990), p307
[19]-J.P.Itié, A.Polian C.Jauberthie-Carillon, E.Dartyge, A.Fontaine, H.Tolentino, G.Tourillon Proceedings of XAFS V Conference, Seattle (1988)
[20]-J.P.Itié, A.Polian, G.Calas, J.Petiau, A.Fontaine,G.Tourillon Phys.Rev.Let.63,4 (1989) 398
[21]–G.E.Ice and C.J.Sparks, N.I.M. 222 121-127 (1984)
[22]-J.McBreen, W.E.O'Grady, G.Tourillon, E.Dartyge, A.Fontaine, K.I.Pandya J.Phys.Chem.1989,**93**, 6308
[23]-H.Tolentino, A.Fontaine, T.Gourieux, G.Krill, J.Y.Henry, J.Rossat-Mignod, A.M.Flank, P.Lagarde, F.Studer IWEPS Kirchberg, Proc. Edited by J.Fink, 1990 H.Tolentino Thesis Nov (1990) Orsay , Université Paris-Sud
[24]-J.Mc Breen, W.E.O'Grady, K.I.Pandya, R.W.Hoffman, D.E.Sayers Langmuir 1987, **3**,428
[25]-D.Guay, J.McBreen, W.E.O'Grady, G.Tourillon et al. in prep.
[26]-P.Kuiper, Ph.D. May 1990 Univ. of Groningen Dep. of App. and Solid St. Ph.
[27]-J.Zaanen, Ph.D. May 1987 Univ. of Groningen Dep. of App. and Solid St. Ph.
[28]-G.Tourillon, G.; In Handbook of conducting polymers; Skotheim, T. A., Ed.Marcel Dekker: New York, 1986; Vol. 1, Chapter 9, p. 293.
[29]-M.Kobayash, N.Colaneri; M.Boyssel, F. Wudl, A.J. Heeger, J. Chem. Phys. 1985, 82, 5717.
[30]-D.Gourier, G.Tourillon, ; J. Phys. Chem. 1986, 90, 5561.
[31]-G.Tourillon, F.Garnier, J. Electroanal. Chem. 1982, 135, 173.
[32]-N.Greenwood, A.Earnshaw, Chemistry of the elements Pergamon Press, New York, 1986.
[33]-N.G.Greaves, P.J.Durham, G.Diakun, P. Quinn, Nature 1981, 294, 139.
[34]-*For the sake of the shortness of this paper we do not include the figures, giving results in an unusual straight forward presentation. But anyone can find them in a more extensive paper*
D.Guay ,G.Tourillon, E.Dartyge, A.Fontaine, H.Tolentino, J.Electroanal.Chem.(1991)305,83.
[35]-L.S.Kau, E.W.Svastits,J.H. Dawson, K.O.Hodgson, Inorg. Chem. 1986, 25, 4307.

DISCUSSION

GOULON - 1. In the high resolution fluorescence detection of XAS described by Dr. J. Hastings, what is the energy resolution needed ?

2. Could you compare the performances of the CCD camera detector that you propose and those of the rotating drone used by Dr. Amemya at the photon factory ?

FONTAINE - 1. The energy resolution of the incoming beam has to match the resolution you aim. As you known we cannot avoid the fundamental square combination of the different components which determine the overall energy resolution.

2. To keep the full benefit of the energy dispersive spectrometer we have to keep the full stability of energy components. This includes keeping the same pixel to collect time-dependent data at the same energy.
Point 2, the dynamics of the image plate is by far too low compared to that one given by silicon-based detector.

DENT - For the $Ce^{IV} \rightarrow Ce^{III}$ how was the reaction coordinate obtained from the XAS spectra shown ?

FONTAINE - The spectra which were displayed were only the XANES part. Full EXAFS have been collected and processed with regular analysis.

JENTYS - Which energy region, from the edge up, are you able to scan with the energy dispersive setup ? Is this region sufficient to obtain data with a comparable accuracy compared to the setup scanning the energy step by step.

FONTAINE - The band pass comes up from the derivative of the Bragg law $\Delta E = E \text{ cotg } \theta d\theta$. We cannot work on E because it is determined by the species you are looking at. cotg θ is made max using Si 111, i.e., using the larger Miller indices. $\Delta\theta$ is given by the curvature. Above 8keV there is no difficulty to get more than 700eV. At the Ti K-edge we got that difficulty to broaden the range. But the dispersive spectrometer is more to be used as a tool for time-resolved experiments. Therefore we are looking at evolution of very well-known systems in their initial state. There are the ideal conditions for sort of "perturbative" approach which have been very often used in Physics, since it is very rare that an ab-initio simulation can give the right answers. Performing high pressure studies we often got only 300-400eV available because of gliches given by the diamond anvils, and we got very nice results (PRL 63, 4, (1989), p.398, Pressure-induced coordination changes in Crystalline and Vitreous GeO_2).

WONG - 1. In your TiO_2 XANES spectrum, what is U_{cd} ?
2. And can one also understand the variation (in intensity) of the same triplet observed in going from rutile → anatase → brookite which have progressive distorsion of the TiO_6 octahedron in the lattice ?

FONTAINE - 1. U_{cd} is an effective U_{cd} since it is really U_{cd} - U_{cp}. The Dipole transition gives $4p^1$ in the final state. The quadripole transition gives 3d1 in the final state. Therefore the core hole gives rise to different relaxations which involve an energy difference of U_{cd} - U_{cp} as it is used to be written. There is, therefore a separation on the final states of the same E_{2g} or e_g band seen through their two channels.

2. What was related in this talk was the rutile spectrum. The evolutions of the "triplet" (which is an overlap of two doublets), came from the evolution of the distorsion of the six-fold coordinated cage which leads to different p-d hybridation.

3. Just for clarity I have to write what I mentionned in the talk. I just reported what has been measured by C. Brouder, E. Beaurepaire, O. Durmeyer and J.P. Kappler from Strasbourg. Unfortunately, I was not involved in this nice piece of work.

TIME-RESOLVED ENERGY - DISPERSIVE DIFFRACTION STUDIES OF DYNAMIC PROCESSES IN MATERIALS SCIENCE

P.Barnes
Industrial Materials Group, Dept. Crystallography, Birkbeck College (University of London), Malet Street, London WC1E 7HX, U.K.; and D.R.S., Daresbury Laboratory, Warringhton WA4 4AD, U.K.

ABSTRACT

Time-resolved energy-dispersive X-ray diffraction, combined with a synchrotron - hard X-ray source, is a versatile technique for studying the synthesis of many modern materials and assessing their in-service performance. Such aims however also imply that a wide variety of demanding physical and chemical conditions can be imposed on the materials under study. This paper discusses such prospects, with illustrations taken from the field of ceramics, cements, zeolites, drugs and amorphous metals.

INTRODUCTION

The pursuit, of collecting X-ray diffraction data within increasingly smaller time intervals, has branched into several technical directions. For monitoring continuous processes, the main contenders are the Laue method for single crystals, position- and area-sensitive detector systems for both single and polycrystalline material, and energy-dispersive diffraction. Allied to synchrotron radiation sources, these techniques are easily capable of yielding sub-second diffraction/scattering patterns, though one should note that information transfer with area detector methods can be a limiting factor when studying continuous processes. In this paper the particular advantages of time-resolved energy dispersive diffraction are considered in relation to dynamic processes in materials science. Quite apart from the goal of achieving the shortest possible diffraction time intervals, there is considerable scope within materials science for examining material on 1 second to 5 minutes time intervals and under a variety of environmental conditions (temperature, pressure, gas-environment). This paper will demonstrate that for many problems in materials science, involving both synthesis and subsequent in-service performance, our understanding can be greatly enhanced by such a diffraction capability. Previous examples will be reviewed and recent (unreported) cases will also be given.

ENERGY DISPERSIVE X-RAY DIFFRACTION (EDXRD)

Following its first demonstration in 1968[1,2], energy dispersive X-ray diffraction has been developed into several modes and even used for crystal structure refinement studies[3,4]. In this paper we will consider synchrotron radiation - energy dispersive diffraction (SR-EDD) in just its rapid-low resolution mode: that is with a white polychromatic incident X-ray beam, from a synchrotron radiation (SR) source and a fixed angle energy-dispersive detector system. Such a system, based on the facility at Daresbury Laboratory (U.K.[5]), is illustrated in Fig.1. The beam incident on the sample can be varied in cross sectional area from diameters of 0.1 mm upwards. The post-sample detector collimation system consists of a single or multiple set of molybdenum foils, 0.5 m in length and separated by 0.1 mm (normally, for diffraction 2θ-angles above 5°) or 0.05 mm spacers (for lower 2θ-angles): the normal setting defines the 2θ-angle to within $\pm 0.1^{\circ}$, this making a negligble contribution to the diffraction peak broadening which is largely governed by the germanium solid state detector resolution (~190 eV FWHM at 5.9 keV). Buras et al.[6] has shown that, for most practical cases, small 2θ-angles ($<10^{\circ}$) should be used for optimum resolution. Hausermann and Barnes[7] have given detailed prescriptions for matching the X-ray geometric optics to the material under study. Of particular importance are the following considerations:

(1) choice of reflection or transmission mode;
(2) choice of single or multiple foil detector collimation systems;
(3) sample dimensions;
(4) definition of the diffracting lozenge (Fig.1) in relation to the **active** sample volume;
(5) the range of useful d-spacings (expressed in effective energy units) in relation to the incident beam energy profile and sample absorption effects.

Often the eventual choice has to be a compromise: For example, the choice of diffraction angle, 2θ, impacts on both the diffracting lozenge dimensions (point (4)) and the energy-range of reflections (point (5)), as well as the peak widths at low 2θ-angles (e.g. $2\theta \lesssim 3^{\circ}$). For the remainder of this paper we will assume that the EDD system has been optimised for the given study and in relation to the performance of the ED-detector and material under study.

SR-EDD ENVIRONMENTAL ENCLOSURES

Materials science studies require a versatile collection of environmental enclosures that can deliver a range of temperatures (from say to -170 to +1500°C), pressures (upto 500 kbars or higher), and gaseous environments around the sample. There is an increasing desire now, in both materials and mineral physics, to be able to study materials under the **combined** action of steady high pressures and temperatures. This is desirable in mineral physics for studying phase diagrams and equations of state for material under conditions akin to those in the Earth's mantle and large planetary interiors; The needs in materials science are many, but particularly in high pressure synthesis, and the performance of lubricants and ceramics under extreme conditions. Based on the experiences at Daresbury (SRS) Laboratory, the most useful combinations so far have been as follows:

(1) The hydration cell: This cell (Fig.1) was first developed by Clark[8] to study the hydration of bulk cement samples under steady state conditions. It has since been modified by Munn et al.[9] to operate over the much wider temperature range of ~5 to 250°C. The sample tubes can be conventional glass test-tubes or made from aluminium alloy when they require a protective PTFE inner sleeve to sustain alkaline environments as required for the hydrothermal synthesis of zeolites and other aluminosilicate structures. Other features include a sealable autoclave cell with thermocouple probe, and various novel remote-control feed and mix systems. The parallel beam geometry permits accurate definition and positioning of the diffracting volume (termed a "lozenge") within the sample (see Fig.1b).

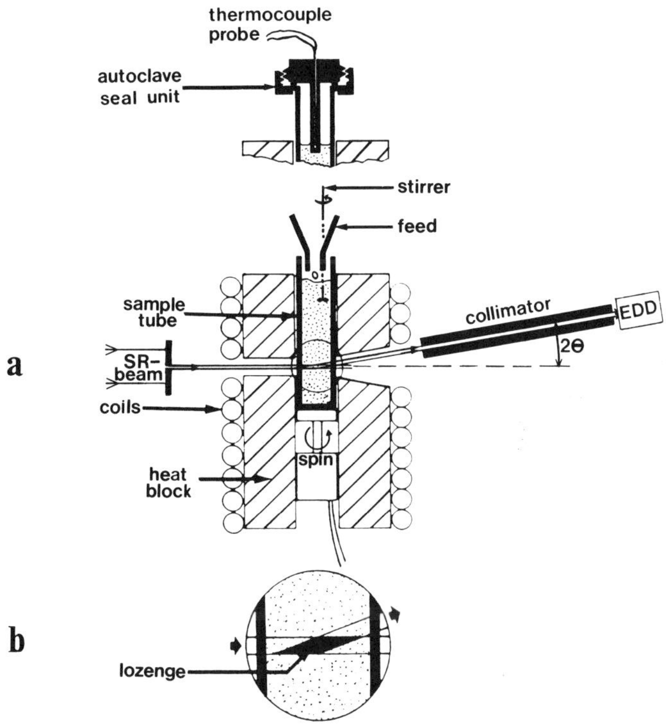

Fig.1a: Schematic of the SR-EDD geometry used (not to scale) and hydration cell with various features (autoclave seal, remote control feed and mix);
1b: The enlarged region indicates the diffracting lozenge.

(2) The furnace stage: With minor modifications to an optical microscope furnace stage, one can obtain sample temperatures between -170 and 1500°C with a fast temperature response: the samples must be small (<10 mm diameter and less than 1 mm in thickness).

(3) Diamond Anvil Cells: A variety of conventional diamond anvil cells are used to deliver pressures of up to 600 kbars within small (0.1 to 0.2 mm diameter) volume samples.

(4) Drickamer-type combination cells: Combined steady state high temperatures and pressures can be achieved over larger sample volumes using variations (Hausermann & Sherman[10]) of the Drickamer-type pressure cell. These can deliver **combined** 500°C + 100 kbars steady conditions with relatively small modificatons to the basic pressure cell design.

SR-EDD STUDIES IN MATERIALS SCIENCE

There are many problems in materials science concerned with the elucidation of structural and dynamic characteristics of a given pathway. These may involve the synthesis or in-service performance of a material within or beyond its normal operating limits. The three most fundmental questions usually asked are:

(1) what intermediate structures, if any, exist between the parent structure and product?,
(2) what are the kinetics of transformation between the various structures involved?, and
(3) can these characteristics ((1) & (2)) be altered by design?

Although "static crystallography" has served materials science well through the solving, refining and identifying of structures of stable phases, basic gaps still remain in our knowledge of the kinetic aspects of "dynamic crystallography". We will now consider some examples taken from the areas of ceramics, cement chemistry, zeolites and other functional materials.

SYNTHESIS OF CERAMICS

The high temperature furnace has been used to study ceramic synthesis up to top temperatures of above 1,400°C. This has involved identification[11] of phases in the kaolinite → mullite systems, as required, for example with decorative ceramics; and the kinetics of transformation in forming high performance ceramics based on the partially-stabilised tetragonal/monoclinic zirconia system. The latter case involves the following transformation sequence:

$$\text{amorphous Zr-hydroxide} \xrightarrow[\text{calcination}]{400\to1{,}300^{\circ}C} \text{tetragonal zirconia} \xrightarrow[\text{conversion}]{1{,}300\to20^{\circ}C} \text{monoclinic zirconia}$$

Mamott et al.[12] have shown that both the rates of calcination and conversion depend on the final pH of the prepared hydroxide. The tetragonal:monoclinic proportions of the final oxide are found to depend (Mamott et al.[12]; Turrillas et al.[13]) on four factors:

(1) final pH of prepared hydroxide;
(2) top (calcination) temperature (e.g. 900→1,300°C);
(3) holding time at top temperature (e.g. 0→7 hours);
(4) cooling rates (e.g. 5→30°C.min^{-1}).

Collecting dynamic diffraction patterns over this synthesis cycle, and covering the 4 variables outlined above, represents a non-trivial problem in data processing. Turrillas[13] has designed a novel intensity contour map to interpret visually the data dispersed over energy-temperature space. Fig.2 shows both the conventional

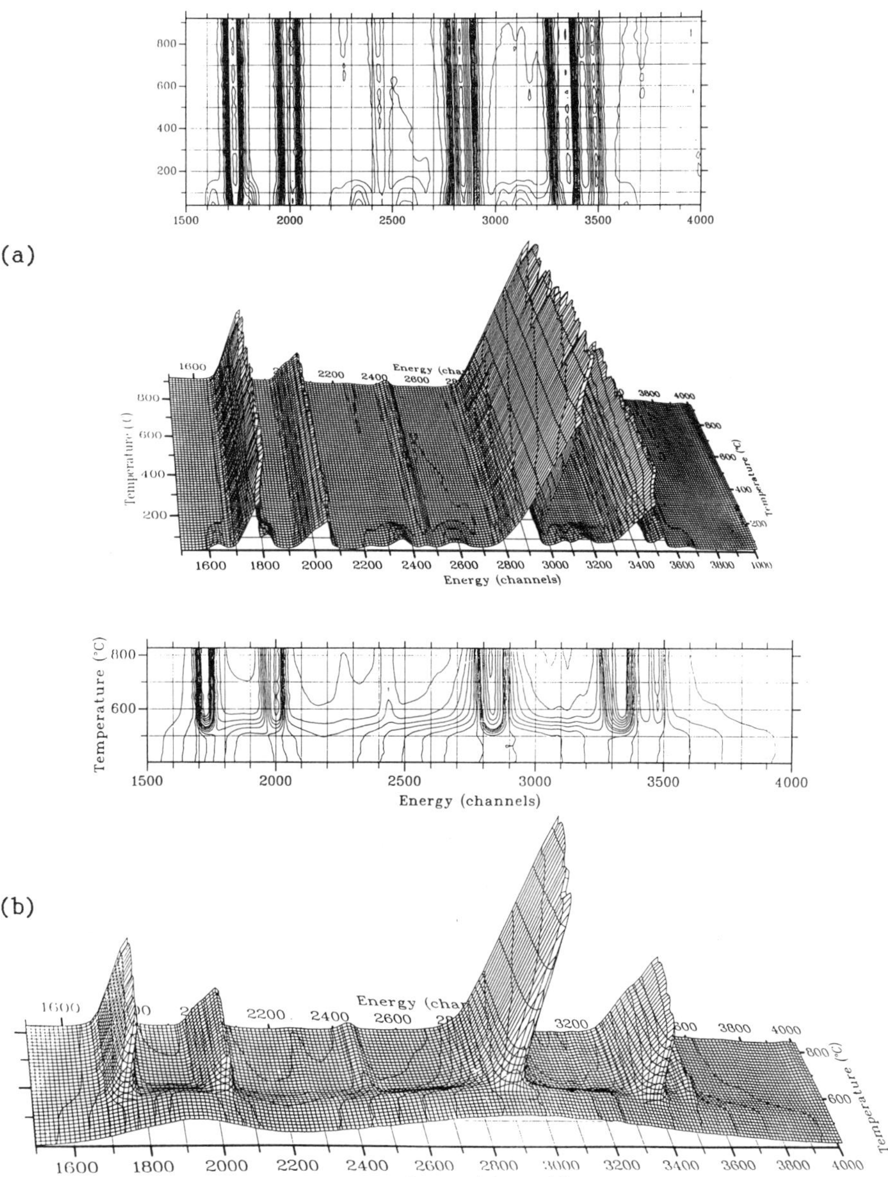

Fig.2: Time-resolved 3-dimensional plots and 2-dimensional contour maps for (a) the calcination, and (b) conversion stages of zirconia synthesis (energy in arbitrary channel units).

3-dimensional and 2-dimensional contour plots for calcination and conversion stages, starting from a pH=10.35 hydroxide. The contour plots pinpoint the onset of crystallisation (~500°C) in the former case and the late conversion (~150°C) in the latter due to a low top temperature (900°C) used.

By contrast, Fig.3 presents a very different picture: even though the starting hydroxide is from a low pH=8.37 batch and the top temperature is maintained for only 1 minute, the conversion commences early at a high temperature (~900°C) due to the high top temperature (1,300°C) used.

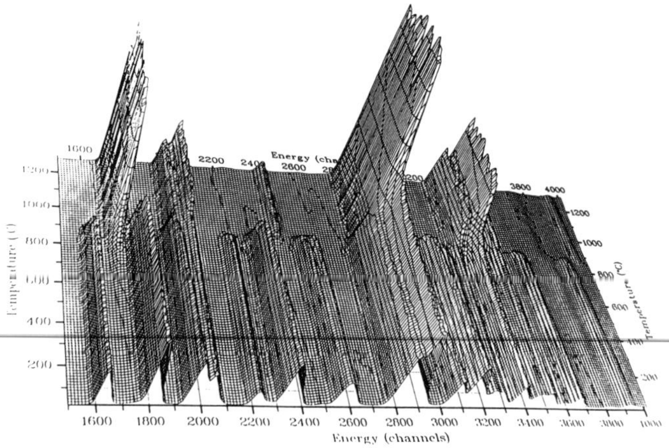

Fig.3: Time-resolved SR-EDD patterns illustrating an early (900°C) tetragonal→monoclinic conversion from a top temperature of 1,300°C, and from a low pH=8.37 batch.

A compilation of results[13], of the types shown in Figs.2 and 3, confirm the conversion dependency on the chemical (1) and physical (2)-(4) variables listed above and also demonstrates the persistence of the pH-effect: i.e. that both calcination and conversion rates increase with higher pH's of the starting hydroxide. This effect persists even upto 1,300°C, though whereas 1 minute of "annealing" at 1,300°C effectively removes the "pH-memory", it is not lost after even 7 hours at 900°C.

Such a profound and persistent effect almost demands some conjectural explanation. One approach suggested[12] is that nuclei for calcination, and subsequently conversion, exist as "embryos" right back in the amorphous hydroxide phase. The pH-effect is believed to result from differing polymerisations of the hydroxide units. These units, believed to be probably tetrameric, polymerise by olation in a one-, two-, or three-dimensional manner with rates dependent on the hydroxide solution pH. Subsequent oxolation of the OH-bridges during calcination then produces oxide aggregates, which resemble those of the original polymerised hydroxide, and which then go on to act as nucleation sites for crystallisation of the tetragonal phase.

HYDRATION OF CEMENTS

The hydration cell, described in Fig.1, has been used to study the hydration of calcium silicate[14,15], calcium aluminate[15,16], and special cements[16]. Calcium aluminate systems pose an interesting challenge on account of the crystallinity of most of the calcium aluminate hydrates and rapidity of some of the chemical sequences. Altogether, a significant array of cement hydration/conversion reactions have now been studied. In particular a rapid hydration to produce a calcium sulpho-aluminate hydrate ($3CaO.Al_2O_3.3CaSO_4.32H_2O$) know as "ettringite" has proved to be particularly interesting in view of its rapid hydration properties and its technological use as a convenient underground mine-filling cement[17]. Whereas most cement hydration scenarios display an early "dormant stage" (e.g. see for example, Skalny et al.[18]), SR-EDD has shown[16] that the following hydration sequence commences more or less immediately on mixing with water:

$$CA + 3C\bar{S} + 2CH + 30H \longrightarrow C_3A.3C\bar{S}.H_{32} \quad - (1)$$

(cement chemistry abbreviations: $C=CaO$, $A=Al_2O_3$, $\bar{S}=SO_3$, $H=H_2O$)

In an attempt to demonstrate how sudden this sequence really is Hausermann et al.[19] devised an in-situ mixing device which can be operated remotely outside the experimental hutch (see Fig.1). By obviating the need to wait for safety searches to be completed, it

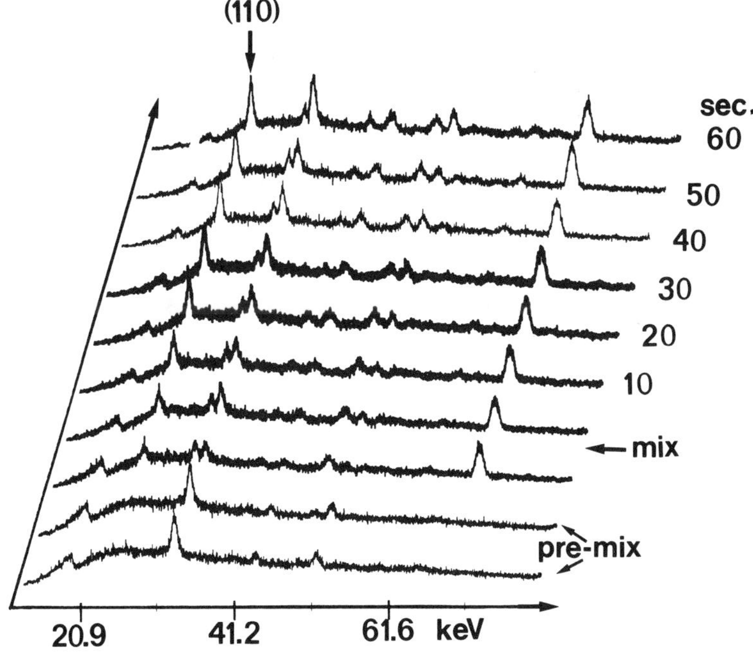

Fig.4: Time-resolved SR-EDD patterns showing the rapid formation of ettringite in ~10 second intervals after initial mixing. The main ettringite (110) peak is marked.

became possible to examine how quickly the ettringite-product really formed. The SR-EDD results show (Fig.4) that this hydration product starts to form within 10 seconds of mixing. The question then remains as to how the calcuim sulphate could have dissolved so quickly to provide sufficient sulphate ions for ettringite crystallisation. SR-EDD appears to have also suggested an answer to this question: an analysis of the ettringite peak positions[16] shows peak-shifts with time that are consistent with the early replacement of sulphate ions by the more available hydroxy/carboxy groups into the ettringite structure. As time further proceeds, the unit cell parameters asymptotically approach the usual values for the fully-sulphated ettringite. The demonstration of a rapid hydration and time-dependent solid solution on a 10-second timescale is no mean achievement for EDXRD.

CONVERSION OF CEMENTS

The instability of calcium aluminate hydrates, under certain high temperature/humidity conditions, was partly responsible for a scare in the mid-seventies (that still remains today) of high alumina cement building structures. The main reaction implicated in this process is:

$$3CAH_{10} \rightarrow C_3AH_6 + 2AH_3 + 18H \qquad -(2)$$

A similar, though slower, reaction is associated with the hydrate, C_2AH_8. These reactions, known as conversion, proceed faster at higher temperatures. An example of such accelerated conversion at ~70°C is given in Fig.5a. The time-resolved sequence shows that here

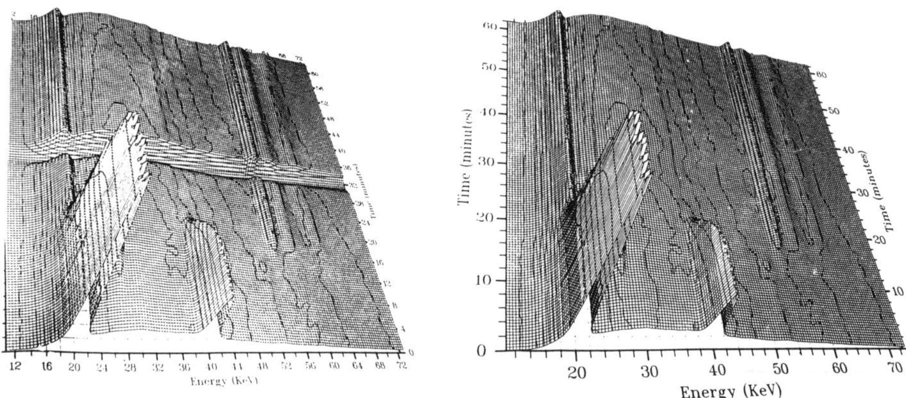

Fig.5a: Time-resolved SR-EDD patterns collected at 2θ=2.6°, showing the conversion at 70°C of hydrate, CAH_{10}, to C_3AH_6. The LHS-figure shows the raw data; the RHS- has been artificially smoothed to cover a necessary interruption of the SR-beam.

the reaction in (2) commences after as little as 12 minutes and is completed after about 20-25 minutes. However these data also confirm another feature that is not immediately evident in Fig.5a, but is much clearer in the 2-dimensional contour map (Fig.5b). This shows that between 8-15 minutes, just before the CAH_{10} phase starts to convert, the secondary hydrate C_2AH_8 appears: this is evident by the

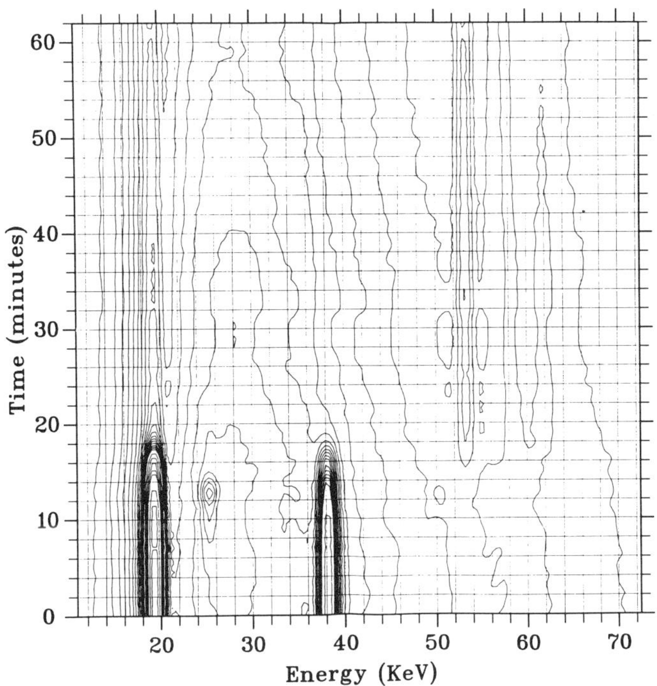

Figure 5b: 2-dimensional time-energy - intensity contour map showing the disappearance of CAH_{10} peaks and emergence of C_3AH_6 peaks between 15-20 minutes, and suprise appearance of C_2AH_8 at 13 minutes.

contours centred around 13 min. and 25 keV, and a very much smaller centre at 50 keV. The temporary appearance of this hydrate is a new feature which will attract much speculation and further research.

HYDROTHERMAL SYNTHESIS OF ZEOLITES

Munn et al.[9] have shown, using modifications to the cell shown in Fig.1, that hydrothermal syntheses can be studied by SR-EDD. EDD is particularly well suited to this problem on 3 accounts:

(1) that a time-resolution, superior to that available with neutron diffraction[20] (~5 minutes) is required;

(2) hard synchrotron (i.e. wiggler) radiation is necessary to penetrate the glass or aluminium/PTFE cell walls and bulk alkaline solutions used in such syntheses;

(3) the X-ray collimation (the lozenge-effect of Fig.1) can effectively remove unwanted diffraction effects from the sample container walls.

The hydrothermal syntheses of both basic (sodalite, zeolite A) and novel (aluminium phophate) zeolite structures, being formed from either silica/alumina gels or from solid material (kaolinite, meta-kaolinite), have now been "captured" using SR-EDD. Assessing the kinetics over a range of synthesis conditions has helped to identify the rate-limiting steps in each case: for example, Munn et al.[9] showed, using kaolinite as the silica/alumina source at ~95oC, that the synthesis could be either rate-limited by the dissolution/break up of the kaolinite-source material or by the assemblage of zeolite framework units, depending on the alkalinity used. These studies have now been extended into the autoclave region (150oC), using an autoclave attachment to the cell (Fig.1), as part of a collaborative study[21] between Cambridge University and Birkbeck College, into the synthesis and stability properties of zeolite structures generally. Fig.6 gives two examples from this study. Fig.6a shows the synthesis of sodalite at 95oC from a gel-mix, but notably with zeolite A appearing as a transitory phase; this behaviour is not typical: both zeolites are usually formed without any crystalline intermediates. Fig.6b shows a transformation of a novel zeolite[21] which becomes unstable once the autoclave seal is broken after synthesis at 150oC. These studies illustrate many of the goals expressed earlier: to characterise the kinetics of growth under alternative pathways; identifying intermediates; and delineating stability regions for the product.

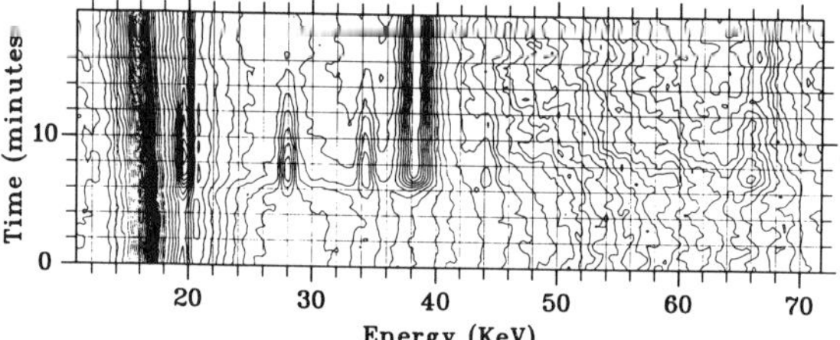

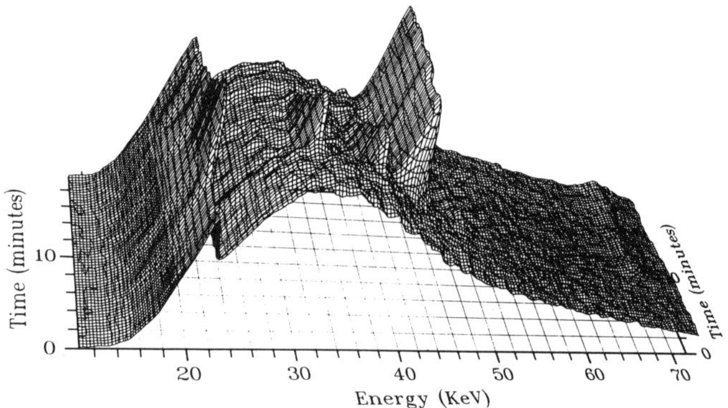

Fig.6a: Time-resolved SR-EDD patterns and 2-dimensional contour map showing the synthesis of zeolite A followed by sodalite from a gel-mix at 95oC. The contour map is particularly effective in showing this, particularly the minor change in the multiple (zeolite A + sodalite) peak at ~38 keV.

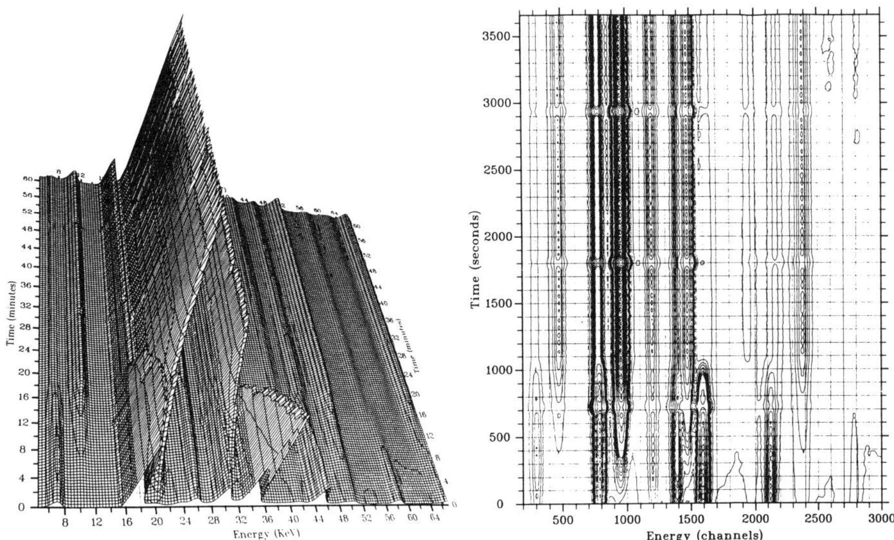

Fig.6b: Time-resolved SR-EDD patterns and 2-dimensional contour map showing an unstable novel zeolite transforming rapidly after synthesis at 150°C.

OTHER MATERIALS

Provided polycrystalline phases are involved, there appears to be a seemingly endless supply of wide-ranging problems suitable for EDXRD analysis. Among other topics that have been studied, albeit in lesser detail than those given above, are the polymorphism of drug materials and crystallisation of amorphous metals. The polymorphism of drugs is important to the drug industry since the choice of a suitable stable polymorph determines its shelf-life (Anwar et al.[22]). Undesirable polymorphic transformations can be induced by changes of temperature (e.g. in storage) and pressure (e.g. in tabletting). Studies of such transformations are being carried out[23] using model polymorphic drug systems, sulphathiazole and sulphamide.

The high temperature crystallisation of amorphous (glassy) metals needs to be understood in order to give working temperature specifications for their use in devices such as low-loss transformer windings. Vrcelj et al.[24] have used SR-EDD to determine the working life of such materials at various operating temperatures, and to identify the products of high temperature crystallisation.

FUTURE DIRECTIONS AND CONCLUSIONS

The belief is that EDXRD will always be an important option when speed of data capture and versatility of environmental cell are important, but clearly also when peak resolution is not a limiting factor. Although the speed of data capture is ultimately linked to detector performance, this does not seem to be a limiting factor for most dynamic processes being studied. We know from image-plate measurements (e.g. Cernik et al.[25]) that there are sampling problems with small sample volumes, though image plate devices can average such texture effects very effectively by integrating around the Debye-Scherrer ring image. Hausermann and Itie[26] however have now shown that this concept can be applied to the EDXRD technique using a small precision-engineered conical detector collimation system and a custom-built annulus-shaped ED-detector crystal. Using such systems they have been able to collect superior EDD patterns on molybdenum disulphide with a ten-fold increase in intensity. Improvements in X-ray sources will impact on EDXRD in those cases where the incident beam intensity profile needs to be matched to a particular d-spacing range under study.

ACKNOWLEDGEMENTS

I would like to express my gratitude to the many colleagues mentioned and referenced, particularly those cases where unpublished work has been included. I also thank the SERC for synchrotron beam time and the SRS personnel for their help and advice.

REFERENCES

1. I.Buras, J.Chwaszczewska, S.Szarras & Z.Szmid, Inst.Nucl.Res. (Warsaw), Rep.No.894/11/PS (1968).
2. B.C.Giessen & G.E.Gordon, Science 159, 973 (1968).
3. A.M.Glazer, M.Hidaka & J.Bordas, J.Appl.Cryst.11, 165 (1978).
4. T.Yamanaka & K.Ogata, J.Appl.Cryst.24, 111 (1991).
5. S.M.Clark, Nucl.Instrum.& Meth. in Phys.Res.A276, 381 (1989).
6. B.Buras, N.Niimura & J.Staun-Olsen, J.Appl.Cryst.11, 137 (1978).
7. D.Hausermann & P.Barnes, subm. to Phase Transitions (1991).
8. S.M.Clark, PhD thesis, University of London, in prep. (1991).
9. J.Munn, P.Barnes, D.Hausermann, S.A.Axon & J.Klinowski, subm. to Phase Transitions (1991).
10. D.Hausermann & W.H.Sherman, private communication (1991).
11. P.Barnes, J.Phys.& Chem.Solids, in press (1991).
12. G.T.Mamott, P.Barnes, S.E.Tarling, S.L.Jones & C.J.Norman, J.Mat.Sci.26, 4054 (1991).
13. X.Turrillas, private communication (1991).
14. P.Barnes, S.M.Clark, S.E.Tarling, E.Polak, D.Hausermann, C.Brennan, S.Doyle, K.J.Roberts, J.N.Sherwood, & R.J.Cernik, Daresbury Laboratory Technical Memorandum DL/SCI/TM55E (1987).
15. P.Barnes,P. S.M.Clark, D.Hausermann, E.Henderson, C.H.Fentiman, S.Rashid, & M.N.Muhamad, subm. to Phase Transitions (1991).
16. M.N.Muhamad, P.Barnes, C.H.Fentiman, D.Hausermann, H.Pollman, & S.Rashid,S, to be subm. to Cement & Concrete Research (1991).
17. S.A.Brooks & J.H.Sharp, in "Calcium Aluminate Cements", ed. by R.J.Mangabhai (Chapman and Hall, London, 1990), p.335.

18. I.Jawed, J.Skalny & J.F.Young, in "Structure and Performance of Cements", ed. by P.Barnes (Applied Science Publishers, London, 1983), p.237.
19. D.Hausermann, P.Barnes, S.Rashid & C.H.Fentiman, unpublished results (1991).
20. E.Polak, J.Munn, P.Barnes, S.E.Tarling & C.Ritter, J.Appl.Cryst. 23, 258 (1990).
21. P.Barnes, J.Munn, J.Klinowski & H.He, unpublished results (1991).
22. J.Anwar, P.Barnes & S.E.Tarling, J.Pharm.Sci.78, 337 (1989).
23. J.Anwar & A.K.Sheradin, unpublished results (1991).
24. R.Vrcelj, D.Hausermann & P.Barnes, unpublished results (1991).
25. R.J.Cernik, S.M.Clark, A.M.Deacon, C.J.Hall, P.Pattison, R.J.Nelmes, P.D.Hatton & M.J.McMahon, Phase Transitions in press (1991).
26. D.Hausermann & J.P Itie, Rev.Sci.Instrum., in press (1991).

DISCUSSION

MILLS - The concept of an annular detector to "smooth-out" the texture effect seems like a good idea ; but now that one is collecting the entire Debye-Scherrer ring you are demanding more of the count-rate capabilities of the detector. Has any thought gone into segmenting the detector or other methods to increase the count rate capability of the energy dispersive detector ?

BARNES - The geometric-optics can be adjusted (e.g. reducing the conical slit width) without sacrificing the concept of averaging around the Debye-Scherrer ring. However you are right to point out that the technique would be ideal for exploitation by an annular array of detector crystals or other segmentations. This would combat texture effects while obtaining better intensity statistics and/or faster collection rates. We do intend to do this: cost is the main obstacle.

WONG - What is the detector that you used for your energy-dispersive diffraction experiments?

BARNES - The results reported here using the standard (non-conical) geometry used a germanium energy-dispersive detector with a resolution of ~190eV at 5.9keV. The energy-dispersive technique is of course totally limited by detector performance: therefore any major developments in detector technology are eagerly awaited.

BIANCONI - What is the indetermination, given by the X-ray fluorescence collimator, on the diffraction angle θ in your experimental set up ?

BARNES - In the standard (non-conical) geometry, 2 flat molybdenum rods (0.5 m length, 100μm separation) provide the final collimation before the detector. This defines the 2θ-collection angle to $\pm 0.01°$. It should be pointed out however that many of the results reported (cements, zeolites) involved large d-spacings (3-13Å) and here it became necessary to use smaller $2\theta,\theta$-angles (e.g. $2\theta = 2 \rightarrow 4°$) to shift the patterns to higher energy, E, to combat absorption of X-rays in the sample and improve peak separation (i.e. reduce $\Delta E/E$). However this ultimately becomes a compromise solution because the well known geometrical "cot θ effect" on peak broadening starts to make a contribution to the peak widths which are otherwise dominated by the detector resolution (see question (2)).

STUDY OF THE PRECIPITATION OF METALLIC PLATINUM INTO IONOMER MEMBRANES BY TIME-RESOLVED DISPERSIVE X-RAY SPECTROSCOPY

P. MILLET
SOPHITECH, 6 rue Henri Poincaré, 91120 Palaiseau, FRANCE

R. DURAND
CREM.GP, BP 75, 38402 Saint-Martin d'Hères, FRANCE

E. DARTYGE, G. TOURILLON and A. FONTAINE
LURE, Bât. 209 D, 91405 Orsay, FRANCE

ABSTRACT

Localized precipitation of metallic platinum in perfluorinated ion-exchange membranes such as Nafion (Dupont de Nemours) allows the preparation of "Solid Polymer Electrolyte" composites which can be used as electrochemical cells in water electrolysers and fuel cells. A theoretical model has been developped to describe the two-step procedure of preparation of those composites, i.e. the ion-exchange and precipitation processes. EXAFS in dispersive mode has been used in LURE (Orsay) to study *in situ* the kinetics of incorporation and precipitation of $[Pt(NH_3)_4\ Cl_2]$ and to check the validity of the model.

INTRODUCTION

Solid Polymer Electrolyte (SPE) technology has been the subject of intensive work during the past decade, especially for water electrolysis[1,2] and hydrogen/oxygen fuel cells[3,4] applications.

In the literature, several methods have been proposed to plate adequat catalytic structures on each side of the membrane sheet. Recently[5], we presented a procedure based on the chemical reduction of cationic noble metal salts into Nafion membranes. The procedure, examplified in ref. 5 with the precipitation of metallic platinum, consists of (Figure 1) : (i) an ion-exchange process during which the membrane, initially in the H^+ form, incorporates a given amount of a cationic "precursor" salt ($[Pt(NH_3)_4]^{2+}$ in the example), (ii) in-situ precipitation of the salt to any solid form by chemical reaction (reduction to metallic platinum by a solution of sodium borohydride in the example). The precipitate predominates near the membrane surfaces and serves as electrode.

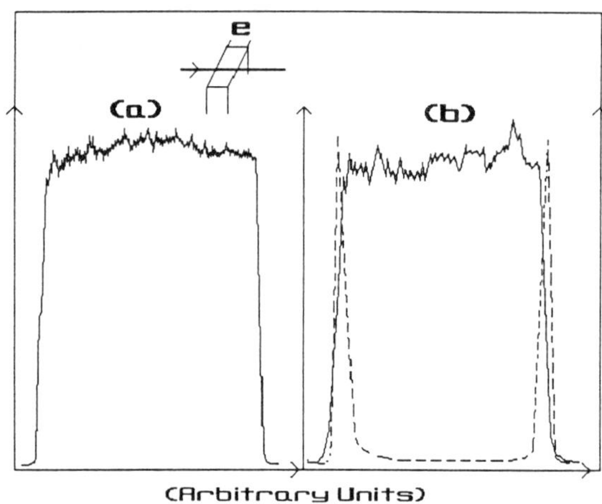

Figure 1 : Electron microprobe concentration profiles across the membrane thickness (e) in arbitrary scale. (a) Nafion-platinum exchanged sample. (b) Nafion-platinum exchanged sample treated with a 15 $g.l^{-1}$ $NaBH_4$ solution for 2 h; sodium (——) and platinum (---) profiles.

That procedure can be extended to the precipitation of various other species to obtain either metal or hydroxide based composites. To determine the optimal conditions of preparation of these composites, we developped a theoretical model which describes the ion-exchange and precipitation processes. Validity of the model was checked in the case of platinum precipitation. EXAFS in dispersive mode was used to follow in-situ the incorporation and precipitation of platinum cations. Concentration profiles in metallic platinum across the membrane thickness were then recorded.

EXPERIMENTAL DETAILS

Perfluorosulfonic polymer membranes (Dupont de Nemours Nafion 117 products) were used for the experiments. $[Pt(NH_3)_4]Cl_2 \cdot H_2O$ (Johnson Matthey) was used as precursor salt for platinum precipitation and $NaBH_4$ (Mercks) was used as chemical reducer.
In situ experiments were performed by EXAFS in dispersive mode at the Laboratoire pour l'Utilisation du Rayonnement Electromagnétique (LURE - Orsay)[6]. Kinetic measurements were performed using the cell shown in Figure 2. Spectra were recorded at the PtL_{III} edge. The ion-echange process was followed by measuring the time variation of the jump height at the L_{III} edge. The jump height was defined as the difference between the base line before the edge and the base line tangent to the first minimum after the absorption maximum[7]. The reduction was followed by measuring the time variation of the white line height.

Concentration profiles in platinum and sodium across the membrane thickness were obtained as previously described[5] using a Camebax electron microprobe analyzer. Diffusion coefficients of the various species have been measured both in membrane and solution[8].

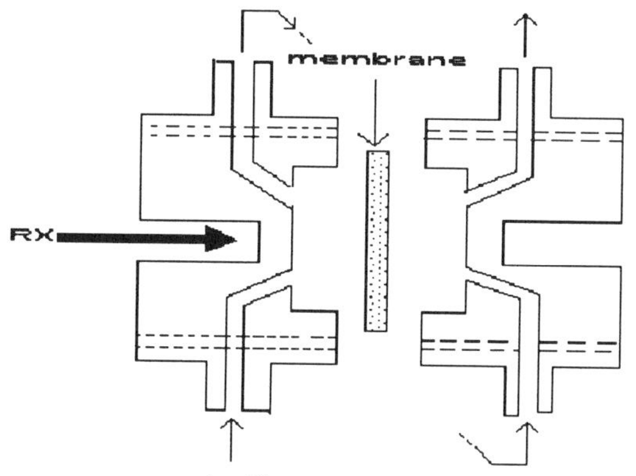

Figure 2 : Schematic diagram of the cell used for kinetic measurements. Data acquisitions were spaced 60 sec. apart. Spectra were recorded in 3.5 sec.

PLATINUM TETRAMINE CHARACTERIZATION

In table I are compiled the nature, number and distance of the nearest neighbors of a given platinum atom in $[Pt(NH_3)_4]\ Cl_2.H_2O$[9].

Because of the somewhat large interatomic distances, EXAFS does not provide information beyond the nearest nitrogen neighbors. However, it is possible to determine changes in interatomic Pt-N distance when the salt is either solid or dissolved or incorporated into Nafion membranes.

Figure 3 shows the radial distribution of platinum tetramine (a) solid, (b) in aqueous solution and (c) incorporated in Nafion membranes. The similitude of the three curves suggests that little change occurs in the structure of the salt during dissolution and incorporation.

Table I : nature, number and distance of the nearest neighbors of a given Pt atom in $[Pt(NH_3)_4]Cl_2.H_2O$.

element	number	distance(Å)
Pt	1	0
N	4	2.04
Pt	2	4.21
Cl	8	4.27
N	8	4.68
H_2O	4	5.21

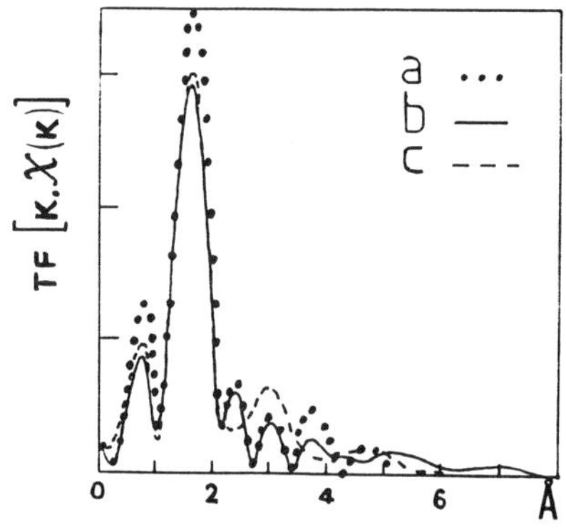

Figure 3 : Radial distribution functions for platinum tetramine. (a) solid; (b) dissolved in water and (c) incorporated into Nafion.

Figure 4-a and 4-b show the K.X(k) vs. K plots (experimental and theoretical) of the main peaks of Figures 3-b and 3-c.

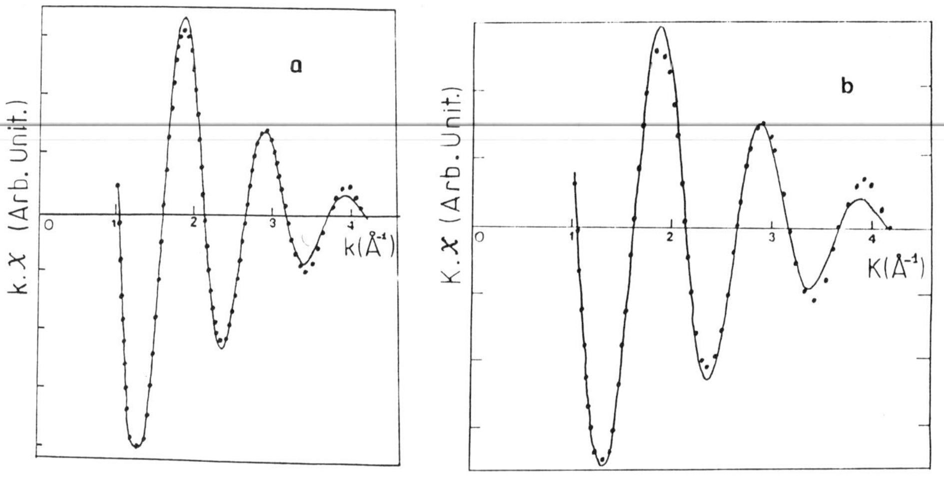

Figure 4 : K.X(k) vs. K plots for the two main peaks of Figure 3-b and 3-c. Experimental (——) and theoretical (...) points. The fit was obtained assuming the existence of 4 nitrogen neighbors at 2.04 A.

The good correlation indicates that the four amine groups remain linked to platinum, even after dissolution in water and incorporation into Nafion membranes; they remain slightly at the same distance (2.04 Å).

Exchange capacity measurements show that one platinum ion is incorporated for each two sulfonate group. All these results suggest that platinum tetramine dichloride dissolves and incorporates Nafion in the form $[Pt(NH_3)_4]^{2+}$.

ION-EXCHANGE PROCESS

We consider here the ion-exchange process that occurs when a Nafion membrane (H^+ form) is soaked in an aqueous solution of platinum tetramine.
The following assumptions were made :
(i) total exclusion of chloride ions;
(ii) platinum concentration $C°$ in bulk solution is sufficiently low to assume that the kinetics is limited by ionic diffusion across the liquid film adjacent to the membrane surfaces;
(iii) concentrations in the bulk solution are supposed constant and invariant (infinite volume and agitation);
(iv) electro-neutrality is respected in the liquid film and we neglect the effect of an eventual electric field ;
(v) the regime of diffusion is assumed to be quasi-stationnary, i.e. concentration gradients of diffusing species across the liquid film are constant at every moment.
Under these assumptions, it can be shown[8] that the system to solve is given by Eqs. (1), (2) and (3).

$$\frac{d_{NPt}}{dt} = \frac{4 \,.\, D_{Pt} \,.\, (C° - C'_{Pt})}{E \,.\, C^- \,.\, l} \qquad (1)$$

$$G = \frac{N_{Pt}}{(1-N_{Pt})^2} \,.\, \frac{4}{K} \,.\, \frac{{D_{Pt}}^2}{{D_H}^2} \qquad (2)$$

$$G \,.\, (C'_{Pt})^2 - (1 + 2.G.C°).\, C'_{Pt} + G.(C°)^2 = 0 \qquad (3)$$

where : E, l = membrane and diffusion layer thicknesses;
C^- = sulfonate concentration in the membrane;
C'_{Pt} = Pt concentration at the interface;
N_{Pt} = ionic fraction of platinum in the membrane;
K = selectivity coefficient at the interface.

The solutions $N_{Pt}(t)$ to the model were obtained numerically. The derivative of order 1 was approximated using the finite difference method.

A comparison of empirical and theoretical results is given in Figure 5 where the relative variation in platinum concentration is plotted vs. time for different platinum tetramine concentrations in solution.

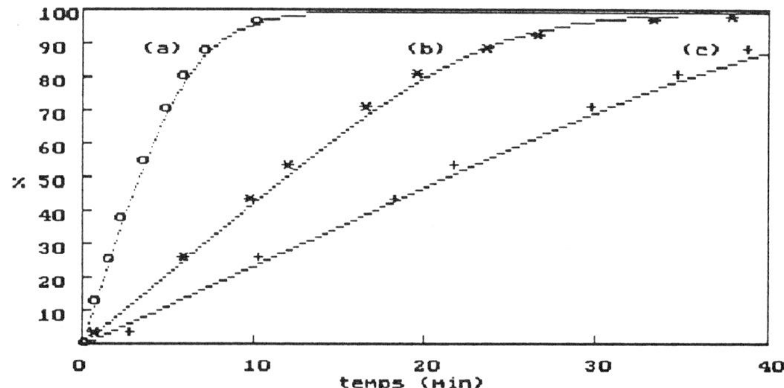

Figure 5 : Kinetics of incorporation of platinum tetramine into Nafion membranes. Comparison of empirical (o,*,+) and theoretical (——) results. (o) $[Pt^{2+}]$ = 3.75 10^{-2} M; (*) $[Pt^{2+}]$ = 1.10^{-2} M; (+) $[Pt^{2+}]$ = 5.5 10^{-3} M.

Theoretical points were obtained assuming a layer thickness of 230 µm, a value which correlate well enough with those deduced from mass transfer correlations[8]. However, under vigourous agitation and when concentrated solutions are used, total exchange can be achieved within a few minutes. In such cases, the kinetics is controlled by ionic diffusion within the membrane.

PLATINUM PRECIPITATION

Platinum tetramine exchanged Nafion membranes were placed in the cell shown in figure 2. Sodium borohydride solutions were then circulated and the time variation of the height of the white line was measured (figure 6). The presence of hydrogen bubbles reduced the precision of the experiments and the fit is not as good as in figure 5. Taking into account BH_4^-, time dependent concentration profiles in metallic platinum were obtained by solving :

$$\Phi_i = - D_i \cdot \frac{\delta C_i}{\delta x} - \frac{D_i \cdot z_i}{R \cdot T} \cdot F \cdot C_i \cdot \frac{\delta \phi}{\delta x} \qquad (4)$$

$$\frac{\delta C_i}{\delta t} = - \frac{\delta \Phi_i}{\delta x} \qquad (5)$$

$$\Sigma_i \, z_i \cdot C_i = 0 \quad \text{(at every x and t)} \qquad (6)$$

where Φ_i is the flux of species i and ϕ the electric potential. Eq. (6) is used as a good approximation of

Poisson's equation[10]. A reaction term r_i was added to the platinum and borohydride continuity equations (5) to account for the chemical reduction :

$$NaBH_4 + 4\ [Pt(NH_3)_4]^{2+} + 8\ OH^- \longrightarrow NaBO_2 + 4Pt^\circ + 16NH_3 + 6H_2O \quad (7)$$

Local variations of metallic platinum were computed at each iteration and added to the values of the precedent iterations. The final $C_{Pt^\circ}(x)$ solutions were obtained after complete reduction of $[Pt(NH_3)_4]^{2+}$. Results obtained are shown in Figure 7.

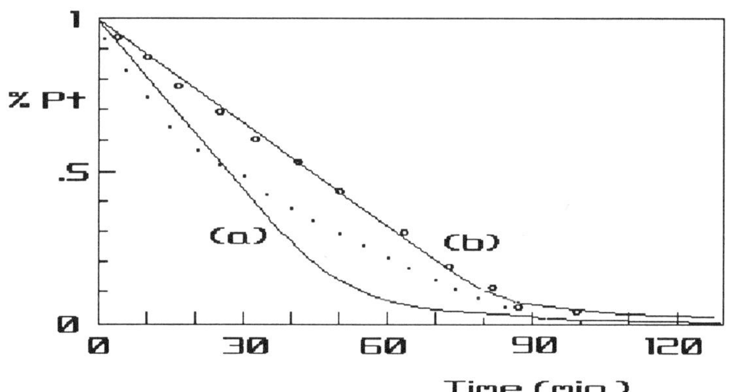

Figure 6 : kinetics of reduction; time variation of $[Pt(NH_3)_4]^{2+}$ concentration within the membrane. (a) $NaBH_4$ = 9 g.l^{-1}; (b) $NaBH_4$ = 3 g.l^{-1}. (.,o) : experimental and (——) theoretical values.

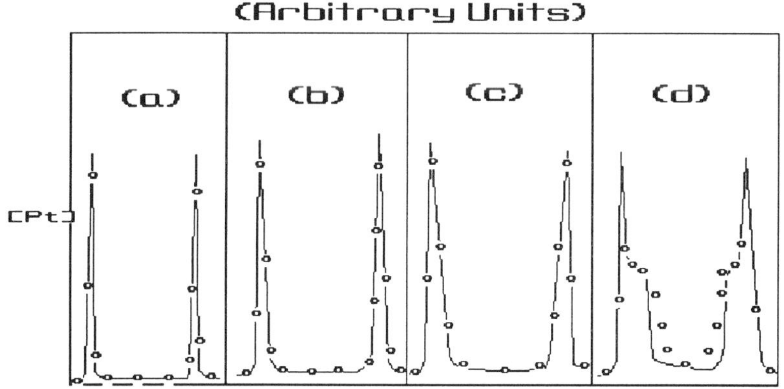

Figure 7 : Experimental (——) and calculated (o) platinum concentration profiles across membrane thickness. Effect of reducer concentration : (a) 1 g.l^{-1}; (b) 15 g.l^{-1}; (c) 40 g.l^{-1} and (d) 80 g.l^{-1}.

CONCLUSIONS

EXAFS in dispersive mode has been used to follow *in situ* the incorporation and precipitation of $[Pt(NH_3)_4 Cl_2]$ within Nafion ionomer membranes. The two steps of incorporation and precipitation have been theoretically described. The model can predict the effect of a change in salt concentration and in reducer concentration. Theoretical results compare well with experimental datas.

While platinum based Nafion composites are interesting for water electrolysis and H_2/O_2 fuel cell applications, the technique of precipitation can be extended to the preparation of various other SPE composites. For example, ruthenium and iridium oxide based electrodes can be used as anodes and cathodes in water electrolysers. Given adequate physical datas (diffusion coefficients of the diffusing species, membrane characteristics), it is possible to optimize the preparative conditions of the composites in order to obtain the desired concentration profiles.

REFERENCES

1. T. Ohta and I. Abe, Proceed. 5th World Hydrogen Energy Conference, Toronto, 1, 47 (1984)

2. R. Oberlin and M. Fischer, Proceed. 6th World Hydrogen Energy Conference, Vienna, 1, 333 (1986)

3. D. Watkins, K. Dircks, D. Epp and A. Harkners, Proceed. 32nd Int. Power Sources Symposium, ECS publication, (1986) p.590

4. S. Srinivasan, E.A. Ticianelli, C.R. Derouin and A. Redondo, J. Power Sources, 22, 359 (1988)

5. P. Millet, R. Durand and M. Pinéri, Int. J. Hydrogen Energy, Vol. 15, n°4. 245 (1990)

6. G. Tourillon, E. Dartyge, A. Fontaine and A. Jucha, Phys. Rev. Let., 57, 603 (1986)

7. R.S. Weber, M. Peuckert, R.A. Dellabetta and M. Boudart, J. Electrochem. Soc., 135, 2535 (1988)

8. P. Millet, Ph.D. Thesis, INPG, Grenoble (1989) FRANCE

9. R.W.G. Wyckhoff, "Inorganic compounds, hydrates and ammoniates", Vol.3 (R.E.W. Krieger, Publishing Company, Malabar, Florida), 1982

10. A.D. Mac Gillivray, J. Chem. Phys., 48 (1968) 2903.

Nuclear Bragg diffraction
a new source for hyperfine spectroscopy, solid state physics, and γ - optics

Rudolf Rüffer
European Synchrotron Radiation Facility, BP 220, F-38043 Grenoble Cedex

Abstract

Nuclear Bragg diffraction with synchrotron radiation as source will become a powerful new X-ray source in the Å-region. This source exceeds by now the brilliance of conventional Mössbauer sources giving hyperfine spectroscopy, solid state physics, and γ - optics further momentum. As example, applications to hyperfine spectroscopy will be discussed.

Introduction

Activities in the field of nuclear Bragg diffraction have got a strong impetus following the pilot experiment which used synchrotron radiation as source[1]. Recent experiments have proved that a precise determination of hyperfine interaction parameters is possible. In the future, with even higher intensities, applications in the field of solid state physics and γ-optics are envisaged.
Nuclear Bragg diffraction experiments with a radioactive source were carried out soon after the discovery of the Mössbauer effect. However, the low brilliance implied very low counting rates and therefore made experiments extremely difficult. Nevertheless, many investigations have been done in the last 25 years which were reviewed by Smirnov[2].
The dynamical theory of nuclear Bragg diffraction was developed by two groups independently, i.e. by Trammell and Hannon[3–6] and Kagan, Afanasev, and Kohn[7–9].
In this paper a brief survey on applications of nuclear Bragg diffraction is given which is followed by a more detailed discussion of the determination of hyperfine interaction parameters. Examples will be given in cases of single crystals of $FeBO_3$ and garnets and polycrystalline Fe foils.

Application on nuclear Bragg diffraction

Let us assume we have a synchrotron radiation beam monochromatized down to an energetic bandwidth of order neV...μeV which is related to the nuclear level width Γ_0. Then a huge variety of very different experiments becomes feasible. There are two main classes of experiments: the *resonant experiments* which make use of the Mössbauer effect in the sample under investigation, and the *nonresonant experiments* which make only use of the high monochromaticity of the γ-ray beam (table I).

The resonant experiments can be subdivided in three groups. The *basic experiments* deal with the proof of the dynamical theory of nuclear Bragg diffraction. Most of the early experiments belonged to this group. The speed-up e.g. was investigated especially in two cases, $FeBO_3$ [10] and yttrium iron garnet (YIG) [11]. The understanding of pure nuclear reflections became more precise by changing the combined hyperfine interaction in the measurements of a YIG single crystal [12].

Table I Examples of resonant and nonresonant experiments

resonant experiments	
basic experiments	hyperfine spectroscopy
speed-up	magnetic field (direction/value)
coherent enhancement	critical phenomena
dynamical beats	EFG (direction,η,value)
quantum beats	isomer shift
suppression of inelastic channel	dynamical processes
optical activity	
interference of elastic processes	
resonant experiments	**nonresonant experiments**
structure analysis	
magnetic/electric structure	Debye-Waller factor
phase problem	thermal diffuse scattering
structure factor	phonon excitation
	interferometry
	shutter (intensity, polarization)

The *hyperfine spectroscopy* in principle covers the field of Mössbauer spectroscopy and related techniques. Those experiments, however, which profit also from the advantages of synchrotron radiation will benefit most from this new technique. For example, scattering experiments become very attractive because they need the very low divergence of synchrotron radiation (in the order of arcsec), which especially applies to the investigation of surfaces and interfaces. Also in case of very small samples (about 1...0.1 mm^2) synchrotron radiation is superior to conventional Mössbauer spectroscopy. In this case high pressure or very low temperature experiments are of interest.

Examples of hyperfine spectroscopy are the measurements of the magnetization curve in case of $FeBO_3$ [24], the determination of hyperfine interaction parameters in case of thulium iron garnet (TIG) for the Tm sites [13], or the polarization analysis in case of Fe_2O_3 [14]. Some of those examples will be discussed in more detail later.

The last subgroup is devoted to *structure analysis*. It is obvious that the electric and magnetic structure including the phases can be determined by the investigation of pure nuclear reflections. Examples are given e.g. for TIG [13] and Fe_3BO_6 [15]. A proposal to determine partial structure factors was made by Faigel [16].

Due to the low counting rate at existing synchrotron radiation sources it is difficult to carry out *nonresonant experiments.* Therefore only one attempt was made so far to search for quasielastic scattering [17]. With the new dedicated synchrotron radiation sources, however, this problem will be overcome. Then experiments will be feasible to determine the Debye-Waller factor [18], search for thermal diffuse scattering [19], or phonon excitation [20].

Moreover, very fast shutters become available as has been shown by Smirnov et al. [21].

Another field will be the γ-interferometrie. Due to the extremely long coherence length, in case of ^{57}Fe up to 30 m, a new thinking has to start.

Hyperfine spectroscopy

In case of Mössbauer spectroscopy the very narrow nuclear level width Γ_0 is used to observe the splitting of nuclear levels. This splitting is caused e.g. by magnetic hyperfine fields or electric field gradients. The energy resolution is $4.5 \cdot 10^{-9}$ eV for the famous ^{57}Fe Mössbauer isotope. In contrast to Mössbauer spectroscopy where the energy differential measurement is the proper technique now the time differential technique becomes superior due to the pulsed structure of synchrotron radiation. The relevant informations show up then in *quantum beats.*

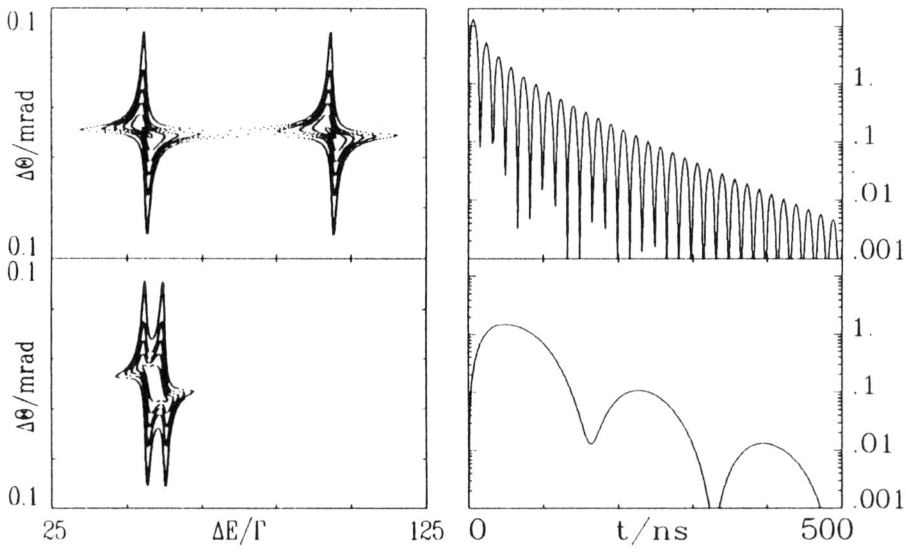

Fig. 1. Contour plot (left) of the frequency response $I(\omega)$ for two resonances with a splitting ΔE of 50Γ (above) and 5Γ (below), and the corresponding time response $I(t)$ (right) integrated over the angle $\Delta\Theta = 0.1$ mrad.

If there are hyperfine splittings of the nuclear levels or line shifts between different sites, quantum beats will appear in the time spectra of the coherently scattered radiation. These beats occur at the difference of frequencies

$$\Omega_B(n, m, \vec{\rho}; n', m', \vec{\rho}') = \omega_{nm}(\vec{\rho}) - \omega_{n'm'}(\vec{\rho}')$$

of all allowed nuclear hyperfine transitions $\omega_{nm}(\vec{\rho})$ from all different nuclear sites $\vec{\rho}$, giving a direct measure of the energy splittings of both the excited and the ground states, as well as of different sites.

To demonstrate the origin of quantum beats, we consider a simple case where only two nuclear resonances of energies E_1 and E_2 ($E_{1,2} = \hbar\omega_0 \pm \frac{1}{2}\Delta E$) and of equal oscillator strength are involved. Considering spatial phases $\varphi_1 = 0$ and $\varphi_2 = \pi$, the frequency response is given by

$$R(\omega) \propto \frac{1}{(E_1 - \hbar\omega - \frac{i}{2}\Gamma_0)} - \frac{1}{(E_2 - \hbar\omega - \frac{i}{2}\Gamma_0)}$$

and the corresponding time response by

$$I(t) \propto |R(t)|^2 \propto e^{-\frac{\Gamma_0 t}{\hbar}} \cdot sin^2(\tfrac{\Delta E \cdot t}{2 \cdot \hbar}).$$

Fig. 1 shows theoretical spectra for different splittings ΔE. The figures on the left show the energy spectra, the figures on the right the corresponding time spectra. The quantum beats show up as the pronounced modulation with frequency $\frac{\Delta E}{2\hbar}$.

The following examples on $FeBO_3$ and garnets were carried out at the high resolution spectrometer at HASYLAB [22]. Pure nuclear reflections were used to overcome the large nonresonant background. These reflections are forbidden for electronic diffraction due to the lower symmetry of the nuclei. This is possible because the nuclei, in contrast to the electrons, are sensitive to the internal fields in the crystal which give rise to another symmetry class. According to experiments and calculations it seems in fact to be quite common that a crystal exhibits pure nuclear reflections [12].

$FeBO_3$ represents a model antiferromagnet scatterer (the weak ferromagnetism which is introduced by the small canting angle between the magnetic fields [23] can be neglected in this context).

In an antiferromagnet the pure nuclear reflections arise entirely from the changing directions of the internal magnetic field. The magnetic unit cell is twice as large as the crystallographic one. In the case of $FeBO_3$, the (n n n)-reflections for odd n (referred to the rhombohedral unit cell) are pure nuclear reflections. The main axis of the electric field gradient (EFG) is perpendicular to the corresponding planes.

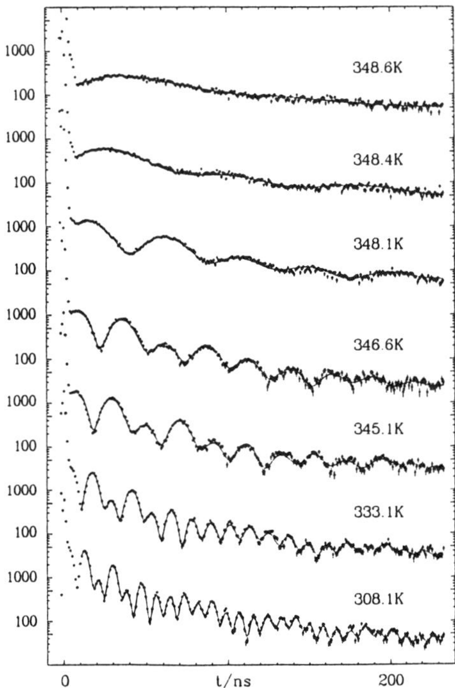

Fig. 2. Time spectra of the decay of the collective nuclear state for the (3 3 3) pure nuclear reflection of $^{57}FeBO_3$ between room temperature and the Néel temperature. The solid lines are the resulting fits by the dynamical theory (from ref. 10).

Measurements were performed from room temperature up to the Néel temperature[24]. Fig. 2 shows the spectra which were measured within about 35 hours. The solid lines are the result of fitting with the dynamical theory. All χ^2 values were below 1.8. The measurements allowed the determination of the magnetic fields with an error of less than 0.02 T. They show the characteristic magnetisation curve of iron borate[23].
Due to the collapse of the pure nuclear reflection at the Néel point the diffracted intensity becomes very low in the vicinity of the Néel temperature[25]. This can be overcome by using 'allowed reflections' which show high diffracted intensity in the whole temperature range. Then, precise determinations of critical exponents are possible.

Garnets are very important in research and applications and they are subject of extended investigations[26]. Their crystal structure is rather complicated and gives still rise to new attempts to refine the structure[27]. In the following YIG and TIG will be described in more detail. Especially the determination of hyperfine parameters and their sensitive dependence on certain parameters will be shown.

The unit cell of YIG contains 160 atoms including 40 iron atoms[28]. They belong to two different crystallographic sites: 16 to the a-site and 24 to the d-site. The iron ions are exposed to a combined magnetic and electric interaction. For both sites independent pure nuclear reflections are predicted.

All ions of the d-site are ferromagnetically coupled and the nuclei experience an identical magnetic interaction strength. They become distinguishable by their electric interaction. While the electric interaction strength is equal for all d-sites, the main axes of the EFG's (V_{zz}) lie along the coordinate axes of the cubic crystal system and vary periodically within the unit cell. One obtains the d_1, d_2, and d_3 subgroups with the V_{zz} axes along the [1 0 0]-, [0 1 0]-, and [0 0 1]-directions, respectively. Only the d_1- and d_2-sites contribute to the (n 0 0) pure nuclear reflections with n = 4m+2 (m=0,1,2,...). Their V_{zz} axes lie within the (0 0 1)-surface. The d_3-site has a vanishing structure factor for these reflections.

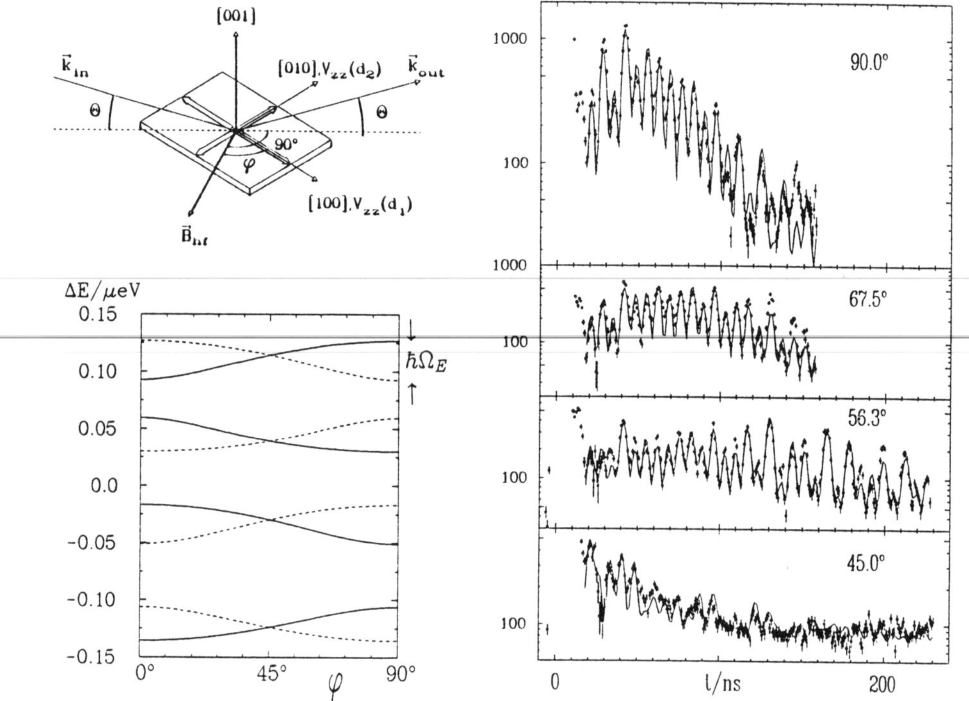

Fig. 3. upper left: Diffraction geometry for the YIG crystal with azimuthal angle φ, scattering plane ($\vec{k}_{in}, \vec{k}_{out}$), and Bragg angle Θ. The direction of the hyperfine field $\vec{B}_{hf}$ and the orientation of the main axis of the EFG within the (001)-surface are shown for the d_1- and d_2-site.
lower left: Splitting of the excited nuclear state for the d_1-site (solid line) and the d_2-site (dashed line) as function of the angle φ between $\vec{B}$ and the [1 0 0]-axis.
right: time spectra of the decay of the collective nuclear state for the (0 0 2) pure nuclear reflection of YIG. The angle φ between $\vec{B}$ and the [1 0 0]-axis is 90 °, 67.5 °, 56.25 °, and 45 °, respectively. The solid lines are the result of a fit using the dynamical theory of nuclear Bragg diffraction (from ref. 12).

As an example the experiments described in ref. 12 on the d-sites were utilized. They show the influence of the relative angle between the hyperfine field $\vec{B}_{hf}$ and the main axis of an EFG.
The experimental situation is shown in fig. 3 (upper left). A YIG crystal, placed in a weak external magnetic field fixed perpendicular to the scattering plane, is rotated by the angle φ. This rotation changes the relative angle between the hyperfine field which is aligned by the external field, and the main axes of the EFG's at the d_1- and d_2-sites which are fixed to the crystallographic axes.

Fig. 3 (right) shows the corresponding diffraction spectra for four different angles φ. The frequencies of the fast beat pattern are the same for all angles φ, i.e. the magnetic interaction stays constant. However, the frequency of the slow overall modulation $sin^2(\frac{1}{2}\Omega_E\, t)$ drastically changes. This can be easily understood from the level splitting of the d_1- and d_2-sites (fig. 3 lower left). For $\varphi = 90\,°$ the splitting of the two sites $\hbar\Omega_E$ is maximal which leads to a relatively 'high' frequency of the modulation. With decreasing angle φ the splitting becomes smaller, and consequently also the frequency decreases. The theoretical calculations additionally show a sensitive dependence of the beat amplitudes on the angle φ.

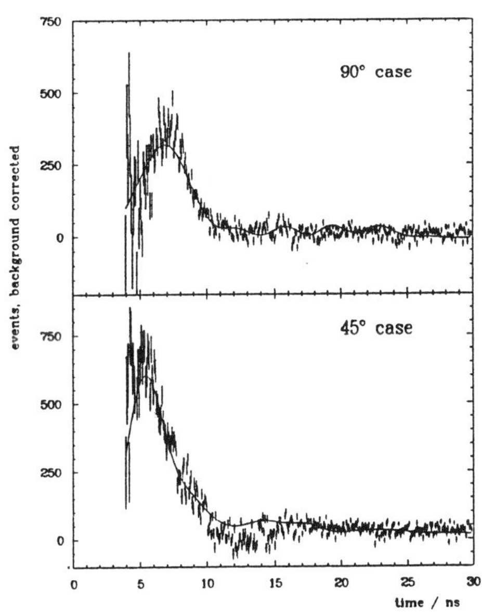

Fig. 4. Time spectra of the (424) reflection of TIG are shown. Two different directions of the external magnetic field, namely 45° and 90° towards the scattering plane were realised. The solid lines are the result of a fit using the dynamical theory of nuclear Bragg diffraction (from ref. 13).

Selecting other reflections, measurements were carried out on the a-sites [29]. To get information about the c-sites, another Mössbauer isotope e.g. ^{169}Tm has to be chosen because this site cannot be occupied by Fe ions.

TIG has the same structure as YIG. The thulium ions replace the 24 yttrium ions on the c-sites. Time differential measurements [13] were performed with two different directions of the external magnetic field, namely 45° and 90° towards the scattering plane (fig. 4). The shape of the quantum beat pattern alters when the direction of the external magnetic field is changed, i.e. the magnetic hyperfine fields are influenced by the external magnetic field which excludes antiferromagnetic coupling of the thulium atoms. Regarding a single nucleus the angle between the main axis of the EFG and the magnetic hyperfine field is different for the 45° and 90° case. Nuclear level splitting and the strength of individual Mössbauer transitions is then different in the above cases thus leads to the observed change in the time spectra. The best simulation of the time spectra was achieved with magnetic hyperfine fields of absolute value of 17 T and 23 T and with a quadrupole splitting ($\frac{1}{2}eqV_{zz}$) of 14.5 mm/s. The two magnetic subgroups are ferromagnetically coupled. For low and high field lattice sites the angles between the main axes of the efg and the [111] direction are 90° and 54.7°, respectively.

The previous examples on hyperfine spectroscopy were performed on pure nuclear reflections. Of course also the investigation of 'allowed' reflections is possible [30]. Another important break-through is the observation of forward scattering of a polycrystalline sample [31]. The same informations, except of course the phases, can be derived from this kind of scattering which is also a collective coherent process as Bragg diffraction.

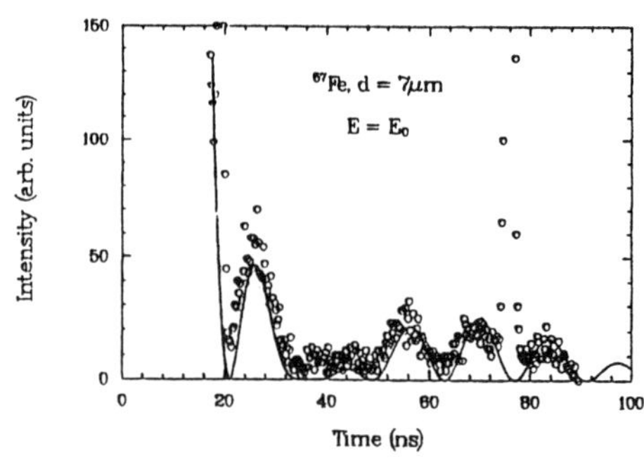

Fig. 5. Time spectra of the forward scattering from a 7 μm polycrystalline-enriched Fe foil (from ref. 31).

The measurements were performed at NSLS using a special monochromator with about 5 meV resolution [32]. A polycrystalline ^{57}Fe foil was put in the monochromatized γ-ray beam in transmission geometry. A vertical magnetic field of approximately 0.1 T aligned the magnetic hyperfine field of the Fe foil perpendicular to the electric polarization vector of the incident beam. Only the $\Delta m = 0$ transitions can be excited in this geometry. The choice of the $\Delta m = 0$ lines is important, since only two such transitions are possible in ^{57}Fe, and each involves a different ground state. Interference between them can only arise if the scattering is coherent. The observation (fig. 5) of the simple beat pattern, coming only from two transitions, demonstrates this interference and hence the coherent nature of the delayed forward scattering radiation. Again from the quantum beat period the magnetic hyperfine field can be derived.

Conclusion

Nuclear Bragg diffraction has the potential to become a powerful technique in hyperfine spectroscopy. Polycrystalline and amorphous samples can be investigated using the forward scattering geometry. In many applications it may be advantageous to apply this technique to single crystals and to carry out measurements in Bragg or Laue geometry. This seems to be realistic because due to the very small beam size only small single crystals are necessary which can be easily grown from many compounds.

With the new dedicated synchrotron radiation sources like ESRF, APS, and SPring-8, this new technique will get further impetus not only in hyperfine spectroscopy but also in the other fields mentioned. With these new sources with their high photon flux further development has to be started on either detectors or monochromators to get rid of the nonresonant background. First attempts have already been made, e.g. single line [25] and absorber filtering [33]. Another promising method to build a nuclear monochromator with bandwidths of up to $200\,\Gamma_0$ around the selected Mössbauer transition energy is the use of grazing incidence antireflecting films (GIAR films) [34,35]. First measurement have recently proved the feasibility [36,37].

References

1. E. Gerdau, R. Rüffer, H. Winkler, W. Tolksdorf, C.P. Klages, J.P. Hannon, Phys.Rev.Lett. 54, 835 (1985)
2. G.V. Smirnov, Hyp.Int. 27, 203 (1986)
3. G.T. Trammell, in *Proceedings of the IAEA Symposium on the Chemical Effects of Nuclear Transformations, Prague, 1960, ed. IAEA, Vienna, 1961, Vol. I, p.75*
4. J.P. Hannon, G.T. Trammell, Phys.Rev. 186, 306 (1969)
5. G.T. Trammell, J.P. Hannon, Phys.Rev. B18, 165 (1978); Phys.Rev. B19, 3835 (1979)
6. J.P. Hannon, G.T. Trammell, Physica B159, 161 (1989)
7. A.M. Afanas'ev, Yu.M. Kagan, Zh.Eksp.Teor.Fiz.Pis.Red. 2, 130 (1965), [JETP Lett. 2, 81 (1965)]
8. Yu.M. Kagan, A.M. Afanas'ev, V.G. Kohn, Phys.Lett. 68A, 339 (1978)
9. Yu.M. Kagan, A.M. Afanas'ev, V.G. Kohn, J.Phys. C12, 615 (1979)
10. U. van Bürck, R.L. Mößbauer, E. Gerdau, R. Rüffer, R. Hollatz, G.V. Smirnov, J.P. Hannon, Phys.Rev.Lett. 59, 355 (1987)
11. J. Arthur, G.S. Brown, D.E. Brown, S.L. Ruby, Phys.Rev.Lett. 63, 1626 (1989)
12. R. Rüffer, E. Gerdau, H.D. Rüter, W. Sturhahn, R. Hollatz, A. Schneider, Phys.Rev.Lett. 63, 2677 (1989)
13. W. Sturhahn, E. Gerdau, R. Hollatz, R. Rüffer, H.D. Rüter, W. Tolksdorf, Europhys.Lett. 14, 821 (1991)
14. D.P. Siddons, J.B. Hastings, G. Faigel, L.E. Berman, P.E. Haustein, J.R. Grover, Phys.Rev.Lett. 62, 1384 (1989)

15. A.I. Chumakov, G.V. Smirnov, M.V. Zelepukhin, U. van Bürck, E. Gerdau, R. Rüffer, H.D. Rüter, Hyp.Int. to be published
16. G. Faigel, Hyp.Int. 61, 1355 (1990)
17. B.C. Larson, J.Z. Tischler, 1991, private communication
18. see e.g. N.M. Butt, D.A. O'Connor, Proc.Phys.Soc. 90, 247 (1967)
19. see e.g. Y. Kashiwase, Y. Kainuma, M. Minoura, J.Phys.Soc.Jap. 51, 937 (1982)
20. see e.g. E. Burkel, B. Dorner, Th. Illini, J. Peisl, Rev.Sci.Instr. 60, 1671 (1989)
21. G.V. Smirnov, Yu.V. Shvyd'ko, O.S. Kolotov, V.A. Pogozhev, Zh.Eksp.Teor.Fiz. 86, 1495 (1984) [JETP 59, 875 (1984)]
22. R. Rüffer, D. Giesenberg, H.D. Rüter, R. Hollatz, E. Gerdau, J. Metge, K. Ruth, W. Sturhahn, M. Grote, R. Röhlsberger, Hyp.Int. 58, 2467 (1990)
23. M. Eibschütz, M.E. Lines, Phys.Rev. B7, 4907 (1973)
24. H.D. Rüter, R. Rüffer, E. Gerdau, R. Hollatz, A.I. Chumakov, M.V. Zelepukhin, G.V. Smirnov, U. van Bürck, Hyp.Int. 58, 2473 (1990)
25. A.I. Chumakov, M.V. Zelepukhin, G.V. Smirnov, U. van Bürck, R. Rüffer, R. Hollatz, H.D. Rüter, E. Gerdau, Phys.Rev. B41, 9545 (1990)
26. see e.g. G. Winkler, Vieweg tracts in pure and applied physics, Vol. 5, Magnetic garnets. Braunschweig/Wiesbaden, Friedr. Vieweg & Sohn 1981
27. J. Dong, K. Lu, Phys.Rev. B43, 8808 (1991)
28. H. Winkler, R. Eisberg, E. Alp, R. Rüffer, E. Gerdau, S. Lauer, A.X. Trautwein, M.Grodzicki, A. Vera, Z.Phys. B49, 331 (1983)
29. R. Rüffer, J. Metge, H.D. Rüter, W. Sturhahn, E.Gerdau, Hyp.Int. to be published
30. J.B. Hastings, D.P. Siddons, G. Faigel, L.E. Berman, P.E. Haustein, J.R.Grover, Phys.Rev.Lett. 63, 2252 (1989)
31. J.B. Hastings, D.P Siddons, U. van Bürck, R. Hollatz, U. Bergmann, Phys.Rev.Lett., 66, 770 (1991)
32. D.P. Siddons, J.B. Hastings, G. Faigel, J.R. Grover, P.E. Haustein, L.E. Berman, Rev.Sci.Instr. 60, 1649 (1989)
33. U. van Bürck, R.L. Mössbauer, E. Gerdau, W. Sturhahn, H.D. Rüter, R. Rüffer, A.I. Chumakov, M.V. Zelepukhin, G.V. Smirnov, Europhys.Lett. 13, 371 (1990)
34. J.P. Hannon, G.T. Trammell, M. Mueller, E. Gerdau, H. Winkler, R. Rüffer Phys.Rev.Lett. 43, 636 (1979)
35. J.P. Hannon, G.T. Trammell, M. Mueller, E. Gerdau, R. Rüffer, H. Winkler, Phys.Rev. B32, 6374 (1985)
36. M. Grote, R. Röhlsberger, M. Dimer, E. Gerdau, R. Hellmich, R. Hollatz, J. Jäschke, E. Lüken, J. Metge, R. Rüffer, H.D. Rüter, W. Sturhahn, E. Witthoff, M. Harsdorff, W. Pfützner, M. Chambers, J.P. Hannon, Europhys.Lett. 14, 707 (1991)
37. R. Röhlsberger, E. Gerdau, O. Leupold, E. Lüken, J. Metge, H.D. Rüter W. Sturhahn, E. Witthoff, R. Rüffer, Europhys.Lett. to be published //

DISCUSSION

GOULON - 1. What would be the ideal filling mode of the machine for your experiment.

2. Polarization effects: anything special to observe with Circular Polarization ?

RUFFER - 1. For experiments with ^{57}Fe which is the most common Mössbauer isotope a bunch separation of about 300ns...500ns would be desirable. That corresponds to a ten bunch mode at ESRF (285ns) at least.

2. In hyperfine spectroscopy I see at the moment no new effect due to incoming circular polarized light. However in some investigations it may be helpful to use such polarization to simplify the response of the sample under investigation.

BARTUNIK - Considering the set-up planned on the ESRF, what is the range in counting rates obtained from nuclear monochromatization ?

RUFFER - The "nuclear" counting rate at ESRF in case of ^{57}Fe, the common Mössbauer isotope, will be about 20kHz in Γ_0 (the nuclear level width $\Gamma_0(^{57}Fe)$ $=5.10^{-9}$eV) for an 1.6m undulator. For other Mössbauer isotopes depending on their Γ_0 the counting rates range from some thousands up to some million counts per second.

HAGELSTEIN - You did show the structural analysis of garnet sites with electronically forbidden reflections only. Is it possible to do such an analysis with electronically allowed reflections ?

RUFFER - There were already some experiments done on electronic and nuclear allowed reflections. The main problem however is the large prompt signal from the electronic diffraction which may overload the detector.

CHEMICAL SHIFTS AND CORE LEVEL LINE SHAPES FOR MONOVALENT, DIVALENT AND TRIVALENT CU COMPOUNDS

O. Gunnarsson, K. Karlsson,* and O. Jepsen
Max-Planck Institut für Festkörperforschung, D-7000 Stuttgart 80, Germany

ABSTRACT

We have studied the chemical shift and the Cu $2p$ core level photoemission spectra of Cu_2O, CuO and $NaCuO_2$, where Cu is formally mono-, di- and trivalent, respectively. We have performed *ab initio* band structure calculations and found that the variation in the net charge of the Cu atom is small. Surprisingly, the smallest net positive charge is found in the trivalent compound. Nevertheless, the calculated chemical shift follows chemical intuition. We provide an explanation using a counting argument. The $2p$ core level line shapes show large variations between the different compounds. For CuO, the leading peak has a large broadening and the largest satellite has a large weight, while for Cu_2O the main peak is narrow and there is only little weight in the satellite region. $NaCuO_2$ is in these respects intermediate between CuO and Cu_2O. These properties are explained using an Anderson impurity model, with parameters obtained from the *ab initio* calculations. Most of the properties can be related to the formal valence of Cu, by using the same counting argument as above.

INTRODUCTION

Cu forms compounds where Cu is formally monovalent, divalent or trivalent. Examples of such compounds are Cu_2O (monovalent), CuO (divalent) and $NaCuO_2$ (trivalent). The question of the valence of Cu has recently attracted much interest in the context of the High T_c superconductors. In some of the undoped parent compounds, Cu is divalent. In the hole doped compounds the formal valence of Cu is increased, while in the electron doped compounds it is reduced. It is then interesting to know how the formal valence of Cu shows up in various experiments, and to what extent it makes sense to think in terms of an increased or reduced Cu valence for the doped High T_c superconductors. Here we study Cu_2O, CuO and $NaCuO_2$ as model compounds and try to relate the chemical shift and spectral properties to the formal valence (or oxidation state). We also study the superconductor $YBa_2Cu_3O_{6.5}$, where one may argue that the three inequivalent Cu atoms have the approximate formal valences one, two and three, respectively.

As the valence increases, one imagines that the net charge on the Cu atom increases. The potential on the Cu site should then become more attractive

*Present address: Physics Department, Chalmers University of Technology, S-412 96 Göteborg, Sweden.

and the binding energy of the core electrons increases. Such a chemical shift is observed for the model compounds mentioned above.[1] We have performed band structure calculations using the density functional formalism in the local density approximation.[2] We find that the net charge on Cu does not follow the expected trend. Thus, we obtain the net positive charges 0.76, 0.87 and 0.56 for the monovalent, divalent and trivalent model compounds, respectively, i.e., Cu has the smallest positive charge in the formally 3+ compound. Our calculated chemical shifts, nevertheless, follow the expected chemical trend. We use a counting argument, where we consider the formal number of holes in the Cu $3d$ and O $2p$ levels, to explain why the chemical shift has a tendency to follow the formal valence. In X-ray photoemission spectroscopy (XPS), the Cu $2p$ spectra show pronounced variations between the model compounds. For instance, the main satellite is very strong for CuO but very weak for Cu_2O, and the main line is broad for CuO but rather narrow for Cu_2O. $NaCuO_2$ is intermediate to Cu_2O and CuO. By using the same counting argument as above we show that most of these features can be related to the formal valence. More extensive accounts of this work have been published elsewhere.[3,4,5]

CHEMICAL SHIFTS AND CHARGE DENSITIES

We have performed linear-muffin-tin[6] band structure calculations for Cu_2O, CuO, $NaCuO_2$ and $YBa_2Cu_3O_{6.5}$ and obtained charge densities for these systems.[3,5] We are in particular interested in the net charge per atom and the distribution of this charge between *s*- *p*- and *d*-electrons. For this purpose, we integrate the charge density inside spheres surrounding the different atoms. This leads to uncertainties, since the results depend on the size of the spheres. To make the results for the different compounds comparable, we therefore use the same radius for a given atom in the different compounds. Even though the absolute values of the charges have a substantial arbitrariness, trends between the different compounds should be relevant. This is particularly true for the Cu $3d$ charges, since more than 96 % of the atomic $3d$ charge falls inside the Cu spheres used here. Therefore small changes in the Cu sphere radius should lead to modest changes in the absolute values of the $3d$ charges, and to even smaller changes in the differences between different systems. We use the radii $R_{Cu} = 2.5\ a_0$ and $R_O = 2.0\ a_0$ for the Cu and O spheres, respectively.

We define the binding energy of the $2p$ core electron relative to the top of the valence band ε_v, and write it as

$$E_{2p} = E_{2p}^{At} + (\varepsilon_v - \Delta V) - E_{rel}, \tag{1}$$

where E_{2p}^{At} is the (positive) atomic binding energy relative to the vacuum level, ΔV is the (ground-state) shift of the $2p$ eigenvalue between the atom and the solid, and E_{rel} is the relaxation energy. We have here separated the changes in the binding energy into an initial state shift $\varepsilon_v - \Delta V$ and a final state relaxation energy E_{rel}. We find that the differences between the three model compounds are mainly due to the initial state shift,[3] and we therefore focus on this shift

Table I. Comparison of theoretical (LDA) and experimental (Expt.) Cu $2p$ core level binding energies. "Shift" shows the change relative to Cu_2O. All energies are in eV and measured relative to the highest occupied level.

	LDA E_{2p}	Shift	Expt.[a] E_{2p}	Shift	Expt.[b] E_{2p}
Cu_2O	930.8	0.0	932.2	0.0	
CuO	931.7	0.9	933.0	0.8	
$NaCuO_2$	932.3	1.5	933.2	1.0	932.7

[a] Ref. 1,7 [b] Ref.8

here. The calculated binding energies are compared with experiment in Table I. The absolute value is correct to within 2 eV, i.e., an error of about 0.2 %. This small error is probably due to the local density approximation. In the following, we focus on the trends between the different compounds, which contain the interesting information. Between the Cu_2O and CuO, the experimental binding energy increases by about 0.8 eV, as found in many independent experiments. Our theoretical increase of 0.9 eV, agrees well with experiment. For the differences between CuO and $NaCuO_2$, Steiner et al[1] find an increase of about 0.2 eV, while Mizokawa et al[8] find a decrease. Our calculation shows an increase of about 0.6 eV, in reasonable agreement with Steiner et al[1] and following chemical intuition. An increase in the binding energy in going from divalent to trivalent Cu has also been found by Allan et al,[9] who studied La_2CuO_4 and $LaCuO_3$.

The results for the calculated charges are shown in Table II. We first consider the $3d$ charges for Cu_2O, CuO and $NaCuO_2$, which one might naively have expected to be 10, 9 and 8, respectively. The calculated variations are much smaller, since such large changes in the $3d$ charge would lead to a large variation in the Cu potential due to the strong Coulomb interaction between the $3d$ electrons, and would be far from a self-consistent solution. This relatively small variation in the number of $3d$ electrons is not surprising and it is consistent with the relatively small measured chemical shifts. It is, however, surprising that the number of $3d$ electrons is essentially the same for the divalent and trivalent model compounds. Even more surprisingly, the net positive charge is smallest for the trivalent model compound. This results from the large $4p$ charge for $NaCuO_2$, which can be understood in terms of the strong O $2p$ to Cu $4p$ interaction, expected for a trivalent system.[3]

The initial state shift $\Delta V - \varepsilon_v$ is shown in Table II. It follows chemical intuition, i.e., the binding energy is smallest for the 1+ compound and largest for the 3+ compound. Table II shows the shift ΔV_{Cu} of the $2p$ level due to the Cu on-site charge. As expected, this shift follows the net Cu charge, and it would tend to make the binding energy smallest for the $NaCuO_2$. However, the

Table II. The 4*s* (n_{4s}), 4*p* (n_{4p}), 3*d* (n_{3d}) and total (n_{Cu}) Cu charges in the ground-state. ΔV is the shift of the core level relative to a free Cu atom, V_{Mad} is the Madelung potential on the Cu site, and $\Delta V_{Cu} \equiv \Delta V - V_{Mad}$ is the shift of the potential on the Cu site due to the on-site Cu charge. The initial state contribution in Eq. (1) to the binding energy is $\varepsilon_v - \Delta V$, where ε_v is the highest occupied state. All energies are in Ryd.

Syst.	n_{4s}	n_{4p}	n_{3d}	n_{Cu}	ΔV	V_{Mad}	ΔV_{Cu}	$\varepsilon_v - \Delta V$
Cu_2O	0.515	0.424	9.296	10.235	0.065	0.702	-0.637	-0.191
CuO	0.454	0.551	9.131	10.136	-0.041	0.795	-0.836	-0.091
$NaCuO_2$	0.578	0.742	9.121	10.441	-0.315	0.230	-0.545	-0.061
Cu"1+"	0.589	0.465	9.281	10.335	0.040	0.543	-0.583	-0.237
Cu"2+"	0.532	0.642	9.296	10.470	-0.128	0.325	-0.453	-0.149
Cu"3+"	0.574	0.721	9.143	10.438	-0.191	0.429	-0.620	-0.086

Madelung potential V_{Mad}, from the other atoms in the solid, reverses this trend. Thus the Madelung potential is much lower for $NaCuO_2$ than for the other two compounds, leading to a larger binding energy for the trivalent compound. Since the Cu Madelung potential is determined by the surroundings of Cu and not by Cu itself, it may seem accidental that the binding energy follows the formal valence. Here, we want to show that this is not the case.

For this purpose, we use a counting argument, and relate the number of electrons to the number of Cu 3*d* and O 2*p* states, which is also the way the formal valence (oxidation state) is determined. For the compounds considered, these states form a complex of bands, which are separated from all other bands by a band gap. For Cu_2O, CuO and $NaCuO_2$ the number of states in these band complexes are 26, 16 and 22, respectively. When the available valence electrons are put into these states, we are left with 0, 1 and 2 holes, respectively, which define the formal valences. We first consider Cu_2O, where the $3d - 2p$ band complex is filled. Due to the hybridization of the Cu 3*d* states with unoccupied states, the 3*d* charge (9.3) is nevertheless smaller than 10. Since Cu is less electronegative than O, the Cu 3*d* level is well (0.14 Ry) above the O 2*p* level. We next consider CuO, which has one hole in the $3d - 2p$ band complex. We first perform a calculation where we use the potentials obtained from the Cu_2O calculation, i.e., we perform one iteration, and fill up the bands completely with 16 electrons (see Table III). The 3*d* charge (9.4) is close to the charge for Cu_2O. We now take into account that CuO has only 15 valence electrons, and remove an electron from the top of the band. Since the Cu 3*d* level is higher than the O 2*p* level in Cu_2O, the top of the band has primarily Cu 3*d* character when the Cu_2O potential is used to calculate the bands. Thus, the removed electron has about 60 % 3*d* character. Because of the large Coulomb interaction between two 3*d* electrons, the removal of 0.6 3*d* electrons leads to a large lowering of

Table III. Charge densities on the Cu atom for a self-consistent Cu_2O calculation, a CuO calculation using the Cu_2O potential and assuming N =16 or 15 electrons per CuO unit and for a self-consistent CuO calculation.

Syst.	Pot.	N	n_{4s}	n_{4p}	n_{3d}	n_{Cu}
Cu_2O	Cu_2O	26	0.515	0.424	9.296	10.235
CuO	Cu_2O	16	0.434	0.579	9.431	10.444
CuO	Cu_2O	15	0.425	0.560	8.855	9.840
CuO	CuO	15	0.454	0.551	9.131	10.136

the $3d$ level, and the charge density obtained is not self-consistent, since such a lowering of the $3d$ level would lead to an almost complete filling of the $3d$ level. To obtain a self-consistent solution, we must have a smaller $3d$ character at the top of the band, i.e., relative to the O $2p$ level, the $3d$ level must be lower in CuO than in Cu_2O. This is achieved by reducing the $3d$ charge (0.17 electrons) in CuO compared with Cu_2O. We can see that this lowering of the $3d$ level is a result of both he counting argument, telling us that there is a hole at the top of the $3d - 2p$ band complex, and a self-consistency argument, telling us that there cannot be very large changes in the $3d$ occupancy, because of the strong $3d - 3d$ Coulomb interaction.

In $NaCuO_2$, there are two holes at the top of the $3d-2p$ band complex. The same argument as above shows that the Cu $3d$ character of the top of the band must then be reduced further compared with CuO, to obtain a self-consistent charge, i.e., the $3d$ level must be lowered further relative to the O $2p$ level. In $NaCuO_2$, this is achieved by the low Cu Madelung potential, and there is no need to reduce the Cu $3d$ charge compared with CuO.

One may ask what would have happened if the Cu Madelung had not been lower in $NaCuO_2$ than in CuO. We have tested this by introducing an artificial external potential which compensates for the lower Madelung potential in $NaCuO_2$. We then find that the system reduces the Cu $3d$ charge by 0.12 in $NaCuO_2$ compared with CuO, so that the Cu $3d$ level is, nevertheless, moved down relative to the O $2p$ level.

Since the net positive Cu charge is smaller than the formal valence for CuO and in particular for $NaCuO_2$, it has been suggested that O becomes increasingly less negative in CuO and $NaCuO_2$ and that the O $1s$ binding energy would then increase in going from Cu_2O to CuO and $NaCuO_2$. Actually the opposite is found experimentally.[10] Within our scheme, this can be easily understood, since the O $2p$ level moves up relative to the Cu $3d$, level as the formal valence of Cu is increased. We find that this relative shift is about 1.9 eV between Cu_2O and CuO, and the same between CuO and $NaCuO_2$. We note that these shifts differ somewhat from the initial state shifts for the O $1s$ level and the Cu $2p$ level, respectively, although they are closely related, since they measure the potentials on the same atoms. To obtain the experimental binding energies, one should, furthermore, add the relaxation energies, which are appreciable, but are found not to change the trends for the Cu $2p$ levels. In spite of these reservations, we find it encouraging that the shifts (1.9 eV) calculated above agree exactly with the experimental results.[10] We also observe that if the Cu $3d$ level had been below the O $2p$ level already in Cu_2O, the arguments above would not have been valid, and it is not clear from these arguments if the Cu potential would then be lowered as the formal Cu valence is increased. It would be interesting to repeat the analysis for Cu_2S and CuS, which apparently show a small chemical shift, and to see if this can be related to the p orbital lying higher for S than for O.

It is interesting to apply this analysis to $YBa_2Cu_3O_{6.5}$, which has 3 inequivalent Cu atoms. In $YBa_2Cu_3O_{7-x}$, there are two CuO_2 plains, where each Cu atom is surrounded by 4 O atoms at distances of $\sim$ 1.94 Å and $\sim$ 1.95 Å,[11] where we have neglected the more distant apex oxygens. This is the same coordination and a similar distance as for CuO, where the Cu-O separations are 1.95 and 1.96 Å. It is therefore tempting to assign the formal valence +2 or somewhat more to these Cu atoms, and we refer to them as Cu$^{"2+"}$ in the following. $YBa_2Cu_3O_{7-x}$ also contains CuO chains. For $x = 0$, all chains are intact, while for $x > 0$ O atoms are missing in some chains. In $YBa_2Cu_3O_{6.5}$, it is believed that every second chain is intact and that in every second chain all the O atoms are missing ("empty chains"). In the intact chains, Cu is surrounded by 4 O atoms at distances of $\sim$ 1.8 Å and $\sim$ 1.94 Å. This is intermediate between the separation for $NaCuO_2$(1.85 Å) and for CuO (1.95 Å and 1.96 Å) and we refer to this as Cu$^{"3+"}$, although one might assign it a somewhat smaller valence. In the empty chains, Cu has two O neighbors at distances of $\sim$ 1.8 Å. Since the two O neighbors in Cu_2O are at about the same distance (1.84 Å), we refer to this as Cu$^{"1+"}$. In Table II we show results for $YBa_2Cu_3O_{6.5}$. First we notice that the *changes* in the initial state shift $\varepsilon_v - \Delta V$ between Cu$^{"1+"}$, Cu$^{"2+"}$ and Cu$^{"3+"}$ are similar to those of the model compounds. This provides a certain justification for our notations "1 + ", "2 + ", and "3 + ". The trends for the $4s$ and $4p$ charges are also similar. The $3d$ charges, however, behave differently for $YBa_2Cu_3O_{6.5}$ and the model compounds. For the "1 + " compounds the $3d$ charges are similar; 9.30 and 9.28. For $YBa_2Cu_3O_{6.5}$ there is, however, a small change in the $3d$ charge between Cu$^{"1+"}$ and Cu$^{"2+"}$ and a rather large reduction between Cu$^{"2+"}$ and Cu$^{"3+"}$, while the model compounds show the opposite trend. The reason for this is that the Cu Madelung potential is substantially lower for Cu$^{"2+"}$ than for Cu$^{"1+"}$ in $YBa_2Cu_3O_{6.5}$. There is therefore no need to reduce the $3d$ charge to obtain the required lowering of the $3d$ level for Cu$^{"2+"}$. Between Cu$^{"2+"}$ and Cu$^{"3+"}$ there is an increase in the Madelung potential. In this case, the $3d$ charge must be reduced to obtain the needed additional lowering of the $3d$ level. We can see that the Madelung potential influences the $3d$ level positions for the different inequivalent Cu atoms in $YBa_2Cu_3O_{6.5}$ in a rather different way than in the model compounds. Nevertheless, the trend of lowering the $3d$ level relative to the O $2p$ level remains, supporting our analysis and our arguments that the systematic downward shift of the $3d$ level is not accidental. We have not tried to determine deviations from the valences "1+", "2+" and "3+" here. This would be particularly interesting in the context of the transport and superconducting properties, and it would then also be interesting to discuss the detailed distribution of the holes between planes and chains.[12]

SHAPE OF THE $2p$ SPECTRUM

In this section we discuss the shape of the $2p$ XPS core level spectra, which have a pronounced leading peak, and for some systems a substantial satellite at about 10 eV higher binding energy. Here, we focus on the width of the main peak and the integrated weight of the satellite. Both of these quantities show substantial variations with the formal valence of Cu. For instance, the weight of the satellite is about half the weight of the main peak for CuO, while for Cu_2O the satellite has little or no weight. Similarly, the broadening of the main peak for CuO is about 3.2 eV, of which about 1.4 eV can be explained as instrumental resolution and life-time broadening, while for Cu_2O there is very little intrinsic broadening apart from the life-time broadening.

The mechanism for the satellite in Cu^{2+} is well understood.[13] The ground-state is assumed to be a mixture of a d^0 and a $d^{10}L^{-1}$ configuration, where L^{-1} is a hole in the valence (ligand) band. When a core hole is created, the strong core hole attraction leads to a large change of the energies of the configurations. Thus the $d^{10}L^{-1}$ is pulled down well below the d^9 configuration. The main peak therefore corresponds to final states which mainly consist of the low-lying $d^{10}L^{-1}$ configuration, while the satellite corresponds to final states with mainly d^9 character. Schematically, we can describe this by introducing the ground-state

$$|E_0(N)>= a^0|d^9> +b^0|d^{10}L^{-1}>, \tag{2}$$

where $|a^0| > |b^0|$ for Cu^{2+} systems and a^0 and b^0 are chosen to have the same sign. The final state corresponding to the main peak is written as

$$|+>= a^f|d^9> +b^f|d^{10}L^{-1}>, \tag{3}$$

where $|b^f| > |a^f|$ and a^f and b^f have the same sign. The state corresponding to the satellite is

$$|->= -b^f|d^9> +a^f|d^{10}L^{-1}>. \tag{4}$$

The weights of the peaks are given by the overlap between the ground-state and the final states,

$$w_+ = |a^0a^f + b^0b^f|^2 \tag{5}$$

for the main peak, and

$$w_- = |-a^0b^f + b^0a^f|^2 \tag{6}$$

for the satellite. If there were no coupling between the configurations in the final state ($a^f = 0$ and $b^f = 1$), the main peak would just reflect the amount ($|b^0|^2$) of d^{10} character of the ground-state, and the satellite the amount ($|a^0|^2$) of d^9 character. If there is a mixing of the two final state configurations ($a^f > 0$), the leading peak grows and the satellite is correspondingly reduced, since the two terms in the expression for $w_\pm$ have the same sign for w_+ but opposite signs for w_-. Thus an increased coupling in the final state leads to a transfer of weight to the leading peak. As a result, the leading peak typically has a larger weight than the satellite, although the d^9 configuration normally has a larger weight than the d^{10} configuration in the ground-state.

For Cu_2O, it is traditionally argued that since this is a d^{10} system, the weight of the satellite should be very small. However, our local density calculations show that there is a relatively small difference between the $3d$ occupancy in Cu_2O ($n_d = 9.30$) and CuO ($n_d = 9.13$). This difference alone cannot explain the large difference in the weights of the satellites. We also notice that the small difference in $3d$ occupancy should not be a defect of our calculation, since a large difference on the order of 1 electron, would have shown up experimentally as a very large chemical shift (on the order of 10 eV). It is therefore necessary to study the difference between these two systems further.

We use the Anderson impurity model, where we explicitly include the $3d$ level on the Cu atom where the core hole is created. This is expected to be a reasonable approximation for core spectra, which sample the local properties of the system. The Anderson model is given by

$$H = \sum_\nu [(\varepsilon_d - U_c(1-n_c))n_\nu + \int \varepsilon n_{\varepsilon\nu} d\varepsilon + \int d\varepsilon (V_\nu(\varepsilon)\psi^\dagger_{\varepsilon\nu}\psi_\nu + \text{h.c.})]$$
$$+U\sum_{\nu<\mu} n_\nu n_\mu + \varepsilon_c n_c, \tag{7}$$

where ε_d is the energy of the $3d$-level on the impurity site. The $3d$ states are labeled by an orbital index m and a spin index σ, and $\nu = (m, \sigma)$ is a combined index. The index ν runs between 1 and $N_d = 10$. If a core hole is created ($n_c = 0$) the $3d$ level is lowered by U_c. The delocalized electrons are described by the second term. The hopping between the conduction states and the $3d$ state is described by the third term, where $V_\nu(\varepsilon)$ is a hopping matrix element. U describes the Coulomb interaction between the $3d$ electrons. All multiplet effects have been neglected. While the multiplets are important for the $3d$ compounds in general, they are not essential for the properties discussed here, namely the width of the main peak and the weight of the dominating satellite. Finally, ε_c gives the energy of the core level, which is assumed to be nondegenerate to simplify the notations.

First we calculate the ground-state wave function. We use a variational approach[14] and write the wave function as

$$|E_0(N)> = \sum_n c_n |n>, \tag{8}$$

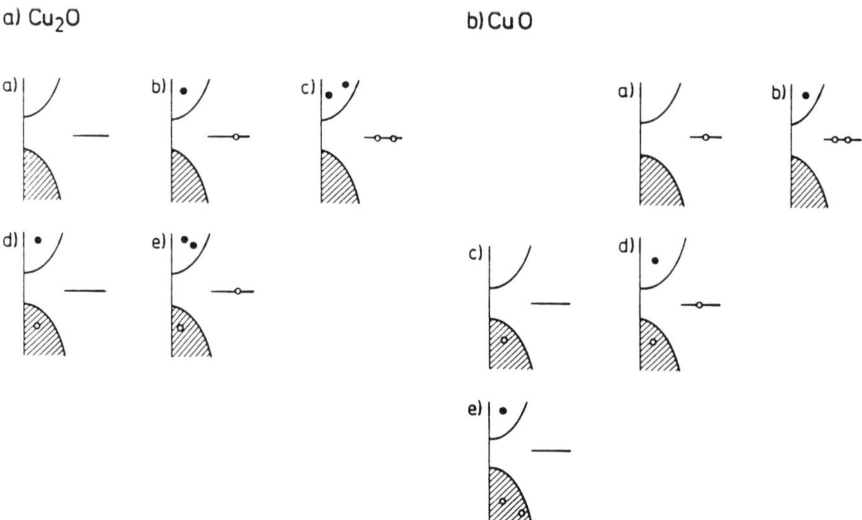

Fig. 1. The many-electron basis states used for Cu_2O (a) and CuO (b). A filled circle represents an electron and an empty circle a hole. The left column shows the d^{10} states, the middle the d^9 states and the right the d^8 states. The states in the upper row are the leading states in a $1/N_d$ expansion.

where $\{|n>\}$ is a many-electron basis set. When choosing the basis set for the three model compounds, we use the same counting argument as above. For Cu_2O there are sufficiently many electrons to fill both the Cu $3d$ level and the O $2p$ level. We therefore introduce a state (a in Fig. 1a) where the valence band and the $3d$ level on the impurity are filled. This state couples to other states, with a hole in the $3d$ level and a conduction electron (b in Fig. 1a). These states couple to additional states, e.g., to states with an electron in the conduction band and a hole in the valence band (d in Fig. 1a). The state c in Fig. 1a is less important, since the d^8 configuration is rather high in energy. The states in the second row are also less important, since the coupling to these states is reduced by $1/N_d$, where N_d is the degeneracy of the $3d$ level.[4,14] Thus the states a and b are the most important ones. For CuO we take into account that there is one electron less than needed to fill up all the O $2p$ and Cu $3d$ levels. Thus we introduce many-electron states with either one hole on the $3d$ level (a in Fig. 1b) or a hole in the valence band (c in Fig. 1b). As for Cu_2O, these states couple to additional states, of which some are shown in Fig. 1b. Here, the most important states are state a and state c. In a similar way, we have to take into account that for $NaCuO_2$ there are two holes in the O $2p$ and Cu $3d$ levels, and we introduce states with two holes on the $3d$ level, states with one hole on the $3d$ level and one hole in the valence band, and states with two holes in the valence band.

In the sudden approximation, the core spectrum is

$$\rho_c(\varepsilon) = \sum_n | < E_n(N-1)|\psi_c|E_0 > |^2 \delta(\varepsilon - E_0 + E_n(N-1)), \quad (9)$$

where $|E_n(N-1)>$ are the final states and ψ_c annihilates a core electron. This result can be rewritten as

$$\rho_c(\varepsilon) = \frac{1}{\pi}\mathrm{Im}g_c(\varepsilon - i0^+), \quad (10)$$

where

$$g_c(z) =< E_0|\psi_c^\dagger \frac{1}{z + H - E_0}\psi_c|E_0 > . \quad (11)$$

The Green's function $g_c(z)$ is calculated by formally assuming that the basis set used for the ground-state calculation is complete. We can then write

$$g_c(z) = \sum_{i,j} < E_0|\psi_c^\dagger|i >< i|\frac{1}{z + H - E_0}|j >< j|\psi_c|E_0 >, \quad (12)$$

where $|i>$ and $|j>$ are the basis states used in the ground-state calculation but with the core level empty. Eq. (12) can be evaluated by using the Lanczos method.[14]

The parameters of the Hamiltonian have been determined from *ab initio* calculations, as discussed elsewhere.[4] Since the $3d$ occupancy plays a particular role in the discussion of the weights of the satellites above, we have, however, adjusted the calculated $3d$ level positions so that the differences between the calculated $3d$ charges in the Anderson model agree with the calculated differences in the local density calculation.[4] Since the theory gives no intrinsic broadening of the main peak in Cu_2O, apart from life-time broadening, as discussed below, we have adjusted the life-time broadening so that the experimental width (1.4 eV) is reproduced by the life-time broadening and experimental resolution. The same broadening is used for CuO and $NaCuO_2$. The results are shown in Fig. 2. First we observe that since we have neglected multiplet effects, the satellites are much narrower than the experimental widths. Otherwise, there is good agreement with experiment. The widths of the main peaks are 3.0 and 1.6 eV for CuO and $NaCuO_2$, respectively, in good agreement with the observed widths of 3.2 and 1.6 eV, respectively. The theory gives a very small satellite for Cu_2O, in agreement with experiment. For CuO, the calculated weight of the satellite is 0.52 of the main peak weight compared with an experimental result of about 0.55. Finally, for $NaCuO_2$ the satellite weight is intermediate between these two extremes, both according to theory and experiment. Because of the problems of subtracting the background, it is hard to give a quantitative weight for the experimental satellites in Cu_2O and $NaCuO_2$.

We now discuss the origin of the large variations in the line shape with variations in the formal valence, and start with the width of the leading peak. First we note that the leading peak corresponds to final states with d^{10} configurations. For CuO the main contribution comes from state c in Fig. 1b, which has a d^{10}

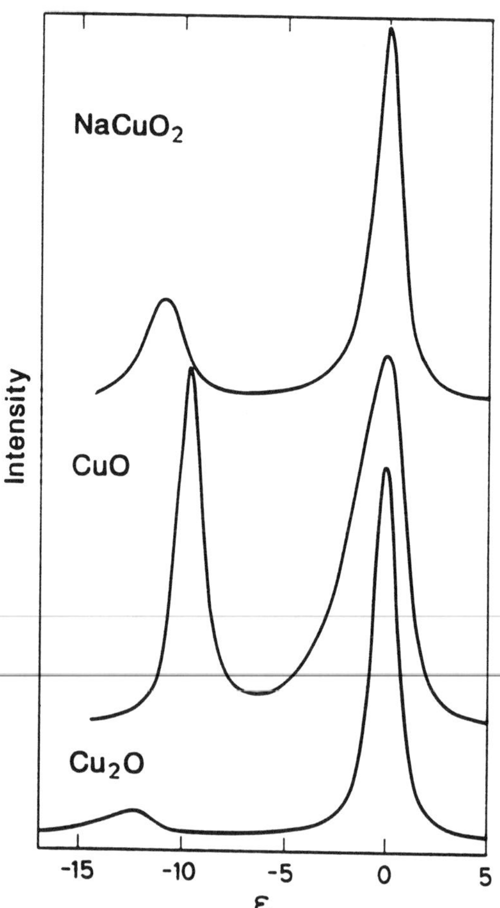

Fig. 2. The calculated $2p$ spectra for Cu_2O, CuO and $NaCuO_2$. The absolute value of the energy is arbitrary and the main peaks have been lined up at zero energy. We have used the parameters $U = 8$ eV and $U_c = 10$ eV. The energy separations between the d^9 and d^{10} configurations are 2.3 eV (Cu_2O), 1.1 eV (CuO) or 2.4 eV ($NaCuO_2$). We have introduced a Gaussian broadening of 0.7 eV full width at half maximum (FWHM) to simulate the instrumental resolution and a Lorentzian broadening of 1.1 eV FWHM to simulate the lifetime broadening.

configuration and a hole in the valence band. The energy of this state depends on the energy of the hole; the further down in energy the hole is, the higher the energy of the state is. For Cu_2O, on the other hand, the main contribution comes from state a in Fig. 1a, which has a fixed energy, since there is no valence hole or conduction electron involved in this state. This difference between CuO and Cu_2O is a direct consequence of the difference in formal valence. The theory therefore predicts that for Cu_2O, the main peak has no broadening, apart from life-time broadening and instrumental resolution. For CuO, on the other hand, the main peak has a width, since the system can end up in many different final d^{10} states with different energies, depending on the energy of the valence hole. Since the weights of these states depend on their coupling to the ground-state, we first write down the ground-state wave function (compare Eq. (2) above)

$$|E_0(N)>= a^0|d^9> + \int d\varepsilon b^0(\varepsilon)|d^{10}\underline{\varepsilon}>, \tag{13}$$

where ε is the energy of the hole, and $b^0(\varepsilon)$ describes the amplitude of states with a d^{10} configuration and a valence hole at ε. For the arguments here, we neglect the coupling between the d^9 and d^{10} configurations in the final state, which is partly justified by the relatively large separation of the d^9 and $d^{10}\underline{\varepsilon}$ configurations in the final state. The state $|d^{10}\underline{\varepsilon}>$ (with a core hole) is then an eigenstate. The spectrum would then have the form

$$\rho_c(\varepsilon + E_0^0) = |b^0(\varepsilon)|^2\Theta(-\varepsilon), \tag{14}$$

where E_0^0 is a constant which is unimportant for the discussion, and $\Theta(-\varepsilon)$ is 0 if $\varepsilon > 0$, and otherwise 1. Here we have put the energy zero at the top of the valence band, which leads to the cut off of the spectrum at $\varepsilon = 0$ by the Θ function. In the ground-state, the states with the hole close to the top of the band tend to have a larger weight, since these states have a low energy. The coefficient $|b^0(\varepsilon)|$ therefore decreases as ε becomes more negative. The unbroadened spectrum predicted by this theory is therefore nonsymmetric, with a tail towards higher binding energy. Such an asymmetry appears to be seen for most Cu^{2+} compounds.[15] We note that the asymmetry discussed here is different from the infra-red singularity,[16] since the present mechanism also works for insulators. We furthermore observe that in Cu_2O, the state b with a d^9 configuration also has a conduction electron. This leads to a broadening of the satellite, while for CuO there is no broadening of the satellite of this type. As a consequence of the explanation above, one may expect the main peak of $NaCuO_2$ to be particularly broad, since the d^{10} configuration in this case has 2 valence holes. For $NaCuO_2$, however, the hopping matrix elements to the top of the valence band are particularly strong, so that the holes are energetically very close to the top of the valence band and therefore cause a small broadening of the peak.

We next discuss the weight of the satellite. We first notice that the satellite in Cu_2O is not quite as weak as it appears. The reason is that for Cu_2O there is a large broadening of the satellite, making it appear smaller than it is, while for CuO there is a broadening of the main peak, making the satellite appear larger than it is. Thus the calculated ratio of the satellite weight to the main peak weight is 0.19, while the corresponding ratio for CuO is 0.52. We next observe that an increase in the weight of the d^{10} configuration in the ground-state leads to an increase in the weight of the leading peak. Since Cu_2O has a larger $3d$ occupancy than CuO, we first have to ask if this can explain the difference in the weights of the two satellites. We have therefore performed a calculation for Cu_2O, where the $3d$ level was raised until the $3d$ occupation became the same as in CuO. This increased the weight of the satellite (from 0.19 to 0.38 of the main peak weight) and the difference between Cu_2O and CuO is therefore partly due to this difference in the $3d$ occupation in the initial state. However, even in this calculation, the weight of the satellite remains smaller for Cu_2O (0.38) than for CuO (0.52). From the discussion below Eq. (6), it follows that weight is transferred from the satellite to the main peak due to the coupling between the final state configurations. This coupling must then be stronger for Cu_2O than for CuO. This may suggest that the one-particle hopping matrix elements $V_\nu(\varepsilon)$ are larger for Cu_2O than for CuO. However, our *ab initio* calculation of $V_\nu(\varepsilon)$ shows that the opposite is true, which is also expected from the formal valence. In CuO there is a hole in an antibonding state at the top of the $3d-2p$ band complex and therefore a contribution to the bonding from the $3d-2p$ coupling. One would therefore expect the geometry of CuO to adjust so that $V_\nu(\varepsilon)$ is larger in CuO than in Cu_2O, which does not have a hole in the $3d-2p$ band complex. In spite of this, the final state coupling is larger in Cu_2O than in CuO. This follows if degeneracy effects in the important basis states in Fig. 1 are considered. The dominating states for Cu_2O are the states a and b in Fig. 1a, and for CuO the states a and c in Fig. 1b. Thus, for Cu_2O, any of the $3d$ electrons can hop into the conduction band, providing 10 different hopping channels, due to the degeneracy of the $3d$ level. For CuO, on the other hand, hopping is only possible if the valence electron, hopping to the $3d$ orbital, has the same quantum number as the hole in the $3d$ level. In CuO there is therefore only one hopping channel. This effect is more than enough to compensate for the smaller one-particle hopping matrix elements $V_\nu(\varepsilon)$ in Cu_2O, and it leads to a substantially stronger final-state coupling in Cu_2O. This results in a larger transfer of weight from the satellite to the main peak in Cu_2O, and it is an important part of the explanation for the weak satellite in Cu_2O. We notice that this degeneracy effect is a direct consequence of the basis states used for Cu_2O and CuO, which are determined by the formal valences.

For $NaCuO_2$ the weight of the satellite is smaller than for CuO. For $NaCuO_2$ the degeneracy effect discussed above leads to the same result as for CuO, i.e., there is hopping in only one channel. This effect therefore cannot explain the difference between $NaCuO_2$ and CuO. The $3d$ occupations in the ground-states are also similar, and cannot explain the differences in satellite weights. However, the one particle hopping matrix elements are larger for $NaCuO_2$ than for CuO. In both compounds Cu is surrounded by 4 O atoms, but in $NaCuO_2$ the separation is smaller, leading to a stronger coupling. This is consistent with the fact that since $NaCuO_2$ is trivalent, there are two holes at the top of $3d-2p$ band complex instead of one hole in CuO, which should make a shorter separation for $NaCuO_2$ favorable. Because of the larger $V_\nu(\varepsilon)$ in $NaCuO_2$, there is a stronger final state coupling, and a larger transfer of weight from the satellite to the main peak.

CONCLUSIONS

We have studied the chemical shift and the XPS line shape of the $2p$ core level in Cu_2O, CuO and $NaCuO_2$, where Cu has the formal valences 1, 2 and 3, respectively. We have also studied $YBa_2Cu_3O_{6.5}$, where the inequivalent Cu atoms can be argued to have the approximate formal valence 1, 2 and 3, respectively. We find that the variations in the net Cu charge with the formal valence are small (a few tenths of an electron) and, surprisingly, that the net charge is less positive for the 3+ compound $NaCuO_2$ than for CuO (2+) and Cu_2O (1+). Nevertheless, our calculations give a chemical shift which follows the formal valence. This happens because the Madelung potential gives an important contribution to the chemical shift, suggesting that the agreement with the formal valence is accidental. We show, however, by using a counting argument, that there is a strong tendency for the Cu $3d$ level to move down relative to the O $2p$ level as electrons are removed from the $3d - 2p$ band complex, i.e., as the Cu valence is increased. This picture is consistent with the measured binding energies for the Cu $2p$ and O $1s$ levels. The $2p$ XPS line shape shows strong variations of main peak width and the weight of the satellite with variations in the valence. By using the same counting argument as for the chemical shift, we can also understand most aspects of the line shape. The large width of the main peak in CuO is found to be due to the presence of a hole in the valence band. The distribution of this hole over a finite energy range gives the broadening. The small weight of the satellite in Cu_2O is partly due to the larger $3d$ occupancy in Cu_2O and partly due to the stronger final state coupling, caused by degeneracy effects, which transfers weight from the satellite to the main peak. All these effects are related to the Cu valence. Finally, we note that the relation between the formal valence and a given physical property has to be studied specifically for that property. For instance, the effective charge deduced from the phonon dispersion in CuO is apparently of the order 1,[17] i.e., much smaller than the formal valence 2+. Thus it should be important to reconsider the usefulness of the formal chemical valence when new properties are discussed.

We want to thank S. Hüfner, J.C. Fuggle, P. Adler, A. Simon and G.A. Sawatzky for useful discussions.

REFERENCES

1. P. Steiner, V. Kinsinger, I. Sander, B. Siegwart, S. Hüfner, C. Politis, R. Hoppe, and H.P. Müller, Z. Phys. B **67**, 497 (1987).
2. P. Hohenberg and W. Kohn, Phys. Rev. **136**, B864 (1964); W. Kohn and L.J. Sham, Phys. Rev. **140**, A1133 (1965); For a recent review see, e.g., R.O. Jones and O. Gunnarsson, Rev. Mod. Phys. **61**, 689 (1989).

3. K. Karlsson, O. Gunnarsson, and O. Jepsen, "Chemical shifts for monovalent, divalent and trivalent Cu compounds", submitted to Phys. Rev. B

4. K. Karlsson, O. Gunnarsson, and O. Jepsen, "Shape of the Cu 2*p* core level photoemission spectrum for monovalent, divalent and trivalent Cu compounds", submitted to Phys. Rev. B

5. K. Karlsson, O. Gunnarsson, and O. Jepsen, "Cu 2*p* chemical shifts for $YBa_2Cu_3O_{6.5}$: What is the valence of the Cu atoms?", to be published.

6. O.K. Andersen, Phys. Rev. B **12**, 3060 (1975); O.K. Andersen and O. Jepsen, Phys. Rev. Lett. **53**, 2571 (1984); O.K. Andersen, Z. Pawlowska, and O. Jepsen, Phys. Rev. B **34**, 5253 (1986); O.K. Andersen, O. Jepsen, and D. Glötzel, in *Highlights of Condensed-Matter Theory*, edited by F. Bassani, F. Fumi, and M.P. Tosi (North-Holland, New York, 1985).

7. Ref. 1 gives the binding energies relative to the Fermi energy, and the results were therefore corrected by the estimated separations between the Fermi energy and the top of the valence band obtained from S. Hüfner (priv. commun.).

8. T. Mizokawa, H. Namatame, A. Fujimori, K. Akeyama, H. Kondoh, H. Kuroda, and N. Kosugi, (to be publ.)

9. K. Allan, A. Champion, J. Zhou, and J.B. Goodenough, Phys. Rev. B **41**, 11572 (1990).

10. I. Sander, PhD thesis, Saarbrücken, Germany, 1990 (unpublished).

11. J.D. Jorgensen, B.W. Veal, A.P. Paulikas, L.J. Nowicki, G.W. Crabtree, H. Claus, and W.K. Kwok, Phys. Rev. B **41**, 1863, (1990).

12. J. Zaanen, A.T. Paxton, O. Jepsen, and O.K. Andersen, Phys. Rev. Lett. **60**, 2685 (1988).

13. A. Kotani and Y. Toyozawa, J. Phys. Soc. Jpn. **35**, 1073; **37**, 912 (1974).

14. O. Gunnarsson and K. Schönhammer, Phys. Rev. B **28**, 4315 (1983); in *Handbook on the Physics and Chemistry of the Rare Earths*, edited by K.A. Gschneider, Jr., L. Eyring, and S. Hüfner, (Elsevier, Amsterdam, 1987), p. 103.

15. See, e.g., P. Adler, H. Buchkremer-Hermanns, and A. Simon, Z. Phys. B **81**, 355 (1990).

16. P. Nozieres and C.T. deDominicis, Phys. Rev. **178**, 1097 (1969).

17. W. Reichardt, F. Gompf, M. Ain, and B.M. Wanklyn, (unpubl.)

DISCUSSION

FONTAINE - How does the real structure, corner sharing (CuO) versus edge-sharing ($NaCuO_2$) affect the Madelung selective role in the game of the position of Cu3d with respect to the O 2p band.

GUNNARSSON - Let V^{Cu}_{Mad} and V^{O}_{Mad} be the Madelung potentials on the Cu and O sites, respectively, and write

$$V^{i}_{Mad} = \sum_j V_{ij}(n_{ij} - Z_j) \qquad (1)$$

where n_j and Z_j are the electron charge and nuclear charge, respectively, on atoms of type j. Then the large lowering of V^{Cu}_{Mad} - V^{O}_{Mad} in going from CuO to $NaCuO_2$ can be traced to a reduction in the absolute value of the negative $V_{Cu,Cu}$. The Madelung constants V_{ij} can be written as

$$V_{ij} = \frac{v_{ij}}{R} \qquad (2)$$

where R is the radius of a sphere which has the same volume as the unit cell and v_{ij} is a structure dependent constant. The largest difference between CuO and $NaCuO_2$ is that R is smaller for CuO than for NaCu O_2 (R_{CuO}/R_{NaCuO_2}= 0.76). Furthermore $v_{Cu,Cu}$ is larger for CuO than for $NaCuO_2$, which may be related to the fact that the nearest neighbor Cu-Cu distance is smaller for $NaCuO_2$ than for CuO, although the unit cell is larger for $NaCuO_2$.

The difference in the Cu-O networks between CuO and $NaCuO_2$ plays a role for the hopping integrals between Cu and O. Each Cu x^2- y^2 orbital couples strongly to a linear combination of one 2p orbital on each of the 4 nearest neighbor O atoms. The coupling between such O 2p linear combinations belonging to different Cu atoms is weaker in $NaCuO_2$ than in CuO, due to differences in the Cu-O network for these two compounds. This is an important reason for the very strong hopping between the Cu x^2- y^2 orbital and the top of the O 2p band for $NaCuo_2$, which leads to the small width of the main peak in the Cu 2p core XPS spectrum.

DYNAMIC PHENOMENA STUDIED WITH A CCD DETECTOR

C. M. Brizard, B. G. Rodricks and E. E. Alp
Advanced Photon Source, Argonne National Laboratory
Argonne, IL 60439

R. MacHarrie
AT&T Bell Laboratories, Murray Hill, NJ 07974

ABSTRACT

A new programmable charge coupled device (CCD) detector based on the CAMAC (Computer Automated Measurement and Control) modular system and coupled to a MicroVax III computer has been developed for time-resolved synchrotron experiments. The programmability of the electronics allows one to use many kinds of CCD chip. Moreover, different detector modes can be chosen according to the time scale of the experiment. Various time-resolved x-ray scattering experiments have already been performed at NSLS and CHESS with this imaging system. For example, a real-time study of the early stages of crystallization of the amorphous metallic alloy $Fe_{80}B_{20}$ was carried out at the X6 beamline at NSLS. Here a spin melt ribbon of the amorphous metal was resistively heated in stages to 600°C and the crystallization observed on the CCD. The detector angular acceptance of 3° allowed for the observation of the evolution of the α-Fe, Fe_3B and the Fe_2B phases simultaneously on a minute time scale.

INTRODUCTION

Some extremely powerful synchrotron radiation sources, namely, the 6-GeV European Synchrotron Radiation Facilities (ESRF), the 7-GeV Advanced Photon Source (APS), and the 8-GeV Super Positron Ring (SPring-8) are currently under construction. These high brilliance machines coupled with fast position sensitive detectors leads to some exciting time-resolved experiments.[1,2,3,4]

One such exciting field is the in-situ crystallization of amorphous metallic alloys. Metallic glasses, regardless of the way in which they are prepared are not in configurational equilibrium but are relaxing slowly by a homogeneous process towards an "ideal" metastable amorphous state of lower energy. The amorphous state inherently possesses the possibility of transforming into a more stable crystalline state. However, the most promising properties of metallic glasses, e.g., the excellent magnetic behavior or the high hardness and strength combined with ductility and high corrosion resistance, have been found to deteriorate drastically during crystallization. Understanding the micromechanisms of crystallization to impede or control crystallization is, therefore, a prerequisite for most applications, as the stability against crystallization determines their effective work limits. On the other hand, controlled crystallization of metallic glasses can be used for designing very special partially or fully crystallized microstructures that cannot be obtained from the liquid or crystalline states. In addition to technological interest, crystallization behavior is attracting an increasing scientific interest.[5]

As one of the instrumentation projects of the APS, an entirely new CCD-based programmable detector for x-ray experiments has been developed. The detector with its dedicated CAMAC crate based microprocessor can perform real-time image display and manipulation. The microprocessor coupled with the programmability of the arbitrary waveform generators, that are necessary to read out a CCD, enables one to use most commercially available CCD chips as the detector active element.[6] This detector has the option of being used with direct x-rays or having a phosphor front end coupled to a focussing element.

CCD (CHARGE COUPLED DEVICE) DETECTOR

The control electronics for the detector is a CAMAC-based system, which is flexible enough to satisfy the readout sequences necessary for most commercially available CCDs. It consists of a CAMAC crate which houses the arbitrary waveform generators, master clock, 12 bit ADC module, 12 x 1M memory module, display driver, and LSI-11 dedicated microprocessor. Global control is accomplished by a MicroVax III. Comunication between it and the LSI is through encoded interrupt bit patterns in the form of Look-at-me (LAMs) received by the interrupt register. The LAMs are transmitted through an output register that is accessed by both computers. The main functions of the LSI are:

1) Controlling waveforms sequencing necessary for normal readout, fast clearing, and integration period.

2) Performing displays on a high resolution color monitor, such as full color contour maps, row slice through a particular section of the CCD, data histogramming, etc. All display is carried out through CAMAC with the LSI reading data from the memory module. This feature has the advantage of being an efficient diagnostic tool as the data do not have to be transferred to the MicroVax III.

3) Informing the MicroVax when the LSI's operation is completed so that the MicroVax can decide what function should be done next.

4) Resetting of all modules when necessary.

The heart of the flexibility lies in the interaction between the LSI and the arbitrary waveform generators. Each generator has two 4095 array data banks. The informations written in them correspond to the waveform shape and time required to readout a single row of data from the CCD, to clear or integrate. The data bank does not have to be written completely. These modules are driven by a clock pulse from a master clock generator. The master clock module is programmed to output a burst of square waveforms, whose number corresponds exactly to the number of memory cells written in the waveform generator. During normal readout, a single burst corresponds to the reading of a single row of data. At the end of each burst, it sends a LAM to the LSI, which decreases a counter corresponding to the number of rows in that particular CCD and restarts the burst to read another row. This continues until all the rows are readout, at the end of which, it informs the MicroVax that the CCD data are stored in memory. The MicroVax has the option of reading the data onto the hard disk or requesting the LSI to display it. The LSI is responsible for switching banks to obtain the necessary waveforms that are required during the clear and integration cycles. Another clock module acts as the convert pulse to digitize the data to 12 bits in the ADC module. This module is also triggered by the clock burst from the master clock, and hence, is enabled only during serial readout. Because the LSI controls the transfer, a delay of about 20 μs between rows results.

This system, in which the LSI keeps track of the number of rows the CCD possesses and the arbitrary waveform generators have the information required to readout the CCD, allows one to program the system for different chips. All outputs from CAMAC go to a connector strip that has a 16-pin cable connected to the CCD vacuum chamber. Depending on the CCD in use, the outputs are connected to the strip such that the 16 pins supply the correct pins on the CCD.

The software currently supports the following three chips, namely, TI 4849, TC 215, manufactured by Texas Instruments; and TH 7883, manufactured by Thomson-CSF. The TI 4849 is a 390 x 584 pixels device, with each pixel being 22.4-μm^2, the TC 215 is a 1024 x 1024 pixel device, with each pixel being12-μm^2, and the TH 7883 is a 384 x 576 pixel device with each pixel 23-μm^2. The chip currently used for this time-resolved x-ray experiment was the TI 4849.

The CCD is Peltier-cooled (-40°C) and placed in a vacuum chamber to prevent condensation on the sensitive face of the CCD chip. The front end of this chamber can be either a quartz window connected to an optical system[7] or a beryllium window for direct x-ray applications (Fig. 1). For the first case, the x-ray photons are converted into visible light through a phosphor. A variety of phosphor screens have been used: Gd_2O_2S:Tb (Trimax 2 - 3M), LaOBr:Tm combined with Gd_2O_2S:Tb (Quanta V - Du Pont), and $CaWO_4$ (Hi-Plus - Du Pont). Overall efficiencies of conversion for x-ray energy to light energy within the phosphor are typically 10-20%. For an x-ray energy of 7 KeV (1.77Å), a light wavelength of 0.5 μm (2.5 eV), and a conversion efficiency of 10%, 280 optical photons are produced for a single x-ray photon. Assuming Lambertian emission in the phosphor, the efficiency with which a lens conveys light is approximately:[8]

$$C=[M/2f(1+M)]^2$$

where M is the image to object magnification and f is the lens "f" number. This equation indicates that lens coupling is inherently inefficient and rarely exceeds a few percent (Fig. 2). It has been estimated that each x-ray photon gives rise to two optical photons at the CCD with the coupling of the two Nikon lenses 50 mm/F1.2 and 200 mm/F 4.[9] Estimating a quantum efficiency of 0.5 for the CCD chip at a 0.5 μm wavelength, one photoelectron is generated for two visible photons. Thus, one photoelectron is generated per x-ray photon striking the phosphor screen.

The efficiency of the TI 4849 has also been determined directly with x-rays of 7 KeV and found to be one photoelectron generated per incident x-ray photon. Direct x-ray imaging is exclusively used for fast time-resolved experiments because of practical details like positioning of the slits and decay time of the phosphor. The previous comparaison shows that even for second or minute experiments, the use of direct x-ray imaging is attractive. The same global yield is obtained. For x-ray experiments that did not require any magnification or demagnification, the x-ray imaging setup has been used whatever the time scale, because of its simplicity.

Three modes used for the detector have been set up.[10] The first mode consists in the reading of the full CCD frame in, typically, 1 s. For the second mode, a slit, which is a few rows wide, is physically put in front of the CCD chip. The exposed lines are shifted down into the serial register. This mode allows us to increase the statisitics without significant increase in data collection time. The time resolution in this case is 20 ms, coming from the data transfer time between the buffer memory and the hard disk. For the third mode, a slit of one row wide is put at the top of the chip. The row exposed is shifted down towards the output register. When the top row arrives in this register, the entire chip is read out in the frame mode. The time resolution is 2 μs, the limit coming from the high capacitance of the CCD row.

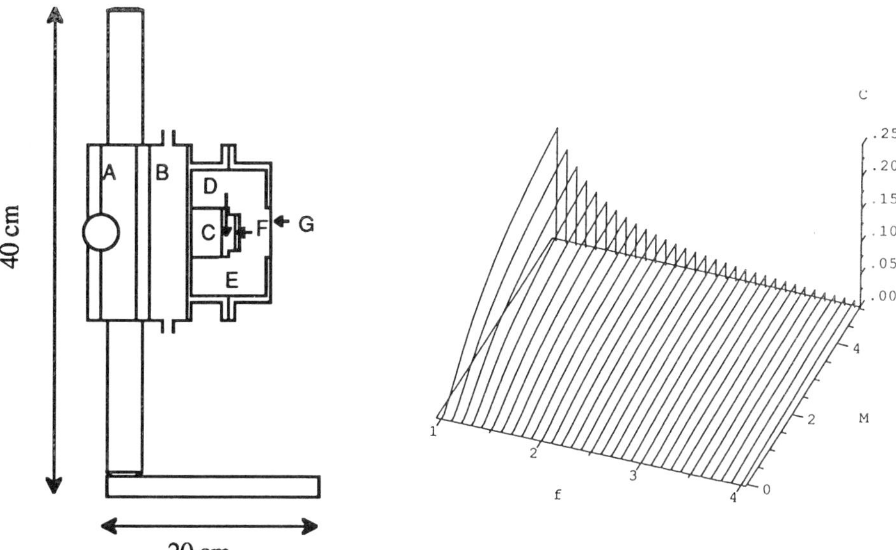

Fig. 1: CCD detector: A) vertical height mounting assembly, B) water-cooled base, C) heat sink, D) Peltier cooler, vacuum chamber, F) CCD chip, G) Beryllium window chamber

Fig. 2: 3D plot of the efficiency of a lens with a Lambertian emission in the phosphor. The 2 parameters are the "f" number of the lens and the image to object magnification, M.

CRYSTALLIZATION OF METALLIC GLASSES

The early stages of crystallization of the compound $Fe_{80}B_{20}$ have been studied for the first time at this time scale (a few minutes) and in this temperature range (from room temperature to 600°C). Table I[11,12,13,14,15] is a summary of the previous x-ray work done on this compound: the time scale is usually in the hour range and/or the temperature range is 600°C-1000°C.

The beamline used for these experiments has a flat double-crystal Si (220) monochromator that allows one to tune the photon energy. The energy was chosen to be 7 keV (1.77 Å), which is just below the Fe fluorescence edge. A rocking curve of Si (100) using the (400) reflection gives an energy spread of 1×10^{-3}. The flux is 5×10^{9} ph/s/mm^2 for an electron beam intensity of 200 mA.

The *in-situ* crystallization was performed in the vertical geometry, and the detector was mounted on the 2–theta arm of the diffractometer. The diffraction patterns of a polycrystalline sample are cones of semi-apex angle 2θ.[16] The main α–Fe reflection is located at a 2θ angle of 51.83°. At the location of the CCD chip (17 cm from the sample), the diffracted pattern corresponds to a large circle of radius 216 mm. The 13-mm long chip intercepts 0.06 rad (3.5°). The maximum distance between the intercepted arc and its chord is 60 μm on the CCD chip plane (3 pixels). This distance is much smaller than the intrinsic variation of the peak position, typically 20 pixels (corresponding to $2.6x10^{-3}$ rad), due to inhomogeneities (crystallites size). So, the diffracted peaks can be assumed to be parallel to the columns of the CCD chip. Hence, the rows can be summed in order to improve the statistics (Fig. 3). If the phenomenum to be studied is faster than a few seconds, the rows would be summed in the serial register of the chip itself. On the other hand, if the time constant of the phenomenum is larger than a few seconds, frames would be recorded, typically every few seconds, and rows can be added later by the software. As the kinetics of the crystallization of $Fe_{80}B_{20}$ are quite slow, the second method was used.

Table I: X-ray diffraction for crystallization studies of $Fe_{80}B_{20}$ amorphous alloys.

Reference	Heat treatment
[11]	T=780°C, Δt=1s-8h
[12]	T=300°C-900°C, ΔT/Δt=100°C/h, Δt=1h.
[13]	T=400°C-1000°C, ΔT/Δt=15°C/h
[14]	T=300°C-900°C, Δt=1h-2h
[15]	T=300°C-600°C, Δt=1h

T is the temperature range at which data were recorded. ΔT/Δt was the heating rate. Δt is the period during which a sample was heated at a constant temperature.

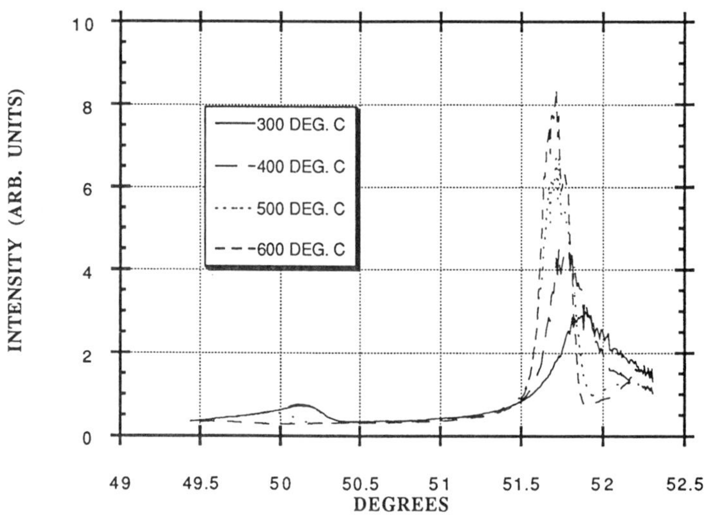

Fig. 3: CCD row sums of the diffracted peaks of $Fe_{80}B_{20}$.

The $Fe_{80}B_{20}$ ribbons are resistively heated (I≈ a few Amps, V≈ a few Volts) in a small vacuum chamber in order to prevent the oxidation of this Fe compound. Isothermal x-ray diffraction spectra are recorded at temperatures from 300°C to 600°C. The field of view of the CCD chip allows one to record simultaneously several diffracted peaks. The Fe_3B (321), α-Fe (110), and Fe_2B (211) peaks are observed respectively at $2\theta=50.20^o$, 51.83^o, and 52.21^o. The fastest crystallization is detected in 11 min as the sample is heated progressively from room temperature to 380°C. This is the first time that the $Fe_{80}B_{20}$ crystallization is detected at such an early stage at low temperature.

A summary of the results[17] (Fig. 4) includes: 1) the eutectic crystallization of Fe_3B and α-Fe appears approximately at 300°C. 2) the metastable Fe_3B phase disappears around 500°C and, simultaneously, the primitive tetragonal Fe_2B (211) peak appears. 3) The average grain size of the crystallites increases from 200 Å (300°C) to 1000 Å (600°C). 4) The main peak shift has been related to an intrinsic physical behavior related to the increase of the d-spacing. (Note that the position of the small peak [Fe_3B] does not change at all; the displacement of the main peak doesn't come from a displacement of the sample. Moreover, the thermal dilatation contribution represents only 10% of this shift.) The d-spacing variation comes from the existence of a supersaturated solution of α-Fe(B). This behavior has already been found in some Fe-B systems.[11] It is emphasized that this finding of the existence of a single phase of α-Fe(B) just after the crystallization is important, because no report on this point has been found in the literature for $Fe_{80}B_{20}$. The B migration out of the α-Fe matrix takes time following the crystallization and phase separation process. With the CCD camera, we are able to observe, for the first time, α-Fe supersaturated with B. This observation is not possible with traditional techniques like Mössbauer spectroscopy, which takes a few days, or even with interrupted quench techniques coupled to x-ray diffraction.

CONCLUSION

A new programmable CCD detector has been developed for time-resolved synchrotron experiments. The versatility of this CAMAC-based system allows one to use many different kinds of commercially available CCD chips. This device can monitor rapid thermal annealings, strain relaxations, crystallization kinetics,etc., by recording x-ray movies on a time scale from minutes to 2 microseconds. The innovative technique used for the experiments decribed here opens new exciting fields to study transient behaviors in matter as they occur.

The fastest time-resolved x-ray experiments will be those using the pulsed nature of the synchrotron beam. At CHESS, for example, the single bunch frequency is 2.56 μs, the x-ray burst duration is 160 ps (FWHM), and a typical number of monochromatic x-ray photons per burst is 5×10^4. With a synchronization between the x-ray bunch arrival and the CCD detector, the time resolution of the detector (2 μs) would allow one to record successive bunch profiles of the synchrotron radiation beam. Also, time-resolved experiments for high flux setups could be made at this speed. For "photon-hungry" experiments, various stroboscopy experiments could be carried out with the three different CCD detector modes. The success of such an experiment relies on the synchronization between the coming x-ray bunch, the probe, and the read-out of the CCD chip according to the mode used. The photons are accumulated on the same CCD area at different stroboscopic times until a good statistics has been reached. This CCD area, which can be a frame, a few rows, or one row according to the mode used, is transferred. The delay time between the x-ray bunch arrival and the probe can be increased and another recording can be made. The resolution of this system is as high as a few nanoseconds, the time interval between the different delay times.

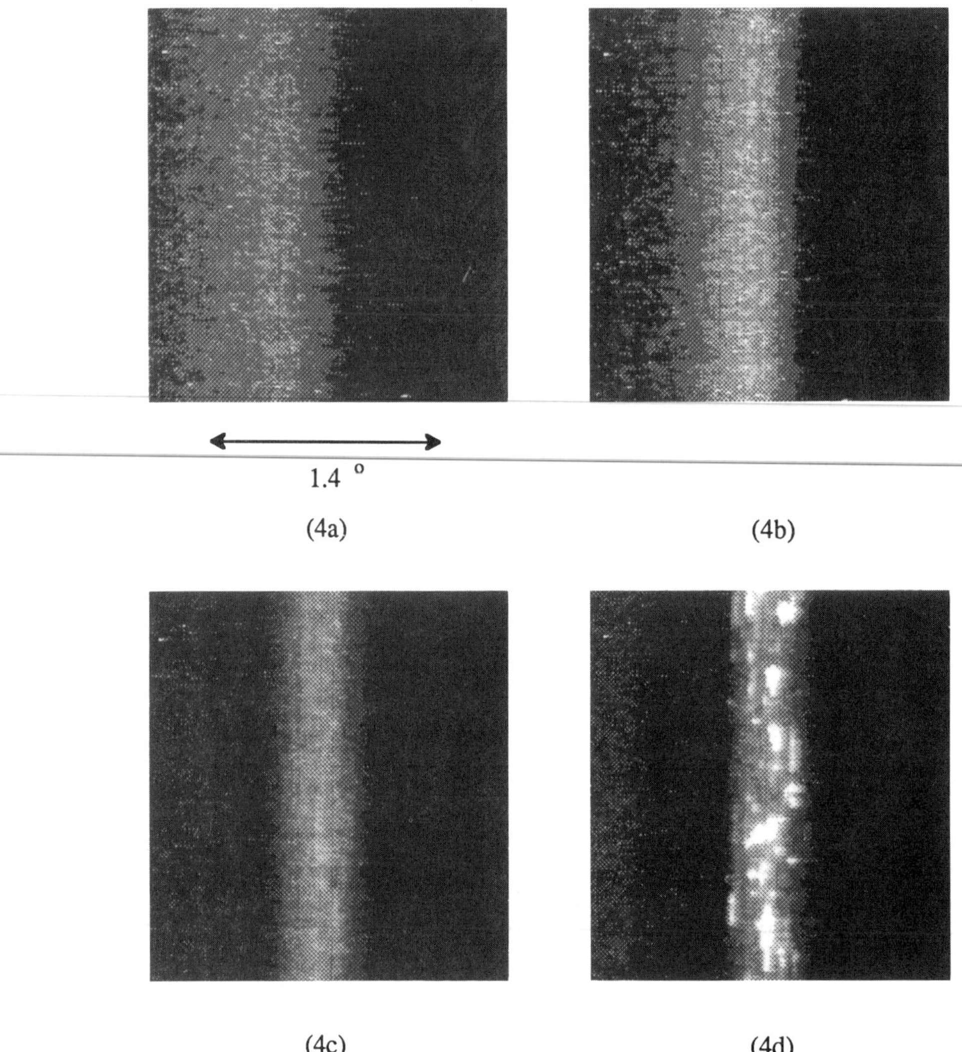

Fig. 4: Full x-ray diffraction pictures obtained with the CCD detector at different temperatures: 300°C (4a), 400°C (4b), 500°C (4c), 600°C (4d). The peaks become more intense and sharper as the temperature increases. At 600°C, very intense spots reveal the presence of large crystallite sizes.

ACKNOWLEDGEMENTS

The authors wish to thank Gopal Shenoy for his great support and Dennis Mills for representing us at the "Synchrotron Radiation and Dynamic Phenomena" Conference in Grenoble; Joe Arko and Ron Hopf deserve thanks for their invaluable help assisting us. It is a pleasure to thank Pedro Montano and Marc Engbretson for the use of X6B at NSLS. Roy Clarke and Walter Lowe are also thanked for their helpful discussions. This work is supported by U.S. Dept. of Energy, BES-Materials Science, under grant contract #W-31-109-ENG-38.

REFERENCES

1. D.M. Mills, "Time-Resolved Studies" in *Handbook on Synchrotron Radiation*, Vol.3, ed. D. Moncton and G. Brown (1991).

2. R. Clarke, W. Passos, W. Lowe, B. Rodricks, and C. Brizard, "Real-Time X-ray Studies of Strain Kinetics in $In_xGa_{1-x}As$ Quantum Well Structures," Physical Review Letters, 66, N.3, (1991)

3. R. Clarke, W. Dos Passos, W. Lowe, B. Rodricks, and C. Brizard, "Real Time X-Ray Studies of Interface Kinetics in Epitaxial Strained Layers," MRS Spring Meeting, Anaheim (1991).

4. W. Lowe, R.A. MacHarrie, R. Clarke, W. Dos Passos, C. Brizard, and B. Rodricks, "Real-Time X-Ray Diffraction Observation of a Pin-Slip Mechanism in Ge_xSi_{1-x} Strained Layers," submitted to Physical Review Letters.

5. U. Koster and U. Herold, *Glassy Metals I*, Ed. H.J. Guntherodt and H. Beck, (Springer-Verlag, 1981) p.225.

6. B. Rodricks and C. Brizard, "A Programmable Imaging System for Synchrotron Studies," accepted for publication to Nucl. Instr. and Meth.

7. C. Brizard and B. Rodricks, "Programmable CCD Imaging System," submitted to Optical Engineering.

8. S.M. Gruner, Rev. Sci. Instrum. 60(7) 1545 (1989).

9. B. Rodricks, R. Clarke, R. Smither, and A. Fontaine, Rev. Sci. Instrum. 60 (8) 2586 (1989).

10. C. Brizard and B. Rodricks, "Programmable CCD Imaging System For Synchrotron Radiation Studies," to be published in Rev. Sci. Instrum.

11. O.T. Inal, L. Keller, F.G. Yost, J. Mater. Sci. 15, 1947-1961 (1980).

12. M. Takahashi, M. Koshimura, and T. Abuzuka, Jpn. J. Appl. Phys. 20, 1821-1832 (1981).

13. Y. Khan, and M. Sostarich, Z. Metallkde. 72, 256 (1981).

14 J.A. Cusido, A. Isalgue, and J. Tejada, Phys. Stat. Sol. (a) 87, 169 (1985).

15. P. Tlomak, S.J. Pierz, L.J. Paulson, and W.E. Brower, Jr. Mater. Sc. and Engineer. 97, 369-372 (1988).

16. L.H. Schwartz, J.B. Cohen, *Diffraction from Materials*, (second edition, MRE).

17. C. Brizard, B.Rodricks, E. Alp, and R. MacHarrie, "In situ x-ray Studies of the early stages of $Fe_{80}B_{20}$ crystallization," submitted to Journal of Materials Science.

TIME RESOLVED X-RAY ABSORPTION SPECTROSCOPY STUDIES OF THE OXIDATION STATE AND THE STRUCTURAL ENVIRONMENT OF COPPER IN ZEOLITE CuNaY DURING HYDROGEN TREATMENT

Michael Hagelstein*, Sabine Cunis, Peter Rabe
Fachhochschule Ostfriesland, D-2970 Emden

Rolf Piffer
Institute of Physical Chemistry, University of Hamburg, D-2000 Hamburg 13

Ronald Frahm
Hamburger Synchrotronstrahlungslabor HASYLAB at DESY, D-2000 Hamburg 52

ABSTRACT

The copper exchanged zeolite CuNaY, a highly active catalyst for the cyclodimerisation reaction of butadiene to vinylcyclohexene, has been investigated by energy dispersive x-ray absorption spectroscopy. This is a new, element specific method for studying the oxidation state and the local structural environment of the copper atoms on a time scale of several hundred milliseconds. In this work the time dependence of heterogeneous reactions of the dehydrated zeolite CuNaY with hydrogen and a gaseous mixture of hydrogen and water vapour is presented.

Admitting pure hydrogen leads to the reduction to Cu^+ but not to metallic copper due to the strong interaction between copper cations and the alumosilicate framework. However, a two step reduction via Cu^+ to small metallic copper clusters is observed after co-admission of hydrogen and water vapour. This demonstrates the importance of the formation and migration of copper complexes in the reduction mechanism.

INTRODUCTION

Microporous zeolites are used in many technical applications. The copper exchanged zeolite CuNaY is a highly active catalyst for the cyclodimerisation reaction of butadiene to cyclohexene[1], an important precursor material for the polymer industry[2]. It crystallizes in the faujasite structure, a stable and rigid alumosilicate framework with a pore volume of 48 Vol% and a large aperture three dimensional channel system[3].

The pretreatment of the zeolite to the active catalyst normally consists of three steps. At first, the sodium ions are partly exchanged to the copper ions in aqueous solution. Then the zeolite is dehydrated and the copper ions are reduced. Hydrogen is a strong reducing agent at high temperatures and the molecules can access even the

*Present Address: ESRF, BP 220, F-38043 Grenoble Cedex

small aperture sodalite cages.

The final oxidation state of the copper ions after the hydrogen treatment depends strongly on the dehydration condition or additional treatment with gases like ammonia and carbon monoxide. The reduction of the Cu^{2+} ions to the metal[4] or the reduction to Cu^{+} ions[5] has been observed. The precise characterisation of the processes determining the oxidation state are of great interest. Another important factor determining the activity and selectivity of zeolitic material is the structure of the lattice and the active complexes.

Time resolved x-ray absorption spectroscopy[6] has been applied to characterize the copper complexes in the alumosilicate framework. The local geometrical structure around the copper ions or atoms is derived from the extended x-ray absorption fine structure (EXAFS) and the oxidation state from the K-edge threshold energy. Kinetic parameters have been deduced from time resolved measurements with the

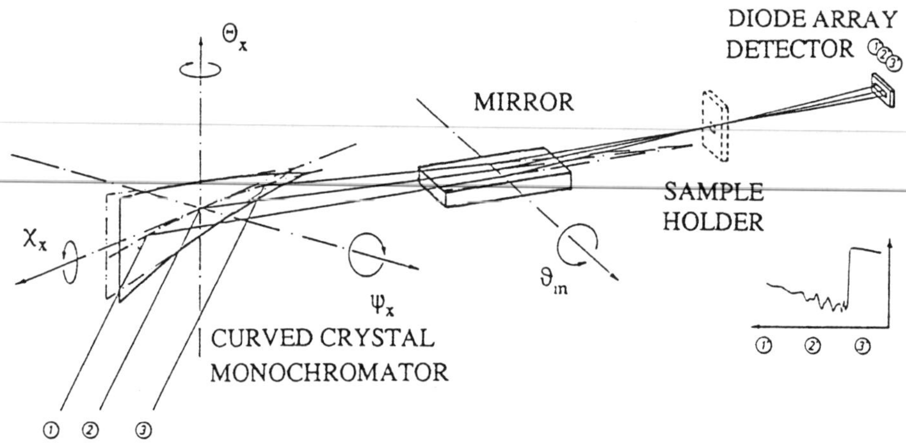

Fig. 1 The schematic representation of the energy dispersive x-ray absorption spectrometer DEXAFS. The quasiparallel synchrotron radiation is focused onto the sample by the curved crystal monochromator. The variation of the Bragg angle Θ along the crystal leads to a position-energy correlation. A complete energy region of x-ray absorption fine structure can be recorded simultaneously.

energy dispersive spectrometer DEXAFS[7] at the synchrotron radiation source DORIS II (Hamburger Synchrotronstrahlungslabor HASYLAB) (fig. 1).

EXPERIMENTAL

Copper(II)-exchanged zeolites were prepared by ion exchange of zeolite NaY with an aqueous solution of $Cu(NO_3)_2$ at room temperature. The composition determined by atomic absorption spectroscopy was $Cu_{17.4}Na_{21.2}Y$. The dried powder was pressed into self supported wafers (80 mg/cm^2). Additionally, a copper free zeolite sample has been prepared. These two specimen were introduced into an all metal vacuum chamber with Be windows.

The DEXAFS spectrometer has been equipped with either a Si(111) crystal and a 1024 photodiodes array for large bandpass spectra or Si(400) and a 512 diodes array for high resolution edge spectroscopy. The higher harmonics were suppressed using a quartz-glass mirror in total reflection geometry. The I_0 and the I_1 signal were measured consecutively with the copper free zeolite and the copper exchanged specimen. The data acquisition of a single frame was set from 100 ms to 300 ms, depending on the synchrotron radiation intensity. Up to 30 frames were accumulated for a single spectrum.

DEHYDRATION OF THE ZEOLITE

Copper-water complexes were present in the large super cages of the zeolite Y framework after the ion exchange[8]. When the samples were dehydrated at high temperatures (510-630 K) and for several hours (up to 51 h) in high vacuum, an EXAFS-signal of the second and third coordination shell has been observed. Close similarities to the EXAFS of CuO clusters with a diameter smaller than 8 Å[8] have been found. It is assumed, that the copper ions loose their water ligands and are bound to zeolite oxygen within the small sodalite cages[2].

REACTION WITH HYDROGEN ON THE WATERFREE ZEOLITE

A $Cu_{17.4}Na_{21.2}Y$ sample has been stored in vacuum for 20 hours at a temperature of 630 K. Complete dehydration can be assumed and the Cu ions are supposed to migrate into the small sodalite cages[2].

An absorption spectrum at the Cu K-edge has been recorded every 20 s during hydrogen admission (T = 570 K, $p_{H_2} = 3 \cdot 10^3$ Pa). The edge shifted to smaller photon energies in the beginning of the reaction (fig. 2). A much slower variation of the characteristic structures at 8982 eV and 8987 eV was observed.

The variation of the normalized absorption relative to the first measurement $\Delta\mu/\mu_1$ ($\Delta\mu = \mu - \mu_1$, μ_1 - absorption for t = 0) for the photon energies 8982 eV and 8987 eV has been plotted against time (fig. 3). The variation of the absorption can be fitted with a sum of two exponentials $f_i(t)$ (1).

$$\Delta\mu/\mu_1(t) = A \sum_{i=1}^{2} (\Delta\mu_\infty/\mu_1)_i \, (1-e^{-t/\tau_i}) \qquad (1)$$

A - constant

$\Delta\mu_\infty/\mu_1$ - normalized absorption for $t = \infty$

τ_i - time constant

The time constant τ_1 for the first fast step was 100 s and for the following process $\tau_2 = 2 \cdot 10^3$ s (table 1). The additional free parameters A and $(\Delta\mu_\infty/\mu_1)_i$ have been fitted to the data.

The biggest shift of the threshold occured during the first three minutes. Thus the fast process with the time constant τ_1 can be correlated with the reduction of the Cu^{2+} ions. The formation of metallic copper species in the zeolite can be excluded from a qualitative comparison of the spectra of the zeolite in the final state (fig. 2,

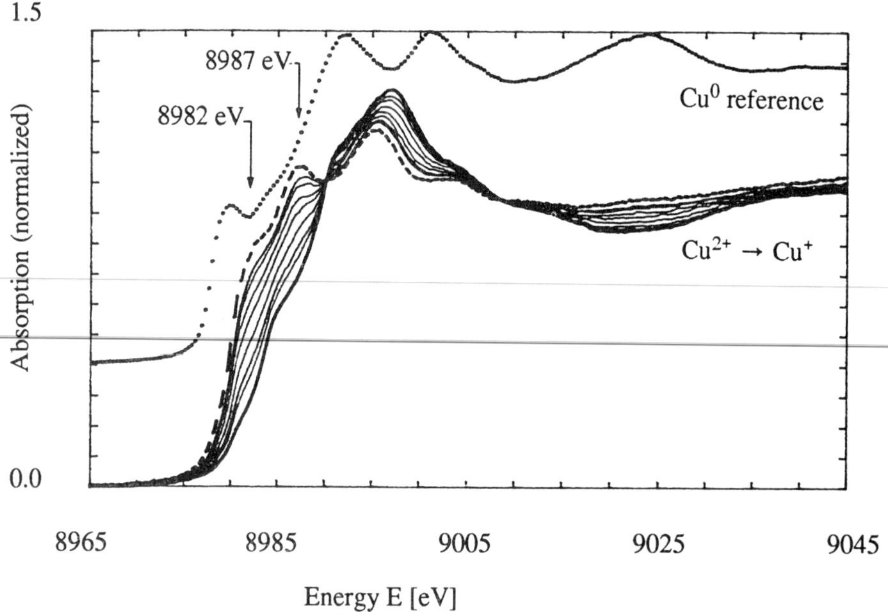

Fig. 2 Time resolved measurement during H_2 admission. Initial state: bold line; final state: bold dashed; reference Cu metal foil: bold dotted.

dashed) with the Cu metal foil spectrum (fig. 2, dotted). A redox reaction (2) of first order kinetics has been formulated.

$$Y \cdot Cu^{2+} + 1/2H_2 \xrightarrow{570K} Cu^{+} \cdot Y \cdot H^{+} \qquad (2)$$

Y - alumosilicate lattice

The reduction to metallic copper has not been observed even after a following oxidation reaction with oxygen and a subsequent reduction with hydrogen. The second process with the larger time constant τ_2 is presumably due to geometrical reordering. A Cu-O distance of 1.98 Å and a coordination number of 2 has been deduced from EXAFS investigations[9] of the zeolite after hydrogen reduction. This leads to

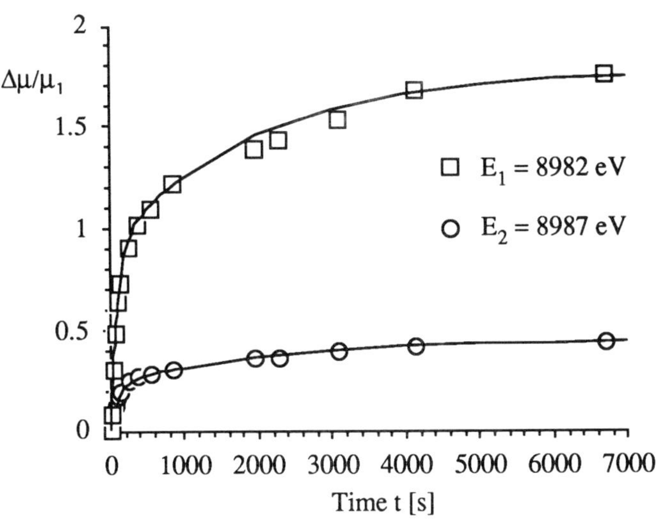

Fig. 3 Variation of the normalized absorption during the redox reaction with hydrogen only. Exponentials have been fitted to the measured points.

Table 1 Constants used to fit the observations with the exponential function.

E [keV]	8982.0	8987.0
A	4.0	1.0

$(\Delta\mu_\infty/\mu_1)_1$	0.22
$(\Delta\mu_\infty/\mu_1)_2$	0.21
τ_1 [s]	100
τ_2 [s]	2000

the assumption that Cu^+ ions are linearly coordinated by two oxygen atoms similar to the geometry in the cuprite structure[10].

REACTION WITH HYDROGEN ON THE REHYDRATED ZEOLITE

A new specimen of $Cu_{17.4}Na_{21.2}Y$ has been stored in vacuum for 8 hours at a temperature of 570 K. The Cu^{2+} ions loose their water coordination and migrate into the small sodalite cages. The sample was rehydrated at room temperature (p_{H2O} = $4.4 \cdot 10^3$ Pa) and heated to 500 K within 30 minutes. Additionally to the water partial pressure hydrogen has been admitted (p_{H_2} = $1.36 \cdot 10^4$ Pa). Both XANES and the EXAFS (a total spectral range of 628 eV) have been recorded with a Si(111) monochromator. The near edge part of the spectrum is shown in fig. 4.

Interesting conclusions can be deduced from the variation of the absorption during this reaction, although for $500\ s < t < 3500\ s$ no spectra could be recorded because of technical reasons. The threshold shifted to smaller photon energies. At the photon energy of 8987 eV the absorption increased for $t < 500$ s (fig. 5) and decreased afterwards. The spectra for $t = 3540$ s and $t = 7020$ s are characterized by a smaller

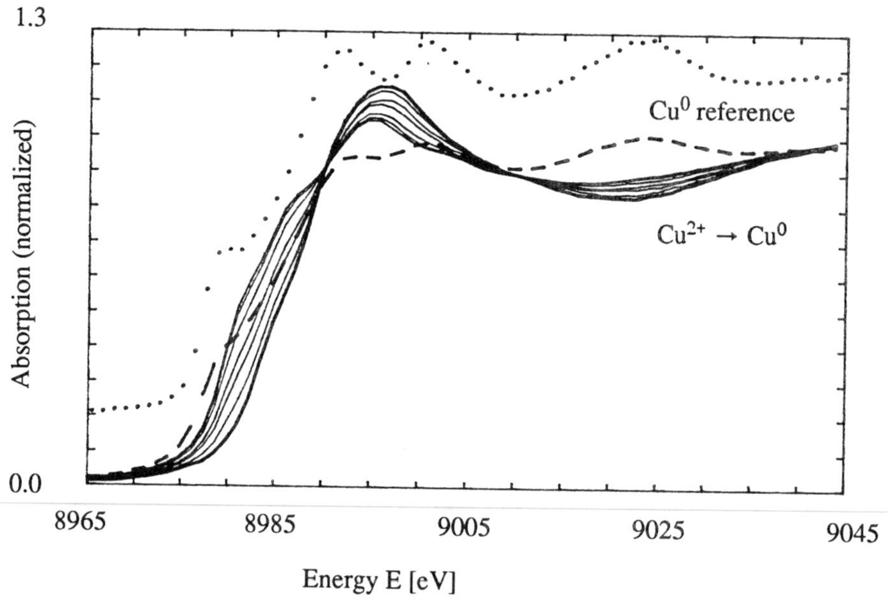

Fig. 4 Time resolved measurement during H_2 admission onto a rehydrated zeolite sample. Initial state: bold line; final state: bold dashed; reference Cu metal foil: bold dotted.

absorption $\Delta\mu/\mu_1$. These final state spectra were completely different from the resulting spectrum after hydrogen treatment of the waterfree zeolite sample. Clearly, the evolution of a metallic phase can be deduced. The redox reaction proceeded in two steps.

1. $Cu^{2+} + 1/2\, H_2 \xrightarrow{500\,K} Cu^{+} + H^{+} \quad (t < 500\ s)$ (4)

2. $Cu^{+} + 1/2\, H_2 \xrightarrow{500\,K} Cu^{0} + H^{+} \quad (t > 3500\ s)$ (5)

The intermediate product of the redox reaction was Cu^+. For $t > 3500$ s only the Cu metal signal was present. Again, the variation of the absorption at the photon energy of 8987 eV and for $t < 500$ s has been fitted with a sum of two exponentials. The first redox process to Cu^+ seemed to split into a pure reduction with the small time constant $\tau_1 = 22$ s and a slower geometrical reordering process with a time constant $\tau_2 = 145$ s.

The variation of the radial distribution around copper atoms has been deduced from the EXAFS of k = 2.5 - 7.5 Å^{-1}. The Fourier transforms F(R') are shown in fig. 6. The coordination number $N_{CU\text{-}O}$ decreased from 4.9 to 3.7 after H_2 admission and the Cu-O distance decreased from 1.99 Å to 1.96 Å. The Fourier transform at 3540 s

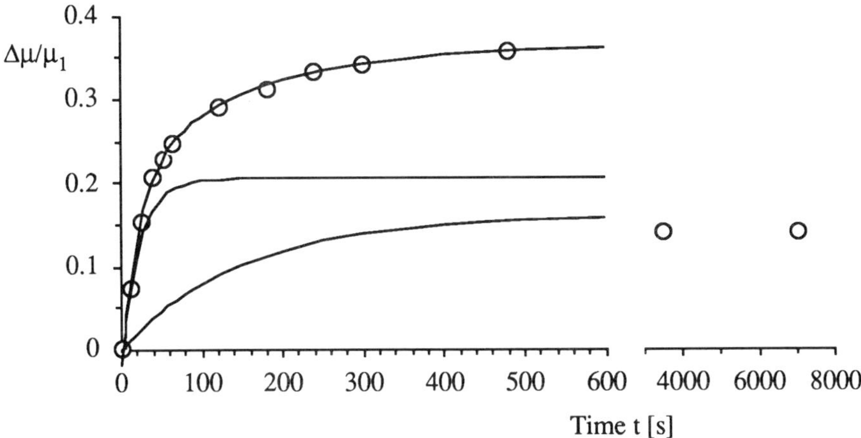

Fig. 5 Variation of the normalized absorption at the photon energy of 8987 eV during the redox reaction with hydrogen on the rehydrated zeolite. A sum of two exponential functions has been used to fit the measured points for t < 500 s.

Table 2 Parameters used for the fit

$(\Delta\mu_\infty/\mu_1)_1$	0.20
$(\Delta\mu_\infty/\mu_1)_2$	0.16
τ_1 [s]	22
τ_2 [s]	145

is dominated by the Cu metal signal with a Cu - O shoulder of the first Cu - Cu peak. The last spectrum at t = 7020 s resemble the Cu metal spectrum of the reference foil with a strong reduction of the amplitude. A detailed analysis[11] showed that this amplitude reduction is consistent with small Cu metal clusters crystallized within the large supercages.

CONCLUSIONS

The stability of the bonds between copper ions and the alumosilicate framework prevents the reduction to the metallic state in the case of completely waterfree zeolite samples. The admission of H_2 lead to the reduction to Cu^+ only. The presence of water during the redox process with H_2 lead finally to the reduction of the copper ions to the metal. A model for this reaction after rehydration can be deduced from the kinetics. The first step is the reduction to Cu^+. The water molecules weaken the Cu^+-zeolite bonds and $[Cu(H_2O)_x]^+$-complexes migrate through the channel system. Finally, small copper metal clusters within the channel system are formed.

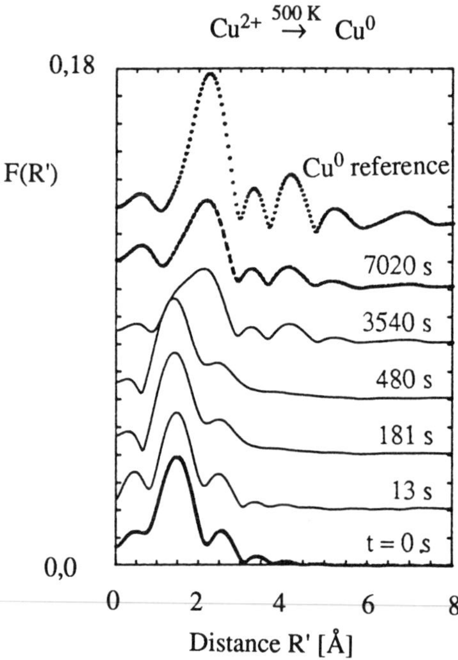

Fig. 6 Fourier transform F(R') of the EXAFS $\chi(k)$ ($k = 2.5 - 7.5$ Å^{-1}) during the H_2, H_2O redox process. An amplitude reduction of the first Cu-O shell and a small decrease of the Cu-O distance is observed after H_2 admission. The final state spectrum resemble the copper metal reference foil.

Acknowledgement

This work is supported by the BMFT under contract No. 05419 DAI and FH Ostfriesland.

REFERENCES

1. H. Reimlinger, U. Krüerke, E. de Ruiter, Chem. Ber. **103**, 2317 (1970)
2. I. E. Maxwell, Adv. Catal. **31**, 2 (1982)
3. D. W. Breck, *Zeolite Molecular Sieves,* Wiley & Sons, New York (1974)
4. S. Tanabe, H. Matsumoto, Bull. Chem. Soc. Jpn., **63**, 192 (1990)
5. P. A. Jacobs, H. K. Beyer, J. Phys. Chem. **83**, 9, 1174 (1979)
6. M. Hagelstein, S. Cunis, P. Rabe, R. Frahm, W. Niemann, R. Piffer, *X-Ray Absorption Fine Structure,* edited by S. S. Hasnain, Ellis Horwood, New York (1991) pp. 546-548
7. M. Hagelstein, S. Cunis, R. Frahm, W. Niemann, P. Rabe, in *Proceedings of the 2nd European Conference on Progress in X-Ray Synchrotron Radiation,* edited by A. Balerna, E. Bernieri, S. Mobilio (SIF, Bologna, 1990), Vol. 25, pp. 407-410
8. R. Piffer, H. Förster, W. Niemann, Catalysis Today, accepted for publication
9. R. Piffer, private communication
10. Z. G. Pinsker, R. M. Imamov, Sov. Phys.-Cryst. **9**, 334 (1964)
11. M. Hagelstein, Thesis (in German), Kiel University (1991)

Pt/Ni CATALYST PREPARATION STUDIED BY *IN SITU* X-RAY ABSORPTION SPECTROSCOPY

Andreas Jentys and Johannes A. Lercher
Institut für Physikalische Chemie, Technische Universität Wien
Getreidemarkt 9, A-1060 Vienna, AUSTRIA

ABSTRACT

The reduction of bimetallic silica supported Pt/Ni catalysts was followed monitoring the changes in the oxidation state of Ni and Pt and the formation of phases during t.p.r.by means of *in situ* X-ray absorption spectroscopy. EXAFS analysis revealed the formation of phases, while XANES was used to monitor the changes in the oxidation state of Ni and Pt, respectively.

The reduction to the final material occurred in two steps: (i) the reduction of Pt and Ni close to Pt and (ii) the reduction of the remaining Ni atoms. The formation of the metallic phases occurred immediately after the reduction process.

INTRODUCTION

During the reduction of chlorine precursors of multimetallic catalysts a complex sequence of reaction steps takes place that determines the structure and the properties of the final material. X-ray absorption spectroscopy offers the possibility to follow the redox processes and the formation of phases element specific. With the XANES (X-Ray absorption near edge structure) the density of vacant states near the Fermi level of the absorber atom are monitored[1]. By comparing the observed features with those of reference compounds of known oxidation state (e.g. bulk metals or ionic compounds), changes in the XANES region can be interpreted as changes in the oxidation state of the absorber atoms. The EXAFS (Extended X-Ray absorption fine structure) provides information about the geometric properties of the neighbors of the absorber atoms[2]. From this the composition and the concentration of phases present in the samples can be estimated by using assumptions for the number and kind of phases present in the material.

It should be especially emphasized that these experiments can be performed *in situ* and provide, thus, a more realistic reflection of the physicochemical properties of the catalysts under reaction conditions than *ex situ* electron spectroscopic methods.

EXPERIMENTAL

CATALYSTS

Silica supported Pt/Ni bimetallic catalyst with Pt : Ni ratios of 9:1, 7:3, 5:5, 3:7, 2:8 and 1:9 were prepared using the incipient wetness technique[3]. The metal loading was constant for all samples $4x10^{-4}mol.g^{-1}$. The compositions of the reduced samples are compiled in Table 1.

X-RAY ABSORPTION SPECTROSCOPY

The chlorine precursors of the catalysts were pressed into self supporting wafers and placed inside a stainless steel cell which allowed to collect XAS during temperature programmed reduction (t.p.r.). After drying the precursors in He at 373 K for one hour t.p.r. was carried out in pure H_2 (5 $ml.s^{-1}$ at NTP) with a temperature increment of 10 $K.min^{-1}$ up to 723 K. The reduction was followed *in situ* by collecting XAS every 25 K. In a second series of experiments the X-ray absorption spectra were collected after intermitting the t.p.r. at 573 K and 723 K by quenching to liquid nitrogen temperature. In order to optimize the signal to noise ratio, the weight of each sample was selected to achieve values of 2.5 for the absorption (μx) of the reduced samples[4]. To compare the XANES of the different samples, the XAS were normalized to the mass areal loading of the edge metal. The XAS were aligned to the energy scale by positioning the first inflection point of the edge to the threshold energy of the bulk metal. The height of the peak above the absorption edge was used to compare the oxidation state of the metal investigated.

Table 1. Compositions of the Ni/Pt catalysts

at% Ni total metal	wt% Ni	wt% Pt	wt% total metal	Accessible metal atoms per gram[#]	Dispersion [%]
100	2.30	0.00	2.30	$3.83*10^{19}$	16
90	2.05	0.86	2.92	$3.86*10^{19}$	16
80	1.81	1.71	3.53	$4.75*10^{19}$	20
70	1.58	2.55	4.13	$4.92*10^{19}$	21
50	1.11	4.20	5.31	$4.70*10^{19}$	21
30	0.66	5.81	6.47	$3.08*10^{19}$	14
10	0.22	7.38	7.59	$3.94*10^{19}$	18
0	0.00	8.15	8.15	$4.84*10^{19}$	22

Deteremined by H_2 chemisorption

The oscillatory part of the X-ray absorption spectra (EXAFS) was obtained by subtracting a polynomial background function from the X-ray absorption spectra and normalized by dividing by the mass areal loading of the edge metal. The contributions of the different coordination shells were isolated by Fourier transformation of the k^2 weighted oscillations. The inverse of the Fourier transformation over a selected range of the radial distribution function yielded the EXAFS for a single coordination shell. From these data the structural parameters were calculated using backscattering amplitude and phase shift functions obtained from bulk metal references in the case of Ni-Ni, Ni-Pt, Pt-Ni and Pt-Pt and from $PtCl_4$ and $NiCl_2.6H_2O$ for Pt-Cl and Ni-Cl respectively[5].

The X-ray absorption spectra were measured at the beamline X18-B at the NSLS, Brookhaven National Laboratory, N.Y.

RESULTS

The XANES of the reference compounds $PtCl_4$ and bulk Pt are plotted in Fig. 1., those of $NiCl_2.6H_2O$ and bulk Ni in Fig. 2. The XANES of $PtCl_4$ and $NiCl_2$ were dominated by a very intense peak above the absorption edge (white line). The intensity of the white line was lower for bulk Ni and Pt than for the chlorides $NiCl_2$ and $PtCl_4$. This difference between the oxidized and the metallic state is explained by a higher density of states above the Fermi level for the oxidized elements[6].

The XANES of both precursors show an intense peak above the absorption edge, very similar to that observed for the reference compounds, $PtCl_4$ and $NiCl_2$, respectively. At the reduction temperature of about 450 K the intensity of these peaks decreased in the case of Ni to half of its initial value, in the case of Pt, to an extent comparable to bulk Pt. With increasing reduction temperature the XANES of Pt did not show further changes, while for Ni a further decrease of the height of the white line was observed around 550 K which was attributed to a second reduction step. After t.p.r. up to 723 K both XANES were comparable to those of the bulk metal references.

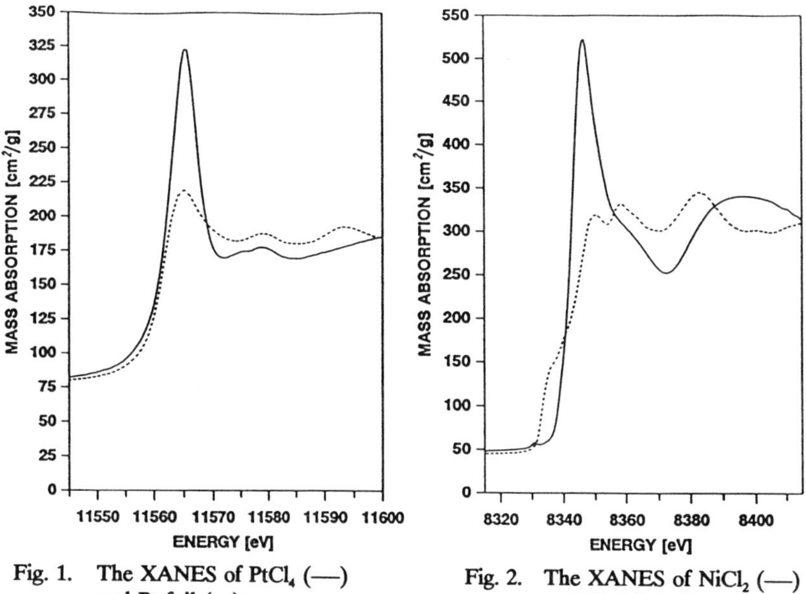

Fig. 1. The XANES of $PtCl_4$ (—) and Pt-foil (---)

Fig. 2. The XANES of $NiCl_2$ (—) and Ni-foil (---)

The XANES of the catalyst containing 80 at% Ni during t.p.r. for the Pt-L_{III} and for the Ni-K edge are plotted in Figs 3. and 4.

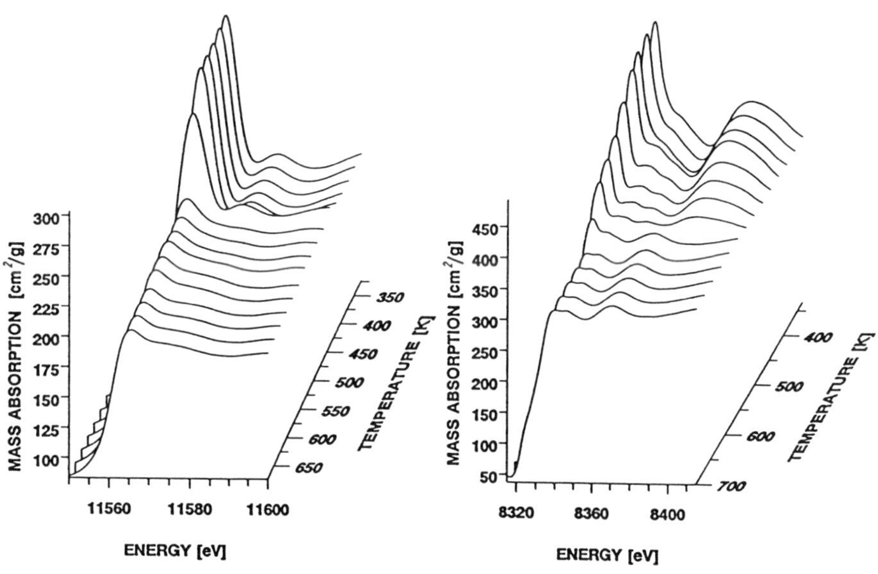

Fig. 3. Variations of the XANES during t.p.r. observed at the Pt-L_{III} edge (80 at% Ni)

Fig. 4. Variations of the XANES during t.p.r. observed at the Ni-K edge (80 at% Ni)

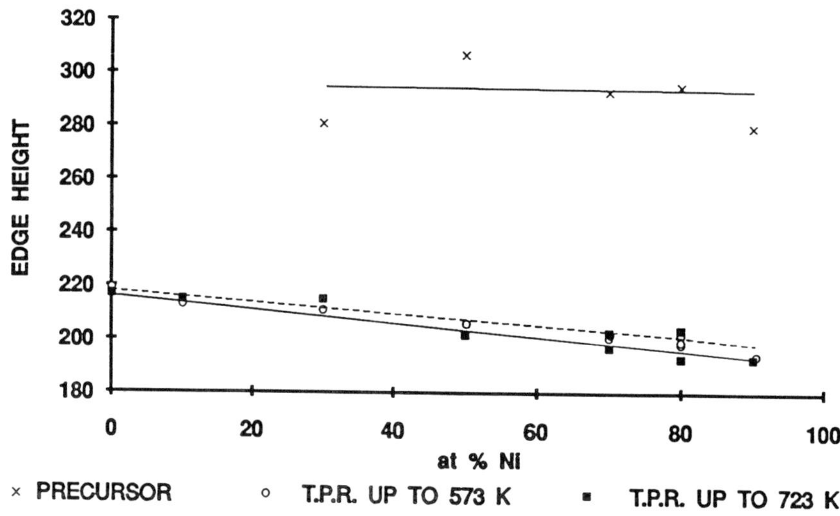

Fig. 5a. The height of the peak above the Pt-L_{III} absorption edge

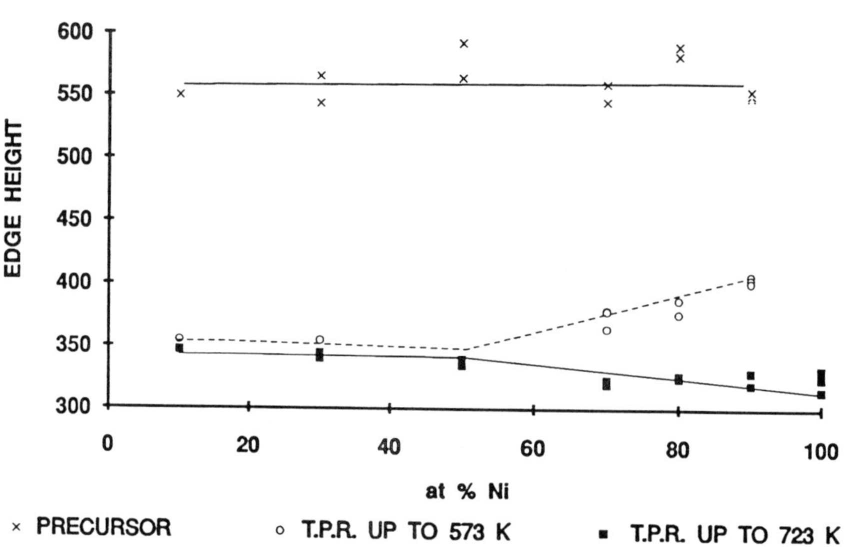

Fig. 5b. The height of the peak above the Ni-K absorption edge

The compiled heights of the peaks above the absorption edge observed for the precursors, after t.p.r. up to 573 K and after t.p.r. up to 723 K are plotted in Fig. 5. For all samples investigated, a significant more intense white line was observed for the precursors compared to the reduced catalysts. For the Pt-L_{III} edges of all catalysts investigated, the height of the peak above the absorption edge after t.p.r. up to 573 K and up to 723 K was equal. For Ni, the reduction behavior changed as a function of the chemical composition. For samples containing 50 at% Ni or less the height of the absorption edge did not change between t.p.r. up to 573 K and 723 K. For samples containing more than 50 at% Ni the height of the peak above the absorption edge was significantly larger after t.p.r. to 573 K than after t.p.r. to 723 K. Assuming additive behavior of the absorber atoms to the XANES, these results indicate that the oxidation state of the Ni atoms changed during t.p.r. between 573 K and 723 K. For all reduced samples the height of the peak above the absorption edge for the Ni-edge as well as for the Pt-edge decreased with increasing Ni content of the catalysts.

The concentration of phases present in the precursors, after t.p.r. up to 573 K and in the reduced catalysts estimated from the EXAFS analysis are plotted in Fig. 6.

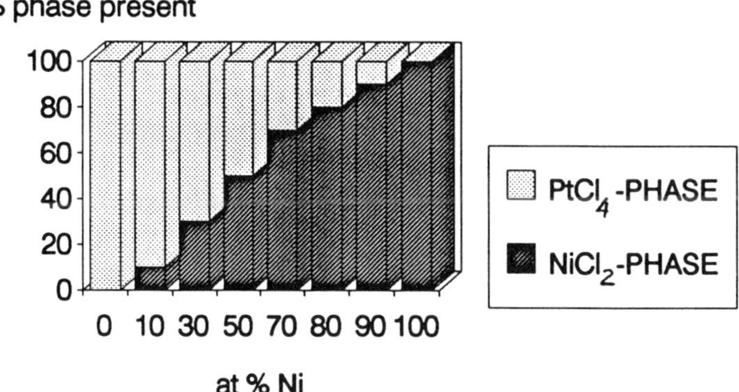

Fig.6a. Concentration of phases present in the precursor

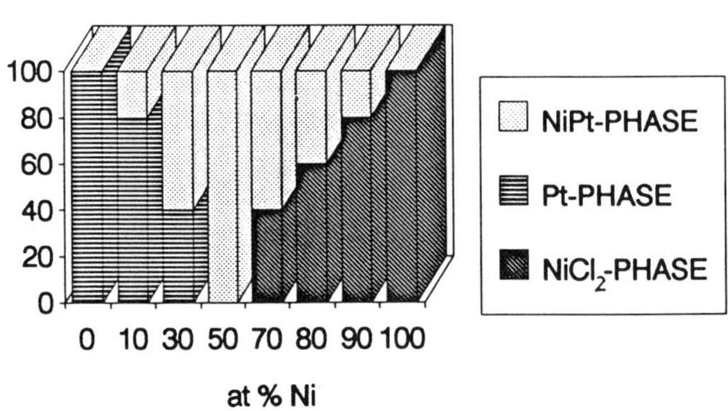

Fig.6b. Concentration of phases present after t.p.r. up to 573 K.

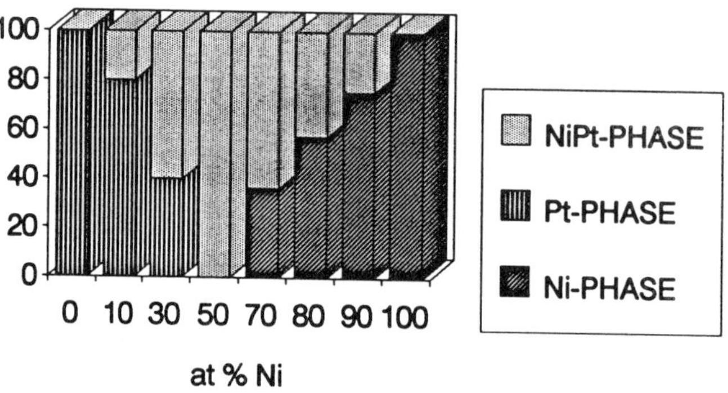

Fig.6c. Concentration of phases present after t.p.r. up to 723 K.

For the precursors of the catalysts two (isolated) phases of $PtCl_4$ and $NiCl_2$ were observed. After t.p.r. up to 573 K the formation of a bimetallic phase was observed for all bimetallic catalysts. Beside this bimetallic phase pure Pt was observed for catalysts containing less than 50 at% Ni, while for samples with a higher Ni content unreduced Ni-species were still observed. After t.p.r. up to 723 K the metals were completely reduced and consisted of a bimetallic NiPt phase and either of a Pt phase (catalysts with less than 50 at% Ni) or of a pure Ni phase (catalysts with more than 50 at% Ni).

DISCUSSION

XANES can be used to follow the oxidation state of Pt and Ni *in situ* during t.p.r.. Because of requiring a temperature resolution of 25 K while using a temperature increment of 10 $K.min^{-1}$ during t.p.r. the time for collecting the XAS is about 150 s. Using energy dispersive systems this time may be enough to collect scans from the absorption edge up to 200 eV (scanning time 10-15 s)[7] or 500 eV (time resolution 60 s)[8]. For the first case it was impossible, as the authors also mentioned, to calculate structural parameters form the obtained EXAFS, while for the second case structural parameters were obtained. For collecting XAS at higher temperatures one further point should be mentioned: While the XANES is almost independent of the sample temperature, the thermal disorder (thermal vibrations) of the atoms cause an additional decay to the observed EXAFS[9].

The oxidation state of the sample can be determined from the XANES in the following ways: (i) from the edge height, (ii) from the area under the peaks above the edge after subtracting the contributions due to the excitation to non discrete final states (continuum step) and (iii) from the position of the absorption edge.

For the system Pt/Ni, as our experiments demonstrated, the first two possibilities lead to the same results with respect to the reduction kinetics. Without using an internal standard for the energy calibration during the experiments it was impossible to determine the position of the edge accurately. It should be noted that, because of the limited resolution of X-ray monochromators, the change of the edge position can only be successfully utilized, if the shift in the energy is in the range of ~1 eV as expected for systems with a large change in the oxidation state.

The reduction behavior of bimetallic Pt/Ni strongly depend on the atomic ratio between

Ni and Pt. For catalysts with 50 at% Ni or less, all metal ions were reduced below 573 K during t.p.r.. The reduced Ni atoms with the appropriate number of (reduced) Pt atoms form immediately a bimetallic, stoichiometric NiPt-phase. The Pt atoms, which could not be incorporated into this phase, constitute an isolated Pt-phase.

Catalysts containing more than 50 at% Ni were not completely reduced below 573 K during t.p.r.. Because of the different height of the white line of the Ni-K edge after t.p.r. to 573 K and 723 K respectively, we conclude that some Ni is present in oxidized form after t.p.r. to 573 K. In contrast, the equal height of the white line for Pt indicated that all Pt atoms were reduced below 573 K. The EXAFS analysis indicated that beside the $NiCl_2$ phase a stoichiometric NiPt phase was formed below 573 K. All Pt (the minor constituent in these samples) was concluded to be in that ordered alloy phase. This indicates that only Ni atoms close to Pt, i.e., those which form the NiPt phase are reduced below 573 K. The other Ni atoms are reduced at higher temperatures and form a separate Ni phase. It is interesting to note that although a mixed chloride precursor was not observed[10] , only those Ni^{2+} cations in close proximity are reduced below 573 K.

The decrease of the height of the white line with increasing Ni concentration after t.p.r. up to 723 K suggests that a partial negative charge transfer from Ni to Pt occurred in the final material[11].

CONCLUSIONS

It was shown that the variation in the height of the white line could be used to measure subtle changes of the oxidation state.

During temperature programmed reduction of silica supported bimetallic catalysts three different processes were observed: (i) the reduction of Pt^{4+} which was independent of its environment, (ii) the reduction of Ni^{2+} in close distance to Pt in parallel to the instantaneous formation of a bimetallic phase and (iii) the reduction of Ni^{2+} at a larger distance to Pt. The processes (i) and (ii) occurred below 573 K during t.p.r., while the third occurred between 573 K and 723 K.

ACKNOWLEDGEMENTS

The work was supported by the "Fonds zur Förderung der Wissenschaftlichen Forschung" under project FWF 6912 CHE (NSF-FWF cooperative program). Research was carried out at the National Synchrotron Light Source (Beamline X18B), Brookhaven National Laboratory, which is supported by the U.S. Department of Energy, Division of Materials Sciences and Division of Chemical Sciences.

REFERENCES

1. J.C.J. Bart, **Advances in Catalysis** (Academic Press, New York, 1986), Vol.34 p.203.
2. E.A. Stern, **Phys. Rev. B**, 10 (8), 3027 (1974).
3. Ch.G. Raab, J.A. Lercher, J.G. Goodwin and J.Z. Shyu, **J. Catal.**, 122, 406 (1990).
4. P.A. Lee, P.H. Citrin, P. Eisenberger and P.M. Kincaid, **Reviews of Modern Physics**, 53 (4), 769 (1981).
5. D.C. Koningsberger and R. Prins, **Principles, Applications, Techniques of EXAFS, SEXAFS and XANES**, (John Wiley & Sons, New York / Chichester / Brisbane / Toronto / Singapore, 1988).
6. A.N. Mansour, J.W. Cook Jr. and D.E. Sayers, **J. Phys. Chem.**, 88, 2330 (1984).
7. D. Bazin, H. Dexbert, J.P. Bouronville and J. Lynch, **J. Catal.**, 123, 86 (1990).

8. J.W. Couves, J.H. Thomas, C.R.A. Catlow, G.N. Greaves, G. Baker and A.J. Dent, **J. Phys. Chem.**, 94 (17), 6517 (1990).
9. R.B. Greegor and F.W. Lytle, **Phys. Rev. B**, 20 (12), 4902 (1979).
10. A. Jentys, G.L. Haller and J.A. Lercher, to be published.
11. A. Jentys, B. J. McHugh, G.L. Haller and J.A. Lercher, **J.Phys. Chem.**, submitted (1991).

MECHANISM OF STRUCTURE REARRANGEMENT DURING DEHYDRATION OF SOME CRYSTALHYDRATES

Yu.A.Gaponov, N.Z.Lyakhov, B.P.Tolochko
Institute of Solid State Chemistry, 630091, Novosibirsk, USSR

M.A.Sheromov
Institute of Nuclear Physics, 630090, Novosibirsk, USSR

ABSTRACT

The experimental results on the structure and morphology of the substance in the reaction zone (reaction interface) during the dehydration of some crystalhydrates were obtained by the worked out X-ray diffraction methods using synchrotron radiation (SR) (in the Siberian SR Center, Novosibirsk). The mechanism of the structure rearrangement in the reaction zone is multistage and considerably depends on the crystalline structure of the initial reagent. The notion of the vacancy structure - a crystalline lattice of the initial reagent, which is stressed and deformed, with the vacancies and pores instead of the molecules of the gaseous product removed into gaseous phase - is a foundation of the suggested model of the arrangement of the reaction zone.

INTRODUCTION

The reactions of the thermal decomposition of the type:

$$AB(sol.) \overset{P,T}{\rightleftarrows} A(sol.) + B(gas.), \qquad (1)$$

in particular the reactions of thermal dehydration of crystalhydrates, proceed, as a rule, topochemically, trough the formation of the reaction zone (reaction interface) separating the solid product from the solid reagent. In the reaction zone (Fig.1) the chemical stage of leaving the molecules of the gaseous product takes place which influences, in some way, the rearrangement of the crystalline substance structure. The mechanism of rearrangement of the substance structure in the reaction zone characterizes the mechanism of the process (1) in general.

The hypothesis on the existence of the vacancy structure - a crystalline lattice of the initial reagent with the vacancies and pores left by the removed into gaseous phase molecules of the gaseous product - is a foundation of the modern knowledge of the arrangement of the reaction zone. This knowledge is realized more completely in the diffusional-kinetic model of the arrangement of the reaction zone taking into account both the process of breaking the chemical bonds of the removed by diffusion molecules of the gaseous product and the processes of crack and pore formation[1,2]. However, none of the physical-chemical models of the arrangement of the reaction zone is directly confirmed, because the method of decreasing the weight of the decomposed reagent is

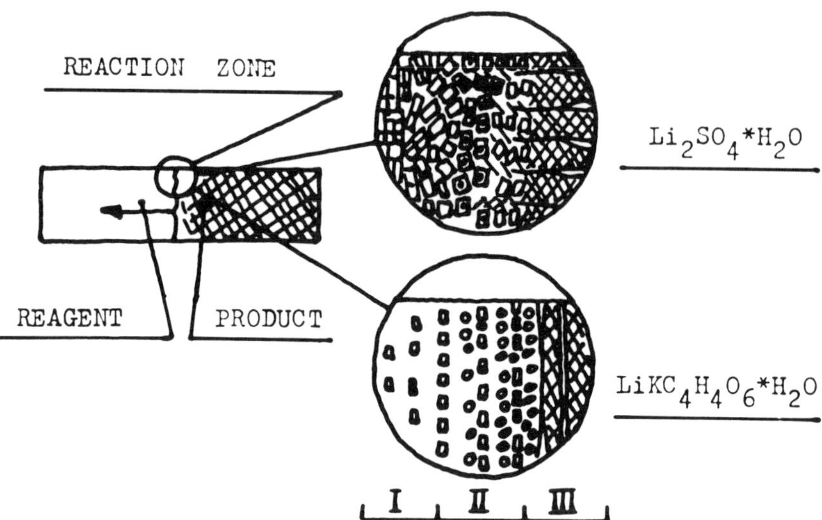

Fig .1. The models of the arrangement of the reaction zone.

the main method for the investigation of reactions of such type. The application of the direct methods for the investigation of the substance arrangement in the reaction zone, in particular X-ray diffractometry using the traditional X-ray sources, is in practice impossible because a number of methodical demands produced by the peculiarities of the reaction zone (the method must have the sufficient time and spatial resolution, the experiments must be carried out "in situ") cannot be satisfied.

The development of SR work at the Siberian SR Center (on the basis of the storage ring VEPP-3, Institute of Nuclear Physics, Novosibirsk) allows one to improve considerably the time and spatial characteristics of the X-ray diffraction methods. So, the method of the local diffractometry was realized on the basis of a high time resolution diffractometer at the station intended to study the dynamics of the structure changes using SR [3]. The spatial extention of the reaction zone (100-200 μm) during the dehydration of some crystalhydrates was confirmed experimentally [4,5]. The method was then realized at the station "Laue diffractometry". On the example of the gypsum (calcium sulphate monohydrate, $CaSO_4*2H_2O$) dehydration of the substance in the reaction zone has been shown to be deformed by the bend of the crystalline lattice of the initial reagent [6,7]. Moreover, it was shown on the example of the dehydration of the lithium potassium tartrate monohydrate ($LiKC_4H_4O_6*H_2O$) that the dehydration proceeds through the formation of the intermediate structure whose crystalline structure continuously changes during the dehydration [7,8].

A method of the simulation (modelling) of the reaction zone was realized with a view to comprehend the mechanism of the rearranging the substance structure in the reaction zone [9]

THE METHOD OF SIMULATING THE REACTION ZONE

The essence of the method intended to simulate the reaction zone is to make the conformity between the changes in time of the crystalline structure of the sample under study (a thin plate with a thickness much smaller than the depth of the reaction zone) and the spatial changes in the structure of the substance in the reaction zone. The thin plate simulates the thin layer of the massive sample across which the reaction zone moves during the topochemical process (Fig.2): in this case the thin plate for each

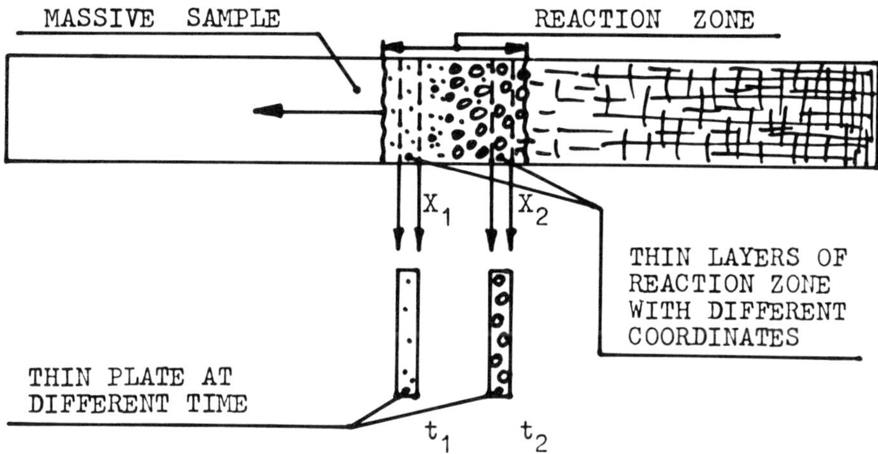

Fig.2. The scheme which explains the essence of the method of simulating the reaction zone.

time moment during the process simulates the different layered cross sections of the propagated reaction zone. The kinetic dependences of the changes of the structural characteristics allow one to simulate the spatial arrangement of the reaction zone. The thickness of the plate must be much smaller than the depth of the reaction zone so as to satisfy the condition of the relative homogeneity of the substance in the reaction zone - about 10 μm for the crystalhydrates. Besides, the reaction conditions must be such that the time needed to register a diffraction picture from the thin plate of the sample would be smaller than the time needed for the reaction zone to pass the distance equal to the depth of this plate.

The time resolution of the station "Laue diffraction" is about 1-2 s. This allows one to investigate the dehydration of single crystalline plates (with 10-20 μm thickness and smaller) of the crystalhydrates of lithium sulphate monohydrate ($Li_2SO_4{*}H_2O$, Y-cut) and lithium potassium tartrate monohydrate

($LiKC_4H_4O_6*H_2O$, Z-cut) under P=5 Pa and at T=350-400 K.

EXPERIMENTAL RESULTS AND DISCUSSION

The multistage character is a typical peculiarity of the kinetic dependences of the structure arrangement: the initial stage of a relatively long induction period and the stage of intensive rearrangement of the structure.

Figs.3.-a,b show typical diagrams of the kinetic dependences of the integral characteristics (intensity and width) of the different fragments of the experimental Laue diffraction patterns obtained during the dehydration of the investigated crystalhydrates. Curve ① corresponds to the reflection of the Laue diffraction pattern of the initial reagent, while curve ② corresponds to the fragment of the diffraction pattern from the polycrystalline product. It is clear that at the initial stage the integral intensity of the reflections increases - the changes in the mosaic and block structure of the initial reagent take place and are caused probably by the dehydration in intergrain and interblock regions (the defective regions) and the diffusional removal of the water molecules into the gaseous phase. From the estimate equation

$$l^2 \sim D*t \ , \qquad (2)$$

where l and t are the characteristic length and time of the diffusional process, D is the diffusion coefficient, it follows that the diffusion coefficient of the water molecules is about $D=10^{-9} sm^2/s$ (the characteristic length is the plate thickness: l=10-20 μm, the characteristic time is the duration of the initial stage: t=2000-3000 s), that corresponds (in the order of magnitude) to the diffusion in the solid state. The changes of the integral width of the reflections can be interpreted as the deviation of the structure of the initial reagent from the primary state: the intermediate distorted and deformed structure is formed because of a decrease of the mole volume of the initial reagent during the dehydration. When the mechanical stresses reach the limit of the mechanical durability of the substance, a destruction of the intermediate structure takes place. This facilitates the removal of the molecules of the gaseous product from the reaction zone and, as a consequence, accelerates the dehydration (the second stage) of the left fragments of the intermediate structure on the law of the compressed volume and gives rise to the formation of the polycrystalline product.

In the case of the dehydration of $Li_2SO_4*H_2O$ single crystal the polycrystalline product already begins to form at the initial stage. Therefore, one can consider that the intermediate structure and the final product are formed simultaneously. The analogous picture is observed during the dehydration of $LiKC_4H_4O_6*H_2O$ single crystal with some difference: forming of both the intermediate structure and the final product takes place successively. Besides, during the dehydration of this crystalhydrate a conside-

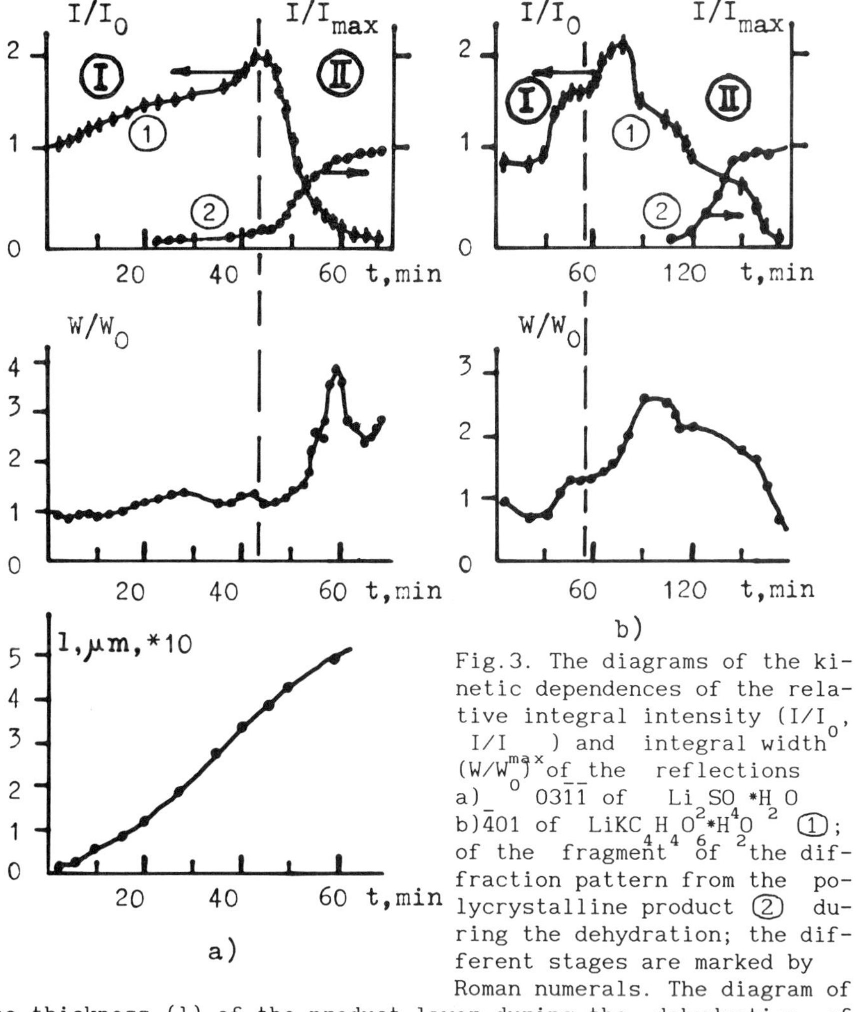

Fig.3. The diagrams of the kinetic dependences of the relative integral intensity (I/I_0, I/I_{max}) and integral width (W/W_0) of the reflections a) $03\bar{1}\bar{1}$ of $Li_2SO_4{*}H_2O$ b) $\bar{4}01$ of $LiKC_4H_4O_6{*}H_2O$ ①; of the fragment of the diffraction pattern from the polycrystalline product ② during the dehydration; the different stages are marked by Roman numerals. The diagram of the thickness (l) of the product layer during the dehydration of $Li_2SO_4{*}H_2O$ single crystal in analogous conditions.

rable broadening of all the reflections of the Laue diffraction pattern is observed. It can be explained by the formation of a substructure in the crystalline lattice of the initial reagent: the concentration of the water molecules is modulated in the lattice so that the intermediate structure formed during the dehydration is distributed in a layers with a period of about 1000-3000 Å which are parallel to the surface of the plane plate of the investigated sample. During the dehydration in the initial stage the thickness of such layers increases but the thickness of the layers of the initial reagent between the layers of the intermediate structure decreases down to 20-40 Å. In the crystalline structure of this crystalhydrate there are several particular planes that apparently determine the observed anisotropy of the dehydration. It was shown that the dehydration of gypsum proceeds through the layer lamination of individual crystallites of the single crystalline sample with the following relaxation of the bend deformation distortions (through the polygonization). On the other hand, the rearrangement of the structure during the dehydration of $Li_2SO_4*H_2O$ the crystalline lattice of which has no any anisotropic properties proceeds without any geometrical regularities of the distribution of the intermediate structure in the volume of the reaction zone. Therefore, it can be concluded that the mechanism of the structure rearrangement during the dehydration depends on the crystalline structure of the initial reagent.

Fig.1 shows the schemes of the arrangement of the reaction zone for the investigated crystalhydrates. The made above analysis of the changes of the structural characteristics allows one to distinguish several regions in the reaction zone. The first region is the region where the initial stage occurs. The dehydration here proceeds on the defect regions of the crystalline lattice: water molecules diffuses into gaseous phase through the layer of intermediate structure. The comparative analysis of the obtained experimental data and the data obtained during the thermogravity analysis allows one to conclude that the intermediate structure obtained in our experiments is the vacancy structure: the crystalline lattice of the initial reagent from what the molecules are removed partially or completely. Fig.3-a shows the diagram of the time dependence of the thickness of the product layer during the decomposition of $Li_2SO_4*H_2O$ single crystal in the analogous conditions [10]. It is clear that in the induction period the sample with thickness of about 10-20 μm must be decomposed completely. But, as it follows from the structural data, the crystalline structure of the initial reagent is not yet destroyed. The second region is a region where the decomposition of the blocks and grains of single crystalline fragments of the reagent proceeds by the law of the compressed volume. Distortion of the intermediate structure gives rise to the formation of the intermediate polycrystalline product followed by the spreading of the crack net accelerating the dehydration in general. The third region is a region of the recrystallization of the intermediate polycrystalline product. The Roman numerals on Fig.1 mark the different regions in the reaction zone.

CONCLUSION

The performed experimental work enables several conclusion to be drown.

Firstly, it was established that the kinetics of the rearrangement of the substance structure in the reaction zone during the dehydration of some crystalhydrates is multistage. The initial stage is the induction period. The second stage is an intensive change of the substance structure. The initial stage is assumed to be a process of the diffusion of the water molecules from the defect regions of the reaction zone and a formation of the intermediate structure. The dehydration of the fragments of the reaction zone which is caused by processes of the crack formation, distortion and destruction of the formed intermediate structure with the following formation of the polycrystalline product is the assumed explanation for the second stage.

The second conclusion is that the mechanism of rearranging the substance structure in the reaction zone during the dehydration of some crystalhydrates depends considerably on the peculiarities of the crystalline structure of the initial reagent.

And thirdly, the suggested models of the arrangement of the reaction zone during the dehydration of the investigated crystalhydrates have the common characteristic peculiarity: the intermediate structure formed during the dehydration is the vacancy structure. The characteristic feature of the models is the fact that different stages in the reaction zone occurs simultaneously in the case of dehydration of $Li_2SO_4*H_2O$, but successively in the case of the dehydration of $LiKC_4H_4O_6*H_2O$.

ACKNOWLEDGEMENT

The authors thank Dr.V.B.Okhotnikov for his help in preparing the single crystalline crystalhydrate samples, Dr.A.A.Sidelnikov for useful remarks during the discussions of the experimental results, Dr.G.A.Savinov for his help in adjustment of the detection system on the experimental station, the storage ring VEPP-3 team for the provision with the needed working parameters of the SR source.

REFERENCES

1. B.I.Yakobson et al., Izv. Sib. Otdel. Nauk SSSR, Ser. Khim. 1 (1985) 20.
2. E.L.Goldberg et al., Izv. Sib. Otdel. Nauk SSSR, Ser. Khim. 1 (1985) 14.
3. N.A.Mezentsev et al., Nucl. Instr. and Meth. A246 (1986) 604.
4. Yu.A.Gaponov et.al.,Izv. Sib. Otdel. Nauk SSSR, Ser. Khim. 3 (1985) 22.
5. V.V.Boldyrev et al., Nucl. Instr. and Meth. A261 (1987) 192.
6. Yu.A.Gaponov et al., Nucl. Instr. and Meth. A282 (1989) 695.
7. Yu.A.Gaponov et al., Rev. Sci. Instrum. 60(7) (1989) 2429.
8. Yu.A.Gaponov et al., Nucl. Instr. and Meth. A282 (1989) 698.
9. Yu.A.Gaponov et al., Preprint INP 90-23 (Novosibirsk. 1990).
10. V.B.Okhotnikov et al., React. Kinet. Catal. Lett. 39(2) (1989) 345.

RAPID TIME-RESOLVED DIFFRACTION STUDIES OF PROTEIN STRUCTURES USING SYNCHROTRON RADIATION

Hans D. Bartunik and Lesley J. Bartunik

Max-Planck-Society, Research Unit for Structural Molecular Biology, c/o DESY, Notkestrasse 85, 2000 Hamburg 52, Germany

ABSTRACT

The crystal structure of intermediate states in biological reactions of proteins or multi-protein complexes may be studied by time-resolved X-ray diffraction techniques which make use of the high spectral brilliance, continuous wavelength distribution and pulsed time structure of synchrotron radiation. Laue diffraction methods provide a means of investigating intermediate structures with lifetimes in the millisecond time range at presently operational facilities. Third-generation storage rings which are under construction may permit to reach a time resolution of one microsecond for non-cyclic and one nanosecond for cyclic reactions. The number of individual exposures required for exploring reciprocal space and hence the total time scale strongly depend on the lattice order that may be affected, e.g., by conformational changes. Time-resolved experiments require high population of a specific intermediate which has to be homogeneous over the crystal volume. A number of external excitation techniques have been developed including in-situ liberation of active metabolites by laser pulse photolysis of photolabile inactive precursors. First applications to crystal structure analysis of catalytic intermediates of enzymes demonstrate the potential of time-resolved protein crystallography.

INTRODUCTION

Many proteins and multi-protein complexes involve rapid structural changes in their biological functioning. In the example of light-driven electron transfer reactions in photosynthetic membrane proteins, transient states have time constants of a few hundred picoseconds up to a few milliseconds. Catalytic reactions of enzymes, as another example, proceed via intermediates with typical lifetimes in the range of 1-10 milliseconds. Transitions between intermediate states with time constants of nanoseconds or longer generally involve changes in the protein conformation. Such structural transitions may be limited to local conformational changes like a reorientation of an individual side chain; in the other extreme, entire molecular domains comprising several hundred atoms may undergo collective large-amplitude motions. Knowledge of the three-dimensional structures of intermediate states would be of considerable importance for a better understanding of structure-function relationships in biological systems. In order to obtain ab-initio structural information, single crystal diffraction experiments on short time scales are required.

The high flux and spectral brilliance of synchrotron radiation sources provide a basis for X-ray diffraction studies of short-lived structural states. The feasibility of rapid time-resolved X-ray diffraction data collection on submillisecond time scales and of external triggering of a biological reaction in a protein crystal was first demonstrated in a study of carbonmonoxy ligand rebinding to sperm whale myoglobin following laser pulse photolysis of the ligand[1,2]. The reaction was cycled, and monochromatic radiation was employed. The use of polychromatic radiation[3] extended the range of possible applications to reactions which may not be cyclicly repeated, at least not in a crystalline environment; these include in particular enzymatic reactions. Laue diffraction methods and external excitation techniques have in the meantime been further developed[4-8]. As a first application of such techniques leading to new structural information, the crystal structure of a catalytic intermediate of an enzymatic reaction of ras p21 was recently determined[9]. This reaction involved comparatively long time constants in the range of 40 minutes. At presently operating 2nd generation storage rings, much shorter time scales in a sub-millisecond range are accessible. The present paper describes methods and techniques of time-resolved single crystal X-ray diffraction, with particular emphasis on polychromatic methods, and applications to studies of enzyme catalysis.

BIOLOGICAL REACTIONS IN CRYSTALLINE SAMPLES

The molecular packing of proteins and multi-protein complexes in crystals leaves space for solvent which fills typically about 50% of the entire crystal volume. The protein-water interface on the molecular surface appears to be essentially the same as for the protein molecule in solution. The solvent channels in the crystal are often so wide that reactant molecules with molecular masses up to a few hundred Dalton may diffuse through. The molecular packing may in favourable cases permit the molecules to undergo conformational changes without disruption of the lattice periodicity. Hence, a number of biological reactions including catalytic enzyme-substrate interactions which involve changes in the tertiary or quarternary structures may proceed in single crystals and be followed by X-ray diffraction.

The crystalline environment may affect the reaction under study in a number of ways, even if the conformation of the protein molecule is essentially the same as in solution. Differences in the relative arrangement of subunits and interactions with neighbouring molecules in contact regions may cause changes in the reaction pathway and rates. In the example of yeast hexokinase, the catalytic turnover rate in the crystalline enzyme is by five orders of magnitudes slower than in solution[10]. The lattice forces in the crystal may in other cases prevent complete turnover, e.g., if only part of the molecules in the unit cell can undergo conformational changes necessary for certain steps of the pathway. In the example of chicken heart aspartate aminotransferase, diffusion of the substrate PPL-Asp into the crystal leads to complete conformational change in one subunit of the enzyme dimer, whereas the movements in the other subunit are hindered by contacts to a neigbouring molecule[11]. In favourable cases, the protein may be crystallized in another crystal form with different space group or cell dimensions and hence different molecular packing.

Degradation in the lattice order, as it may occur when changes in the conformation affect the relative orientation of neighbouring molecules, may limit the applicability of polychromatic diffraction methods and hence affect the time scale of experimental studies; this is further discussed below.

POPULATION OF INTERMEDIATE STATES

Diffraction studies of biological structural kinetics require high population (well above 50%) of a specific intermediate state. Examples of rapid perturbation techniques which may be employed for external triggering are listed in Table I. The time scale of 10 ns for direct laser excitation refer to the typical pulse length and jitter of excimer lasers. The time scale (1 ms) of excitations involving caged compounds[12] is defined by the lifetime of an intermediate in the photolytic activation. Caged compounds are precursors of metabolites like ATP, GTP or Ca^{2+} which are inactivated by a photolabile 'cage'. Removal of the cage with a laser pulse produces a jump in concentration of the active metabolite or in pH. Dyes like proflavin with broad absorption bands in the visible may be used to generate laser-induced temperature jumps in crystals which would otherwise be transparent to visible light. Such dyes may be diffused into the sample at low concentrations so that only a negligible percentage of all protein molecules in the crystal will be affected by possible binding of the dye, e.g., in the vicinity of active sites. In this way, 5-10°C jumps in the temperature may be achieved. Other rapid excitation techniques may involve pressure jumps or external AC electric fields.

Table I Rapid external triggering techniques

Reaction type	Excitation technique and time scale	
Enzymatic reaction	T-jump (laser)	10 ns
	pH-jump (caged protons, laser)	1 ms
	C-jump (caged metabolites, laser)	1 ms
Photosynthesis	Optical pumping (laser)	10 ns
Ligand rebinding	Photolysis (laser)	10 ns

A number of reactions may be cyclicly triggered. In the example of carbonmonoxy myoglobin, the ligand may be debound from the heme iron by laser pulse photolysis; at room temperature, the ligand rebinds on a ms time scale. This reaction may be initiated in the crystal many thousand times; the repetition rate has to be limited to about 10 s^{-1} in order to avoid heating of the sample by more than a few degrees. Techniques for cyclic stimulation of enzymatic reactions in crystals have not yet been developed.

Cryo-conditions may be useful in a number of applications. The lifetimes of catalytic intermediates of enzymatic reactions may be prolonged by several orders of magnitude by cooling to temperatures in the range of typically 170-220 K. Further, cooling may help to enhance the population of a specific intermediate by separating energetically neighboured states along the reaction pathway. Effects of heating and radiation damage in the incident X-ray beam may be significantly reduced at temperatures below about 210 K. In flow-cell experiments involving diffusion of substrate molecules into enzyme crystals at subzero temperatures, suitable cryosolvents[13] are required which prevent the formation of crystalline ice in the sample; for example[14], aqueous methanol solutions may be used at temperatures down to about 100 K. In other experiments, shock-freezing techniques may be applied for temperatures between 80 and 150 K.

Conformational transitions between intermediate states with extremely short time constants < 10-100 ns may probably not be investigated by single crystal diffraction, at least not at high resolution. This is a consequence of the loss in coherence over the crystal volume during relaxation of the conformation to a new stationary-state intermediate. Rearrangement of an extended part of a protein molecule will involve time scales of 10-100 ns. Time-resolved diffraction studies with shorter exposure times may be of interest in the case of reactions which involve only local changes in the conformation. For example, the geminate state in heme proteins formed upon debinding of a CO ligand from the heme iron by laser pulse photolysis may possibly be studied in this way; discrimation against the subsequent readaptation of the protein matrix has to be achieved through ns time-resolution, e.g., in a single-bunch exposure.

DIFFRACTION METHODS

Crystal structure analysis in general requires measurement of a sufficiently complete set of structure amplitudes containing, e.g., 40-50% or more of all independent reflection intensities within a given resolution sphere. This may be achieved by monochromatic crystal rotation techniques or by Laue diffraction methods. Depending on the diffraction method and the exciting wavelength bandwidth, full data collection will involve a different number of subsequent exposures and different time scales. The following discussion emphasizes polychromatic methods, because of their high potential for rapid time-resolved studies.

Crystallographic phases for structure analysis of short-lived intermediates will in general be known to acceptable approximation from previous studies of stable states, e.g. of the native enzyme. This assumption should be valid at least in the case of reactions involving small-amplitude conformational changes.

Broad-bandpass Laue diffraction methods

Polychromatic diffraction methods make use of the continuous distribution of synchrotron radiation over a broad wavelength range, in particular on wiggler or bending-magnet beamlines. The entire volume of reciprocal space (Fig. 1) contained between the limiting Ewald spheres corresponding to the minimum and maximum incident wavelengths of the bandpass, respectively, gives rise to simultaneously excited reflections. An estimate for the number of reflections, N_L, which may be recorded simultaneously on a two-dimensional detector screen may be obtained from a comparison in in reciprocal space of the volume defined by the limiting Ewald spheres and a maximum take-off angle $2\theta_{max}$ to the volume of a sphere with radius $(1/2)\cdot(V_{RUC})^{-1/3}$. V_{RUC} is the volume of the reciprocal unit cell.

$$N_L \approx (8/V_{RUC}) \cdot \sin^4\theta_{max} \cdot (2\cdot\cos 2\theta_{max}+3) \cdot [(1/\lambda_{min})^3 - (1/\lambda_{max})^3]$$

In the example of data collection to 2.5 Å resolution from an orthorhombic structure with a unit cell volume of 10^6 Å^3, using wavelengths in the range from 0.6 to 1.8 Å, about 37.000 reflections are simultaneously excited in a single diffraction exposure.

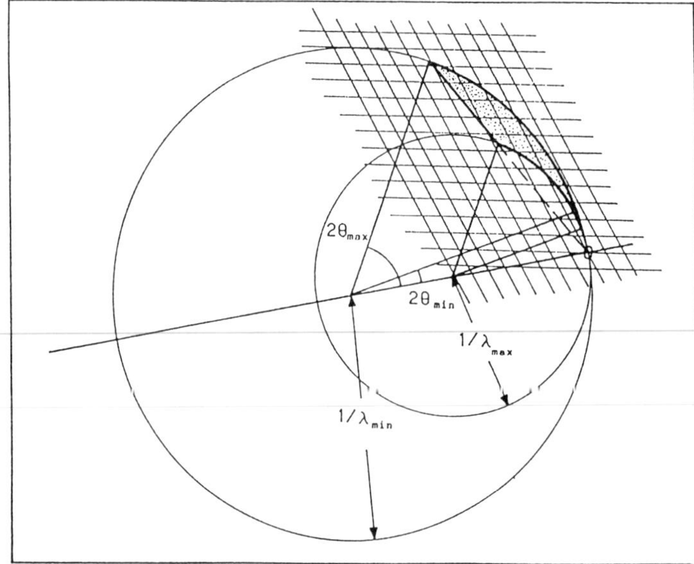

Fig. 1: Laue diffraction geometry in reciprocal space

Structure amplitudes |F(hkl)| may be derived[15] from the integrated reflection intensities I(hkl).

$$I(hkl) = I_o \cdot Q \cdot dV$$
$$Q = (e^2/mc^2 VUC)^2 \cdot P \cdot L \cdot |F(hkl)|^2$$
$$L = \lambda^4/\sin^2\theta$$

Laue processing software packages which have been developed by different groups[4,5,7] predict the spatial coordinates and exciting wavelength ranges for a given Laue pattern on the basis of a refined orientation matrix. After spot integration, structure factors are calculated and scaled by comparing equivalent reflections in different wavelength bins. The average precision in structure amplitudes derived from Laue diffraction exposures tends to be poorer than from monochromatic crystal rotation measurements; but it is in general adequate for calculations of meaningful electron density maps from a sufficiently complete data set, even at high resolution.

The question whether a single Laue exposure will provide enough information for crystal structure analysis is of importance for an estimate of the time resolution which may be reached in the experiment. The answer first of all depends on the space group symmetry. In the frequent case of orthorhombic space groups, a maximum of about 40% of all possible independent reflections may be recorded simultaneously from a favourable crystal orientation[16]; such a percentage may just be sufficient for structural studies at high resolution. The completeness reaches values of up to more than 90% for higher symmetries[17]. In general, the completeness in Laue data sets is quite poor at low resolution corresponding to d-spacings above ca. 4 Å. It may be seen from Fig. 1 that the volume between the limiting Ewald spheres gets very thin near the origin. This inefficiency of Laue methods to explore reciprocal space at low resolution affects the contrast in electron density maps[16]. Taking the example of $2F_o$-F_c difference Fourier maps (Fig. 2) calculated for orthorhombic bovine pancreatic trypsin on the basis of Laue structure amplitudes and model phases, the map still permits to follow the polypeptide chain unambiguously. However, the density is interrupted at locations along the main chain, and the significance level of side chain density is considerably lower on average than in maps based on monochromatic structure amplitudes. The contrast in these Laue maps could be improved significantly by including a small number of model structure amplitudes randomly chosen in the resolution range, e.g., from 4-7 Å. Such a procedure may be adequate if the intermediate structure remains close to a known structure of a stable state. It may fail in the case of large-amplitude conformational changes. An alternative solution to the problem posed by incomplete sampling of reciprocal space might be provided by the use of direct methods, e.g. maximum-entropy methods[18].

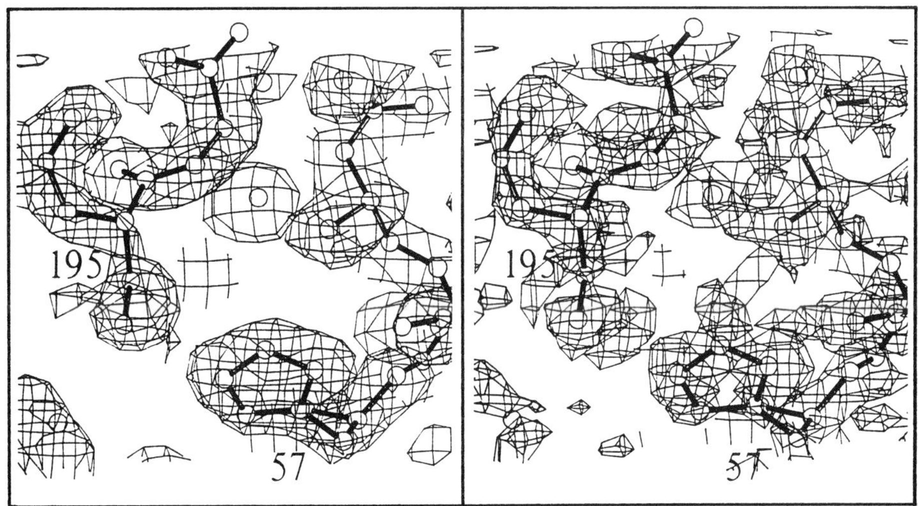

Fig. 2: Comparison of electron density maps of the active site of BP trypsin calculated on the basis of (a) monochromatic data (b) Laue data

The number of Laue structure amplitudes which may be derived from an exposure in practise is further reduced by a number of experimental factors. Incoherent scattering from the sample and its environment causes a poor average signal-to-noise ratio, since the incident wavelength bandwidth is broad as compared to the excitation range of individual reflections. As a consequence, weak reflections may not be measurable, and the apparent diffraction limit is shifted to larger d-spacings. A number of reflections overlap energetically or spatially. Whereas the percentage of wavelength overlaps tends to be relatively small[19], spatial overlaps of neighbouring reflection spots may often affect a substantial part of the diffraction pattern. The percentage of spatial overlaps is strongly enhanced by lattice disorder that may be induced, e.g., by conformational changes during an enzymatic reaction in the crystal. A slight broadening in the crystal mosaic spread by a few tenths of a degree may already cause streaks to occur in the Laue pattern[7]. In theory, structure amplitudes may be derived even from spatially overlapping reflection spots; this requires, however, reliable models for the resolution functions in reciprocal space which will in general not be available.

Thus, derivation of a sufficiently complete Laue data set from a single exposure is only possible under the conditions of a high-symmetry space group and well-preserved lattice periodicity. In the more general case, a series of exposures have to be recorded subsequently. A relatively slight decrease in the crystalline order may already make it necessary to apply reduced-bandpass Laue or even monochromatic diffraction methods on longer total time scales.

Reduced-bandpass Laue and monochromatic diffraction methods

The bandwidth in the incident beam has to be reduced with increasing crystal disorder. Then, exploration of reciprocal space requires an increasing number of subsequent diffraction exposures which may be based on Laue methods using a reduced wavelength bandpass of 5-10% or on monochromatic crystal rotation methods, typically with a 0.1% bandpass.

Reduced-bandpass Laue diffraction[7,20] data may be collected by two different techniques. In the one case, still exposures are recorded from a number of different stationary crystal orientations relative to the incident beam. A considerable number of all reflections is only partly excited; relative scaling of the Laue structure amplitudes therefore requires a reliable estimate of the degree of partiality. With an alternative technique, a series of diffraction patterns is recorded from the crystal in one fixed orientation during a wavelength scan. Each of these "Scanning-Laue" patterns corresponds to a well-defined section within the total (broad) wavelength range. Scanning-Laue methods offer a number of advantages. They lead to fully excited reflection intensities; a smaller number of exposures has to be taken; a single stationary crystal orientation facilitates the use of cooling and external excitation techniques.

Monochromatic techniques provide an optimum signal-to-noise ratio and diffraction to the highest possible resolution. Crystal rotation data collection will in general yield structure amplitudes of higher accuracy than Laue methods. On the other hand, exploration of reciprocal space requires a substantially higher number of exposures around different crystal orientations, due to the use of a narrow bandwidth.

In Table II, the different diffraction methods are compared with respect to the number of exposures and the time scales of data collection. The estimates are based on the example of a medium-sized orthorhombic protein structure with a unit cell volume of $5\cdot10^5$ Å^3 and data collection to 2.0 Å resolution. The exposure times refer to the use of a double-focussing wiggler station (BW6) at DORIS III. The experimental parameters are discussed below.

TABLE II

diffraction method	number of exposures	total amount of data	minimum exposure time per image/total	transfer time to data store per image/total	time needed for data proc.$ (100.000 refl.)
WB-L	2	4 MB	1 ms / 2 ms	20 ms / 40 ms (V)	4.5 h
SC-L	10	60 MB	10 ms / 0.1 s	0.1 s / 1 s (H)	4.5 h
RB-L	30	60 MB	10 ms / 0.3 s	0.1 s / 3 s (H)	4.5 h
M-1.0	90	180 MB	1 s / 1.5 min	0.1 s / 9 s (H) 2 s / 3 min (D)	4.5 h 4.5 h
M-0.1	900	1.8 GB	0.4 s / 6 min	2 s / 30 min (D)	4.5 h

$ using MADNES[21] on DECstation 5000 / ULTRIX
WB-L: white-beam Laue method; bandwidth ca. 1.5 Å
SC-L: Scanning-Laue; 5% bandwidth
RB-L: reduced-bandwidth Laue; 5% bandwidth
M-1.0: monochromatic crystal rotation; 0.1% bandwidth; 1.0° rotation
M-0.1: monochromatic crystal rotation; 0.1% bandwidth; 0.1° rotation
V: VideoRAM H: histogramming memory 20 MB/s D: dump onto disk

The number of exposures required vary between 2-3 for white-beam Laue and a minimum of 90 for monochromatic crystal rotation techniques. A comparable amount of structural information could be obtained from about 10 Scanning-Laue exposures each corresponding to a scan interval of, on average, 0.15°.

DATA ACQUISITION TECHNIQUES AND TIME SCALES

Fig. 3 shows a scheme of a typical experimental set-up used in time-resolved diffraction data acquisition[8]. Synchrotron radiation is focussed onto the sample crystal by total reflection from a (e.g., toroidal) mirror. For broad-band ("white-beam") conditions on a bending-magnet or wiggler station, all wavelengths above the cut-off wavelength of the mirror are present; absorption in windows along the incident beam path, sample and sample environment strongly weakens the components in the incident beam corresponding to long wavelengths above ca. 1.8 Å. As a simple alternative, a pin-hole geometry without a focussing mirror may be employed; it may be adequate on high-brilliance insertion-device beamlines. For reduced-bandwidth applications, a monochromator may be inserted. A (Gaussian) wavelength distribution with $\Delta\lambda/\lambda \approx$ 5-10% may be defined with a double mosaic-crystal or multilayer monochromator[20];

alternatively, undulator radiation may be used without the need for a monochromator. A well-defined bandpass may be produced with a scanning-monochromator system[7].

The time scales of data acquisition depend on the diffraction method. The exposure times indicated in Table II have been estimated for a wiggler beamline (BW6) with double-focussing optics at DORIS III, and use of an area detector with high detection quantum efficiency (DQE > 50%). For conventional monochromatic data collection involving a 0.1% bandpass and 1° rotation exposures, a total time scale of 1-2 minutes may be reached, assuming that the time required for reading out the area detector and data storage is negligible. With a 5% bandpass and reduced-bandpass Laue diffraction, the number of exposures at different crystal orientations decreases by a factor of about three, and the exposure time per image drops from 1 s to about 10 ms; hence total time scales of a few hundred milliseconds may be reached. Scanning-Laue data acquisition involving a scan through an entire wavelength bandwidth of 1.5 Å with read-out of separate diffraction images for each adjacent 0.15 Å section reduces the total exposure time by a further factor of three. Use of white-beam Laue methods pushes the time-resolution to the range of a few milliseconds. These estimates refer to studies under non-cyclic conditions.

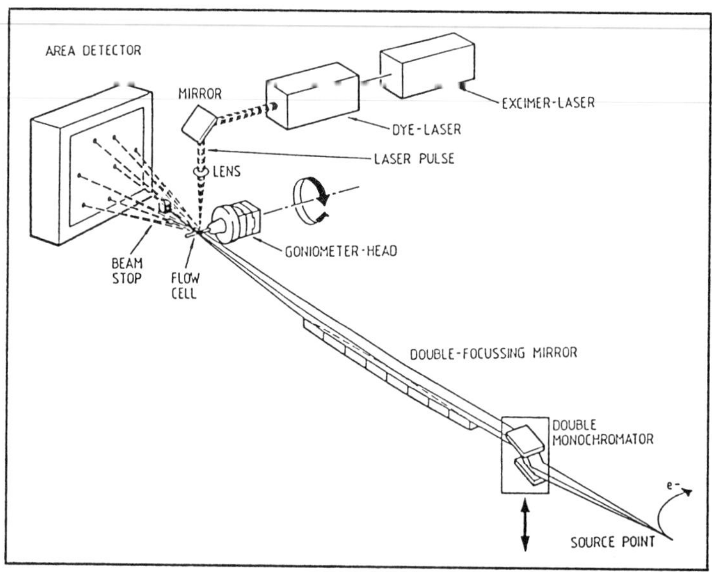

Fig. 3: Scheme of experimental set-up for diffraction data collection with external stimulation of the sample

As an illustration to the feasibility of rapid diffraction data collection on 2nd generation storage rings, Fig. 4 shows a white-beam Laue pattern of orthorhombic bovine pancreatic trypsin which was recorded on the 10-pole wiggler beamline W2 at DORIS II. Despite pin-hole geometry without focussing X-ray optics and the use of photographic film (with a DQE ≈ 5%), an exposure time of 50 ms was sufficient in order to measure diffraction out to very high (1.5 Å) resolution. With double-focussing geometry and a detector system with a high DQE, e.g. an image plate, the exposure time may be reduced by more than three orders of magnitude. Hence, in this case of a very strongly diffracting protein crystal, even a time scale of less than 100 μs may eventually be reached at DORIS. This example indicates that the exposure times given in Table II correspond to rather conservative estimates. On 3rd generation storage rings like the ESRF or APS, white-beam Laue studies of trypsin and other well-ordered structures will only require sub-microsecond exposure times. For small proteins, exposure to a single bunch may then be sufficient, hence a sub-nanosecond time scale may be reached.

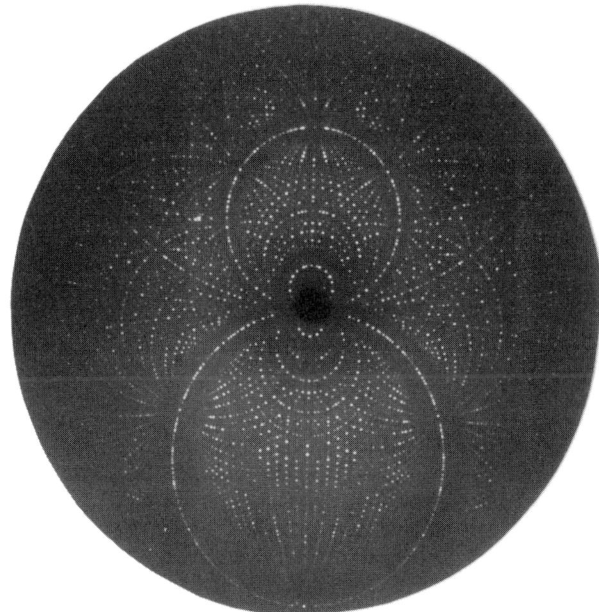

Fig. 4:
White-beam Laue diffraction exposure of BP trypsin on photographic film - exposure time 50 ms

Exposure times in the range > 10 μs may be defined with a rather simple chopper in the incident beam, used in combination with a conventional electromagnetic shutter in order to avoid frame overlap. Extraction of single synchrotron bunches may be achieved with a rapidly rotating chopper[22]. In such ns-time resolved applications, a trigger pulse for synchronizing the external excitation of a reaction in the sample to the X-ray bunch may be obtained from the master clock of the storage ring; such a technique was previously described and applied in ns time-resolved diffraction studies of low-molecular weight structures[2].

At present, photographic film is still frequently used in Laue data collection. Image plates offer a much higher DQE, in particular at short wavelengths, and a broad dynamical range over five orders of magnitude[23]. With integrating detector systems like image plates or film, the overall time resolution is essentially defined by the time needed for cassette exchange, if the experiment requires the recording of a series of subsequent exposures, e.g., at different crystal orientations. Instead, a streak-camera concept may be applied which involves translation of the plate during an exposure[24]. In view of the dense packing of reflection spots in white-beam Laue patterns, such a technique requires large film or image plates at long distance from the sample. For many applications involving more than one exposure, an on-line area detector system with rapid read-out would be ideally suited, since it would permit to carry out experiments in an interactive way. Such detector systems, in particular TV camera systems based on the use of CCDs, are being developed in a number of different laboratories. The data storage times included in Table II refer to the use of an on-line area detector with a read-out time of 20 ms per image. In this case, the total time scale of the experiment is defined for most diffraction methods by the time required for data storage, with the possible exception of monochromatic 1° rotation data collection (M-1.0). The data storage media and speeds assumed in Table II correspond to state-of-the-art technology. Interactivity in the diffraction experiment further requires rapid processing of data. About 4-5 hours are presently needed on a powerful workstation for reduction and processing of a data set comprising about 100.00 reflections (Table II). Developments of parallel computing schemes for speeding up data processing by about two orders of magnitude are in progress.

STUDIES OF ENZYMATIC REACTION INTERMEDIATES

The biological functioning of enzymes involves transient changes in their conformations which are essential for the specificity, direction and rates of biocatalytic reactions. After the initial step of substrate binding to the enzyme under formation of a Michaelis complex, one or more intermediate states are formed, until the products are released and the enzyme returns to its native conformation. In transition states, the enzyme may adopt distorted conformations which are difficult to predict theoretically by molecular modelling, even if the conformations of the enzyme in its native state or in stable complexes with inhibitors ("transition-state analogues") are known. Low-temperature or time-resolved X-ray diffraction are presently the only techniques which may provide ab-initio three-dimensional structural information for intermediates with short lifetimes at ambient temperatures. Time-resolved studies are being pursued by an increasing number of groups. In the following, applications to three different enzymes including elastase, hexokinase and ras p21 are described. These studies serve as examples for the problems and techniques involved in such experiments, and for the structural information which may be derived.

Serine proteases hydrolyze specific peptide bonds of their substrates under formation of an acyl-enzyme intermediate. The catalytic reaction involves relatively small conformational changes which are limited to the substrate binding pocket and the active site of the enzyme. In solution and at room temperature, complete turnover occurs typically on time scales of several milliseconds. The lifetime of the intermediate state may be greatly prolonged by cryoenzymological techniques. However, the formation of the intermediate is difficult to follow in the crystal by spectroscopic techniques. In the case of catalytic acylation of the substrate tBoc-Pro-Ala-Ala-OMe by porcine elastase, time-resolved Laue X-ray diffraction was applied[25] in order to monitor the occurence of conformational changes through their effect on the crystalline order. The substrate was diffused into the enzyme crystal at a temperature of 200 K. Laue exposures were recorded with an on-line area detector (FAST/ENRAF-NONIUS) following rapid changes in the temperature by steps of 5-10 K. At temperatures near 240 K, suddenly streaks appeared in the pattern indicating disorder. Subsequent diffraction data collection to 2.0 Å resolution with crystal rotation techniques, due to the broadened cyrstal mosaicity, provided the basis for crystal structure analysis of a productive acyl-enzyme intermediate. The enzyme structure was refined, and the substrate was located in electron density in difference Fourier maps; on this basis, the interaction between elastase and substrate in an acyl-enzyme intermediate state was modelled. This represents the first successful application of cryo-techniques to the study of enzymatic reaction intermediates at high resolution.

Yeast hexokinase is an example for an enzyme undergoing large-amplitude conformational changes in catalytic reactions. A first step in the reaction involves binding of glucose to hexokinase. Previous determinations of the crystal structures of native yeast hexokinase P2 (in form of a "compact dimer") and of a complex of yeast hexokinase P1 (monomer) with glucose provided evidence for relative motions of the two domains of hexokinase upon binding of glucose[26]. Fig. 5 shows a model of the enzyme-glucose complex. In a second step, ATP binds to hexokinase inducing further conformational changes preceding catalytic phosphorylation of glucose.

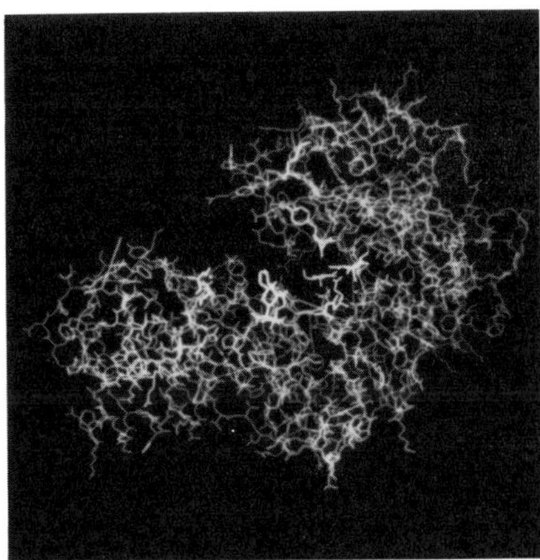

Fig. 5: Molecular model of hexokinase P1 with glucose bound to the cleft between the two domains

A series of X-ray diffraction experiments have been undertaken on hexokinase P2 (HK-P2) with the aim to investigate the interaction with the sugar substrate in the crystalline environment, and to study an intermediate ternary complex formed by the enzyme, glucose and ATP. An orthorhombic crystal form was used containg HK-P2 as elongated dimers[27]. Due to lattice constraints, the reaction rates are slowed down from about 200 s^{-1} to 0.06 min^{-1}, and complete turnover is prevented in the crystal[9]. White-beam Laue patterns recorded from native HK-P2 on the non-focussing wiggler station W2 at DORIS with an exposure time of 1 s show diffraction to about 2.0 Å

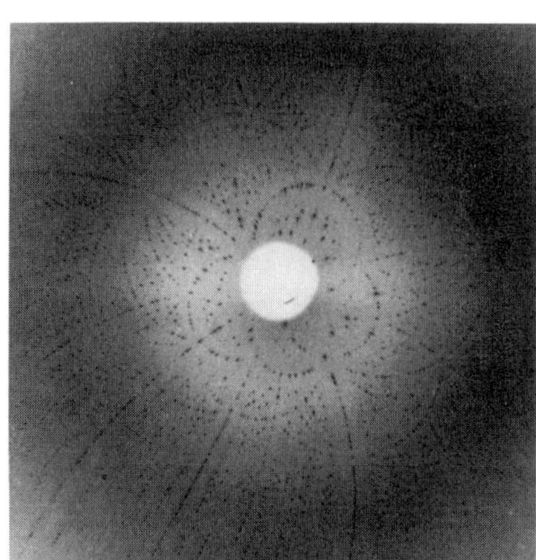

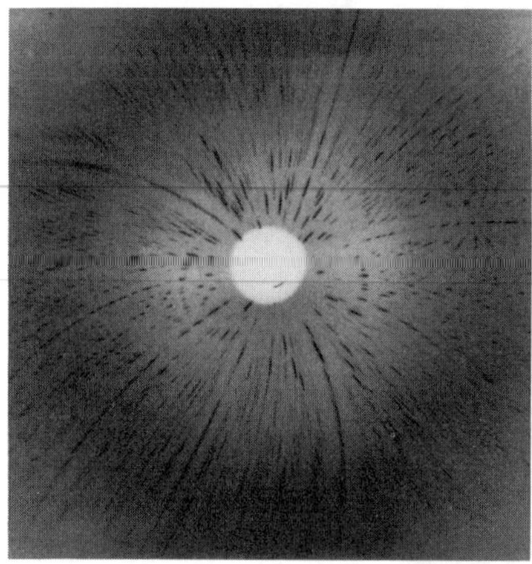

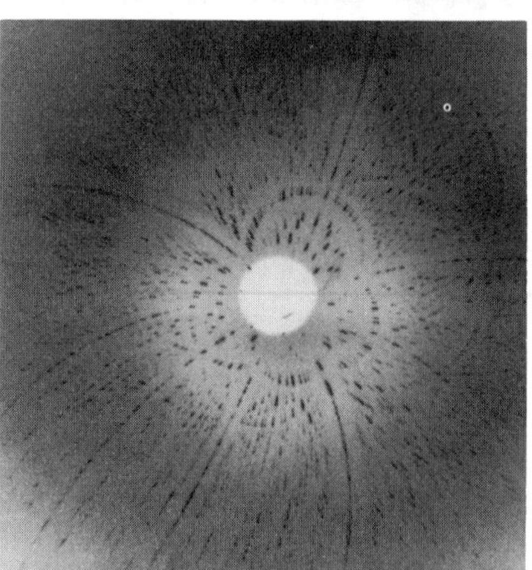

Fig. 6:
Changes in lattice order upon binding of glucose to hexokinase P2 followed by Laue methods.
(a) Native enzyme;
(b) 1 min,
(c) 5 min after rapid solvent exchange.

resolution (Fig. 6a). Binding of glucose to the enzyme was followed in a series of white-beam Laue exposures (Figs. 6a-c). Using a flow cell and optical monitoring of the solvent exchange, glucose-containing solvent was rapidly flushed over the enzyme crystal. A Laue exposure taken 0.5 s after the glucose-containing solvent reached the crystal did not show any diffraction indicating a severe loss in lattice periodicity. However, the lattice reordered on a time scale of minutes. After one minute, the lattice order is partly restored, (Fig. 6b). After five minutes, the initial lattice order has greatly recovered; two sublattices with different cell dimensions are now present (Fig. 6c). The lattice disorder results from binding of the sugar to the enzyme molecule inducing conformational changes. In the presence of a phosphate buffer, diffusion of glucose in to the hexokinase crystal does not affect the lattice order; phosphate is known to inhibit the enzyme. In further experiments, binding of glucose to HK-P2 was followed by diffusion of caged ATP into the crystal; a full account of these studies will be given elsewhere. A stoichiometric amount of active ATP was liberated in-situ with a single (10 ns) pulse at 307 nm from a Xe*Cl excimer laser; the activation process in the crystal was monitored by time-resolved optical microspectrometry at 410 nm. A white-beam Laue exposure taken immediately after laser activation shows quite high crystalline order; structural analysis is in progress.

Flash photolysis and a combination of Laue and monochromatic diffraction techniques have been applied in a study[8] of GTP hydrolysis catalyzed by a fragment of c-Ha-ras p21. A single crystal of a complex of p21 with caged GTP was illuminated with several flashes from a xenon flash lamp in order to convert the (inactive) caged GTP into active GTP. Thus, hydrolysis of GTP to GDP was initiated in the crystal. This reaction took place on a time scale of about 40 minutes. The rather long time scale of the experiment may have favoured a possible reordering of the crystal lattice after initiation of the catalytic reaction. Nevertheless, the lattice order of some crystals was affected, but it was possible to apply Laue diffraction methods and to determine the crystal structure of the intermediate formed upon binding of active GTP. Difference Fourier maps showed substantial conformational changes in an extended area of the enzyme structure around the GTP binding site. This represents the first study of an enzymatic reaction intermediate by Laue methods which provided new structural information.

CONCLUSIONS

The development of Laue diffraction methods for use in protein crystallography has stimulated considerable interest in applications to structural analysis of intermediate states. This is based on the possibility to explore reciprocal space by measuring reflection intensities at a small number - in favourable cases only one - of different stationary orientations of the sample crystal relative to the incident polychromatic beam. The high flux and spectral brilliance which is already available at operational 2nd generation storage rings make exposures on a millisecond time range feasible, at least in the case of strongly diffracting structures. It has recently been demonstrated on an undulator station at CHESS, that a diffraction pattern (containing only a few hundred reflections, due to a quite narrow bandpass) from a medium-sized protein structure, lysozyme, can be obtained from an exposure to a few synchrotron bunches of 120 ps duration each[20]. At 3rd generation facilities which are under construction, e.g., at the ESRF and APS, microsecond exposure times may be sufficient for full data collection including measurement of in the order of 100.000 reflections extending to high resolution. In the case of cyclic reactions, the time resolution may be pushed even into the nanosecond range.

Laue methods are efficient in the case of high-symmetry space groups. For tetragonal, hexagonal or cubic space groups, a sufficiently high percentage of all independent reflections may possibly be measured from a suitable single orientation of the crystal. In the most frequent cases of orthorhombic or monoclinic protein structures, a few or several exposures at different crystal orientations will be required. The total time scale of the experiment is then much longer, possibly in the range of seconds. The number of exposures needed at different orientations further depends on the lattice order. Unfortunately, transient lattice disorder appears to be often induced by conformational changes; a possible reordering occurs on a rather long time scale of minutes. Even a relatively slight broadening in the crystal mosaicity makes it necessary to substantially reduce the incident wavelength bandwidth. Using Scanning-Laue techniques, in the order of ten exposures are then required at one stationary crystal orientation; the total time scale may be in the range of hundred milliseconds. With an increasing number of subsequent exposures, the overall time resolution is more and more defined by the speed of detector read-out and data storage rather than the total exposure time.

The microsecond to second time range is probably of highest interest for most applications in time-resolved protein crystallography. Collective motions provoking structural changes in extended parts of the conformation will involve time constants longer than about hundred nanoseconds. Lattice constraints in the crystalline environment or cooling to moderately low temperatures may greatly prolong the time required for relaxation of the protein matrix to a new (intermediate) structure with sufficiently high lattice periodicity. Another crucial and difficult step in time-resolved experiments is the external initiation of biological reactions so that high and homogeneous population of a specific state is achieched in the entire crystal volume; further development of techniques will be essential for many possible applications.

Potentially important applications include in particular studies of enzymatic reactions. Three-dimensional structural information on catalytic intermediates would be of considerable importance for investigations of structure-function relationships in a great number of different enzymes; furthermore, it may provide a basis for a more efficient development of transition state analogues, e.g. inhibitors, in pharmaceutical drug design. In addition to enzyme structure research, diffraction studies of conformational changes are of interest for a number of different classes of biological structures including, among others, transport proteins, contractile proteins and viruses. Time-resolved single crystal diffraction techniques provide a unique tool for structural biology whose full power has yet to be developed.

REFERENCES

1. Bartunik, H.D. (1983). Nucl. Instr. Meth. 208, 523.
2. Bartunik, H.D. (1984). Rev. Phys. Appl. 19, 671.
3. Moffat, K., Szebenyi, D. and Bilderback, D. (1984). Science 233, 1423.
4. Moffat, K. (1989). Ann. Rev. Biophys. Biophys. Chem. 18, 309.
5. Helliwell, J.R., Habash, J., Cruickshank, D.W., Harding, M.M., Greenhough, T.J., Campbell, J.W., Clifton, I.J., Elder, M., Machin, P.A., Papiz, M.Z. and Zurek, S. (1989). J. Appl. Cryst. 22, 483.
6. Hajdu, J., Acharya, K.R., Stuart, D.I., Barford, D. and Johnson, L.N. (1988). Trends Biol. Sci. 13, 104.
7. Bartunik, H.D. and Borchert, T. (1989). Acta Cryst. A45, 718.
8. Bartunik, H.D. (1991). Crystal structure analysis of biological macromolecules by synchrotron radiation diffraction. In: Handbook on Synchrotron Radiation, Vol. 4, eds. E.Rubenstein, S.Ebashi and M.Koch. Amsterdam: Elsevier Science Publ.
9. Schlichting, I., Almo, S.C., Rapp, G., Wilson, K., Petratos, K., Lentfer, A., Wittinghofer, A., Kabsch, W., Pai, E.F., Petsko, G.A. and Goody, R.S. (1990). Nature (London) 345, 309.
10. Wilkinson, K.D. and Rose, I.A. (1980). J. Biol. Chem. 255, 7569.
11. Kirsch, J.F., Eichele, G., Ford, G.C., Vincent, M.G., Jansonius, J.N., Gehring, H. and Christen, P. (1984). J. Mol. Biol. 174, 497.
12. McCray, J.A. and Trentham, D.R. (1989). Ann. Rev. Biophys. Chem. 18, 239.
13. Douzou, P. and Petsko, G.A. (1984). Adv. Protein Chem. 36, 246.
14. Walter, J., Steigemann, W., Singh, T.P., Bartunik, H.D., Bode, W. and Huber, R. (1982). Acta Cryst. B38, 1462.
15. Zachariasen, W.H. (1945). Theory of X-Ray Diffraction in Crystals. New York: Wiley
16. Bartunik, H.D., Bartsch, H.H. and Huang Qichen (1991). Acta Cryst A, in press.
17. Clifton, I.J., Elder, M. and Hajdu, J. (1991). Acta Cryst. 24, 267.
18. Gilmore, C.J., Bricogne, G. and Bannister, C. (1990). Acta Cryst. A46, 297.
19. Cruickshank, D.W., Helliwell, J.R. and Moffat, K. (1987). Acta Cryst. A43, 656.
20. Bartsch, H.H., Bartunik, H.D., Hohlwein, D. and Zeiske, T. (1990). Acta Cryst. A46, Suppl. C-21.
21. Messerschmidt, A. and Pflugrath, J.W. (1987). J. Appl. Cryst. 20, 306.
22. LeGrand, A.D., Schildkamp, W. and Blank, B. (1989). Nucl. Instr. Meth. A 275, 442.
23. Amemiya, Y., Matsushita, T., Nagakawa, A., Satow, Y., Miyahara, J. and Chikawa, J. (1988). Nucl. Instr. Meth. A 266, 645.
24. Watanabe, N. and Sakabe, N. (1990). KEK Progress Report 90-3, p. 89.
25. Bartsch, H.H., Bartunik, H.D., Summers, L.J., Meyer, E.F. and Powers, J.C. (1989). Z. Kristallogr. 186, 14.
26. Bennett, W.S. and Steitz, T.A. (1980). J. Mol. Biol. 140, 211.
27. Steitz, T.A., Fletterick, R.J. and Hwang, K.J. (1973). J. Mol. Biol. 78, 551.

DISCUSSION

GRATTON - Do you think there is a possibility to distinguish between large domain motions and local motions of side chains ?

BARTUNIK - Domain motions will involve changes in the coordinates both of main and side chain atoms over an extended part of the molecule. Such motions should be clearly distinguishable from local side chain motions, even at medium resolution.

BARNES - Could you elaborate on the situations where you (temporarily) lost your diffraction pattern during activation.

BARTUNIK - The hexokinase-P2/glucose binding experiment was carried out in order to investigate the transition from the open conformation of the native enzyme to the closed of the complex. Of particular interest was whether the transition involved intermediate steps in the crystalline environment. The diffraction pattern disappeared within less than 0.5s after the glucose-containing solvent reached the native enzyme crystal. Taking the incident wavelength bandwidth and the cell dimensions of HK-P2 into account, the absence of a diffraction pattern may be explained by a (temporary) broadening of the crystal mosaic spread from about 0.1° to more than 1.5°. Since glucose actually binds to the HK-P2 molecule in the crystal (in the absence of inihibiting phosphate), the lattice disorder appears to result from large conformational changes which affect the relative orientation of the subunits within the dimer and the packing of dimers.

CAFFREY - Please comment on the possibility of osmotic shock to the crystal upon contact with glucose (susbstrate) containing solution. I'm concerned about the possibility of "non-enzymatic" effects of the composition-jump.

BARTUNIK - The lattice disordering upon diffusion of glucose into the hexokinase crystal does not occur at all, if the enzyme is inhibited, e.g., by phosphate. This indicates that non-enzymatic effects, in particular an osmotic shock, may be excluded.

SOLID COMBUSTION REACTIONS CHARACTERIZED BY TIME-RESOLVED X-RAY ABSORPTION SPECTROSCOPY

R. Frahm*, Joe Wong*#, J.B. Holt#, E.M. Larson#, B. Rupp# and P.A. Waide#

*Hamburger Synchrotronstrahlungslabor HASYLAB at DESY
Notkestr. 85, D-2000 Hamburg 52, GERMANY
and
#Lawrence Livermore National Laboratory, P.O. Box 808,
Livermore, CA 94551, USA

ABSTRACT

A QEXAFS (quick-scanning EXAFS) technique has been employed to monitor the site specific chemistry and local coordination changes about selected reactants in a class of highly exothermic SHS (Self-propagating High temperature Synthesis) reactions. Real time EXAFS measurements during the reaction were made in the time frame of a few seconds. By tuning the monochromator to a specific energy at which maximum changes occur in an EXAFS feature of an element from the reactant phase to the product, a time resolution down to 20 ms was achieved. The Ni + Al -> NiAl reaction has been investigated in some detail in light of a possible intermediate phase in the so-called "after-burn" region.

INTRODUCTION

SHS is a class of fascinating high temperature reactions in which at least one of the reactants is a solid. The underlying basis is the ability of highly exothermic reactions to sustain themselves in the form of a reaction or combustion front. The temperature of the combustion front can be extremely high (~4000 K) and the rate of propagation can be relatively rapid (~100 mm/s). Such SHS process permits an opportunity to study reactions at extreme thermal gradients (10^5 K/cm) and under conditions such that adiabatic conditions are conveniently invoked in theoretical analysis. Because of the high rate of combustion and extreme thermal conditions, the chemical kinetics and dynamics of phase transformations at the combustion front are not well understood[1,2]. In this paper, we have applied a recently developed QEXAFS method[3,4] to investigate the site-specific chemistry and local coordination changes about the Ni reactant in the Ni + Al -> NiAl combustion reaction.

EXPERIMENTAL

The experiments were performed at SSRL (Stanford Synchrotron Radiation Laboratory) at the wiggler beamline 10-2[5]. The electron storage ring SPEAR was operated at 3.0 GeV with injection currents of about 50 mA.The 31 pole wiggler was operated at a magnetic field of 1.45 T. The synchrotron beam was monochromatized by two Si(111) crystals mounted on a single goniometer. The spectrometer resolution was ~1 eV at the Ni K-edge at 8333 eV. The flux of the monochromatic beam was of the order of 10^{11} photons/s.

The samples, in the form of a cube 19 mm on edge, were pressed from dry-mixed 50-50 atomic mixtures of Ni and Al powders having a green density of ~55% of theoretical value. Average particle sizes for Ni and Al were 5 and 20 microns respectively. The QEXAFS data collection was first initiated at ambient temperature, and within 20-30 s the sample was ignited electrically from the top by a W coil. A W/W+26%Re micro-thermocouple was used to record synchronously the temperature-time profile with the QEXAFS data collection.

RESULTS AND DISCUSSION

In the QEXAFS mode, the fluorescence spectra were recorded from -40 to +160 eV with respect to the Ni K edge at 8333 eV. Each spectrum took ~3 s to record and some ~3 s delay was used to store and display the data on the computer giving a total of 6 s between consecutive scans. For the Ni + Al reaction it was found consistently that the fcc Ni reactant is transformed (reacted) to a Ni-Al phase within one QEXAFS frame of 6 s. The temperature profile vs. time is shown in Fig. 1. t = 0 corresponds to triggering of the QEXAFS data collection, the sharp rise from A to B marks the arrival of the combustion front at the region of the sample illuminated by the synchrotron beam.

The EXAFS signals at points A, B, C, D and E along the temperature-time profile of Fig. 1 were extracted. Fourier transforms of the EXAFS curves at B, C, D and E are plotted in Fig.2. The first peak at ~2 Å corresponds to the first and second coordination shells about the Ni centre in NiAl which is known to have a CsCl bcc structure. The Fourier peaks at ~3.5 Å and ~4.5 Å are those of the third and fourth shells. The first radial structural feature is the only dominant feature at the maximum reaction temperature at 1190 °C and increases in magnitude as the reacted specimen cools showing a reduction in the Debye-Waller factor in the measured EXAFS signal. At 730 °C and some 30 s after the passage of the combustion front, the fourth shell becomes apparent and increases in magnitude with further temperature decrease. This instance marks the low temperature limit for the formation of longer range order above the Ni coordination sphere as the NiAl product is being formed. At 410 °C, the third shell appears as a shoulder and becomes resolved at 110 °C. This series of QEXAFS measurements reveals the time events of coordination changes about the Ni center in the so-called "after-burn" region well after the combustion front has swept through the specimen.

The velocity of combustion front for the Ni + Al reaction was measured to be >20 mm/s. Therefore, within 1 s the combustion front has swept through the whole length of the 19 mm long sample. In order to better time resolve the dynamical event at the combustion zone, we employed a constant energy scan mode by tuning the monochromator to an energy which exhibits a sizable variation of an EXAFS feature from reactant to product. The normalized Ni XANES transmission spectra for fcc Ni metal and for bcc NiAl show a change of EXAFS amplitude from a maximum in Ni metal to a minimum in NiAl compound at 48 eV above the Ni K-edge. The variation of x-ray fluorescence signal was then measured at 20 ms time frame in the course of the combustion[6]. The intensity vs. time curve consists of a sharp decay during the first few seconds in which time the system attains its maximum temperture upon passage of the combustion front. This is followed by a slow decay as the temperature of the system falls from its maximum value to end of the measurement.

Correlating with our recent TR-XRD findings on this combustion system[7] that the final NiAl product appears 10 s or so after the passage of the combustion front,the initial sharp decay in the constant energy experiments[6] may be understood in terms of a phase transformation from the fcc Ni metal to some Ni-Al phase in the period when the system reaches its maximum temperature. The bcc NiAl final product is observed (and formed) at a lower temperature a few seconds later in accordance with Fourier transforms of the QEXAFS data at 730 ^oC and lower shown in Fig. 2. A slow decay beyond 20 s may be attributed to a reduction in the Debye-Waller factor of the Ni EXAFS in NiAl with temperature decrease, which in turn increases the downward movement of the minimum feature in the NiAl EXAFS at 8381 eV. This intensity-decay feature at constant energy is reproducible at both 8381 eV and at 8347 eV.

ACKNOWLEDGEMENTS

This work is supported under the auspices of the U.S. Department of Energy (DOE) by the Lawrence Livermore National Laboratory under contract W-7405-ENG-48. R.F. is grateful to LLNL for hospitality and support during an extended stay. J.W. is grateful to the Alexander von Humboldt Stiftung for a 1991 senior U.S. scientists award to continue this work at HASYLAB.

REFERENCES

1. Z.A. Munir and U. Anselmi-Tamburini, Mater. Sci. Rep. **3**, 277 (1989).
2. A.G. Merzhanov, in *Combustion and Plasma Synthesis of High-temperature Materials,* eds Z.A. Munir and J.B. Holt, VCH publishers, New York (1990), pp. 1-53.
3. R. Frahm, Nucl. Instrum. Meth. **A270**, 578 (1988).
4. R. Frahm, Rev. Sci. Instrum. **60**, 2515 (1989).
5. V. Karpenko J.H. Kinney, S. Kulkarni, K. Neufeld, C. Poppe, K.G. Tirsell, J. Wong, J. Cerino, T. Troxel, J. Yang, E. Hoyer, M. Green, D. Humphries, S. Marks and D. Plate, Rev. Sci. Instrum. **60**, 1451 (1989).
6. R. Frahm, Joe Wong, J.B. Holt, E.M. Larson, R. Rupp and P.A. Waide, (1991) to be published.
7. Joe Wong, E.M. Larson, J.B. Holt, P.A. Waide, B. Rupp and R. Frahm, Science, **249**, 1406(1990)

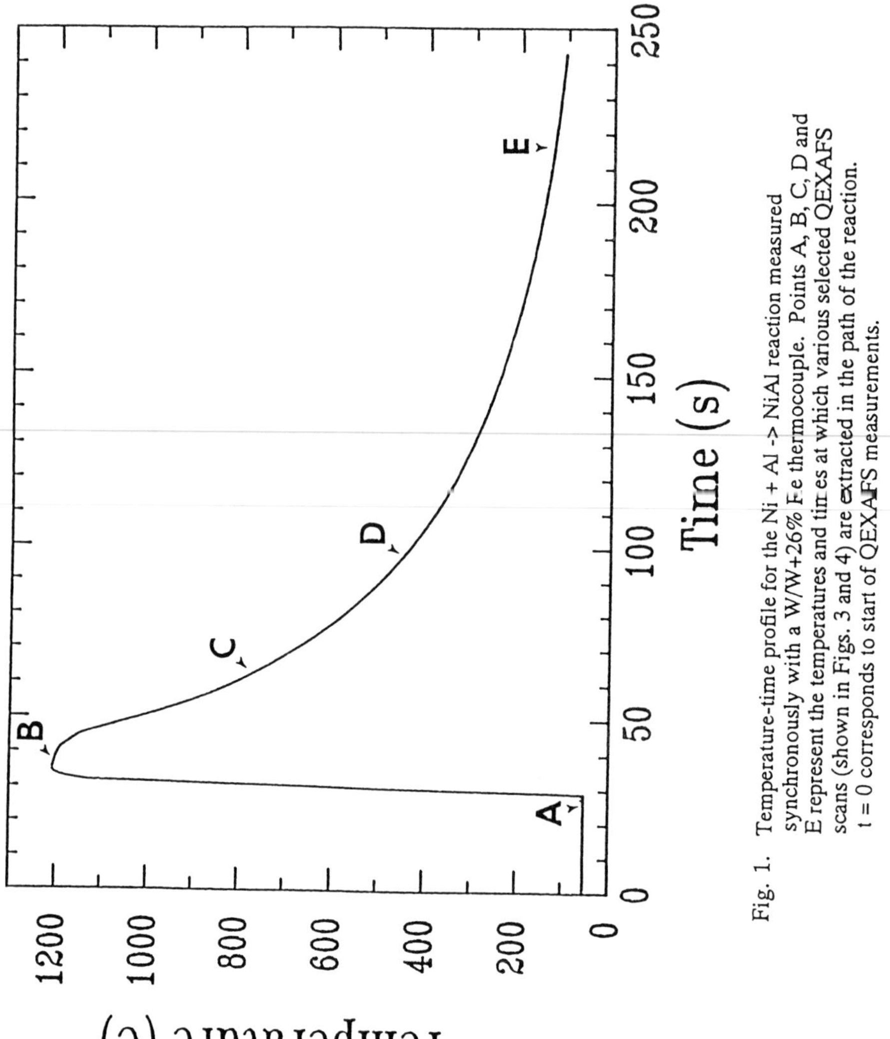

Fig. 1. Temperature-time profile for the Ni + Al -> NiAl reaction measured synchronously with a W/W+26% Fe thermocouple. Points A, B, C, D and E represent the temperatures and times at which various selected QEXAFS scans (shown in Figs. 3 and 4) are extracted in the path of the reaction. t = 0 corresponds to start of QEXAFS measurements.

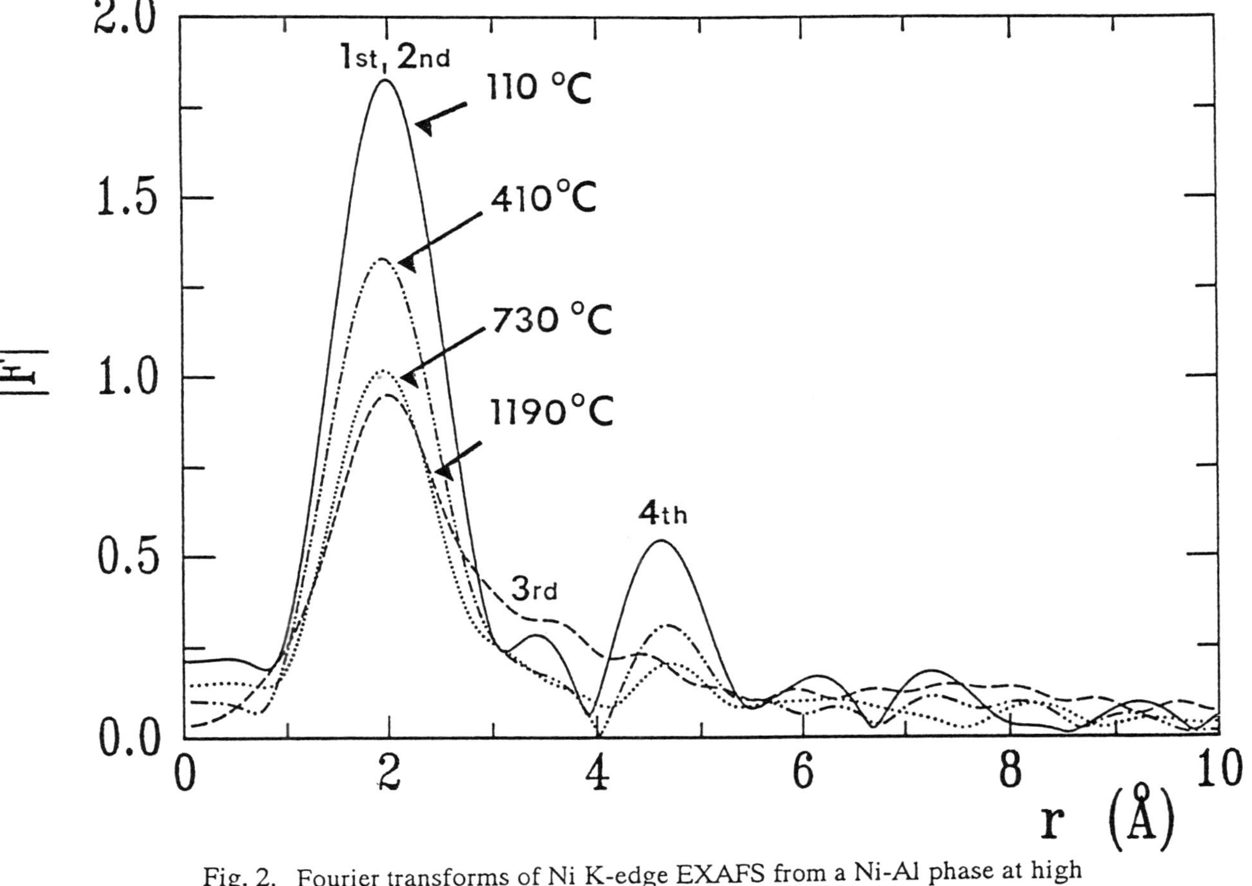

Fig. 2. Fourier transforms of Ni K-edge EXAFS from a Ni-Al phase at high temperature and from NiAl formed in the after-burn region.

DISCUSSION

GOULON - I certainly agree with your qualitative interpretation. I still have difficulty to follow you for the quantitative analyses of EXAFS:
- In a reactive or time evolving system the number of different sites, or, let say, the disorder effects are increasing dramatically.
Don't you think that this should reduce a lot of structural content of the EXAFS spectra and that you may degrade further the situation by reducing the k-range that you investigated ?
On the other hand, I agree with the strategy to use "finger prints" in the XANES to identify the presence of some species.

WONG - I agree with your points above.

JENTYS - 1. Are the Fourier transformed EXAFS of Ni corrected to the temperature of the sample ?
2. Did you try to heat up the sample after the reaction to see if you were observing changes in the local environment of Ni while the sample was cooling down or only the effect of the decreasing temperature in the EXAFS ?

WONG - 1. No
2. Yes, we did heat up the reacted specimen and obtained the same EXAFS spectrum as that at the same temperature on cooling right after the reaction.

FAST SCANNING EXAFS : AN USEFUL TOOL IN TIME-RESOLVED STUDIES OF CHEMICAL PROCESSES

C. Prieto, V. Briois, Ph. Parent, F. Villain, P. Lagarde, H. Dexpert, B. Fourman
LURE, Université de Paris-Sud, 91405 Orsay, France

A. Michalowicz
LURE and Laboratoire de physicochimie structurale, Université Paris-Val de Marne, 91000 Créteil, France

M. Verdaguer
LURE and Laboratoire de chimie des métaux de transition, Université P. et M. Curie, 75252 Paris Cedex 05, France

ABSTRACT

The X-ray absorption spectroscopy station EXAFS III (of the D1 line of the D.C.I. ring at LURE) has been modified to record data in the fast scanning mode. After a brief description of the experimental set-up, results of selected kinetics experiments are presented. Interests and limitations are shortly discussed.

INTRODUCTION

X ray absorption fine structures, X.A.F.S., when properly handled, can be used since the availability of synchrotron radiation, as a local probe of structure and electronic vacant states for practically all elements. Useful in any case, the technique becomes invaluable when applied to systems with no long range order. A continuous endeavour in instrumentation allows to extend its applicability to new systems (*in situ* experiments, time-resolved studies, dilute samples...).

The step by step X.A.F.S. acquisition was the first mode to be used and remains the most spread. It allows to record spectra with all the available detection geometries, transmission, fluorescence, electron detection. Its main limitations lie in the mechanical noise induced by the step-by-step mode and the important acquisition time (≈ 10-20 mn for a 1 keV spectrum at 10 keV). Dispersive EXAFS is a clever mean to overcome these limits. Acquisition times at the ms scale are available. Long accumulation times allow access to diluted samples. Nevertheless, neither fluorescence nor electron detection can be used. A large energy range is forbidden or at least difficult to cover by crystal's curvature. Fast EXAFS fills the gap between the two above techniques[1-4] : the high photon flux conditions of synchrotron radiation allows in principle to decrease the counting time - thanks to a quasi-continuous rotation of the stepping motor -, without significant loss of the signal/noise ratio. All the detection possibilities and flexibility of the classical step by step mode are preserved but an appreciable gain in time recording per point thus per spectrum becomes possible with two consequences : structural kinetics can be followed in the minute range of time and there is a non negligible gain of beam time.

A brief description of the fast EXAFS spectrometer and its attached equipments is given before the presentation of some examples of kinetic and static experiments.

EXPERIMENTAL

The EXAFS III station is an all-purposes EXAFS station modified for fast scanning X-ray absorption measurements. Figure 1 presents a schematic description of the experimental set-up.

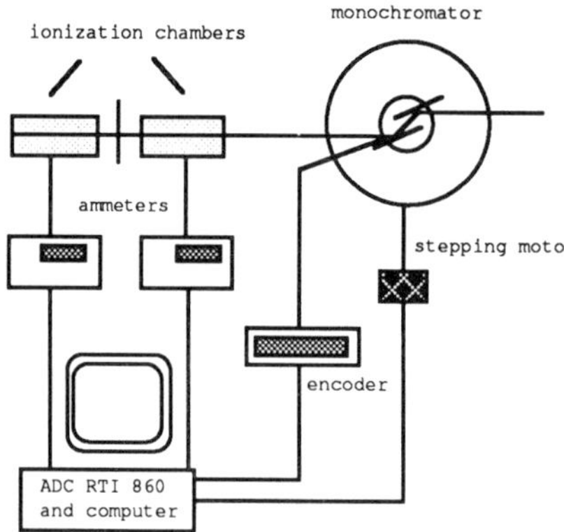

Fig. 1 : Schematic description of the station

Description

The monochromator[5] is made up of a double crystal in the asymmetric Bragg-Bragg geometry ; a translation of the second crystal, governed by a cam[6,7], provides a fixed height of the monochromatic beam during the rotation. The angle sweeping (12°-72°) is controlled by a stepping motor (0.72 arcsec/step) and the angular position is recorded by an encoder Heidenhain ROD 700 with a 1 arcsec accuracy. During the fast EXAFS acquisition, the monochromator continuously sweeps the energy domain at a programmable constant speed (from 10 steps/s to 1000 steps/s). The incident and transmitted beam intensities are measured by two ionization chambers linked with two picoammeters Keithley model 617. While on the conventional step-by-step mode the picoammeter outputs are digitalized through a voltage to frequency converter, they are sampled here by an ultra-fast analog to digital converter board during the acquisition (Analog Device model RTI 860 ; two channels are used to register each 60000 informations (I_0 and I). The ADC board is controlled by a Compac 286 computer which also drives the stepping motor of the monochromator and the angle encoder via an IEEE-488 interface.

An experimental spectrum is made of 500 points, each one resulting of the average of 120 measurements with a tunable sampling period, depending on the angular domain swept and on the rotation speed of the monochromator. In the fast EXAFS acquisition mode, a 1000 eV spectrum at the copper K-edge (8979 eV) can be obtained in about 30 sec with a data collection time of 60 msec / point. Thanks to the high photon flux provided by the synchrotron light (about 10^8 detected photons at 9 keV), the signal to noise ratio is very close to that of a step-by-step experiment (about 3.10^{10} detected photons / s), which typically takes 15 mn[2,4]. To illustrate this point, figure 2 compares the absorption spectrum of ZnO at the zinc K-edge (9659 eV) recorded (a) in the fast EXAFS and (b) in the step by step(1s/point) modes ; it is clear that the signal/noise ratio is similar in the two experiments.

Another example is given in figure 3, which presents the radial distribution function of In_2O_3 at the In K-edge (27940 eV), also obtained in both modes. This shows that even at high energy, where the photon flux is weaker than at the Zn K-edge by about

2 orders of magnitude, fast EXAFS recording can be made successfully, even when a single scan is achieved.

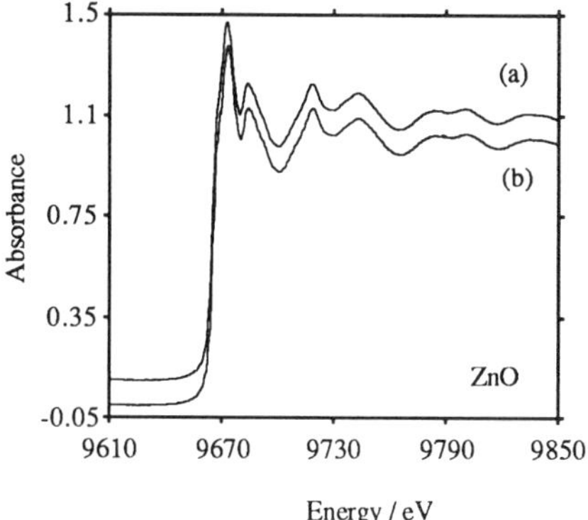

Fig. 2 : Absorption spectra of ZnO. (a) fast EXAFS (b) step by step EXAFS.

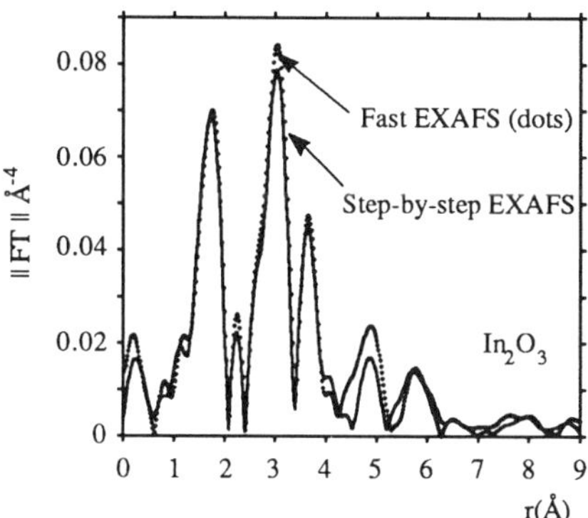

Fig. 3 : Comparison between the radial distribution functions of In_2O_3 at the indium K-edge (27940 eV), obtained in the fast EXAFS and step-by-step modes.The Fourier transform is given uncorrected for phase shift. The different peaks represent the successive oxygen and indium shells around the excited indium atoms

Attached equipments

To allow high quality *in situ* studies of kinetic experiments under various constraints (temperature, pression, controlled atmosphere, etc.), various equipments are available. For experiments on solid compounds with thermal variation (catalysts oxidation or reduction, phase transitions, etc.), we dispose of programmable furnaces (ambiant to 600°C) under controlled gaz flow and also liquid nitrogen or helium flow cryostats. For kinetic studies on liquid compounds requiring constraining cautions due to reactivity with environment, we dispose of a Teflon cell equipped with suitable windows (pyrolytic graphite, Kapton, Mylar), variable thickness and thermostatic temperature control (0°-150°C). When the chemical reaction proceeds out of the cell, the cell content can be quickly renewed by a peristaltic pump at the different steps of the reaction. Some experiments using these equipments are given below to illustrate our to-day possibilities.

To-day limitations

At the present state, the limiting parameter for a further possible gain in time in our experiment is not the photon flux but either the motor's speed at low energy or, at higher energy, the settling time (1-99%) of the electrometer used to amplify the current output of the ionization chambers (Keithley 617 : 12.5 ms in the 0-2 nA range). Figure 4 compares the absorption spectrum of $ZnSiO_4$ (willemite) at the zinc K-edge, recorded with two different acquisition time ((a) : 14 ms / point ; (b) : 330 ms / point) (from A. Ramos). It demonstrates that with a too fast acquisition, one observes a loss of resolution specially in the case of high dynamic phenomena as white lines or pre-edges transitions. The EXAFS signal, which slowly oscillates around an average value, is not affected by this effect. The limit comes when the interval time between two experimental points is of the same order of magnitude as the settling time, when the absorbance is quickly varying with energy. Highly dynamic picoammeters with shorter settling times or more efficient detectors, as silicon diodes, can be used to solve the problem.

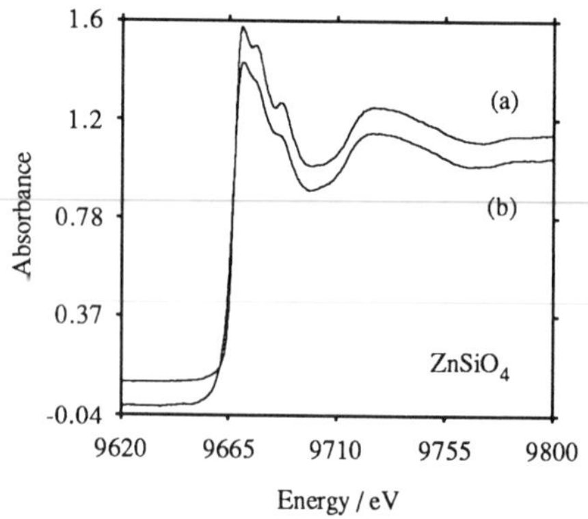

Fig. 4 : Effect of the monochromator speed on the resolution : $ZnSiO_4$ absorption spectra recorded at (a) 1.2 eV/s and (b) 28.1 eV/s.

EXPERIMENTS

Are shown some applications illustrating the specific results that the fast collection mode has brought to the understanding of different questions in reactive chemistry.

A. Redox behaviour of Cerium

1. Nature of the oxidation state

Cerium is a widely used lanthanide, particularly in ceramics materials, phosphors, etc.... It displays well-defined L_{III} and L_{II} edges, with strong intensity.

Figure 5 displays the results of a simple titration of aqueous Fe^{2+}_{aq} by aqueous Ce^{4+}_{aq} (0.1 mol l^{-1} solutions), recorded at the L_{III} edge in the above quoted cell, which allows a quick running of the sample solution from the reactor to the cell. The spectrum x = 0 (x, titrated fraction of Fe^{2+}_{aq}) is a pure Ce^{4+}_{aq} spectrum. The one with x = 1 a pure Ce^{3+}_{aq} spectrum. The large difference between the two spectra allows to follow redox kinetics with cerium ions. Furthermore, this trivial but quantitative titration experiment (realized in about 1 hour) demonstrates simply that the spectrum of Ce^{4+}_{aq} is not an actual mixture of Ce^{4+}_{aq} and Ce^{3+}_{aq}, as first proposed by Sham[8], but that the two peaks are intrinsic peaks of Ce^{4+}_{aq} since each of the experimental spectra ($0 < x < 1$) can be exactly reproduced from linear combination of initial and final spectra. A possible interpretation of the two peaks has been given by Bianconi *et al.*[9].

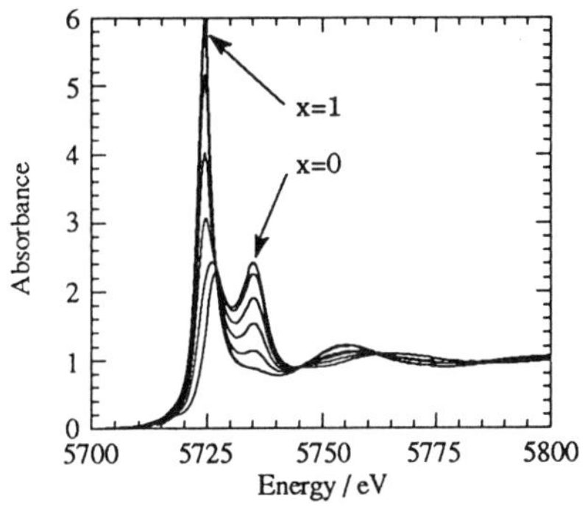

Figure 5 : Titration of Fe^{2+}_{aq} by Ce^{4+}_{aq} (0.1 mol l^{-1} solutions)

2. Continuous structural changes during ethanol oxidation by Ce^{4+}_{aq}

The second order kinetics oxidation of alcohols by aqueous cerium^{4+} can be followed by edge (Figure 6) and EXAFS measurements. The detailed analysis of the EXAFS data[10], shows the lengthening of the Ce-O bond lengths (+ 0.09 Å from 2.48 to 2.57Å) and the increase of the number of oxygen neighbours around the cerium (from 8 to 12) during the reduction process of the cerium.

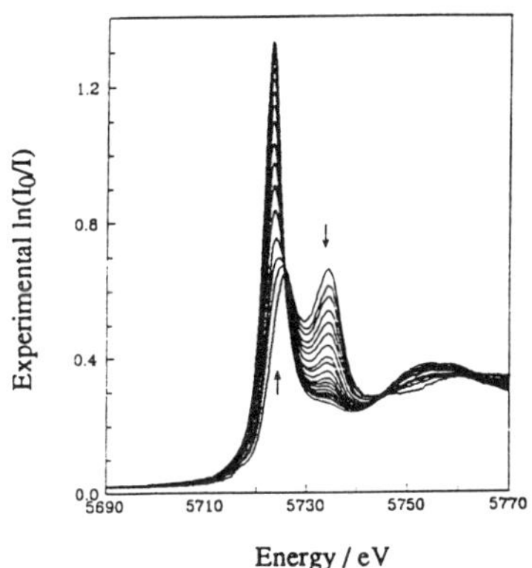

Figure 6 : Experimental spectra of the reduction of Ce^{4+}_{aq} by ethanol

B. Dioxygen uptake and release by cobalt complexes

Square planar complexes of cobalt(II), $Co^{II}L$ (Scheme 1), are known to easily bind two pyridine molecules in axial positions (Scheme 2). In both geometries, the complexes are insensitive to dioxygen. When removing by heating one of the pyridine molecule from the bis(pyridine) adduct, under dioxygen atmosphere, dioxygen uptake occurs and leads to a black diamagnetic binuclear compound, proposed to be a peroxo-Co(III) complex (Scheme 3). With further heating the Co^{II} square planar complex is obtained.

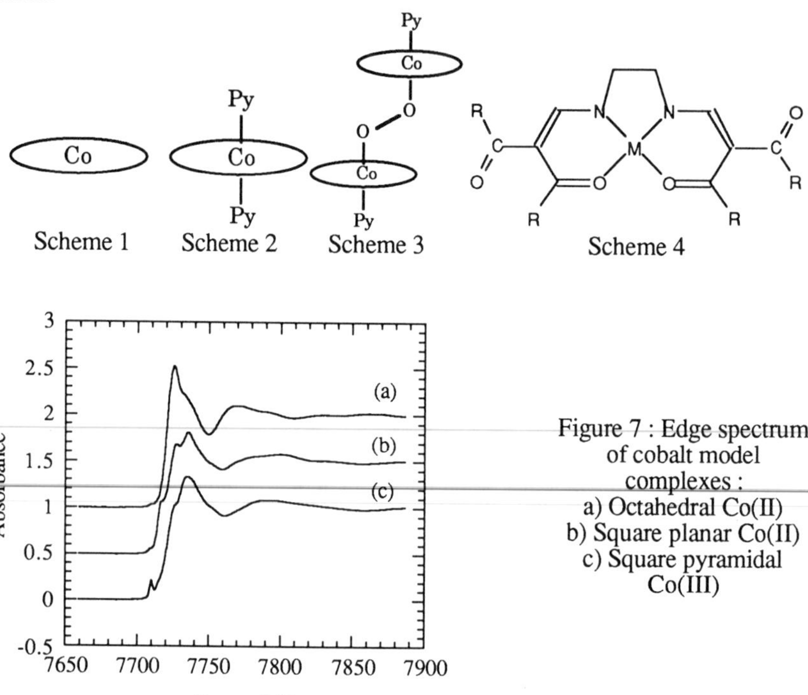

Figure 7 : Edge spectrum of cobalt model complexes :
a) Octahedral Co(II)
b) Square planar Co(II)
c) Square pyramidal Co(III)

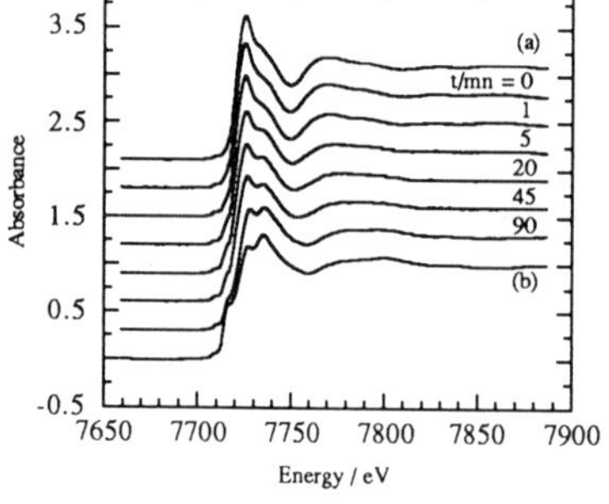

Figure 8 : Kinetics of dioxygen uptake by the cobalt complex. Are shown the octahedral (a, top) and the square planar (b, bottom) cobalt(II) complexes, together with the time-resolved spectra (time t/mn)

Fast scanning EXAFS was used to follow the edge changes during the reaction of dioxygen uptake and release by a CoL complex designed by Jäger (L = ligand of scheme 4, with R= OC_2H_5)[11], using a furnace in the 25-150°C range, at a temperature heating rate of about 1.5 °C / mn, under air stream.The total time of the experiment was 2 hours. The results are shown in Figures 7 and 8. Figure 7 displays three models (octahedral bis(pyridine) $Co^{II}L$ (0.1 mol l^{-1}) (a), solid state square planar $Co^{II}L$ (b), and solid state square pyramidal $Co^{III}L(-CH_3)$ (c). Figure 8 shows the kinetics results at the edge. An interesting feature of the edge spectra of this first experiment is that the III oxidation state (Figure 7, (c)) is non observable at any step of the kinetics, contrary to what is generally postulated by the Py-$\{Co^{III}L\}(O_2^{2-})\{Co^{III}L\}$-Py formulation.

C. Bimetallic Pt / Re Catalysts under dihydrogen and hydrocarbon flow

Reforming is a heavy industry process, where bimetallic modified catalysts may be used. Two particular points of interest are the preparation of the catalysts and their coking by hydrocarbons. A feasibility experiment to follow the reaction *in situ*, under gaseous reactants flow was recently performed both at platinum and rhenium edges[12]. We present in Figure 9 the results obtained at two stages of the experiment, *i.e.* the preparation of the catalyst from a bimetallic Pt(1%)-Re(1%) chlorinated deposit supported on γ-alumina under dihydrogen flow and different pressures (1 and 3 bars) , using a furnace in the 25°C-500°C range. Total experimental time is 3 hours.

The absorption spectra were recorded continuously during the experiment and only a few selected data at the platinum edge are shown in Figure 9. The Fourier transform clearly displays that at this stage, the bulk of oxidized platinum is reduced to disordered metallic platinum. In a second step, gaseous *n*-heptane was flowed with dihydrogen. The appearance in the Fourier Transform of a Pt-C shell at low R demonstrates that coking occurs after 2 hours at 300°C (but not at 460°C).

This example shows that, in this kind of time range, fast EXAFS allows to follow "continuously" *in situ* kinetics processes with rather dilute samples, using the simple surroundings of step by step EXAFS, without the time limitations intrinsic to such a technique.

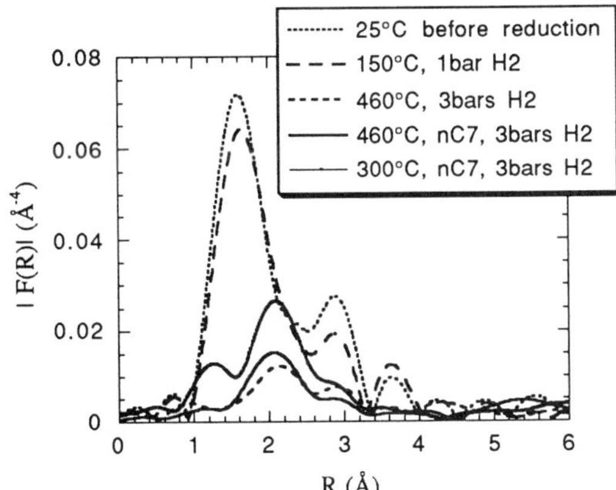

Figure 9 : Fourier transforms at the Platinum L_{III}-edge of a bimetallic Pt-Re catalyst studied *in situ* at different stages of the chemical process : reduction and coking

CONCLUSION

The conventional E.X.A.F.S. station was transformed to record structural kinetics. Experiments at different energies demonstrate that this scanning mode can bring specific results non allowed when using the conventionnal step-by-step mode. The limits of this collecting mode can be pushed further (see for example Ref 2 and Frahm *et al.*[13,14].

Instrumental developments are currently in progress at LURE to reduce recording time not only to enhance the available range of chemical rates, but also to allow a more efficient use of the beam. As a matter of fact, besides the described time-resolved experiments, and the ones collected in our station reported in this conference[15,16], one of the most useful applications of such a station is the possibility for users to investigate more samples in a given time or to get higher quality data, by increasing the counting statistics.

REFERENCES

1. R. Frahm, *Nucl. Inst. Meth.*, **A270**, 578 (1988)
2. P. Lagarde, M. Lemmonier and H. Dexpert, *Physica B* **158,** 337 (1989)
3. R. Frahm, *X-ray absorption fine structure*, S.S. Hasnain Ed., Ellis Horwood, New York, p. 732 (1990)
4. C. Prieto, P. Parent, F. Le Normand, P. Lagarde and H. Dexpert, Conference Proceedings Vol. 25, *"2nd European Conference on Progress in X-Ray Synchrotron Radiation Research"*, A. Balerna, E. Bernieri and S. Mobilio (eds.), SIF, Bologna, 441 (1990)]
5. J. M. Dubuisson, J. M. Dauvergne, C. Depautex, P. Vachette and C. E. Williams, *Nucl. Inst. Meth.*, **A246**, 636 (1986).
6. M. Lemonnier, O. Collet, C. Depautex, J.M. Esteva and D. Raoux, *Nucl. Instr. and Meth.* **152**, 109 (1978)
7. J. Goulon, M. Lemonnier, R. Cortes, A. Retournard, D. Raoux, *Nucl. Instr. and Meth.* **208**, 625 (1983)
8. T.K. Sham, a) *J. Chem. Phys.*, **79,** 1116 (1983) ; b) *Phys Rev. B*, **40**, 6045, (1989)
9. A. Bianconi, A. Marcelli, H. Dexpert, R. Karrnatak, A. Kotani, T. Jo, J. Petiau, *Phys. Rev. B*, **35,** 806 (1987)
10. C. Prieto C., P. Lagarde, H. Dexpert,V. Briois , F. Villain, M. Verdaguer
a) *Meas. Sci. technol.* **2**, (1991), under press
b) *J. Chem. Phys. Solids*, under press
11. a) E.G. Jäger, M. Rost, M. Rudolph, *Z. Chem.*, **29**, 415 (1989) ; b) E. Jäger, F. Villain, M. Verdaguer, work in progress
12. N. Guyot-Sionnest, F. Villain, D. Bazin, H. Dexpert, F. Le Peltier, J. Lynch, J.P. Bournonville, *Catal. Lett.*, under press
13. R. Frahm, T.W. Barbee, W. Warburton, *Phys. Rev. B*, **44**, 2822 (1991)
14. J., Wong, .E.M. Larson, J.B. Holt, P.A. Waide, B. Rupp, R. Frahm, G. Nutt, *this conference*
15. M. Weibel, J. El Fallah, F. Villain, F. Le Normand, *this conference*
16. C. Cartier, E. Prouzet, A. Tranchant, R. Messina, F. Villain, H. Dexpert, *this conference*

QUESTIONS

D. Mills : Why does the quick EXAFS technique has less noise than the standard stepping technique ?
Answer : Everything being equal, the electronic noise is the same. The mechanical noise is decreased due the microstepping technique.
J. Wong : How many steps by degrees of Bragg angle is your set-up at LURE for QEXAFS ?
Answer : 5000 steps per degree.
S. Leach : Have you correlated your results of time-resolved chemical process with those obtained by other techniques, for examples by rapid Raman spectroscopy ?
Answer : No, but it could be done when the chemical system at hand deserves such a comparison.

DISCUSSION

MILLS - Why does the quick EXAFS technique have less noise than the standard stepping technique ?

VERDAGUER - Everything being equal, the electronic noise is the same. The mechanical noise is decreased due to the microstepping technique.

WONG - How many steps by degrees of Bragg angle is your set-up at LURE for QEXAFS ?

VERDAGUER - 5000 steps per degree.

LEACH - Have you correlated your results of time-resolved chemical process with those obtained by other techniques, for examples by rapid Raman spectroscopy ?

VERDAGUER - No, but it could be done when the chemical system at hand deserves such a comparison.

COMBINED X-RAY ABSORPTION SPECTROSCOPY AND X-RAY POWDER DIFFRACTION

Andrew J Dent and G Neville Greaves
SERC Daresbury Laboratory
Warrington WA4 4AD, UK

John W Couves and John M Thomas
The Royal Institution of Great Britain
21 Albermarle Street, London W1X, UK

ABSTRACT

The complementary nature of x-ray absorption fine structure (XAFS) spectroscopy and x-ray diffraction (XRD) is described. In particular XAFS records the local structure whilst XRD detects the long range crystallinity enabling the heterogeneity of materials like single phase catalysts to be explored. Both measurements can be combined to facilitate novel *in situ* experiments. We have used a horizontal energy dispersed x-ray beam and a photodiode array to detect transmission XAFS and a curved position sensitive detector positioned in the vertical plane to record XRD. This arrangement has been used to follow the formation of mixed oxide catalysts.

INTRODUCTION

The structure of a heterogeneous material is usually characterised through independent measurements. X-ray absorption fine structure (XAFS) spectroscopy is increasingly used as it is chemically selective.[1] Although the atomic environments of particular elements can be obtained by tuning to the corresponding x-ray absorption thresholds, each radial distribution function is restricted to neighbours generally within 5 or 6 Å of the excited atom. X-ray diffraction (XRD) on the other hand detects the average crystallinity of the material. High resolution powder diffraction techniques are now providing solutions for a variety of crystal structures.[2] The specific properties of disordered materials like single phase catalysts, however, require a knowledge of both short and long range order. In this case separate measurements of XAFS and XRD can be made and the two analysed in conjunction. Recently this approach has been used with success in structurally characterising an ion exchanged zeolite.[3] Scanning monochromator and scanning angle arrangements were used on specimens prepared *in situ* in different experiments. The reproducibility of *in situ* conditions for separate measurements, however, is not easy and the desirability of detecting XAFS and XRD on the same specimen under the same conditions is self-evident. Since a Debye-Scherrer pattern exists whenever an XAFS measurement is made a combined experiment in principal is trivial. For some time now energy dispersed x-ray beams have been used to study the structural chemistry of materials in the time domain.[4] Transmission spectra can be readily obtained using a suitably adapted photodiode array (PDA) system.[5] As all wavelengths are collected at once dynamic measurements can be made. A variety of curved position sensitive detectors (PSD) are available for x-ray powder diffraction. As the XRD pattern is detected in angle dispersed mode this type of detector is also well-suited to *in situ* measurements. In this paper we will describe how PDA and curved PSD systems can be brought together in a combined XAFS-XRD experiment.

IN SITU ENERGY DISPERSIVE XAFS

In Fig. 1 a fixed arrangement is illustrated which has been devised for *in situ* experiments. This is based on an energy dispersed geometry developed for dynamic XAFS experiments [6] but has been augmented with a curved PSD to enable powder diffraction patterns to measured at the same time. The combined data acquisition system is shown schematically in Fig. 3. The dispersive XAFS part will be described and illustrated first.

The bent triangular crystal selects a broad bandpass, ΔE, from the incoming white x-ray beam which is geometrically dispersed and horizontally focussed. It subtends an angle, $\Delta\theta$, at the sample given by

$$\Delta\theta = \tan\theta \frac{\Delta E}{E} \qquad (1)$$

where θ is the Bragg angle of the monochromator. For a a bandpass of 250 eV at 8.3 keV, for instance, the focussing angle, $\Delta\theta$, of a Si [111] monchromator (θ = 13.8^0) is 0.43^0. ΔE is chosen to straddle the absorption threshold, E, of the selected element and the transmission spectrum is recorded on the PDA placed after the sample stage. The spectral resolution is also given by eqn. 1. in this case $\Delta\theta$ is the convolution of the Darwin width of the monochromator (3.4 10^{-5} at θ = 13.8^0) convoluted with the angle subtended by the source (1.8 10^{-3} for station 7.4 at the SRS) and angle subtended by a 25 μm pixel (typically 3 10^{-5}). For a Si [111] monochromator the corresponding resolution at 8.3 keV is 2 eV. Time resolved XAFS measured in this way is illustrated in Fig. 2 where changes in the metal environment of a heterogeneous catalyst are recorded. The Ni K-edge of nickel oxide supported on alpha alumina is shown during thermal reduction in a stream of hydrogen.[8] The reduction of nickel takes place over a period of 15 min. as the temperature reaches 450^0 C. Analysis of the reduced fine structure just prior to this closely matches the structure of Ni in crystalline NiO, with 6 nearest neighbour oxygens surrounded by 12 nickels. Once reduction is complete XAFS spectra are consistent with a disordered form of Ni metal. The dramatic change in fine structure oscillations can clearly be seen in Fig.2, along with the chemical shift in the x-ray absorption threshold.

The PDA used to detect the polychromatic XAFS shown in Fig. 2 is a cooled RL1024S Reticon array operating at 220 K.[7] The data acquisition system for the PDA is VME based and interfaces with the user through the main control computer, a Microvax II. The image of the transmission spectrum is read out serially at a single channel readout rate of 200kHz resulting in a response time of approximately 2 mS over 512 pixels. In order to obtain sufficient statistics in the absorption signal to resolve the fine structure much longer accumulation times are generally required for all but the most concentrated systems. The total exposure time for a particular frame might be around 1s and the duration between frames might be anything from seconds to minutes. The start and stop parameters for the PDA set by the user are administered by the instruction sequencer and recorded spectra are stored in the VME memory as indicated in Fig. 3. A major improvement in operational flexibility and also in quality of data has been obtained by separating the experimental operations defined by the user from those affecting the front-end electronics of the PDA. These are fixed once the

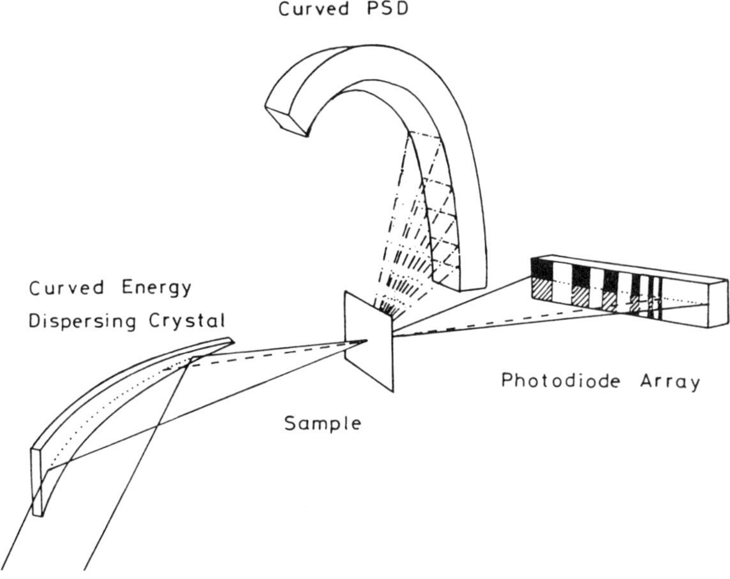

Fig. 1. Schematic diagram showing the arrangement for recording *in situ* energy dispersive XAFS and powder XRD.

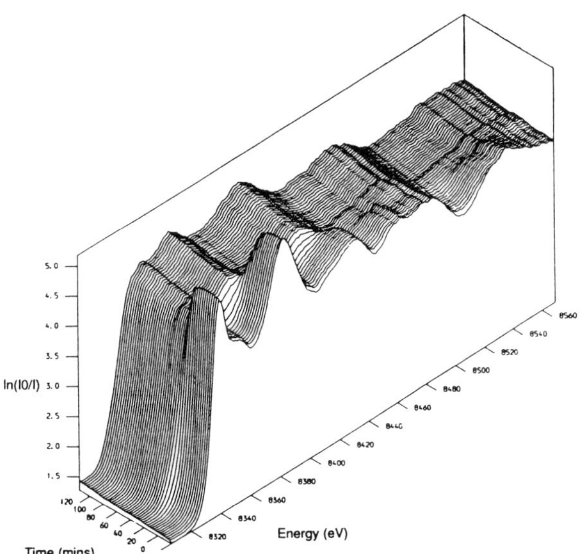

Fig. 2. Raw Ni K-edge data following the reduction of NiO on alpha alumina in a 10%/90% hydrogen/nitrogen mix.The temperature was raised from 200°C to 500°C in 60 mins and then held for a further 60 mins.

sequence and readout parameters have been optimised and a local sequencer facilitates these operations interfacing with the PDA via level translators. It also provides registration signals to the ADC (16 bits at 500 kHz). The local sequencer and level translator units together with the preprocessing electronics are placed in a screened environment adjacent to the detector. The detailed operation of the PDA is controlled by a C program resident in the VME system which gives the correct timing sequence to the PDA and also ensures that the resulting data is stored in the VME memory. A fuller description of the design and operation of this PDA system are given elsewhere.[9]

COMBINED XAFS AND POWDER XRD ARRANGEMENT

The x-ray diffraction from a powdered sample can be measured alongside energy dispersive XAFS using a curved position sensitive detector (PSD). This has been placed vertically above the sample stage at the monochromator focus in Figs. 1 and 3. Although the XRD and XAFS patterns can in principal be collected simultaneously, in practice the large bandpass, ΔE,required for XAFS broadens the diffraction lines and the also causes fluorescence adding considerably to the XRD background. These effects can be eliminated by making the measurements in sequence with different ΔE's. A movable slit has therefore been included which opens when the PDA is operating and narrows when the curved PSD is operating. In particular a bandpass of 700 eV has been used for the XAFS measurements reported here with a reduction to 30 eV for XRD. The reduced bandpass for XRD results in a focussing angle (eqn. 1) of ~ 0.05^0 or $9\ 10^{-4}$. Whilst this exceeds the monochromator Darwin width and source size contributions to the angular dispersion discussed above, it is well-matched to the angular resolution of the curved PSD. An INEL detector has been used offering an angular range of 120^0 and an angular precision of ~0.04^0 or $7\ 10^{-4}$.

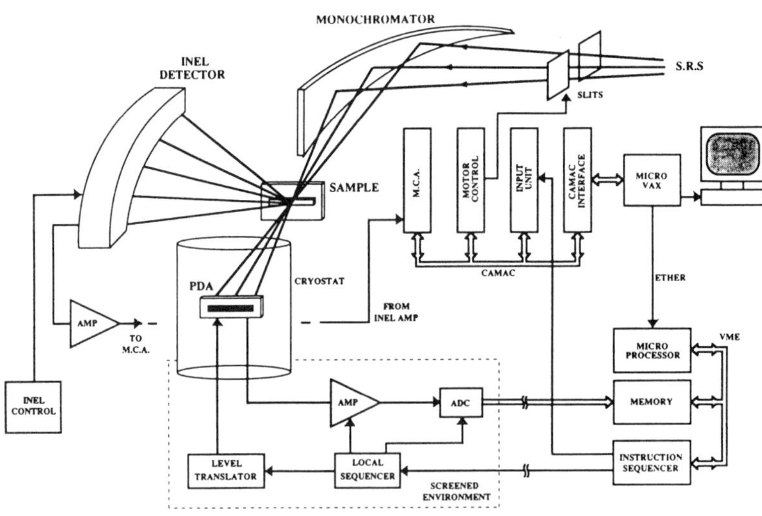

Fig. 3. Schematic showing the control of the combined XAFS-XRD experiment.[10]

Returning to Fig. 3 the output of the INEL curved PSD is fed to an MCA. This and the motor control for the monochromator entrance slits are CAMAC driven. XRD powder patterns from the INEL are stored in the Microvax II memory. These are appropriately registered together with the PDA files stored in the VME memory so that XRD and XAFS data can be considered in the correct timing sequence. Once the sequence and particular time frames for the PDA and PSD have been chosen, the operation of the experiment is controlled by the VME microprocessor. This provides the PDA with its timing sequence and routes the data to the VME memory. A signal is then sent to the Microvax via CAMAC to 'stop down' the horizontal beam incident on the monochromator. Data acquisition from the INEL commences utilising the MCA and when this is finished it is stored in the Microvax II memory. The motorised slits are then returned to their original position and the routine pauses, if necessary, before the VME system recommences data collection with the PDA. Further details of the instrumentation for combined XAFS-XRD are given in ref 10.

XAFS AND XRD OF MIXED OXIDE CATALYSTS

We illustrate the utility of the dual measurement of XAFS and XRD by examining the generation of two mixed oxide catalysts: Cu - Mn oxide and Cu - Zn oxide.

The copper manganese oxide system is a powerful oxidant even at room temperature. The precursor material, is a monophasic carbonate in which the two metals are intimately mixed. This can be decomposed by stepwise heating in air. By 470^0 C the structure is almost completely amorphous and the material is an active catalyst (Hopcalite). At higher temperatures it recrystallises to a spinel phase of $Cu\,Mn_2\,O_4$ in which it is believed some Cu^{2+} is reduced to Cu^+ residing on tetrahedral sites, leaving Mn^{3+} and Mn^{4+} ions in octahedral sites. The catalytic activity of MnO_2 promoted by the incorporation of copper then may well be due to the redox couple:

$$Cu^{2+} + Mn^{3+} = Cu^{+} + Mn^{4+}$$

The Cu K-edges are shown in Fig. 4 following part of the decomposition of the carbonate from 300^0 C to 350^0C. An edge shift to lower energies can just be discerned suggesting the formation of Cu^+ ions. There are some small changes in the shape of the first minimum, but more obviously there is an increase in the period of the initial fine structure oscillation indicative of a decrease in the Cu - O bond length. This would be expected if Cu^{2+} ions, initially occupying octahedral sites were reducing to Cu^+ ions and began to moving to tetrahedral sites.

The corresponding powder patterns for the decomposition of the sample of $CuMn_2(CO_3)_3$ are plotted in Fig. 5. The signal quality is poor and this is due to the substantial fluorescence from Mn excited at 8.89 keV. None the less the diffraction lines significantly decrease over this temperature range, clearly anticipating transformation to the amorphous Hopcalite phase.

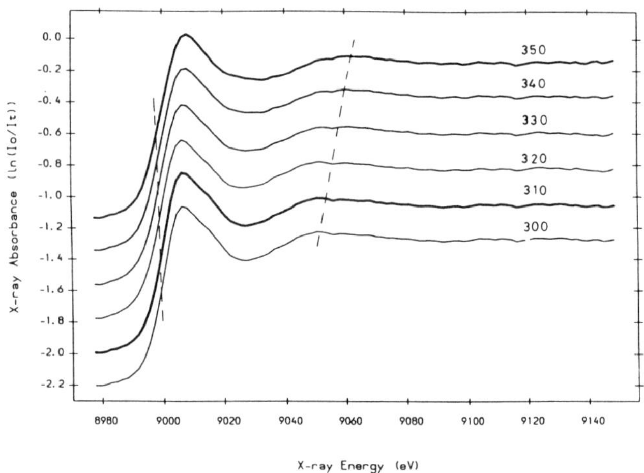

Fig. 4 Cu K-edges measured during the decomposition of $CuMn_2(CO_3)_3$ in air at the temperatures in deg C shown. The dashed lines indicate the change in the edge position and the shift in the position of the first XAFS peak following the white line.

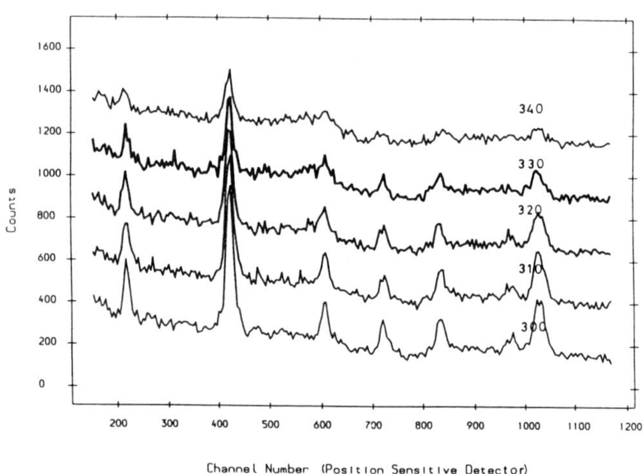

Fig. 5. Powder patterns following the decomposition of $Cu\ Mn_2\ (CO_3)_3$ taken with 1.3885 Å x-rays. The 2 theta scale is uncalibrated. The XRD profiles were obtained alongside the Cu K-edge XAFS shown in Fig. 4 using the arrangement shown in Fig. 1. The large background signal is due to fluorscence from Mn excited by 8.99 keV radiation.

Cu - Zn oxides are catalysts for methanol synthesis. Aurichalcite, $(Cu_{1-x} Zn_x)(CO_3)_2 (OH)_6$, is a suitable precursor to generate a highly dispersed catalyst. It is a layered hydroxycarbonate containing a homogeneous distribution of Cu and Zn. Aurichalcite for which Cu/Zn = 5/95 was first calcined in dry air and then reduced in a 10% H_2/N_2 mix. Cu K-edge XAFS and XRD at 1.3885 Å were measured together and results are shown in Figs. 6 and 7. The temperature was ramped at 2^0C/min from ambient. At each temperature XAFS scans were accumulated for 1 min. and XRD patterns for 5 mins. A gap of 8 mins was left between each combined measurement. Three paired results obtained during decomposition are plotted in Fig. 6. The Cu K-edge shows little change during calcination whilst the XRD profile changes dramatically, the initial complex aurichalcite structure giving way to the simpler metal oxide formed at 450^0 C. As in the formation of Hopcalite, aurichalcite is calcined at first into a highly disordered phase around 300^0 C, at which stage most of the interlayer carbonate and hydroxyl groups have decomposed. During this metamorphosis in long range crystalline order the immediate environment of Cu recorded by the XAFS is clearly largely unaffected. A wide bandpass was used which just included the Zn threshold in addition to the Cu XAFS and little or no changes were observed for the major cation either.

When the calcined material was further heated to 500^0 C in the reducing atmosphere, significant changes took place both in the XRD and in the Cu K-edge XAFS as shown in Fig. 7. The initial XRD scans are dominated by the powder pattern of ZnO. The 2 theta axis was calibrated against an Al standard and the diffraction lines matched with values from the JCPDS database. Remaining lines are due at first to CuO. However, when reduction takes place the CuO diffraction lines disappear to be replaced by those of fcc Cu metal. These are marked in Fig. 7. The persisting ZnO pattern can be clearly seen. The changes taking place on reduction evident in the powder patterns plotted in Fig. 7, can also be seen in the Cu K-edge XAFS. The initial resonance typifying CuO vanishes to be replaced by the familiar two-peaked threshold of metallic Cu. There is a chemical shift in the threshold to lower energies and substantial changes in the XAFS period occurs, similar to the changes occurring in Ni O dispersed on alpha alumina referred to earlier (see Fig. 2).

CONCLUSIONS

The combined detection of an x-ray absorption spectrum and an x-ray diffraction pattern has obvious advantages for studying structural changesin the short and long range order in materials. The fixed geometry described here is particularly suited to following these dynamically *in situ.* Other dual detection systems can be envisaged using a similar combination of PSD's. Small and large angle scattering could be measured for instance using a PDA and a curved PSD respectively. Whilst the XAFS-XRD technique has been illustrated with reference to the preparation of heterogeneous catalysts, it would evidently also be useful in studying devitrification of glasses for example or the preparation of solids from gels. By employing glancing angles of incidence surface corrosion processes could be examined *in situ.*

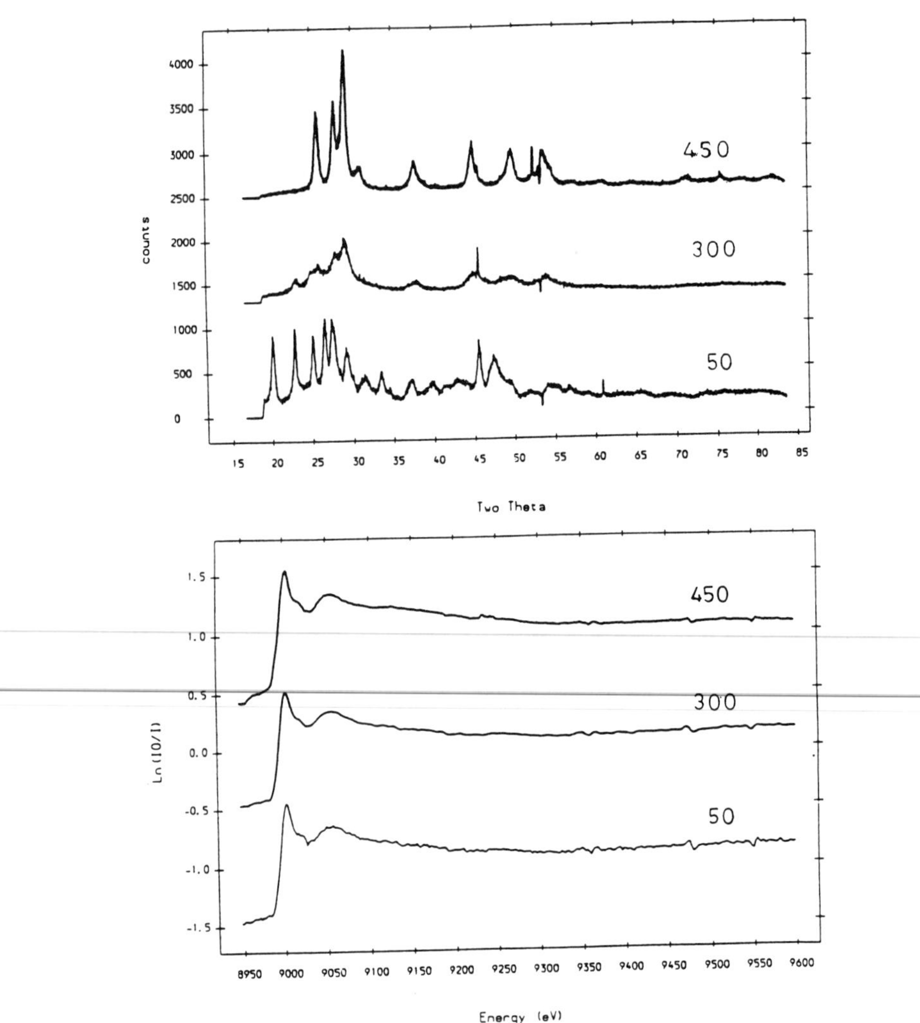

Fig. 6 *In situ* combined XAFS-XRD measurements of the decomposition of aurichalcite ($Cu_{1-x}Zn_x)(CO_3)_2(OH)_6$ to the Zn - Cu oxide catalyst. Powder patterns are shown above (1.3885 Å) and Cu K-edge XAFS below. The temperatures corresponding to the scans are in deg C.

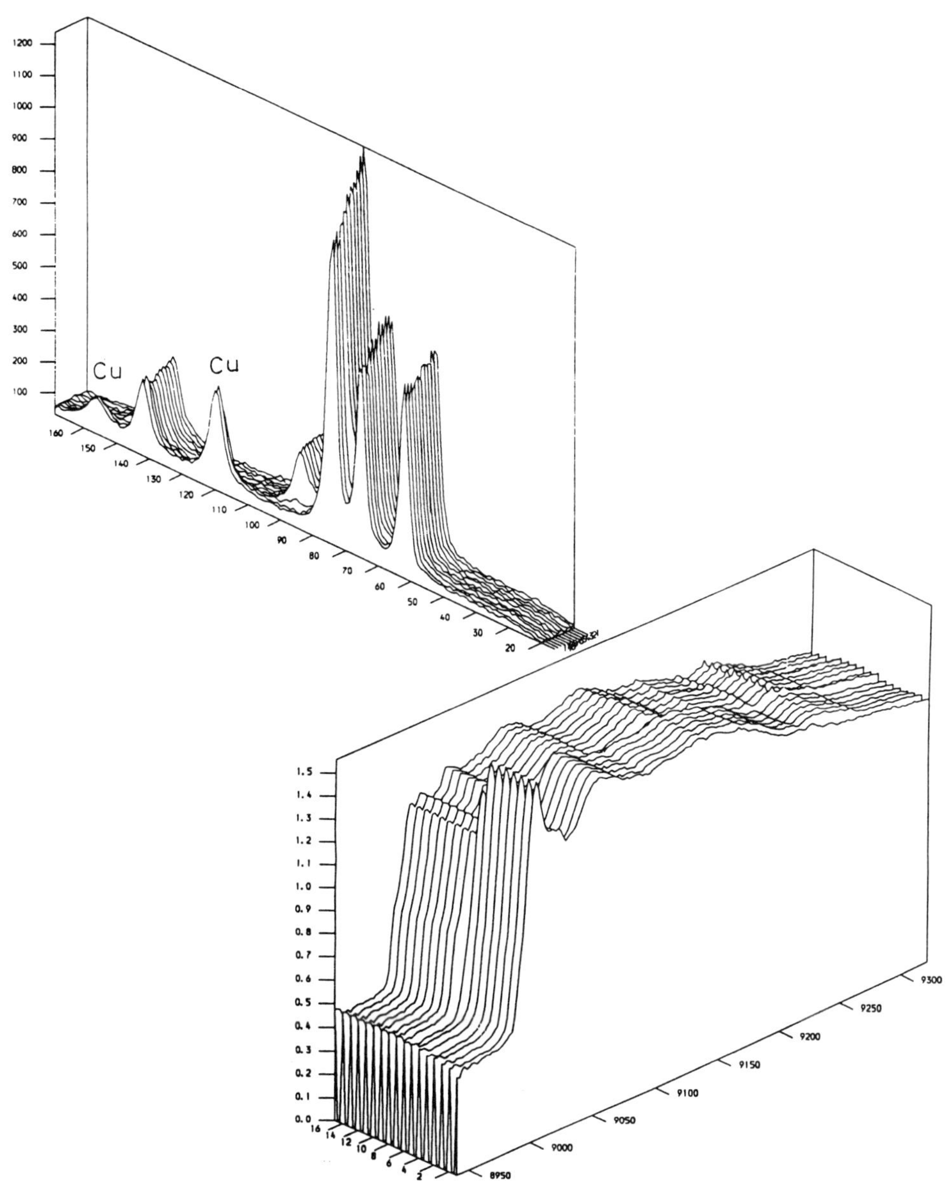

Fig. 7 In situ combined XAFS - XRD measurements following the thermal reduction of a 5/95 Cu-Zn mixed oxide in a 10% H_2 /N_2 mixture. The diffraction lines for metallic Cu are indicated.

ACKNOWLEDGEMENTS

We would like to thank SERC for provision of synchrotron radiation facilities and many of our colleagues for their cooperation, particularly Gareth Baker, Gareth Derbyshire, Richard Farrow, Richard Jones, Colin Morrell, Srinivasan Natarajan, Christine Ramsdale, Gopinathan Sankar, Matthew Wells and Paul Wright.

REFERENCES

1. C.R.A. Catlow and G.N. Greaves, *Applications of Synchrotron Radiation*, Blackie and Son Ltd, Glasgow and London, 1990, Chaps. 6, 7 and 8.

2. C.R.A. Catlow and G.N. Greaves, *Applications of Synchrotron Radiation* , Blackie and Son Ltd, Glasgow and London, 1990, Chap. 2.

3. E. Dooryhee, G.N. Greaves, A.T. Steel, R.P. Townsend, S.W. Carr, J.M. Thomas and C.R.A. Catlow, "Structural studies of High-area zeolitic Absorbents and Catalysts by a Combination of High Resolution X-Ray powder Diffraction and X-ray Absorption Spectroscopy," *Faraday Discuss. Chem. Soc.* 89, 119-136, 1990.

4. E. Dartyge, C. Depautex, J.M. Dubuisson, A. Fontaine, A. Jucha, P. Leboucher and G. Tourillon, "X-ray Absorption in Dispersive Mode; A New spectrometer and a Data Acquisition System for Fast Kinetics," *Nucl. Instr. and Meth.*, A246, 452-460, 1986.

5. G.E. Derbyshire, W.I. Helsby, A.J. Dent, S.A. Wright, R.C. Farrow, G.N.Greaves, C. Morrell and G.J. Baker, "Current and Future Energy dispersive EXAFS Detector Systems," *Adv. X-Ray Analysis* 34 in press 1991.

6. R.P. Phizackerly, Z.V. Rek, G.V. Stephenson, S.D. Conradson, K.O. Hodgson, T. Matsushita and H. Oyanagi, "An Energy-Dispersive Spectrometer for the Rapid Measurement of X-ray Absorption Spectra Using Synchrotron Radiation," *J. Appl. Phys*,161, 220-232, 1983.

7. N.M. Allinson, G. Baker, G.N. Greaves and J.K. Nicoll, "PDA System for Energy dispersive EXAFS," *Nucl. Instr. and Meth.*, A266, 592-597, 1988.

8. G. Baker, A.J. Dent, G.E. Derbyshire, G.N. Greaves, C.R.A. Catlow, J.W. Couves and J.M. Thomas, "Time Resolved Structural Studies of Nickel Exchanged Zeolite Y and Nickel Oxide Using Energy dispersive EXAFS," *X-Ray Absorption Fine Structure*, Ellis Horwood, London,1991, p738-741.

9. D. Bogg, A.J. Dent, G.E. Derbyshire, R.C. Farrow, G.N. Greaves, W.I. Helsby, C. Morrell, C.A. Ramsdale and M.P. Wells, "A VME Based Photodiode Array System for Energy Dispersive EXAFS," 4th *Int. Conf. on Synchrotron Radiation Instruementation*, Chester , July 15/19, 1991.

10. A.J. Dent, G. Bushnell-Wye, J.W. Couves, R. Cernik, G.N. Greaves and J.M. Thomas, "Combined Energy Dispersive X-ray Absorption Spectroscopy (XAS) and X-ray Powder Diffraction," 4th *Int. conf. on Synchrotron Radiation Instruementation*, Chester, July 15/19 1991.

DISCUSSION

ELAND - Could you explain how the position of arrival of X-rays is encoded in your 120° position-sensitive detector ?

DENT - The INEL detector consists of a sharp blade at 9600V in a 6.4 bar atmosphere of He/Me. The detector works in the "avalanche" region of a Geiger-Müller tube and so when an X-ray enters the detector an avalanche of electron are produced which travel down a delay line. The arrival time of the pulse is measured at each end of the delay line and hence the position of the X-ray can be found. The accuracy is ± 0.5mm corresponding to a 0.1° diffraction.

WONG - What is the angular range of your detector for TR-XRD experiments ?

DENT - The angular range of the XRD detector is 120°. The range of measurement ~18-100°; the lower limit due to having the detector "high" enough for the direct beam to pass beneath for the XAS measurement, the upper limit being due to the furnace geometry.

IN SITU EVOLUTION OF LITHIUM INTERCALATION WITHIN V_2O_5 IN FUNCTION OF TIME

C. CARTIER(1), E. PROUZET(2), A. TRANCHANT(3), R. MESSINA(3), F. VILLAIN(1), H.DEXPERT(1)

(1) LURE (C.N.R.S., M.E.N., C.E.A.), Université de Paris-Sud, 91405 Orsay, France
(2) Institut des matériaux, 2 rue de la Houssinière, 44072 Nantes cedex 03, France
(3) LECSO, UM 28, C.N.R.S., BP 28, 2, Rue H. Dunant 94320 Thiais, France

Abstract

With a dedicated cell which allows us to investigate on line the structural modifications occuring during different electrochemical reactions by X-ray absorption spectroscopy, we have been able to follow directly the successive steps of the inter/de-intercalation process of lithium within the V_2O_5 matrix. The in-situ measurements performed on $Li_xV_2O_5$ compounds in the range ($0<x<0.8$) allows to point out small and significant modifications of both electronic state and atomic surroundings of vanadium. Combining the X-ray absorption results, we suggest a model for the evolution of the cathode material.

Introduction

Systems based on alkali metal anodes (such as lithium) and solid cathodes show considerable promises for high energy density storage batteries. Especially, the solid cathodes based on the insertion of the alkali cation in the host lattice, together with this of the electron, insertions which occur without producing new phases or disrupting this lattice as long as the lithium content stays inside a certain homogeneous range are very attractive since such a behaviour enhances the reversibility of the electrochemical processes.

Among the solid cathodes, V_2O_5 is of interest since it can reversibly incorporate, at high potentials (upper to 3V vs Li) one lithium per mole of V_2O_5 [1 - 4]. However, a deeper intercalation of lithium (x>1 lithium for V_2O_5) or repeated intercalations during charge/discharge cycles cause serious structural and morphological damages, and a significant decrease of the reversibility of the electrochemical processes is then simultaneously registered [5-6].

Given the considerable interest in the effect of the structural changes on the electrochemical behaviour of the material cathodes, especially in the case of V_2O_5, the interest of the X-ray absorption spextroscopy (XAS) becomes evident since the technique brings stereochemical informations around an absorbing atom. So, we have already used it for studying V_2O_5 and $Li_xV_2O_5$ compounds generated during its electroreduction [7].

However, ex-situ absorption measurements require numerous samples (one for each lithium content) and consequently, in-situ X-ray absorption experiments appears to suit best for our studies. Major advantages of our in-situ X-ray measurements are (i) to follow precisely the structural and electronic changes, during the electrochemical treatment on the same sample and (ii) to avoid the problems of eventual structural rearrangements, upon removal of the sample from the electrochemical cell.

Consequentely, we designed a cell which allows us to investigate on line the structural changes occuring during different electrochemical reactions by XAS. We have been able to follow directly the successive steps of the inter/de-intercalation process of the lithium within the V_2O_5 matrix. Ex-situ measurements previously performed on the electrogenerated $Li_xV_2O_5$ compounds [7-8] are used as references and compared to these in-situ XAS results.

Experimental section

XAFS measurements

X-ray absorption spectra have been recorded at the vanadium K absorption edge on the fast EXAFS station (EXAFS III) at LURE (DCI ring using 1.85 GeV positrons with an average intensity of about 250 mA). This station is equipped with a two crystals Si(311) monochromator for harmonics rejection Helium-Neon and air filled ionization chambers were respectively used to mesure the flux intensity before and after the sample.

The XANES spectra were recorded from 5440 to 5600 eV, with one second accumulation time per point every 0.30 eV. The spectrum of a 6 µm thick vanadium metallic foil was recorded before or after each unknown XANES spectrum. This insures an energy accuracy determination of 0.30 eV. The EXAFS spectra were recorded over 1000 eV.

Due to the signal to noïse ratio quality, we have operated in an intermediate range of data scanning, which correponds to 5 minits per spectrum (100 eV range). We add two spectra before Fourier transforming. Even if this period of time seems to be large, it's however small henought to correspond to the relaxation time of our system. So that we are sure to get a reliable study between structural evolutions and physical property behaviour.

Electrochemical cell

Our cell has been designed in order to allow X-ray absorption transmission mode recording during the electroinsertion process. The electrochemical system (Figure 1) is inserted between two stainless steel plates (60x60x4 mm) which act also as current collectors, separated by a Viton joint. The positive holds 3.10^{-5} moles of V_2O_5 mixed with 180 wt% of amorphous carbon and PTFE solution (SOREFLON) to obtain a latex, the negative is a 100 µm thick lithium metallic foil.

Kapton windows (for X-ray transparency) avoid the latex to compres the latex. A Viledon disk is used as an electrolyte container [$LiAsF_6$ 1M in PC/DME 50/50] and a molecular Celgard foil prohibites electrical short circuit. The intercalation has been performed by successive 5mn discharge under constant current (I=90 µA) followed by a 15 mn relaxation (time necessary to obtain a constant value for the potential) step cycles.

X-ray absorption : data processing

All the XANES spectra have been treated in the same way : after an energy calibration, the background has been substracted through a linear function determined by a least-square fitting of the pre-edge experimental points. All the spectra have been normalized by taking as unit the absorbance value at 5600 eV.

The EXAFS analysis was performed following a classical way already decribed [8-9]. The whole procedure used the EXAFS chain programs written by A. Michalowicz [10].

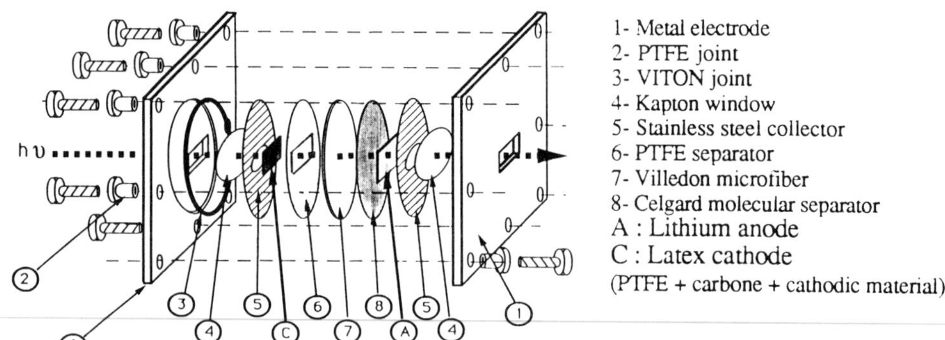

Figure 1. In-situ cell for X-ray absorption study of lithium electrochemical intercalation.

Results of the ex situ study

All the electroreduction domain ($Li_xV_2O_5$ compounds with $0<x<2.4$) has been investigated.

XANES

Near edge spectrum gives informations on the electronic structure of the absorbing atom. Vanadium K edge XANES spectra obtained for different lithium contents are reported on figure 2.

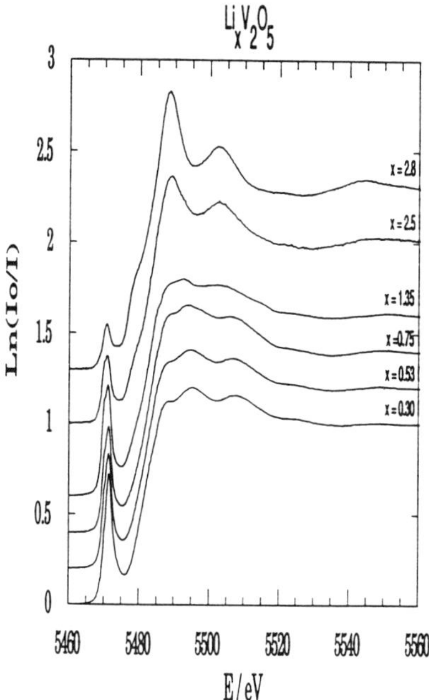

Figure 2. Evolution of the vanadium K edge XANES with the lithium content.

We can distinguish two regions : for $x \leq 1$ (before the important drop of the potential), the evolutions are weak ; beyond, they become stronger.

The first peak (called pre-edge) corresponds to the transition to the first energy level non completely filled (Molecular Orbital essentially $3d_M$, in the case of vanadium K edge). I would like to discuss first the evolution of the pre edge position and its intensity which occurs during the V_2O_5 electroreduction.

Two prominent modifications are obtained as electroreduction proceeds :

- first, a slight change in the pre-edge position. A 1.0 eV shift to lower energy is obtained for the most reduced compounds (x>2). This shift is consistent with the transformation of V(V) into V(IV) already suggested by electrochemical yields. The 'double band' pre-edge structure for partially reduced compounds (0<x<2) indicates the simultaneous presence of the two oxidation states V(V) and V(IV).

- second, a progressive decrease of the pre-edge intensity. The origin of this decrease may be either a modification of the vanadium coordination sphere which becomes less distorted as electroreduction proceeds (decrease of the $4p_V/3d_V$ ratio on the molecular orbital final excited state [11]) and an increase of the V-O bond lengths which induces a decrease of the $2p_O/3d_V$ ratio (smaller effect) [12].

The other part of the edge is also affected since the beginning of the reaction, particularly the first absorption maximum. Multiple scattering calculations performed on V_2O_5 [13] and polarized X-ray absorption experiments on oriented V_2O_5 gel [13] have shown a z polarization for this transition. Thus, changes of this band is the signature of modifications occuring on the z axis. Combining this evolution with the pre-edge examination, we can suggest that the long distance oxygen (2.78 Å) is moving into the vanadium first coordination sphere. (VO_5 --->VO_6 chromophore) : the event on the z axis and the variation of the stereochemistry are then explained.

EXAFS

Quantitative evolutions of the structural parameters (bond distances, number of neighbours and Debye-Waller factor), obtained from EXAFS data analysis, are the following :

- FIRST SHELL V-O : The variation of the Debye-Waller factor (which represents the thermal and static disorder) is weak. The regular increase of the V-O distances (they are converging to 2.00 Å for $x > 2$) is consistent with the electroreduction of V(V) into V(IV). The number of oxygen first neighbours, quasi constant for $x<1$, increases for higher lithium contents (it goes from ≈ 5 to ≈ 6). This is certainly due to the entrance of the oxygen atom at 2.78 Å in the vanadium first coordination sphere.

- SECOND SHELL V-V : Before $x = 1$, the evolution of the structural parameters is weak. Beyond $x = 1$, we observe (i) a significant decrease of the V-V distances and (ii) a strong decrease of the V-V correlation number (4 V atoms second neighbours in V_2O_5 , 2 V atoms for $x = 2.4$). These results show that the intercalation process induces local disorder in the second shell surrounding the vanadium atoms : this phenomenon begins for 0.9 lithium per mole of V_2O_5, and becomes then more important.

Results of the in situ study

We have seen that in the reversible domain, modifications of the X-ray absorption spectra are weak. To describe this domain and to obtain reliable and precise results, it is necessary to built a cell which allows us to perform simultaneously electroinsertion and X-ray absorption measurements. The discharge curves are reported on figure 3 : the results obtained with our in-situ cell are close from those obtained with a classical electochemical cell. Using the system described in figure 1, X-ray absorption measurements were made all along the reversible domain in between $0.12 \leq x \leq 0.76$ (one spectrum every 0.03 lithium intercalated).

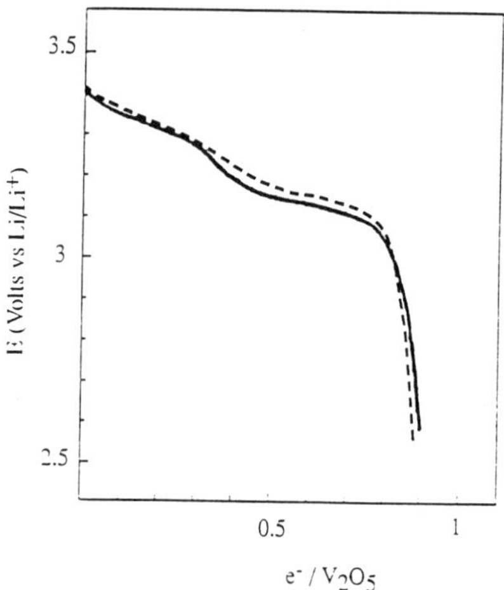

Figure 3. Comparison of the V_2O_5 discharge curve obtained with our in situ cell (---) with this obtained using a classical electrochemical cell (——).

EXAFS

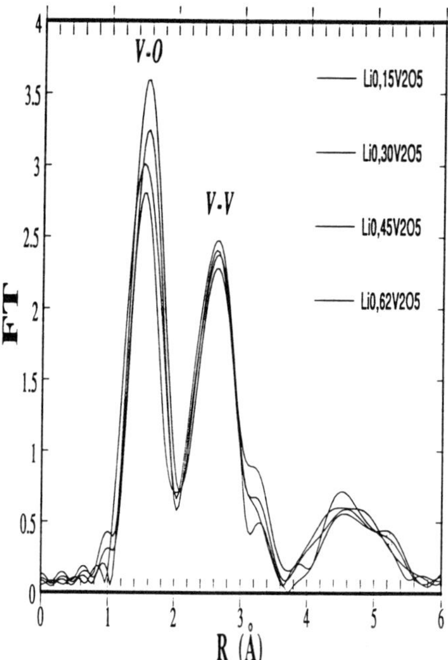

Figure 4. Modulus of the Fourier Transform for the $Li_xV_2O_5$ compounds : x = 0.15, 0.30, 0.45, 0.62.

As we can see on figure 4, in-situ analysis allows to point out small modifications which are not observed during ex-situ measurements. These modifications are now reliable since X-ray absorption experiments were performed on the same sample as electroreduction proceeds. Qualitatively, in this domain, the variations in the first shell are more important than those in the second one.

The results of the fitting procedure for the first (V-O) and the second (V-V) shells are reported on figure 5.

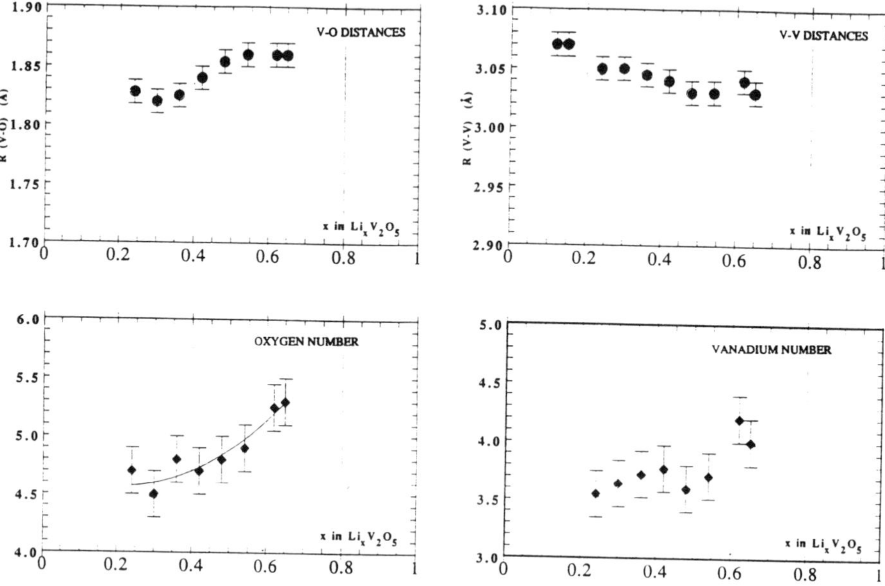

Figure 5. Variation of the structural parameters of the first and second shell around vanadium with the lithium content.

The number of oxygen atoms in the first shell around the vanadium increases significantly (the Debye-Waller factor was fixed in the fitting procedure). At the same time, the mean V-O distances increase (from 1.83 to 1.87 Å), corresponding to the V(V) ---> V(IV) transformation. The lithium insertion modified strongly each polyhedron.

Concerning the second shell of vanadium atoms, the shortening of the V-V distances shows however that these polyhedron are differently linked, with an increase of the compactness. There is a slight increase of the number of vanadium neighbours.

The evolutions of the second shell are quantitatively half of those relative to the first one. So it's obvious that the kinetic of the kinetic of the modifications of the structural parameters is different for the chromophore and for the lattice.

Conclusion

The in situ X-ray absorption study of the lithium electroinsertion intoV_2O_5 ($Li_xV_2O_5$ compounds) performed between $0 \leq x \leq 0.76$ showed weak but significant modifications of the host lattice, and we can suggest a model for the evolution of the cathodic material :

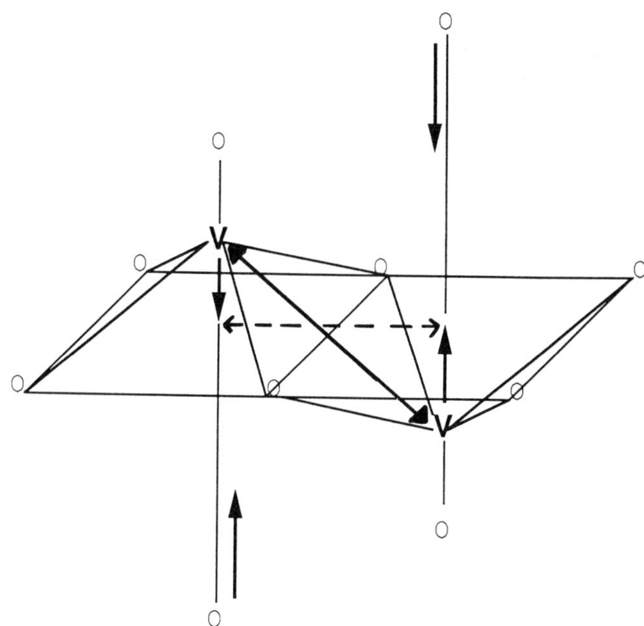

The pyramidal shape of the starting polyhedron becomes more symmetric as the oxygen at long distance is entering in the first coordination sphere. The consequence is that the vanadium atom is moving to the center of the square, and it's clear that the V-V distance is shortened. Around the strong potential jump ($x \approx 1$), this shortening becomes more important which induces a strong compactness of the lattice : finally, there is a rupture of the polyhedron series.

The structural evolutions and electrochemical results are in line with this model. This results was difficult to obtain without in-situ X-ray absorption experiments.

Acknowledgements : this work was supported by contracts from DRET. n° 86108 and 891210.

References

1 A. Tranchant, R. Messina and J. Perichon, J. Electroanal. Chem. 113 (1980) 225.
2 J.P. Pereira-Ramos, R. Messina and J. Perichon, J. Power Sources 16 (1985) 193.
3 M.S. Whittingham, J. Electrochem. Soc. 123-2 (1976) 315.
4 C.R. Walk and J.S. Gore, J. Electrochem. Soc. 122 (1975) 68c.
5 N. Kumagai, K. Tanno, T. Nakajima and N. Watanabe, J. Electrochimica Acta 28-1 (1983) 17.
6 S. Hub, A. Tranchant and R. Messina, J. Electrochimica Acta 33-7 (1988) 997.

7 C. Cartier, A. Tranchant, M. Verdaguer, R. Messina and H. Dexpert, J. Electrochimica Acta 35-5 (1990) 889.
8 C. Cartier, M. Verdaguer, H. Dexpert, A. Tranchant and R. Messina, Proceedings of the 6th International Conference on X-ray Absorption Fine Structure, S.S. Hasnain Ed., Ellis Horwood, New-York (1991) 276.
9 B.K. Teo, EXAFS : Basic principles and Analysis, Springer, Berlin, 1986.
10 A. Michalowicz in M. Verdaguer and H. Dexpert (eds.), Structures fines d'absorption en chimie, Vol. 3 - N° 1, Ecole du CNRS, Garchy, 1988.
11 C. Cartier and M. Verdaguer, J. Chim. Phys. 86 (1989) 1607.
12 F. Babonneau, S. Doeuff, A. Leaustic, C. Sanchez, C. Cartier and M. Verdaguer, Inorg. Chem. 27 (1988) 3166.
13 S. Stizza, G. Mancini, M. Benfatto, C.R. Natoli, J. Garcia and A. Bianconi, Phys. Rev. B 40 (1989) 12229.

DISCUSSION

WONG - 1. What amplitude function was used to derive the V-V coordination in the $Li_xV_2O_5$ compounds ?

2. Are the crystal structures determined (known) for these $Li_xV_2O_5$ compounds ?

CARTIER - 1. Theoretical amplitudes and phase shifts functions (Mc Kale parameters) were used in the fitting procedure of the second (V-V) and the first (V-O) shell.

2. The lattice parameters for some $Li_xV_2O_5$ compounds are known only for a few x values. We studied in-situ the first part of the discharge curve by means of X-ray powder diffraction. The results indicated changes of the crystal structure during the reaction (C. Cartier et al., Electrochimica Acta 35 (1990) 889). After the strong potential jump, the cathodic material becomes rapidely amorphous.

Time-Resolved Diffraction Studies of Fast Solid Combustion Reactions at High Temperature

Joe Wong, E.M. Larson, J.B. Holt, P.A. Waide, B. Rupp, R. Frahm* and G. Nutt

Lawrence Livermore National Laboratory, University of California, P.O. Box 808, Livermore CA 94551, USA
and
*Hamburger Synchrotronstrahlungslabor HASYLAB at DESY, Notkestrasse 85, D-2000 Hamburg 52, Germany

ABSTRACT

Real time synchrotron diffraction has been used to monitor the phase transformations of highly exothermic, fast self-propagating solid combustion reactions on a subsecond time scale down to 50 milliseconds. A specially designed reaction chamber was constructed to enable simultaneous in-situ diffraction and IR thermal imaging to be measured. A position-sensitive photodiode array detector capable of a full scan of 1024 pixels in 4 ms was used to record the time-resolved diffraction patterns from a fixed location of the specimen during passage of the combustion wave front. The detector was triggered by an inframetric camera or a thermocouple placed upstream from the area illuminated by the x-rays. The phase transformations and chemical changes in a number of simple binary systems of the type A + B -> AB are reported here. These include Ti + C -> TiC, Ni + Al -> NiAl, Ta + C -> TaC, and 2Ta + C -> Ta_2C. This new experimental approach can be used to study the chemical dynamics of high-temperature solid-state phenomena and to provide the needed database to test various models for solid combustion process.

INTRODUCTION

The combustion of gaseous reactants has been actively studied for many years[1]. However, there is a class of combustion, where at least one of the reactants is a solid, which has received little attention from materials scientists thus far. The products of these "solid flames" are technologically important materials such as ceramics, intermetallics, and composites[2]. This class of solid combustion reactions is universally accompanied by the release of a large amount of heat. Once initiated with an external source such as an electrically heated tungsten coil or a laser, these reactions become self-sustaining and propagate to completion within seconds. These so-called Self-propagating High temperature Synthesis (SHS) reactions are characterized by a fast moving combustion front (1-100 mm/s) and a high self-generated reaction temperature (1000 - 4000K).

Materials produced by SHS are finding increasing use in such applications as cutting tools, high-tech electronic ceramics, high-performance structural materials, high-temperature superconductors and heating elements for furnaces. When used in material manufacturing, SHS yields solid products that required little postproduction machining and generate less waste materials than conventional furnace processes. In addition, the SHS process is much more efficient, both in terms of energy and time, with products produced in seconds instead fo the hours or days required in conventional processes.

Although, in principle, the basic concepts of this method of materials synthesis are relatively easy to apply, the chemical reactions and dynamics of phase transformations at the combustion front are not well understood[3]. This is true even of the simplest A + B -> AB class of combustion reactions. Until recently[4], these reactions were difficult to investigate because of their high rates of combustion and extreme thermal conditions. Burn front velocities and temperature profiles could be measured as functions of time and sample position, and the product phase(s) and microstructure could be examined at the conclusion of the reaction. However, no conventional technique allowed *in situ* examination of phase transformations and chemical kinetics at the combustion front.

EXPERIMENTAL

A schematic of the apparatus used is shown in Fig. 1. Silicon photodiode array detectors manufactured by Princeton Instruments were used as position-sensitive detectors to record the time-resolved x-ray diffraction (TR-XRD) patterns from the specimen ignited in a specially designed reaction chamber-diffractometer with a vertical $\theta - 2\theta$ geometry. Each photodiode array is 25 mm long and covers a 6 deg. window in 2θ-space. The detector is capable of recording a full scan of 1024 pixels in 4 ms. The sample holder and detectors were each positioned with a high-precision motor-driven stage to enable independent θ and 2θ motions. The sample holder was made of stainless steel and lined with a grafoil sheet to protect the holder from corrosion by the hot specimen during combustion. The reaction was ignited by the passage of a current through a tungsten coil adjacent to the pressed sample block.

The thermal profile of the burn front was recorded concurrently by using an imaging infrared camera (Inframetrics, Inc. Model 600). The Inframetrics camera sends a temperature value for a pre-selected pixel every 50-60 msec through an RS-232 port, which is connected to the serial port of an IBM-AT computer which records the temperature profile. The sample temperature is measured at a position ~1 cm upstream from the position of the x-ray beam. The temperature is compared to pre-set triggering temperature. As soon as the measured value reaches or exceeds this temperature, a pulse is sent to the array detector controller and the diffraction data collection is activated thus establishing a time zero for the scans. The diffractions peaks were recorded at a fixed scan rate (e.g. 100 ms per scan) for a total number of scans (e.g. 500 scans) with a second IBM AT computer.
Another method used was to trigger the TR-XRD recording with a thermocouple[4]. Details of the design and construction of this high-speed time-resolved diffractometer reaction chamber are described elsewhere[5,6].

Time-resolved diffraction measurements were performed on beam line X-11A at the Brookhaven National Synchrotron Light Source (NSLS) with the x-ray storage ring operating at an electron energy of 2.528 GeV and an injection current of ~200 mA. The synchrotron beam, 0.7 mrad (unfocused) or 2 mrad (focused), passed through a 1 mm vertical entrance slit and was monochromatized with a double Si(111) crystal at 8 keV. The estimated photon flux for the unfocused beam was ~10^{10} photons/s at the sample. Prior to ignition, the chamber was pumped down and back-filled with a He partial pressure of ~0.5 bar to avoid oxidation of the metal particles during combustion and to minimize x-ray scattering by air. Diffraction specimens in the form of a 19 mm cube, or cylinder 19 mm in diameter and 19 mm long were pressed from dry-mixed weighed out atomic mixtures of elemental powders. The average particle size of the Ti, Al, Ni, and Ta powders were 10, 20, 5 and 2 microns, respectively. Submircon amorphous carbon procured from CABOT corporation as a Monarch 900 carbon black was used. The density of the pressed pellets was ~55-60% of the theoretical density.

RESULTS AND DISCUSSION

Formation of TiC

The TR-XRD results for the Ti + C -> TiC reaction is shown in Fig. 2. To monitor the formation of TiC, the detector was appropriately positioned to cover the strongest diffraction peaks of both the Ti reactant and the TiC product, ie. the Ti(101) and TiC(200) peaks. For this reaction, 500 scans were made each taking 200ms, for a total scan time of 100 s. From the 200 scans, seven scans are selected to demonstrate the sequence of critical events for this combustion reaction. At 0 s, the detector was triggered and recorded the Ti(101) line at ambient temperature. This reactant line persists for just over 1 s and decrease abruptly at 1.2 s. This instant marks the arrival of the wavefront at the region of the sample illuminated by the synchrotron beam. At 1.4 s, the Ti(101) line has disappeared, with simultaneous appearance of the TiC(200) line at a higher 2θ angle. From 1.4 s on, the TiC peak grows until at 1.8 s formation of the TiC phase is completed. The burn front velocity of this reaction is ~5mm/s. It is easily deduced that formation of the final TiC product takes place at the combustion front. Subsequent scans to 100 s show a continuous shift of the TiC(200) peak to higher 2θ values, indicating lattice contraction caused by cooling of the reacted specimen. From this sequence of TR-XRD scans, we learn that the first step in the combustion process is the melting of Ti(m.p.=1660 C). Subsequently, within the same 200 ms period, the molten Ti reacts with the solid carbon particles to form TiC. The total reaction time for the complete formation of TiC is within 0.4± 0.2 s.

Formation of NiAl

Ni and Al are stronger x-ray scatterers than are Ti and C. With this system, we achieved time resolution of 100 ms and 10 ms using a focussed beam. The wavefront velocity of this combustion reaction is much higher (>20mm/s) than that of the TiC rection. The TR-XRD results for this system is shown in Fig. 3. Again the detector is triggered at 0 s and recorded the Ni(111) and Al(200) diffraction peaks at ambient temperature. At 0.1 s, the intensity of both reactant peaks decrease, marking the arrival of the combustion wavefront in the region of the sample illuminated by the synchrotron beam. At 0.2 s, only the high-temperature Ni peak persists with increased intensity, indicating that the combustion front has completely swept the region illuminated by the x-ray beam. At 0.3 s, the broad Ni diffraction peak is replaced by a series of sharp peaks centered at about the same 2θ region. These sharp lines persist for more than three seconds and may arise from Ni and/or an unidentified Ni-Al alloy phase with various grain orientations at high temperature. At 3.6 s, the diffraction intensity from the reacting system falls to a minimum, after which another set of sharp line appear, centerd at 44.1 deg. and exhibiting a very high intensity peak at 5.4 s but decreasing sharply between 5.6 and 10 s. At 10.3 s, another cluster of sharp lines centered at ~44.5 deg. appears, their relative intensities varying non-systematically with time. In principle, these sharp lines and rapid intensity changes could be due to intermediate phases, grain orientation , grain motion at high temperature or a combination of these effects. This cluster of lines assignable to the (110) reflection of the NiAl product remains for the duration of the diffraction experiment, to 190 s, shifting to high 2θ value only due to lattice contraction as the specimen cools.

Since the wavefront velocity of this reaction was measured to be 40 mm/s, the combustion front has therefore passed throught the 19 mm specimen within 0.5 s of ignition. Therefore, the Bragg peaks observed after 3.6 s must arise from phases formed after passage of the wavefront, that is, in the "after burn" region. Fig. 4 plots the integrated intensity in the region 43-46 deg. as a function of time and summarized the time-resolved diffraction events in the Ni + Al system. The sharp high intensity peak A at 5.4 s is flanked on both sides by several low-intensity peaks. After ~20 s, broader peaks, like B, are observed. Beyond 50 s, the plot becomes almost featureless as the NiAl(110) product peak shifts to high 2θ due to thermal contraction. The sharp feature at 5.4 s is most likely due to an intermediate phase formed before formation of the final NiAl product.

Formation of TaC

TR-XRD results for the Ta + C -> TaC reaction is shown in Fig. 5. Two detectors were used simultaneously, one centered at 36.5 deg. and a second one at 72 deg. to collect the Bragg scattering of the Ta (110) and (211) peaks at the start of the reaction. Fortunately, both TaC and Ta2C have their major peaks in these windows. Each TR-XRD scan was recorded in a 100 ms time frame. The two Ta metal peaks are prominent until 5.6 s have passed when thermal effects broaden and diminish their intensities due to heating. The Ta(110) peak splits at 6.7 s and another peak is observed at ~35 deg. The Ta(211) has almost disappeared and only a hint of scattering is observed in the center of the higher angle detector. Both TaC and Ta_2C are clearly evident at 7.5 s in both detectors. The burn fornt velocity for this reaction is ~2.0mm/s and with a beam size of 1.2 mm the burn front would pass through the x-ray spot in 0.5 s. After 30 s, most of the scattering from Ta_2C has disappeared leaving only TaC as the final product of the reaction.

In Fig. 6 the integrated peak intensites of the reactants and products in the Ta + C combustion are plotted. The reduction in intensity of the Ta peak due to thermal broadening from heating of the sample is largely responsible for the slope of the metal peak between 3 - 6 s. An intersection of the Ta and TaC lines shows that TaC forms immediately upon passage of the burn front. But the TaC intensity does not reach its maximum until the Ta_2C peak has grown and then diminished. The appearance and disappearance of the Ta_2C phase clearly indicates that it is an intermediate phase of the Ta + C reaction.

Formation of Ta_2C

The burn front velocity for the 2Ta + C -> Ta_2C reaction is about a factor of 2 faster than that for the formation of TaC discussed in the previous section. We monitored the sub-carbide reaction at a 50 ms time scale. The TR-XRD data for this reaction is shown in Fig. 7. It is seen that within a time frame of 50 ms, the Ta peaks begin to decrease at 1.6 s and the major Ta_2C(101) peak appears at 1.65 s. In the next frame at 1.7 s all the subcarbide peaks are evident in both detectors with no indication of any other species all the way to the end of reaction. The apparent shift to low angle in the last scan at 7.5 s is due to a drop in the height of the sample surface with respect to the beam as the whole sample contract upon cooling. In Fig. 8, the integrated intensities of the Ta reactant and Ta_2C product are plotted as a function of time. As can be seen, the slope of the curve showing the formation of product is much steeper than that of the TaC reaction.. Almost as soon as Ta disappears, the Ta_2C product reaches it maximum intensity.

The adiabatic temperatures of this reaction and of the TaC formation described in the previous section are calculated to be 2376 C and 2452 C, respectively. These maximum reaction temperatures are well below the melting points of Ta (3020 C), carbon (~3800 C), the corresponding products Ta_2C (`3300 C) and TaC (~3825 C) and eutectic temperatures

(2825 C and 3375 C) known in the binary Ta-C phase diagram. Hence there is no formation of a liquid phase and the synthesis of these two carbides are truly solid-state combustion.

CONCLUDING REMARKS

The series of experiments reported here demonstrated that TR-XRD using synchrotron radiation is indeed a very powerful and perhaps unique method of following the phase transformations and chemical dynamics of solid combustion reactions *in situ* at high temperature. When combustion front velocity and temperature profile are measured synchronously and correlated with the TR-XRD scans, all participating phases can be identified as a function of time and temperature. Intrinsic, real-time kinetic data of this sort are much needed to permit testing of existing theoretical models of solid combustion and to provide the basis for developing news theories. Furthermore, using higher flux synchrotron sources currently available at wiggler beamlines (10^{12} photons/s) or brighter third generation sources to be available at ALS at Berkeley, ESRF at Grenoble, APS at Argonne and SPring-8 in Japan, we can achieve higher spatial resolution for a closer look at the combustion front in this interesting class of high temperature solid combustions.

ACKNOWLEDGMENTS

This work is supported under the auspices of the U.S. Department of Energy (DOE) by the Lawrence Livermore National Laboratory under contract W-7405-ENG-48. We are also grateful for the support of DOE, Division of Materials Sciences, under contract DE-AS05-80-ER10742, for its role in the development and operation of beam line X-11 at the NSLS. The NSLS is supported by DOE, Divisions of Materials Sciences and Chemical Sciences, under contract DE-AC02-76CH00016. One of us (JW) is grateful to the Alexander von Humboldt Stiftung for a 1991 senior U.S. award to continue this work at HASYLAB, Germany.

References

1. See for example the volume of articles published in *Combustion and Flame* since 1957.
2. Z.A. Munir, Ceram. Bull. **27**, 342 (1988).
3. Z.A. Munir and U. Anselmi-Tamburini, Mater. Sci. Rept. **3**, 277 (1989).
4. Joe Wong, E.M. Larson, J.B. Holt, P.A. Waide, B. Rupp and R. Frahm, Science, **249**, 1406 (1990)
5. E.M. Larson, P.A. Waide and Joe Wong, Rev. Sci. Instrum., **62** 53, (1991).
6. E.M. Larson, J.B. Holt, Joe Wong, P.A. Waide, B. Rupp and L. Terminello, J. Mater. Res. (1991), submitted.

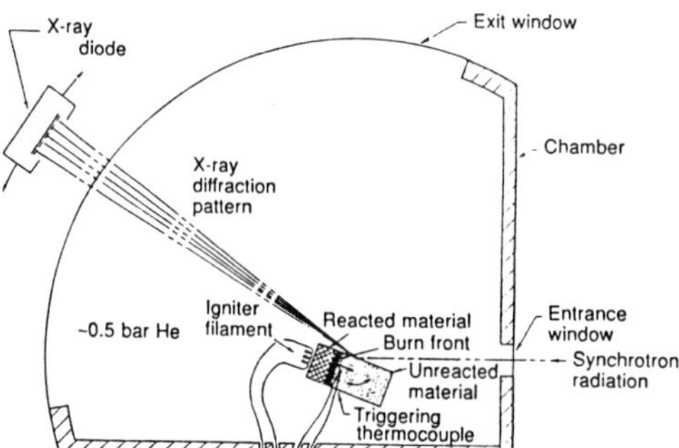

Fig. 1 Schematics of high speed TR-XRD reaction chamber used to study solid combustions.

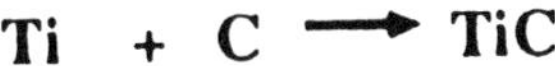

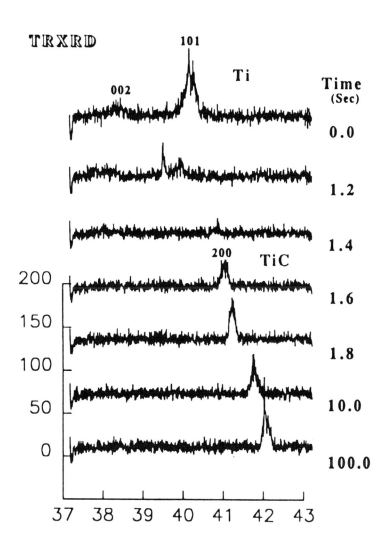

Fig. 2 Selected TR diffraction scans of the Ti + C -> TiC reaction.

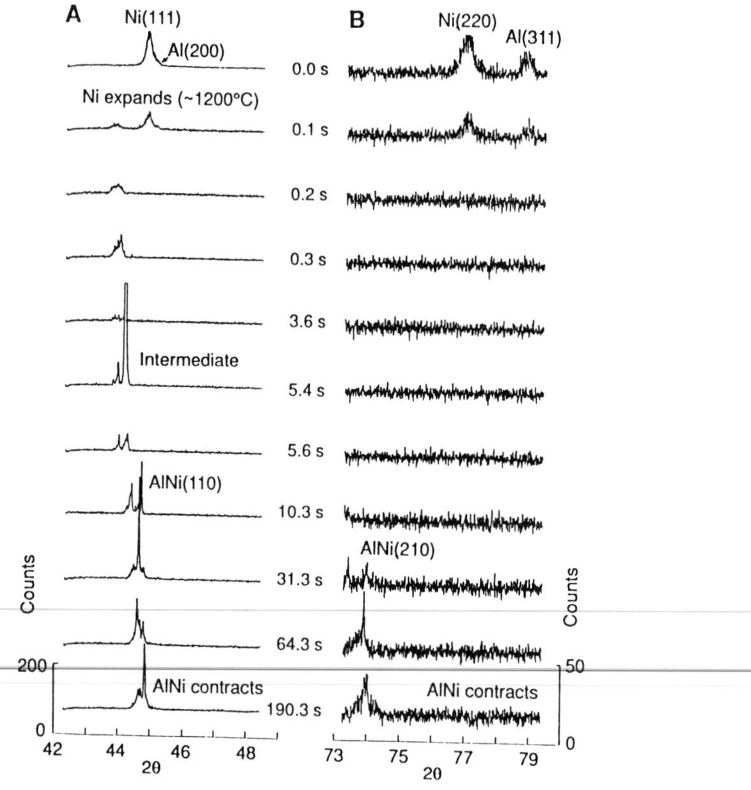

Fig. 3. Selected TR diffraction scans of the Ni + Al -> reaction.

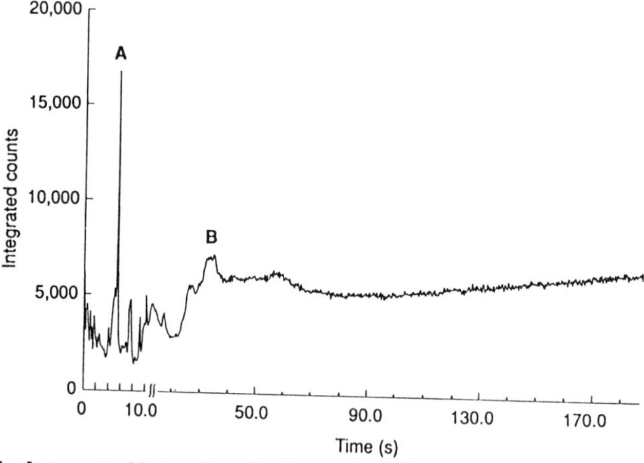

Fig. 4. Integrated intensity plot from the TR scans shown in Fig. 3 vs,time.

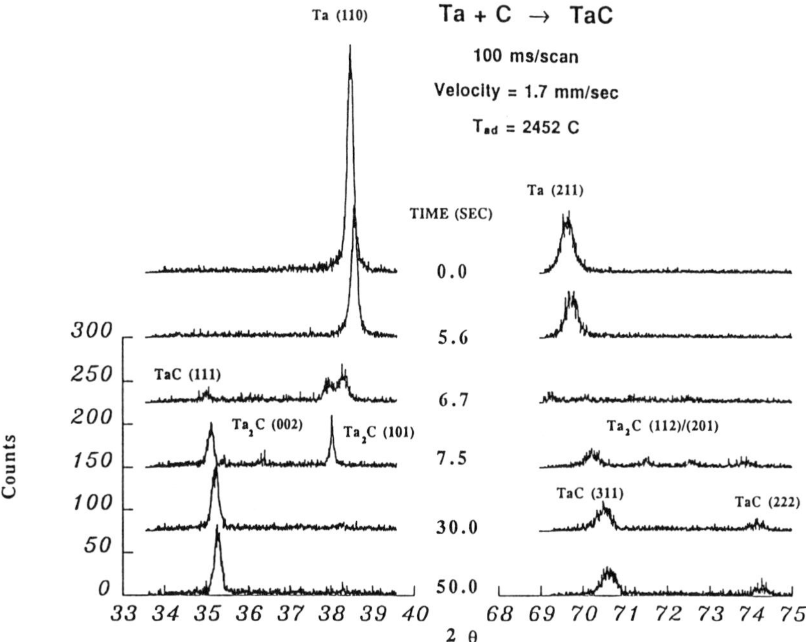

Fig. 5. Selected TR diffractions scans of the Ta + C -> TaC reaction.

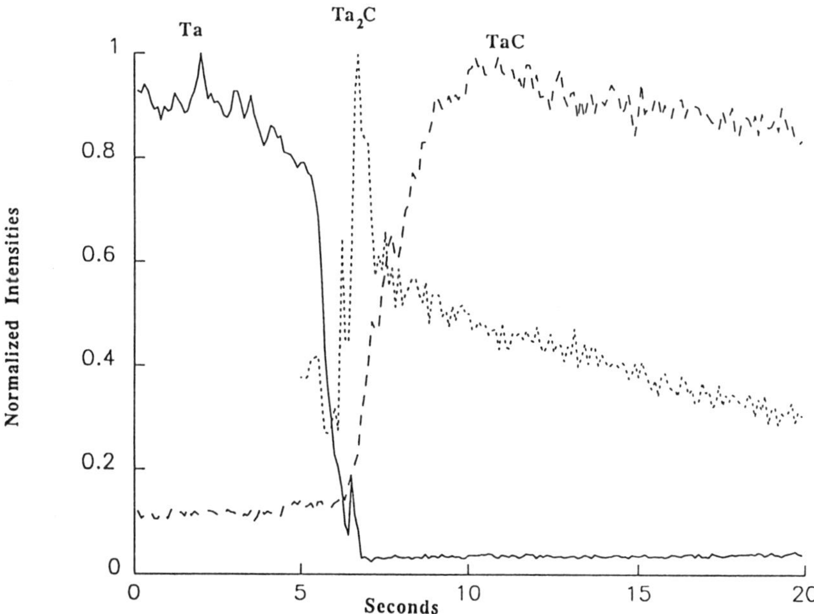

Fig. 6. Integrated intensity plots from the TR scans shown in Fig. 5 vs.time.

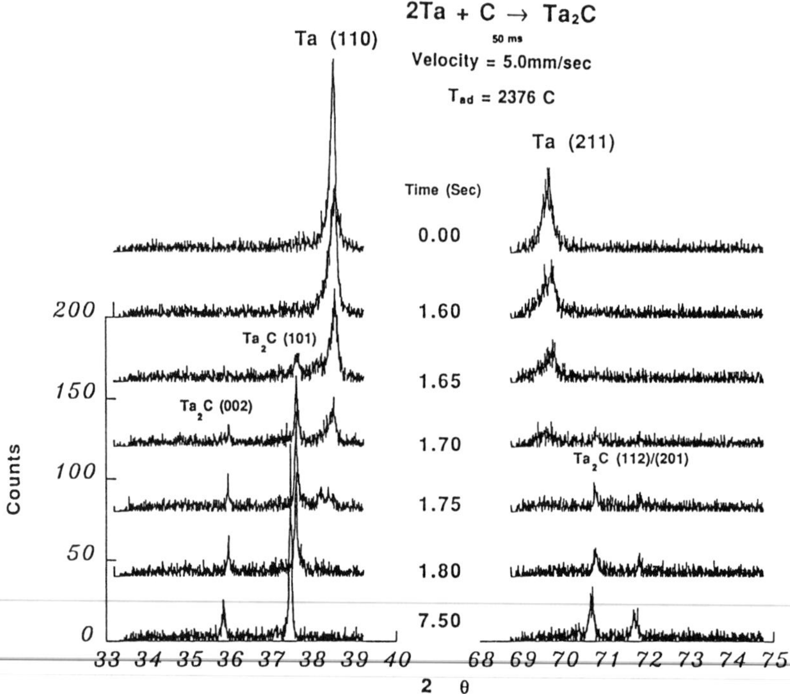

Fig. 7. Selected TR diffraction scan of the 2Ta + C -> Ta_2C reaction.

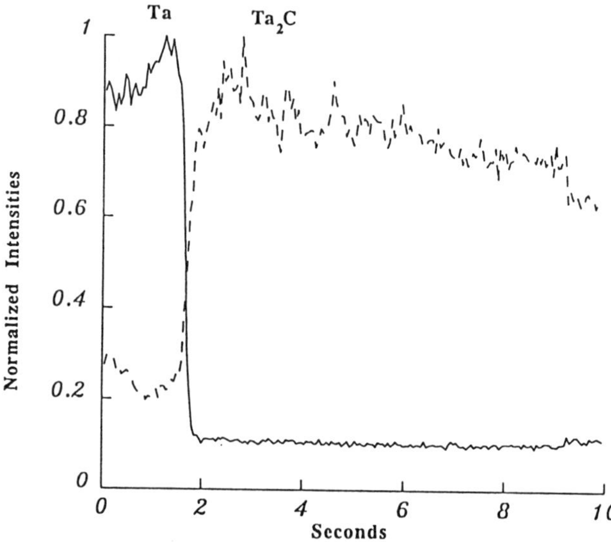

Fig. 8. Integrated intensity plots from the TR scans shown in Fig. 7 vs. time.

DISCUSSION

LEACH - 1. Are there many cases where you see vaporisation at the base edge of the sample undergoing combustion ?

2. Carborundum that has been vaporized contains not only SiC but also SiC_2 and Si_2C in equilibrium, at temperatures somewhat above your so-called adiabatic temperature for the Si + C combustion system. Did you see any evidence for SiC_2 or Si_2C being formed in your solid state combustion case ?

WONG - 1. No, we have not noticed or particularly looked for such vaporization. We do see gaseous evolution during combustion in certain systems, e.g. like "carbon"-containing systems, but this is mainly due to moisture in the raw reactants.

2. We have no experience with the combustion if Si+C $\rightarrow$ SiC.

BARNES - What are the spatial dimensions of material being dynamically studied in relation to the movement of the reaction wavefront during one collection time interval ?

WONG - For the Ni+Aℓ $\rightarrow$ NiAℓ reaction, the measured velocity of the burn front is 20mm/sec. For a scan time of 100ms/scan the burn front travels 2mm during a scan. Now the vertical height of beam is 1mm, and at an angle of 25° incident to sample surface, beam size on sample is ~2.6mm. Thus the beam size on sample is larger than the distance swept by the burn front in the time frame. Also spatial temperature profile indicates a burn front width of ~0.5mm. For the Ta+C$\rightarrow$TaC reaction, burn front velocity = 2mm/sec, scan time was 100ms/scan, vertical slit = 25mm, angle of incidence = 30°, the burn front travels 0.2mm during a scan and beam size is 0.5mm on sample. So again, we are **not** spatially resolving the burn front.

V. THIRD GENERATION SR SOURCES & ROUND TABLE DISCUSSION

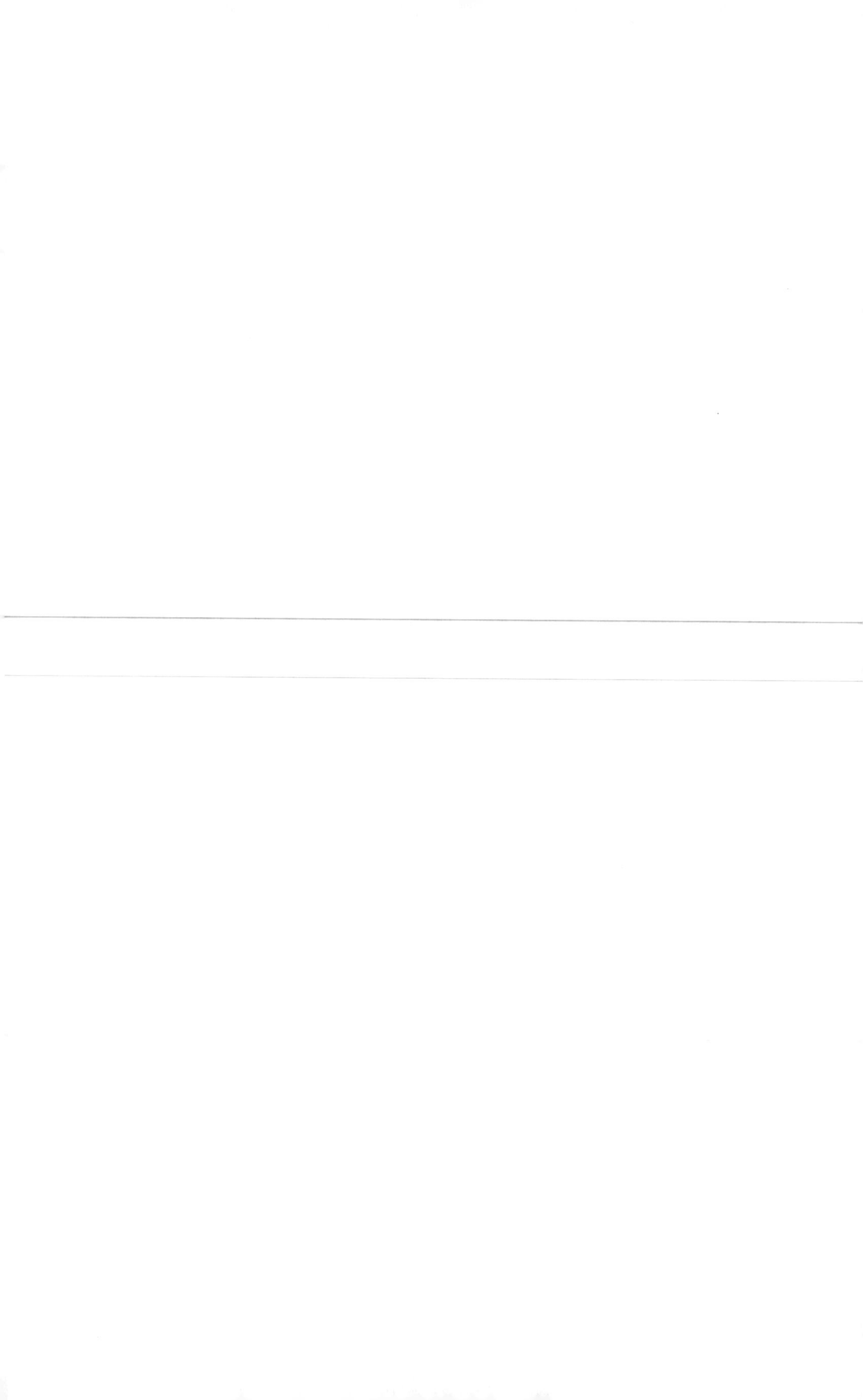

ESRF AND THE THIRD GENERATION OF SYNCHROTRON RADIATION SOURCES

R. Haensel
ESRF, BP220, F-38043 Grenoble, France

ABSTRACT

The European Synchrotron Radiation Facility ESRF is the first member of the third generation of Hard X-ray Synchrotron Radiation sources, in which the electron/ positron beam emittance and the total length of insertion devices (wavelength shifters, wigglers, undulators) are pushed to their present limits. This results in an optimized photon flux and brightness (the number of photons emitted per second per unit source area, per unit solid angle, within a given bandwidth). The high photon flux allows geometrical structures of condensed matter with extremely high time resolution to be investigated and thus to follow physical, chemical or macromolecular reactions and structural changes (rearrangements) in real time. The high brilliance makes it possible to study extremely small samples (micrometer range) and larger hetereogeneous samples with extremely high spatial resolution.

To reach and maintain this performance the 0.1 mm thick electron/positron beam must be kept stable in its lateral and angular position in the insertion devices all around the 850 meter circumference storage ring over a typical period of 10 hours. This constitutes a real challenge for the suppression of vibrations and differential settlements of the buildings housing the radiation source and the several dozens of beamlines with a typical length of 75 meters (and longer in some cases).

Another challenge to be met is the high radiation load on the optical elements with their atomic scale surface flatnesses and the enormous data flow to be handled by the 0-, 1-, and 2-dimensional X-ray detectors.

INTRODUCTION

Since their discovery, nearly a hundred years ago, X-rays have played an extremely important role in the investigation of the structure of matter. While many of the fundamental discoveries with X-rays have been made with classical laboratory 'bremsstrahlung' X-ray tubes during the first half of this century, the field of X-ray applications got a tremendous boost thirty years ago by the 'discovery' that electron synchrotrons, very much to the dismay of High Energy Physicists, are extremely intensive X-ray sources. The origin of the 'Synchrotron Radiation' is the transversal acceleration of charged particles in a magnetic field (magneto-bremsstrahlung). The first observations of Synchrotron Radiation were made in the bending magnets of an electron synchrotron (hence the name), but transversal acceleration of charged particles in magnetic fields and hence production of Synchrotron Radiation can also occur in betatrons, in bending magnets of storage rings, in periodic assemblies of magnets (called wavelength shifters, wigglers or undulators, depending on their configuration) which are inserted into the straight sections of accelerators and storage rings, and, last but not least, in magnetic fields in cosmic space.

The first experiments with Synchrotron Radiation (around 1960 in Washington/DC and Frascati) were performed using, in parasitic mode, accelerators and (lateron) storage rings, that were initially built as part of nuclear or high-energy physics programmes. Many of these first generation machines were eventually converted into dedicated Synchrotron Radiation sources (SURF in Gaithersburg, DCI in Orsay, SPEAR in Stanford) but some do still run in a time-sharing mode as high-energy machines and partially dedicated Synchrotron Radiation sources (DORIS in Hamburg, ADONE in Frascati, CESR in Ithaca, the different VEPP rings in Novosibirsk, and the accumulator ring of Tristan in Tsukuba). Since these machines were designed for high-energy physics programmes with a limited number of interaction regions for high-energy experiments, the emittance is not optimized all around the machines for the installation of Synchrotron Radiation beamlines.

This and the limited capacity of existing machines, which were not able to meet the increasing demands of Synchrotron Radiation users, called from 1975 on for a new second generation of machines which were optimally designed for their use as Synchrotron Radiation sources with the radiation mainly coming from the bending magnets. In this category we find more than a dozen operating machines in Europe, the United States and in Asia.

To meet the needs for more brightness several improvements have been made in machines of the first and second generations by incorporating a limited number of insertion devices into existing straight sections(e.g. SPEAR and PEP in Stanford, BESSY in Berlin, the Photon Factory in Tsukuba), by inserting by-passes (DORIS), and by converting the magnetic lattice into a High Brightness Lattice (SRS in Daresbury).

The third generation of Synchrotron Radiation Sources addresses the rising demand for high brightness by categorically lowering the emittance of the ring with the insertion of numerous quadrupole and sextupole magnets and by providing more and longer straight sections for insertion device installation. These measures of course considerably increase the circumference of the machines. In the low energy regime (below 2 GeV) one machine (SuperACO in Orsay) is in operation and a dozen more are under construction. In the Hard X-ray regime three projects with energies of 6 GeV and above are under way (ESRF in Grenoble, APS at the Argonne National Laboratory and SPring8 in Harima Science Garden City). Although the wigglers and undulators constitute the principal sources of Synchrotron Radiation for these machines, their bending magnets are by no means unimportant: their brightness is still superior to that of many insertion devices in sources of the first and second generations.

THE DEVELOPMENT OF THE ESRF PROJECT

When around 1975 it became obvious that the Synchrotron Radiation sources of the first generation were no longer able to meet the increasing demand of Synchrotron Radiation users, both qualitatively and quantitatively, first attempts were started to build dedicated Synchrotron Radiation sources in several countries. While in Europe these attempts were successful for machines in the 1-2 GeV range (SRS in Daresbury, BESSY in Berlin), it very soon became clear that for a Hard X-ray machine with an energy of 5 GeV or above, in view of its large size, both the financing and the exploitation could only be envisaged in a European framework.

The European Science Foundation ESF took up this idea and installed a 'Working Group on Synchrotron Radiation' under the chairmanship of H. Maier-Leibnitz. The group published a first report in 1977. In the following years an ad hoc Committee on Synchrotron Radiation of the ESF chaired by Y. Farge worked out 'The Feasibility Study' with three supplements on 'The Scientific Case', 'The Machine' and the 'Instrumentation'. This work eventually led to the creation of the European Synchrotron Radiation Project ESRP, located at the European Center of Nuclear Physics CERN in Geneva. Under the leadership of B. Buras and S. Tazzari and in close collaboration with the machine experts of the LEP project and numerous Synchrotron Radiation experts in all European countries a report was published in 1984, which finally gave the starting signal for serious discussions between the governments of those European countries, which had started the initiative through the ESF.

In 1985 the governments of France, Germany, Italy, Great Britain and Spain signed a 'Memorandum of Understanding' and started the ESRF activities in Grenoble/France. The choice of Grenoble ideally fulfilled the conditions of an excellent scientific environment as previously formulated by the ESRP Project Group:

- Several universities,
- local branches of the national research organizations CNRS (Centre National de la Recherche Scientifique) with laboratories for crystallography, low temperatures, high magnetic fields (the latter jointly operated with the German Max-Planck-Gesellschaft) and the
- CEA (Commissariat à l'Energie Atomique) with many institutes including the Division d'Electronique, de Technologie et d'Instrumentation LETI and ASTEC, an office for Technology Transfer,
- the Institut Laue-Langevin,
- the outstation of the European Molecular Biology Laboratory EMBL,
- the Institute for Millimetric Radioastronomy IRAM (jointly operated by the CNRS and the Max-Planck-Gesellschaft),
- the Centre National d'Etudes des Télécommunication CNET, and
- several national and international industrial companies (Air Liquide, Bull, Hewlett&Packard, Pechiney to mention only a few)

represent the high standard of Grenoble as a city of Science and industry related Research & Development.

A point of special attraction for the ESRF to go Grenoble was the immediate proximity of the Institute Laue-Langevin ILL, which successfully operates the High Flux Neutron Reactor HFR for the European Neutron Users' Community for more than 20 years. The complementarity of ESRF and ILL sharing a joint site and a number of services will provide the European scientists with unique possibilities for investigations in the fields of solid state physics, chemistry, materials science, biology, and medicine.

The ESRF Foundation Phase started in 1986 with the installation of the ESRF team in Grenoble. During 12 months the technical parameters of the machine, the scientific case, the final location of all buildings and the budget for the next 11 years (6.5 years for the construction of the machine and the first set of beamlines followed by the first 4.5 years of operation with the completion of all 30 beamlines), and legal documents were established, allowing in 1988 the start of the construction and the signature of the Convention by 11 governments (to the 5 above mentioned came also Switzerland, Belgium and 4 Nordic Countries: Denmark, Finland, Norway and Sweden), to which at the end of 1991 came The Netherlands as 12th member country.

ESRF SOURCE CHARACTERISTICS

The main parameters of the source are summarized in the following Table I:

Nominal beam energy	6 GeV
Circumference	844.39 m
Horizontal emittance	7 nm*rad
Vertical emittance	0.6 nm*rad
Number of straight sections	32 (29 suitable for insertion devices)
Number of bending magnets	64 (suitable for 29 beamlines)
Length of straight sections	6.34 m (for 3 modules)
Radiation sources	Undulators (max. 14.4 keV in the fundamental)
	Wigglers
	Wavelength shifters
	Bending magnets (at 10 and 20 keV)
Number of bunches	1 - 992
Beam current	>100 mA in multi-bunch mode
	7.5 mA in single-bunch mode
Estimated halflife-time	10 hours

Fig. 1 shows an overview of the accelerator buildings (status 1990).

The electrons/positrons are generated and preaccelerated in a 200 MeV linear accelerator and brought to the final energy of 6 GeV in a booster synchrotron of 330 meter circumference.

The storage ring has a total circumference of 850 meters, 200 meters of which are occupied by the insertion devices; another 200 meters are used for the 64 bending magnets, and the remaining 450 meters for the 320 quadrupole and 224 sextupole magnets and all other ancillary devices (such as beam position monitors, pumping ducts, valves etc.). The large number of quadrupole and sextupole magnets are needed for the extremely small horizontal beam emittance with resulting beam sizes of 100-400 micrometers in different straight sections.

The expected current is 100 mA for multi-(ie.992)bunch operation with a pulse distance of 2.8 nsec and 5 mA for single-bunch operation with a pulse distance of 2.8

microsec (the time for one orbit). It is however expected that the full 100 mA can also be obtained with 20 bunches in the machine. This would allow to obtain bunch distances of 140 nsec, giving a good compromise between the users aiming for maximum current irrespective of the time structure and those interested in a good time structure for time resolved experiments. The bunch length will be appr. 15 nsec.

The 29 straight sections for insertion devices will be equipped with a large variety of wavelength shifters, wigglers, undulators allowing to cover a wide photon energy range with X-rays of different polarization characteristics (linear, circular, elliptic).

STATUS OF THE MACHINE

The preinjector was purchased from industry and delivered at the beginning of 1991. The commissioning of this accelerator was finished in June 1991.

The commissioning phase of the booster synchrotron was started in September 1991. On 12 November for the first time electrons were accelerated to the final energy of 6 GeV. Before the end of the year the commissioning was successfully finished with the first extraction of the beam into the transfer line connecting the booster synchrotron and the storage ring.

The storage ring commissioning phase will start on 15 February 1992 and will last until the end of 1992. Fig. 2 shows the storage ring in the final state of assembly. In 1993 the first set of at least 7 beamlines will be commissioned. The first external users can start their experiments at the beginning of 1994.

EXPERIMENTAL PROGRAMME

The goal of ESRF is to provide 30 beamlines, providing the optimum exploitation of the high brightness, the high photon flux and also the spectral distribution of the photons towards high energies, and aiming at a good balance of access for the different fields of applications (physics, chemistry, biology, medicine, materials science, etc.). At least 7 beamlines will be ready for the starting phase of the users' operation at the beginning of 1994. A total of 18 beamlines (see Table II) will be available one year later (beginning of 1995) with the remaining beamlines to be put into operation until the end of 1998.

The 30 EXAFS beamlines are constructed and operated under the responsibility of the ESRF and occupy practically all insertion devices. On the other hand, only a small fraction of the available 29 bending magnet beamlines will be used for the ESRF programme. The remaining ones are offered to interested groups (so called Collaborating Research Groups: CRG) from the member countries, which will be able to provide their own budget for the investment and recurrent costs and which are allowed to exploit two thirds of the available time on their beam line for their own scientific programme with the remaining one third to be given to the general ESRF users. Arrangements with 4 CRGs have been concluded until now.

Table II: ESRF beamline programme (status end of 1991

Beamline	Scientific Goals	Source
1. Microfocus	Micro-Diffraction Small-Angle Scattering High Pressure	Undulator 4-15 keV
2. Multipole Wiggler Materials Science	Small Molecule Crystallography Magnetic Scattering	Wiggler 4-60 keV
3. Multipole Wiggler White beam	Laue Protein Crystallography High Pressure ED Monochromatic Option	Wiggler 4-60 keV
4. High Brilliance	Real Time Small-Angle Scattering Monochr. Protein Crystallogr.	Undulator tunable around 10 keV
5. High Energy X-ray Scattering	Gamma-ray Diffraction Small-Angle Scattering Compton Scattering	Wavelength Shifter
6. Circular Polarization	Dichroism in EXAFS, SEXAFS Spin-dependent Photoemission Microscopy at 2.5 keV	Helical Undulator $E \leq 4$ keV
7. Surface Diffraction	Surface Structural Studies Phase Transitions Growth Mechanisms Liquid Surface Diffraction	Undulator $k_{max} = 1.85$
8. Dispersive EXAFS	Time-resolved Structural Studies	Wiggler or Tapered Undulator
9. "Open" Undulator	Temporary use by external groups	
10. "Open" Bending Magnet	Temporary use by external groups	
11. Mössbauer High Resolution	Nuclear Bragg Scattering High Resolution (5-100 meV) Inelastic Scattering at 0-5 eV Energy Transfer, Electronic and Vibrational Excitations	Undulator 14 keV
12. Assymetric Wiggler	Magnetic Scattering	Assymetric Wiggler
13. Surface Science	SEXAFS, Standing Waves	Undulator
14. High Energy Wiggler	Microtomography, possibly Angiography	Wiggler
15. Powder Diffraction	Structure Determination	Bending Magnet, later: Undulator
16. Wiggler Long Beamline	Topography	Multipole Wiggler
17. Anomalous Scattering	Materials Science	Undulator 4-15 keV
18. EXAFS	2 EXAFS stations	Bending Magnet

Photo : A.M. Freund / A. Childéric

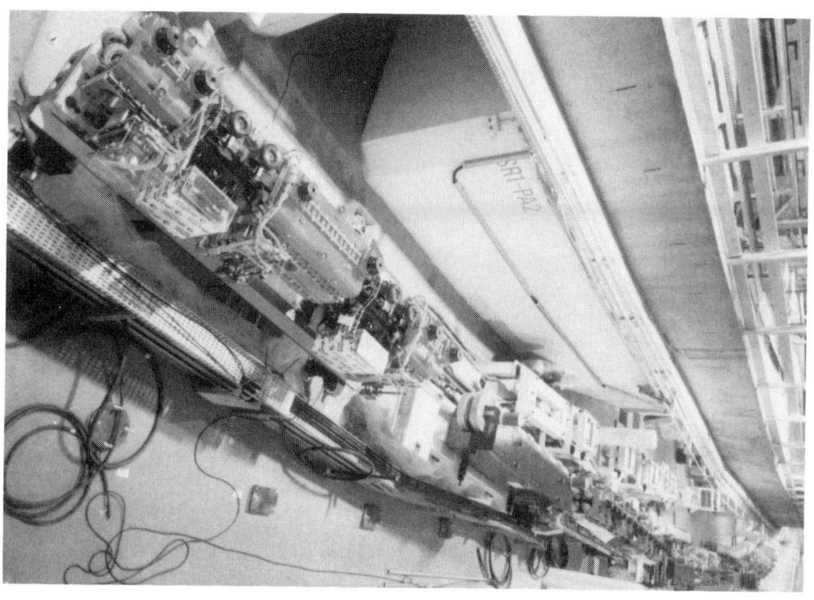

DISCUSSION

<u>NENNER</u> — This conference deals wih dynamic phenomena. Could you comment on the applications along these lines ?

Do you intend to operate ESRF in a one bunch operation or, generally speaking, a low repetition rate.

<u>HAENSEL</u> — Given the high photon flux and brilliance on the one hand and the time structure on the other the ESRF beam line program will provide all means for time-dependent measurements.

The mode of operation will follow the beam line applications. A good compromise will be I_{max} = 100mA with 20 bunches, and from time to time single bunch mode will be provided.

<u>WONG</u> — What is the critical energy of the bending magnet radiation at ESRF ? What is the corresponding photon flux say at 10 keV ?

<u>HAENSEL</u> — E_c = 20 keV in the main part and 10 keV in the soft ends
10^{13} ph/sec.mrad 0.1%.

<u>BAUMGARTEL</u> — What is the bunch length ? Can it be changed ?

<u>HAENSEL</u> — The bunch length is around 15 ps. It can be changed with additional cavities but this is a very complicated issue.

ROUND TABLE: "TIME-DEPENDENT EXPERIMENTS : PRESENT AND FUTURE"

Discussion leader : Irène Nenner

CEA - DSM/DRECAM, Service des Photons, Atomes et Molécules, Bâtiment 522, Centre d'Etudes de Saclay, 91191 Gif sur Yvette cedex, France

and

LURE, Laboratoire mixte CNRS-CEA-MENJS, Bâtiment 209D, Centre Universitaire de Paris sud, 91405 Orsay cedex, France

I - Introduction

This round table showed that the rates of physical phenomena studied by synchrotron radiation (SR) cover a very large time scale from the second to the picosecond range and that time-resolved experiments apply to numerous fields such as material sciences, surface science, chemistry, biology, molecular physics. This is illustrated by the diagram of Table 1, extracted from the review of Mills (1991). One distinguishes two classes of phenomena i) those which require experiments using the pulsed nature of SR ii) those which take SR as a continuous source. The latter are generally "flux hungry" experiments. The limit between these two groups lies around 10 μs, which is the very maximum of the interpulse period obtained with existing machines.

The present status of time-resolved experiments shows that the limitations are very different in each time domain, as are the challenges. Consequences of the future use of third generation machines are discussed. In addition, certains general aspects of the relations of the concerned scientific community with those of other scientific areas are considered.

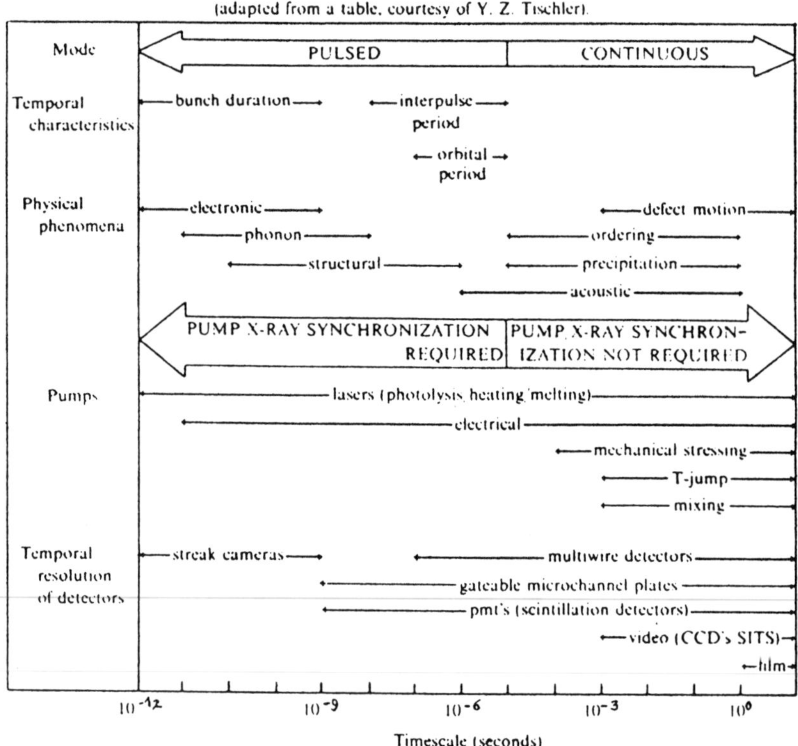

Table 1 : Physical phenomena, temporal characteristics of SR, temporal resolution of detectors, from Mills 1991.

II- "Flux hungry" X-ray and soft X-ray experiments

Flux hungry experiments, which are relevant for time-dependent investigations in materials science (Mills 1991) are : i) time resolved X-ray diffraction (TR-XRD), ii) time-resolved small X-ray scattering (TR-SAXS) and iii) time-resolved X-ray absorption spectroscopy (TR-XAS) iv) time-resolved X-ray photoemission spectroscopy (TR-XPS) or time-resolved Auger electron spectroscopy (TR-AES). TR-XRD probes crystal structure, phase transformation and evolution with time. It is a long-range order probe of the unit cell x translational vector. TR-SAXS probes time-resolved structure in the

10-1000 Å range, e.g. pores. TR-SAXS probes the time variation of local atomic and electronic (chemical bonding) structure in the range to about 10 Å. TR-XAS includes both energy dispersive EXAFS (Frank et al 1983) and quick-scanning EXAFS (Frahm 1988). TR-XPS and TR-AES probe the kinetics at surfaces of processes such as catalytic reactions, interface formation, epitaxy.

Present capabilities of the X-ray tools for time dependent measurements are limited by both detector speed and photon flux to a temporal resolution of tenths of ms (Mills 1991). Third generation storage rings such as ESRF (Grenoble), APS (Argonne) and SPring-8 (Japan) with high intensity and high brightness photon sources, remove the "flux" limitation to attain us temporal resolution. This is simply due to the fact that if 10^{11} photons/s allows ms resolution, 10^{14} photons/s allows μs resolution. Then the speed of the detector response becomes the remaining factor.

Thus, since the construction of various third generation sources is now under way, future prospects of time dependent experiments beyond the ms regime lies in the development of **faster detectors**. Furthermore, the need of such detectors ought to be driven by the "science" need with incorporation of the necessary hardware and software designs in order to use these high flux sources effectively for time-resolved studies. Two dimensional CCD detectors with 584x390 pixels are available (Clarke and Lowe 1991) for real time X-ray scattering with a recovering time of 10 μsec. The high brightness of these new photon sources also enable spatial resolution under to the micron range, so that we may then have tools such as TR-microXRD, TR-microSAXS, TR-microXAS, TR-microXPS and TR-microAES in our structural tool box.

In order to reach a better time resolution in the subnanosecond range, one could envision (Clarke and Lowe 1991) to use the time structure of SR. Third generation machines opearated with a long interpulse period (a few microseconds) and short pulse length (100 psec typically) would permit to investigate laser processing, growth, annealing, ion implantation activation, recrystallization studies.

Besides the present (limited ?) efforts to develop fast and high resolution detectors in existing facilities, it is worth to notice that there are a number of technical developments made for Astrophysics and High Energy Physics and which could be profitably transferred to SR (e.g. fully depleted CCD cameras that are radiation hard). The main problem is the necessary allocated budget : to start such a program, one needs 1 to 10 millions FF a year ! Finally it should be mentioned that a European workshop on "X-ray Detectors for Synchrotron Radiation Sources" organized by A.H. Valenta, J. Morse and D. Raoux, has treated extensively this subject in Aussois, France (September 30-October 4, 1991)

III- Direct time-resolved experiments in the UV, VUV and X-ray ranges

The physical phenomena occuring below 10 μs, only briefly described in Table 1, are presented in more details in the diagram of Table 2. They are relevant to molecular physics (gas phase molecules, adsorbed molecules on surfaces, condensed molecules or molecules in matrices), and to biology. In molecular physics, they are related to electronic spectroscopy and dynamics of the decay of excited states (Nenner and Beswick 1987). In Biophysics, there are also related to rotatonal dynamics of fluorescent macromolecules in solution (Brochon 1980, Brochon et al 1991). On the short time side, solvent relaxation, fast autoionization, vibrational motions, direct ionization and dissociation belong to the subpicosecond range and are inaccessible by SR pumps. On the "long" time side, indirect phenomena such as predissociation, slow autoionization, fluorescence are readily accessible. This is because the pulse width, ΔT, lies currently in the subnanosecond range and the interpulse period T approaches the microsecond range. In these experiments, which do not require high incidence SR flux, the detection is based on high efficiency photon, electron or ions counting techniques. This is especially true for coïncidence techniques (Electron-ions, fluorescence photon-ions, electron-ion-ion etc...) for which the number of investigated events should be low enough to avoid false coïncidences and paralysis (Baer 1991, Eland 1991). The detection can also work in the frequency domain in measuring phase change and demodulation (Gratton 1984).

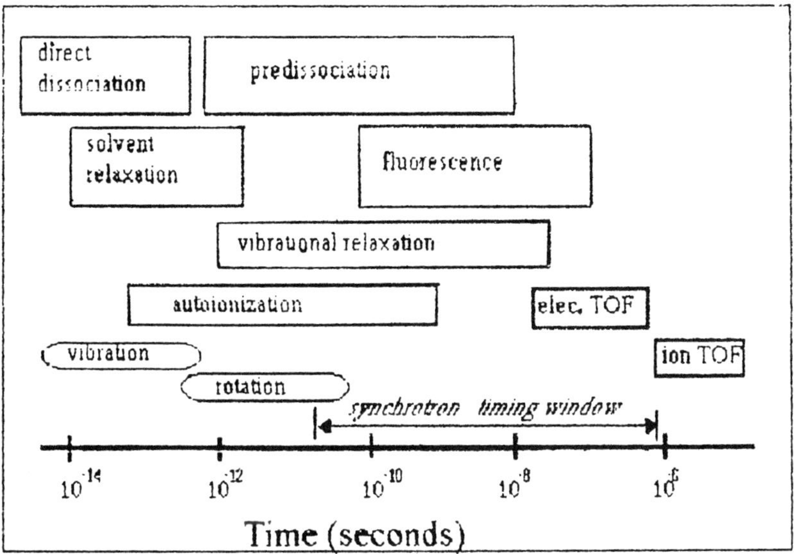

Table 2 : Detailed physical phenomena in the submicrosecond range and time domain for particle detection

A - Long interpulse periods and possible extensions

The present facilities in which the pulsed structure of SR is used, operate the machine with a number of bunches low enough to obtain an interpulse period, T, long enough compared to the pulse width, ΔT. On the other hand, for a given machine, the ΔT/T typically ranges from 100 to 10000 (for long length machines) with T in the 1 to 0.1 microsecond range. Third generation facilities tend to operate the ring with many bunches because of high brillance requirements. Even though the ΔT reduced to 10 ps, the interpulse period is also reduced to a few ns, thus keeping ΔT/T constant. On the one hand, new fast phenomena like fast (pre)dissociation and autoionization can be accessible, if the challenge of developing fast particle detectors is achieved. One the other hand, the 100 to 1000 ns interpulse range, optimized for fluorescence and electron time of flight detection, requires a **single bunch operation** of the machine with T= 100 ns, at the expense of the intensity per pulse and of the total intensity available for other users. This immediately suggests that **dedicated shifts** should be planned for such time-resolved studies, as long as the number of these exotic users is sufficient compared to the total and the total number of shifts is large enough to allow a significant length of time for them.

Extension of the interpulse period with a given machine can be achieved with **choppers**. It appears that HASYLAB at Hambourg (Germany) has a five years experience of mechanical choppers with a maximum frequency of 1.4 kHz in connection with SR in the VUV spectral range (Möller and Zimmerer 1987). Firstly, the storage ring DORIS has itself an excellent time structure. The length (FWHM) of one bunch is approximately 130 ps and the interpulse period is either 240 ns (HASYLAB dedicated beam time with 4 bunches) or 960 ns (high energy shifts with one bunch). The maximum current in the storage ring is obtained with 4 or 8 bunches instead of 480 bunches (which is the maximum number). Secondly, the chopper installed at the experimental station Superlumi (Time -resolved fluorescence for long lived species in the solid phase, like N_2 in a matrix), extends the interpulse period up to the ms range, whereas the intensity of the photon beam is decreased by a factor twelve. This offers the possibility to perform time-resolved measurements with a large dynamic range in a multi-bunch (n=8) mode with a high current.

Another way of extending the interpulse period is to operate the machine in filling a fraction of bunches, in a multibunch mode. This has been achieved with the Bessy machine, at Berlin (Germany) in filling only about 60 out of the 104 buckets. A pseudo long bunch of 110 ns has allowed (Waterstradt et al 1991) to produce a zero kinetic energy spectrum (ZEKE) with an adapted extraction field technique. This promising method does not reduce significantly the total light intensity as compared to those obtained with all bunches filled. This could avoid the one bunch operation and dedicated shifts. Further improvements should be planned with third generation machines having different multibunch modes.

B - Reduction of the SR pulse length. Competition with lasers ?

The goal of reducing the pulse length down to the picosecond time range is relevant in molecules (or biological species), as seen in Table 2. The challenge is not to develop fast photon detectors because they are available, (Table 1), but rather fast electron detectors for time of flight measurements and a number of coïncidence techniques combining electrons/ions, or photon/ion or electron/photon for spectroscopy. The obvious limits of ΔT (bunch length) is given by the minimum intensity per bunch. It should be noticed that the divergence of SR, after the transport beam line, induces a broadening of the photon pulse. Typically it will be difficult to go below 10 ps, (1 ps =0.3 mm) unless one develops shortening methods similar to those used with high intensity laser radiation. SR with 10 ps pulses remains advantageous because the pulse shape does not change with wavelength.

Below this value of 10 ps, the coherent radiation offered by laser harmonic generation (ΔE/E = 0.005 %) overcomes largely the possibilities of SR (ΔE/E = 0.1 %) because such radiation can be available in highly specialized laboratories from the UV to the VUV range, beyond 100 eV, in the femtosecond time domain with intensities higher than SR (ondulators) by several orders of magnitude (L'Huillier 1991). Notice that the apparent superiority of these laser sources does not hold when one considers the tunability in a wide range e.g. over thousands of cm^{-1}. It is well known that the requirements for scanning over a large wavelength range cannot be readily met by lasers. The time consuming operations such as the changing of mirrors, dyes, pump lasers, as well as the inevitable realignment requirements, make such scans impractical. It is thus evident that time resolved SR capabilities are complementary to those of lasers.

C - Synchronization of SR with lasers

The combination of lasers with VUV and X-ray SR is clearly developing, in atomic and molecular physics (Wuilleumier et al,

Sonntag et al, Nahon et al 1991) and in surface physics, materials sciences and Biophysics (Mills 1991). The first experiments were performed with CW lasers, taking SR as a continuous source. More advanced ones, take advantage of the pulsed nature of SR with an association of pulsed lasers (Mills 1991). Different types of pump (laser)-probe(SR) experiments are possible for a large range of excited state lifetimes. If it large compared to the interpulse period T, one can use conventional pulsed lasers as long as their repetition rate is sufficient to create a large enough density of excited states. This has been realized with the CHESS storage ring (T=2.5 μs) with a frequency doubled Nd:Yag lase at 5300 A, 20 Hz and X-ray SR (7.08 to 7.2 keV). The delay between the pump mand probe beams were varied from the micro to millisecond time range. If the excited state lifetime is shorter than T and longer than the pulse length ΔT, then high repetition rate lasers are necessary. This has been realised by Mitani (1988) with the UVSOR ring in Japan in a time resolved fluorescence and modulation spectroscopies.

Other types of experiments in which SR is the pump are under way, especially in molecular physics for which interesting phenomena occur often in the nanosecond range or below (see Table 2). A single bunch operation with an interpulse period long enough to adjust the time delay between the pump and probe pulses, is required. Depending on the physical phenomena, the interpulse period must be of several tenths of nanoseconds. Such experiments require the synchronization of laser pulses by the RF cavity of the ring and the minimization of the jitter. Notice that high repetition rate (several tenths of kilohertz) lasers should be highly desirable to approach those of SR (MHz range).

Beyond the combination of conventional or advanced laboratory laser sources, it should be noticed that many two photon experiments require a laser in the infra-red, among which free electron lasers produced in linear accelerators are especially attractive because of their tunability, the pulse width in the picosecond domain and a repetition rate which could be adjusted to a SR source (Ortega et al 1989).

IV - General comments and conclusions

The scientific community interested in time-resolved experiments with SR is still small compared to the total number of SR users. Because of the large variety of problems found in many scientific fields, this community is scattered and its various components do not interact naturally. The present conference is the first occasion to our knowledge, to trigger such an interaction. Suggestions have been made to encourage the present users to act as a group.

*The X-ray users, like Astronomers, could combine their efforts to satisfy their needs. This is true for "flux hungry" X-ray experiments which obviously need **fast detectors**.
*Users of the UV and VUV clearly demand two distinct modes of operations of the machine :
i) short single bunch operation
ii) large interpulse period with a long single (or pseudo) pulse.
Such modes of operations require that such potential users interact strongly with machine builders, very early in the conception time of a new machine. An alternative would be to build a specifically tailored machine for this community.

Acknowledgements

Special thanks are due to T. Baer, J.C. Brochon, M. Daniels, J. Goulon, A. Hopkirk, S. Leach, D. Menzel, M. Möller, J. Wong, to help me gathering the main streams of the panel discussion.

References

T. Baer, J. Booze and K.M. Weitzel "Photoelectron Photoion Coïncidence studies of ion disociation dynamics" in Vacuum Ultra-Violet Photoionization and Photodissociation of Molecules and Clusters, C.Y. Ng editor. World Scientific Publ. Co. Singapore (1991) pp259-296

J.C. Brochon, (1980), "Protein Dynamics and Energy transduction" in Proceedings of the 6th Tanigushi International Symposium, pp 163-169 (Ishiwata S. Editor), Tokyo. J.C. Brochon, F. Merola, and A.K. Livesey (this proceedings).

R. Clarke and W.P. Lowe, Synchrotron Radiation News 4, 25 (1991)

M.Daniels, J.P. Ballini and P. Vigny (these proceedings)

J.H.D. Eland (1991) "Coïncidence studies of multiionized molecules" in Vacuum Ultra-Violet Photoionization and Photodissociation of Molecules and Clusters, Ed. Ng C.Y., World Scientific, Singapore.

R. Frahm, Nucl. Instum. Meth. A270, 528 (1983).

A.M. Frank et al. Nucl. Instrum. Meth. 208, 651 (1983)

M. Glass-Maujean and H. Fröhlich (these proceedings)

E. Gratton et al., Rev. Sci. Instrum. 55, 4686 (1984) and this proceedings

A. L'Huillier, L.A. Lompré, G. Mainfray, C. Manus-Dubosc, Adv. Opt. and Atom. Mol. Physics (in press) ; A. L'Huillier, private communication.

D.D. Mills "Time-resolved studies" in Handbook on SR, ed by G.S. Brown and D. Moncton, North Holland, New York (1991), Chap. 9 p. 291 and refs. therein.

T. Mitani, T. Yamanaka, M. Susuki, T. Horigome, K. Hayakawa, I. Yamazaki, J. Lumi. 39, 313 (1988).

T. Möller and G. Zimmerer, Physica Scripta T17, 177 (1987).

L. Nahon, J. Tremblay, M. Larzillière, L. Duffy, P. Morin, Nucl. Inst. Methods B 47, 72 (1990)

I. Nenner and J.A. Beswick, "Molecular Photodissociation and Photoionization" in Handbook on Synchrotron Radiation, vol 2, G.V. Marr ed. Elsevier (1987), Chap. 6 pp 355-466.

M. Ortega et al, NIM A, 285, p. 97 (1989)

B. F. Sonntag, Physica Scripta, T34, 93 (1991)

E. Waterstradt, R. Jung, G. Reiser, H.J. Dietrich, K. Müller-Dethlefs and E.W. Schlag, (private communication)

F. J. Wuilleumier, J. Phys. (Paris) Colloq. 43 C2-347 (1982)

PARTICIPANTS

ANDREONI W. Ms
IBM Zürich Res. Labs.
CH-8803 RUSCHLIKON
SWITZERLAND

ARTEMIEV A.N.
Kurchatov Inst. Atomic Energy
123182 MOSCOW
USSR

AVALDI L.
I.M.A.I. de C.N.R.
CP 10 / 00016 MONTEROTONDO SCALO
ITALY

BAER T.
Chemistry Dept.
Univ. of North Carolina
CHAPEL HILL
NC 27599-3290 (USA)

BARNES P.
Birkbeck College/Crystallography
Malet street
LONDON WC1E 7HX (UK)

BARRAUD A.
CEA / DRECAM / SCM
F-91191 GIF SUR YVETTE
France

BARTUNIK H.D.
MPG Molecular Biology, at DESY
Notkestrasse 85
D-2000 HAMBURG 52
F.R.G.

BAUMGARTEL H.
Inst. f. Physikalische Chemie
Takustrasse 3
D-1000 BERLIN 33 (FRG)

BEHRET H.
Deutsche Bunsen Gesellschaft
Varrentrappstr. 40
D-6000 FRANKFURT/Main 90
F.R.G.

BESWICK J.A.
LURE / Bât. 209 D
F- 91405 ORSAY
France

BIANCONI A.
Dipartimento di Fisica
Università di Roma "La Sapienza"
P.le Aldo Moro / I-00185 ROMA
Italy

BORGHINI G. Ms
Dipartimento di Fisica
Università di Roma "La Sapienza"
P.le Aldo Moro 5 / I-00185 ROMA
Italy

BOZIO R.
Istituto di Chimica Fisica
via Loredan 2 / I-35131 PADOVA
Italy

BRECHIGNAC C. Ms
Laboratoire Aimé Cotton
F-91405 ORSAY / Bât. 505
France

BRIOIS V. Ms
LURE / Bât. 209 D
F-91405 ORSAY
France

BROCHON J.C.
LURE / Bât. 209 D
F-91405 ORSAY
France

BROCKLEHURST B.
Chemistry Department
University of Sheffield
SHEFFIELD S3 7HF (UK)

BROSOLO M. Ms
Dipart. di Scienze Chimiche
Università di Trieste
via A.Valerio 22
I-34100 TRIESTE TS
Italy

BRUTSCHY B.
Inst. f. Physikalische Chemie
der F.U.Berlin
Takustrasse 3
D-1000 BERLIN 33 (FRG)

CAFFREY M.
Dept. of Chemistry
Ohio State University
COLOMBUS OH 43210
U.S.A.

CARTIER C. Ms
LURE / Bât. 209 D
F-91405 ORSAY
France

CAULETTI C. Ms
Univ. di Roma "La Sapienza"
Dipartimento di Chimica
P.le Aldo Moro / I-00185 ROMA
Italy

CHEREPKOV N.A.
Aviation Instruments Making Inst.
Hertsen St. 67
190000 LENINGRAD
USSR

CLARKE R.
University of Michigan
Randall Laboratory
ANN ARBOR MI 48109 (USA)

DANIELS M.
Oregon State University
Radiation Center
CORVALLIS OR 97331-5903 (USA)

DECLEVA P.
Università di Trieste
Scienze Chimiche
via A.Valerio 22
I-34100 TRIESTE TS (Italy)

DE LANGE C.A.
Physical Chemistry
University of Amsterdam
Nieuwe Achtergracht 127
NL-1018 WS AMSTERDAM
The Netherlands

DELLA LONGA S. Ms
Dip. Medicina Experimentale
Università dell'Aquila
I-67100 L'AQUILA (Italy)

DELWICHE J.
Spectroscopie de photoélectrons
Univ. de Liège / B 6
SART TILMAN 4000 LIEGE
Belgium

DENT J.A.
SERC Daresbury Laboratory
DARESBURY
WARRINGTON WA4 4AO (UK)

DEXPERT H.
LURE / Bât. 209 D
F-91405 ORSAY
France

DUJARDIN G.
Photophysique moléculaire CNRS
Bât. 213
F-91405 ORSAY (France)

DURAND J. Ms
Résidence Albert Ier
47bis rue Albert Ier
F-41000 BLOIS (France)

DURAND R.
CREMGP / ENSEEG
BP n°75
F-38402 St-MARTIN D'HERES
France

ELAND J.H.D.
Physical Chemistry Laboratory
University of Oxford
South Parks Rd / OXFORD OX1 3QZ
U.K.

ERMAN P.
Physics Dept. I
Royal Inst. of Technology
S-100 44 STOCKHOLM (Sweden)

FISONS INSTRUMENTS
85 av. Aristide Briand
F-94110 ARCUEIL (France)

FEYEN B.
RUCA/Dept. Computer Science
Groenenborgerlaan 171
B-2020 ANTWERPEN (Belgium)

FONTAINE A.
LURE / Bât. 209 D
F-91405 ORSAY (France)

FOX K. Ms
Unilerver Res./ Port Sunlight
Quarry Rd. East
Bebington Wirral
MERSEYSIDE L63 3JW (UK)

FRONZONI G. Ms
Dip. Scienze Chimiche
Università di Trieste
via A.Valerio 22
I-34127 TRIESTE (Italy)

GAPONOV Yu.A.
Institute Nuclear Physics
Synchrotron Radiation Center
630090 NOVOSIBIRSK 5USSR)

GAVEAU M.A.
DRECAM / SPAM / Bât. 522
CEN de Saclay
F-91191 GIF SUR YVETTE (France)

GEISSLER E.
Spectrométrie physique
BP n° 87
F-38402 St-MARTIN D'HERES
France

GLASS-MAUJEAN M. Ms
ENS / Spectroscopie hertzienne
4 pl.Jussieu
F-75252 PARIS Cedex 05
France

GOULON J.
E.S.R.F. / BP n° 220
F-38043 GRENOBLE Cedex (France)

GRATTON E.
Univ. of Illinois
LUMIS Lab./Physics Bldg
1110 W.Green Street
URBANA IL 61801 (USA)

GUNNARSSON O.
MPI Festkörperforschung
Postfach 80 06 65
D-7000 STUTTGART 80 (FRG)

HAENSEL R.
E.S.R.F / BP n° 220
F-38043 GRENOBLE Cedex (France)

HAGELSTEIN M.
E.S.R.F. / BP n° 220
F-38043 GRENOBLE Cedex (France)

HERLIN N. Ms
CEN De Saclay
DRECAM / SPAM
F-91191 GIF SUR Yvette (France)

HERTEL I.V.
Fak. f. Physik / Univ. Freiburg
Hermann-Herderstr. 3
D-7800 FREIBURG i.Br.(FRG)

HOPKIRK A.
SERC Daresbury Lab.
WARRINGTON Cheshire WA4 4AD (UK)

JENTYS A.
Inst. f. Physikalische Chemie
der T.U.Wien
Getreidemarkt 9
A-1060 WIEN (Austria)

KALLNE E. Ms
Dept. of Physics I
Royal Inst. of Technology
S-100 44 STOCKHOLM (Sweden)

KILCOYNE A.L.
Fritz Haber Inst. der MPG
Faraday Weg 4-6
D-1000 BERLIN 33 (FRG)

LABARTHE B.
MECA 2000
37 rue St-Léger
F-78540 VERNOUILLET (France)

LAJZEROWICZ J.
Spectrométrie physique
BP n° 87
F-38402 ST-MARTIN D'HERES(France)

LEACH S.
Photophysique moléculaire CNRS
Bât. 213
F-91405 ORSAY (France)

LE NORMAND F.
IPCMS
Groupe Surfaces-Interfaces
4 rue Blaise Pascal
F-67070 STRASBOURG (France)

LISINI A. Ms
Dip. di Scienze Chimiche
Università di trieste
via A.Valerio 22
34127 TRIESTE (Italy)

MANCEAU M.
Minéralogie-Cristallographie
UPMC/4 pl.Jussieu/Tour 16
F-75252 PARIS Cedex 05 (France)

MARCELLI A.
I.N.F.N. / L.N.F.
P.O.Box 13
I-00044 FRASCATI (Italy)

MARECHAL J.L.
CEN de Saclay
DRECAM / SPAM
F-91191 GIF SUR YVETTE (France)

MENZEL D.
Physik Abt. der T.U.München
D-8046 GARCHING (FRG)

MEYER M.
LURE / Bât. 209 D
F-91405 ORSAY (france)

MILLIE Ph.
CEN de Saclay
DRECAM / SPAM / Bât.522
F-91191 GIF SUR YVETTE (France)

MILLS D.
APS/Argonne National Lab.
ARGONNE IL 60439 (USA)

MOLLER T.
II-Inst. f. Exp. Physik
Universität Hamburg
Luruper Chaussee 149
D-2000 HAMBURG 50 (FRG)

MORAWECK B.
I.R.C. / C.N.R.S.
2 av. Albert Einstein
F-69626 VILLEURBANNE (France)

MORIN P.
LURE / Bât. 209 D
F-91405 ORSAY (France)

MOTTE-TOLLET F. Ms
Univ. de Liège/Chimie physique
SART TILMAN Bât. B6
B-LIEGE 4000 (Belgium)

NAHON L.
LURE / Bât. 209 D
F-91405 ORSAY (France)

NENNER I. Ms
CEN de Saclay
DRECAM / SPAM / Bât. 522
F-91191 GIF SUR YVETTE (France)

PANNETIER J.
Inst. Laue-Langevin
156 X
F-38042 GRENOBLE Cedex (France)

PARENT Ph.
LURE / Bât. 209 D
F-91405 ORSAY (France)

PIANCASTELLI M.N. Ms
Dip. Scienze Chimiche
Univ. di Roma/Tor Vergata
via O.Raimondo
I-00173 ROMA (Italy)

RANDALL K.
Fritz Haber Inst. der MPG
Faradayweg 4-6
D-1000 BERLIN 33 (FRG)

RIEKEL C.
E.S.R.F / BP n° 220
F-38043 GRENOBLE Cedex (France)

RODNYI P.A.
Leningrad State Technical Univ.
LENINGRAD 195251 (USSR)

RUFFER R.
E.S.R.F. / BP n° 220
F-38043 GRENOBLE Cedex (France)

RUHL E.
Inst. f.Physikalsiche Chemie
der F.U.Berlin
Takustrasse 3
D-1000 BERLIN 33 (FRG)

SCHIRMER J.
Theoretische Chemie
Physikalisch Chemisches Inst.
der Universität Heidelberg
Im Neuenheimer Feld 253
D-6900 HEIDELBERG (FRG)

SCHMELZ H.C. Ms
CEN de Saclay
DRECAM / SPAM / Bât. 522
F-91191 GIF SUR YVETTE (France)

SCHMIDBAUER M.
Fritz Haber Inst. der MPG
Faradayweg 4-6
D-1000 BERLIN 33 (FRG)

SCHROEDER J.
Inst. f. Physikalische Chemie
der Universität Göttingen

Tammannstr. 6
D-3400 GOTTINGEN (FRG)

SIMONS J.P.
Dept. of Chemistry
University Park
NOTTINGHAM NG7 2RD (UK)

TANAKA K.
Photon Factory/ Nat.l Lab.
for High Energy Physics (KEK)
TSUKUBA, IBARAKI 305
Japan

TEREKHIN M.
Kurchatov Inst. Atomic Energy
Division Synchrotron Radiation
123 182 MOSCOW (USSR)

THISSEN R.
Spectroscopie de photoélectrons
SART TILMAN / Bât. B6
B-4000 LIEGE (Belgium)

TOBITA S.
Gunma College of Technology
580 TORIBA, maebashi
GUNMA (Japan)

TROYANOWSKY C.
SFC/Division de Chimie physique
10 rue Vauquelin
F-75005 PARIS (France)

TWESTEN I. Ms
FU Berlin, Fb Physik
AG Schwentner
Arnimallee 14
D-1000 BERLIN 33 (FRG)

UNDERHILL A.E.
Univ. College of N.Wales
BANGOR Gwynedd L57 2DG (UK)

VALETTE B.
MECA 2000
37 rue St-Léger
F-78540 VERNOUILLET (France)

VARIAN VACUUM PRODUCTS
via Fratelli Varian 54
I-10040 LEINI (TO) (Italy)

VERDAGUER M.
Chimie des métaux de transition
UPMC / Bât.F / 4e étage
4 pl.Jussieu
F-75252 PARIS Cedex 05 (France)

VILLAIN F. Ms
LURE / Bât. 209 D
F-91405 ORSAY (France)

WEITZEL K.M.
Inst. f. Physikalische Chemie
Takustrasse 3
D-1000 BERLIN 33 (FRG)

WONG J.
Hasylab at DESY
Notkestrasse 85
D-2000 HAMBURG 52 (FRG)

Author Index

AIP Conference Proceedings

		L.C. Number	ISBN
No. 106	Predictability of Fluid Motions (La Jolla Institute, 1983)	83-73641	0-88318-305-6
No. 107	Physics and Chemistry of Porous Media (Schlumberger-Doll Research, 1983)	83-73640	0-88318-306-4
No. 108	The Time Projection Chamber (TRIUMF, Vancouver, 1983)	83-83445	0-88318-307-2
No. 109	Random Walks and Their Applications in the Physical and Biological Sciences (NBS/La Jolla Institute, 1982)	84-70208	0-88318-308-0
No. 110	Hadron Substructure in Nuclear Physics (Indiana University, 1983)	84-70165	0-88318-309-9
No. 111	Production and Neutralization of Negative Ions and Beams (3rd Int'l Symposium) (Brookhaven, NY, 1983)	84-70379	0-88318-310-2
No. 112	Particles and Fields – 1983 (APS/DPF, Blacksburg, VA)	84-70378	0-88318-311-0
No. 113	Experimental Meson Spectroscopy – 1983 (7th International Conference, Brookhaven, NY)	84-70910	0-88318-312-9
No. 114	Low Energy Tests of Conservation Laws in Particle Physics (Blacksburg, VA, 1983)	84-71157	0-88318-313-7
No. 115	High Energy Transients in Astrophysics (Santa Cruz, CA, 1983)	84-71205	0-88318-314-5
No. 116	Problems in Unification and Supergravity (La Jolla Institute, 1983)	84-71246	0-88318-315-3
No. 117	Polarized Proton Ion Sources (TRIUMF, Vancouver, 1983)	84-71235	0-88318-316-1
No. 118	Free Electron Generation of Extreme Ultraviolet Coherent Radiation (Brookhaven/OSA, 1983)	84-71539	0-88318-317-X
No. 119	Laser Techniques in the Extreme Ultraviolet (OSA, Boulder, CO, 1984)	84-72128	0-88318-318-8
No. 120	Optical Effects in Amorphous Semiconductors (Snowbird, UT, 1984)	84-72419	0-88318-319-6
No. 121	High Energy e^+e^- Interactions (Vanderbilt, 1984)	84-72632	0-88318-320-X
No. 122	The Physics of VLSI (Xerox, Palo Alto, CA, 1984)	84-72729	0-88318-321-8
No. 123	Intersections Between Particle and Nuclear Physics (Steamboat Springs, CO, 1984)	84-72790	0-88318-322-6
No. 124	Neutron-Nucleus Collisions: A Probe of Nuclear Structure (Burr Oak State Park, 1984)	84-73216	0-88318-323-4

No. 125	Capture Gamma-Ray Spectroscopy and Related Topics – 1984 (Int'l Symposium, Knoxville, TN)	84-73303	0-88318-324-2
No. 126	Solar Neutrinos and Neutrino Astronomy (Homestake, 1984)	84-63143	0-88318-325-0
No. 127	Physics of High Energy Particle Accelerators (BNL/SUNY Summer School, 1983)	85-70057	0-88318-326-9
No. 128	Nuclear Physics with Stored, Cooled Beams (McCormick's Creek State Park, IN, 1984)	85-71167	0-88318-327-7
No. 129	Radiofrequency Plasma Heating (Sixth Topical Conference) (Callaway Gardens, GA, 1985)	85-48027	0-88318-328-5
No. 130	Laser Acceleration of Particles (Malibu, CA, 1985)	85-48028	0-88318-329-3
No. 131	Workshop on Polarized ^{3}He Beams and Targets (Princeton, NJ, 1984)	85-48026	0-88318-330-7
No. 132	Hadron Spectroscopy–1985 (International Conference, Univ. of Maryland)	85-72537	0-88318-331-5
No. 133	Hadronic Probes and Nuclear Interactions (Arizona State University, 1985)	85-72638	0-88318-332-3
No. 134	The State of High Energy Physics (BNL/SUNY Summer School, 1983)	85-73170	0-88318-333-1
No. 135	Energy Sources: Conservation and Renewables (APS, Washington, DC, 1985)	85-73019	0-88318-334-X
No. 136	Atomic Theory Workshop on Relativistic and QED Effects in Heavy Atoms (Gaithersburg, MD, 1985)	85-73790	0-88318-335-8
No. 137	Polymer-Flow Interaction (La Jolla Institute, 1985)	85-73915	0-88318-336-6
No. 138	Frontiers in Electronic Materials and Processing (Houston, TX, 1985)	86-70108	0-88318-337-4
No. 139	High-Current, High-Brightness, and High-Duty Factor Ion Injectors (La Jolla Institute, 1985)	86-70245	0-88318-338-2
No. 140	Boron-Rich Solids (Albuquerque, NM, 1985)	86-70246	0-88318-339-0
No. 141	Gamma-Ray Bursts (Stanford, CA, 1984)	86-70761	0-88318-340-4
No. 142	Nuclear Structure at High Spin, Excitation, and Momentum Transfer (Indiana University, 1985)	86-70837	0-88318-341-2
No. 143	Mexican School of Particles and Fields (Oaxtepec, México, 1984)	86-81187	0-88318-342-0
No. 144	Magnetospheric Phenomena in Astrophysics (Los Alamos, NM, 1984)	86-71149	0-88318-343-9
No. 145	Polarized Beams at SSC & Polarized Antiprotons (Ann Arbor, MI & Bodega Bay, CA, 1985)	86-71343	0-88318-344-7
No. 146	Advances in Laser Science–I (Dallas, TX, 1985)	86-71536	0-88318-345-5
No. 147	Short Wavelength Coherent Radiation: Generation and Applications (Monterey, CA, 1986)	86-71674	0-88318-346-3

No. 148	Space Colonization: Technology and The Liberal Arts (Geneva, NY, 1985)	86-71675	0-88318-347-1
No. 149	Physics and Chemistry of Protective Coatings (Universal City, CA, 1985)	86-72019	0-88318-348-X
No. 150	Intersections Between Particle and Nuclear Physics (Lake Louise, Canada, 1986)	86-72018	0-88318-349-8
No. 151	Neural Networks for Computing (Snowbird, UT, 1986)	86-72481	0-88318-351-X
No. 152	Heavy Ion Inertial Fusion (Washington, DC, 1986)	86-73185	0-88318-352-8
No. 153	Physics of Particle Accelerators (SLAC Summer School, 1985) (Fermilab Summer School, 1984)	87-70103	0-88318-353-6
No. 154	Physics and Chemistry of Porous Media—II (Ridge Field, CT, 1986)	83-73640	0-88318-354-4
No. 155	The Galactic Center: Proceedings of the Symposium Honoring C. H. Townes (Berkeley, CA, 1986)	86-73186	0-88318-355-2
No. 156	Advanced Accelerator Concepts (Madison, WI, 1986)	87-70635	0-88318-358-0
No. 157	Stability of Amorphous Silicon Alloy Materials and Devices (Palo Alto, CA, 1987)	87-70990	0-88318-359-9
No. 158	Production and Neutralization of Negative Ions and Beams (Brookhaven, NY, 1986)	87-71695	0-88318-358-7
No. 159	Applications of Radio-Frequency Power to Plasma: Seventh Topical Conference (Kissimmee, FL, 1987)	87-71812	0-88318-359-5
No. 160	Advances in Laser Science–II (Seattle, WA, 1986)	87-71962	0-88318-360-9
No. 161	Electron Scattering in Nuclear and Particle Science: In Commemoration of the 35th Anniversary of the Lyman-Hanson-Scott Experiment (Urbana, IL, 1986)	87-72403	0-88318-361-7
No. 162	Few-Body Systems and Multiparticle Dynamics (Crystal City, VA, 1987)	87-72594	0-88318-362-5
No. 163	Pion–Nucleus Physics: Future Directions and New Facilities at LAMPF (Los Alamos, NM, 1987)	87-72961	0-88318-363-3
No. 164	Nuclei Far from Stability: Fifth International Conference (Rosseau Lake, ON, 1987)	87-73214	0-88318-364-1
No. 165	Thin Film Processing and Characterization of High-Temperature Superconductors (Anaheim, CA, 1987)	87-73420	0-88318-365-X
No. 166	Photovoltaic Safety (Denver, CO, 1988)	88-42854	0-88318-366-8
No. 167	Deposition and Growth: Limits for Microelectronics (Anaheim, CA, 1987)	88-71432	0-88318-367-6
No. 168	Atomic Processes in Plasmas (Santa Fe, NM, 1987)	88-71273	0-88318-368-4

No. 169	Modern Physics in America: A Michelson-Morley Centennial Symposium (Cleveland, OH, 1987)	88-71348	0-88318-369-2
No. 170	Nuclear Spectroscopy of Astrophysical Sources (Washington, DC, 1987)	88-71625	0-88318-370-6
No. 171	Vacuum Design of Advanced and Compact Synchrotron Light Sources (Upton, NY, 1988)	88-71824	0-88318-371-4
No. 172	Advances in Laser Science–III: Proceedings of the International Laser Science Conference (Atlantic City, NJ, 1987)	88-71879	0-88318-372-2
No. 173	Cooperative Networks in Physics Education (Oaxtepec, Mexico, 1987)	88-72091	0-88318-373-0
No. 174	Radio Wave Scattering in the Interstellar Medium (San Diego, CA, 1988)	88-72092	0-88318-374-9
No. 175	Non-neutral Plasma Physics (Washington, DC, 1988)	88-72275	0-88318-375-7
No. 176	Intersections Between Particle and Nuclear Physics (Third International Conference) (Rockport, ME, 1988)	88-62535	0-88318-376-5
No. 177	Linear Accelerator and Beam Optics Codes (La Jolla, CA, 1988)	88-46074	0-88318-377-3
No. 178	Nuclear Arms Technologies in the 1990s (Washington, DC, 1988)	88-83262	0-88318-378-1
No. 179	The Michelson Era in American Science: 1870–1930 (Cleveland, OH, 1987)	88-83369	0-88318-379-X
No. 180	Frontiers in Science: International Symposium (Urbana, IL, 1987)	88-83526	0-88318-380-3
No. 181	Muon-Catalyzed Fusion (Sanibel Island, FL, 1988)	88-83636	0-88318-381-1
No. 182	High T_c Superconducting Thin Films, Devices, and Application (Atlanta, GA, 1988)	88-03947	0-88318-382-X
No. 183	Cosmic Abundances of Matter (Minneapolis, MN, 1988)	89-80147	0-88318-383-8
No. 184	Physics of Particle Accelerators (Ithaca, NY, 1988)	89-83575	0-88318-384-6
No. 185	Glueballs, Hybrids, and Exotic Hadrons (Upton, NY, 1988)	89-83513	0-88318-385-4
No. 186	High-Energy Radiation Background in Space (Sanibel Island, FL, 1987)	89-83833	0-88318-386-2
No. 187	High-Energy Spin Physics (Minneapolis, MN, 1988)	89-83948	0-88318-387-0
No. 188	International Symposium on Electron Beam Ion Sources and their Applications (Upton, NY, 1988)	89-84343	0-88318-388-9
No. 189	Relativistic, Quantum Electrodynamic, and Weak Interaction Effects in Atoms (Santa Barbara, CA, 1988)	89-84431	0-88318-389-7
No. 190	Radio-frequency Power in Plasmas (Irvine, CA, 1989)	89-45805	0-88318-397-8

No. 191	Advances in Laser Science–IV (Atlanta, GA, 1988)	89-85595	0-88318-391-9
No. 192	Vacuum Mechatronics (First International Workshop) (Santa Barbara, CA, 1989)	89-45905	0-88318-394-3
No. 193	Advanced Accelerator Concepts (Lake Arrowhead, CA, 1989)	89-45914	0-88318-393-5
No. 194	Quantum Fluids and Solids—1989 (Gainesville, FL, 1989)	89-81079	0-88318-395-1
No. 195	Dense *Z*-Pinches (Laguna Beach, CA, 1989)	89-46212	0-88318-396-X
No. 196	Heavy Quark Physics (Ithaca, NY, 1989)	89-81583	0-88318-644-6
No. 197	Drops and Bubbles (Monterey, CA, 1988)	89-46360	0-88318-392-7
No. 198	Astrophysics in Antarctica (Newark, DE, 1989)	89-46421	0-88318-398-6
No. 199	Surface Conditioning of Vacuum Systems (Los Angeles, CA, 1989)	89-82542	0-88318-756-6
No. 200	High T_c Superconducting Thin Films: Processing, Characterization, and Applications (Boston, MA, 1989)	90-80006	0-88318-759-0
No. 201	QED Stucture Functions (Ann Arbor, MI, 1989)	90-80229	0-88318-671-3
No. 202	NASA Workshop on Physics From a Lunar Base (Stanford, CA, 1989)	90-55073	0-88318-646-2
No. 203	Particle Astrophysics: The NASA Cosmic Ray Program for the 1990s and Beyond (Greenbelt, MD, 1989)	90-55077	0-88318-763-9
No. 204	Aspects of Electron–Molecule Scattering and Photoionization (New Haven, CT, 1989)	90-55175	0-88318-764-7
No. 205	The Physics of Electronic and Atomic Collisions (XVI International Conference) (New York, NY, 1989)	90-53183	0-88318-390-0
No. 206	Atomic Processes in Plasmas (Gaithersburg, MD, 1989)	90-55265	0-88318-769-8
No. 207	Astrophysics from the Moon (Annapolis, MD, 1990)	90-55582	0-88318-770-1
No. 208	Current Topics in Shock Waves (Bethlehem, PA, 1989)	90-55617	0-88318-776-0
No. 209	Computing for High Luminosity and High Intensity Facilities (Santa Fe, NM, 1990)	90-55634	0-88318-786-8
No. 210	Production and Neutralization of Negative Ions and Beams (Brookhaven, NY, 1990)	90-55316	0-88318-786-8
No. 211	High-Energy Astrophysics in the 21st Century (Taos, NM, 1989)	90-55644	0-88318-803-1
No. 212	Accelerator Instrumentation (Brookhaven, NY, 1989)	90-55838	0-88318-645-4
No. 213	Frontiers in Condensed Matter Theory (New York, NY, 1989)	90-6421	0-88318-771-X 0-88318-772-8 (pbk.)

No. 214	Beam Dynamics Issues of High-Luminosity Asymmetric Collider Rings (Berkeley, CA, 1990)	90-55857	0-88318-767-1
No. 215	X-Ray and Inner-Shell Processes (Knoxville, TN, 1990)	90-84700	0-88318-790-6
No. 216	Spectral Line Shapes, Vol. 6 (Austin, TX, 1990)	90-06278	0-88318-791-4
No. 217	Space Nuclear Power Systems (Albuquerque, NM, 1991)	90-56220	0-88318-838-4
No. 218	Positron Beams for Solids and Surfaces (London, Canada, 1990)	90-56407	0-88318-842-2
No. 219	Superconductivity and Its Applications (Buffalo, NY, 1990)	91-55020	0-88318-835-X
No. 220	High Energy Gamma-Ray Astronomy (Ann Arbor, MI, 1990)	91-70876	0-88318-812-0
No. 221	Particle Production Near Threshold (Nashville, IN, 1990)	91-55134	0-88318-829-5
No. 222	After the First Three Minutes (College Park, MD, 1990)	91-55214	0-88318-828-7
No. 223	Polarized Collider Workshop (University Park, PA, 1990)	91-71303	0-88318-826-0
No. 224	LAMPF Workshop on (π, K) Physics (Los Alamos, NM, 1990)	91-71304	0-88318-825-2
No. 225	Half Collision Resonance Phenomena in Molecules (Caracus, Venezuela, 1990)	91-55210	0-88318-840-6
No. 226	The Living Cell in Four Dimensions (Gif sur Yvette, France, 1990)	91-55209	0-88318-794-9
No. 227	Advanced Processing and Characterization Technologies (Clearwater, FL, 1991)	91-55194	0-88318-910-0
No. 228	Anomalous Nuclear Effects in Deuterium/Solid Systems (Provo, UT, 1990)	91-55245	0-88318-833-3
No. 229	Accelerator Instrumentation (Batavia, IL, 1990)	91-55347	0-88318-832-1
No. 230	Nonlinear Dynamics and Particle Acceleration (Tsukuba, Japan, 1990)	91-55348	0-88318-824-4
No. 231	Boron-Rich Solids (Albuquerque, NM, 1990)	91-53024	0-88318-793-4
No. 232	Gamma-Ray Line Astrophysics (Paris–Saclay, France, 1990)	91-55492	0-88318-875-9
No. 233	Atomic Physics 12 (Ann Arbor, MI, 1990)	91-55595	0-88318-811-2
No. 234	Amorphous Silicon Materials and Solar Cells (Denver, CO, 1991)	91-55575	0-88318-831-7
No. 235	Physics and Chemistry of MCT and Novel IR Detector Materials (San Francisco, CA, 1990)	91-55493	0-88318-931-3
No. 236	Vacuum Design of Synchrotron Light Sources (Argonne, IL, 1990)	91-55527	0-88318-873-2

No. 237	Kent M. Terwilliger Memorial Symposium (Ann Arbor, MI, 1989)	91-55576	0-88318-788-4
No. 238	Capture Gamma-Ray Spectroscopy (Pacific Grove, CA, 1990)	91-57923	0-88318-830-9
No. 239	Advances in Biomolecular Simulations (Obernai, France, 1991)	91-58106	0-88318-940-2
No. 240	Joint Soviet-American Workshop on the Physics of Semiconductor Lasers (Leningrad, USSR, 1991)	91-58537	0-88318-936-4
No. 241	Scanned Probe Microscopy (Santa Barbara, CA, 1991)	91-76758	0-88318-816-3
No. 242	Strong, Weak, and Electromagnetic Interactions in Nuclei, Atoms, and Astrophysics: A Workshop in Honor of Stewart D. Bloom's Retirement (Livermore, CA, 1991)	91-76876	0-88318-943-7
No. 243	Intersections Between Particle and Nuclear Physics (Tucson, AZ, 1991)	91-77580	0-88318-950-X
No. 244	Radio Frequency Power in Plasmas (Charleston, SC, 1991)	91-77853	0-88318-937-2
No. 245	Basic Space Science (Bangalore, India, 1991)	91-78379	0-88318-951-8
No. 246	Space Nuclear Power Systems (Albuquerque, NM, 1992)	91-58793	1-56396-027-3 1-56396-026-5 (pbk.)
No. 247	Global Warming: Physics and Facts (Washington, DC, 1991)	91-78423	0-88318-932-1
No. 248	Computer-Aided Statistical Physics (Taipei, Taiwan, 1991)	91-78378	0-88318-942-9
No. 249	The Physics of Particle Accelerators (Upton, NY, 1989, 1990)	92-52843	0-88318-789-2
No. 250	Towards a Unified Picture of Nuclear Dynamics (Nikko, Japan, 1991)	92-70143	0-88318-951-8
No. 251	Superconductivity and its Applications (Buffalo, NY, 1991)	92-52726	1-56396-016-8
No. 252	Accelerator Instrumentation (Newport News, VA, 1991)	92-70356	0-88318-934-8
No. 253	High-Brightness Beams for Advanced Accelerator Applications (College Park, MD, 1991)	92-52705	0-88318-947-X
No. 254	Testing the AGN Paradigm (College Park, MD, 1991)	92-52780	1-56396-009-5
No. 255	Advanced Beam Dynamics Workshop on Effects of Errors in Accelerators, Their Diagnosis and Corrections (Corpus Christi, TX, 1991)	92-52842	1-56396-006-0
No. 256	Slow Dynamics in Condensed Matter (Fukuoka, Japan 1991)	92-53120	0-88318-938-0
No. 257	Atomic Processes in Plasmas (Portland, ME, 1991)	91-08105	0-88318-939-9